IVB C

Animal Models *in* Orthopaedic Research

edited by

Yuehuei H. An
Richard J. Friedman

CRC Press
Boca Raton   London   New York   Washington, D.C.

Library of Congress Cataloging-in-Publication Data

Catalog record is available from the Library of Congress

This book contains information obtained from authentic and highly regarded sources. Reprinted material is quoted with permission, and sources are indicated. A wide variety of references are listed. Reasonable efforts have been made to publish reliable data and information, but the author and the publisher cannot assume responsibility for the validity of all materials or for the consequences of their use.

Neither this book nor any part may be reproduced or transmitted in any form or by any means, electronic or mechanical, including photocopying, microfilming, and recording, or by any information storage or retrieval system, without prior permission in writing from the publisher.

The consent of CRC Press LLC does not extend to copying for general distribution, for promotion, for creating new works, or for resale. Specific permission must be obtained in writing from CRC Press LLC for such copying.

Direct all inquiries to CRC Press LLC, 2000 Corporate Blvd., N.W., Boca Raton, FL 33431.

Trademark Notice: Product or corporate names may be trademarks or registered trademarks, and are only used for identification and explanation, without intent to infringe.

© 1999 by CRC Press LLC

No claim to original U.S. Government works
International Standard Book Number 0-8493-2115-8
Printed in the United States of America 2 3 4 5 6 7 8 9 0
Printed on acid-free paper

*To Kay Q. Kang, M.D.
Without her love, inspiration, and support,
this book would not have been possible.*

Yuehuei H. An, M.D.

*To my wife Vivian, and my daughters Arielle and Leah,
for their patience, understanding, love and support.*

Richard J. Friedman, M.D., FRCS(C)

The Editors

Yuehuei H. An, M.D. is Associate Professor and Co-director of the Orthopaedic Research Laboratory, Department of Orthopaedic Surgery, Medical University of South Carolina, Charleston, SC, and Adjunct Assistant Professor of Bioengineering, Department of Bioengineering, Clemson University, Clemson, SC.

Dr. An graduated from the Harbin Medical University, Harbin, China, in 1983 and was trained in orthopaedic surgery at Ji Shui Tan Hospital (residency), Beijing, China, and in hand surgery at Sydney Hospital (Clinical Fellow), Sydney, Australia.

In 1991, Dr. An joined with Dr. Richard J. Friedman in the Department of Orthopaedic Surgery at the Medical University of South Carolina to establish an Orthopaedic Research Laboratory. Since then, he has been directing and supervising graduate students, research fellows, and residents on numerous orthopaedic research projects. Dr. An has published more than 50 scientific papers and book chapters and presented more than 60 research abstracts. He is an active member of six academic societies, including the Orthopaedic Research Society (USA), Society for Biomaterials (USA), American Society of Biomechanics, and Tissue Engineering Society (USA).

Dr. An's current research interests include repair of bone and cartilage defects using tissue engineering techniques, improving bone or soft tissue ingrowth to implant surfaces, prevention of bacterial adhesion and prosthetic infection, and development of animal models for orthopaedic applications.

He enjoys art and created most of the drawings used in his chapters.

Richard J. Friedman, M.D., FRCS(C), is Professor and Director of the Orthopaedic Research Laboratory, Department of Orthopaedic Surgery, Medical University of South Carolina, Charleston, SC, and Adjunct Professor of Bioengineering, Department of Bioengineering, Clemson University, Clemson, SC.

He graduated from the University of Toronto School of Medicine in 1980 and after that was trained in orthopaedic surgery at Johns Hopkins University and Harvard University.

Dr. Friedman is the founder and director of the Orthopaedic Research Laboratory at the Medical University of South Carolina. He has published more than 100 scientific papers and book chapters and presented more than 200 research abstracts. In 1994, he edited *Arthroplasty of the Shoulder* published by Thieme Medical Publishers, New York, a reference book in human shoulder arthroplasty. He has been invited as a guest speaker for many national and international academic meetings. He has served as a chairman or member of more than 30 academic or professional committees or meetings and is a member of more than twenty societies or associations.

Dr. Friedman's current research interests include shoulder and elbow surgery, bone and cartilage repair using tissue engineering techniques, bone ingrowth, total joint replacement, orthopaedic biomaterials, and development of animal models for orthopaedic applications.

Contributors

Yuehuei H. An, M.D., Associate Professor and Co-director, Orthopaedic Research Laboratory, Department of Orthopaedic Surgery, Medical University of South Carolina, Charleston, SC and Adjunct Assistant Professor, Department of Bioengineering, Clemson University, Clemson, SC

Todd D. Bell, M.D., Orthopaedic Resident, Department of Orthopaedic Surgery, Medical University of South Carolina, Charleston, SC

Sarah A. Bingel, V.M.D., Ph.D., Assistant Professor, Department of Comparative Medicine and Department of Pathology, Medical University of South Carolina, Charleston, SC

Earl R. Bogoch, M.D., FRCSC, Associate Professor, Division of Orthopaedic Surgery, Department of Surgery, Director, Wellesley Orthopaedic Research Laboratory, University of Toronto, Toronto, ON, Canada

Robert C. Bray, M.D., FRCSC, Orthopaedic Surgeon, University of Calgary, Department of Orthopaedic Surgery, Joint Injury and Arthritis Research Group, Calgary, AB, Canada

Frank P. Cammisa, Jr., M.D., Chief, Spine Surgery Service, Division of Orthopaedic Surgery, Hospital for Special Surgery, Cornell Medical College, New York, NY

Edgar G. Dawson, M.D., Clinical Professor, Department of Orthopaedic Surgery, UCLA School of Medicine, Los Angeles, CA

Masao Deguchi, M.D., Chief of Spine Surgery, Department of Orthopaedic Surgery, Nagano Red Cross Hospital, Nagano, Japan

Robert A. Draughn, D.Sc., Professor and Chairman, Department of Materials Science, College of Dental Medicine, Medical University of South Carolina, Charleston, SC

Richard T. Fosse, M.R.C.V.S., Associate Director LAR, Rhone Poulenc Rorer, Dagenham Research Centre, Dagenham, Essex, UK.

Richard J. Friedman, M.D., FRCSC, Professor and Director, Orthopaedic Research Laboratory, Department of Orthopaedic Surgery, Medical University of South Carolina, Charleston, SC, and Adjunct Professor, Department of Bioengineering, Clemson University, Clemson, SC

Federico Girardi, M.D., Spine Surgery Fellow, Division of Orthopaedic Surgery, Hospital for Special Surgery, Cornell Medical College, New York, NY

Stuart B. Goodman, M.D., Professor, Stanford University Medical Center, Division of Orthopaedic Surgery, Stanford, CA

Helen E. Gruber, Ph.D., Senior Scientist and Director of Research Histology, Orthopaedic Research Laboratory, Carolina Medical Center, Charlotte, NC

John A. Jansen, D.D.S., Ph.D., Professor and Head, Department of Biomaterials, University of Nijmegen Dental School, Nijmegen, The Netherlands

Tokumi Kanemura, M.D., Orthopaedic Surgeon, Department of Orthopaedic Surgery, JR Tokai Hospital, Nagoya University School of Medicine, Nagoya, Japan

Linda E. A. Kanim, M.A., Research Associate, University Spine Associates, UCLA School of Medicine, Los Angeles, CA

Noriaki Kawakami, M.D., Chief of Orthopaedic Surgery, Department of Orthopaedic Surgery, Meijou Hospital, Nagoya University School of Medicine, Nagoya, Japan

Donald B. Kimmel, Ph.D., Senior Scientist and Director, Department of *In vivo* Bone Biology, Merck & Co., Inc., West Point, PA

Jan Klompmaker, M.D., Registrar, Department of Orthopaedic Surgery, University Hospital Nijmegen, Institute of Orthopaedics, Nijmegen, The Netherlands

Martine LaBerge, Ph.D., Associate Professor, Department of Bioengineering, Clemson University, Clemson, SC

Martin Lind, M.D., Ph.D., Research Fellow, Orthopaedic Research Laboratory, Aarhus University Hospital Noerrebrogade, Aarhus, Denmark

Theodore T. Manson, B.A., Graduate Student, Musculoskeletal Research Center, Department of Orthopaedic Surgery, University of Pittsburgh, Pittsburgh, PA

Kensaku Masuhara, M.D., Associate Professor, Department of Orthopaedic Surgery, Osaka University Medical School, Osaka, Japan

Hiromi Matsuzaki, M.D., Associate Professor, Chief of Spine Division, Department of Orthopaedic Surgery, Nihon University Surugadai Hospital, Tokyo, Japan

Minoru Matui, M.D., Assistant Professor, Department of Orthopaedic Surgery, Osaka University Medical School, Osaka, Japan

Jason J. McDougall, Ph.D., Postdoctoral Fellow, University of Calgary, Joint Injury and Arthritis Research Group, Calgary, AB, Canada

Shimpei Miyamoto, M.D., Ph.D., Assistant Professor, Department of Orthopaedic Surgery, Osaka University Medical School, Osaka, Japan

Erica L. Moran, B.Sc., Research Associate, Wellesley Orthopaedic Research Laboratory, University of Toronto, Toronto, ON, Canada

Katsuya Nakata, M.D., Graduate student, Department of Orthopaedic Surgery, Osaka University Medical School, Osaka, Japan

Theodore R. Oegema, Jr., Ph.D., Professor, Department of Orthopaedic Surgery, University of Minnesota, Minneapolis, MN

Keiro Ono, M.D., Professor Emeritus, Department of Orthopaedic Surgery, Osaka University School of Medicine, Director, Osaka Kosei-nenkin Hospital, Osaka, Japan

Donald L. Pruitt, M.D., Hand Surgeon, Alabama Orthopaedic Clinics, Mobile, Alabama, USA

Warren K. Ramp, Ph.D., Senior Scientist, Orthopaedic Research Laboratory, Carolina Medical Center, Charlotte, NC.

Harvinder S. Sandhu, M.D., Division of Orthopaedic Surgery, Hospital for Special Surgery, Cornell Medical College, New York, NY

Alison C. Smith, D.V.M., Associate Professor, Department of Comparative Medicine and Division of Cardiology, Medical University of South Carolina, Charleston, SC

Yong Song, M.D., Research Assistant, Stanford University Medical Center, Division of Orthopaedic Surgery, Stanford, CA

Audrey A. Stasky, B.A., Senior Laboratory Technician, Orthopaedic Research Laboratory, Carolina Medical Center, Charlotte, NC

Dale R. Sumner, Ph.D., Professor and Chairman, Department of Anatomy, Professor, Department of Orthopaedic Surgery, Rush Medical College, Rush–Presbyterian–St. Luke's Medical Center, Chicago, IL

M. Michael Swindle, D.V.M., Professor and Chairman, Department of Comparative Medicine and Department of Surgery, Medical University of South Carolina, Charleston, SC

Thomas M. Turner, D.V.M., Assistant Professor, Department of Orthopaedic Surgery, Rush Medical College, Rush–Presbyterian–St. Luke's Medical Center, Chicago, IL

Robert M. Urban, Research Administrator, Department of Orthopaedic Surgery, Rush Medical College, Rush–Presbyterian–St. Luke's Medical Center, Chicago, IL

René P. H. Veth, M.D., Professor and Chairman, Department of Orthopaedic Surgery, University Hospital Nijmegen, Institute of Orthopaedics, Nijmegen, The Netherlands

Denise M. Visco, Ph.D., Research Fellow, Merck Research Laboratories, Department of Pharmacology, Merck & Co., Inc., Rahway, NJ

Tracy M. Vogrin, B.Sc., Graduate Student, Musculoskeletal Research Center, Department of Orthopaedic Surgery, University of Pittsburgh, Pittsburgh, PA

Ken Wakabayashi, M.D., Assistant Professor, Department of Orthopaedic Surgery, Nihon University Surugadai Hospital, Tokyo, Japan

Savio L.–Y. Woo, Ph.D., A. B. Ferguson Professor and Director, Musculoskeletal Research Center, Department of Orthopaedic Surgery, University of Pittsburgh, Pittsburgh, PA

Kazuo Yonenobu, M.D., Associate Professor, Department of Orthopaedic Surgery, Osaka University Medical School, Osaka, Japan

Franklin A. Young, Jr., D.Sc., Professor, Department of Materials Science, College of Dental Medicine, Medical University of South Carolina, Charleston, SC

Preface

Research using animal models provides important knowledge of pathological conditions that eventually can lead to the development of more effective clinical treatment of diseases in both humans and animals. This book covers most of the major animal models used in studies of biomaterials and orthopaedic disorders. It is to be used as a reference book and is primarily directed towards surgeons, investigators, research fellows, graduate students, or anyone working in the field of orthopaedic or biomaterial research. It is intended to serve as a basis for a literature search before embarking on a detailed research project. It is possible that other scientists or physicians in areas unrelated to orthopaedic or biomaterial research may also find this material useful as a source of reference, or as a tool to expedite their research.

This book is an outgrowth of the editors' own quest for information about animal research methodology in orthopaedic and biomaterial research and, more importantly, represents tremendous support from the orthopaedic and biomaterial research communities. The 11 chapters written by the editors are a combination of knowledge gained from personal experience and the research literature. The remaining 20 chapters are contributions from 45 well-known experts in their fields of interest from throughout the world.

The book has 31 chapters and is divided into eight major parts. Part I is a general discussion about the care and use of laboratory animals and experimental designs in orthopaedic research. Part II describes the most commonly used evaluation methods in orthopaedic animal research. Detailed descriptions of common animal models used in orthopaedic research are given in Parts III–VIII.

The book is designed to be concise as well as inclusive and more practical than theoretical. The text is simple and straightforward. Appropriate numbers of tables, diagrams, line figures, and photographs are used to make the contents more vivid. The appendices include a list of periodicals and publications related to orthopaedic research, laboratory animals, and procurement sources. Full bibliographies at the end of each chapter guide readers to more detailed information on the subject. A book of this length cannot possibly discuss every animal model that has ever been produced in orthopaedic research, but it is felt that the major models and their applications have been included.

Yuehuei H. An
Richard J. Friedman

Foreword

Research is at the heart of progress in orthopaedic surgery as it is in all other fields of medicine. The products of research have changed the face of orthopaedics and have provided for millions of human beings the chance of enjoying active and productive lives.

Research encompasses a broad range of activities, but it should always begin with a burning question that will lead to the development of one or more hypotheses and the need to test them experimentally.

For that purpose, there are a number of methods available to the investigator. They include *in vitro* experiments, the use of cadaveric material, computer models, physical models, and clinical databases.

Animal models are an integral part of the process and, as such, are frequently used in orthopaedic research. However, prior to their application there are some fundamental issues that must be addressed. Is the animal experiment truly necessary or can the questions be answered using other methods?

Experimental work requiring the use of living creatures can never be taken lightly. An enormous controversy surrounds us regarding the ethical issues involved. Thus, the use of animal models for subjects that are important to mankind can only be justified when there is no other viable alternative.

Experiments in animals when justified and properly planned and executed have been essential in the acquisition of new knowledge.

Animal models have allowed us to understand the natural history of disease, to develop new and improved surgical techniques, and to predict the effect of a given treatment or surgical procedure. They have been critical in the development and in the evaluation of implants, one of the basic elements of modern orthopaedics. Animal models play a crucial role in biocompatibility evaluation, which is the most fundamental basis of knowledge necessary in the biomaterials field. Tissue engineering and the use of new technologies based on molecular biology developments require animal experiments. And the list can goes on and on. In every aspect of orthopaedic research, the use of animal models constitutes an essential step that leads to the eventual application of newly acquired information to the human patient.

This work then addresses a very important subject and one that to my knowledge has not been covered in such a comprehensive manner in any other book or publication.

There are a number of features that make this book unique in addition to the subject and the depth in which it has been addressed. The editors have played a very active role in the conception and in the execution of the project. Of the 31 chapters that comprise the book, 11 were written by the editors themselves. This is unusual and reflects on the one hand their expertise and knowledge of the subject and, on the other, the level of their commitment.

The book is divided into eight parts. The first two deal with principles of detailed methodology. The other six address the use of models for specific purposes. Given the nature of the musculoskeletal system, a broad picture needs to be considered. Bone, cartilage, joint replacement, ligaments and tendons, spinal conditions and microsurgical techniques are included with chapters that address in detail the use of animal models in most areas of related research.

I particularly enjoyed Part One of the book, including all its five chapters, which by themselves represent in my judgment a major contribution. The ethical issues involved are presented in a very objective light, taking into account the concerns of the orthopaedic research community as well as those of animal rights advocates. This is a topic with which all investigators should be very familiar.

There is no research without appropriate experimental design. The young investigator will find important information to guide him from the conception of the basic ideas, to the development and execution of the experiment, and to the eventual publication of results in a scientific journal.

My own perception is that it will be among our young trainees and investigators that the book will have its major impact. The seasoned researcher and the basic scientist will find this book very valuable as well, and an excellent source of reference given the breadth with which the subject has been approached. This book has a great deal to offer to everyone involved in orthopaedic research.

Jorge O. Galante, M.D.

Contents

PART I — GENERAL CONSIDERATIONS OF USING LABORATORY ANIMALS

1 Ethics and Regulations for the Care and Use of Laboratory Animals 3
   Alison C. Smith, Richard T. Fosse, and M. Michael Swindle

2 Experimental Design, Evaluation Methods, Data Analysis, Publication, and Research Ethics 15
   Yuehuei H. An and Todd D. Bell

3 Animal Selections in Orthopaedic Research 39
   Yuehuei H. An and Richard J. Friedman

4 Surgical Facilities, Peri-operative Care, Anesthesia, and Surgical Techniques 59
   Alison C. Smith and M. Michael Swindle

5 Euthanasia and Necropsy 71
   Sarah A. Bingel

PART II — EVALUATION METHODS IN ORTHOPAEDIC ANIMAL RESEARCH

6 Methods of Evaluation in Orthopaedic Animal Research 85
   Yuehuei H. An

7 Histological Study in Orthopaedic Animal Research 115
   Helen E. Gruber and Audrey A. Stasky

8 Mechanical Properties and Testing Methods of Bone 139
   Yuehuei H. An and Robert A. Draughn

9 Mechanical Testing of Cartilage 165
   Martine LaBerge

10 Mechanical Testing of Ligaments and Tendons 175
   Savio L.-Y. Woo, Theodore T. Manson, and Tracy M. Vogrin

PART III — ANIMAL MODELS OF BONE CONDITIONS

11 Animal Models of Fracture or Osteotomy 197
   Yuehuei H. An, Richard J. Friedman, and Robert A. Draughn

12 Animal Models for Testing Bioabsorbable Materials 219
   Yuehuei H. An and Richard J. Friedman

13 Animal Models of Bone Defect Repair 241
   Yuehuei H. An and Richard J. Friedman

14 Animal Models of Osteonecrosis 261
   Kensaku Masuhara, Minoru Matui, Katsuya Nakata, and Keiro Ono

15 Animal Models of Osteopenia or Osteoporosis 279
   Donald B. Kimmel, Erica L. Moran, and Earl R. Bogoch

PART IV — ANIMAL MODELS OF ARTICULAR CARTILAGE AND JOINT CONDITIONS

16 Animal Models of Articular Cartilage Defect 309
   Yuehuei H. An and Richard J. Friedman

17 Animal Models of Meniscal Repair 327
   Jan Klompmaker and René P. H. Veth

18 Animal Models of Osteoarthritis 349
   Theodore R. Oegema, Jr. and Denise M. Visco

19 Animal Models of Rheumatoid Arthritis 369
   Erica L. Moran and Earl R. Bogoch

PART V — ANIMAL MODELS OF JOINT REPLACEMENT AND RELATED CONDITIONS

20 Animal Models for Studying Soft Tissue Biocompatibility of Biomaterials 393
   John A. Jansen

21 Animal Models for Studying Bone Ingrowth and Joint Replacement 407
   Dale R. Sumner, Thomas M. Turner, and Robert M. Urban

22 Animal Models for Investigations of Biomaterial Debris 427
   Martin Lind, Yong Song, and Stuart B. Goodman

23 Animal Models of Orthopaedic Prosthetic Infection 443
   Yuehuei H. An and Richard J. Friedman

PART VI — ANIMAL MODELS FOR THE STUDY OF LIGAMENTS AND TENDONS

24 Animal Models of Ligament Repair 461
   Jason J. McDougall and Robert C. Bray

25 Animal Models of Tendon Repair 477
   Donald L. Pruitt

26 Animal Models for Ligament and Tendon Fixation 491
   Franklin A. Young, Jr. and Yuehuei H. An

PART VII — ANIMAL MODELS OF SPINAL CONDITIONS

27 Animal Models of Spinal Instability and Spinal Fusion 505
Harvinder S. Sandhu, Linda E. A. Kanim, Federico Girardi,
Frank P. Cammisa, Jr., and Edgar G. Dawson

28 Animal Models of Spinal Cord Compression 527
Shimpei Miyamoto, Kazuo Yonenobu, and Keiro Ono

29 Animal Models for Reconstruction of Vertebral Column and Intervertebral Disc 539
Hiromi Matsuzaki and Ken Wakabayashi

30 Animal Models of Scoliosis 549
Noriaki Kawakami, Masao Deguchi, and Tokumi Kanemura

PART VIII — MICROSURGICAL TECHNIQUE

31 Microsurgery and Orthopaedic Animal Models 567
Yuehuei H. An

Appendix 1   Abbreviations 583

Appendix 2   Journals and Publications Related to Orthopaedic Research
and Laboratory Animals 585

Appendix 3   Major Sources of Laboratory Animals 587

Index 589

Part I

General Considerations of Using Laboratory Animals

1 Ethics and Regulations for the Care and Use of Laboratory Animals

Alison C. Smith, Richard T. Fosse, Warren K. Ramp, and M. Michael Swindle

CONTENTS

I. Introduction 3
II. Animal Research and Ethics 4
   A. Importance of Animal Research 4
   B. The 'Three Rs' and Alternatives to Animals in Orthopaedic Research 5
III. Legislation and Guidelines 5
   A. U.S. Legislation and Guidelines 5
      1. In General 5
      2. Definitions 6
      3. Institutional Policies and Responsibilities 7
      4. Protocol Review 7
   B. European Legislation 9
      1. In General 9
      2. European Directive 88/609 9
      3. CoE Convention ETS 123 9
   C. Legislation and Guidelines of Other Countries 10
IV. Obtaining Approval from the IACUC 11
V. Good Laboratory Practice 11
References 13

I. INTRODUCTION

The use of animals has played a vital role in the numerous medical advances in the field of orthopaedic research. This is primarily because of the inherent limitations of alternatives to animal experimentation that preclude the study of the interactions of the various tissues and organ systems in the intact organism. A postition statement issued by the American Academy of Orthopaedic Surgeons has endorsed the appropriate use of animals in orthopaedic research in addition to acknowledging the importance of increasingly refined alternative approaches. The group believes that the continued use of animals is justified based on the human and animal benefits to quality of life issues.[1] Other professional organizations engaged in biomedical research also have position statements regarding the use of animals.[2,3]

The current climate for the conduct of biomedical research which involves animal use in general is stormy. Viewpoints are frequently polarized between the biomedical research community and animal protection groups. At their most extreme, these attitudes have resulted in the unfair characterization of scientists as uncaring individuals who view animals as a means to an end and animal advocates as anti-intellectuals. The reality is that there are many caring scientists who approach the use of animals responsibly and many sensible animal advocates. In the end, research animals' best interests will be served when both groups find a middle ground from which a reasonable dialogue can ensue.

Although public attitudes have shifted towards support of animal welfare and animal rights viewpoints over the last 20 years, public opinion continues to strongly support animal experimentation. Recent polls indicate that 60–80% of the population accepts the use of animals in biomedical research; however, support for animal use varies with the species of animal used and the type of research.[4,5] In general, the public is more concerned with the reduction and elimination of pain and suffering for research animals as opposed to experiments that require animals to be killed.

With that said, it is the intent of this chapter to briefly describe some of the ethical issues involved in animal research and to summarize the current regulatory standards under which biomedical research using animals, including orthopaedic research, is conducted. The regulations and guidelines for research animals are living documents and continue to be revised. Salient issues included in these documents involve minimizing animal pain and distress, the use of alternatives when appropriate, and justification of the number and species of animals selected.

II. ANIMAL RESEARCH AND ETHICS

A. Importance of Animal Research

Public controversy regarding animal experimentation can be traced back two centuries with the establishment of the use of animal experimentation in scientific methodology. The antivivisection movement was established in response to the new field of experimental physiology, in which key discoveries were made possible by the use of animals.[4] Many of the ethical issues born during this period are still intensely debated today. It is beyond the scope of this chapter to give a complete overview of the current major philosophical arguments regarding animal experimentation; however, the reader is referred to some of the many recent publications for further reading.[4–12]

Defenders of animal experimentation generally argue in favor of the benefits of research to both humans and animals. Philosophically, this argument is in the utilitarian tradition wherein the consequences of a particular action are weighed in terms of benefit and harm to all those affected. The utilitarian or consequentialist approach suggests that the considerable benefits attained through animal experimentation easily outweigh the cost in terms of animal pain and distress.

Another approach in support of animal research is made by identifying distinguishing features that are morally significant and which separate man from animals. One of the most frequently debated issues is whether animals have moral agency. Carl Cohen of the University of Michigan, a supporter of animal research, argues that animals lack rights. He has written that rights are claims, or potential claims, that entail obligations among members of society. Since animals lack the ability to comprehend the rules that come with duties and do not have obligations, they have no rights.[4]

Philosophers have been some of the key voices and founding intellectuals in the modern animal movement. Peter Singer, the author of the seminal animal rights book, *Animal Liberation*, uses the utilitarian argument to attack animal research on the basis that moral status should be conferred on animals based on their ability to experience pain and suffering. Singer identifies the capacity to suffer with sentience and, therefore, with right to moral status. He argues that the benefits of animal experimentation are trivial compared to the cost of animal suffering. Singer does not confer moral equality between animals and man; instead he argues that equal consideration be given to animals when their interests (such as being hurt) are equal.[5]

The leading proponent of the position that animals should be accorded rights is philosopher Tom Regan. He asserts that animal life has inherent value which should accord animals the right to not be used as a means to an end. He believes that no matter how much good results, animal experimentation is morally wrong because it violates animals' basic rights.[4]

B. The 'Three Rs' and Alternatives to Animals in Orthopaedic Research

The English scientists William Russell and Rex Burch, in their 1959 publication of *The Principles of Humane Experimental Technique*, were the first to voice scientists' concern for experimental animals.[13] They described the principles of the three "Rs" which promoted goals for research scientists: (1) replacement of animals by use of *in vitro* methods or by using animals that are phylogenetically lower; (2) reduction of the number of animals required; and (3) refinement of experimental methods to reduce the ethical costs in terms of painful or stressful procedures. Some decades later, the "three Rs" have been adopted by the scientific community and have served as the definition of the current search for alternatives.

In orthopaedic research, the state of the art is such that replacement alternatives are not developed to the point of eliminating all animal use. At this time, it is likely that most orthopaedic researchers would say that replacement alternatives will never be developed. However, all "three Rs" have current applications including the use of phylogenetically lower animals and the use of techniques such as cell/tissue cultures, benchtop experiments, and computer simulations. Cell and organ cultures can be used for toxicity and biocompatibility testing of drugs and biomaterials, as well as for evaluating effects of hormones, growth factors, and environmental factors on bone metabolism. This approach can give insight into direct effects of these influences on biological processes involved in bone diseases that would otherwise be difficult or impossible to obtain from whole animals. In addition, invertebrate animals and microorganisms are utilized for studies of mechanisms of biological mineralization and pathogenesis of osteomyelitis. Physical and computer models are also used and may be appropriate for studies of joint mechanics, prosthesis wear, and complex physiological systems such as calcium homeostasis, while cadaveric materials can be applied to development of surgical techniques and constructs. Computerized patient registries serve as excellent epidemiological databases for retrospective clinical studies and for outcome assessment. Compared to animal models, the advantages of these systems are that they are usually faster and less expensive, produce less pain and distress in vertebrate animals, and reduce the number of vertebrate animals used in research. Some alternative methods used in orthopaedic animal research are summarized in Table 1.

An issue that is likely to remain a source of controversy for orthopaedic researchers who use animals and for individuals who care for animals is whether surgical procedures of extremities should be performed unilaterally or bilaterally. From the scientific standpoint as well as in the interest of using fewer animals, the case is often made that a bilateral model is preferred since studies can be designed to allow an animal to be its own control. Unilateral models have the advantage of producing less pain and distress to the individual animal. Clearly, protocols of this kind must be carefully evaluated and the scientific merits weighed against the potential for pain and distress to the study animals.

III. LEGISLATION AND GUIDELINES

A. U.S. Legislation and Guidelines

1. In General

The conduct of biomedical research using animals carries with it the imperative that the care and use of animals is appropriate and their treatment is humane. In the United States, this imperative is in the form of two federal laws which govern the use of all animals used in research, testing, and education. The Animal Welfare Act (AWA) of 1966 (P.L. 89-544) and its subsequent amendments are administered by the United States Department of Agriculture (USDA) and implemented by USDA's Animal and Plant Health Inspection Service (APHIS). The regulations which promulgate the AWA are published in Title 9 of the Code of Federal Regulations and are commonly referred to as the Animal Welfare Regulations.[36] The AWA regulates appropriate care and treatment for

TABLE 1
Examples of Alternative Testing Methods

Method	Material	Use[Ref.]
Cell cultures	Osteoblast	Osteopetrosis pathogenesis[14]
		Poly-L-lactate biocompatibility[15]
		Particle toxicity[16]
	Fibroblast, Myoblast	Phenotypic expression[17]
		Titanium biocompatibility[18]
Organ cultures	Tibia	Antibiotic toxicity[19]
	Tibia, Calvaria	pH effects[20]
	Parietal bone	Estrogen effects[21]
Bacteria	*Bacterionema matruchotti*	Calcification mechanism[22]
	Staphylococcus aureus	Osteomyelitis pathogenesis[23]
Invertebrates	Mollusk shell	Mineral nucleation[24]
	Crustacean shell	Calcification mechanism[25]
Cadaveric materials	Wrist	Fracture fixation[26]
	Spine	Spine stabilization[27]
	Knee	Ligament reconstruction[28]
Computer models	Knee	Prosthesis development[29]
	Total hip prosthesis	Implant positioning[30]
	Joint prosthesis	Wolff's law[31]
Physical models	Joint simulator	Wear behavior[32]
	Joint loading model	Exercise physiology[33]
Epidemiologic databases	Patient registry	Retrospective clinical studies[34]
	Clinical and outcome data	Testing treatment hypotheses[35]

animals used in research and includes provisions regarding their sale, shipping, purchase, housing, and veterinary care. The *Public Health Service Policy on Humane Care and Use of Laboratory Animals of 1986* (PHS Policy) was published by the Public Health Service (PHS) to implement The Health Research Extension Act of 1985.[37]

In addition to the above laws, there are several federal documents that promulgate guidelines for institutions that receive federal funding. The PHS Policy endorses and supplements the *U.S. Government Principles for the Utilization and Care of Vertebrate Animals Used in Testing, Research, and Training*, a set of guidelines developed by the Interagency Research Animal Committee. The PHS Policy is applicable to all PHS-conducted or supported activities involving animals, regardless of the institution or country in which the work is conducted. The NIH *Guide for the Care and Use of Laboratory Animals* also endorses the U.S. government principles and its use by institutions as the basis for the development and implementation of animal care programs is required by the PHS. The guide contains guidelines concerning institutional policies and responsibilities; animal environment, housing, and management; veterinary medical care; and physical plant and is widely recognized as a primary reference on animal care.[38] These regulations will be summarized as they pertain to the field of orthopaedic research using animals, specifically issues concerning surgery and minimizing pain and distress.

2. Definitions

Part 1 of the AWA defines terms used in parts 2 and 3, which outline the law's requirements. While the guide lacks a list of definitions, it does define certain issues related to surgery in the body of the text. Both the USDA regulations and the NIH guide define a major operative procedure as any surgical procedure that penetrates and exposes a body cavity or produces permanent impairment of physical or physiological functions. Both the USDA regulations and the guide prohibit multiple major survival surgery on a single animal unless it has been scientifically justified by the principal investigator and approved by the Institutional Animal Care and Use Committee (IACUC) or unless it is part of routine veterinary care required to protect the health of the animal.

From a veterinary care and animal welfare perspective, a prevailing issue is the ability to conduct animal research while minimizing pain and distress. The USDA regulations define a painful procedure as any procedure that could reasonably be expected to cause more that momentary or slight pain or distress in humans in excess of that of an injection. Although neither the guide nor the PHS policy include such a definition, the fourth Government Principle states that procedures that cause pain or distress in humans should be considered to cause pain or distress in animals.

Another term that is defined by the USDA regulations and that is relevant to the issue of pain and distress is that of a paralytic drug. The USDA defines a paralytic drug as one which produces partial or complete loss of muscle contraction and which has no anesthetic or analgesic properties, so that the animal cannot move, but is completely aware of its surroundings and can feel pain. The guide specifically states that paralytic drugs are not anesthetic or analgesic.

3. Institutional Policies and Responsibilities

The USDA regulations, the guide, and the PHS policy call for the appointment by the chief executive officer at each research institution of an Institutional Animal Care and Use Committee (IACUC). IACUC members must be qualified through experience and expertise to oversee and evaluate the institution's animal program, procedures, and facilities. The composition of the IACUC varies depending upon which of the regulatory documents are cited, but all call for membership to include a doctor of veterinary medicine with training or experience in laboratory animal science and medicine and one public member not affiliated with the institution in any way other than as a member of the IACUC. The PHS policy and the USDA regulations both state that the veterinary member have direct or delegated program authority and responsibility for activities involving animals at the research facility. The PHS policy and the guide state that one member should be a practicing scientist experienced in research involving animals.

Common to all three regulatory documents are the following IACUC functions. The IACUC must conduct a complete review of an institution's program of animal care and animal facilities every six months. It then must prepare and submit reports of its evaluations to the institutional official. The committee is also charged with the review of all protocols that involve the use of animals. The committee has the authority to review and approve, require modifications to secure approval, or withhold approval of proposed procedures involving animals and proposed significant changes in ongoing protocols.

4. Protocol Review

Review and approval of the animal use components of a research proposal is one of the principal charges of the IACUC, and approval must be obtained before beginning any portions of the research project that involve animal use. The regulations differ somewhat regarding some specific aspects of protocol review. The USDA regulations require identification of the species and approximate number of animals requested. The USDA regulations and the guide require that the principal investigator provide a rationale for using animals and justification for the number of animals to be used. The guide further specifies that the number of animals requested should be justified statistically whenever possible. The third U.S. government principle makes a broader statement regarding the justification of animal use by stating that animals should be selected "of an appropriate species and quality and the minimum number required to obtain valid results." The USDA regulations require a complete description of the proposed use of the animals and the procedures that will be used to limit discomfort and pain to that which is unavoidable to conduct the study.

The USDA regulations and the PHS policy state that procedures which involve animals should be designed to avoid or minimize discomfort, pain, and distress. The USDA regulations further specify that a principal investigator must have considered alternatives to procedures that may cause more than momentary pain or distress and that documentation be provided regarding the methods used to determine that alternatives were unavailable. Consistent with the idea of reducing and replacing the use of animals, both the guide and the third U.S. Government Principle recommend the use of *in vitro* biological systems or computer simulation when appropriate. Unnecessary duplication of previous experiments is discouraged by both the guide and the USDA regulations, and a written assurance to that effect is required by the USDA regulations.

All the regulatory documents require the appropriate use of sedatives, analgesics, or anesthetics for procedures that may cause more than momentary or slight pain or distress unless the withholding is justified for scientific reasons in writing by the investigator. In addition, they also use similar wording to address the issue of severe or chronic pain or distress that cannot be relieved by stating that those animals should be euthanized at the end of the procedure or, if appropriate, during the procedure. The USDA regulations require veterinary consultation during protocol planning for any procedures that may cause more than momentary or slight pain or distress.

Use of paralytic drugs without anesthesia for surgical or other painful procedures is absolutely contraindicated by the guide, the USDA regulations, and U.S. Government Principles. Because they lack either analgesic or anesthetic properties, the concern is that animals which are immobilized but still conscious could experience pain or distress. The guide makes specific recommendations regarding their usage when combined with anesthetics for surgical procedures. The guide also recognizes that their use as a sole agent may be appropriate for certain nonpainful, well-controlled studies. However, it also definitively states that any proposed activities of this sort receive a careful evaluation by the IACUC to ensure that animal well-being is safeguarded.

All the regulatory documents use similar wording regarding qualifications of personnel conducting procedures on animals. The USDA regulations, the guide, and the PHS policy specifically charge the committee to evaluate and determine that personnel conducting procedures on animals have received adequate training in the procedures used. Assessment of personnel qualifications is required regardless of professional degree since it is implicit in the requirements that specific knowledge regarding animal anesthesia, surgery, or other experimental techniques is necessary for performing these tasks in a humane and scientifically acceptable manner.

Provision of adequate veterinary care is a requirement in all the regulations and guidelines. The USDA regulations and the guide also require appropriate provision of pre- and post-operative care for animals undergoing surgery. The guide makes specific recommendations regarding preoperative planning, monitoring during surgery, and provision of post-operative care.

The USDA regulations and the guide address issues specific to protocols which involve surgery. The USDA regulations require that all survival surgery be performed using aseptic procedures and techniques. They further specify that major operative procedures on non-rodent species may be performed only in dedicated facilities maintained under aseptic conditions. Minor surgical procedures and survival surgery on rodents do not require dedicated facilities, but must be performed using aseptic procedures. The guide makes similar recommendations, but recognizes that modifications to aseptic procedures may be appropriate for certain surgical procedures in rodents. The USDA regulations describe aseptic procedures to include the use of surgical gloves, masks, sterile instruments, and aseptic techniques. The guide gives a similar description of aseptic technique but elaborates specific practices involved in its application.

Whereas the USDA regulations specify only that survival surgery on nonrodent species be performed in a dedicated facility maintained and operated using aseptic procedures, the guide provides more specific recommendations on facility design. The guide divides the functional components of an aseptic surgery suite into surgical support, animal preparation, surgeon's scrub, operating room, and post-operative recovery. Design features that help minimize traffic flow and contamination are recommended.

Despite the fact that more than one set of regulations governs the use of animals in orthopaedic research, there is general accord among the pertinent laws and guidelines. The implementation of the regulations will be unique to each research institution based on its available facilities and research program. Legal authority for upholding the applicable regulations rests with the IACUC; its responsibilities for oversight of an animal care program are numerous and specific. Ultimately, a successful animal care program is based on a marriage of the professional standards described by the regulations and their implementation using professional judgment.

B. European Legislation

1. In General

Countries in Europe, led by the United Kingdom, have a long history of legislative regulation of biomedical research with animals. The British Anti-Cruelty Act was introduced by Richard Martin in 1822. He later founded the Royal Society for the Prevention of Cruelty to Animals. The British Cruelty to Animals Act regulating animal experimentation was introduced in 1876. Animal protection in France can be traced to the Grammont law of 1850. France passed legislation specifically applied to animal experimentation in 1963. Similar laws have been passed in most European countries.

Europe has been in the process of harmonizing several of the regulations that cover many activities, including experimentation with animals. Central to European legislation are the "three Rs" proposed by Russell and Burch.[13] The concepts of reduction, replacement and refinement are the basic concepts that have formed the foundation of the main sets of European laws and regulations.

Currently in Europe there are two collaborative blocs that have led to common legislation regulating the use of animals in biomedical research. Experimentation using animals is regulated in Europe by two sets of transnational laws and treaties, the European Union Directive 88/609,[39] and the Council of Europe Convention for the Protection of Vertebrate Animals used for Experimentation and other Scientific Purposes, CoE ETS 123.[40]

2. European Directive 88/609

The European Union (EU) comprises 15 member states that have agreed to cooperate in a number of areas that primarily regulate free flow of goods and economic activity in the member state areas. The process — commonly called harmonization — is designed to establish a free European economic zone with a common set of minimal laws.

The EU adopts legislation that has different levels of application. Regulations serve as binding laws in the member countries. Directives set minimum requirements designed to achieve a desired result but allow some degree of discretion that can be exercised by each member state; decisions are binding only to certain parties. The directive 88/609 is an example of legislation that allows some degree of discretion that can be exercised by each member state. Common to all the member states' legislation is a minimum requirement that serves to coordinate legislation.

3. CoE Convention ETS 123

The Council of Europe (CoE) is an international organization currently comprising 40 member states, including the 15 member states of the European Union. Treaties and conventions adopted by the CoE members are not laws that are applicable in all the member states. The decisions of the CoE become binding only after a certain number of member states ratify a convention. The CoE convention became binding after three states ratified the convention. Unlike the EU which has an economic base for its legislative focus, the CoE is concerned with issues of democracy, human rights and the adoption of common practices, in this case common minimum rules for the use of animals in biomedical research.

The two sets of laws are very similar and for the purposes of this chapter can be viewed as being identical. Therefore, the CoE convention will be used as the legislative system to describe how biological experiments with animals in the majority of European states are regulated. The convention opened for signature on March 18, 1986, and had as its aim the protection of vertebrate animals used for experimental and other scientific purposes. The convention has defined sets of criteria for the treatment of animals that apply in all the 40 member states. These cover guidelines for housing, breeding, sources of animals, and the competence of persons who have responsibility for care of animals, or the performance of procedures, whether these be planning or practical hands-on procedures. The convention also allows for the introduction of quality standards. It is reviewed at multilateral negotiations held in Strasbourg every third year. These meetings are designed to review the application of the convention and suggest modifications that may be needed at any time.

The most significant area that has been addressed is that of competence. The convention states that all persons engaged in research with animals shall be deemed competent (Article 26 CoE and Article 14 EU). No person in the CoE/EU area shall be allowed to perform or plan procedures involving animals, or assume responsibility for planning procedures on animals, unless he/she has taken part in training specified by the CoE/EU. The implication of this is that uniform competence requirements are in the process of being adopted in all the participating countries. The EU established an expert committee under the auspices of the Federation of European Laboratory Animal Science Association with a mandate to suggest curricula for training and teaching personnel in Europe.[41] The committee defined four categories of personnel: caretakers (Group A), laboratory animal technicians (Group B), researchers (Group C), and laboratory animal specialists (Group D). Similar recommendations have been proposed for euthanasia,[42,43] and microbiologic quality.[44] The convention states that stray or ownerless animals shall not be used, and that all cats and dogs shall be purposely bred. The same applies to all other species with the exception of farm animals and wild caught primates. Minimum requirements for housing, cage sizes, and for environmental enrichment are specified, but it is up to each member country to exceed these specifications.

The convention does not require the formation of institutional animal care committees at each user establishment. Some form of peer review authority is required, but this is left to the discretion of the member country. This has led to the development of several systems ranging from local ethical committees in several countries, e.g., Sweden, to decentralized responsible persons who act as extended forms of a central departmental level, e.g., Norway. Other countries have instituted systems of inspectors who certify the use of animals and monitor the procedures and institutions, e.g., the U.K., the Netherlands. There are varying degrees of lay person involvement. All countries require that the researcher obtain written permission prior to performing an experiment. Permission is granted following review by an official authority, and on condition that the applicant is competent to perform the procedures that are specified in the application. Each application shall demonstrate that the procedure cannot be performed without the use of animals, and that a suitable animal-free alternative is not available.

C. Legislation and Guidelines of Other Countries

To the authors' knowledge, similar legislation, guidelines, and policies on the care and use of laboratory animals exist in several other countries, reflecting a worldwide concern for animal welfare. These countries include Canada, Australia, Japan, and Taiwan. In Canada, federal legislation which governs animal welfare is under Section 446 of the Criminal Code. In addition, numerous provincial acts pertain specifically to the use of experimental animals in research, teaching and testing. The *Guide to the Care and Use of Experimental Animals* contains guidelines established by the Canadian Council on Animal Care to evaluate animal care and use in Canadian universities and government and commercial laboratories.[45] The *Australian Code of Practice for the Care and Use of Animals for Scientific Purposes* and its subsequent revisions cover all live vertebrate species used for scientific purposes in research, agriculture, biology, industry, and teaching. The code descibes the responsibilities of institutions and individuals that use animals as well as requirements for their procurement and care.[46] In Japan, research using animals is closely regulated by the government under the *Standards Relating to the Care and Management of Experimental Animals*, which is part of the Law Concerning the Protection and Control of Animals. The Japanese regulations are comprehensive and include policies related to animal transport, animal health, public health, waste disposal, and breeding.[47] Researchers in the above-mentioned countries or other countries should observe the national or local legislation, regulations, or guidelines on the care and use of laboratory animals and help to promote them.

IV. OBTAINING APPROVAL FROM THE IACUC

In order to fulfill its responsibilities concerning protocol review, each IACUC at a research institution requires the submission by the principal investigator of a written protocol which completely describes all aspects of animal use involved in a research protocol. The protocol must address the specific federal requirements described above as well as any applicable state, local or institutional requirements. If an institution receives federal funding or is accredited by AAALAC International, it is compelled to meet the requirements described in the PHS Policy and the guide as well as to comply with the legally binding standards in the USDA regulations. This is the case for the majority of the institutions in the United States which utilize animals to conduct orthopaedic research.

The actual format of the form used by the IACUC for protocol review is generally individualized by each institution. This also applies to the way in which the principal investigator is asked to provide the relevant information. For example, some forms will ask a series of very specific questions, with additional elaboration prompted by response to a particular question or section. Other formats may be less structured and simply ask for a narrative description to questions or information requested. Regardless of the format for animal protocol review used by a particular IACUC, the principal investigator should be aware that the information which she/he is requested to provide is based on the current applicable regulations and guidelines. Responses should be concise but not so abbreviated as to omit relevant and required information, e.g., dosages for drugs, their routes of administration, and dosing schedule.

The perception that more "paperwork" is now necessary for approval of animal protocols compared to past requirements is widespread and is actually true to some extent. This is due to the fact that revisions continue to be made in the pertinent regulations and guidelines. Impetus for change has come from both public concern regarding the treatment of animals used in biomedical research as well as from the scientific community's commitment to animal welfare and the need to formulate standards in step with current scientific knowledge and technology. It should be remembered, however, that responsibility for the ethical and humane use of animals used in research begins with the investigator's decision to use animals. The required process of protocol submission, review, with appropriate modifications when necessary, and approval is the mechanism by which investigators can best justify and describe the necessity to use animals. By providing appropriate responses to the information required, the investigator not only expedites the entire review process, but also satisfies the societal demands for accountability that are represented by the applicable regulations and guidelines.

V. GOOD LABORATORY PRACTICE

Products such as medical devices intended for human use, human and animal drugs, and biological products are regulated by the Food and Drug Administration (FDA) and must undergo preclinical testing prior to clinical testing. Standards exist for testing such FDA regulated products that support applications for research or marketing permits in the document entitled *Good Laboratory Practice (GLP) for Nonclinical Laboratory Studies*.[48] Adherence to the practices described in the document is intended to assure the quality and integrity of the safety data filed in accordance with applicable sections of the Food, Drug, and Cosmetic Act and the Public Health Service Act. The regulations concerning GLP studies are comprehensive and include standards for all personnel, facilities, equipment, test articles, and records involved in a nonclinical laboratory study.

Any facility which conducts a GLP study is subject to inspection by the FDA. The FDA must be permitted access to inspect the facility and any records or specimens maintained as part of a study. Copying of records must also be permitted. Any facility which does not permit inspection by the FDA will not have its nonclinical laboratory study considered by the FDA for support of an application for research or marketing permit.

All personnel involved in conducting a GLP study must have the qualifications to perform their assigned functions. Qualifications can include education, training, and experience or a combination thereof, and must be documented for all individuals involved. A study director with the appropriate professional background must be designated for each GLP study and has overall responsibility for the technical conduct of the study as well as the interpretation and reporting of results.

A testing facility must have a quality assurance unit which is independent of the personnel who conduct the study. The quality assurance unit is responsible for monitoring each study to provide assurance that the facilities, equipment, personnel, methods, practices, records, and controls conform with the regulations. The quality assurance unit is charged with performing and documenting periodic inspections of a study. Any problems encountered during the inspection must be immediately identified to the study director and management.

The regulations contain specifications regarding animal facilities, support areas, and laboratory space. Animal facility design must assure separation of species or test systems, separate individual projects, provide for quarantine of animals, and accommodate routine or specialized housing. In addition, there must be sufficient space to archive all raw data and specimens from completed studies with access limited to authorized personnel.

Equipment used during a GLP study to collect and assess data must be adequately tested and calibrated. Documentation of these procedures is required, as written provisions in the event of equipment malfunction.

Standard operating procedures (SOPs) must be in place to assure the quality and integrity of the data collected during the course of a study, and any deviation from the written protocol must be documented as part of the raw data. SOPs should include at a minimum animal care procedures; animal facility procedures; methods used to receive, identify, store, handle, mix, and sample control and test articles; animal observations; laboratory tests; procedures for handling animals found moribund or dead during a study; necropsy procedures, collection and identification of specimens; histopathology; data handling, storage, and retrieval; maintenance and calibration of equipment; and animal transfer, placement, and identification. Special emphasis is given to having procedures in place that could affect the outcome of the study, such as inadvertent animal misidentification or exposures to test or control articles. Food and water provided to animals must be analyzed periodically during the course of the study for contaminants that might reasonably be expected to be present and affect study results.

The entire GLP study must be conducted in written accordance with the study protocol, appropriately documented by the individuals involved, and all deviations recorded. At the conclusion of a study, a comprehensive final report, signed by the study director, must be prepared that includes the objectives of the study, all methodologies involved in data collection and analysis, a description of the animals used, identification of all personnel involved in the study, and a description of where all raw data and specimens will be stored.

Finally, the regulations set forth conditions under which testing facilities may be disqualified for failing to comply with the requirements. If a facility has been disqualified, it can seek to be reinstated by providing evidence to the Commissioner of the FDA that it has taken appropriate corrective actions to assure compliance with the regulations.

REFERENCES

1. *A Position Statement: Animals in Biomedical Research,* American Academy of Orthopaedic Surgeons, Park Ridge, IL, 1987.
2. *Report of the American College of Laboratory Animal Medicine on Adequate Veterinary Care in Research, Testing and Teaching,* American College of Laboratory Animal Medicine, 1996.
3. *Use of Animals in Biomedical Research: The Challenge and Response,* American Medical Association, White Paper, 1992.
4. Orlans, F. B., *In the Name of Science: Issues in Responsible Animal Experimentation,* Oxford University Press, Oxford, U.K., 1993.
5. Rowan, A. N., Loew, F. M., and Weer, J. C., *The Animal Research Controversy: Protest, Process and Public Policy,* Tufts University School of Veterinary Medicine, Tufts University, Boston, 1995.
6. Cheyney, D. I., Seyfarth, R. M., *How Monkeys — See the World-Inside the Mind of Another Species,* University of Chicago Press, Chicago, 1990.
7. DeGrazia, D. D., *Taking Animals Seriously — Mental Life and Moral Status,* Cambridge University Press, 1996.
8. Dennett, D. C., *Consciousness Explained,* Little Brown, Philadelphia, 1991.
9. Griffin, D. R., *Animal Minds,* University of Chicago Press, 1992.
10. Mitchell, R. W. and Thompson, N. S., *Deception — Perspectives on Human and Nonhuman Deceit,* SUNY Press, Albany, NY, 1986.
11. Moussaieff, J. M., *When Elephants Weep — The Emotional Lives of Animals,* Dell Publishing, New York, 1995.
12. Mukerjee, M., "Trends in animal research," *Sci. Am.,* 276, 86, 1997.
13. Russell, W. M. S. and Burch, R. L., *The Principles of Humane Experimental Technique,* UFAW, Potters Bar, 1992.
14. Jackson, M. E., Sundquist, K. T., and Marks, S. C., Jr., "Use of bone cell cultures to study skeletal pathology," *Microsc. Res. Tech.,* 33, 232, 1996.
15. Otto, T. E., Nulend, J. K., Patka, P., Burger, E. H., and Haarman, H. J., "Effect of (poly)-L-lactic acid on the proliferation and differentiation of primary bone cells *in vitro*," *J. Biomed. Mater. Res.,* 32, 513, 1996.
16. Sun, J.-S., Tsuang, Y.-H., Liao, C.-J., Liu, H.-C., Hang, Y.-S., and Lin, F.-H., "The effects of calcium phosphate particles on the growth of osteoblasts," *J. Biomed. Mater. Res.,* 37, 324, 1997.
17. Glowacki, J., "Cellular reactions to bone-derived material," *Clin. Orthop.,* 324, 47, 1996.
18. Vinall, R. L., Gasser, B., and Richards, R. G., "Investigation of cell compatibility of titanium test surfaces to fibroblasts," *Injury,* 26 (Suppl. 1), S-A21, 1995.
19. Murakami, T., Murakami, H., Ramp, W. K., and Hanley, E. N., Jr., "Interaction of tobramycin and pH in cultured chick tibiae," *J. Orthop. Res.,* 14, 742, 1996.
20. Ramp, W. K., Lenz, L. G., and Kaysinger, K. K., "Medium pH modulates matrix, mineral, and energy metabolism in cultured chick bones and osteoblast-like cells," *Bone Miner.,* 24, 59, 1994.
21. Sato, K., Nohtomi, K., Shizume, K., Demura, H., Kanatani, H., et al., "17 beta-estradiol increases calcium content in fetal mouse parietal bones cultured in serum-free medium only at physiological concentrations," *Bone,* 19, 213, 1996.
22. Swain, L. D., Renthal, R. D., and Boyan, B. D., "Resolution of ion translocating proteolipid subclasses active in bacterial calcification," *J. Dent. Res.,* 68, 1094, 1989.
23. Hudson, M. C., Ramp, W. K., Nicholson, N. C., Williams, A. S., and Nousiainen, M. T., "Internalization of *Staphylococcus aureus* by cultured osteoblasts," *Microb. Pathogen.,* 19, 409, 1995.
24. Halloran, B. A. and Donachy, J. E., "Characterization of organic matrix macromolecules from the shells of the Antarctic scallop, *Adamussium colbecki*," *Comp. Biochem. Physiol. [B] Biochem. Mol. Biol.,* 111, 221, 1995.
25. Strus, J. and Compere, P., "Ultrastructural analysis of the integument during the moult cycle in *Ligia italica* (Crustacea, Isopoda)," *Pflugers Arch.,* 431 (Suppl. 2), R251, 1996.
26. Moore, M. S., Popovic, N. A., Daniel, J. N., Boyea, S. R., and Polly, D. W., Jr., "The effects of a wrist brace on injury patterns in experimentally produced distal radial fractures in a cadaveric model," *Am. J. Sports Med.,* 25, 394, 1997.

27. Smith, M. E., Cibischino, M., Langrana, N. A., Lee, C. K., and Parsons, J. R., "A biomechanical study of a cervical spine stabilization device: Roy-Camille plates," *Spine,* 22, 38, 1997.
28. Lintner, D. M., Dewitt, S. E., and Moseley, J. B., "Radiographic evaluation of native anterior cruciate ligament attachments and graft placement for reconstruction. A cadaveric study," *Am. J. Sports Med.,* 24, 72, 1996.
29. Walker, P. S. and Garg, A., "Range of motion in total knee arthroplasty. A computer analysis," *Clin. Orthop.,* 262, 227, 1991.
30. Robinson, R. P., Simonian, P. T., Gradisar, I. M., and Ching, R. P., "Joint motion and surface contact area related to component position in total hip arthroplasty," *J. Bone Joint Surg.,* 79B, 140, 1997.
31. Prendergast, P. J. and Huiskes, R., "The biomechanics of Wolff's law: recent advances," *Ir. J. Med. Sci.,* 164, 152, 1995.
32. McKellop, H. A. and Rostlund, T. V., "The wear behavior of ion-implanted Ti-6Al-4V against UHMW polyethylene," *J. Biomed. Mater. Res.,* 24, 1413, 1990.
33. Gruber, K., Denoth, J., Ruder, H., and Stussi, E., "Zur Mechanik der Gelenkbelastung," *Z. Orthop. Ihre Grensgeb.,* 129, 260, 1991.
34. Brand, D. A., Krag, M. H., Hausman, M. R., Trainor, K. F., Akelman, E., et al., "A patient registry for orthopaedic surgery," *Clin. Orthop.,* 252, 262, 1990.
35. Westberg, E. E., Mann, N. H., III, and Spengler, D. M., "Integrating and presenting clinical and treatment outcome data for cost-effective case management," *Comput. Biol. Med.,* 27, 31, 1997.
36. *Code of Federal Regulation, Title 9, Chapter 1, Subchapter A-Animal Welfare, Part 2*, U.S. Government Printing Office, Washington, DC, 1966.
37. *Public Health Service Policy on Humane Care and Use of Laboratory Animals,* NIH, 1986.
38. *Guide for the Care and Use of Laboratory Animals,* National Academy Press, 1996.
39. Council Directive of Nov. 24, 1986 on the approximation of laws, regulations, and administrative provisions of the member states regarding the protection of animals used for experimental and other scientific purposes. Commision of the European Communiites, Brussels, 1986.
40. *Explanatory Report on the European Convention for the Protection of Vertebrate Animals used for Experimental and other Scientific Purposes.* Committee of Ministers of the Council of Europe, Strasbourg, 1986.
41. "FELASA recommendations on the education and training of persons working with laboratory animals: categories A and C," *Lab. Anim.,* 29, 121, 1995.
42. Close, B., Banister, K., Baumans, V., Bernoth, E. M., Bromage, N., et al., "Recommendations for euthanasia of experimental animals: Part 2. DGXT of the European Commission," *Lab. Anim.,* 31, 1997.
43. Close, B., Banister, K., Baumans, V., Bernoth, E. M., Bromage, N., et al., "Recommendations for euthanasia of experimental animals: Part 1. DGXI of the European Commission," *Lab. Anim.,* 30, 293, 1996.
44. Rehbinder, C., Baneux, P., Forbes, D., van Herck, H., Nicklas, W., et al., "FELASA recommendations for the health monitoring of mouse, rat, hamster, gerbil, guinea pig and rabbit experimental units," *Lab. Anim.,* 30, 193, 1996.
45. Olfert, E. D., Cross, B. M., and McWilliam, A. A., *Guide to the Care and Use of Experimental Animals*, 2nd edition, Canadian Council on Animal Care, Ottawa, 1993.
46. Larkin, R. A. and Brooks, R. M., "Laboratory animal welfare around the globe: Australia: New South Wales," *Lab Animal,* 25, 24, 1996.
47. Nasto, B., "Laboratory animal welfare around the globe: Japan," *Lab Animal* , 25, 32, 1996.
48. *Good Laboratory Practice for Nonclinical Laboratory Studies,* Department of Health and Human Services, Washington, DC, 1992.

2 Experimental Design, Evaluation Methods, Data Analysis, Publication, and Research Ethics

Yuehui H. An and Todd D. Bell

CONTENTS

I. Question, Hypothesis and Experimental Purpose 16
II. Writing A Research Proposal 18
III. Experimental Design 19
    A. Number of Animals Required (Sample Size) 19
        1. Review of Similar Previous Studies 19
        2. Preliminary Data 19
        3. Animal Variance 20
        4. Evaluation Methods 20
    B. Randomization and Sampling Errors 20
    C. Controls 21
        1. Commonly Used Control Groups 22
        2. Unilateral and Bilateral Models 22
        3. Unicortical and Bicortical Plug Models 23
IV. Evaluation Methods 24
    A. Selection of Evaluation Methods 26
    B. Common Sources of Error 26
        1. Procedural Errors 26
        2. Systemic Errors of Testing Systems 27
        3. Data Collection Error 27
        4. Data Entry and Processing Error 27
        5. Personal Error 27
V. Data Analysis 27
    A. Descriptive Statistics 28
    B. One Sample Analysis 29
    C. Unpaired Comparisons 29
    D. Paired Comparisons 29
    E. Analysis of Variance (ANOVA Test) 29
    F. Correlation and Regression Analysis 30
    G. Nonparametric Data 31
VI. Publication 31
    A. Title Page 31
    B. Abstract 31
    C. Introduction 32
    D. Materials and Methods 32

  E. Results 32
  F. Discussion 33
  G. References 33
 VII. Research Ethics 33
  A. Ethical Use of Laboratory Animals 33
  B. Obligations of Researchers 33
  C. Authorship and Acknowledgment 34
  D. Intellectual Property 34
  E. Integrity 35
  F. Never Duplicate Publication 35
Acknowledgments 35
References 35

I. QUESTION, HYPOTHESIS AND EXPERIMENTAL PURPOSE

The first step in generating a research project is the formulation of a new idea or question to be investigated. Orthopaedic surgeons often formulate research ideas when they encounter problems during their routine diagnostic, treatment and follow-up procedures. Ph.D. researchers or others who work exclusively in the laboratory may derive research ideas from conversations with orthopaedic surgeons or by reading the biomedical literature. It is often difficult to determine the most appropriate manner in which to investigate the idea. Although surgeons are the original source of many research ideas, researchers are usually more informed on the biomedical literature. The literature is one of the most important tools in helping researchers to evaluate the originality and feasibility of testing a particular idea. Occasionally, what had seemed to be a "new idea" to a surgeon may already have been tested and published in the literature. So, whether the primary investigator is a surgeon or a laboratory researcher, a thorough search of the literature is necessary in order to assess the originality, necessity, and the feasibility of carrying out the project. A review of previous studies allows the researcher to refine and reformulate the hypotheses, designs, methods and tools employed previously. The new internet technology has made medical and research literature much more accessible. However, one must still go to a library and obtain the original papers because abstracts are too brief to give all of the necessary information.

  The hypothesis is an unproven theory which is tentatively put forward in order to be tested. It is a more specific, logical and scientific form of a thought or idea and consists of the main question or questions to be investigated. Several hypotheses may be involved in the verification of one idea.

  The experimental purpose is a statement of the general goal of the study. It states exactly what the investigator plans to accomplish and it must be based on established experimental designs and evaluation methods. It is impossible to conduct a good, relevant research project without a specific question, clear hypothesis, or a well-defined purpose.

  The following examples may help readers understand the meanings of and the relationships between the idea or question, the hypothesis, and the experimental purpose. Example 3 also demonstrates the logical progression of experimental projects which may arise from a single idea.

Example 1 Mechanical Symmetry of Rabbit Bones[1]

a. Brief Review and Questions Bilateral animal models are used extensively for evaluating biocompatibility and biofunctionality of biomaterials in bones. Implicit in such an approach is the assumption that, in the same animal, a pair of corresponding right and left bones have similar mechanical properties. Only a few published articles, however, verify this assumption. The rabbit is commonly used in orthopaedic animal research and the mechanical properties of rabbit bones

TABLE 1
Comparison of Long Bone Mechanical Properties between Normal Control, Contralateral Control, and Treatment Groups*

Bone	Treatment	Side	Strength (MPa)	Elastic modulus (GPa)
Femur	Normal control	Both femur	130 ± 5	13.6 ± 0.4
	Arthritis model	Left (control)	97 ± 21	8.3 ± 1.5
		Right (arthritis)	80 ± 16	7.1 ± 1.4
Tibia	Normal control	Both tibia	195 ± 6	21.3 ± 0.7
	Arthritis model	Left (control)	186 ± 26	15.9 ± 2.9
		Right (arthritis)	173 ± 19	14.9 ± 2.2
Humerus	Normal control	Both humerus	167 ± 5	13.3 ± 0.6
	Arthritis model	Left (control)	158 ± 25	11.9 ± 2.7
		Right (arthritis)	157 ± 26	12.4 ± 2.4

* The values were generated using a three-point bending test.[2,3]

are important parameters in many experimental designs. Thus, studies are needed to evaluate the mechanical symmetry of rabbit long bones. This will allow better experimental designs to be created.

b. Hypothesis and Purpose The hypothesis of this study is that the mechanical properties of rabbit long bones are symmetrical. The experimental purpose of the study is to verify the hypothesis by examining the bending and indentation parameters of rabbit long bones on each side. The results demonstrated no significant differences between the right and left femur, tibia, or humerus for any of the bending or indentation parameters (Table 1).

Example 2   Bone Ingrowth to Implant Surfaces in Osteopenic Bone[4]

a. Brief Review and Questions Total joint replacements are a mainstay of treatment for debilitating arthritis. Many of these procedures are performed on patients with osteopenic bone due to inflammatory arthritis or osteoporotic bone conditions. Further research is needed to determine whether the osteopenia has detrimental effects on bone ingrowth into the implant surfaces (Question 1). It is also known that under normal conditions bone ingrowth varies with different implant surfaces (this includes the material composition of the implant and the surface texture of the implant). Is this also true in osteopenic bone (Question 2)? Intra-articular injection of carrageenan has been shown, in previous studies, to induce an inflammatory arthropathy and peri-articular osteopenia in rabbits (Review for method selection). This provides a useful and appropriate model for the investigation.

b. Hypothesis and Purpose The hypotheses of this study are that bone ingrowth into an implant surface is different in osteopenic bone (Hypothesis 1) and that the ingrowth of osteopenic bone is altered by different implant surfaces (Hypothesis 2). The experimental purpose of this research project is to use a carrageenan-induced arthritis model and three different implant surface textures to evaluate the bone-implant interface and surrounding cancellous bone in the distal femur of adult rabbits. The contralateral femurs will serve as the treatment controls. They will have identical implants placed, but will not receive carrageenan injections. Both histomorphometric analysis and mechanical testing will be performed to determine the quality and quantity of bone ingrowth, as well as the mechanical strength of the interface. The results demonstrated that carrageenan-induced arthritis influences the quality of the adjacent bone and subsequently affects the bone ingrowth into different implant surfaces.

Example 3 Effects of Albumin Coating on Bacterial Adhesion and Implant Site Infection Rate

a. Study 1[5] One of the most devastating complications in the surgical treatment of arthritis is infection. It may lead to complete failure of a joint replacement, occasionally necessitating amputation of the extremity. It has been demonstrated that adhesion of bacteria to a biomaterial surface is the initial step in the development of prosthetic infection. This adhesion may occur during the surgical procedure (through the air or by direct contact) or after implantation (by direct seeding from the bloodstream). It is also known that serum proteins, such as albumin, have inhibitory effects on bacterial adhesion to biomaterial surfaces.

A hypothesis is formulated to address the question of whether or not albumin-coating will effectively inhibit bacterial adhesion to implant surfaces (Question 1). The hypothesis is that albumin coating will reduce the number of *Staphylococcus epidermidis* which adhere to a titanium surface (Hypothesis 1). The experimental purpose is to use a computerized epifluorescent bacterial counting method to quantify the number of bacteria adhering to albumin-coated and non-albumin coated titanium surfaces, *in vitro* (Purpose 1). The results demonstrated that serum albumin coating inhibits *S. epidermidis* adhesion to titanium surfaces by more than 95 percent.

b. Study 2[2] After demonstrating the inhibitory effects of albumin coating with respect to bacterial adhesion, a subsequent question arose. How long would the albumin coating persist under physiologic conditions (Question 2)? Titanium implant surfaces were coated with bovine serum albumin (BSA) using a cross-linking agent. The implants were then placed in a solution of phosphate buffered saline at 37°C with agitation to simulate a physiologic environment. The inhibitory effect on bacterial adhesion was then evaluated at different time periods. The results revealed that only 10% of the coated bovine serum albumin (BSA) decayed off the surface during a 20-day incubation period and the inhibitory effect of the albumin coating on bacterial adherence remained high (more than 85 percent) throughout the length of the experiment (20 days).

c. Study 3[6] The knowledge accumulated in the above studies established the potential effectiveness of albumin coating in preventing prosthetic infection. However, these were *in vitro* experiments. Could the results be replicated *in vivo* (Question 3)? The new hypothesis is that the cross-linked albumin coating will effectively reduce the frequency of prosthetic infection *in vivo* (Hypothesis 3). The purpose of this study is to use a rabbit model to evaluate the *in vivo* effects of a serum protein coating on the implant infection. The results of the experiment demonstrated a much lower infection rate (27%) in rabbits with albumin coated implants vs. those with uncoated implants (62%).

II. WRITING A RESEARCH PROPOSAL

The research proposal is a document which reflects the extensive investigation into the validity and originality of a particular idea and describes exactly how the experiment will be carried out. It normally consists of background information on the subject, the specific aims of the study, an experimental plan, the materials and methods which will be used, the significance of the project, and appropriate references.

The research proposal is essential for the following reasons: (1) it facilitates the formulation of a detailed experimental protocol which precisely defines the hypothesis, the purpose, the experimental plan, the evaluation methods, the potential difficulties and possible solutions, the expected results and the appropriate statistical methods for analyzing the data; (2) it provides written documentation that the idea is original and viable by referencing related work already completed on the subject; (3) it provides a tool for communicating with other researchers in the field, thus allowing further verification of the idea and possibly eliciting useful suggestions before the project is started; (4) it provides a detailed outline of the project, which serves to motivate all personnel involved, defining the role of each individual and the exact manner in which the goals of the project will be accomplished; and finally, (5) with certain modifications, it may serve as a grant application for gaining financial support from the local institution or other research foundations.

Details on how to plan a new project and how to write a research proposal can be found in the books by Ingle,[7] Leedy,[8] Hawkins and Sorgi,[9] and Manly,[10] or the review by Mendenhall.[11] Several important aspects in the development of a research proposal in orthopaedic animal research will be discussed, including determination of the number of animals (sample size) to be used, randomized sampling, designing control groups, and common experimental designs based on statistical principles.

III. EXPERIMENTAL DESIGN

A. Number of Animals Required (Sample Size)

The number of animals needed for a particular study can be determined based on information acquired from the literature or from the results of preliminary studies. The required number depends on the intrinsic variability among the animals being used, the consistency of the surgical procedure which will be performed, the accuracy of the evaluation methods, and the statistical techniques which will be used to analyze the data.

Statistically, sample size is related to power, effect size, and significance level.[12] It is important to realize that sample size calculations will always be approximate and that it is impossible to predict the exact outcome of any particular experiment.[13] Many competent researchers lack sufficient statistical training to determine the appropriate sample size, the validity of statistical principles employed in an experimental design, and the statistical methods for data analysis. This is particularly true when sophisticated analysis is required. Consultation with a trained statistician is strongly advised.

1. Review of Similar Previous Studies

It is very helpful if information from previous studies with similar designs is available to the researcher. According to the standard deviation or coefficient of variance, the required number of animals can easily be estimated. For example, the number of rats (precise specimens) needed to study fracture healing for one time period using one evaluation method (histomorphometry or mechanical pushout test) is 8–12; the number of rabbits (precise specimens) required to study bone ingrowth in a femoral condyle using one evaluation model is 8–10; and the number of rabbits needed for the evaluation of different infection rates in two or more implants is 10–15.[6,14]

These numbers may seem empirical. However, based on the belief that the majority of materials in the literature are statistically sound, one can assume that the numbers adapted from the literature were obtained by precise statistical analysis. A statistician should be consulted, however, to verify the accuracy of the estimation before the experiment is started or before a manuscript is submitted for publication.

2. Preliminary Data

If a suitable number cannot be determined after consulting other investigators and extensively searching the literature, a separate preliminary study should be designed. This pilot study should be designed such that only a few animals or specimens are evaluated. Usually one or two standardized evaluation methods may be used to determine the appropriate number of animals to be used in the project. For example, 6–8 animals or specimens can be used in a preliminary study with histomorphometrical analysis or mechanical testing as evaluation methods. Based on this preliminary study, the standard deviation (SD) of the mean (of a specified parameter, such as bone density, bone mechanical strength, or cartilage thickness), coefficient of variation, and mean difference among groups (effect size) can be determined.

Sample size is one parameter which helps to determine the power, effect size (or magnitude of the effect) and level of significance of a study.[12] The power of a study is the likelihood of rejecting the null hypothesis. An 80% level is generally viewed as adequate. Effect size is a measure of the difference among the groups. Cohen[15] defines a small effect as 0.2 of a standard deviation, a moderate effect as 0.5 of a SD, and a large effect as 0.8. It is more difficult to detect a small effect of the independent variable than it is to detect a large effect. So, if a small difference is expected between the control and treatment group, a relatively large sample size is necessary. The significance level is the probability of rejecting a true null hypothesis; it is often set at 0.05. In the book by Cohen,[15] both power and sample size tables can be found. When planning a study, the researcher should determine the desired power, acceptable significance level, and expected effect size and use these three parameters to determine the necessary sample size.[12]

3. Animal Variance

In order to reduce the inter-animal variance and thus limit the number of animals needed, the study group should be as homogenous as possible. Ideally, the animals should be of the same strain, sex, age, weight and similar serum antibody status if applicable. Several experimental animal strains can be obtained which, due to extensive inbreeding, have very similar or identical genetic makeups. This is the case for SD rats, NZW rabbits and some strains of minipigs. Dogs and cats, on the other hand, are relatively heterogeneous, with respect to strain, age and body weight. Therefore, in a similarly designed study employing cats and dogs, the required number will be larger than with the other animals.

Two strategies which can markedly reduce the total number of animals needed are the use of paired designs and the use of multiple specimens in a single animal. In a paired design, the contralateral limb in each animal acts as the control. For example, four pairs of cortical plugs (each pair consisting of an experimental plug and a control plug) can be implanted into the femurs of a single animal. Each of the experimental plugs can be compared to a control plug which was implanted in a similar location in the contralateral femur. This arrangement provides nearly identical environments for the experimental and control groups which increases the power of the study. Note that, when using a cortical plug model, unicortical implantation is recommended. The reasons for this will be discussed later in the chapter. If two implant surfaces need to be compared for one time period, four dogs are normally enough for histological analysis (two animals) and mechanical testing (two animals, pushout test in most cases). According to previous studies, 6–8 pairs of specimens are enough for histomorphometrical evaluation or mechanical testing (pushout test) using paired student t-test for data analysis.

4. Evaluation Methods

The number of animals needed also depends on the design of the experiment, the evaluation methods and the statistical test for data analysis. For qualitative methods, such as descriptive histology, 2–4 animals (or specimens) are enough. For quantitative analyses, such as histomorphometrical or biomechanical evaluations, analyzed with a paired student t-test or repeated measures ANOVA, the number of animals (or specimens) in each group should be at least 6–8. Analysis of data with an unpaired student t-test normally requires a sample size of 8–12. Cohen's book[15] provides tables that may be used to help to determine the appropriate sample size for a particular statistical test.

B. Randomization and Sampling Errors

Randomization involves dividing a population into two or more groups such that every individual has an equal chance of being placed into each group. Unless study subjects are completely homogenous, they should be randomized. This is essential before a valid conclusion can be drawn about a causal relationship between a treatment and the observed effect. This strategy attempts to equally distribute characteristics across experimental groups and thereby eliminate selection bias.[16]

Randomization of animals or specimens into different groups can be performed by using a random permutation table, throwing dice, drawing in a lottery or by using computer-generated random permutations. The latter is presented in detail by Martin et al.[17]

The random assignment of subjects to groups does not guarantee the equivalent distribution of all extraneous variables in the group. There must be a sufficient number of animals or specimens in each group for randomization to have a high probability of distributing the extraneous variables similarly across all groups. The distribution of identifiable variables, such as weight, age and sex should be evaluated after randomization is complete to determine whether they have been equally distributed.

When using multifactorial designs (e.g., three treatment groups and three time periods) or designs requiring a large number of animals, animal facilities may not be capable of housing all the animals at the same time. Thus the animals will need to be delivered on two or more occasions. All animals need to be randomized for study participation, but this situation is complicated by the potential differences among the animals delivered at different times (i.e., age, body weight, or susceptibility to disease). To eliminate this source of variation, all treatments should be equally represented in the animals from each delivery.

The basic objective of sampling is that a sample should be chosen to represent its population. An estimate of a population parameter that is determined from a random sample will generally differ from the true value to some extent. This difference is referred to as the sampling error. Sampling error reflects the inherent uncertainty of conclusions about a population based solely on information gained from sample data (a subset of the population). The magnitude of sampling error is a function of sample size. This is due to the large amount of inherent variation in estimates based on small samples as compared to the smaller inherent variation seen with large sample sizes. The amount of statistical uncertainty associated with a particular study can be expressed in the form of confidence intervals. These intervals are largely determined by the sample size. Intervals based on small samples are relatively wide, reflecting a relatively large sampling error. Intervals based on adequate sample sizes are more narrow, reflecting a smaller sampling error.

Two sampling procedures which are commonly used in biomedical research are simple random sampling and stratified sampling. A simple random sample is one that is drawn from a population (a larger animal group or tissue specimen) such that each element of the population has an equal probability of being included in the sample.[10,18] Theoretically, it allows the knowledge gained by close examination of the sample group to be extrapolated to the larger population. Simple random sampling is useful in many situations; however, it does have some disadvantages. One problem encountered with simple random sampling is that, by design, there is no control over how the sample items are distributed in the population. This becomes an issue in cases where it is necessary to have the sample items spread out evenly within the population. For example, when viewing histologic sections of a tissue with a heterogeneic or anisotropic structure (such as cancellous bone), it is important to view sample sections which are distributed relatively evenly throughout the region of interest. The larger population, in this case, would be the thousands of potential sections from a block of tissue specimen. Stratified sampling procedures are often used to aid in solving this problem.[10] In stratified sampling, the specimen is divided into a series of non-overlapping strata (for example, a 10-mm thick bone block might be divided into consecutive 1-mm thick bone layers) and specimens are then chosen from each stratum by simple random sampling.

C. Controls

There are several different types of control groups, including normal controls, treatment controls and time controls. They are designed to act as comparisons for the experimental groups. A normal control is composed of a group of animals which does not receive the experimental treatment. Ideally, all characteristics, such as the strain, age and body weight, should be consistent between the normal control and treatment groups. They should also have the same living conditions and feeding patterns. The objective is to vary only the factor or factors under investigation so that a valid comparison can be made between the two groups. Data obtained in the past or data from the literature can also serve as a normal control, but only if every factor involved is consistent with the current experiment.

1. Commonly Used Control Groups

For widely used species a considerable amount of normal control data can be obtained from the literature, such as the mechanical, histological, or biochemical characteristics of bone or cartilage. In bilateral experimental designs, the contralateral limb can be used as a treatment control but does not necessarily represent a normal control. This distinction is important. It is due to the fact that any abnormality created in the treated limb may potentially have indirect effects on the contralateral limb. An example of this is seen in a study by Frank et al.[19] They demonstrated the abnormalities created in the contralateral knee of rabbits after unilateral transection of the medial collateral ligament. It has been hypothesized that the injury causes a shift of the weight-bearing burden to the contralateral limb. Another example is the systemic effect which is seen in the treatment control limb of carrageenan induced knee arthritis in the rabbit model (Table 1). The inflammatory arthritis not only causes reduced mechanical strength and elastic modulus of the long bones of the treated limb but also affects the contralateral limb (though less severely) and even the front limbs. The literature shows that treatment of one limb creates changes in the contralateral limb in a variety of different tissues, including: bone, cartilage, ligament, and muscle.[19] The strength of a treatment control is that, by being a part of the same animal, it reduces the amount of unintended variation between the experimental and control groups. A contralateral (treatment) control is sufficient in many situations, such as: testing the effects of chemically induced arthritis or electrical stimulation on bone or cartilage tissue; comparing the efficacy of a new fracture fixation method to a standard method applied to the contralateral limb; comparing bone ingrowth into two different implant surfaces (one in each limb); and for examining the effects of immobilization on osseous and cartilaginous tissues.

Finally, many studies utilize a time control. In this type of design, each experimental subject serves as its own control. Measurements are made prior to administration of the experimental element and at consecutive intervals during the course of the study. Such would be the case in the study of a new drug which is intended to cause an increase in bone mineral density (BMD). Measurements of each individual's BMD might be made before the drug is given (baseline), and taken after 2, 6, and 12 months of treatment. At the conclusion of the trial comparisons can be made to the pretreatment levels. Alternatively, comparisons can be made between different subjects after various periods of treatment. For example, an examination of bone ingrowth into prosthetic implants may be undertaken. If a total of 30 experimental subjects are utilized, comparisons of the pushout strengths of the implants could be made after six, 12 and 24 weeks by sacrificing 10 of the subjects for each time period.

For some experiments, such as the first successful reconstruction of an amputated arm, controls are unnecessary or are impossible to design. There are also some situations which do not permit control of all variables. In these cases, the best available design should be applied with awareness of the design defects.

2. Unilateral and Bilateral Models

Bilateral animal models, also referred to as "paired designs," are used extensively (in 95% of the studies) for evaluating the biocompatibility and biofunctionality of foreign materials in bones.[20] In this type of design, one leg is used as the control side while the contralateral leg acts as the experimental side. One of the major advantages of using bilateral models is the more efficient (higher power for statistical analysis) comparison between control and experimental groups. The variation for any given long bone in different animals is greater than the variation between paired bones in the same animal.[2] Therefore, a "paired left-right" comparison in a single animal is a closer comparison than between two different animals. "Paired" designs are based on the assumption that certain properties between bones of the left and right limbs are symmetrical. It has been previously established that the structure and mechanical properties of certain human and animal long bones are symmetrical. If conditions are well controlled, the number of animals in each group can be reduced significantly. When using an animal for which symmetry has not been experimentally established, a different approach is necessary. In this case the best design is to randomly designate the control and experimental side of the animal. By doing this, a potentially existing asymmetry between the two sides can be eliminated. The control and experimental side can then be compared using a paired Student t-test.

In some cases bilateral models are not appropriate. When major procedures are performed, especially those involving a joint, they create some degree of disability. If using a bilateral model, this may cause ethically unacceptable disability to the experimental animal.[21] The situation is better tolerated if the animal has an untreated contralateral limb. This will allow the animal to heal over time and return to its previous level of function. In this type of situation, a unilateral model should be designed. Unlike human subjects, animals are not cooperative and unexpected complications occur. It is always wise to be a little conservative if potential risks have been foreseen. Common unilateral and bilateral animal models used in orthopaedic research are listed in Table 2, including published models and potential models.

3. Unicortical and Bicortical Plug Models

Special attention should be given to the use of transcortical plug models in dogs for studies of bone ingrowth into implants. Theoretically, a bicortical plug model can double the number of specimens for testing. By cutting the implant in the middle, one half can be used for histological examination and the other for mechanical testing. It is a good design if there are no major complications, such as fracture. However, the drill hole through two cortices reduces bone strength significantly, increasing the risk of fracture. This potential risk has been brought into serious consideration recently by our group. We recently had a catastrophic failure of experimental method resulting in two dogs being sacrificed at the beginning of a new project when femoral bicortical plug implantation led to bone fractures (Figure 1). The result was the loss of experimental animals without gaining significant scientific data. In addition, we had to deal with serious criticism and doubt from both the animal caretakers and the IACUC (Institutional Animal Care and Use Committee) although we have been qualified animal researchers for many years. A research group in Europe is known to have experienced a similar occurrence. On re-evaluation, the procedure may not be valid because it was only published by one research group and no details of the animals used (body weight, age) or animal casualties (any fractures due to the procedure?) were reported in any of the original articles.[22–25] Therefore, a quick investigation has been performed on the effect of drill holes on the strength of the canine femur. The results show that a unicortical, 5-mm diameter drill hole in a canine femur creates a 38% reduction in bending strength, while a bicortical drill hole of the same size causes a reduction of over 50%. These results are basically in accordance with those of others (Table 3). A 50% reduction in bone strength is cause for concern, and thus the safety of the bicortical bone plug model should be carefully considered. The risk of fracture probably could have been minimized by using larger dogs, smaller diameter drill holes in areas of thick bone (i.e. close to the proximal or distal metaphyseal areas) and restrictive caging of the animals (limiting the cage space to prevent potential animal jumping). Carefully designed, a bicortical model has several advantages. It reduces the number of animals which are needed for an experiment and also has a greater statistical power. On the other hand, a bilateral unicortical femoral plug model as described by Bobyn et al.[28] appears to be safe although it does not have the statistical power of a

TABLE 2
Unilateral and Bilateral Animal Models Can Be Used in Orthopaedic Research*

Category	Model	Location	Animals Used
Unilateral model	Joint replacement	Hip	Rabbit, dog, sheep, goat, monkey
		Knee	Rabbit, sheep, goat
	Transarticular osteotomies	Knee	Rabbit, dog, sheep, goat, monkey
	Osteomyelitis	Tibia, femur	Rabbit, dog
	Prosthetic infection	Diaphyseal tibia	Rabbit
		Diaphyseal femur	Dog, rabbit
		Femoral condyles	Rabbit
	Segmental bone defect	Femur	Rat, rabbit, dog
		Tibia	Dog, sheep
		Radius	Rabbit, dog, rat
	Bone lengthening	Tibia	Dog, sheep
Arthritis (osteo-, inflam., infectious)	Knee		Dog, rabbit
	Articular cartilage defect	Femoral condyles	Rabbit, dog
	Meniscal repair		Dog, rabbit, goat, sheep, monkey
	ACL reconstruction		Dog, rabbit, goat, sheep, monkey
	Knee collateral ligament repair	Med. collateral lig.	Rabbit, dog, goat, sheep, rat
	Tendon repair	Flexor tendon	Chicken, Rabbit, dog
	Nerve repair	Sciatic nerve	Mouse, rat, rabbit, monkey (median nerve)
Bilateral model	Fracture healing	Femur	Rat, rabbit, dog
		Tibia	Rat, rabbit, dog
		Fibulae	Rat, rabbit, dog
		Radius	Rat, rabbit, dog
	Bone defect repair	Femur	Rat, dog
		Radius	Rabbit, dog, rat
		Ulna	Rabbit, dog
Bone ingrowth to implant surfaces	Femoral condyles		Rabbit, dog, sheep, pig, rat, Guinea pig
	(plug, screw, intramedul. rod)	Upper tibia	Rabbit, dog, goat, sheep, pig, rat
		Diaphyseal femur	Dog,† sheep, rabbit, rat, pig, Guinea pig
		Diaphyseal tibia	Dog, sheep, goat, rabbit, rat, pig
		Diaphyseal humerus	Dog, rabbit, goat, sheep, pig
		Ulna	Turkey

* Including published models and potential models.
† Trans-bicortical plug is not recommended.

bicortical model. Our intent is to notify researchers who may be considering the use of bicortical models that the risk of fracture is significant.

IV. EVALUATION METHODS

Chapter 6 summarizes some of the more useful evaluation methods in orthopaedic research, including clinical evaluation, necropsy, morphological or structural analysis, biochemical evaluation, mechanical testing, and the use of specialized devices or equipment. The most commonly used methods in orthopaedic animal research are: clinical observation, radiography, macro-observation at necropsy, histological evaluation, and mechanical testing. More sophisticated methods, such as electron microscopy, CT, and MRI, have many advantages but are expensive and often unnecessary.

Experimental Design, Evaluation Methods, Data Analysis, Publication, and Research Ethics

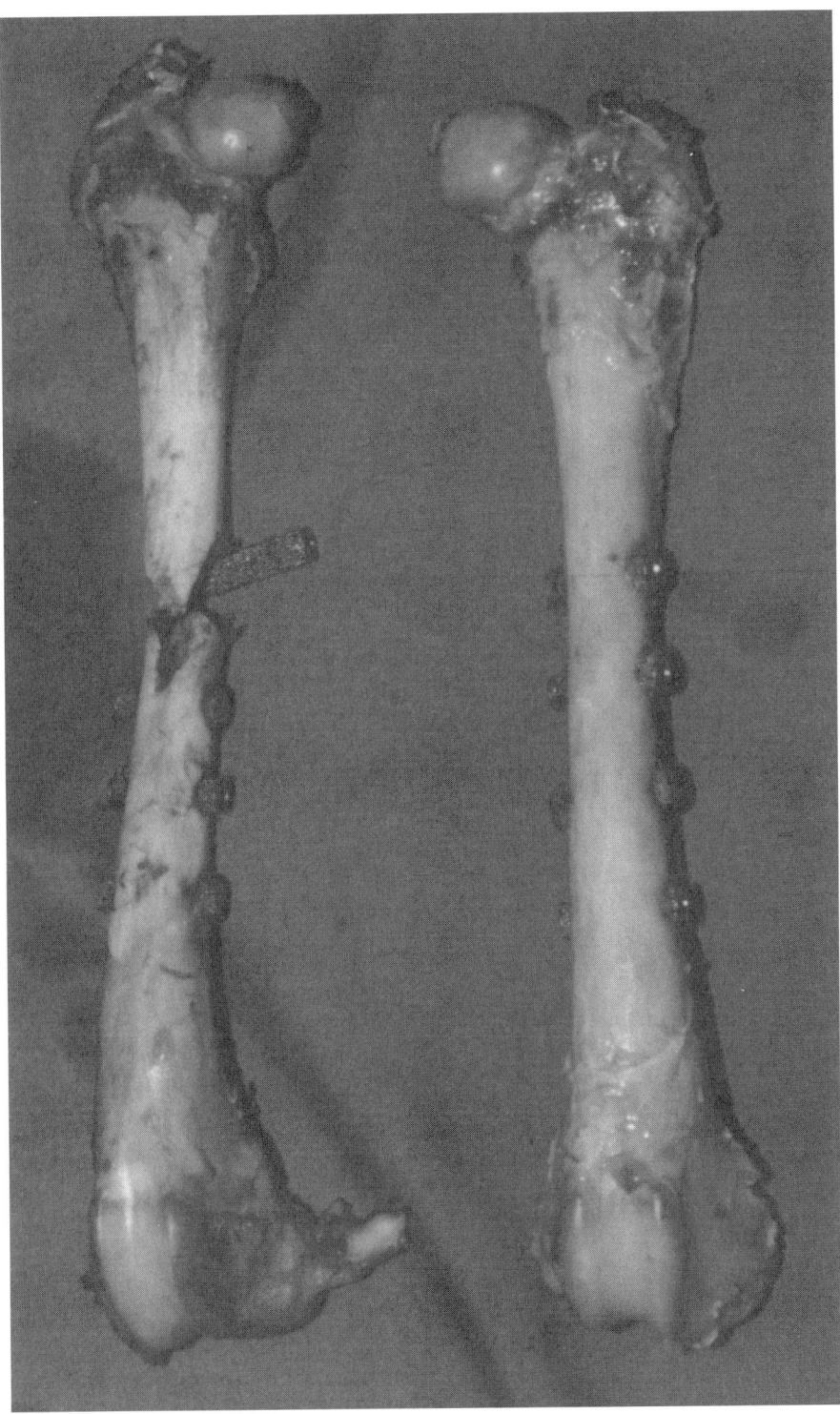

FIGURE 1. Photograph showing a bilateral canine femoral fracture at the sites of bicortical plugs.

TABLE 3
Effect of Transcortical Drill Hole on the Mechanical Strength of Diaphyseal Bone

Subject	Bone	Bone diam. (mm)	Hole diam. (mm)	Hole diam./ outer diam. of bone	Hole type	Mechanical test	Strength reduction (%)	1st author[Ref.]
Sheep	Femur	?	?	20%	Unicortical	Torsional	34%	Edgerton[26]
		19.6	?	50%	Unicortical	Torsional	60%	Hipp[27]
Dog	Femur	13.7	6	43.8%	Unicortical	4-pt. bending	37.6%	Authors' lab.
		13.7	6	43.8%	Bicortical	4-pt. bending	50.8%	Authors' lab.

A. SELECTION OF EVALUATION METHODS

The methods of evaluation should be carefully selected. Although many sophisticated and expensive testing methods have been described, the most useful techniques in animal studies are those which are simple, reliable, valid, efficient and economical. Most importantly, the test method must be valid, it must be capable of effectively measuring the parameter which it is intended to measure. The reliability of a method represents its ability to accurately and precisely measure the intended property. The accuracy of a measurement is the degree to which it represents the actual value of the measured parameter. Precision is a measure of the internal consistency of a measuring device, and represents the reproducibility of a measurement. A method can be extremely precise but inaccurate. For example, when evaluating a measuring caliper, if an object with a known thickness of 500 micrometers is measured and is consistently read as having a thickness of 600 micrometers, then the caliper is highly precise but is not accurate. An evaluation method must be both accurate and precise to be considered highly reliable.

The development of new evaluation methods requires evaluation of its accuracy. This involves comparison of the new test with a standard, which is usually a verified, well-established method. It is also very helpful to use at least two different techniques to demonstrate the accuracy of the results.

In order to test the precision of a method, repeated measurements of a known parameter can be made and then compared. Alternatively, the same procedure may be carried out by several different laboratory groups under standardized conditions, these results can then be compared.[7] The results can then be analyzed for precision using Student's t-test or by correlation analysis.

B. COMMON SOURCES OF ERROR

Besides sampling error, inaccurate data can be the result of improper procedures during the preparation and conduction of an experiment (inadequate surgical procedures or specimen preparation), systemic error of testing systems, data collection error and personal error.

1. Procedural Errors

Procedural errors are those which result from inadequate experimental procedures, such as rough or inaccurate surgical procedures, insufficient or harsh decalcification of osseous tissues, or inappropriate specimen preparation for mechanical testing. For example, when cylindrical implants are to be implanted, the size of drill holes should be consistent. Large deviations in drill hole size will influence the amount of bone ingrowth. Inadequate preparation of tissues for histomorphological study will cause changed morphology or deviated quantitative parameters. Bone surfaces which are not perpendicular to the long axis of the cylindrical implant will result in altered mechanical values and inaccurate load-displacement curves.

2. Systemic Errors of Testing Systems

Systemic error is the inherent error of a measuring device, due to imperfections in its manufacture or modifications. This inherent error creates a relatively consistent difference between the obtained value and the true value. The obtained value is always on one side of the true value, either greater or smaller. All measurements are subject to some degree of error. For example, a significant machine compliance of a mechanical testing system will cause underestimation of the specimen stiffness.[1] Systemic error of a testing system can be corrected by careful calibration and zeroing. Also, it is important to realize that the inherent error of an evaluation method is almost always greater than that described by its inventors.

3. Data Collection Error

Missing data in a research project may pose many problems for investigators. In animal research, one common reason for loss of data is the reduced number of subjects caused by unexpected animal complications or deaths. Careless labeling of specimen or incorrect recording of data during an experiment are also potential sources of error.

4. Data Entry and Processing Error

Errors may also occur during the process of raw data entry into the computer, especially when many numbers or many categories are involved. Careful data verification procedures must be implemented during and after the input process. Data should be entered into the computer and verified by two different people. The use of computer software designed specifically for data entry may also reduce the chance for error by incorporating range checks, skip patterns and rekey verification into the data entry process.

5. Personal Error

Like the systemic error of a machine, some degree of personal error is inevitable. Most personal errors lie outside of the individual's awareness. Even the most conscientious, well-trained and ethical investigator is not completely free of error. Having an evaluation repeated by a second trained individual provides necessary verification and may also help detect the presence of bias. This is especially important when laboratory personnel have knowledge of the treatment of the animal and the expected results. Whenever possible, observations should be made by trained personnel blind to the treatment applications in an additional attempt to protect against bias and deliberate fraud.[16]

Reliable operator performance is the result of careful selection and training of personnel and time-sampling checks of the reliability of performance.[7] It is the director's responsibility to test the reliability of each new technical assistant, preferably without the knowledge of the individual. The performance of any laboratory group has a greater chance of success when the team members place checks on each other and report all errors when they are detected or suspected.

V. DATA ANALYSIS

With the flourishing developments in computer technology, many software programs are available for statistical analysis.[29] State of the art computer software, highly tested for accuracy and efficiency, is now readily available for biomedical research and is cost effective. Statistica, StatView for Macintosh, and SAS (which is compatible with both Macintosh, Windows and mainframe systems) are some of the more efficient and widely accepted programs. Simpler programs such as StatWork and Excel are straight-forward and also adequate in many cases. Though these programs certainly make statistical analysis less laborious, they are not foolproof. If the appropriate tests are not properly applied, the results will be misleading. Errors may also occur in the analysis phase due to incorrect calculations or data transformations.

The keys to successful statistical analysis are to have an effective experimental plan and an appropriate method for analysis. It is important to understand the strengths, weaknesses and underlying assumptions for each test. Inadequate or ineffective use of statistical methods is seen often, even in major medical journals, such as *J. Bone Joint Surg. (Br),*[30] *Brit. Med. J.,*[31] and *J. Bone Joint Surg. (Am).*[32] Due to the complexity of applying statistical testing techniques, it is strongly recommended that a statistician be consulted during the process of experimental planning, data analysis, and manuscript preparation.

Differences between experimental and control data may serve to support or refute an experimental hypothesis. Thus it is essential to determine whether the observed differences between groups are statistically significant or simply due to chance variation. There are a number of different statistical methods which can be used to determine statistical significance; each is appropriate in specific situations.

A. Descriptive Statistics

A descriptive statistic is a number which is intended to summarize some characteristic of a larger source of data. Examples of commonly used descriptive statistical terms include: measures of central tendency (mean, median and mode) and measures of variability (minimum, maximum, range, standard deviation, standard error of the mean, and coefficient of variation).[33]

The mean is the sum of the observations divided by the number of observations. Each observation plays a part in the calculation of the mean, so difficulties can arise if there are outliers in the data. Often, outliers are important parts of the data. Occasionally, however, the correction or disposal of outlying data can be justified, but only when obvious error is found in data collection or transfer. For example, an unusually weak pushout strength value may be omitted if it is found that the implant was improperly positioned during surgery (positioned mainly in medullary canal but not in cancellous bone bed as it is supposed to be).

The median is the central value in a set of ordered numbers. Unlike the mean, every number in a set does not enter into the computation of the median. The strength of the median is that it is not sensitive to the extreme scores in a set of data. It is appropriately used for analysis of data with a skewed distribution, such as the "time to recurrence." The mode is simply the most frequently seen value in a set of data. It is most useful for demonstrating the clustering of values in a set of data.[12] The minimum value is the smallest value in a set of data and the maximum value is the largest one. The range is the absolute value of the difference between the minimum and the maximum values. These are simplified expressions of the variability of a set of numbers. Often they do not accurately represent the rest of the dataset, but they are easy to document and interpret.

A more precise calculation of variability is the standard deviation (SD). It represents the average deviation from the mean of the individual observations. SD is valid when the data falls in a normal or Gaussian distribution. The standard error of the mean (SEM) is a statistic that estimates the central tendency of the data or the variability of the sample mean in the population (when repeated samples of the same size are taken from the population). It is calculated by dividing SD of the observations by the square root of the number of observations. SD is often used when the purpose of an observation is to compare two means such as the means of mechanical strength of bone in control and treated groups. SEM is more useful when the goal is to determine the range of a normal value in a population such as white blood cell count in a human population or the mechanical strength of human bone.

The coefficient of variation (CV) is a unitless expression of variability. It is calculated by dividing the sample standard deviation by the sample mean. It is especially useful when comparing the variability of several different measurements, or when measurements are made in different units. Normally, if a CV value is under 0.2, it indicates a relatively low variability. Values less than 0.1 indicate a very small deviation or they may indicate a sufficiently large sample size. The limitation of CV arises when the mean is very small (close to 0). In this situation the CV value will be quite large (since the mean is in the denominator of the equation) and may not accurately represent the variation of the dataset.

B. One Sample Analysis

One sample analysis is used when it is necessary to compare the average of some parameter in an experimental group to a generally accepted or hypothesized parameter (rather than comparing it to a control group). The most commonly used method for one sample analysis is the one sample t-test. It compares an experimentally determined sample mean to a hypothesized mean for the population, and determines the likelihood that the observed differences between the two means occurred by chance alone. The results of such a t-test are reported in the form of a p-value, which ranges from 0 to 1. A p-value approaching one indicates a high likelihood that any difference between the sample mean and the hypothesized mean is purely due to chance and is not considered significant. On the other hand, a low p-value, such as 0.05 (a commonly used limit) or smaller, indicates that the difference between the two means is significant and is probably not due to chance alone.

C. Unpaired Comparisons

The unpaired student t-test is used to assess the difference between the average measurements of two separate groups. This might include comparisons between an experimental and a control group or between two different experimental groups. For example, one might want to determine whether a new drug causes an increase in bone mineral density. An experiment might be arranged such that an experimental group is treated with the new drug, while the control group receives a placebo. After a sufficient length of time, the BMD of each group is measured. The student t-test would be applied to statistically test for a difference in the means of the two groups. The results are expressed in p-values, as previously described.

D. Paired Comparisons

Paired comparisons serve to compare two sets of data which are naturally paired in some way. The most common scenario is a comparison of two separate sets of data taken from the same experimental group at different times or under different conditions. An example of this type of experiment would be the evaluation of the effectiveness of a muscle stimulator in improving strength. Measurements of strength might be taken before application of the muscle stimulator (baseline), after six weeks of use and after 12 weeks of use. Another typical situation is a study which utilizes a bilateral experimental model to evaluate the ingrowth of osteopenic bone into a prosthetic implant, and compares it to the "normal" bone ingrowth in the contralateral femur. Paired comparisons are generally more powerful then unpaired ones, since each experimental subject serves as its own control. The results can be analyzed using the paired t-test or, for two or more repeated measurements, the repeated measures ANOVA. The results are again reported as p-values.

E. Analysis of Variance (ANOVA Test)

ANOVA is a statistical tool which may be used to evaluate the impact of one or more nominal independent variables on a continuous dependent variable. A detailed discussion of the theoretical basis of ANOVA is beyond the scope of this book. In simplistic terms, ANOVA makes comparisons of the variations within and between experimental groups, and by doing this it determines whether group means differ from each other (Figure 2). Groups with similar means tend to produce a

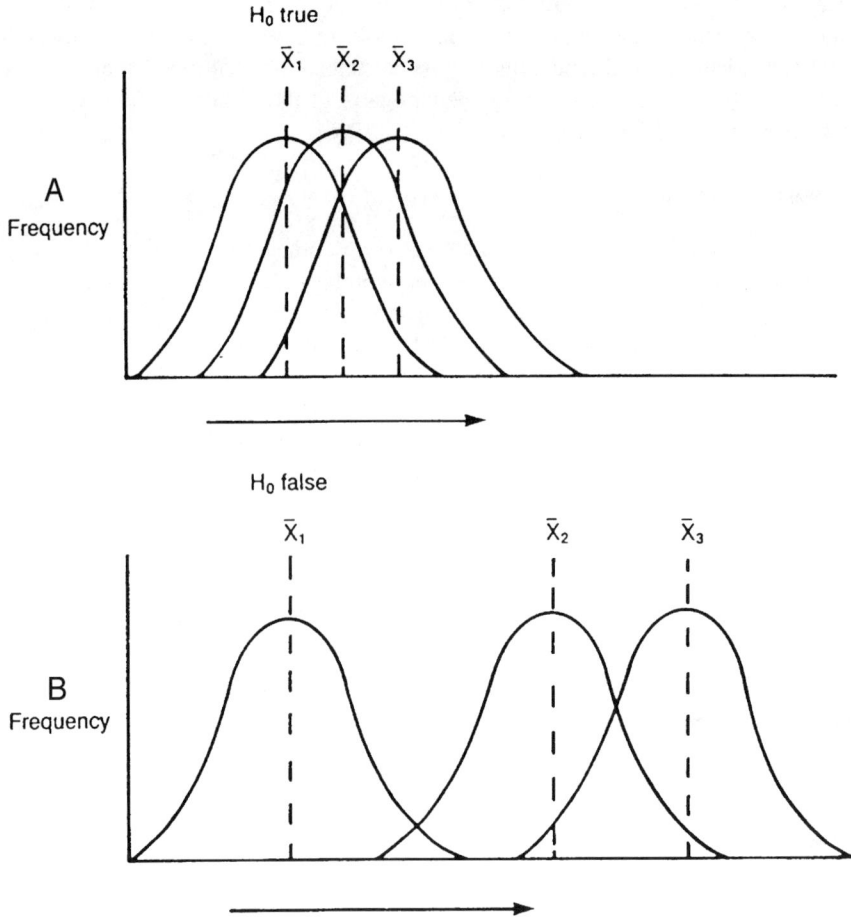

FIGURE 2. Figure A illustrates the relatively small amount of intergroup variance which is present when group means are not significantly different. In Figure B the group means are significantly different, consequently there is a greater intergroup variance. (From Reference 12, with permission)

relatively small intergroup variance (variance among all groups), whereas groups with significantly different means create a relatively large intergroup variance. One strength of ANOVA is its ability to simultaneously compare multiple independent variables. For example, a study might be undertaken to determine whether or not the region of the country in which someone lives (nominal independent variable) has an effect on their risk of osteoporosis (dependent variable). ANOVA allows one to compare the rates of osteoporosis of individuals in the Northeast, South, Midwest and West Coast regions to determine if there are differences based on location. If differences are observed using ANOVA, post-hoc methods such as Fisher's PLSD, Scheffé's F and Bonferroni/Dunn procedures can then be applied to examine differences between various pairs of data.

F. Correlation and Regression Analysis

Correlation analysis is a tool used to indicate the degree of linear relationship between two variables. This relationship is expressed as a correlation coefficient, with a value ranging from -1 to $+1$. A correlation coefficient of 0 means that no linear relationship exists between the two variables. A value of $+1$ indicates a strong positive correlation. This reflects a linear relationship between variables such that when one variable increases, the other increases to a proportional degree, and when one decreases, the other decreases to a proportional degree. A correlation coefficient of –1, while still indicating a linear relationship, connotes a strong negative correlation. When one variable increases the other decreases to a proportional degree, and vice versa. Correlation coefficients might be used to describe the relationship between bone density and mechanical strength.[3] As BMD increases, the mechanical strength of the bone increases to a proportional degree. This indicates a strong positive correlation between the two. Correlation analysis measures only the linear relationship between two variables and it should not be used if the relationship is nonlinear. A scattergram may be used to determine whether or not a relationship is linear.

Regression analysis is used to predict the value of a dependent variable based on the value of one or more independent variables. It can be applied only when: (1) there is a linear relationship between the independent and dependent variable, (2) all variables are continuous, and (3) all values are independent of each other. In reference to the previous example, regression analysis might be used to predict the mechanical strength of a bone (dependent variable) based on its BMD (independent variable).

G. Nonparametric Data

Nonparametric tests are statistical techniques which can be applied when there is no assumption of a Gaussion distribution within the population. Thus they are often referred to as distribution-free tests. While not as powerful or flexible as their parametric counterparts (i.e. t-test and ANOVA), nonparametric tests can be applied in situations where parametric tests are not valid. There are a number of nonparametric tests. Chi-square is the most commonly used and it requires nominal data. Others, which require ordinal data, include the Mann-Whitney U test, the Wilcoxon Signed Rank test, and the Kruskal Wallis test. The Mann-Whitney U test is used to test the hypothesis that the distributions of two different sets of data are equal. It is the nonparametric equivalent of an unpaired t-test. Similarly, the Wilcoxon Signed Rank test is the nonparametric equivalent of the paired t-test. The Kruskal Wallis test is analogous to ANOVA and tests whether two or more sets of data come from the same distribution or from different distributions. Further discussion of their uses may be found in many statistical texts such as those by Abacus Concepts, Inc.,[33] Munro et al.,[12] and Forthofer and Lee.[18]

VI. PUBLICATION

The functions of a scientific paper are to present new research methodologies, the results of a scientific project, or new theories derived from the results. How well these functions are carried out depends on their presentation in the manuscript. Detailed information on how to write a scientific article can be found in books by Leedy,[8] Hawkins and Sorgi,[9] Bay,[34] O'Connor,[35] Garb,[36] and Whimster,[37] or in papers by Morris[30] and the International Committee of Medical Journal Editors.[38] The following are some brief guidelines for writing a successful paper.

A. Title Page

The title page should include the following information: (1) the title, which should be both concise and informative; (2) full names and academic degrees of all authors; (3) the affiliation and location where the project was conducted; (4) information about the corresponding author, including address and contact numbers; and (5) a short running title (no more than 40 characters).

B. Abstract

An abstract is a condensed form of the paper. Included in it are the research hypothesis, the experimental purpose and plan, the methods, the results, and conclusions. Most importantly, an abstract should be self-contained and should give readers a concise overview of the entire paper.

To ensure a self-contained quality, no unspelled abbreviations should be used. The appropriate number of words for an abstract is 250–350. If the abstract is oversimplified it is virtually useless, while an excessively lengthy abstract makes it difficult for the reader to grasp the main points of the article. According to the requirements of publishers, 3–6 key words should be supplied at the end of the abstract to assist indexers in cross-referencing the article. This indexing information may also be published with the article.

C. Introduction

A clear statement of the question to be answered, the hypothesis, and the experimental purpose should be given in the introductory section. Often, a brief review of the history and a discussion of any recent developments in the field are given so that readers can easily understand the questions. For projects without a clear hypothesis, a clearly stated question and purpose are essential.

D. Materials and Methods

The materials used and experimental protocol should be discussed in detail. Information concerning the animal species used should be given, including strain, sex, age, body weight, number of animals (or specimens), housing conditions and state of health.[39] Failure to include this information makes comparisons between studies very difficult or impossible. Examples include comparisons of mechanical strength of bone or cartilage.

Any new materials, drugs, devices or methods used in the study should be presented in detail. The sources, including the names and locations of the companies, of any major supplies or materials used in the experiment should be indicated. For previously described methods, properly published references should be cited. Meeting abstracts, local proceeding papers, or papers from a rare book are inappropriate sources to reference since they are not widely available. After a reference for a method is cited, a brief description of the method should be given in order to fully relate the content of the study. A detailed description of the surgical approach, technique, surgical procedure and performance of it must be included in surgical studies. Also, the sampling procedure, sample size determination, the measures used for reducing errors, and the statistical methods used for data analysis should be included in this section of the manuscript.

E. Results

In the results section, a detailed description of the findings should be given in the form of text, tables, graphs, line drawings or photographs. A combination of the above-mentioned components can be used depending on the nature of the data. The precision of the measurement techniques and observed results (confidence intervals, standard deviation or standard error) should accompany the presentation.

For newly developed animal models, any casualties or complications associated with the procedure should be indicated. It is uncommon for a new model to be developed without any casualties or complications. These casualty reports are frequently left out of the papers. In a recent review of the methods sections of biomedical research papers, 30% failed to mention the total number of animals used, and animal deaths were not always recorded.[39]

Some reviewers have suggested that for a single set of data either a table or a graph should be used, but not both. The authors do not necessarily support this view. If the values in the data set are relative or standardized numbers, only a graph is capable of effectively showing the difference or trend. On the other hand, although a graph is good for catching the reader's eye, a table containing the actual values is more informative. A table of actual values allows easier comparison among studies from different groups. When only relative values are included in a study, as in the correlation between mechanical properties and structural or biochemical parameters, comparisons between studies are difficult if not impossible.

F. Discussion

The discussion section should include an explanation of the results obtained from the current study and the conclusions or proposed theories which are drawn from the results. Results may be discussed and compared with other particularly relevant studies and topics. Opinions on the validity of the data and the reliability of the testing methods should be given here. Existing data in the literature should be used for comparison and support for the new data or theory. If applicable, this section may also mention the of clinical relevance of the results. The discussion is normally completed with a summary of the major conclusions. In some journals the conclusions are given in a separate section. Most authors feel that conclusions should not be stated, especially in the abstract, unless they are completely justified by the results.[30] Lengthy discussion is usually unnecessary and makes it difficult for readers to focus on the salient points of the article. Most readers prefer a simple, straightforward, and relevant discussion section.

G. References

The references should be relevant, current, and complete. Excessive use of references is not appropriate for an original paper. More references should be used for review papers but they also should be representative, relevant, and current. Outdated references are difficult to find and frequently are useless. Older references should only be used to bring in original theories or methods or for supporting the theory that the authors are going to propose. Included in the reference should be: the authors (up to three or five), editors, title, journal name, book title, volume, edition, page number, publisher, location of the publisher and year of publication. Every reference needs to be verified with the original paper or by a MedLine search before the paper is submitted for publication.

VII. RESEARCH ETHICS

A. Ethical Use of Laboratory Animals

Due to public concern for animals' rights and welfare, Russell and Burch, two English scientists, introduced *The Principles of Humane Experimental Technique* in 1959.[40] They described the principle of the "three Rs": (1) replacement of animals in biomedical research, by using *in vitro* methods or by using animals that are phylogenetically more primitive; (2) reduction of the number of animals required; and (3) refinement of experimental methods to reduce the ethical costs in terms of painful or stressful procedures. Today, in North America, most European countries, and several other developed countries, specific regulations and guidelines have been established for the proper use and care of laboratory animals. Researchers should comply with all regulations and guidelines available in their own countries and should promote the development and perfection of them. See Chapters 1, 2, and 5 for more information on the ethical use and care of laboratory animals.

B. Obligations of Researchers

In a 1992 paper, Manly summarized the four obligations of researchers: obligations to the society, obligations to sponsors or employers, obligations to colleagues, and obligations to human subjects.[10] Although the obligations to society and human subjects are not directly applied to animal research, achievements in animal research will eventually serve human patients. Therefore, researchers are expected to deliver the highest quality of work to the public. The investigator should follow a particular code of ethics for the design, analysis and reporting of a study. In certain cases, project information should be kept confidential and not exposed to other researchers without permission from the principal investigator or sponsor. The sponsor should be informed of any significant changes in an ongoing project and any major problems should be reported, as well. The investigators should have a realistic idea of their level of ability and the probability that a project can be carried out successfully. Most importantly, investigators should be truthful about their research results and shortcomings so that other researchers can learn from them and ultimately promote continued development in the field. Fabricated experimental results and incorrect information are detrimental and may mislead others' studies, resulting in the waste of research funds and scientific impurity.

C. Authorship and Acknowledgment

The senior investigator is responsible for ensuring that the intellectual contribution (conception, design, improvement of methodology, analysis and interpretation of data, and the final approval of the paper) and the amount of effort (conducting experiment, collecting data, and drafting illustrations or methodology section) by co-authors are reflected in the authors' designation or sequence in the authors' list. It is very important to ensure that all authors desire to be included as such. Corresponding or senior authors are those who contribute original ideas, execute the experimental design, and draw conclusions or theories from the results. General supervision of the research group or participation solely in the acquisition of funds does not justify authorship. Corresponding or senior authors are also responsible for the correct spelling of co-authors' names. All authors should also be willing to assume public responsibility for its content, although for some journals only corresponding or senior authors are held accountable.

Acknowledgments of appreciation to those who helped on the project should be included. The names of the financial supporters should be also be clearly stated to demonstrate the gratitude of the authors and to prevent any potential conflicts of interest.

D. Intellectual Property

The term "intellectual property" was formally defined in the 1967 Stockholm Convention, which established the World Intellectual Property Organization. Article 2 (viii) of the convention provides that "intellectual property" shall include the rights relating to: literary, artistic and scientific works; performances of performing artists, phonograms and broadcasts; inventions in all fields of human endeavor; scientific discoveries; industrial designs; trademarks, service marks and commercial names and designations; protection against unfair competition; and all other rights resulting from intellectual activity in the industrial, scientific, literary or artistic fields. Protection of intellectual property is the subject of many international agreements.

The development of new techniques in the field of biotechnology has led to the emanation of such treaties as the *Treaty on Intellectual Property in Respect of Integrated Circuits*, signed in Washington in 1989 and the *Budapest Treaty on the International Recognition of the Deposit of Microorganisms for the Purposes of Patent Procedure*, signed in Budapest in 1977. Increases in the trade of intellectual property over the past decade has made it apparent that there is a compelling need for further international cooperation in this area.

Protection of international property rights is now the subject of much interest and debate on the local and world stage. International agreements concerning intellectual property protection afford inventors only limited patent protection in certain circumstances. It has become apparent that more comprehensive and formal agreements are needed in the area of biomedical research. When an article is published in the literature, the copyright is generally transferred from the author(s) to the publisher for a certain period of time.

Authors should respect the originality of others' theories and methods by referencing their original papers or by explicitly indicating the sources (such as personal communications or ideas from journal manuscripts or grant proposals reviewed by the author). It is unethical for investigators to take original ideas from others' manuscripts or grant proposals and present them as their own. Original theories or methods which impact science must be distinguished from common knowledge, and the former must always be cited.

E. INTEGRITY

All data with informative detail should be clearly presented to the readers. Although outright fraud is a rare occurrence, simply avoiding mention of important facts may also lead to disastrous consequences when others attempt to repeat an experiment or use the described procedures. A typical example is the lack of data on animal casualties during the process of developing a new model.[39] Even an occasional complication may indicate significant risks for a proposed model. The weaknesses of a project should be addressed openly. It is also important to discuss any factors which may have adversely affected the results. Mention of them should allow the problems to be avoided in the future.

F. NEVER DUPLICATE PUBLICATION

It is unethical to submit the same article to more than one peer-reviewed journal simultaneously and it is even worse to have the same paper published in two different journals. This type of unethical practice in publication can be easily recognized with the computerized literature searches (such as MedLine) which are now available. It is acceptable to submit the paper to a second journal only after it has been rejected by the first one. It is also unethical to publish a study based on a certain set of data, and then to publish a followup article without adding a significant amount of new data or information. Finally, one should not publish two different versions of the same paper in order to publish in more than one place.

ACKNOWLEDGMENTS

The authors want to thank Dr. Subrata Saha, Ph.D., Department of Bioengineering, Clemson University, for his assistance in the mechanical testing of canine femurs (The data are included in the section on Controls) and Linda E. A. Kanim for her time spent reviewing the chapter and her numerous suggestions for improvement.

REFERENCES

1. An, Y. H., Kang, Q., and Friedman, R. J., "The mechanical symmetry of rabbit long bones studied by bending and indentation test," *Am. J. Vet. Res.*, 57, 1786, 1996.
2. An, Y. H., Stuart, G. W., McDowell, S. J., McDaniel, S. E., Kang, Q., and Friedman, R. J., "Prevention of bacterial adherence to implant surfaces with a cross-linked albumin coating *in vitro*," *J. Orthop. Res.*, 14, 846, 1996.
3. Kang, Q., An, Y. H., and Friedman, R. J., "The mechanical properties and bone densities of canine cancellous bones," *J. Mater. Sci. Mater. Med.*, 9, 263, 1998.
4. Friedman, R. J., An, Y. H., Jiang, M., Draughn, R. A., and Bauer, T. W., "*In vivo* mechanical and histological evaluation of bone ingrowth and apposition to metal implants of different surface textures in the rabbit femur," *J. Orthop. Res.*, 14, 455, 1996.,
5. An, Y. H., Friedman, R. J., Draughn, R. A., Smith, E. A., Nicholson, J., and John, J. F., "Rapid quantification of staphylococci adhered to titanium surfaces using image analyzed epifluorescence microscopy," *J. Microbiol. Meth.*, 24, 29, 1995.
6. An, Y. H., Bradley, J., Powers, D. L., and Friedman, R. J., "An *in vivo* study of preventing prosthetic infection using crosslinked albumin coating," *J. Bone. Joint. Surg.*, 79B, 816, 1997.
7. Ingle, D. J., *Principles of Research in Biology and Medicine*, Lippincott, Philadelphia, 1958.
8. Leedy, P. D., *Practical Research,* Macmillan, New York, 1980.
9. Hawkins, C. and Sorgi, M., *Research, How to Plan, Speak and Write About It,* Springer-Verlag, Berlin, 1985.
10. Manly, B. J., *The Design and Analysis of Research Studies,* Cambridge University Press, Cambridge, 1992.

11. Mendenhall, H. V., "Surgical principles of biomaterial implantation," in *Encyclopedia Handbook of Biomaterials and Bioengineering*, Part 1, Vol. 1, Wise, D. L., et al., Eds., Dekker, New York, 1995, 3.
12. Munro, B. H., Jacobson, B. J., and Braitman, L. E., "Introduction to inferential statistics and hypothesis testing," in *Statistical Methods for Health Care Research*, 2nd edition, Munro, B. H. and Page, E. B., Eds., Lippincott, Philadelphia, 1993.
13. Matthews, D. E. and Farewell, V. T., *Using and Understanding Medical Statistics*, 3rd edition, Karger, Basel, 1996, Chapter 15.
14. Petty, W., Spanier, S., Shuster, J. J., and Silverthorne, C., "The influence of skeletal implants on incidence of infection," *J. Bone Joint Surg.*, 67A, 1236, 1985.
15. Cohen, J., *Statistical Power Analysis for the Behavioral Sciences*, Lawrence Erlbaum Associates, Hillsdale, NJ, 1987.
16. Vølund, A., "Experimental design and statistical evaluation," in *Handbook of Laboratory Animal Science*, Svendsen, P. and Hau, J., Eds., CRC Press, Boca Raton, FL, 1994, Chapter 15.
17. Martin, R. A., Daly, A., and DiFonzo, C. J., "Randomization of animals by computer program for toxicity studies," *J. Environ. Pathol. Toxicol. Oncol.*, 6, 143, 1986.
18. Forthofer, R. N. and Lee, E. S., *Introduction to Biostatistics. A Guide to Design, Analysis, and Discovery*, Academic Press, San Diego, 1995.
19. Frank, C. B., Loitz, B., Bray, R., Chimich, D., King, G., and Shrive, N., "Abnormality of the contralateral ligament after injuries of the medial collateral ligament. An experimental study in rabbits," *J. Bone. Joint. Surg.*, 76A, 403, 1994.
20. LaBerge, M. and Powers, D. L., "Scientific basis for bilateral animal models in orthopaedics," *J. Invest. Surg.*, 4, 109, 1991.
21. An, Y. H., Friedman, R. J., Latour, R. A., Draughn, R. A., and Powers, D. L., "Bioabsorbable screw fixation of osteotomies in the dog femur," *Clin. Orthop.*, 355, 300, 1998.
22. Anderson, R. C., Cook, S. D., Weinstein, A. M., and Haddad, R. J., Jr., "An evaluation of skeletal attachment to LTI pyrolytic carbon, porous titanium, and carbon-coated porous titanium implants," *Clin. Orthop.*, 182, 242, 1984.
23. Thomas, K. A., Cook, S. D., Renz, E. A., Anderson, R. C., Haddad, R. J., Jr., et al., "The effect of surface treatments on the interface mechanics of LTI pyrolytic carbon implants," *J. Biomed. Mater. Res.*, 19, 145, 1985.
24. Thomas, K. A. and Cook, S. D., "An evaluation of variables influencing implant fixation by direct bone apposition," *J. Biomed. Mater. Res.*, 19, 875, 1985.
25. Cook, S. D., Walsh, K. A., and Haddad, R. J., Jr., "Interface mechanics and bone growth into porous Co-Cr-Mo alloy implants," *Clin. Orthop.*, 193, 271, 1985.
26. Edgerton, B. C., An, K. N., and Morrey, B. F., "Torsional strength reduction due to cortical defects in bone," *J. Orthop. Res.*, 8, 851, 1990.
27. Hipp, J. A., Edgerton, B. C., An, K. N., and Hayes, W. C., "Structural consequences of transcortical holes in long bones loaded in torsion," *J. Biomech.*, 23, 1261, 1990.
28. Bobyn, J. D., Pilliar, R. M., Cameron, H. U., and Weatherly, G. C., "The optimum pore size for the fixation of porous-surfaced metal implants by the ingrowth of bone," *Clin. Orthop.*, 150, 263, 1980.
29. Brown, R. A. and Beck, J. S., "Medical statistics on microcomputers," *Br. Med. J.*, London, 1990.
30. Morris, R. W., "A statistical study of papers in the *Journal of Bone and Joint Surgery (Br) 1984*," *J. Bone Joint Surg.*, 70B, 242, 1988.
31. Gardner, M. J., Altman, D. G., Jones, D. R., and Machin, D., "Is the statistical assessment of papers submitted to the 'British Medical Journal' effective?" *Brit. Med. J.*, 286, 1485, 1993.
32. Senghas, R. E., "Statistics in the *Journal of Bone and Joint Surgery*: Suggestions for authors," *J. Bone Joint Surg.*, 74A, 319, 1992.
33. *StatView*, Abacus Concepts, Inc., Berkeley, CA, 1992.
34. Bay, R. A., *How to Write and Publish a Scientific Paper*, 3rd edition, Cambridge University Press, Cambridge, 1989.
35. O'Connor, M. and Woodford, E. P., *Writing Successfully in Science*, Harper Collins Academic, London, 1991.
36. Garb, J. L., *Understanding Medical Research*, Little & Brown, Boston, 1996.
37. Whimster, W. F., *Biomedical Research, How to Plan, Publish and Present It*, Springer, London, 1997.

38. International Committee Medical Journal Editors, "Uniform requirements for manuscripts submitted to biomedical journals," *J. Am. Med. Assoc.,* 269, 2282, 1993.
39. Smith, J. A., Birke, L., and Sadler, D., "Reporting animal use in scientific papers," *Lab. Anim.,* 31, 312, 1997.
40. Russel, W. M. S. and Burch, R. L., *The Principles of Humane Experimental Technique,* UFAW, Potters Bar, 1992.

3 Animal Selections in Orthopaedic Research

Yuehuei H. An and Richard J. Friedman

CONTENTS

I. Why Use Animal Models in Orthopaedic Research? 39
II. How to Choose Animal Models in Orthopaedic Research 40
    A. Ethical and General Considerations..40
        1. Ethics..40
        2. Availability 40
        3. Housing Requirements 41
        4. Ease of Handling...41
        5. Cost..42
        6. Susceptibility to Disease 42
    B. Available Background Data of the Animal..42
    C. Commonly Used Animals in Orthopaedic Research..42
        1. Rabbits...42
        2. Rats..44
        3. Dogs...45
        4. Goats..45
        5. Sheep 45
        6. Mice...46
        7. Primates...46
        8. Pigs 46
        9. Horses..46
        10. Other Animals 46
    D. Commonly Used Animal Models in Orthopaedic Research 46
    E. FDA Guidelines..50
III. Concluding Remarks 50
References 50

I. WHY USE ANIMAL MODELS IN ORTHOPAEDIC RESEARCH?

Animal models provide important knowledge of pathological conditions that can eventually lead to the development of more effective clinical treatment of diseases in both humans and animals. Research using animal models acts as the bridge between *in vitro* studies (such as studies of protein adsorption, cell adhesion and toxicological tests) and human clinical trials. It is an essential research tool which is applicable for many biomedical projects.

When making a decision on whether or not an animal model should be used for a particular study, the following questions should be raised: (1) Can the knowledge gained from the animal model be extrapolated to human conditions? (2) Are there any alternative methods which can verify the hypothesis? (3) Does the procedure cause extraordinary pain or disability to the animals? An *in vivo* study using an animal model may be considered necessary and appropriate if the procedure does not cause extraordinary pain or disability to the animals, the knowledge gained can be applied for the benefit of humans or animals, and there are no other *in vitro* alternatives.

For example, in the development of a new bioabsorbable material, measurements of its mechanical strength and degradation rate in a saline environment must be tested. Its biocompatibility in cell culture must also be assessed (Figure 1). After these *in vitro* studies, the material cannot immediately be tested in humans because it may have undetected toxic effects in human tissues. In the past, some products containing polylactide acid (PLA) and polyglycolic acid (PGA) have caused tissue lysis in human subjects leading to aseptic abscesses. Will the new absorbable material cause a similar problem? This question has to be answered before clinical trials are undertaken. Therefore, an animal model (using lower level animals such as rats) can be used to test the biocompatibility and degradation rate of the material, *in vivo*. Tests such as these are normally accomplished by subcutaneous and intraosseous implantation. If the experiment does not reveal any significant toxic effects, a second animal model (using higher level animals such as rabbits) can be used to evaluate the potential applications of the material, such as fixation of fractures or osteotomies or repair of ligaments or cartilage defects. At the same time, the process of material degradation and replacement by host tissues could investigated. If the material functions well enough for fracture fixation or ligament repair in all animal models, then consideration may be given to a cautious, well controlled human trial (Figure 1).[1]

II. HOW TO CHOOSE ANIMAL MODELS IN ORTHOPAEDIC RESEARCH

A. ETHICAL AND GENERAL CONSIDERATIONS

1. Ethics

Scientists today must realize that the use of animals in biomedical research has become an ethical issue. Although a small number of people feel that animal experiments should be stopped altogether, most have more moderate views. They recognize the need for animal research but feel that certain limitations and regulations are necessary. In general, the use of phylogenetically primitive animal species is more readily accepted by the public, so research with invertebrates is preferred over the use of vertebrates. Similarly, the use of pigs, rabbits and goats is preferred over the use of animals such as dogs or cats, to which humans are emotionally attached. Those studies which are directly applicable to human patients are also more well-accepted than studies with primarily basic scientific value. The amount of discomfort experienced by the animals is another important element. While it may be possible to justify experiments which cause slight harm or distress to the animals, those which create severe pain and suffering are usually not acceptable. A limited number of animals should also be used. Because of these factors, the choice of animal species in any given experiment is often the result of compromise. Ideally, scientific judgment should be the deciding factor; however, public concerns are also important and sometimes outweigh the scientific merit of a study. It is the authors' opinion that rats, rabbits, goats and sheep should be considered first in a new project. For most orthopaedic animal studies, there is no specific reason for dogs to be used when goats and sheep are also available.

2. Availability

Sometimes, the process of selecting animal species for a study could be further influenced by the availability of animals, boarding requirements, and cost. Generally speaking, securing animals of a particular species, strain, type, or age is very important in the selection of experiment animals. In some developed countries such as the United States, the availability of commonly used animals in orthopaedic research is not a problem. You can obtain rats, rabbits, dogs, goats or sheep any time during the year. You may want to contact the animal resources early enough if an animal

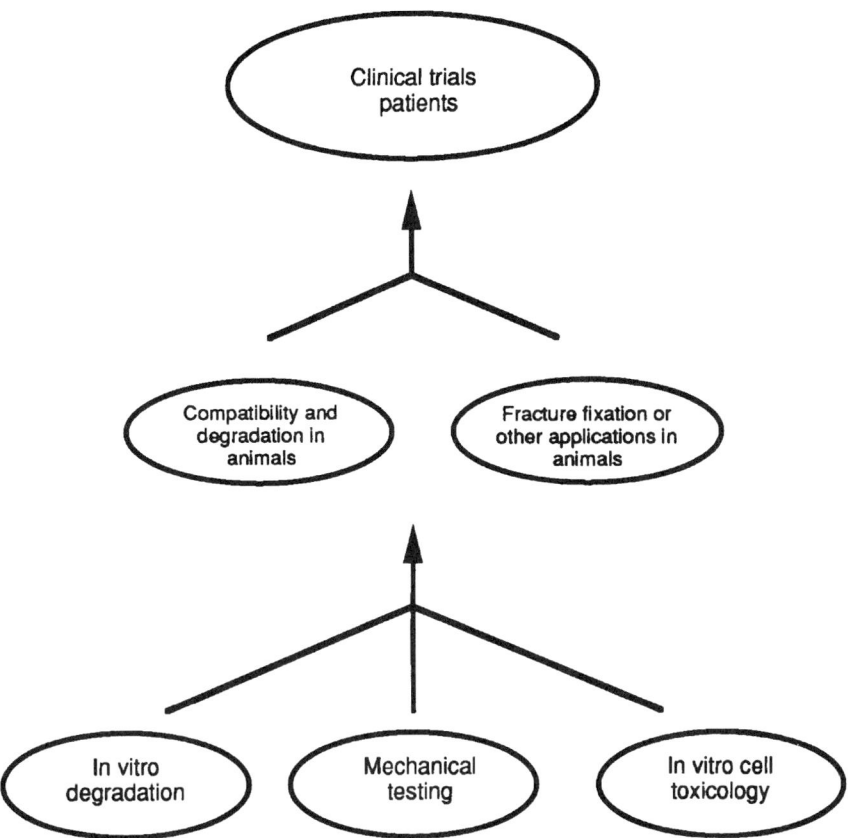

FIGURE 1. Schematic drawing of the steps in the development of a bioabsorbable material for clinical application.

species of a particular age or a rare species such as a primate is required. In many developing countries, availability is still a major problem because of the lack of standardized laboratory animal resources.

3. Housing Requirements

At certain institutions, the animal research facility may not have the capacity to house certain animal species. Then, the researcher may want to select another animal species if he or she does not want to travel to other facilities. More often than not, the latter is not really a better idea than choosing another species instead if it is acceptable for the project. Often, the researchers do not have a chance to inspect and treat the animals, leading to less reliable data. Also, doing a project at a commercial facility or one with a long traveling distance is very costly.

4. Ease of Handling

One reason more rats or rabbits than goats or sheep are selected in biomedical research is the easy handling of these animals. Large animals are good for certain investigations requiring large sized bones or other tissues, such as ACL reconstruction using ligament prosthesis designed for the human knee. Large animals create more difficulties than small animals for transportation, housing, peri-operative care, specimen handling, and disposal.

5. Cost

The costs of animal purchasing, transportation, quarantine time, housing, unexpected loss, surgical supply or services, or special equipment such as CT or MRI should not be decisive factors in selecting experimental animals; however, the costs of these items should be carefully calculated before embarking on the project (Table 1). Also, financial limitations such as the size of the grant or the definite amount of money available for the project are important factors in selection of animals. The bottom line is that you can use a cheaper animal as long as you believe the data derived from that animal species are valid to verify the hypothesis.

6. Susceptibility to Disease

Spontaneous diseases in animals during experiment can seriously compromise the experimental plan, confuse research data, and raise the cost (Table 1). Researchers should avoid using animal species known to have high incidence of a particular disease. For example, conventional rats are susceptible to chronic respiratory disease and conventional rabbits are likely to have *Pasteurella multocida* infections. Because of their low costs they are often used for acute procedures. If the study is chronic and long survival (more than a month), SPF (virus and antibody free) animals should selected.

B. AVAILABLE BACKGROUND DATA OF THE ANIMAL

A brief survey of the animal models used in the research papers published in *J. Orthop. Res.* in 1992 to 1996 was conducted to show the preference of researchers on the use of animal subjects in orthopaedic research (Figure 2, Table 2). The results showed that the most commonly selected animals are rabbits, rats, dogs, and goats. One should be aware of the limitations of this survey (one particular time period for only one journal).

First, the existing research data in the literature for an animal species, such as anatomy, physiological features, or the responses to drugs and surgical procedures, especially the information relevant to orthopaedic research, are very important for the selection of research animals. From Table 1, it is very clear there are much more background data for rabbits, dogs, rats, and goats than for the other animals, which reflect the value and suitability of these animals. Second, the existing data help researchers to repeat or extend previous work because they need the information of the species, strain, and basic biological data of the animal and the corresponding surgical procedures employed. Furthermore, the basic data help veterinary doctors and animal caretakers to better maintain the animal during the research period. These considerations have been influential in the fact of wide usage of only the few species shown in Figure 2. However, it is encouraging now that goats and mice are becoming valuable research subjects in the field. From an ethical point of view, it is easier for people to accept animals other than dogs for use in research.

C. COMMONLY USED ANIMALS IN ORTHOPAEDIC RESEARCH

1. Rabbits

The rabbit is one of the most commonly used animals in orthopaedic research. The above-mentioned *J. Orthop. Res.* survey showed that rabbits were used in 26% of the total animal studies (45 out of 171). Also based on the relevant publications (through the MedLine search), rabbits are more suitable for the studies of articular cartilage repair, ACL or medial collateral ligament (MCL) reconstruction,[2] fracture or osteotomy, bone ingrowth,[3] bone defect repair, steroid-induced osteonecrosis,[4,5] or osteoarthritis.[6] With the increasing interest on articular cartilage repair using tissue engineering techniques, the rabbit femoral articular defect model has become more and more popular in the last five years.[7-9] Although not shown clearly in the JOR survey (Table 2), rabbit

TABLE 1
Common Animal Species Used in Orthopaedic Research

Animal	Lifespan (Years)	Common disease	Average weight (kg)	Cost/purchase ($*/per animal)	Cost/housing ($/day/animal)	Housing requirements	Ease of handling
Rat (conventional)	2–3	Multiple infections	0.2–0.4	15–25/each	0.20	Nothing specific	Easy
Rat (VAF† or SPF‡)	2–3	None	0.2–0.4	15–25/each	0.20	Temperature/moisture controlled room	Easy
Rabbit (conventional)	7–8	Pasteurellosis	3–5	25–35/each	1–2	Temperature/moisture	OK, one person
Rabbit (SPF)	7–8	None	3–5	35–55/each	1–2	Temperature/moisture controlled room	OK, one person
Dog	10–12	Heart worm Canine hepatitis	10–25	300/each	4–5	Large cage	OK, two persons
Goat	10–15	Bacterial or viral pneumonia	50–70	250/each	4–5	Nothing specific	Difficult, two persons
Sheep	10–15	Bacterial or viral pneumonia	50–70	250/each	4–5	Nothing specific	Difficult, two persons

* $: US dollar; † VAF: Virus antigen free; ‡SPF: Specific pathogen free.

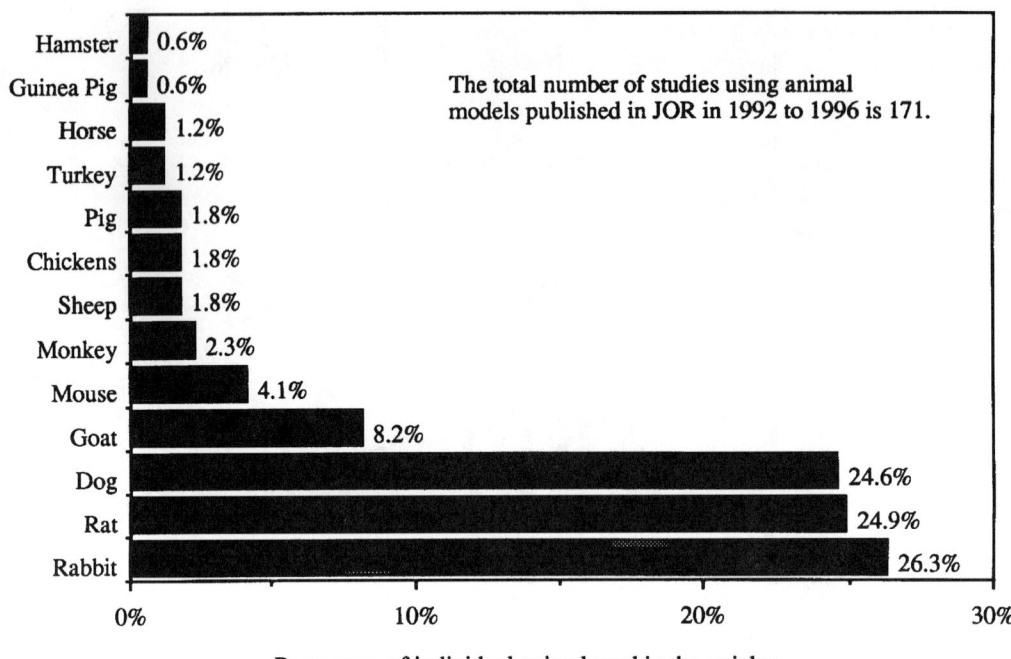

FIGURE 2. The animal subjects used in the research papers published in *J. Orthop. Res.* in 1992 to 1996.

bone defect models, such as radial[10,11] or cranial defects,[12,13] have been well established. Many researchers used rabbits for the studies of inflammatory arthritis and local osteopenia.[14–16] Rabbits are also successful subjects for the studies of osteomyelitis,[17] septic arthritis,[18] and foreign body or prosthetic infection[19] since they seem to be more susceptible to bacterial infection than other animals such as rats. In the authors' laboratory, the rabbit distal femur plug model has been used successfully for the studies of bone ingrowth to implants with different surface coatings or textures.[16,20,21]

Above all, the rabbit is a relatively high level vertebrate, having a good size which enables easy surgical operations and convenient radiographic, histologic and mechanical analysis and it is relatively economical compared to dogs. One shortcoming of using rabbits is they seem to be more fragile than rats and dogs. There are more unexpected deaths caused by complications or diseases in rabbits.

2. Rats

The rat, a rodent, has a mean healthy lifespan of 21-24 months. Bone elongation ceases by age 6–9 months, an age after which considerable useful experimental lifespan remains.[22] Although the rat is a lower level vertebrate compared to the rabbit, dog and goat, it is another most popular animal subject in orthopaedic research for its low cost and easy handling. It plays a very important role in the field, such as the studies of fracture,[23-25] bone defect repair,[26-28] bone ingrowth,[29,30] bone or joint infections,[31,32] osteoporosis,[22,33,34] osteomyelitis,[35] bone circulation,[36] prosthetic debris,[37,38] biocompatibility (subcutaneous or intramuscular implantation of biomaterials),[39,40] hemocompatibility of vascular prosthesis,[41] or nerve repair.[42,43] It seems to be a universal animal subject since rats have been used everywhere in orthopaedic research. Due to all of its advantages it is not unreasonable to consider the rat first as the animal subject for a new project.

TABLE 2
The Animal Models Used in the Research Papers Published in the *Journal of Orthopaedic Research* from 1992 to 1996 (Total: 171 Articles)

Tissues	Models	Rabbits	Dogs	Rats	Goats	Mice	Monkeys	Sheep	Chickens
Bone	Fracture/osteotomy	4	1	5	1		2		
	Bone defect/bone substitute	3	8	4					
	Subcutaneous bone substitute			5					
	Bone ingrowth	2	7	2	1		1		
	Bone biology/growth	1	2	3					
	Infections of bone, joint, or implant	3		3					
	Prosthetic debris			2					
	Limb lengthening/bone distraction	3			1				
	Electrical property/electrical stimulation	1	1						
	Bone tumor			2		2			
Joint	Osteoarthritis		4				2	1	
	Inflammatory arthritis	2	1	2		2			
Cartilage	Cartilage biology or repair	7	4	1	1	1			
Ligament	ACL reconstruction	6			9		1		
& tendon	MCL reconstruction	7		2					
	Tendon repair	2	3	3					3
	Soft tissue subcutaneous biocompatibility			1					
	Blood vessel/blood flow	2	5		1				
	Nerve repair	1		2		1			
	Muscle	1	1	2					
Others			5	2		1			
Total		45	42	41	14	7	4	3	3

3. Dogs

The dog is a higher level vertebrate. It has probably the closest *in vivo* condition to the human except for nonhuman primates. Dogs have played a dominant role in orthopaedic research. This animal has been used extensively and successfully in the studies of fracture,[44,45] bone defect repair,[46] bone ingrowth,[47,48] prosthetic infections,[49,50] osteomyelitis,[51] prosthetic debris,[52] osteonecrosis,[53] osteoporosis,[22,33] osteoarthritis,[6] ACL reconstruction,[2] meniscal repair and reconstruction,[54,55] cartilage biology or repair, spinal procedures, or bone vasculature and blood flow.

4. Goats

Goats are becoming popular and valid animal subjects in orthopaedic research in recent years, especially for their role in the research of ACL[2,56] and bone ingrowth.[57,58] They also have been used for the studies of biocompatibility,[59] joint replacement,[60] fracture,[61] limb lengthening,[62] meniscal repair,[63,64] or cartilage biology or repair.[65,66] It is a newcomer compared to dogs and lacks basic data. Due to their nature as a higher level vertebrate and non-pet status compared to dogs, goats will play a significant role in orthopaedic research in the future. The shortcomings of using goats include difficulty of handling and requirement for large housing space.

5. Sheep

The sheep is a large animal similar to a goat, which has both growing and adult skeletal phases, but the age of peak bone mass is not clear. There is less information about sheep as a subject for orthopaedic research compared to dogs and goats, but the available data are increasing in recent years. It is common to notice sheep as the selected animal in the literature such as bone defect repair,[67,68] osteoporosis (questionable),[69] meniscal repair,[70] cartilage defect repair,[71] osteoarthritis,[72] ACL reconstruction,[73,74] vessel graft,[75] nerve repair[76] or limb lengthening.[77]

6. Mice

The mouse is a small rodent, which has become more and more popular in skeletal research because of the ease with which its genome can be manipulated and investigated. Both regular mice or nude mice have been widely used in the studies of osteogenesis in bone or soft tissues,[78-80] chondrogenesis of potential materials in subcutaneous tissue,[81,82] osteoporosis, inflammatory arthritis, bone tumor, or nerve repair. The most common use of the mouse is the screening of potential substances for osteogenesis or chondrogenesis.

7. Primates

Nonhuman primates have both growing and adult skeletal phases, and the peak bone mass occurs at age 10–11 years. From a scientific point of view, primates are ideal for biomedical research because among animals they are the closest to humans. They have been used in the studies of osteoporosis,[22,33,34] bone ingrowth,[83] bone repair,[84] cartilage repair,[85] or osteoarthritis.[86] Because of the lack of availability and high cost, the use of primates has been limited to the research projects for which they are definitely necessary, such as the evaluation of new potential drugs for osteoporosis.[22,33,34]

8. Pigs

The pig has both growing and adult skeletal phases. It has been reported as a subject for the studies of effects of exercise on the skeleton,[87] osteoporosis (more work needed for its validity),[22] post-traumatic osteonecrosis of the femoral head,[88] fractures of cartilage and bone,[89] and bone ingrowth in the metaphyseal plug model.[90]

9. Horses

The horse is the largest animal used as an experimental subject. It has been used mainly for studies of cartilage or joint conditions since there is rich cartilage tissue in the horse. It has been reported as an experimental animal for articular defect repair,[91,92] experimental synovitis,[93] and septic arthritis.[94]

10. Other Animals

Guinea pigs are very popular for studies of osteoarthritis.[95,96] They are also used for a model of post-traumatic osteomyelitis.[97] Other animals are sometimes used, such as cats for osteoarticular transplantation,[98] hamsters for implant infection,[99] chickens for studies of scoliosis[100] or tendon repair,[101] or turkeys for bone remodeling.[102]

D. COMMONLY USED ANIMAL MODELS IN ORTHOPAEDIC RESEARCH

According to the literature and the authors' experience the suggested animal models for common orthopaedic studies are listed in Table 3. The animal models cited in the table are selected representatives with the authors' preference. Most of them are recent publications and are not necessarily the original models. Also, they are not inclusive.

TABLE 3
Author's Preferences of Animal Models for Common Orthopaedic Studies

System & Studies	Animal	Model	First author, year[Ref.]
Bone			
Osteogenesis in soft tissue	Nude mice/Rat	Subcutaneous pocket on the back	Hara 1996[80]/Nathan 1988[103]
	Rat/Rabbit	Intramuscular implantation	Aspenberg 1989[104]/Ripamonti 1996[105]
Fracture or osteotomy	Rat	Tibial/femoral fracture	Bak 1988[23]/An1994[24]/Bonnarens 1984[25]
	Rabbit	Radial/tibial fracture or osteotomy	Bruce 1987[106]/Braten 1990[107]/Murray 1996[108]
	Dog	Radial fracture	Bellah 1987[44]/Glennon 1994[45]
		Femoral condyle osteotomy	An 1997[109]
Bone defect or bone graft	Rabbit/Rat	Calvarial defect	Schmitz 1986[12]/Ashby 1996[13]/Kobayashi 1995[27]/Sweeney 1995[28]
	Rabbit/Dog/Rat	Radial defect	Niedzwiedzki1993[10]/Nyman 1995[11]/Johnson 1996[46]/Solheim 1992[26]
	Dog/Sheep/Rat	Large femoral defect	de Pablos 1994[110]/Ehrnberg 1993[111]/Puelacher 1996[112]
Bone ingrowth	Dog/Goat	Transcortical plug	Cameron 1976[47]/Bobyn 1980[113]/Verheyen[114]
		Femoral condyle plug	Søbelle1991[115]
	Rabbit	Distal femur plug (transverse)	Friedman 1995[20], 1996a[21], 1996b[16]
		Distal femur plug (longitudinal)	Feighan 1995[3]
	Rat	Tibial/femoral intramedullary nail	Ducheyne 1992[29]/Hazen 1993[30]
Bioabsorbable implant	Rat/Rabbit	Subcutaneous implantation	Pistner 1993[116]/Törmälä 1991[117]
	Rabbit	Femoral/tibial osteotomy	Bostman 1992[118]/Matsusue 1991[119]
	Dog	Femoral condyle osteotomy	An 1997[109]
		Femoral diaphyseal osteotomy	An 1997[109]/Leggon 1988[120]
Osteonecrosis	Dog	Femoral head/freezing	Malizos 1993[53]
	Rabbit	Femoral head, steroid/Shwatzman reaction	Matsui 1992[4]/Yamamoto 1995[5]
	Pig	Femoral head, posttraumatic	Seiler 1996[88]
	Rat	Femoral head, hypertensive	Kataoka 1992[121]
Osteoporosis or osteopenia	Rat/Primate	Ovariectomy	Cesnjaj 1991[33]/Thompson 1995[34]/Kimmel 1996[22]
Bone lengthening	Rabbit/Dog/Goat	Tibia	Nakamura 1995[122]/Danial 1994[123]/Lin 1996[124]
	Rat	Tibia (Ilizanov type)	Aronson 1997[125]
Inhibition of physeal growth	Rabbit	Upper tibial physeal stapling, ablation	Ross1997[126]
Prosthetic infection	Rabbit	Femoral head/knee replacement	Southwood 1985[127]/Blomgren 1981[128]/Belmatoug 1996[29]
		Femoral condyle metal plug	An 1997[19]/Isiklar 1996[130]
	Dog	Femoral/tibial intramedullary nailing	Petty 1985[49]/1988[50]/Isiklar 1993[131]

TABLE 3 (continued)
Author's Preferences of Animal Models for Common Orthopaedic Studies

System & Studies	Animal	Model	First author, year[Ref.]
Osteomyelitis	Rabbit/Rat	Tibia	Scheman 1941[132]/Mayberry-Carson 1992[17]/Rissing 1985[31]/Spagnolo 1993[35]
	Dog	Tibia/femur	Fitzgerald 1983[51]/Philipov 1995[133]
Joint			
Septic arthritis	Mouse/Rat	Multiarthritis, hematogenous *S. aureus*	Bremell 1991[134]/Bremell 1994[32]
	Rabbit	Knee, intraarticular injection of *S. aureus*	Stricker 1996[18]
Osteoarthritis	Dog/Rabbit	Knee	Moskowitz 1992[6]
	Guinea pig	Knee	de Bri 1995[95]/Whatson 1996[96]
Inflammatory arthritis	Rabbit	Knee injection of carrageenan/antigen	Bogoch 1988[14]/Kang 1997[15]/Beesley 1992[135]
	Rat	Collagen-induced arthritis	Knoerzer 1995[136]
Joint replacement	Dog	Total hip replacement	Kraemer 1995[52]/Finkelstein 1995[137]/Dowd 1995[138]
	Rabbit	Femoral head replacement	Southwood 1985[127]
Biomaterial debris	Rat	Sucutaneous air pouch	Gelb 1994[37]/Naidu 1996[139]
		Intraarticular administration	Lewis 1995[38]
	Rabbit	Proximal tibia implantation of debris	Gooman 1991[140]/1994[141]
	Dog	Total hip replacement	Bobyn 1995[142]/Kraemer 1995[52]
Cartilage			
Subcutaneous chondrogenesis	Nude mice	Subcutaneous pocket on the back	Paige 1996[81]/Fujisatao 1996[82]
Articular cartilage defect	Rabbit/Dog	Distal femoral joint defect	Freed 1994[7]/Kandel 1995[8]/Brittberg 1996[9]/Shortkroff 1996[143]
Meniscus repair and grafting	Rabbit/Dog/Goat	Repair	Huang 1989[144]/Arnoczky 1994[54]/Miller 1995[145]
	Rabbit/Dog/Sheep	Grafting or reconstruction	Messner 1994[146]/de Groot 1996[55]/Edwards 1996[70]
Ligament and tendon			
MCL repir	Rabbit/Rat	Complete or partial leceration of CML	Frank 1995[147]/Schreck 1995[148]/Litke 1994[149]/Batten 1996[150]
MCL reconstruction	Rabbit	Autograft or xenograft	Milthorpe 1994[151]/King 1995[152]
ACL reconstruction	Rabbit/Goat/Dog	ACL reconstruction	Arnoczky 1990[2]
Artificial ACL anchor	Goat	ACL reconstruction	Young 1995[56]
Tendon repair	Chicken/Rabbit	Laceration of flexor profundus tendons	Kugota 1996[101]

Soft tissues

Biomaterial biocompatibility	Rat/Mice	Intramuscular implant	McNamara 1981[40]/McGeachie1992[153]
	Rat/Dog	Subcutaneous implant	An 1997[39]/Picha 1996[154]/Hunt1996[155]/Campbell 1989[156]
Small diameter artificial vessel	Rat	Abdominal aorta	Bartels 1988[41]/Okoshi 1993[157]
	Rabbit	Carotid artery, infrarenal aorta	Cassel 1989[158]/Greisler 1991[159]
	Dog	Femoral artery/carotid artery/aorta	Matsuumoto 1973[160]/Sandusky 1995[161]/Kito 1996[162]
Nerve repair/graft	Rat/Rabbit	Ischiatic/sciatic nerve defect/grafting	Hall 1994[42]/Giandino 1995[43]/Amillo 1995[163]

Spine

Spinal instability	Rabbit/Pig	Lumbar spine	Stokes 1989[164]/Kaigle 1995[165]
Spinal fusion	Rabbit/Dog/Goat	Lumbar spine	Boden 1995[166]/Sandhu 1997[167]/Brantigan 1994[168]
Spinal cord compression	Mouse/Rabbit	Lumbar spine	Miyamoto 1992[169]/Saito 1992[170]
Verteb. column graft/replacement	Dog	Thoracic spine defects	Olson 1991[171]
Interverteb. disc graft/repacement	Dog	Lumbar spine	Frick 1994[172]/Matsuzaki 1996[173]
Scoliosis	Chicken/Rabbit	Pinealectomy/rib-growth stimulation	Kanemura 1997[100]/Agadir 1988[174]

E. FDA GUIDELINES

For certain animal models, the animal selection is relatively easy since there are guidelines established by professional authorities. In the United States, guidelines have been established by the FDA for certain preclinical and clinical studies such as preclinical testing of new drugs. For example, the recent *FDA Guidelines for Preclinical and Clinical Evaluation of Agents Used in the Treatment or Prevention of Postmenopausal Osteoporosis (1994)* delineate specific preclinical models to demonstrate the efficacy and safety of new, potential agents for osteoporosis therapy.[175] The guidelines recommend that agents be evaluated in two animal species, including a ovariectomized rodent model (rat) and a second non-rodent model (large remodeling animals, dogs, or preferably ovariectomized primates). Although there are some controversies about individual guidelines,[34] most of the guidelines are scientifically sound.

III. CONCLUDING REMARKS

The selection of an animal model sounds easy, but it is not. Many factors have to be considered before the decision is made, including the appropriateness of the model to the human condition, the available background data of the animal, the availability, housing requirement, cost, and the ease of experimental manipulation of the animal, as well as the ethical implications of using research animals.

REFERENCES

1. Navia, J. M., *Animal Models in Dental Research,* University of Alabama Press, 1977, 79.
2. Arnoczky, S. P., "Animal models for knee ligament research," in *Knee Ligaments: Structure, Function, Injury,* Daniel, D., Akeson, W. H., and O'Connor, J. J., Eds., Raven Press, New York, 1990, 401.
3. Feighan, J. E., Goldberg, V. M., Davy, D., Parr, J. A., and Stevenson, S., "The influence of surface-blasting on the incorporation of titanium-alloy implants in a rabbit intramedullary model," *J. Bone Joint Surg.,* 77A, 1380, 1995.
4. Matsui, M., Saito, S., Ohzono, K., Sugano, N., Saito, M., et al., "Experimental steroid-induced osteonecrosis in adult rabbits with hypersensitivity vasculitis," *Clin. Orthop.,* 277, 61, 1992.
5. Yamamoto, T., Hirano, K., Tsutsui, H., Sugioka, Y., and Sueishi, K., "Corticosteroid enhances the experimental induction of osteonecrosis in rabbits with Shwartzman reaction," *Clin. Orthop.,* 316, 235, 1995.
6. Moskowitz, R. W., "Experimental models of osteoarthritis," in *Osteoarthritis: Diagnosis and Medical/Surgical Management,* 2nd edition, Moskowitz, R. W., Ed., Saunders, Philadelphia, 1992, 213.
7. Freed, L. E., Grande, D. A., Nohria, A., Emmanural, J., Mikos, A. G., and Langer, R., "Joint resurfacing using chondrocytes and synthetic biodegradable polymers," *J. Biomed. Mater. Res.,* 28, 891, 1994.
8. Kandel, R. A., Chen, H., Clark, J., and Renlund, R., "Transplantation of cartilagenous tissue generated *in vitro* into articular joint defects," *Artif. Cells Blood Substit. Immobil. Biotechnol.,* 23, 565, 1995.
9. Brittberg, M., Nilsson. A., Lindahl, A., Ohlsson, C., and Peterson, L., "Rabbit articular cartilage defects treated with autologous cultured chondrocytes," *Clin. Orthop.,* 326, 270, 1996.
10. Niedzwiedzki, T., Dabrowski, Z., Miszta, H., and Pawlikowski, M., "Bone healing after bone marrow stromal cell transplantation to the bone defect," *Biomaterials,* 14, 115, 1993.
11. Nyman, R., Magnusson, M., Sennerby, L., Nyman, S., and Lundgren, D., "Membrane-guided bone regeneration. Segmental radius defects studied in the rabbit," *Acta Orthop. Scand.,* 66, 169, 1995.
12. Schmitz, J. P. and Hollinger, J. O., "The critical size defect as an experimental model for craniomandibulofacial nonunions," *Clin. Orthop.,* 205, 299, 1986.
13. Ashby, E. R., Rudkin, G. H., Ishida K., and Miller, T. A., "Evaluation of a novel osteogenic factor, bone cell stimulating substance, in a rabbit cranial defect model," *Plast. Reconstr. Surg.,* 98, 420, 1996.
14. Bogoch, E., Gschwend, N., Bogoch, B., Rahn B., and Perren, S., "Juxtaarticular bone loss in experimental inflammatory arthritis," *J. Orthop. Res.,* 6, 648, 1988.

15. Kang, Q., An, Y. H., Butehorn, H. F., and Friedman, R. J., "Morphological and mechanical study of the effects of experimentally induced inflammatory knee arthritis on rabbit long bones." *J. Mater. Sci. Mater. Med.,* 9, 463, 1998.
16. Friedman, R. J., An, Y. H., Jiang, M., Butehorn, H. F., Draughn, R. A., and Bauer, T. W., *Bone ingrowth into porous and HA coated titanium implants in experimental inflammatory arthritis.* Presented at the 5th World Biomaterials Congress, Toronto, May 28–June 2, 1996, 567.
17. Mayberry-Carson, K. J., Tober-Meyer, B., Gill, L. R., Lambe, D. W., Jr., and Costerton, J. W., "Osteomyelitis experimentally induced with *Bacteroides thetaiotaomicron* and *Staphylococcus epidermidis.* Influence of a foreign-body implant," *Clin. Orthop.,* 280, 289, 1992.
18. Stricker, S. J., Lozman, P. R., Makowski, A. L., and Gunja-Smith, Z., "Chondroprotective effect of betamethasone in lapine pyogenic arthritis," *J. Pediatr. Orthop.,* 16, 231, 1996.
19. An, Y. H., Bradley, J., Powers, D. L., and Friedman, R. J., "An *in vivo* study of preventing prosthetic infection using crosslinked albumin coating," *J. Bone Joint Surg.,* 79B, 816, 1997.
20. Friedman, R. J., Bauer, T. W., Garg, K., Jiang, M., An, Y. H., and Draughn, R. A., "Histological and mechanical comparison of hydroxyapatite coated cobalt-chrome and titanium implants," *J. Appl. Biomater.,* 6, 231, 1995.
21. Friedman, R. J., An, Y. H., Jiang, M., Draughn, R. A., and Bauer, T. W., "*In vivo* mechanical and histological evaluation of bone ingrowth and apposition to metal implants of different surface textures in the rabbit femur," *J. Orthop. Res.,* 14, 455, 1996.
22. Kimmel, D. B., "Animal models for *in vivo* experimentation in osteoporosis research," in *Osteoporosis,* Marcus, R., Feldman, D., and Kelsey, J., Eds., Academic Press, San Diego, 1996, 671.
23. Bak, B. and Andreassen, T. T., "Reduced energy absorption of healed fracture in the rat," *Acta Orthop. Scand.,* 59, 548, 1988.
24. An, Y. H., Parent, T., Friedman, R. J., and Draughn, R. A., "Production of a standard closed fracture in the rat tibia," *J. Orthop. Trauma,* 8, 111, 1994.
25. Bonnarens, F. and Einhorn. A., "Production of a standard closed fracture in laboratory animal bone," *J. Orthop. Res.,* 2, 97, 1984.
26. Solheim, E., Pinholt, E. M., Andersen, R., Bang, G., Sudmann, B., and Sudmann, E., "The effect of a composite of polyorthoester and demineralized bone on the healing of large segmental defects of the radius in rats," *J. Bone. Joint. Surg.,* 74A, 1456, 1992.
27. Kobayashi, K., Agrawal, K., Jackson, I. T., and Vega, J. B., "The effect of insulin-like growth factor 1 on craniofacial bone healing," *Plast. Reconstr. Surg.,* 97, 1129, 1996.
28. Sweeney, T. M., Opperman, L. A., Persing, J. A., and Ogle, R. C., "Repair of critical size rat calvarial defects using extracellular matrix protein gels," *J. Neurosurg.,* 83, 710, 1995.
29. Ducheyne, P., Ellis, L. Y., Pollack, S. R., Pienkowski, D., and Cuckler, J. M., "Field distributions in the rat tibia with and without a porous implant during electrical stimulation: a parametric modeling," *IEEE Trans. Biomed. Eng.,* 39, 1168, 1992.
30. Hazan, R., Brener, R., and Oron, U., "Bone growth to metal implants is regulated by their surface chemical properties," *Biomaterials,* 14, 570, 1993.
31. Rissing, J. P., Buxton, T. B., Weinstein, R. S., and Shockley, R. K., "Model of experimental chronic osteomyelitis in rats." *Infect. Immun.,* 47, 581, 1985.
32. Bremell, T., Lange, S., Holmdahl, R., Ryden, C., Hansson, G. K., and Tarkowski, A., "Immunopathological features of rat *Staphylococcus aureus* arthritis," *Infect. Immun.,* 62, 2234, 1994.
33. Cesnjaj, M., Stavljenic, A., and Vukicevic, S., "*In vivo* models in the study of osteopenias," *Eur. J. Clin. Chem. Clin. Biochem.,* 29, 211, 1991.
34. Thompson, D. D., Simmons, H. A., Piries, C. M., and Ke, H. Z., "FDA guidelines and animal models for osteoporosis," *Bone,* 17, 1255, 1995.
35. Spagnolo, N., Greco, F., Rossi, A., Ciolli, L., Teti, A., and Posteraro, P., "Chronic staphylococcal osteomyelitis: a new experimental rat model," *Infect. Immun.,* 61, 5225, 1993.
36. Grundnes, O. and Reikeras, O., "Acute effects of intramedullary reaming on bone blood flow in rats," *Acta Orthop. Scand.,* 64, 203, 1993.
37. Gelb, H., Schumacher, H. R., Cuckler, J., Ducheyne, P., and Baker, D. G., "*In vivo* inflammatory response to polymethylmethacrylate particulate debris: effect of size, morphology, and surface area," *J. Orthop. Res.,* 12, 83, 1994.

38. Lewis, C. G., Belniak, R. M., Plowman, M. C., et al., "Intraarticular carcinogenesis bioassays of CoCrMo and TiAlV alloys in rats," *J. Arthroplasty*, 10, 75, 1995.
39. An, Y. H., Trilokekar, N., Tian, Y., Friedman, R. J., von Recum, A., and Powers, D. L., *Evaluation of biocompatibility of sulfonated polylactide*, Presented at the Ann. Meet. Acad. Surg. Res., San Antonio, Texas, Sept. 4–6, 1997.
40. McNamara, A. and Williams, D. F., "The response to the intramuscular implantation of pure metals," *Biomaterials*, 2, 33, 1981.
41. Bartels, H. L. and van der Lei, B., "Small-calibre vascular grafting into the rat abdominal aorta with biodegradable prostheses," *Lab. Anim.*, 22, 122, 1988.
42. Hall, G. D. and van Way, C. W., III, "A comparison of nerve grafting and tissue expansion techniques in the rat," *Microsurgery*, 15, 439, 1994.
43. Giardino, R., Nicoli Aldini, N., Perego, G., Cella, G., Maltarello, M. C., et al., "Biological and synthetic conduits in peripheral nerve repair: a comparative experimental study," *Int. J. Artif. Organs*, 18, 225, 1995.
44. Bellah, J. R., "Use of a double hook plate for treatment of a distal radial fracture in a dog," *Vet. Surg.*, 16, 278, 1987.
45. Glennon, J. C., Flanders, J. A., Beck, K. A., Trotter, E. J., and Erb, H. N., "The effect of long-term bone plate application for fixation of radial fractures in dogs," *Vet. Surg.*, 23, 40, 1994.
46. Johnson, K. D., August, A., Sciadini, M. F., and Smith, C., "Evaluation of ground cortical autograft as a bone graft material in a new canine bilateral segmental long bone defect model," *J. Orthop. Trauma.*, 10, 28, 1996.
47. Cameron, H. U., Pilliar, R. M., and Macnab, I., "The rate of bone ingrowth into porous metal," *J. Biomed. Mater. Res.*, 10, 295, 1976.
49. Petty, W., Spanier, S., Shuster, J. J., and Silverthorne, C., "The influence of skeletal implants on incidence of infection," *J. Bone Joint Surg.*, 67A, 1236, 1985.
50. Petty, W., Spanier, S., and Shuster, J. J., "Prevention of infection after total joint replacement. Experiments with a canine model," *J. Bone Joint Surg.*, 70A, 536, 1988.
51. Fitzgerald, R. H., "Experimental osteomyelitis: Description of a canine model and the role of depot administration of antibiotics in the prevention and treatment of sepsis," *J. Bone Joint Surg.*, 65A, 371, 1983.
52. Kraemer, W. J., Maistrelli, G. L., Fornasier, V., Binnington, A., and Zhao, J. F., "Migration of polyethylene wear debris in hip arthroplasties: a canine model," *J. Appl. Biomater.*, 6, 225, 1995.
53. Malizos, K. N., Quarles, L. D., Seaber, A. V., Rizk, W. S., and Urbaniak, J. R., "An experimental canine model of osteonecrosis: characterization of the repair process," *J. Orthop. Res.*, 11, 350, 1993.
54. Arnoczky, S. P., Cooper. T. G., Stadelmaier, D. M., and Hannafin, J. A., "Magnetic resonance signals in healing menisci: an experimental study in dogs," *Arthroscopy*, 10, 552, 1994.
55. de Groot, J. H., de Vrijer, R., Pennings, A. J., Klompmaker, J., Veth, R. P., and Jansen, H. W., "Use of porous polyurethanes for meniscal reconstruction and meniscal prostheses," *Biomaterials*, 17, 163, 1996.
56. Young, F. A. and An, Y. H., "A new artificial ACL anchor," *Mater. Res. Soc. Symp. Proc.*, 394, 31, 1995.
57. Callahan, B. C., Lisecki, E. J., Banks, R. E., Dalton, J. E., Cook, S. D., and Wolff, J. D., "The effect of warfarin on the attachment of bone to hydroxyapatite-coated and uncoated porous implants," *J. Bone Joint Surg.*, 77A, 225, 1995.
58. Radder, A. M., Leenders, H., and van Blitterswijk, C. A., "Application of porous PEO/PBT copolymers for bone replacement," *J. Biomed. Mater. Res.*, 30, 341, 1996.
59. Gangjee, T., Colaizzo, R., and von Recum, A. F., "Species-related differences in percutaneous wound healing," *Ann. Biomed. Eng.*, 13, 451, 1985.
60. Oates, K. M., Barrera, D. L., Tucker, W. N., Chau, C. C., Bugbee, W. D., and Convery, F. R., "*In vivo* effect of pressurization of polymethyl methacrylate bone-cement. Biomechanical and histologic analysis," *J. Arthroplasty*, 10, 373, 1995.
61. Curtis, M. J., Brown, P. R., Dick, J. D., and Jinnah, R. H., "Contaminated fractures of the tibia: a comparison of treatment modalities in an animal model," *J. Orthop. Res.*, 13, 286, 1995.
62. Hu, J. and Zheng, X. Z., "Biomechanical study on bone healing in experimental limb lengthening," *Chung Hua Wai Ko Tsa Chih*, 32, 249, 1994.
63. Zhang, Z., Arnold, J. A., Williams, T., and McCann, B., "Repairs by trephination and suturing of longitudinal injuries in the avascular area of the meniscus in goats," *Am. J. Sports Med.*, 23, 35, 1995.

64. Port, J., Jackson, D. W., Lee, T. Q., and Simon, T. M., "Meniscal repair supplemented with exogenous fibrin clot and autogenous cultured marrow cells in the goat model," *Am. J. Sports Med.*, 24, 547, 1996.
65. Jackson, D. W., Halbrecht, J., Proctor, C., van Sickle, D. and Simon, T. M., "Assessment of donor cell and matrix survival in fresh articular cartilage allografts in a goat model," *J. Orthop. Res.*, 14, 255, 1996.
66. Butnariu-Ephrat, M., Robinson, D., Mendes, D. G., Halperin, N., and Nevo Z., "Resurfacing of goat articular cartilage by chondrocytes derived from bone marrow," *Clin. Orthop.*, 330, 234, 1996.
67. Hallfeldt, K. K., Stutzle, H., Puhlmann, M., Kessler, S., and Schweiberer, L., "Sterilization of partially demineralized bone matrix: the effects of different sterilization techniques on osteogenetic properties," *J. Surg. Res.*, 59, 614, 1995.
68. Viljanen, V. V., Gao, T. J., Lindholm, T. C., Lindholm, T. S., and Kommonen, B., "Xenogeneic moose (*Alces alces*) bone morphogenetic protein (mBMP)-induced repair of critical-size skull defects in sheep," *Int. J. Oral Maxillofac. Surg.*, 25, 217, 1996.
69. Newman, E., Turner, A. S., and Wark, J. D., "The potential of sheep for the study of osteopenia: current status and comparison with other animal models," *Bone,* 16 (Suppl), 277S, 1995.
70. Edwards, D. J., Whittle, S. L., Nissen, M. J., Cohen, B., Oakeshott, R. D., and Keene, G. C., "Radiographic changes in the knee after meniscal transplantation. An experimental study in a sheep model," *Am. J. Sports Med.*, 24, 222, 1996.
71. Homminga, G. N., Bulstra, S. K., Kuijer, R., and van der Linden, A. J., "Repair of sheep articular cartilage defects with a rabbit costal perichondrial graft," *Acta. Orthop. Scand.*, 62, 415, 1991.
72. Ghosh, P., Read, R., Armstrong, S., Wilson, D., Marshall, R., and McNair, P., "The effects of intraarticular administration of hyaluronan in a model of early osteoarthritis in sheep. I. Gait analysis and radiological and morphological studies," *Semin. Arthritis Rheum.*, 22 (Suppl 1), 18, 1993.
73. Bolton, C. W. and Bruchman, W. C., "The GORE-TEX expanded polytetrafluoroethylene prosthetic ligament," *Clin. Orthop.*, 196, 202, 1985.
74. Amis, A. A., Camburn, M., Kempson, S. A., Radford, W. J., and Stead, A. C., "Anterior cruciate ligament replacement with polyester fibre. A long-term study of tissue reactions and joint stability in sheep," *J. Bone Joint Surg.*, 74B, 605, 1992.
75. Simoni, G., Galleano, R., Civalleri, D., et al., "Pharmacological control of intimal hyperplasia in small diameter polytetrafluoroethylene grafts. An experimental study," *Int. Angiol.* 15, 50, 1996.
76. Lawson, G. M. and Glasby, M. A., "A comparison of immediate and delayed nerve repair using autologous freeze-thawed muscle grafts in a large animal model. The simple injury," *J. Hand Surg.*, 20B, 663, 1995.
77. Steen, H., Fjeld, T. O., Miller, J. A., and Ludvigsen, P., "Biomechanical factors in the metaphyseal- and diaphyseal-lengthening osteotomy. An experimental and theoretic analysis in the ovine tibia," *Clin. Orthop.*, 259, 282, 1990.
78. Sun, Y., Lu, Y., Hu, Y., Ma, F., and Chen, W., "Induction of osteogenesis by bovine platelet transforming growth factor-beta (TGF-beta) in adult mouse femur," *Chin. Med. J.*, 108, 914, 1995.
79. Miyazawa, K., Kawai, T. and Urist, M. R., "Bone morphogenetic protein-induced heterotopic bone in osteopetrosis," *Clin. Orthop.*, 324, 259, 1996.
80. Hara, A., Ikeda, T., Nomura, S., Yagita, H., Okumura, K., and Yamauchi, Y. "*In vivo* implantation of human osteosarcoma cells in nude mice induces bones with human-derived osteoblasts and mouse-derived osteocytes," *Lab. Invest.*, 75, 707, 1996.
81. Paige, K. T., Cima, L. G., Yaremchuk, M. J., Schloo, B. L., Vacanti, J. P., and Vacanti, C. A., "*De novo* cartilage generation using calcium alginate-chondrocyte constructs," *Plast. Reconstr. Surg.*, 97, 168, 1996.
82. Fujisato, T., Sajiki, T., Liu, Q., and Ikada, Y. "Effect of basic fibroblast growth factor on cartilage regeneration in chondrocyte-seeded collagen sponge scaffold," *Biomaterials,* 17, 155, 1996.
83. Shaw, J. A., Wilson, S. C., Bruno, A., and Paul, E. M., "Comparison of primate and canine models for bone ingrowth experimentation, with reference to the effect of ovarian function on bone ingrowth potential," *J. Orthop. Res.*, 12, 268, 1994.
84. Ripamonti, U., Bosch, C., van den Heever, B., Duneas, N., Melsen, B., and Ebner, R., "Limited chondro-osteogenesis by recombinant human transforming growth factor-beta 1 in calvarial defects of adult baboons (Papio ursinus)," *J. Bone Miner. Res.*, 11, 938, 1996.
85. Robinson. P. D., "Histologic study of articular cartilage repair in the marmoset condyle," *J. Oral Maxillofac. Surg.*, 51, 1088, 1993.

86. Carlson, C. S., Loeser, R. F., Jayo, M. J., Weaver, D. S., Adams, M. R., and Jerome, C. P., "Osteoarthritis in cynomolgus macaques: a primate model of naturally occurring disease," *J. Orthop. Res.*, 12, 331, 1994.
87. Raab, D. M., Crenshaw, T. D., Kimmel, D. B., and Smith, E. L., "A histomorphometric study of cortical bone activity during increased weight-bearing exercise," *J. Bone Miner. Res.*, 6, 741, 1991.
88. Seiler, J. G., III, Kregor, P. J., Conrad, E. U., III, and Swiontkowski, M. F., "Posttraumatic osteonecrosis in a swine model. Correlation of blood cell flux, MRI and histology," *Acta. Orthop. Scand.*, 67, 249, 1996.
89. Tomatsu, T., Imai, N., Takeuchi, N., Takahashi, K., and Kimura, N., "Experimentally produced fractures of articular cartilage and bone. The effects of shear forces on the pig knee," *J. Bone Joint Surg.*, 74A, 457, 1992.
90. Buser, D., Schenk, R. K., Steinemann, S., Fiorellini, J. P., Fox, C. H., and Stich, H., "Influence of surface characteristics on bone integration of titanium implants. A histomorphometric study in miniature pigs," *J. Biomed. Mater. Res.*, 25, 889, 1991.
91. Shamis, L. D., Bramlage, L. R., Gabel, A. A., and Weisbrode, S., "Effect of subchondral drilling on repair of partial-thickness cartilage defects of third carpal bones in horses," *Am. J. Vet. Res.*, 50, 290, 1989.
92. Pullin, J. G., Collier, M. A., Das, P., Smith, R. L., de Bault, L. E., et al., "Effects of holmium: YAG laser energy on cartilage metabolism, healing, and biochemical properties of lesional and perilesional tissue in a weight-bearing model," *Arthroscopy*, 12, 15, 1996.
93. Palmer, J. L., Bertone, A. L., Malemud, C. J., and Mansour, J., "Biochemical and biomechanical alterations in equine articular cartilage following an experimentally-induced synovitis," *Osteoarthritis Cartilage*, 4, 127, 1996.
94. Whithair, K. J., Bowersock, T. L., Blevins, W. E., Fessler, J. F., White, M. R., and van Sickle, D. C., "Regional limb perfusion for antibiotic treatment of experimentally induced septic arthritis," *Vet. Surg.*, 21, 367, 1992.
95. de Bri, E., Reinholt, F. P., and Svensson, O., "Primary osteoarthrosis in guinea pigs: a stereological study," *J. Orthop. Res.*, 13, 769, 1995.
96. Watson, P. J., Hall, L. D., Malcolm, A., and Tyler, J. A., "Degenerative joint disease in the guinea pig. Use of magnetic resonance imaging to monitor progression of bone pathology," *Arthritis Rheum.*, 39, 1327, 1996.
97. Passl, R., Muller, C., Zielinski, C. C., and Eibl, M. M., "A model of experimental post-traumatic osteomyelitis in guinea pigs," *J. Trauma*, 24, 323, 1984.
98. Henry, W. B., Jr., Schachar, N. S., Wadsworth, P. L., Castronovo, F. P., Jr., and Mankin, H. J., "Feline model for the study of frozen osteoarticular hemijoint transplantation: qualitative and quantitative assessment of bone healing," *Am. J. Vet. Res.*, 46, 1714, 1985.
99. Nakamoto, D. A., Haaga, J. R., Bove, P., Merritt, K., and Rowland, D. Y., "Use of fibrinolytic agents to coat wire implants to decrease infection. An animal model," *Invest. Radiol.*, 30, 341, 1995.
100. Kanemura, T., Kawakami, N., Deguchi, M., Mimatsu, K., and Iwata, H., "Natural course of experimental scoliosis in pinealectomized chickens," *Spine*, 22, 1563, 1997.
101. Kubota, H., Manske, P. R., Aoki, M., Pruitt, D. L., and Larson, B. J., "Effect of motion and tension on injured flexor tendons in chickens," *J. Hand. Surg.*, 21A, 456, 1996.
102. Rubin, C., Gross, T., Qin, Y. X., Fritton, S., Guilak, F., and McLeod, K., "Differentiation of the bone-tissue remodeling response to axial and torsional loading in the turkey ulna," *J. Bone Joint Surg.*, 78A, 1523, 1996.
103. Nathan, R. M., Bentz, H., Armstrong, R. M., Piez, K. A., Smestad, T. L. and Ellingsworth, L. R., "Osteogenesis in rats with an inductive bovine composite," *J. Orthop. Res.*, 6, 324, 1988.
104. Aspenberg, P. and Lohmander, L. S., "Fibroblast growth factor stimulates bone formation. Bone induction studied in rats," *Acta. Orthop. Scand.*, 60, 473, 1989.
105. Ripamonti, U., "Osteoinduction in porous hydroxyapatite implanted in heterotopic sites of different animal models," *Biomaterials*, 17, 31, 1996.
106. Bruce, G. K., Howlett., C. R., and Huckstep, R. L., "Effect of a static magnetic field on fracture healing in a rabbit radius. Preliminary results," *Clin. Orthop.*, 222, 300, 1987.
107. Braten, M., Terjesen, T., Svenningsen, S., and Kibsgaard, L., "Effects of medullary reaming on fracture healing. Tibial osteotomies in rabbits," *Acta Orthop. Scand.*, 61, 327, 1990.

108. Murray, D. W., Wilson-MacDonald, J., Morscher, E., Rahn, B. A., and Kaslin, M., "Bone growth and remodelling after fracture," *J. Bone Joint Surg.*, 78B, 42, 1996.
109. An, Y. H., Friedman, R. J., Latour, R. A., Draughn, R. A., and Powers, D. L., "Bioabsorbable screw fixation of osteotomies in the dog femur," *Clin. Orthop.*, 355, 300, 1998.
110. de Pablos, J., Barrios, C., Alfaro, C., and Canadell, J., "Large experimental segmental bone defects treated by bone transportation with monolateral external distractors," *Clin. Orthop.*, 298, 259, 1994.
111. Ehrnberg, A., De Pablos, J., Martinez-Lotti, G., Kreicbergs, A., and Nilsson, O., "Comparison of demineralized allogeneic bone matrix grafting (the Urist procedure) and the Ilizarov procedure in large diaphyseal defects in sheep," *J. Orthop. Res.*, 11, 438, 1993.
112. Puelacher, W. C., Vacanti, J. P., Ferraro, N. F., Schloo, B., and Vacanti, C. A., "Femoral shaft reconstruction using tissue-engineered growth of bone," *Int. J. Oral Maxillofac. Surg.*, 25, 223, 1996.
113. Bobyn, J. D., Pilliar, R. M., Cameron, H. U., and Weatherly, G. C., "The optimum pore size for the fixation of porous-surfaced metal implants by the ingrowth of bone," *Clin. Orthop.*, 150, 263, 980.
114. Verheyen, C. C., de Wijn, J. R., van Blitterswijk, C. A., de Groot, K., and Rozing, P. M., "Hydroxylapatite/poly(L-lactide) composites: an animal study on push-out strengths and interface histology," *J. Biomed. Mater. Res.*, 27, 433, 1993.
115. Søballe, K., Hansen, E. S., Rasmussen, H. B., Jorgensen, P. H., and Bünger, C., "Bone graft incorporation around titanium-alloy- and hydroxyapatite-coated implants in dogs," *Clin. Orthop.*, 272, 282, 1991.
116. Pistner, H., Bendix, D. R., Mühling, J., and Reuther, J. F., "Poly (L-lactide): a long-term degradation study *in vivo*. Part III. Analytical characterization," *Biomaterials*, 14, 291, 1993.
117. Törmälä, P., Vasenius, J., Böstman, O., Vainiopää, S., Laiho, J., et al., "Ultra-high-strength absorbable self-reinforced polyglicolide (SR-PGA) composite rods for internal fixation of bone fractures: *In vitro* and *in vivo* study," *J. Biomed. Mater. Res.*, 25, 1, 1991.
118. Böstman, O. M., Päivärinta, U., Partio, E., et al., "The tissue-implant interface during degradation of absorbable polyglycolide fracture fixation screws in the rabbit femur," *Clin. Orthop.*, 285, 263, 1992.
119. Matsusue, Y., Yamamuro, T., Yoshii, S., Oka, M., Ikada, Y., Hyon, S.-H., and Shikinami, Y., "Biodegradable screw fixation of rabbit tibia proximal osteotomies," *J. Appl. Biomater.*, 2, 1, 1991.
120. Leggon, R. E., Lindsey, R. W., and Panjabi, M. M., "Strength reduction and the effects of treatment of long bones with diaphyseal defects involving 50% of the cortex," *J. Orthop. Res.*, 6, 540, 1988.
121. Kataoka, Y., Hasegawa, Y., Iwata, H., Matsuda, T., Genda, E., et al., "Effect of hyperbaric oxygenation on femoral head osteonecrosis in spontaneously hypertensive rats," *Acta Orthop. Scand.*, 63, 527, 1992.
122. Nakamura, E., Mizuta, H., and Takagi, K., "Knee cartilage injury after tibial lengthening. Radiographic and histological studies in rabbits after 3–6 months," *Acta. Orthop. Scand.*, 66, 313, 1995.
123. Daniel, B. L., Waanders, N. A., Zhang, Y., et al., "The use of ultrasound mean acoustic attenuation to quantify bone formation during distraction osteogenesis performed by the Ilizarov method. Preliminary results in five dogs," *Invest. Radiol.*, 29, 933, 1994.
124. Lin, C. C; Huang, S. C., Liu, T. K., and Chapman, M. W., "Limb lengthening over an intramedullary nail. An animal study and clinical report," *Clin. Orthop.*, 330, 208, 1996.
125. Aronson, J., Shen, X. C., Skinner, R. A., Hogue, W. R., Badger, T. M., and Lumpkin, C. K., Jr., "A rat model of distraction osteogenesis," *J. Orthop. Res.*, 15, 221, 1997.
126. Ross, T. K. and Zionts, L. E., "Comparison of different methods used to inhibit physeal growth in a rabbit model," *Clin. Orthop.*, 340, 236, 1997.
127. Southwood. R. T., Rice, J. L., McDonald, P. J., Hakendorf, P. H., and Rozenbilds, M. A., "Infection in experimental hip arthroplasties," *J. Bone Joint Surg.*, 67B, 229, 1985.
128. Blomgren, G., "Hematogenous infection of total joint replacement. An experimental study in the rabbit," *Acta Orthop. Scand. Suppl.*, 187, 1, 1981.
129. Belmatoug, N., Cremieux, A. C., Bleton, R., et al., "A new model of experimental prosthetic joint infection due to methicillin-resistant *Staphylococcus aureus*: a microbiologic, histopathologic, and magnetic resonance imaging characterization," *J. Infect. Dis.*, 174, 414, 1996.
130. Isiklar, Z. U., Landon, G. C., Daruiche, R., Fernau, R., and Musher, D., "Penetration of vancomycin into biofilm: An *in vivo* orthopaedic implant infection model," *Trans. Orthop. Res. Soc.*, 18, 458, 1993.
131. Isiklar, Z. U., Darouiche, R. O., Landon, G. C., and Beck, T., "Efficacy of antibiotics alone for orthopaedic device related infections," *Clin. Orthop.*, 332, 184, 1996.

132. Scheman, L., Janota, M., and Lewin, P., "The production of experimental osteomyelitis," *J. Am. Med. Assoc.,* 117, 1525, 1941.
133. Philipov, J. P., Pascalev, M. D., Aminkov, B. Y., and Grosev, C. D., "Changes in serum carboxyterminal telopeptide of type I collagen in an experimental model of canine osteomyelitis," *Calcif. Tissue Int.,* 57, 152, 1995.
134. Bremell, T., Lange, S., Yacoub, A., Ryden, C., and Tarkowski, A., "Experimental *Staphylococcus aureus* arthritis in mice," *Infect. Immun.,* 59, 2615, 1991.
135. Beesley, J. E., Jessup, E., Pettipher, R., and Henderson, B., "Microbiochemical analysis of changes in proteoglycan and collagen in joint tissues during the development of antigen-induced arthritis in the rabbit," *Matrix,* 12, 189, 1992.
136. Knoerzer, D. B., Karr, R. W., Schwartz, B. D., and Mengle-Gaw, L. J., "Collagen-induced arthritis in the BB rat. Prevention of disease by treatment with CTLA-4-Ig," *J. Clin. Invest.,* 96, 987, 1995.
137. Finkelstein, J. A., Anderson, G. I., Waddell, J. P., Richards, R. R., and Humeniuk, B., "A Madreporic-surfaced femoral component in a canine total hip arthroplasty model: bone remodelling response at 6 and 24 months," *Can. J. Surg.,* 38, 501, 1995.
138. Dowd, J. E., Schwendeman, L. J., Macaulay, W., et al., "Aseptic loosening in uncemented total hip arthroplasty in a canine model," *Clin. Orthop.,* 319, 106, 1995.
139. Naidu, S. H., Beredjiklian, P., Adler, L., Bora, F. W., Jr., and Baker, D. G., "*In vivo* inflammatory response to silicone elastomer particulate debris," *J. Hand. Surg.,* 21A, 496, 1996.
140. Goodman, S. B., Fornasier, V. L., and Kei, J., "Quantitative comparison of the histological effects of particulate polymethylmethacrylate versus polyethylene in the rabbit tibia," *Arch. Orthop. Trauma Surg.,* 110, 123, 1991.
141. Goodman, S. B., "The effects of micromotion and particulate materials on tissue differentiation. Bone chamber studies in rabbits," *Acta. Orthop. Scand. Suppl.,* 258, 1, 1994.
142. Bobyn, J. D., Jacobs, J. J., Tanzer, M., et al., "The susceptibility of smooth implant surfaces to periimplant fibrosis and migration of polyethylene wear debris," *Clin. Orthop.,* 311, 21, 1995.
143. Shortkroff, S., Barone, L., Hsu, H. P., et al., "Healing of chondral and osteochondral defects in a canine model: the role of cultured chondrocytes in regeneration of articular cartilage," *Biomaterials,* 17, 147, 1996.
144. Huang, T. L., Lin, G. T., O'Connor, S., Chen, D. Y., and Barmada, R., "Healing potential of experimental meniscal tears in the rabbit. Preliminary results," *Clin. Orthop.,* 267, 299, 1991.
145. Miller, M. D., Ritchie, J. R., Gomez, B. A., Royster, R. M., and de Lee, J. C., "Meniscal repair. An experimental study in the goat," *Am. J. Sports Med.,* 23, 124, 1995.
146. Messner, K., "Meniscal substitution with a Teflon-periosteal composite graft: a rabbit experiment," *Biomaterials,* 15, 223, 1994.
147. Frank, C. B., Loitz, B. J., and Shrive, N. G., "Injury location affects ligament healing. A morphologic and mechanical study of the healing rabbit medial collateral ligament," *Acta Orthop. Scand.,* 66, 455, 1995.
148. Schreck, P. J., Kitabayashi, L. R., Amiel, D., Akeson, W. H., and Woods, V. L., Jr., "Integrin display increases in the wounded rabbit medial collateral ligament but not the wounded anterior cruciate ligament," *J. Orthop. Res.,* 13, 174, 1995.
149. Litke, D. S. and Dahners, L. E., "Effects of different levels of direct current on early ligament healing in a rat model," *J. Orthop. Res.,* 12, 683, 1994.
150. Batten, M. L., Hansen, J. C., and Dahners, L. E., "Influence of dosage and timing of application of platelet-derived growth factor on early healing of the rat medial collateral ligament," *J. Orthop. Res.,* 14, 736, 1996.
151. Milthorpe, B. K., "Xenografts for tendon and ligament repair," *Biomaterials,* 15, 745, 1994.
152. King, G. J., Edwards, P., Brant, R. F., Shrive, N. G., and Frank, C. B., "Intraoperative graft tensioning alters viscoelastic but not failure behaviours of rabbit medial collateral ligament autografts," *J. Orthop. Res.,* 13, 915, 1995.
153. McGeachie, J., Smith, E., Roberts, P., and Grounds, M., "Reaction of skeletal muscle to small implants of titanium or stainless steel: a quantitative histological and autoradiographic study," *Biomaterials,* 13, 562, 1992.
154. Picha, G. J. and Drake, R. F., "Pillared-surface microstructure and soft-tissue implants: effect of implant site and fixation," *J. Biomed. Mater. Res.,* 30, 305, 1996.

155. Hunt, J. A., Flanagan, B. F., McLaughlin, P. J., Strickland, I., and Williams, D. F., "Effect of biomaterial surface charge on the inflammatory response: evaluation of cellular infiltration and TNF alpha production," *J. Biomed. Mater. Res.,* 31, 139, 1996.
156. Campbell, C. E. and von Recum A. F., "Microtopography and soft tissue response," *J. Invest. Surg.,* 2, 51, 1989.
157. Okoshi, T., Soldani, G., Goddard, M., and Galletti, P. M., "Very small-diameter polyurethane vascular prostheses with rapid endothelialization for coronary artery bypass grafting," *J. Thorac. Cardiovasc. Surg.,* 105, 791, 1993.
158. Cassel, W. S., Mason, R. A. Campbell, R., Newton, G. B., Hui, J. C., and Giron, F., "An animal model for small-diameter arterial grafts," *J. Invest. Surg.,* 2, 181, 1989.
159. Greisler, H. P., Cabusao, E. B., Lam, T. M., Murchan, P. M., Ellinger, J., and Kim, D. U., "Kinetics of collagen deposition within bioresorbable and nonresorbable vascular prostheses," *ASAIO Transections,* 37, M472, 1991.
160. Matsumoto, H., Hasegawa, T., Fuse, K., Yamamoto, M., and Saigusa, M., "A new vascular prosthesis for a small caliber artery," *Surgery,* 74, 519, 1973.
161. Sandusky, G. E., Lantz, G. C., and Badylak, S. F., "Healing comparison of small intestine submucosa and ePTFE grafts in the canine carotid artery," *J. Surg. Res.,* 58, 415, 1995.
162. Kito, H. and Matsuda, T., "Biocompatible coatings for luminal and outer surfaces of small-caliber artificial grafts," *J. Biomed. Mater. Res.,* 30, 321, 1996.
163. Amillo, S., Yanez, R., and Barrios, R. H., "Nerve regeneration in different types of grafts: experimental study in rabbits," *Microsurgery,* 16, 621, 1995.
164. Stokes, I. A., Counts, D. F., and Frymoyer, J. W., "Experimental instability in the rabbit lumbar spine," *Spine,* 14, 68, 1989.
165. Kaigle, A. M., Holm, S. H., and Hansson, T. H., "Experimental instability in the lumbar spine," *Spine,* 20, 421, 1995.
166. Boden, S. D., Schimandle, J. H., and Hutton, W. C., "An experimental lumbar intertransverse process spinal fusion model. Radiographic, histologic, and, biomechanical healing characteristics," *Spine*, 20, 412, 1995.
167. Sandhu, H. S., Kanim, L. E., Toth, J. M., et al., "Experimental spinal fusion with recombinant human bone morphogenetic protein-2 without decortication of osseous elements," *Spine,* 22, 1171, 1997.
168. Brantigan, J. W., McAfee, P. C., Cunningham, B. W., Wang, H., and Orbegoso, C. M., "Interbody lumbar fusion using a carbon fiber cage implant versus allograft bone. An investigational study in the Spanish goat," *Spine*, 19, 1436, 1994.
169. Miyamoto, S., Takaoka, K., Yonenobu, K., and Ono, K., "Ossification of the ligamentum flavum induced by bone morphogenetic protein. An experimental study in mice," *J. Bone Joint Surg.,* 74B, 279, 1992.
170. Saito, H., Mimatsu, K., Sato, K., and Hashizume Y., "Histopathologic and morphometric study of spinal cord lesion in a chronic cord compression model using bone morphogenetic protein in rabbits," *Spine*, 17, 1368, 1992.
171. Olson, E. J., Hanley, E. N., Jr., Rudert, M. J., and Baratz, M. E., "Vertebral column allografts for the treatment of segmental spine defects. An experimental investigation in dogs," *Spine*, 16, 1081, 1991.
172. Frick, S. L., Hanley, E. N., Jr., Meyer, R. A. Jr., Ramp, W. K., and Chapman, T. M., "Lumbar intervertebral disc transfer. A canine study," *Spine*, 19, 1826, 1994.
173. Matsuzaki, H., Wakabayashi, K., Ishihara, K., Ishikawa, H., and Ohkawa, A., "Allografting intervertebral discs in dogs: a possible clinical application," *Spine*, 21, 178, 1996.
174. Agadir, M., Sevastik, B., Sevastik, J. A., Persson, A., and Isberg, B., "Induction of scoliosis in the growing rabbit by unilateral rib-growth stimulation," *Spine*, 13, 1065, 1988.
175. *Guidelines for Preclinical and Clinical Evaluation of Agents Used in the treatment or Prevention of Postmenopausal Osteoporosis*, Division of Metabolism and Endocrine Drug Products: Food and Drug Administration, 1994.

4 Surgical Facilities, Peri-Operative Care, Anesthesia, and Surgical Techniques

Alison C. Smith and M. Michael Swindle

CONTENTS

I. Introduction 59
II. Surgical Facilities and Equipment 60
III. Preoperative Care 61
IV. Anesthesia And Analgesia 62
    A. Principles of Anesthesia 62
    B. Preoperative Agents 62
    C. Injectable Anesthetic Protocols 63
    D. Inhalational Anesthesia 64
    E. Analgesia 65
V. Surgical Technique 65
VI. Post-Operative Care 67
References 68

I. INTRODUCTION

Successful outcomes in the use of animal models in orthopaedic research are best achieved by an interdisciplinary approach among researchers, surgeons, veterinarians, and animal care staff. This approach also generally results in optimization of the ethical and welfare aspects for the animals involved. A team approach involves coordination of all facets of an animal protocol, beginning with preoperative planning to identify any special needs related to the anesthetic or operative regimens as well as requirements for postoperative care. Careful protocol preparation and IACUC review will necessitate that these protocol issues be described; however, some degree of firsthand interaction among key personnel can contribute greatly to the smooth initiation of a new study or the successful modification of one with unanticipated experimental difficulties. Each of the various professional and technical backgrounds have different but unique perspectives and areas of expertise to bring to bear on any given project and those different areas of proficiency should be used to advantage.

    As in any survival surgical procedure, but especially in orthopaedic surgery, meticulous attention to aseptic technique and appropriate tissue handling is necessary to minimize the possibility of infection as well as the experimental variables associated with exaggerated tissue responses to trauma. In many cases it may be necessary for the MD surgeon to become familiar with species-specific anatomical variations for optimal technical results. In addition, MDs may need to be educated regarding the species-specific differences that dictate which anesthetic regimens are appropriate as well as which postoperative medications and practices are suitable. Guidelines for training of personnel who perform surgery on research animals have been published and should be consulted.[1,2]

The design of the surgical and support areas should be such that all aspects of providing appropriate medical and surgical practices are facilitated. Facilities and equipment should be suitable for the species of animals utilized and the complexity of the procedures performed. Regulatory and legal guidelines must also be met. This chapter will provide an overview of the issues related to animal anesthesia and surgery.

II. SURGICAL FACILITIES AND EQUIPMENT

The principal issue regarding surgical facilities in the Animal Welfare Act is more of a programmatic issue rather than an engineering standard. The USDA regulations state that major survival surgical procedures in nonrodent species must be performed in a dedicated facility which is maintained under aseptic conditions. Aseptic procedures include the use of surgical attire, sterile instruments and aseptic techniques. Minor operative procedures and survival surgery in nonrodent species do not require a dedicated facility but must still be performed using aseptic procedures.[3]

The guide makes similar programmatic recommendations but is more specific regarding what are considered suitable design features for aseptic surgical facilities.[4] Emphasis is placed on the integration of different functional areas and features that help minimize contamination. Design features that are described are general recommendations. What is important is that the facility be appropriate for the species used, the complexity of the procedures, and the magnitude of the surgical program.

The guide recommends the following functional elements for most survival surgical facilities: surgical support, animal preparation area, surgeon's scrub area, operating rooms, and postoperative recovery. In some cases, with particularly low volume programs using only a few species, it may be appropriate to combine certain functional areas such as the animal preparation and postoperative recovery areas. Some factors which would influence the suitability of this arrangement would include whether compatible species were recovered simultaneously and if noise could be minimized for animals recovering from anesthesia. In general, the guide recommends that the entire surgical area be physically separate from the rest of the animal facility, but that careful consideration be given to its physical relationship to animal housing areas, radiology facilities, diagnostic laboratories, and office areas.

Design features to reduce the potential for contamination should be utilized. Use of solid, nonporous materials for interior surfaces and minimizing the amount of fixed equipment in the operating room facilitate sanitization. Minimizing traffic flow in the operating rooms and appropriate ventilation systems are also critical.

Other design features should include availability of adequate lighting, scavenging of waste anesthetic gases, and sufficient electrical outlets for support equipment. The size of the operating room should be determined by which species are used, the amount of support equipment necessary, and the number of staff involved in procedures. For a more detailed description of issues related to the design of surgical facilities, the reader is referred to Ruys.[5]

Practices should be in place to maintain the operating room as free from contaminants as possible. All floors, lights and furniture should be wiped down with a suitable disinfectant prior to and after each use, as well as at the end of the day. Personnel should don surgical scrub suits, caps, masks, and shoe covers prior to entering the operating room.

All surgical instruments and bioimplants must be sterilized either by autoclaving or ethylene oxide sterilization. Instrument packs should be prepared to allow adequate air removal and steam penetration of the package during autoclaving. They should be double wrapped and external chemical indicators should be used on all packs, regardless of sterilization method. Efficacy of specific steam sterilization cycles should be monitored using biological indicators.

The type of surgical procedure will dictate the specific instrument needs, however, purchase of high quality surgical instruments will be most economical in the long run. Cleaning and decontamination of instruments should be performed as soon as possible after use since effective decontamination is impaired if debris is allowed to dry on surfaces. Instruments should be dry prior to wrapping in order to prevent the potential for wet instrument packs after autoclaving. Guidelines for sterilization procedures have been published and should be consulted for appropriate practices.[6]

Proximity of the operating suite to radiology facilities is important for orthopaedic procedures. Radiographic equipment should be able to accommodate species of diverse sizes and be periodically maintained and calibrated. In addition, the room should be spacious enough to accommodate a portable anesthetic machine and a gurney for taking intra- or immediate postoperative films. Automatic processors are convenient for programs that perform large numbers of radiographs. It may also be useful to have view boxes mounted in the operating room in a location easily viewed by the surgeon.

A well-appointed surgical suite will have an array of equipment related to the maintenance and monitoring of anesthesia in addition to equipment used by the surgeon. Gas anesthesia machines, ventilators and ECG monitoring devices are routinely used for most nonrodent species undergoing anesthesia. Similarly, circulating water blankets and electronic temperature probes are also routinely employed. Blood pressure monitoring by direct or indirect methods may be indicated for particularly lengthy procedures. Infusion pumps provide accurate delivery of balanced electrolyte solutions and may also be utilized for anesthetic protocols that involve continuous infusions of parenteral agents. Electrocautery and suction may also greatly facilitate surgical manipulations. More specialized equipment may be necessary for projects that involve arthroscopic techniques.

III. PREOPERATIVE CARE

Preoperative care procedures and health monitoring will vary with the species utilized and the requirements of the experiment. All protocols should require the use of healthy and relatively young animals. Knowledge of the times of closure of various growth plates for the species utilized is often a determining factor for selecting the age range for the species to be ordered for orthopaedic research.[7-10]

Reputable commercial vendors that supply specific pathogen free animals should be used as the source for rodents and rabbits. These suppliers routinely perform intensive health monitoring to document quality assurance and freedom from endogenous pathogens in their animals. Presence of endogenous viral or bacterial pathogens can be an important source of experimental variation in terms of an animal's ability to undergo anesthesia, initiate and sustain normal tissue response to surgery, as well being free of a reservoir of bacteria that can potentially seed any bioimplants. Vendor health monitoring reports can be supplemented with in-house laboratory tests if vendor surveillance is desirable, however, generally no preoperative work-up is performed prior to surgery for these species.

Because of their extremely high metabolic rate, it is not necessary or desirable to withhold food or water from rodents preoperatively. An overnight fast is recommended for rabbits; however, they should be allowed free access to water until just prior to surgery. For these species, to assure a good surgical outcome, emphasis should be placed on intraoperative support, appropriate anesthetic regimens, limiting operating time to as brief a period as possible and providing suitable analgesics postoperatively when necessary.

Dogs expressly bred for research with comprehensive health profiles can be obtained from commercial vendors and are the most expensive to purchase. Alternatively, dogs can be obtained from USDA licensed Class B dealers or, in some areas, from local animal shelters. Dogs purchased from Class B dealers, while not purpose-bred for research, are healthy animals released for sale after undergoing a health-screening and "conditioning" regimen. If animals are available from local shelters, their low purchase price is frequently offset or negated by their uncertain or poor health status. These animals must undergo a comprehensive health-screening and conditioning regimen before they can be released for research protocols.

Sources of agricultural species used in research are more variable. Swine and some ruminant species can be obtained from commercial sources. However, it is also common to purchase farm animals directly from nearby farms. Farm source animals, particularly some ruminant species, should be routinely tested for species specific zoonotic diseases.

As a minimum, routine preoperative testing of dogs and other large animal species generally involves documentation that the animals are parasite-free and that they have a normal hematological profile. Other tests, such as blood chemistry or coagulation profiles, should be performed based on study requirements. Preoperatively large animals should be fasted overnight, although access to water should be allowed.

IV. ANESTHESIA AND ANALGESIA

The diverse species of laboratory animals used in orthopaedic research precludes a detailed discussion of all of the anesthetic agents that can be used for these protocols. Species specific physiologic requirements and biological parameters should be discussed with a veterinarian prior to initiating the protocol. Principles of anesthesia and general anesthetic recommendations will be discussed in this selection. Veterinary textbooks on anesthesia and analgesia should be used as a resource if more specific information is required.[11-16]

A. Principles of Anesthesia

Orthopaedic procedures are generally invasive major surgeries requiring deep surgical anesthesia. The anesthetic protocol should provide anesthesia, analgesia and muscle relaxation while maintaining homeostasis. The physiologic effects of the anesthetic protocol should be carefully considered, especially if physiologic measurements are to be made while under anesthesia.[14,17,18] The selection of the protocol will also depend upon the equipment available and the expertise of the personnel. In general, inhalant anesthesia is the preferred method for general anesthesia in most species. Dosages of agents for species likely to be used in research are in the charts, however, the reader should also refer to the text for discussion of indications and contraindications. Postoperative analgesia is indicated for orthopaedic protocols unless it must be withheld for scientific reasons.

B. Preoperative Agents

Preoperative agents are used to sedate the animal to prevent anxiety, to lower the dosage of general anesthetics administered and to prevent undesirable physiologic reactions, such as vagal stimulation. Preoperative agents fall into the classes of anticholinergics, tranquilizers, dissociative agents and a-adrenergic agonists and antagonists.[11-16]

Anticholinergic agents such as atropine are useful to prevent bradycardia during intubation and to dry bronchiole secretions through its vagolytic action. Glycopyrrolate may be used as an atropine substitute. The agents are generally administered 5–15 min. prior to induction under general anesthesia, usually in combination with other preanesthetic agents. Atropine will induce a transient tachycardia and would be contraindicated if that effect is undesirable.[11-16]

Tranquilizers are used to sedate animals and to reduce the dosage of general anesthetic agents. They are generally administered 15–30 min. prior to general anesthesia. The phenothiazine derivative tranquilizers are the most common ones used in veterinary anesthesia, in particular, acepromazine. Other agents such as promazine and chlorpromazine may induce undesirable side effects in some species. The phenothiazine derivatives are mild a-adrenergic blockers and also cause peripheral vasodilation. Their effects generally last approximately 8–12 hours.[11-16]

Benzodiazepine tranquilizers include diazepam and midazolam. These agents provide good sedation with minimal cardiovascular effects. Midazolam is a water soluble agent which is more potent and shorter acting than diazepam.[19] When used as a sole agent it has a duration of action of approximately 20 min. Diazepam is relatively long acting with effects of 4–8 hours in most species. Diazepam is effective to counteract seizure activity. These agents may be combined with other agents to induce surgical anesthesia for minor procedures.

Dissociative agents include ketamine and the combination agent tiletamine-zolazepam (Telazol).[20] These agents may be used to induce animals into a state of dissociative anesthesia causing them to be unaware of their surroundings but with only mild analgesia. Animals will not have muscle relaxation and will have mild clonic/tonic types of muscular activity. These agents are extremely useful for chemical restraint of approximately 20 min. in most species. They may be used as part of a general anesthesia protocol when combined with other agents. Their use as general anesthetics is discussed under injectable agents. Ketamine causes a transient tachycardia with minimal hemodynamic effects. The cardiovascular effects of Telazol may be more pronounced in some species.[11–16]

Alpha adrenergic agonists and antagonists include xylazine and medetomidine,[21] which are the two most commonly used agents in this class. They are best utilized in combination with other agents to provide general anesthesia. Xylazine causes bradycardia, heart block, peripheral vasodilation and nausea in many species. These side effects can be counteracted with atropine. Medetomidine has less of the undesirable effects of xylazine. These agents have mild analgesic activity, which may be very transient in some species. They generally have activity for 20 min.[11–16]

C. Injectable Anesthetic Protocols

When injectable agents are utilized for general anesthesia, they should be administered as continuous iv infusions following induction. The use of infusion protocols enables the anesthetist to provide a stable level of physiologic effects instead of the unpredictable fluctuations in baseline activities that may occur when repeated bolus injections are administered. Combining agents also causes variations of the effects of the individual agents to occur depending upon the species and the dosages. Combinations may either potentiate or nullify the undesirable effects of some agents making it difficult to predict the physiologic effects of the protocol.[14,16] Commonly recommended injectable anesthetic protocols are listed in Table 1.

Dissociative agents have been widely combined with other agents to induce anesthesia. As sole agents, they do not provide sufficient analgesia for general surgery. The two most commonly used agents are ketamine and the combination agent tiletamine/zolazepam (Telazol).[20] They are usually combined with tranquilizers or a-adrenergic agents. Commonly used combinations are: ketamine/acepromazine, ketamine/diazepam, ketamine/midazolam, ketamine/xylazine, ketamine/medetomidine, ketamine/Telazol, ketamine/Telazol/xylazine. None of these combinations can be recommended for all species and protocols. Their individual effects are very species dependent. Most of the combinations provide 20–30 min. of general anesthesia in most species, which is not sufficient for prolonged or highly invasive surgeries.[11–16]

Barbiturate anesthesia is used in all species. However, it requires iv access for most species. They may be administered as an ip injection for rodents. The barbiturates are potent cardiorespiratory depressants which are dose dependent in their activity. In large animals, they are best administered as iv infusion protocols. Pentobarbital is relatively long acting in most animals and may provide 30–45 min. of anesthetic activity following a single iv injection. It requires hepatic metabolism and excretion for biodegradation. Thiobarbiturates, such as thiopental and thiamylal, are primarily excreted by the kidneys and have much shorter anesthetic times of 5-20 min. The dosage of barbiturates are reduced 1/3–1/2 by administration of tranquilizers or other preanesthetic agents.[11–16]

Propofol is a steroidal anesthetic that must be administered by continuous iv infusion. The agent may be profoundly hypotensive in some species. It may be administered as a sole agent for general anesthesia in some species.[14–16]

Other agents such as detomidine, etomidate and alpha chloralose are generally not reliable for general anesthesia in most species.[14–16] They may be indicated as part of a general anesthetic

TABLE 1
Commonly Recommended Injectable Anesthetic Protocols*

Animal	Agent[Ref.]	Dosage	Route
Mouse or Rat	Fentanyl/fluanisone+midazolam†[11]	10 ml/kg for mice, 2.7 ml/kg for rats	IP
	Pentobarbital[14,15,16,17]	40–50 mg/kg	IP
	Ketamine/xylazine[11,14,15,16,17]	100 mg/kg + 10 mg/kg	IP
Guinea Pig	Fentanyl/fluanisone+midazolam†[11]	8 ml/kg	IP
	Pentobarbital[11,14,15,16,17]	35–40 mg/kg	IP
	Ketamine/xylazine[11,14,15,16,17]	50 mg/kg + 2 mg/kg	IP
Rabbit	Fentanyl/fluanisone+midazolam†[11]	0.3 ml/kg (IM) + 2 ml/kg (IV)	IM+IV
	Pentobarbital[11,14,15,16,18]	40–45 mg/kg	IV
	Ketamine/xylazine[11,14,15,16,18]	35 mg/kg + 5 mg/kg	IM
	Ketamine/medetomidine[11,15,16,18]	25 mg/kg + 0.5 mg/kg	IM
	Ketamine/acepromazine[11,14,15,16,18]	50 mg/kg + 1 mg/kg	IM
Cat	Ketamine/acepromazine[11,3,5,14,15,19]	10–20 mg/kg + 1 mg/kg	IM
	Ketamine/xylazine[11,3,5,14,15]	15 mg/kg + 1 mg/kg	IM
	Ketamine/medetomidine[11,15]	7 mg/kg + 0.08 mg/kg	IM
	Tiletamine/zolazepam[15]	6–12 mg/kg	IM
	Pentobarbital[11,14,15,19]	20–30 mg/kg	IV
	Thiopental[11,14,15,19]	8–12 mg/kg	IV
Dog	Fentanyl/fluanisone[11]	0.1–0.2 mg/kg	IM
	Ketamine/midazolam[11,15]	10 mg/kg + 0.5 mg/kg	IM
	Tiletamine/zolazepam[14,15,19]	6–12 mg/kg	IM
	Pentobarbital[11,14,15]	20–30 mg/kg	IV
	Thiopental[11,14,15]	8–12 mg/kg	IV
Goat or sheep	Ketamine/xylazine[12,14,15]	2.2–7.5 mg/kg+0.1 mg/kg	IV
	Ketamine/medetomidine[12,15]	0.5 mg/kg + 0.02 mg/kg	IV
	Pentobarbital[12,15]	20–30 mg/kg	IV
	Thiopental[12,14,15]	25 mg/kg	IV
Swine	Ketamine/acepromazine[11,12,14,15,16]	33 mg/kg + 1.1 mg/kg	IM
	Ketamine/medetomidine[14,15,16,21]	10 mg/kg + 0.2 mg/kg	IM
	Tiletamine/zolazepam[14,15,16]	2–8 mg/kg	IM
	Pentobarbital[11,12,14,15]	20–40 mg/kg	IV
	Thiopental[6,11,12,14,15,16]	6.6–25 mg/kg	IV
Baboon	Ketamine[14,15,24]	5–10 mg/kg	IM
	Ketamine/diazepam[14,15,24]	10 mg/kg + 0.2 mg/kg	IM
Macaques	Ketamine[14,15,24]	5–20 mg/kg	IM
	Ketamine/xylazine[14,15,24]	10 mg/kg + 0.25 mg/kg	IM

* These injectable agents provide approximately 20–30 minutes of anesthesia in most species. For prolonged procedures it is best to maintain general anesthesia after induction with an inhalational anesthetic (see text).
† Mixture of 1 part fentany/fluanisone + 2 parts water for injection + 1 part midazolam.

protocol in special circumstances. The combination agent fentanyl/fluanisone (Hypnorm) is commonly used in rodents.[11]

D. INHALATIONAL ANESTHESIA

Inhalational anesthesia should be the primary choice for general anesthesia unless it is contraindicated by the scientific protocol or the inability of laboratory personnel to administer it properly. The most commonly used inhalant anesthetics in veterinary anesthesia are isoflurane, halothane and methoxyflurane. Other newer agents include desflurane, sevoflurane and enflurane.

Older agents such as ether and chloroform should not be considered for general anesthesia. The agent that has the most widely indicated applications is isoflurane. It provides surgical anesthetic levels at concentrations ranging from 0.5-2.0% in most species. Nitrous oxide does not provide sufficient analgesia as a sole agent for general anesthesia in animals and should only be used in combination with other agents to reduce the level of the inhalant required.[11-16,22]

Nitrous oxide, halothane and methoxyflurane have biosafety considerations when used in the laboratory. In susceptible individuals, they may cause hepatic and renal complications. All inhalant anesthetics, and especially these agents, require gas scavenging systems.[11-16]

All inhalational anesthetics are best utilized with equipment designed for their delivery, including a vaporizer in a closed or semi-closed system. In rodents they are sometimes delivered as open agents on cotton balls, but this use should be limited to fume hoods or the equivalent. Equipment and hoses should be checked for leaks and soda lime cannisters for absorption of carbon dioxide should be routinely cleaned and changed.

Inhalant agents are delivered in oxygen, nitrous oxide, air or a combination after flowing the delivery gas through a vaporizer. Flow rates for most species range from 10–15 ml/kg. The number of respirations required per minute is highly species and age dependent. If positive ventilation is used, most species require 18-22 cm H_2O airway pressure. The most common protocols for general anesthesia are isoflurane 0.5-2% delivered in oxygen or oxygen:nitrous oxide 2:1.[11-16]

E. ANALGESIA

Postoperative analgesia should be required for all orthopaedic protocols unless a specific scientific justification exists for witholding them. Intraoperative or preoperative administration of these agents has been shown to reduce the pain reflex and to reduce the postopertive recovery times of some species. Animals should at least receive the first injection of an analgesic prior to recovering from general anesthesia. The length of administration depends upon the clinical condition of the animal and requires professional judgment.[23]

The most commonly used agents in veterinary analgesia are the opioids. Specifically, buprenorphine and butorphanol are the most commonly used agents for most species. Some opioids, such as morphine, are associated with a high incidence of untoward reactions in some species.[14,16,17,23,24] Common analgesics are listed in Table 2.

Nonsteroidal antiinflammatory drugs (NSAIDs) may be used in combination with opioids to enhance analgesia and antiinflammatory activity. The most widely used agent in animals is phenylbutazone. It does not provide sufficient analgesia postopertively for major surgery, however, it may be used to reduce the dosage of opioids. Other agents, such as aspirin and acetaminophen, are less likely to be potent enough to be used in orthopaedic surgery and have hepatic and renal toxicity in some species. Newer NSAID agents, such as ketaprophen, may be sufficient to provide postopertive analgesia in some species. Their use in many species has not been validated.[14-16]

Injections of local anesthetics may also be used to enhance analgesia. They may be given as dorsal nerve root injections to provide regional analgesia or as infiltrations along the incision line.[14-16]

V. SURGICAL TECHNIQUE

Strict adherence to aseptic procedures and principles of careful tissue handling are necessary for the evaluation of biomaterials. The goal of the surgeon should be to minimize all factors that have the potential to impede normal wound healing, thereby making the interpretation of data most meaningful by minimizing experimental variability. The precepts that are outlined below should be applied to all species undergoing orthopaedic procedures.

Preparation of the animal involves decontamination of the incision site and enough of the surrounding surgical area to prevent wound contamination. Hair removal and initial skin cleansing

TABLE 2
Commonly Recomended Analgesics

Agent	Animal[Ref.]	Dosage (mg/kg)	Route	Duration of Action (hrs)
Buprenorphine	Mouse[11,14,23,24,25]	0.5–2.0	SC	12
	Rat[11,14,23,24,25]	0.1–0.5	SC	12
	Guinea Pig[11,14,23,24,25]	0.5	SC	8–12
	Rabbit[11,14,23,24,25]	0.02–0.05	SC	8–12
	Cat[11,13,14,15]	0.005–0.01	SC	12
	Dog[11,13,14,15]	0.01–0.02	SC	12
	Sheep[11,12,13,14,15]	0.005–0.01	IM, SC	6
	Goat[11,12,13,14,15]	0.005	IM, SC	6–12
	Swine[11,12,13,14,15,16]	0.05–0.1	IM	12
	Primates[11,12,13,14,15,24]	0.01	IM	6–8
Butorphanol	Mouse[11,14,23,24,25]	1–5	SC	4–6
	Rat[11,14,23,24,25]	2	SC	6
	Rabbit[11,14,23,24,25]	0.1–0.5	SC	6
	Cat[11,13,14,15]	0.4	SC	6
	Dog[11,13,14,15]	0.2–0.4	SC	6
	Sheep[11,12,13,14,15,20]	0.5	SC, IM	4–6
	Swine[11,12,13,14,15,16]	0.1–0.3	SC, IM	4–6
Phenylbutazone	Mouse[14,15]	30	PO	12
	Rat[14,15]	20	PO	12
	Guinea Pig[14,15,17]	40	PO	12
	Cat[14,15]	10	PO	12
	Dog[14,15]	6–22	PO	8
	Goat or sheep[12,14,15]	4–8	PO	12
	Swine[14,15,16]	10–20	PO	12
	Primate[14,15,24]	50	PO	12
Ketorolac	Dog[14]	1–1.5	PO	6
	Swine[16]	1	IM, PO	12
	Primates[24]	1	IM	12

should be performed outside the operating suite in the case of rabbits and large animals and in a site separate from the operating area for rodents. Hair should be removed as close to the skin as possible while simultaneously maintaining the integrity of the epithelium. Loose hair should be thoroughly removed from the surgical area followed by application of a surgical antiseptic beginning over the incision site and working outward to avoid contamination from the periphery.

Following transport to the operating room or table, a "sterile" scrub is performed prior to draping. Three successive scrubs with a surgical disinfectant are performed using sterile gauze sponges, sponge forceps and surgical gloves using the same outward pattern. An application of 70% alcohol follows a suitable contact time for the disinfectant used, then the entire area is dried. It is then preferable to apply a plastic, incisable drape over the entire area just prepared. This type of drape is a better bacterial barrier than cloth drapes because it is impermeable to fluids and prevents wicking of moisture from the periphery into the incision.

Draping proceeds by closely surrounding the incisional area with four drapes followed by covering the entire animal and back table with one large drape. For large animals separation of the sterile field from the anesthetist at the head of the table is accomplished by using two IV poles positioned on opposite sides of the table to suspend the large drape with towel clamps. The end result, even for rodents and rabbits, should be the formation of one entire sterile field that includes the completely draped animal, operating table, and adjacent back table with instruments. This method of draping greatly facilitates the maintenance of a sterile field and instruments during surgery.

Use of good surgical technique is critical to produce the minimal amount of tissue damage necessary to the procedure. Gentle tissue handling, meticulous hemostasis, and closure of dead space all contribute to minimizing the intensity and duration of the inflammatory response and preventing hematoma or seroma formation that favor bacterial growth. Incisional margins should be carefully apposed to avoid excessive tension which can create irritation postoperatively and result in self-mutilation. Suture material should be selected to minimize the inflammatory response, making synthetic absorbable suture materials preferable to catgut and silk. In species with an adequate amount of subcutaneous tissue, skin closure is best accomplished with a buried suture line using a subcuticular pattern. Monofilament nylon or stainless steel wound clips can be used for skin closure of rodents; however, rodents may succeed in chewing out sutures or staples if tissue irritation occurs. Routine wound care for many procedures usually only requires observation of the incisional area for normal healing. Veterinary surgical textbooks should be consulted for species-specific techniques and practices.[16,25-29]

VI. POST-OPERATIVE CARE

The postoperative period can be divided into two important phases, recovery from anesthesia and the period of tissue healing. All species should be carefully and frequently observed during recovery from anesthesia, which is characterized by presence of normal vital signs and return of righting reflexes. Intensity of postoperative monitoring will vary with the species; however, return to normothermia and attention to cardiorespiratory function are important in all species. Circulating water jackets can be safely used in all species to maintain body temperature without running the risk of tissue injury. Heat lamps and heating pads can also be used but in such a way as to prevent overheating or tissue damage. Intraperitoneal administration of warmed electrolyte solutions is an easy method of aiding rewarming and providing cardiovascular support to rodents. For rodents monitoring respiratory rate and pattern as well as mucous membrane color is advisable. Rodents should not be allowed to recover on wood shavings since they can occlude the nose and stick to the mouth and eyes. Soft cloths or disposable pads should be used instead and can aid in maintaining body temperature.

Monitoring vital signs, mucous membrane color, and surgical incisions at regular intervals should be routine for rabbits and large animals. ECG monitoring and use of pulse oximetry is also useful, but may not be necessary for all procedures. In addition to routine monitoring during postanesthetic recovery, resources should also be available to handle potential complications. This generally involves support of the cardiorespiratory system by the administration of fluids and emergency drugs as well as providing airway support and supplemental oxygen.

For many procedures, animals may be returned to their home cages following anesthetic recovery. However, in some cases it may be necessary to rely on cage confinement to limit activity for a certain period of time postoperatively. These situations require the use of professional judgment and should be worked out in advance among the researcher, veterinarian, and animal care staff. Postoperative monitoring should be performed at least once daily for a specified time frame and include measurement of vital signs; assessment for resumption of normal physiologic functions, species-specific behavior and activity levels; and healing of surgical incisions. In addition, a careful assessment should be made regarding the degree of postoperative pain and distress and the need for analgesia. This topic has received detailed discussion elsewhere to which the reader is referred for further information.[30,31] Other precepts of good nursing care should be followed and include offering palatable foods to stimulate appetite, use of bandaging to provide padding for sore extremities, and providing a comfortable, quiet environment.

For orthopaedic procedures, it is especially important to evaluate the animal's ability to ambulate and perform normal body movements during the immediate postoperative period and at intervals until study completion. Animals with musculoskeletal pain that cannot be adequately controlled with analgesics or animals that develop clinical musculoskeletal signs should be routinely radiographed, if necessary, under anesthesia in order to obtain diagnostic films. Animals with uncontrollable pain or distress should be euthanized.

For clinical diagnostic reasons as well as to meet certain study requirements, it is also necessary to have the means to perform various microbiological, hematological, and clinical chemistry tests. Meaningful information can only be obtained if appropriate procedures are followed during sample collection, processing, and transport. If in-house laboratory facilities are limited or unavailable, arrangements should be made in advance with commercial laboratories that can competently handle veterinary specimens.

REFERENCES

1. Brown, M. J., Pearson, P. T., and Tomson, F. N., "Guidelines for animal surgery in research and teaching," *Am. J. Vet. Res.,* 54, 1544, 1993.
2. Academy of Surgical Research, "Guidelines for training in surgical research in animals," *J. Invest. Surg.,* 2, 263, 1989.
3. *Code of Federal Regulation, Title 9, Chapter 1, Subchapter A — Animal Welfare, Part 2,* U.S. Government Printing Office, Washington, DC, 1966.
4. *Guide for the Care and Use of Laboratory Animals,* National Academy Press, 1996, 1.
5. Ruys, T., *Handbook of Facilities Planning. Laboratory Animal Facilities,* Van Nostrand Reinhold, New York, 1991.
6. "Good hospital practice: steam sterilization and sterility assurance," *Assoc. Adv. Med. Instr.,* 1988, 115.
7. Foster, H. L., Small, J. D., and Fox, J. G., *The Mouse in Biomedical Research,* Academic Press, New York, 1982.
8. Weisbroth, S. H., Flatt, R. E., and Kraus, A. L., *The Biology of the Laboratory Rabbit,* Academic Press, New York, 1974.
9. Baker, H. J., Lindsey, J. R., and Weisbroth, S. H., *The Laboratory Rat,* Academic Press, New York, 1979.
10. Getty, R., *Sisson and Grossman's The Anatomy of the Domestic Animals,* W. B. Saunders, Philadelphia, 1975.
11. Flecknell, P. A., *Laboratory Animal Anaesthesia,* 2nd Ed., Academic Press, New York, 1996.
12. Riebold, T. W., Goble, D. O., and Geiser, D. R., *Large Animal Anesthesia, Principles and Techniques,* 2nd Ed., Iowa State University Press, Ames, IA, 1995.
13. Short, C. E., "Inhalant Anesthetics," in *Principles and Practice of Veterinary Anesthesia,* Williams & Wilkins, Baltimore, 1987.
14. Kohn, D. H., Wixson, S. K., White, W. J., and Benson, G. J., *Anesthesia and Analgesia in Laboratory Animals,* Academic Press, New York, 1997.
15. Thurmon, J. C., Tranquilli, W. J., and Benson, G. J. *Lumb and Jones' Veterinary Anesthesia,* 3rd ed., Williams & Wilkins, Baltimore, 1996.
16. Swindle, M. M., *Surgery, Anesthesia & Experimental Techniques in Swine,* Iowa State University Press, Ames, IA, 1998.
17. Jenkins, W. L., "Pharmacologic aspects of analgesic drugs in animals: an overview," *J. Am. Vet. Med. Assoc.,* 191, 1231, 1987.
18. Thurmon, J. C. and Benson, G. J., "Pharmacologic consideration in selection of anesthetics for animals," *J. Am. Vet. Med. Assoc.,* 191, 1245, 1987.
19. Smith, A. C., Zellner, J. L., Spinale, F. G., and Swindle, M. M., "Sedative and cardiovascular effects of midazolam in swine," *Lab. Anim. Sci.,* 41, 157, 1991.
20. Ko, J. C. H., Williams, B. L., McGrath, C. J., Short, C. E., and Rogers, E. R., "Comparison of anesthetic effects of telazol-xylazine-xylazine, telazol-xylazine-butorphanol and telazol-xylazine-azaperone combinations in swine," *Contemp. Topics Lab. Anim. Sci.,* 35, 71, 1997.
21. Flecknell, P. J., "Medetomidine and antipamezole: potential uses in laboratory animals," *Lab. Anim.,* 26, 21, 1997.

22. Steffey, E. P. and Eger, E. I., "Nitrous oxide in veterinary practice and animal research," in *Nitrous Oxide*, Eger, E. I., Ed., Elsevier, New York, 1985, 305.
23. Flecknell, P. A., "The relief of pain in laboratory animals," *Lab. Anim.,* 18, 147, 1984.
24. Blum, J. R., "Laboratory animal anesthesia," in *Experimental Surgery and Physiology: Induced Animal Models of Human Disease*, Swindle, M. M. and Adams, R. J., Eds., Williams & Wilkins, Baltimore, 1988, 329.
25. Laber-Laird, K., Swindle, M. M., and Flecknell, P., *Handbook of Rodent and Rabbit Medicine*, Pergamon, New York, 1996.
26. Bojrab, M. J., *Current Techniques in Small Animal Surgery,* Lea & Febiger, Philadelphia, 1983.
27. Swindle, M. M., *Basic Surgical Exercises Using Swine,* Praeger, New York, 1983.
28. Dougherty, R. W., *Experimental Surgery in Farm Animals,* Iowa State University Press, Ames, IA 1981.
29. von Recum, A. F., *Handbook of Biomaterials Evaluation*, Macmillan, New York, 1986.
30. Aronson, A. L., Clark, J. D., Dubner, R., Gebhart, G. F., Hughes, H. C., et al., *Recognition and Alleviation of Pain and Distress in Laboratory Animals,* National Academy Press, Washington, DC, 1992.
31. Thurmon, J. C., Tranquilli, W. J., and Benson, G. J., "Perioperative Pain and Distress," in *Lumb and Jones' Veterinary Anesthesia*, 3rd Ed., Williams & Wilkins, Baltimore, 1996.

5 Euthanasia and Necropsy

Sarah A. Bingel

CONTENTS

I. Introduction 71
II. Euthanasia Methods 72
    A. Decapitation or Cervical Dislocation 73
    B. Carbon Dioxide 73
    C. Gas Anesthetic Overdose 74
    D. T61 74
    E. KCl or Exsanguination 74
    F. Barbiturate Overdose 74
III. Necropsy 75
IV. Common Sampling Procedures 76
    A. Sampling for Histopathological Evaluation 76
    B. Sampling for Mechanical Testing 77
    C. Joint Fluid Collection 77
    D. Sampling for Serology 77
    E. Sampling for Cultivation 78
        1. Bacteria 78
        2. Fungi 78
        3. Viruses 79
References 79

I. INTRODUCTION

The term *euthanasia* literally translated from the Greek means good death. Euthanasia techniques which result in rapid loss of consciousness with minimal stress or anxiety and are followed by cardiac and respiratory arrest and cessation of brain function are considered humane euthanasia methods.[1] Ideally, these processes rapidly disrupt all afferent sensory pathways from peripheral pain receptors as well as blocking central processing of these pathways in the thalamus, cerebral cortex and subcortical centers.[1]

Investigators who need to euthanize an animal as part of a project must do so in compliance with relevant guidelines set forth in the *Public Health Service (PHS) Guide for the Humane Care and Use of Laboratory Animals*,[2] the *Recommendations of the AVMA Panel on Euthanasia (1993)*[1] in the United States, or the *Recommendations for Euthanasia of Experimental Animals (1997)* of the European Commission.[3] Euthanasia may also become necessary as a way of alleviating pain or distress that cannot be eliminated by the use of analgesics. For animals on protocols where pain or distress may occur before such studies begin, it is essential that the researcher and the veterinarian agree on a point at which the animal will be terminated from the study for humane reasons and euthanized.[4]

It is essential that persons performing euthanasia be adequately trained in the methods of restraint of the animal as well as the technical skills required to perform the act so as to minimize pain, stress and anxiety in the animals. In the United States, the PHS guide requires that all methods of euthanasia proposed by researchers be reviewed by an Institutional Animal Care and Use Committee (IACUC).[2]

Once euthanasia has been done, it is essential that the animal be examined for loss of vital signs.[1,4] If there is any uncertantity, one may also decapitate the animal or create a bilateral thoracotomy or pneumothorax or exsanguinate it to confirm death. It is preferable not to have other animals in the room where the euthanasia is being performed as powerful pheromones are frequently emitted by the animals being euthanized or sounds may be made that are distressful to other animals.[1,2,4]

II. EUTHANASIA METHODS

The choice of methods is governed by many variables such as the species, age, physical condition, number of animals that need to be done at the same time, cost, availability of equipment or controlled substances, whether one needs tissues for metabolic or other studies that must be free of chemical residues and the skills of the person performing the act. Euthanasia methods are generally classified as physical techniques, chemical agents or inhaled gases.[1] Physical methods have a high potential for causing pain or stress prior to the loss of consciousness if not carried out properly. Therefore, all physical methods must be justified by researchers in their grants and protocols and they must be approved by the IACUC.[2] Physical method should only be performed by trained personnel so that it results in a quick and painless death with minimal stress to both the animal and the person performing it.[1,2] Animals should always be handled as gently as possible to minimize anxiety and stress. As a rule, the non-explosive inhaled gases, carbon dioxide and barbiturates are preferred over the physical methods. A summary of the acceptable euthanasia agents for each species is summarized in Table 1.

TABLE 1
Euthanasia Methods Recommended by AVMA

Species	Decapitation	Cervical Dislocation	CO2†	Barbiturate Overdose	Gas Anesthetic Overdose (in hood)	KCl or Exsanguination (under anesthesia)	T-61‡ (IV only)
Cat				A (IV, IP)	A	A	A
Chicken	A§	A	A	A (IV, IP)	A	A	A
Dog				A (IV, IP)	A	A	A
Goat				A (IV)		A	A
Guinea Pig	A		A	A (IV)	A	A	A
Hamster	A	A	A	A (IP)	A	A	A
Mouse	A	A	A	A (IP)	A	A	A
Primate				A (IV, IP)		A	
Rabbit	A	A*	A	A (IV, IP)	A	A	A
Rat	A	A*	A	A (IP)	A	A	A
Sheep				A (IV)		A	A
Swine				A (IV)		A	A

* Acceptable for immature animals only (rats & guinea pigs < 200g, rabbits , 1kg).[1]
† Prolonged time is required for immature animals.
‡ Animals must be sedated prior to use.[3]
§ A = acceptable.

Animal caretakers, students and some researchers may find it psychologically disturbing to have to repeatedly euthanize animals.[4–6] This is especially true when they have become emotionally attached to the animals. In such cases it is prudent to have someone else euthanize them. Skilled caretakers and technicians who can handle the animals compassionately at the time of euthanasia can do a lot toward preventing distress in the animals.

A. Decapitation or Cervical Dislocation

Decapitation is acceptable for rodents weighing less than 200 grams and for small rabbits less than one kg.[1] The major advantage to this technique is that it provides a rapid loss of consciousness and is a quick way to obtain tissues with minimal metabolic artifacts or chemical residues.[1,7] It is generally performed with a guillotine. Mikeska and Kiemm[8] believed that decapitation produces a powerful arousal stimulus and they demonstrated the presence of low voltage fast activity in the decapitated head for 13–14 sec. However, both they and Derr[9] documented that the time required for the oxygen tension in the brain to drop low enough for the brain to lose consciousness from hypoxia is 2.7 seconds. In addition, Vanderwolf et al.[10] have shown that this EEG activity is not indicative of a conscious state. It has been clearly shown that sectioning of the spinal cord at any level results in immediate loss of pain perception below the level of the section. Sensory fibers for the head and scalp enter the spinal cord at the levels of C2-C3.[11] It is for that reason that decapitation should be performed at the atlanto–occipital joint or at C1-C2. If done in this manner, decapitation is a humane method.[1,10,12] The major disadvantages to this technique are that it is potentially dangerous to the person doing it and that it does require some handling and restraint of the animals.

Cervical dislocation is performed by placing a rod immediately behind the base of the skull and pressing down while at the same time pulling quickly on the tail or hindlimbs with the thumb and index finger until you feel the vertebrae separate. High cervical dislocation at the atlanto–occipital joint results in crushing and severing of the spinal cord.[1] Unlike decapitation, dislocation does not result in an almost immediate loss of consciousness as a result of loss of blood flow. Holson[12] believed that for this reason, this technique may cause more pain perception than decapitation. Death results from paralysis of respiration and subsequent anoxia. This technique is approved for rodents weighing less than 200 grams as well as poultry and rabbits less than 1 kg.[1] The person doing it must be very skilled at the technique in order to do it humanely. The advantages are the same as those for decapitation. This technique may result in pulmonary artifacts like blood in alveoli and vascular congestion.[13]

B. Carbon Dioxide

Carbon dioxide is a safe, inexpensive, non-flammable agent for euthanisia of rodents. Carbon dioxide can be delivered either in the form of dry ice or as a compressed gas in cylinders. If it is used in the form of dry ice, there must be no physical contact between the ice and the animal being euthanized.[1,4] Carbon dioxide generated by any other source is not acceptable.[1–4] Neonatal animals are resistant to this agent so prolonged exposure is necessary or barbiturates could be used as they work well in neonates. Carbon dioxide results in rapid lowering of the pH of the blood and cerebrospinal fluid. Hypoxia and death are the result of direct depression of the cerebral cortex, subcortical structures and the respiratory center and cerebral vasodilatation as well as a direct effect on the myocardium.[13] If a euthanasia chamber is used, care must be taken to avoid overcrowding.

Butler et al.[14] have shown that carbon dioxide alters arachidonic acid metabolism and smooth muscle responses to acetylcholine. Carbon dioxide had also been documented to alter lymphoproliferation and cell mediated lympholysis.[15] This has always been considered a painless agent. However, a recent study using human subjects has shown that high concentrations of carbon dioxide, 80–100%, can be noxious and painful to the nose and throat.[16] This is believed to be due to the formation of carbonic acid on mucous membranes. The authors concluded that rats exposed to the same concentrations may also experience pain and discomfort. They further suggested that carbon dioxide could still be considered a humane agent if the animals were not exposed to concentration greater than 70% until after they had become unconscious. Carbon dioxide does not alter the histologic integrity of most tissues including the brain but it may cause vascular congestion, pulmonary edema or microhemorrhages in the lungs of rodents.[13]

C. Gas Anesthetic Overdose

Inhalant anesthetics include halothane, methoxyflurane, enflurane, isoflurane and ether. Ether is explosive, flammable and very irritating and for these reasons it is not recommended.[17] Gas overdose is an excellent choice for small (<7 kg) animals where venipuncture is technically difficult and in chickens.[1] Halothane, methoxyflurane and isoflurane are nonflammable, nonexplosive and can be delivered either in a closed container like a closed chamber or bell jar or via a face mask. Care must be taken to ensure that there is an adequate supply of air inside the closed chamber and that the animal does not come in direct contact with the liquids as they are irritating. Halothane is the most soluble and most effective for inducing rapid loss of consciousness and, therefore, is the agent of choice.[1] Isoflurane is the least soluble but it has a pungent odor and animals tend to hold their breath. Death is the result of direct suppression of respiration and of the cerebral cortex, and other vital centers.[1] These agents have been reported to cause no alterations in the histology of most tissues except for congestion of alveolar capillaries.[13] Both halothane and methoxyflurane have been shown to alter some metabolic parameters[7,14] and cause lymphocyte proliferation.[15]

D. T61

This is an injectable nonbarbiturate agent that is no longer commercially manufactured or available in the United States. It is a combination of tetrazine HCl, butyramide and a curariform drug.[18,19] According to the European Commission's recommendations for euthanasia, T61 is an acceptable agent if given intravenously and slowly to a sedated animal as it may result in pain if given too quickly to rabbits, rodents, dogs, cats, ferrets and large animals.[3] Dogs frequently become distressed and vocalize and move their legs. There is evidence to suggest that respiratory arrest may occur before the loss of consciousness.[18,19] Hellbreckers et al.[20] recently published data from dogs and rabbits disputing this study's findings. T61 is believed to exert both a narcotic and direct effect on the respiratory centers in addition to the paralytic effect of the curariform drug on the respiratory muscles and the heart leading to circulatory collapse, hypoxia and death.[3,18,19] The side effects of this agent are esthetically unacceptable to many people.

E. KCl or Exsanguination

Exsaguination is an acceptable method of euthanasia in most species only if done in an unconscious or anesthetized animal.[1,3] Hypovolemia can cause extreme distress and anxiety and for this reason it can never be used alone. In a research setting this method is a convenient way of doing terminal blood collection on antibody producing animals in order to get as much serum as possible. Bleeding can be done either by venipuncture or cardiac stick. It is not effective in birds because of their tendency for clot formation[3] or in reptiles because of their lower metabolic rate and high tolerance of hypoxia. Potassium chloride is a rapidly acting cardiotoxic agent. However, it causes seizures, gasping, and vocalizations and cannot be used as a euthanasia agent in awake animals. It is acceptable for use only in unconscious animals.[1,3,19]

F. Barbiturate Overdose

Barbiturates can be used on a wide variety of species and produce a very rapid effect with minimal discomfort to the animal. They result in death by inducing deep anesthesia and unconsciousness with

CNS and respiratory center depression.[1,4,18,19] Although many barbiturates are acceptable agents, sodium pentobarbital is the most commonly used for euthanasia.[4] Administration of the drug should be via the intravenous route and never given by intrapulmonary or intracardiac routes in conscious animals because both routes are stressful and painful.[4] However, the intraperitoneal route is acceptable for rodents, dogs, cats, rabbits and primates. The dosage for rodents is 150-200 mg/kg body weight.[4] Persons using this technique should be adequately trained in proper restraint of the animal and in proper injection technique to minimize anxiety and stress to the animal. One of the disadvantages of barbiturates is that they are controlled substances and accurate records must be kept of their use. Because they cause relaxation of smooth muscles, splenomegaly is a common gross finding. They are not reported to cause any histologic aberrations to tissues other than vascular congestion of the spleen and lungs.[13] However, they have been documented to alter endocrine functions as well as metabolic and lymphocytic functions.[7,14,15]

III. NECROPSY

In order to avoid artifacts caused by autolysis, a necropsy should be done immediately after the animal has died. If this is not possible the carcass should be placed in a leakproof plastic bag and put in the refrigerator until the necropsy is performed. The body should never be frozen if histopathologic examination of tissues is needed as ice crystals form inside the cells and when the body thaws the cells rupture making histologic evaluation difficult or impossible. The bodies of small rodents tend to undergo autolytic changes rapidly.[21] Within 40–60 min. after death the villi in the small intestine are denuded, the bronchial epithelium separates from the underlying lamina propria, there is cytoplasmic alteration in skeletal muscle fibers and there are changes in the adrenals, lymph nodes and parathyroid glands.[22]

There is an old saying that "A necropsy is a message of wisdom from the dead to the living."[23] The degree of enlightenment derived by this quest for wisdom is dependent largely upon the thoroughness with which the necropsy is performed. A systematic standard operating procedure (SOP) for performing a necropsy must be followed if the results are to be reproducible and uniform.[21,24] This is a stage at which nontreatment variability can be introduced and interfere with an otherwise reproducible study.[21] Having a good detailed necropsy form that follows a logical sequence for the recording of gross observations and organ weights is helpful. Accurate morphologic evaluations require that samples be taken from representative regions of each organ.[21,25] This practice should also be followed when trimming tissues.

Within a research setting, necropsies play a major role in detection and diagnosis of diseases present in laboratory animals. Many diseases can cause significant alterations in the metabolic or immune systems of affected animals and thereby interfere with research studies. Necropsies also play a major role in the quality assurance programs for rodents and other animals in institutions and in screening of vendors. They are also required for toxicologic studies and many research protocols involving biomedical implant devices, surgical models and drug trials.

Necropsies should be performed in a dedicated room with restricted access and used only for this purpose because of the high potential for the spread of pathogenic organisms. Disposable waterproof gowns, latex gloves, plastic shoe protectors and protective eye wear should always be worn as well as masks when indicated. Instruments should be autoclavable or able to be thoroughly disinfected. Containers used for collection of tissues or those used to transport specimens to the lab should be leak proof. Care should also be taken not to contaminate the outside of the containers with blood or tissues. If contamination does occur the outside should be thoroughly cleaned or disinfected prior to removal from the necropsy room. In addition, no one should leave the necropsy room wearing contaminated gowns, boots etc. It is essential that there be adherence to the guidelines from the Centers for Disease Control (CDC) for the handling of potentially harmful pathogenic organisms.[26] Biosafety level 2 pathogens are frequently encountered especially with primates and these necropsies should be left to a pathologist or highly trained laboratory personnel.[26,27] There should also never be food or drinks in the necropsy room. Once the necropsy is done, the carcass as well as all contaminated disposable items, should be placed in sealed biohazard bags and either incinerated or autoclaved. The necropsy table, walls, counters sink etc. should be thoroughly disinfected. Instruments should be either disinfected or autoclaved.

IV. COMMON SAMPLING PROCEDURES

A. Sampling for Histopathological Evaluation

For any morphological, quantitative, or stereological study, one of the most efficient and powerful sampling procedures is the "systematic random sampling" protocol. Within the structure of interest (such as bone, cartilage, ligament or muscle), tissue blocks are sampled systematically and the pattern adopted can be varied among animals. Subdivision into tissue slices and then sample blocks is always carried out at regular intervals according to preplanned systematic protocols.[28]

Diagnostic necropsies for the detection of diseases are a science and best left to a pathologist who has been trained in the recognition of diseases and altered morphology as well as how to collect the required samples for virology or toxicology testing, etc.[24,29]

Specimens should be recorded by measuring the size and describing the precise location in a given organ (lung, bone, muscle, or joint), its color, and its consistency. Observations of the gross specimens should be done in purely descriptive terms, not in diagnostic terms.[29] The reason for this is that there is significant opportunity for variability in interpretation of many lesions from one observer to another.[29] Photography of gross specimens is also a useful way to document findings as well as to obtain teaching materials.

In general, slicing of organs or tissues should be done as sharp dissection with a scalpel blade rather than cutting with scissors that tend to crush the margins of the tissue. Samples for histopathology should be equal to or less then 5 mm thick to permit optimal fixation by the fixative.[22] The most commonly used fixative is 10% neutral buffered formaldehyde. Samples for electron microscopy should be 2–3 mm thick. The two most commonly used EM fixatives are gluteraldehyde and Karnovsky's solution both of which are toxic and should only be used in a fume hood.

The methodology for the retrieval and subsequent analysis or testing of orthopaedic implants has been adequately documented.[30–34] The use of a form on which to record all of the clinical history, radiographic findings and gross observations at the time of removal of the implant and that permits easy conversion of this data for statistical analysis is recommended.[30] The gross description should include the appearance of the implant itself, the color, shape, size as well as a description of the fiberous capsule around it and adjacent soft tissues.[31] In addition, photographing the implant at the time of removal adds a permanent visual record to document these findings. Culture swabs for aerobic and anaerobic (and mycobacteria if indicated) cultures should be taken to document the presence or absence of infection.[30,35] The implant should be excised along with the adjacent soft tissues and any capsule. A macroscopic examination for cracks, scratches, corrosion etc. may be done using a stereomicroscope.[36] If either scanning or transmission electron microscopy are to be performed, it is recommended to do whole body perfusion with buffered gluteraldehyde.[37,38] Caution should be observed as this is toxic and should be done in a fume hood while wearing gloves and protective eye ware.

For optimal fixation of tissues, especially large sized specimens, perfusion fixation techniques should be used. The procedure should be performed under general anesthesia. For large bones or joints of dogs, goats, or sheep, perfusion fixation should be done through catheterization of major arteries supplying the limb of interest or aorta and vena cava for lower limb fixation.[39] The blood is then replaced with heparinized saline, followed by formalin. Lungs should be infused with formalin by injecting it down the trachea and then tying it off.[22] Similarly, formalin can be injected into loops of intestine to preserve normal morphology and speed fixation.[36] The ideal ratio of the volume of fixative to tissue is 10:1 for formalin and twice that for alcohol solutions.[24]

B. SAMPLING FOR MECHANICAL TESTING

Specimens of bone, cartilage, tendon and ligament, or other soft tissues to be used for mechanical testing should be harvested with sufficient extra tissues around the area of interest. A hand saw or wire saw is efficient for cutting bone. Soft connective tissue specimens should be placed in saline or PBS or wrapped in saline soaked sponge until testing. Keeping surrounding soft tissue (muscle, fascia, or skin) intact is very helpful for protecting the bone from drying.

C. JOINT FLUID COLLECTION

If it is necessary to collect synovial fluid at the time of necropsy for examining cellular or biochemical components, it should be done first so as not to contaminate the skin with intestinal contents etc. There are a variety of ways to collect synovial fluid depending upon the size and ease of tapping the joint. Sterile techniques always should be followed so as not to contaminate the sample obtained.[40] The skin should be clipped and a Betadine solution permitted to remain for at least 1–2 min followed by 70% alcohol. Sterile gloves should be used and the area could be draped with sterile gauze sponges or small drapes if needed. Alternately if the joint is not easily tapped or is very small, a sterile surgical approach can be made to the joint and fluid obtained by aspiration after opening the capsule. A sterile 18 gauge needle is preferred. Synovial fluid should be collected into EDTA tubes for submission to the clinical lab for analysis.[40]

Routine examination of synovial fluid from animals should include: a total cell count, a differential count and a ratio of WBC:RBC, mucin clot quality which is a measure of the viscosity and quality of the hyaluronic acid, a total protein and an A/G ratio, specific gravity, color, viscosity and glucose concentration.[40] Total cell counts from normal joints in animals vary widely from joint to joint and the range for dogs is 0–2900 cells. The microorganisms most frequently recovered from spontaneous septic arthritis in pigs are *Actinobacillus suis*, *Staphylococcus aureus*, *Haemophilus parasuis*, *Mycoplasma hyorhinis* and *hyosynoviae*, *Corynebacterium pyogenes* and *Erysipelothrix rhusiopathiae*. Spontaneous septic arthritis in rats and mice is most frequently caused by *Staphylococus sp.*, *Mycoplasma arthriditis*, *Corynebacterium kutscheri* and *Streptobacillus moniliformis*.[41] Spontaneous septic arthritis is rare in carnivores except for *Borelia burgdorferi* in dogs.[40] In osteoarthritis or rheumatoid arthritis, many biological makers released can be detected by biochemical or immunochemical assays (see Chapters 6, 18, and 19).

D. SAMPLING FOR SEROLOGY

Blood or serum samples are commonly used for routine serological evaluation, bacterial culturing in infections of bone, joint or an implant, or examining markers of pathological conditions (such as serum sulfated glycosaminoglycans released from arthritic joints, acid phosphatase in progressive osteoporosis or bone resorption, or alkaline phosphatase and osteocalcin during fracture healing or bone formation [see Chapters 6, 15]).

At necropsy blood can be drawn from the hearts or from the axillary regions of anesthetized rodents after severing the axillary vessels. Serum samples from animals other than rodents should be drawn before euthanasia is performed. Serum samples should be collected into tubes containing a serum separator. Blood should not be forced through a small gauge needle or it will hemolyze the red cells. The serum should be separated as soon as possible and stored at –70°C.[42–44]

Serology may be used to diagnose viral infections for viruses that are very difficult to isolate and grow in tissue culture and for viruses with a prolonged course. Multiple samples may be required to demonstrate seroconversion by finding a four-fold increase in viral titer between the acute sample taken during the first two weeks of infection and a convalescent sample taken three weeks later.[42–44] The use of serology to monitor the health status of rodent colonies is standard operating procedure in many institutions. These samples, typically, are obtained by cardiac stick from an anesthetized rodent as a terminal procedure. Its limitation is that there is a variable lag time between exposure to the agent and the presence of a detectable titer and, therefore, it is not helpful in animals that become acutely ill and die. Wherever possible, serologic monitoring should be combined with necropsy of sentinel animals and examination for ecto- and endoparasites.

E. SAMPLING FOR CULTIVATION

1. Bacteria

In order to completely diagnose many pathological conditions, it is essential to culture for growth of microorganisms. Obtaining material for bacterial culture must be done in such a manner as to not contaminate the area being sampled which could result in overgrowth of the pathogen by the contaminating bacteria.[43,44] Ideally, one should try to get needed cultures before the necropsy begins to prevent contamination of the skin lesion or before opening any internal organs. Sterile instruments, gloves and gowns should be used as needed to get the sample.[42] For most aerobic cultures commercially available sterile polyester swabs such as the Culturette (Becton Dickinson Microbiologic Systems) with transport media are adequate.[25,42]

Sterile PBS may also be used for short term transport of some organisms but is not acceptable for Pastuerella.[45] If it is suspected that there are only low numbers of organisms present, it is desirable to submit a piece of tissue which can be ground up and cultured.[25] Urine can be collected by aspiration from the bladder with a sterile syringe and can be transported to the lab in a sterile screw capped container.[25,43] Normally sterile fluids like synovial fluid or pericardial fluid can be aspirated into a sterile syringe and then put into screw capped sterile tubes if the volume is small, or if the volume is large, it can be placed into blood culture tubes.[40,43,44] Blood cultures should be obtained while the animal is alive but if that is not possible, sterile blood can be obtained from the aorta or precava if done under aseptic technique.

Many anaerobic organisms are very fastidious and care should be taken not to expose the samples to oxygen or drying.[42] Anaerobic samples should be collected into a commercially available system like Anaerobic culturette (Becton Dickinson Microbiologic Systems), the Anerobic pack (Difco Labs.) or the Gas Pack Pouch (BBL Microbiology Systems).[27,42] Regular aerobic swabs are a poor choice for recovery of anaerobes.[27]

The use of cotton swabs should be avoided as they contain fatty acids that may retard the growth of bacteria.[42] Swabs are acceptable for samples from mucosal surfaces in the respiratory, urogenital systems and the conjunctiva.[43] Carey Blair transport media should be used for recovery of Shigella, Salmonella, Vibrio and Campylobacter.[42] Abscesses should be sampled by submitting 1–5 ml of pus in a sterile screw capped tube as well as a 1 cm piece of the wall of the abcess as some organisms are not recovered in the pus.[42]

It is important that bacterial cultures be obtained as soon as possible after death as gut organisms as well as others multiply rapidly after death and invade tissues which may result in overgrowth of the pathogen. In addition, the viability of some pathogens including Mycoplasma decreases rapidly after death.[24] Sampling techniques should be done so that a sufficient quantity of the material is obtained in order to optimize successful recovery and cultivation of the pathogen and so that the sample collected is representative of the lesion or disease process. Even if excellent collection techniques are utilized, failure to correctly transport or store the specimen prior to submitting it to the lab could result in a negative culture. It is wise to submit the culture as quickly as possible because the viability of some microorganisms decreases rapidly.[42,43]

2. Fungi

Proper collection of the specimen and how it is transported are of major importance in successful recovery of fungi.[43] Aseptic technique should be observed to try to avoid bacterial contamination of the culture. Scrapings from skin or nail lesions should be placed in a sterile petri dish or collected into mycobiotic or mycosel agar. Pieces of tissue may be submitted in sterile screw capped tubes for mincing or grinding by the lab.[43]

3. Viruses

The diagnosis of viral infections is dependent upon isolation and identification of the virus in cell cultures or tissues and by detection of an antibody titer in the serum.[46] The successful recovery and identification of infectious viruses necessitates that the necropsy be done within two hours of death.[40] In addition, the virus may only be present at very early stages of the disease process. Many viruses are very labile and specimens collected for viral isolation and identification should be kept moist and cold but not frozen.[44,46] If there is a transit delay time of greater than one hour, samples should be refrigerated at 4°C or placed in a container with cold packs. Samples can be stored at 4°C for up to five days.[45] Specimens should be collected into special viral transport media that contain either serum albumin or gelatin to protect the virus and antibiotics to stop the overgrowth of the culture by bacteria. Stuart media, Amies media, Liebovitz-Emory media and Hanks balanced salt solution are commercially available vial transport media.[42] A significant loss of infectious titer occurs when enveloped viruses like *Herpes simplex*, *Varicella zosteer* and Influenza are kept at room temperature or frozen at –20°C. This is not true for nonenveloped viruses like adenoviruses or enteroviruses.[42] Blood for viral isolation should be collected into sterile tubes containing an anticoagulant and refrigerated but not frozen. If it is necessary to store vial cultures longer than five days they should be frozen at –70°C. Coles[40] has published a list of many of the common animal viruses and what tissues should be submitted to the lab. If the samples must be shipped via the mail to the lab it is prudent to contact the lab at the time of necropsy and follow their directions as to how to package the tissues. Safe handling and processing of samples must be done in full compliance with all applicable CDC regulations.[26]

REFERENCES

1. "Report of the AVMA Panel on Euthanasia," *J. Am. Vet. Med. Assoc.,* 202, 231, 1993.
2. Institute of Laboratory Animal Resources, *Guide for the Care and Use of Laboratory Animals*, National Academy Press, Washington, DC, 1996, 65.
3. Close, B., Banister., K., Baumans, V., et al., "Recommendations for euthanasia of experimental animals: Part 2. DGXT of the European Commission," *Lab. Anim.,* 31, 1, 1997.
4. National Research Council, *Laboratory Animal Management — Rodents*, National Academy Press, Washington, DC, 1996, 105.
5. Wolf, T., "Lab animal technicians," *Vet. Clin. North Am.,* 15, 449, 1985.
6. Arluke, A., "Uneasiness among laboratory technicians," *Lab. Anim.,* 19, 20, 1990.
7. Bathena, S. T., "Comparison of effects of decapitation and anesthesia on metabolic and hormonal parameters in Sprague Dawley rats," *Life Sci.,* 50, 1649, 1992.
8. Mikeska, J. A. and Klemm, W. R., "EEG evaluation of asphyxia and decapitation euthanasia of the laboratory rat," *Lab. Anim. Sci.,* 25, 175, 1975.
9. Derr, R. F., "Pain perception in decapitated rat brain," *Life Sci.,* 49, 1399, 1991.
10. Vanderwolf, C. H., Buzsaki, G., Cain, D. P., Cooley, R. K., and Robertson, B., "Neocortical and hippocampal electrical activity following decapitation in the rat," *Brain Res.,* 451, 340, 1988.
11. Hanak, M. and Scott, A., *Spinal Cord Injury — An Illustrated Guide for the Health Care Professional*, Springer, New York, 1983.
12. Holson, R. R., "Euthanasia by decapitation: evidence that this technique produces prompt painless unconsciouness in laboratory rodents," *Neurotoxic. Teratol.,* 4, 233, 1992.
13. Feldman, D. B. and Gupta, B. N., "Histopathologic changes in laboratory animals resulting from various methods of euthanasia," *Lab. Anim. Sci.,* 26, 218, 1976.
14. Butler, M. M., Griffey, S. M., Clubb, F. J., Gerrity, L. W., and Campbell, W. S., "The effects of euthanasia techniques on vascular arachidonic acid metabolism and vascular and intestinal smooth muscle contractility," *Lab. Anim. Sci.,* 40, 277, 1990.

15. Howard, H. L., McLaughlin-Taylor, E., and Hill, R. L., "The effect of mouse euthanasia techniques on subsequent lymphocyte proliferation and cell mediated lymphocyte assays," *Lab. Anim. Sci.,* 40, 510, 1990.
16. Danneman, P. J., Stein, S., and Walshaw, S. O., "Humane and practical implications of using carbon dioxide mixed with oxygen for anesthesia or euthanasia of rats." *Lab. Anim. Sci.,* 47, 376, 1997.
17. Blackshaw, J. K., Fenwick, D. C. , Beattie, A. W., and Allan, D. J., "The behavior of chickens and rats during euthanasia with chloroform, carbon dioxide and ether." *Lab. Anim.,* 22, 67, 1988.
18. Lumb, W. V. Dashi, K., and Scott, R. J., "A comparative study of T61 and pentobarbital euthanasia of dogs," *J. Am. Vet. Med. Assoc.,* 172, 1978, 149.
19. Lumb, W. V., "Euthanasia by noninhalent pharmocologic agents," *J. Am. Vet. Med. Asoc.*, 165, 851, 1974.
20. Hellbreckers, L. J., Baumans, V., Bertens, A. P. M. G., and Hartman, W., "On the use of T61 for euthanasia of domestic and laboratory animals: an ethical evaluation." *Lab. Anim.,* 24, 200, 1990.
21. Bucci, T. J., "Evaluation of altered morphology," in *Handbook of Toxicologic Pathology*, Haschek, W. M. and Rousseaux, C. G., Eds., Academic Press, San Diego, 1991, Chapter 3.
22. Seaman, W. J., *Postmortem Changes in the Rat: A Histologic Characterization*, Iowa State University Press, Ames, 1987.
23. Benbrook, E. B., "The value of the necropsy in veterinary medicine," *J. Am. Vet. Med. Assoc.* 111, 65, 1947.
24. Feinstein, R. E., "Postmortem procedures," in *Handbook of Laboratory Animal Science*, Svendsen, P., Hau, J., Eds., CRC Press, Boca Raton, 1994, Chapter 23.
25. Evans, M. J., Shaw, S. G., and Wells, J. P. "Quantitative techniques for morphological evaluation," in *Handbook of Toxicologic Pathology*, Haschek, W. M. and Rousseaux, C. G., Eds., Academic Press, San Diego, 1991, Chapter 4.
26. Richmond, J. Y. and McKinney, R. W. *Biosafety in Microbiological and Biomedical Laboratories.* U.S. Dept. of Health and Human Services and the Centers for Disease Control, Washington, DC, 1993.
27. Baron, E. J. and Feingold, S. M., *Bailey and Scott's Diagnostic Microbiology*, 8th Ed., Mosby, St. Louis, 1990, 49.
28. Cruz-Orive, L. M., and Hunziker, E. B., "Steriology for anistropic cells: application to growth cartilage," *J. Microsc.,* 143, 47, 1986.
29. Dodd, D. C., "The pathologist in toxicologic testing and evaluation," in *Handbook of Toxicologic Evaluation*, Haschek, M. and Rousseaux, C. G., Eds., Academic Press, San Diego, 1991, Chapter 2.
30. Cook, S. D. and Lavernia, C. J., "Post implantation evaluation of surgical implants," in *Handbook of Biomaterials Evaluation*, von Recum, A. F., Ed., Macmillan, New York, 1986, Chapter 12.
31. Matalaga, B. F., "Tissue preparation," in *Handbook of Biomaterials Evaluation*, von Recum, A. F., Ed., Macmillan, New York, 1986, Chapter 29.
32. Woodward, S. C. and Salthouse, T. N., "The tissue response to implants and its evaluation by light microscopy," in *Handbook of Biomaterials Evaluation*, von Recum, A. F., Ed., Macmillan, New York, 1986, Chapter 30.
33. Murray, G. I. and Ewew, S. W., "A novel method for optimum biopsy specimen preservation for histochemical and immunohistochemical analysis," *Am. J. Clin. Pathol.,* 95, 131, 1991.
34. Salthouse, T. N., "Histochemical and quantitative microscopy," in *Handbook of Biomaterials Evaluation*, von Recum, A. F., Ed., Macmillan, New York, 1986, Chapter 33.
35. French, H. G., Cook, S. D., and Haddad, R. J., Jr., "Correlation of tissue reaction to corrosion in osteosynthetic devices," *J. Biomed. Mater. Res.,* 18, 817, 1984.
36. Fenwick, B. W. and Kruckenberg, S. "Comparison of methods used to collect canine intestinal tissues for histological evaluation," *Am. J. Vet. Res.,* 48, 1276, 1987.
37. Dociu, N. and Bilge, F. H., "Evaluation by scanning electron microscopy," in *Handbook of Biomaterials Evaluation*, von Recum, A. F., Ed., Macmillan, New York, 1986, Chapter 32.
38. Sheffield, W. D. and Matlaga, B. F., "Evaluation by transmission electron microscopy," in *Handbook of Biomaterials Evaluation,* von Recum, A. F., Ed., Macmillan, New York, 1986, Chapter 31.
39. von Recum, A. F., *Handbook of Biomaterials Evaluation*, Macmillan, New York, 1986, 352, 379, and 442.
40. Coles, E. H., *Veterinary Clinical Pathology,* 4th ed., W. B. Saunders, Philadelphia, 1986

41. Carter, G. R., "Isolation and identification of bacterial specimens from clinical specimens," in *Diagnostic Procedures in Veterinary Bacteriology and Mycology*, Charles C. Thomas, Springfield, 1984, 19.
42. Baron, E. J. and Funegold, S. M., *Bailey and Scotts's Diagnostic Microbiology*, 9th Ed., Mosby, St. Louis, 1994, 53, 542.
43. Murray, P. R., Kobayashi, G. S., Pfaller, M. A., and Rosenthal, K. B., *Medical Microbiology*, 2nd Ed., Mosby, St. Louis, 1994, 160, 400, 542.
44. Washington, J. A., *Principles of Diagnosis in Medical Microbiology*, 4th ed., Baron, S., Ed. University of Texas Medical Branch at Galveston, 1996, 153, 284
45. Shimoda, K., Maejima, K., Kuhara, T., and Nakagawa, M., "Stability of pathogenic bacteria from laboratory animals in various transport media," *Lab. Anim.*, 25, 228, 1991.
46. White, D. O. and Fenner, F. J., *Medical Virology*, 4th ed., Academic Press, San Diego, 1994, 191.

Part II

Evaluation Methods in Orthopaedic Animal Research

6 Methods of Evaluation in Orthopaedic Animal Research

Yuehuei H. An

CONTENTS

  I. Introduction 86
 II. Clinical Observation 86
III. Radiography 86
     A. Plain Radiography 86
     B. High Resolution Radiography and Microradiography 87
 IV. Macro Observation at Necropsy 87
  V. Structural and Morphological Evaluation 87
     A. Measurement of Length and Area 87
        1. Direct Measurement 88
        2. Measurement Based on X Ray Images 88
        3. *In vivo* Measurement 88
     B. Dissecting Microscopy 88
     C. Histology and Histomorphometry 89
        1. Paraffin Embedding and Decalcification 89
        2. Plastic Embedding and Sectioning 89
        3. Staining Procedures 90
        4. Histological Evaluation 91
        5. Histomorphometry 92
     D. Electron Microscopy 93
        1. Scanning Electron Microscopy 93
        2. Transmission Electron Microscopy 94
     E. Confocal microscopy 94
 VI. Biochemical Methods 95
     A. Detecting Markers in Body Fluid 95
     B. Detecting Biochemicals in Tissues 95
VII. Mechanical Testing 96
VIII. Special Techniques 96
     A. Autoradiography, Bone Scan, and Scintigraphy 96
     B. Computed Tomography 97
     C. Magnetic Resonance Imaging 97
     D. Single-Photon Absorptiometry, Single X Ray Absorptiometry,
        and Dual-Energy Absorptiometry 98
     E. Ultrasound 98
     F. Arthroscopy 99
     G. Measuring Tissue Blood Flow 99

IX. Molecular Biological Techniques 99
    A. Basic Terminology and Methods in Molecular Biology 99
    B. Etiology and Histopathogenesis..100
    C. Diagnosis and Prognosis 101
    D. Gene Therapy 101
References 102

I. INTRODUCTION

Evaluation methods should be carefully selected. Although many sophisticated techniques have been reported, the methods selected for most animal studies should be valid, reliable, efficient, simple, necessary, available, and economical (see Chapter 3). In orthopaedic animal research, clinical observation, radiography, macro-observation at necropsy, histological evaluation, and mechanical testing are used in most cases.

It is impossible to include a detailed description of every method which has been developed in orthopaedic animal research in a single chapter. This review only outlines the more common methods used in orthopaedic animal research and most of them are only briefly mentioned, with the purpose of helping readers find appropriate methods for their projects. One may refer to other chapters for methods of specific interest, for example, the methods for measuring bone apparent density and ash density in Chapter 8.

II. CLINICAL OBSERVATION

Based on the requirements of the study, animals should be weighed periodically. Unless otherwise indicated, animals should be weighed weekly. Any weight loss or decreased weight gain may indicate that something is wrong with the animal. Body temperature should be measured periodically and usually daily for the first week after surgery. The feeding patterns and the amount of daily feces should be also checked. Some experimental studies result in temporary weight loss from stress and other factors.

A normal and healthy animal looks calm, comfortable, has normal reactions to environmental stimuli, and is well groomed. Unlike humans, animals cannot clearly communicate that they are in pain. Therefore, observations should be made with attention to the following signs of pain:[1] (1) activity change (hyperactivity, inactivity, recumbency, withdrawal); (2) vocalization change (increase or decrease); (3) changes of eating and drinking patterns; (4) aggressive or defensive behavior; (5) changes in behavioral patterns (such as grooming, foraging, exercising, or sleep); and (6) change in body temperature. Any animal showing signs of pain should be examined carefully for bone fracture or wound abnormalities, such as infection.

A normal, healthy wound should appear dry with intact sutures. Swelling may be caused by the granulation tissue involved in wound repair (hard swelling) or may be due to hematoma or bacterial infection (soft swelling). Excessive drainage, erythema or soft swelling are indicative of infection (see Chapter 23). Any swelling, especially soft swelling, should be inspected carefully. Joint range of motion should be without limitation. Nonweight-bearing or a carried leg indicates pain or other potential problems such as fracture.

III. RADIOGRAPHY

A. Plain Radiography

Radiography is the basic method for evaluating fracture healing and bone defect repair (see Chapter 11 and 13). Pre-operative radiographs confirm normal anatomy and demonstrate the size of the bone, which is very helpful for designing or choosing fixation devices and implants of appropriate size or shape. In the operating room, roentgenograms using a C-arm unit are useful in determining the quality of internal fixation and the placement of implants. Radiographs should be taken immediately after surgery to examine the position of the fracture or defect and the quality of fixation. Periodic radiographs of the fracture or defect site are essential for monitoring the process of repair (See Chapters 11 and 13 for semi-quantitative radiographic scoring systems for fracture healing and bone defect repair). In addition, angiography has been used to examine the revascularization of a healing fracture or a vascularized bone graft.[2]

B. HIGH RESOLUTION RADIOGRAPHY AND MICRORADIOGRAPHY

After the animals have been sacrificed, the bone specimens should be radiographed again using a high resolution X ray machine, such as a Faxitron (Hewlett Packard Co., McMinnville, OR). Based on these high resolution radiographs, bone density can be quantitated by measuring the light intensity or illuminance using a digital photometer, such as a Spectron unit (Denver, Co). This technique is commonly called quantitative roentgenographic densitometry. Tiedeman et al.[3] found that X ray density correlated well with the ash density and mechanical properties of bone. They believe that this method is superior to radiographic scoring method and is capable of detecting small differences in mineral content even using standard radiographs.

Microradiography (using a high resolution X ray machine) based on thin bone sections (about 0.5–1.0 mm) provides detailed images of bone structures. It can be used in the quantification of bone apposition and ingrowth into an implant.[4-6] The quality of the bone structure around an implant can also be assessed using this method. A variation called microradiographic videodensitometry is also useful for analyzing bone structures and bone density.[5,7] The vascularity of repaired bone tissues can be evaluated using microradiography (or microangiography). In this technique radio-opaque substances such as Micropaque, $BaSO_4$, or lead oxide are injected into the arterial vasculature prior to specimen processing. This allows visualization of the cross-sectioned vessels on thin sections. Vascular trees can also be demonstrated using larger sections.[8-10]

IV. MACRO OBSERVATION AT NECROPSY

At the necropsy, macromorphology is the essential part of evaluation during surgical or necropsy procedures. The nature, color, and amount of the subjects of interest should be recorded. Regular photography is very important for documentation of what have been found at surgery or necropsy. The specimens containing osseous tissues should be radiographed.

For procedures involving bone, attention should be paid to the size of callus, the alignment of the diaphysis, the positions of the implants or fixation devices, and the surrounding tissues. For conditions involving joints, the morphology of the articular cartilage is the most important consideration. Pathological findings include roughened areas caused by arthritis, cartilage defects, fracture lines, or osteophytes. Also, the size of the joint, joint capsule and synovium; the amount and characteristics of the joint fluid; and the appearance of ligaments, menisci and the soft tissues around the joint should be observed.

Body organs and tissues, such as liver, kidney, lung, or muscle, can be harvested for evaluation of the concentrations of ions released from implanted biomaterials. Using flameless atomic absorption spectroscopy, it has been demonstrated that the lungs of minipigs contain the largest amount of titanium (Ti) ions five months after implantation of Ti screws in the mandible.[11]

V. STRUCTURAL AND MORPHOLOGICAL EVALUATION

A. MEASUREMENT OF LENGTH AND AREA

Many methods have been used for measuring length and area of bone or the dimensions of other organs or tissues. These methods include direct measurement using a ruler or caliper, measurement based on radiographs, or the use of specially designed devices. There may be significant inter-method or inter-observer variability, therefore, several observers may be needed to perform the same procedure independently or more than one method may be used. Caution should be taken when inter-study comparison is made, especially if different measurement techniques are employed.

1. Direct Measurement

Most length and area measurements are accomplished reliably by traditional techniques (a ruler or sliding digital caliper) such as the measurements of long bone dimensions (length, internal and external width).[12,13] A specially designed electronic caliper has also been reported.[14]

2. Measurement Based on X Ray Images

Two dimensional measurements are often made from X ray images using a ruler or caliper.[15–18] The percentage of magnification should be considered when using X ray images for measurements. The amount of magnification depends on the distance between the specimen and the film. A metal bar or strip with known length can be used as a reference. It should be placed at the same distance from the film as that of the subject. A standard goniometer is effective for measuring angles based on radiographic images. A custom computer program based on X ray images has been developed in the author's laboratory. It is capable of evaluating the periostial and endostial dimensions of the upper humerus and glenoid. Parameters which may be evaluated include humeral canal width, shaft width, tuberosity offset, head offset, radius of curvature of the head and glenoid, head diameter, canal flare index, glenoid height and depth, arc of enclosure, radius of curvature and depth of cancellous bone.[19]

The length or perimeter of irregular lines or the area of irregular bone or tissue specimens can be measured using computer image analysis. Most image software has the capacity to measure length and area. Images of interest (photographs, radiographs or prints) can be scanned into the computer, displayed on the screen, outlined and measured. Careful calibration of the software is necessary before accurate measurements can be made.

Measurement of the torsion angle of long bones (in humans) and the femoral angle of inclinaton (in dogs) has been reported with the use of computed tomography (CT),[20] a digital coordinator-goniometer,[21] and a symmetric axis based method.[22]

3. *In vivo* Measurement

In vivo measurement of bone length or limb length is a challenge. Often, soft tissue landmarks are drawn on the skin and the distance between the marks are measured with a tape measure. A technique called kyniklometry has been reported for measuring the distance between soft tissue landmarks on the lower legs of conscious rabbits. It is comparable to X ray stereophotogrammetry.[23] A specially designed goniometer has been reported for the measurement of joint angles in clinical practice.[24] Limb circumference measurements are useful for monitoring the progress of a swollen limb, joint, or the growth of a limb. For the quantification of limb circumference, a tape measure is effective. Methods using an electronic digitizer and a mathematical formula for an ellipse (for fetal head and body circumferences) have also been reported.[25]

B. Dissecting Microscopy

Dissecting microscopy with magnification up to 5X has been used for examining and photographically documenting the surface morphology of articular cartilage, bone, soft tisues, or implant-tissue interface. Wet specimens can be used, which is an advantage over regular, low magnification SEM. The latter may change the morphology of the specimen surface due to the critical point drying procedure. In the author's laboratory, dissecting microscopy has been found to be useful

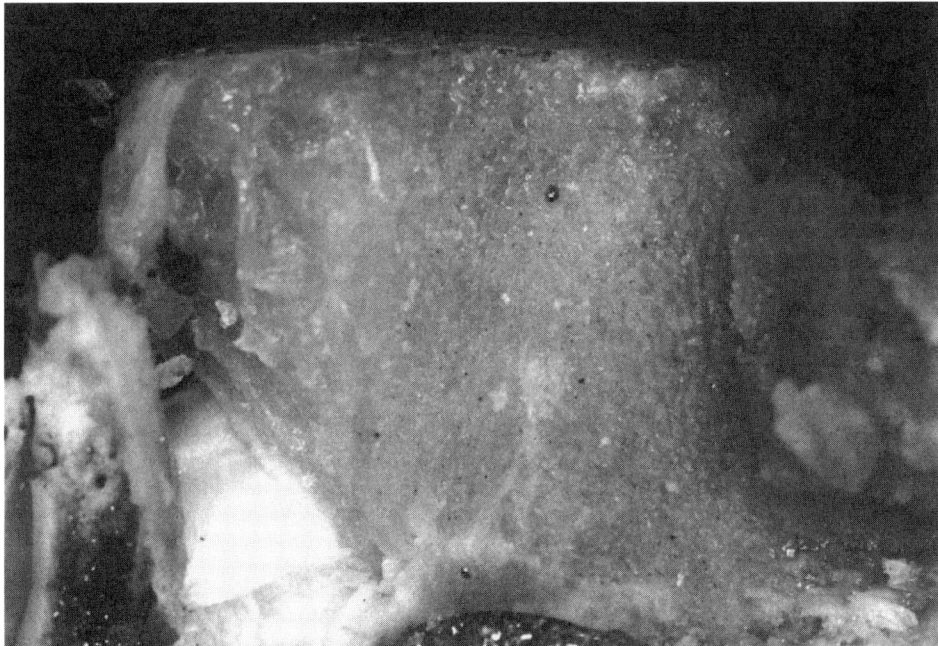

FIGURE 1. Implant debris are shown on the bone surface after the implant is removed.

for observing implant debris in the implant bed after the implant is removed (Figure 1). A grading system for evaluation of the severity of arthritis under dissecting microscope was reported by Sommerlath and Gillqist:[26] normal cartilage (Grade 0), fibrillation (Grade 1), pannus and fibrillation (Grade 2), superficial clefts (Grade 3), deep localized clefts (Grade 4), large defects (Grade 5), and complete loss of cartilage on the weight bearing area (Grade 6).

C. HISTOLOGY AND HISTOMORPHOMETRY

1. Paraffin Embedding and Decalcification

Paraffin embedding and sectioning remains the most common method for histologic study of soft tissues (subcutaneous tissue, muscle, tendon, ligament), cartilage, and decalcified bone specimens. In the decalcification of bone specimens, the goal is to achieve enough decalcification to allow successful sectioning but to avoid over-decalcification so that cellular details remain intact, thus facilitating successful enzyme and immunohistochemical staining. Common solutions for decalcification include nitric acid, HCl, formic acid, and EDTA. See Chapter 7 and the review by Skinner et al.[27] for the details of decalcification methods.

2. Plastic Embedding and Sectioning

Undecalcified preparation and sectioning are specialized procedures for the evaluation of osseous tissues (bone, calcified tissues), dental tissues, and especially specimens containing metal implants (see the review by Sanderson[28] and Chapters 7 and 20 for details). In this technique specimens are embedded in plastic media such as methylmethacrylate or Spurr's resin.

There are three major sectioning methods for plastic embedded specimens: (1) direct sectioning using heavy microtomes, (2) "sawing-grinding," and (3) sawing-only. Small undecalcified bone specimens embedded in glycol or methyl methacrylate can be cut using automatic rotary microtomes (such as the Jung Supercut 265) with a tungsten carbide blade (for methyl methacrylate or glycol methacrylate) or with a large glass blade (for glycol methacrylate). Larger undecalcified specimens can be cut using a sliding microtome (such as the Jung Model K, Heidelberg).

'Sawing-grinding" is the traditional method used with plastic embedded specimens. The specimen is sectioned with a diamond-coated wafering saw (such as the Buehler Isomet 2000, Struers Accutome-5, or Leco VC-50) into 0.2–1.0 mm thick slices. The slices are then glued onto a Plexiglass slide and ground on a grinding machine (such as the Buehler Ecomet 3, Struers Dap-V, or Leco VP–160) to produce 30–100 μm thick sections. In patient and skilled hands, the thickness of the ground sections can be less than 15 μm.[29] Well controlled systems with automatic grinding capacity, such as the Exakt sawing-grinding system (Exakt Apparatebau, Germany),[30] are also available but are costly. The process is tedious and time consuming. Also, because the slices made before grinding are relatively thick, for small specimens successful cuts have to be guaranteed for production of useful sections for evaluation without wasting.

There are two systems available for sawing-only procedures. One is the modified inner circular "sawing" technique (Fijnmetaal Techniek Amsterdam, The Netherlands) reported by van der Lubbe and Klein (see Chapter 20).[31,32] This technique can create 12±5 μm thick sections without grinding. Another sawing-only system is a diamond-coated wire saw unit, Histosaw (Delaware Diamond Knives, Wilmington, DE). According to our experience and others,[33] sections as thin as 30 μm can be created using this method in a matter of minutes. The sections are then readily glued onto regular glass microslides, stained, and coversliped. Because of its simplicity, efficiency, and relatively low cost (15,000 US dollars) it is becoming a popular method for the sectioning of undecalcified or implant-containing specimens. The advantage of these two techniques is that they are capable of sectioning hard tissues or implant-containing specimens without grinding.

Overlapping between the boundaries of implant and tissue may occur due to the porosity of some implant surfaces or because of an imperfect angle between the interface and the sectioning plane (the correct angle is 90 degrees).[29,34] This phenomenon can be conquered by making the thinnest sections possible and orienting the angle between the interface and sectioning plane at 90 degrees. If a cylindrical implant is used the plane of the cut should be perpendicular to the long axis of the implant. This allows multiple, correctly angled cuts to be made without repositioning the implant. If the cut is made longitudinal to the long axis of the implant, wrong angle phenomenon occur because, theoretically, no right angle sections can be made in this orientation since the thickness of the section cannot be zero (Figure 2). Parr et al.[34] demonstrated that interlabel distances were not significantly affected by section thickness. They suggested that the use of microradiographs for histomorphometrical analysis of the implant-bone interface is superior to brightfield analysis because of the low variability of microradiographical data and the added ability to obtain bone mineral density measurements. However, the correct sectioning angle (90°) and the thinnest possible sections always should be obtained in order to utilize the advantages of brightfield observation (visibility of cellular detail and the composition of tissues around the implant).

3. Staining Procedures

Many staining methods are available (see Chapter 7). H&E staining remains the basic and common procedure for most tissues. It can be used for both decalcified and undecalcified specimens. Both Goldner's stain[30,35] and von Kossa's stain[36] allow differentiation of osteoid from mineralized bone matrix. Other common stains for bone sections include Giemsa,[6,36] toluidine blue,[29,30,37] methylene blue/basic fuchsin,[38,39] and Stains-All.[40,41] Tetracycline labeling is a type of staining which is administered before animals are sacrificed. It is based on the propensity for tetracycline to deposit within bone, and is used for examining bone growth and remodeling. It requires undecalcified sectioning.[69] Safranin O (for GAG)-fast green,[43] Alcian blue (for PGs), and periodic acid-Schiff (for chondroitin sulfate and glycoproteins)[44] are commonly used in the evaluation of articular cartilage.

Vascular injections of India ink or other dyes prior to tissue fixation are often used for studying the vascularity of various tissues, such as bone or callus,[45] meniscus, ligament or tendon.[10,46]

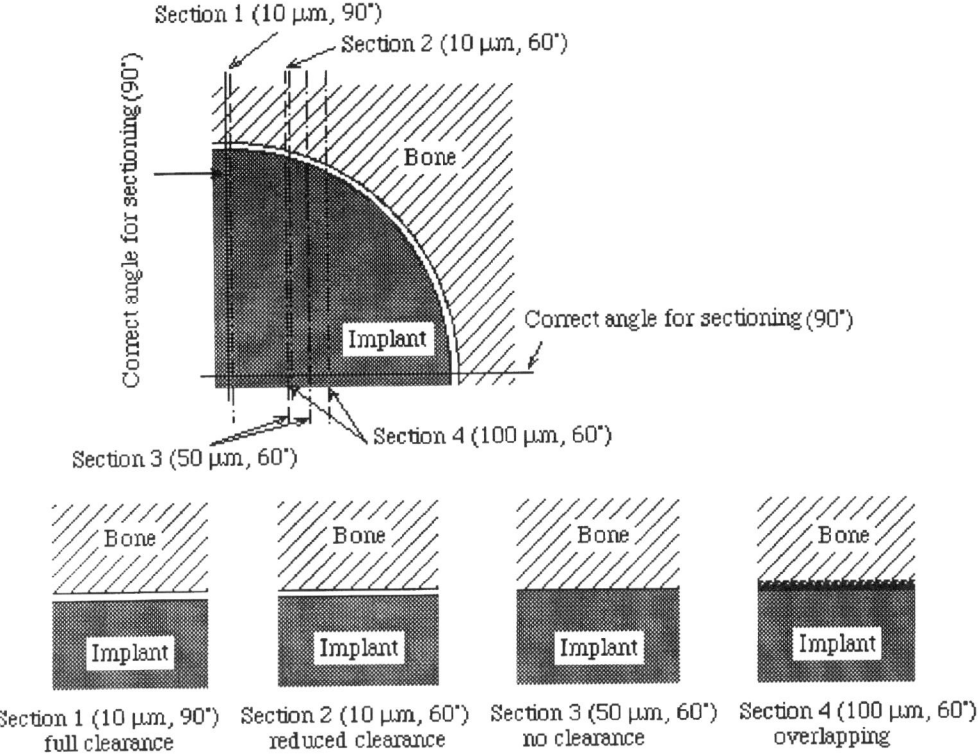

FIGURE 2. Schematic diagram showing the effects of section thickness and the angle of sectioning plane on the appearance of the implant-bone interface (a clearance at the interface is assumed).

Enzyme staining procedures have been developed for localization of alkaline phosphatase (ALP) and acid phosphatase (ACP) in bone, cartilage,[47] and tissues surrounding an implant.[48]

Immunohistochemical staining (IHCS) techniques have been further developed in recent years. IHCS techniques are used for examining biochemicals in cartilage, bone, ligament, tendon, and other tissues. Immunostaining can be used to identify types I, II and IV collagen, glycoproteins, laminin, tenasin, and fibronectin using plastic embedded bone specimens.[49] Common macromolecules such as cartilage matrix protein (CMP),[50] type I, II, and III collagen,[51-53] and PGs[54] have also been successfully localized in cartilage specimens. Other cartilage biochemicals for which IHCS staining methods have been developed include type V, VI, X and XI collagen,[53,55] chondroitin sulfate, keratan sulfate,[56] PGs,[54] stromelysin, tumor necrosis factor-a (TNF-a), TNF receptors,[57] and fibronectin.[58] IHCS techniques have been used to demonstrate the distribution of types I, II, and III collagen at the soft tissue-implant interface,[59] the healing tendon-bone interface,[60] and the ligament to bone attachment.[56] IHCS procedures for substance P, tyrosine hydroxylase, and neurofilament have been used to evaluate nerve regeneration.[61] See Chapter 7 for additional staining techniques.

4. Histological Evaluation

Histologic evaluations of bone, cartilage, ligament, tendon, synovium, and other soft tissues have been used extensively in orthopaedic research (see Chapter 7). Observation and characterization are normally done under a light microscope. Descriptive histology and histomorphometry are the two main types of histological study. Depending on the particular situation, either or both may

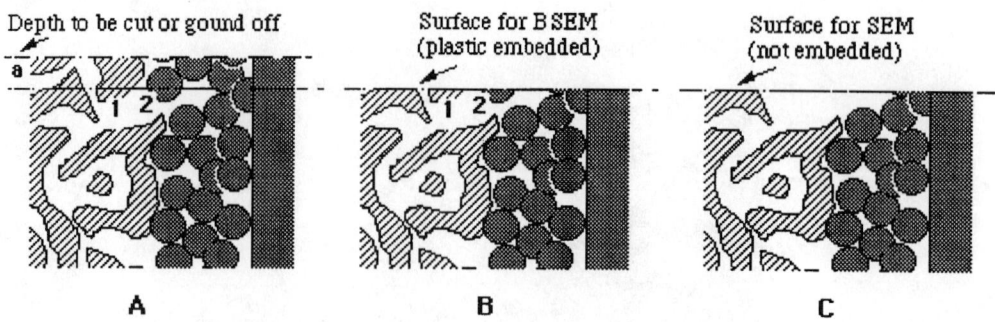

FIGURE 3. Schematic diagram showing the effects of SEM sample preparation on the trabecular bone volume. The upper surface of the tissue block containing implant-bone interface needs to be cut or ground off to create a surface for SEM or BSEM evaluation (A). If the specimen is embedded in plastic media, the end of trabeculae "1" and the lower part of bead "2" will be preserved on the surface (B). If the specimen is not embedded, the end of trabeculae "1" and the lower part of bead "2" may fall off the surface, assuming there are no continuities or connections on the sagittal plane (C).

be used. Descriptive histology is used to give a general picture of the tissue of interest, including the morphology, structure, and arrangement of cells or matrix. Scoring systems are often designed in order to semi-quantify the components of interest. An example of this is the estimation of quantity of new bone formation in a bone defect (Table 3 in Chapter 13). Full bone formation in a defect is scored as 3, moderate bone formation as 2, mild bone formation as 1, and no new bone formation as 0. The data is analyzed using non-parametric analysis of variance. Examples of other such scoring systems can be found in the appropriate chapters. Systems have been developed for evaluation of fracture healing (Table 5 in Chapter 11), bone defect repair (Table 5 in Chapter 13), cartilage defect repair (Table 3 in Chapter 16), and biocompatibility of soft tissue implants (Table 1 in Chapter 20).

5. Histomorphometry

Histomorphometric analysis has been performed using histological sections, microradiographs,[34] and backscattered electron microscopic (BSEM) images (on plastic embedded surfaces).[6,62,63] Standard SEM images of a specimen surface are less favorable for histomorphometric analysis due to the fact that overlying components from adjacent areas are not well demonstrated (Figure 3).

Histomorphometry is a methodology for quantitatively analyzing (1) length (perimeter or boundary), such as the surface perimeter of an implant, (2) distance between points, such as the clearance at implant-tissue interface or distance between the central lines of two trabeculae, (3) area, such as trabecular bone area or repair tissue area, and (4) the number of components of interest, such as trabecular number, vessel number, or cell number.[64] These parameters are the four types of primary measurements which can be made based on two dimensional (2D) images. Three dimensional (3D) parameters or structures can be calculated or reconstructed from 2D measurements according to carefully considered assumptions. Although accurate 3D data is necessary for proper comparison between different specimens (such as treated and control bone structure), it is often very difficult to reconstruct a 3D structure based on a single 2D image because the structures of most biological tissues (such as bone tissue) are anisotropic. This problem has been partially conquered by the introduction of quantitative CT,[65–67] MRI,[67] and confocal laser scanning microscopy, all of which can easily section and reconstruct the specimen.

In spite of its limitations, 2D histomorphometrical analysis remains a common and useful method for analyzing the structural changes in trabecular bone,[64,68] the callus composition in healing fracture sites, the repair tissues of bone or cartilage defects, the pathological changes in arthritis, the bone apposition and ingrowth into implant surfaces (see Chapter 21), and the soft tissue-implant interface (see Chapter 20).

Standard nomenclature, symbols, and units for bone histomorphometry can be found in the review by Parfitt et al.[64] The more commonly used terms for trabecular bone structures include BV (bone volume) or TBA, trabecular bone area, which is the trabecular surface area divided by the total area in mm^2; Tb.Th (trabecular thickness, the average thickness of trabeculae in μm); Tb.Sp (trabecular separation, the average distance between trabeculae, representing the amount of marrow space in μm). Common parameters for trabecular bone spatial connectivity include Tb.N (trabecular number, the average number of continuous trabecular elements encountered per unit area), Ho.N (hole number, the average number of holes per unit area), N.Nd (trabecular node number, nodes: trabecular branch points), N.Tm (trabecular terminus number, termini: trabecular end points), and Nd/Tm ratio. Most of the parameters can be measured using specialized imaging software.

Quantifiable parameters can be used for histomorphometrical analysis of fracture callus,[69] repair tissues of bone defects,[70,71] cartilage defects,[72] ectopic bone formation (soft tissue ossicles),[73,74] mineralized bone, non-mineralized bone, new bone, old bone, chondral tissue, fibrocartilage, hyaline cartilage, and fibrous vascular tissue. Also, parameters for cartilage repair, cartilage thickness and area, degree of attachment, and surface roughness have been developed.[75]

Parameters which are used in the evaluation of experimental arthritis include: articular cartilage thickness and area, synovial cell layer thickness, subchondral bone plate thickness, periarticular bone structure and spatial connectivity.[68,76–79] Other measurements of synovial or cartilage morphology include synovial cell density, chondrocyte and necrotic cell density, the concentration of lipid-containing cells, and mean surface destruction grade.[80]

In the histomorphometrical analysis of implant-bone interface, the useful parameters are: (1) bone apposition (or ongrowth), which is the fractional linear extent of bone apposed to implant surface divided by the total surface perimeter of the implant (i.e. the surface potentially available for apposition)[79,81] and (2) bone ingrowth, which represents the amount of ingrown bone per unit of available surface area, porous space and ingrowth depth.[79,82–84] In the case of bone ingrowth within an osteopenic bone bed, the structure of the bone, represented by TBA, Tb.Th, Tb.N, and Tb.Sp, should be also analyzed.[79,84] Characterization of the soft tissue-implant interfaces produced in percutaneous and subcutaneous implantation may include quantification of epidermal downgrowth, sulcus width, capsule thickness, macrophage density or fibroblast density (see Chapter 20).[85,86]

In examining the scar tissue within a ligament defect, areas of interest which have been analyzed include blood vessels, fat cells, loosely arranged collagen, disorganized collagen, dense cellular infiltrates, and the mixture of these elements.[87]

For vascular repair, Karim et al.[88] reported a histomorphometrical analysis of the number of vascular smooth muscle cells, actin stain positive cells, total cells, and the neointimal collagen area.

Like macro-measurements of bone dimensions, there may be significant intermethod or interobserver variability in histomorphometric analysis. When necessary, several observers may need to perform the same procedure independently or more than one method may be employed for comparison.[89]

D. Electron Microscopy

1. Scanning Electron Microscopy

Scanning electron microscopy (SEM) and backscattered electron microscopy (BSEM) are important methods for evaluation of the structure and morphology of bone structures,[7,68,90] trabecular bone surfaces,[91] osteoclast-bone interfaces,[92] cartilaginous surfaces or structures,[93] implant surfaces,[81,94] wear debris,[95] implant beds,[96] and implant-tissue interfaces.[62,97–99] SEM is also a popular method for examining the vascular structure of various corrosion casted tissues (using Mercox injection), including bone,[46,100] muscle,[101] joint,[102] ligament and tendon.[103] The shortcomings of SEM are that the specimen has to be dried before observation, causing distortion of the original spatial structure and morphology,[93] and in some instruments specimen size is limited. The first problem seems to have been solved by the new low-temperature or cryo-SEM system.[93,104]

2. Transmission Electron Microscopy

Transmission electron microscopy (TEM) is the most powerful method for evaluating the ultrastructure and morphology of large molecules (such as proteoglycans or collagens), subcellular components, cells, and even the implant-tissue interface. Immunolabeled electron microscopy makes it possible to locate biochemicals of interest, such as: PGs within cartilage;[105] collagens within cartilage cells and matrix;[106,107] and osteopontin, fibronectin, and osteocalcin within the cells or matrix adjacent to implant surfaces.[99] A relatively new, high voltage electron microscopic tomography method can be used to view the structural relationships between collagen and mineral in bone.[108] The system also has direct 3D imaging capability.

In routine cases, tissues are embedded in Epon or Spurr's medium[109] and ultrathin sections (50 nm to 2.0 μm) are cut.[39,99,108,110,111] Section preparation of metal implant-tissue interface had been a challenge for some time. In many cases the implants are removed for easier preparation of ultrathin sections.[37,112] However, implant removal inevitably damages the implant-tissue interface. Therefore, several methods for preparing tissue ultrathin sections containing the intact implant-tissue interface have been explored. These include: (1) using a soft cored implant coated with a thin layer of metal,[110,113] (2) removing bulk metal with electrochemical dissolution before embedding,[112,114] and (3) removing bulk metal by sawing-grinding techniques.[39,115] Also, 10 μm sections created by the inner circular "sawing" technique can be used directly for TEM examination.[32]

E. Confocal Microscopy

Developed in the 1980s, confocal laser scanning microscopy (CSLM) has already become a new star among the numerous imaging methods used in biomedical research. It has a wide variety of applications and its use has become widespread in orthopaedic research. The technique utilizes a laser beam which can penetrate tissue to a depth of 300–500 μm and thus reflect images beneath the surface of a specimen. Stored multilayer 2D images can then be reorganized to show 3D or cross-sectional pictures. The advantages of CLSM over the conventional SEM are its ability to view the images within a specimen or cell and the fact that it can be used with wet specimens.

CSLM has been used for viewing the structures at the implant-tissue interface, such as unmineralized bone matrix or mineralized bone.[116,117] Using CSLM, Piattelli et al.[116] found that a layer of unmineralized bone matrix lies at the interface of mineralized bone and titanium screw surface in a rabbit tibial model. Their study revealed that while 40% of the titanium surface contained bone apposition, only 10% of the bone was in direct contact with the screw surface while the other 30% was separated from the surface by an unmineralized tissue layer. Confocal microscopy has also been employed for the examination of cellular survival and proliferation in autogenous flexor tendon grafts in a canine model.[118] In another study, CSLM was used to determine the location of viable chondrocytes in frozen and thawed osteochondral articular cartilage.[119] An additional use of CSLM is for examining cell location and population in cell-seeded porous constructs.[120] It has also been utilized to detect the location of type IX collagen in cartilage[121] and type X and XI collagen in the bovine collateral ligament-bone junction.[122,123]

VI. BIOCHEMICAL METHODS

A. Detecting Markers in Body Fluid

Markers are substances which can be detected in various body fluids (including synovial fluid, serum, and urine) and may reflect pathologic conditions. There are three categories of markers which are commonly used in the study of bone and cartilage. These include: (1) metabolic products of cartilage (chondrocalcin [C-propeptide of type II collagen], pyridinium crosslinks, hyaluronic acid [HA]) and sulfated glycosaminoglycans [S-GAG] such as chondroitin sulfate [CS] and keratan sulfate [KS]) and bone metabolic products (osteocalcin [bone gla-protein], hydroxyproline and deoxypyridinium crosslinks); (2) specific enzymes of cartilage (stromelysin or collagenase) and bone (bone-specific alkaline phosphatase, tartrate-resistant acid phosphatase); and (3) hormonal substances, growth factors, or cytokines relevant to cartilage repair (prostaglandin E2 [PGE2], interleukin–1 [IL–1]) and bone repair (endothelial cell-stimulating angiogenesis factor).

Generally, markers of joint pathology (osteoarthritis, rheumatoid or inflammatory arthritis, or joint injury) include elevated levels of S-GAG (CS and KS), HA, chondrocalcin, stromelysin, collagenase, and decreased levels of IL–1 and TNF-OC in synovial fluid, elevated levels of S-GAG (CS and KS), and HA and a decreased level of osteocalcin (due to suppressed bone formation in RA) in serum, and an elevated level of pyridinium crosslinks (due to bone resorption in RA) in urine.[124–127] Indicators of bone resorption (in osteoporosis) include elevated levels of tartrate-resistant acid phosphatase and pyridinoline and/or pyridinoline-containing peptides in plasma and elevated levels of fasting calcium and hydroxyproline, deoxypyridinium crosslinks and hydroxylysine glycosides in urine.[128–130] Markers for bone formation (in fracture healing, hyperparathyroidism, or hyperthyroidism) include increased levels of osteocalcin, procollagen peptides, and bone-specific alkaline phosphatase in serum.[128–130]

Enzyme linked immunosorbent assay (ELISA), radioimmunoassay (RIA) and high-pressure liquid chromatography (HPLC) are common biochemical methods used for detecting markers in body fluid. Due to their relatively noninvasive nature, these assays are clinically important for monitoring the pathological progress and rehabilitation of certain osseous or cartilaginous conditions. ELISA has been used for evaluating GAG,[131] KS,[132,133] or CS[134] in synovial fluid; osteocalcin,[135,136] GAG,[131] and KS[135,137] in serum; and telopeptide of type I collagen in urine.[136] RIA has been used for detecting KS[138] in synovial fluid; osteocalcin,[139] carboxyterminal propeptide and telopeptide of type I collagen[140] in serum. Immunoassays are rapid, extremely sensitive and require little specialized equipment. HPLC methods have also been shown to be very effective in quantifying some markers, such as the synovial fluid HA,[141] serum PGs,[141,142] and urinary pyridinium crosslinks of collagen.[135,143,144]

B. Detecting Biochemicals in Tissues

Another class of histological stains can be used to localize certain biochemicals in cells or tissues, but they provide only qualitative data. Examples include: safranin O and Alcian blue which specifically stain GAG in cartilage and immunohistochemical stains which can selectively label ALP or ACP in bone and cartilage.

Numerous biochemical tests are available for quantification of biochemical components in bone, cartilage, ligament, tendon, and other tissues. For articular cartilage, DNA content can be evaluated with several different methods,[145–147] the hydroxyproline content (a measure of collagen) with a colorimetric[148,149] or ^{3}H-hydroxyproline method,[150] the ratios of type I and II collagen with SDS-polyacrylamide gel method,[151] PGs or GAG content with a hexosamine method,[152] electrophoresis,[153] immunoblotting,[153] or 1,9-dimethylmethylene blue method,[149,154] PGs synthesis with RIA,[149,155,156] CMP synthesis with a immunoblotting method,[50] and fibronectin content with ELISA.[58] Biochemical methods are also used to evaluate gelatinase activity[149,157,158] and tissue inhibitor of metalloproteinase (TIMP).[149,159] Immunochemical methods in cartilage research were summarized by Hardingham.[160]

TABLE 1
Mechanical Properties of Tissues and Materials Commonly Encountered in Orthopaedic Research

Material	Ultimate strength (MPa)	Elastic modulus (GPa)	First author, year[Ref.]
Biological materials			
Cortical bone	35–283	5–23	An 1998 Chapter 8
Cancellous bone	1.5–38	10–1570 (MPa)	An 1998 Chapter 8
Tendon	10–200	50–1000 (MPa)	Silver 1987[172]
Skin	4–14	6–44 (MPa)	Silver 1987[172]
Hyaline cartilage	—	0.41–0.89 (MPa, compression)	Mow 1997[175]
Arterial wall	0.5–1.72	1 (MPa)	Silver 1987[172]
Biomaterials			
Al_2O_3	—	550	Wright 1990[176]
Cobalt alloy	700	200	Wright 1990[176]
SS	850	180	Wright 1990[176]
Ti alloy	1250	110	Wright 1990[176]
HA	600	19	Heimke 1986[177]
PMMA	35	3	Wright 1990[176]
UHMWPE	27	1	Wright 1990[176]
Synthetic rubber	10–12	4 (MPa)	Black 1988[174]

For bone tissues, assays for DNA, RNA, calcium, ALP, osteocalcin, and collagen content are available.[161–164] Interestingly, structurally intact and functionally active ALP has even been isolated from clavicle fragments of an Egyptian mummy (before 2000 BC) with an ELISA method using a monoclonal antibody.[165]

For ligament and tendon, biochemical methods have been used to evaluate DNA content,[166,167] collagen synthesis and content,[166–169] PGs or GAG synthesis and content,[166–167] collagenase activity,[170] and fibronectin content.[171] Determination of the biochemical composition of vasculature also has been reported, including hydroxyproline (for collagen), total tissue protein, DNA and RNA.[88]

VIII. MECHANICAL TESTING

Mechanical properties are the basic parameters of living tissues[172,173] and biomaterials[89,174] which are especially important in orthopaedic research. Mechanical properties of bone, cartilage, ligament, tendon, and other tissues have been well documented. Values of strength and elastic modulus of common biological materials and biomaterials are listed in Table 1. Mechanical testing techniques for bone tissues include bending, compression, tension, indentation, torsion, screw pullout test, strain gauges and ultrasonic methods (see Chapter 8). Mechanical testing procedures for bone-implant interface include pushout, pullout, screw pullout test, and a torque removal technique (see Chapter 8). The indentation and confined compression tests are common procedures for testing mechanical properties of articular cartilage (see Chapters 9 and 16). For ligament and tendon, a tensile test should be used (see Chapter 10).

IX. SPECIAL TECHNIQUES

A. Autoradiography, Bone Scan, and Scintigraphy

Macro- and microautoradiography can be used to examine fracture healing or other repairing tissue in animal models. Briefly, a radioactive chemical (such as ^{99}Tcm, ^{45}Ca, or ^{3}H) is injected intravenously before the animal is sacrificed.[178,179] For macro-autoradiography, bone slices are cut from the fracture site and the slices (or whole bone) are placed on X ray film in a dark room. The film is developed after 12 hrs of exposure. For micro-autoradiography, paraffin sections or plastic embedded sections are made, emulsion coated, exposed for 1 to 27 days (depending on the radiochemical employed) and then developed.[178,179] The sections are then examined under light microscope. Using micro-autoradiography, DePalma et al.[178] studied the repair process of cartilage defects by quantification of labeled cells in the defects and adjacent marrow areas. Greiff found that radioactivity was localized in the callus and growth plate (the mineralizing regions).[179] Autoradiography has been also used for evaluation of osteoarthritis,[180] inflammatory arthritis (decreased uptake by bone at the early stage of the disease),[181] repair tissues in bone defects (using ^{3}H-thymidine),[182] bone formation around implants,[111] and cellular and collagen distribution within repaired ligaments.[169]

Radioactive isotopes such as ^{99}Tcm, or ^{67}Ca are used for bone scan or scintigraphy. Several hours after injection of the isotope, the radioactivity absorbed by normal bone, necrotic bone, or fracture callus is detected by a scintillation detector. Normally, increased uptake of radioisotope is found at the sites of bone growth or repair with rich blood circulation,[183,184] and decreased or absent uptake is seen in the areas of ischemic osteonecrosis.[185] This technique has been used for evaluation of bone healing,[184,186] infected fracture,[186] inflammatory arthritis,[181] ischemic osteonecrosis,[2,185] bone healing around implant,[187] and ischemic muscle lesions.[186]

B. COMPUTED TOMOGRAPHY

CT has been used for examining bone structure, bone destruction, new bone formation during fracture healing and bone lengthening in animal models.[188–190] Noninvasively, CT also has been used to evaluate the revascularization of femoral head necrosis in a canine model (with Micropaque injection).[2]

Quantitative CT (QCT) is capable of analyzing bone structure, even in small rat bones and is believed to be more sensitive than dual-energy X ray absorptiometry (DEXA).[191] The spatial resolution of CT on cancellous specimens can reach 8–80 μm.[192,193] The recent development of QCT has resulted in images with high 3D resolution, which may be used for 3D reconstruction of cancellous bone.[66,67] Based on a 2D array, Feldkamp developed what has become known as the microCT scanner for 3D reconstruction of bone.[194] Another CT method for 3D reconstruction is the X ray tomographic microscope (XTM), which allows *in vivo* evaluation of cancellous bone.[65]

Quantitative CT has also been used for evaluating the density and mechanical properties of bone. It can be applied *in vivo* or on excised bone specimens.[195] CT numbers, image intensity or CT density values (such as Hounsfield units: HU), are measured in the areas of interest. CT density is based on relative attenuation of X rays by a scanned body as compared with attenuation by water. In general, zero HU equals the density of water and –1,000 HU corresponds to the relative density of air. Cortical bone has CT density greater than +1,000 HU and cancellous bone has values ranging from –25 to 714 HU. An average CT value of water is determined for each scan to adjust the systemic error of the machine.[195] By correlation analysis, power functions between CT density and mechanical values (such as strength or elastic modulus), apparent density and ash density of bone can be formulated. Therefore, mechanical values and densities of bone can be predicted by CT values.[189,195,196] The advantage of QCT is that it can be applied noninvasively and *in vivo*.

C. MAGNETIC RESONANCE IMAGING

MRI provides exceptional soft tissue contrast not afforded by other imaging methods. Therefore, the internal components of synovial joints, i.e., cartilage, menisci, synovium, ligaments, etc. can be seen in exquisite detail.[197] In a canine model, MRI can even distinguish herniated intervertebral disc tissue from scar.[198]

MRI has been used for examining changes in bone and cartilage such as the progression of osteoarthritis[199,200] (see Chapter 18), fracture healing,[201,202] altered bone structure after bone biopsy,[203] healing menisci,[204] cartilage thickness,[205,206] and early osteonecrosis in a variety of animal models.[185,207,208] In a canine model, the necrotic changes of the femoral head are thought to be detectable by MRI within four weeks after the ischemic insult.[207] MRI is also useful for detecting early changes of bone marrow produced by bacterial infection.[209]

Recent studies have shown that MRI also may provide high resolution 3D images of trabecular architecture.[67,210–212] MRI is believed to be superior to CT and ultrasound methods for this purpose due to its ability to distinguish the boundary between muscle and bone and even between the cortical and cancellous regions within the bone.[213] MRI has also been used for evaluating bone mineral density (BMD),[214–216] and predicting bone elastic modulus (needs further study),[217] which is very significant for the diagnosis and monitoring of osteoporosis. The latter topic merits increased research attention.

D. Single-Photon Absorptiometry, Single X ray Absorptiometry, and Dual-Energy Absorptiometry

Single-photon absorptiometry (SPA) and dual energy absorptiometry (DEA) are two noninvasive methods for measuring bone mineral content (BMC) and BMD. They are most commonly applied to the appendicular skeleton.[188] The radioactive sources for SPA are ^{125}I and ^{241}Am. SPA has commonly been used for measuring the mineral density of the distal radius, ulna, calcaneous and femoral neck. Using formulas generated by regression analysis, SPA can be used to estimate the mechanical properties of healing bone.[218,219,201] A new method for measurement of bone mass is single X ray absorptiometry (SXA) reported by Borg et al.[220] The SXA device has an X ray tube which emits X rays at an energy level of 40 kVp and 0.2 mA. It has been used to measure the BMC and BMD of forearm bones and the results have shown a very good correlation with the more traditional SPA method.[220]

DEA can be performed with either radioisotopes or X rays. When the dual-energy source is derived from X rays, the technique is termed dual-energy X ray absorptiometry (DEXA). A high correlation has been found between DEXA and traditional methods for measuring bone density.[221] DEXA has been demonstrated to accurately measure the BMC and BMD of very small areas of interest,[222] such as in rat bone.[223–225] Like SPA, DEXA has been commonly used to evaluate BMC, BMD, and mechanical properties of normal bone,[226] healing bone[188,201,227] and osteoporotic bone[225,228,229] in animal models.

E. Ultrasound

Clinically, ultrasound has significant diagnostic values for the evaluation of joint effusions, cartilage defects and joint capsule thickening.[230] Muscle or tendon tears can also be detected and the healing process monitored. Using ultrasound tomography, correct diagnoses can be made for animals with soft tissue abscesses, foreign bodies, hematomas, and soft tissue tumors. The recently developed high frequency ultrasound backscattered microscope has proven to have the ability to visualize the subsurface structures in immature articular cartilage and some of its developmental changes.[231] Another recent report has demonstrated the use of echography *in vivo* and *in vitro* for the assessment of changes in articular cartilage and subchondral bone in experimental arthritis models.[232,233] Ultrasound has also been used for detecting osteomyelitis (OM) in turkeys.[234] OM lesions were recognized as hyperechoic (bright white) disruptions in cortical bone with a specificity rate of 83%.

Quantitative ultrasound (QUS) parameters, such as broadband ultrasound attenuation, ultrasound velocity, and ultrasound attenuation, can be used to investigate bone structure.[193] In an *in vitro* study on trabecular bone cubes, it was found that ultrasound parameters were significantly associated with bone structural indices, such as Tb.Sp or trabecular connectivity.[193]

QUS is becoming an alternative to photon absorptiometry in assessing bone density. This has been useful in the diagnosis and management of osteoporosis.[235–237] The diagnostic sensitivity of QUS on BMD is similar to that of DEXA, even on small rat bones.[235] It is also very useful for predicting hip fracture risk.

Ultrasound is a very important tool for measuring mechanical properties of bone as well (see Chapter 8). Ultrasonic techniques offer some advantages over direct mechanical tests for measuring the elastic modulus of bone.[238] Specifically, the specimens can be smaller, with less complicated shapes (cylinder or cube) and several anisotropic properties can be tested using one specimen.[239] Recently, with the combination of vibration analysis and ultrasound velocity measurements, whole bone mechanical characteristics have been assessed *in vivo*.[240]

F. Arthroscopy

Diagnostic arthroscopy can be used to visualize the surfaces of most components in the joint (articular cartilage, intra-articular ligaments, menisci and synovium).[241,242] It has also been used for intraarticular surgical procedures, such as ACL transection.[243] The use of arthroscopy is limited to larger animals due to the size of available scopes.

G. Measuring Tissue Blood Flow

Several methods have been employed for measuring tissue blood flow, these include: (1) the microsphere method, in which radioactive beads injected into the bloodstream distribute into various organs in proportion to their blood supply;[244] (2) the indicator-dilution technique, in which clearance of the indicator is a measure of tissue perfusion;[245] and (3) laser-Doppler flowmetry, in which relative perfusion values are determined via detection of blood cell movement. This is done by analyzing the Doppler shift of backscattered light originating from a monochromatic laser light source.[246,247] The latter is a nondestructive method for determining real-time blood flow in a variety of tissues including muscle and bone.[247,248] It has been used to investigate the blood flow in bone under different pathological situations, such as osteonecrosis, fracture healing, or arthritis.

X. MOLECULAR BIOLOGICAL TECHNIQUES

Amazing advances in molecular biological technology (MBT) have had an enormous impact on virtually every aspect of medicine, including orthopaedic surgery. MBT is a powerful tool. It has been used for the isolation and analysis of specific regions of chromosomal DNA which indicate various pathologic conditions. In animal models, gene transfer has proven to be an effective therapeutic method for many musculoskeletal diseases. Clinical trials of gene therapy in humans for rheumatoid arthritis are expected to appear soon. Comprehensive reviews of the use of molecular biological techniques in orthopaedic research have been reported by Bridge,[249] Sandberg,[250] and Shore and Kaplan.[251,252]

A. Basic Terminology and Methods in Molecular Biology

The basic terminology given here is adapted from the reviews by Shore and Kaplan.[251,252] A gene is a unit of heredity, consisting of a segment of chromosomal DNA that is required for production of a functional protein or RNA. The gene contains both coding and regulatory regions. A transgene is a foreign gene which has been spliced into an animals original genomic DNA. mRNA is a type of RNA that contains protein coding information. Nucleotide sequence refers to the order of nucleotides in a given segment of DNA or RNA. Translocation is the transfer of a portion of DNA from one chromosome to another. A probe is a DNA or RNA molecule that is labeled, or tagged, and can then be used to locate a complementary DNA or RNA strand through hybridization. Vectors are DNA molecules that are used as carrier molecules for cloned DNA sequences. They contain information which allows recombinant molecules to be replicated in host bacterial cells. A plasmid is a small circular double-stranded DNA molecule which is found in bacteria and replicates independently of the host chromosome. They are commonly used as vectors in molecular cloning. A recombinant DNA molecule is a DNA molecule containing segments of DNA from different origins, such as a piece of human DNA that has been joined to a plasmid DNA. A clone is a term used to describe identical segmental DNA molecules produced by recombinant DNA technique. Molecular cloning is a process by which a specific segment of DNA is isolated and then numerous identical copies, or clones, of that segment of DNA are generated.

Hybridization is the process of matching complementary strands of DNA or RNA or both to form a double stranded molecule. *In situ* hybridization (ISH) is the hybridization of a DNA or RNA probe to a target molecule that has not been extracted from its original cellular location, within a chromosome or in a fixed tissue section.[253] Immunohistochemistry is analogous to ISH for nucleic acids and is used to detect the distribution of a specific protein within a cell or tissue. In immunohistochemistry, a specific antibody serves as the probe to detect the protein of interest. Gel electrophoresis is a method of separating DNA, RNA, or protein molecules based on their size and electrical charge. This technique makes use of the fact that, under an electrical field small molecules migrate through a gel matrix (agarose or acrylamide) faster than larger molecules. Southern blotting or transfer is a technique which is used to transfer DNA that has been electrophoresed through an agarose gel onto a solid support for hybridization. Northern blotting or transfer is the process of transferring RNA onto a solid filter support for hybridization. Western blotting or transfer is the process of transferring proteins that have been electrophoresed through an acrylamide gel onto a solid filter support for detection of a specific protein by antibody labeling.

B. Etiology and Histopathogenesis

Cloned DNAs can be used to identify the chromosomal location of a gene. This is important in helping to associate a specific gene with a disease. The localization of specific genes has been accomplished by determination of critical, consistent kayotypic breakpoints which are characteristic of specific histological types of mesenchymal neoplasms.[249]

Several methods are available for localization of specific genes, translocated chromosomes, or chromosomal breakpoints, for example: fluorescent *in situ* hybridization (FISH), in which a fluorescently labeled probe is allowed to hybridize to specific chromosomes and thus determines their gene location microscopically.[252] Also, *in situ* hybridization (ISH) is capable of detecting mRNA for a specific gene, using nucleic acid probes which are complementary to and hybridize with the mRNA. Because it is performed on tissue sections, ISH localizes the mRNA to individual cells within the tissue and thereby allows investigation of the spatial distribution and heterogeneity of expression of particular genes within a population of cells. The method described by Hicks et al.[253] is a typical example of ISH. It describes, in detail, a method for performing ISH on skeletal tissue cells (growth plate cells) using synthetic oligonucleotide (a short sequence of nucleotides) probes. This technique is currently being translated into diagnostic practice.[254]

Specific chromosomal abnormalities have been associated with a number of skeletal neoplasms. In Ewing's sarcoma or Askin tumor, a rearrangement of chromosome 11 and 12 t(11;12) has been found. A translocation of chromosomes X and 18 t(X;18) is characteristic of synovial sarcoma of the extremities. In osteosarcoma, chromosomal regions 1p11–13, 1q10–12, 1q21-22, 11p15, 12p13, 17p12–13, 19q13, and 22q11–13 are frequently rearranged.[255] Also, overexpression of the cyclin G1 gene is frequently observed in human osteosarcoma cells.[256] In 50% of solitary lipomas, a structural rearrangement of 12q13–14 has been detected. Many chromosomal translocations are tumor-specific.[249] The identification of clonal chromosome abnormalities has also been reported in association with orthopaedic conditions, such as proliferative fasciitis, Dupuytren contractures and osteochondromatosis. This technique has provided new evidence for a neoplastic origin of these lesions in contrast to previous theories of reactive, developmental, or hormonal etiologies.

Collagens is the basic component found in the extracellular matrix of healthy tissues. Mutations in collagen structures cause a variety of diseases that include osteogenesis imperfecta (mutations in one of the two structural genes for Type I procollagen), chondrodysplasia (mutations in genes for Type II collagen), and possibly, some forms of osteoporosis, osteoarthritis, and aortic aneurysms (defects in Type I, II, or III collagen, respectively).[257,258] Many disease phenotypes have been produced in transgenic mice by introducing mutated collagen genes. Such animal models have proven to be excellent tools for investigating the consequences of mutations in collagen genes and have helped to identify additional diseases caused by collagen defects.[259]

Gene expression during fracture repair has been reviewed by Sandberg et al.[250] They summarized the regulation of genes coding for extracellular matrix components and growth regulatory molecules during fracture healing. The information available focuses on the sequential expression of genes coding for collagens, PGs, and other matrix proteins during secondary (callus) healing. The temporal and spatial distribution of different connective tissue components (mesenchyme, cartilage, and bone) is closely linked to the expression of genes coding for their characteristic constituents. The current and the near future development of MBT will provide answers to some persistent questions in fracture healing.

C. Diagnosis and Prognosis

The examination of specific genes, chromosomal translocations, and gene expression allows direct and accurate diagnosis of many clinical conditions (such as cancer). It also permits characterization of the changes which occur in different stages of tissue repair (cartilage or bone repair). Some especially powerful applications include: diagnosing specific disease processes which may have varying histological characteristics; distinguishing certain neoplasms which have similar histological appearances; and distinguishing certain benign tumors from their malignant counterparts. Chromosome analysis and reverse transcription-polymerase chain reaction (PCR) methods are commonly used for disease diagnosis and follow-up.

Cytogenetic findings obtained after clinical diagnosis have many applications. They can be used to: (1) serve as indicators of disease progress (such as the transformation of a benign tumor into its malignant counterpart or transformation from one tumor into another); (2) indicate the need for a change in treatment plan (as would be the case for a malignant transformed tumor): and (3) monitor the effectiveness of treatment (by detecting reduction or elimination of a specific tumor or monitoring the changes seen in different stages of fracture healing).

D. Gene Therapy

Gene transfer is a procedure in which certain therapeutic genes are administered locally or systemically. This can be accomplished by direct introduction into diseased sites (such as tumor, bone or cartilage defects, or arthritic joints) or by *ex vivo* transfer into cells or tissues which are then transplanted into the diseased location. Systemic administration allows the secreted gene products to enter the circulation and thereby reach the disease sites. A variety of vectors, including retroviruses, adenoviruses, herpes simplex viruses, and liposomes, as well as naked DNA, have been used to deliver genes into living subjects.

Several gene transfer methods aimed at the treatment of osteosarcoma have been reported. Ko et al.[260] reported that the recombinant adenovirus (Ad) vector containing the thymidine kinase (TK) gene driven by the osteocalcin (OC) promoter (Ad-OC-TK), when delivered concurrently with acyclovir, is highly selective in blocking the growth of osteosarcoma in a nude mouse model. Another recent report showed the efficacy of a high-titer antisense cyclin G1 retroviral vector in a nude mouse model of osteosarcoma.[256]

A number of highly specific post-translational enzymes involved in collagen biosynthesis have recently been cloned. There has been increasing interest in the possibility that the unique post-translational enzymes involved in collagen biosynthesis may offer attractive targets for specifically inhibiting excessive fibrotic reactions in a number of diseases. Several investigations also have suggested that it may be possible to inhibit collagen synthesis with oligo-nucleotides or antisense genes.[258]

Gene therapy offers novel possibilities for the treatment of inflammatory or rheumatoid arthritis.[261] Presently, *in vitro* and *in vivo* investigations are directed toward gene transfer in order to allow the delivery of genes whose products possess antiarthritic properties, such as IL–1 receptor antagonist gene or tumor necrosis factor inhibitor (TNFI) gene.[262] Gene transfer using a retrovirus vector has been successful in achieving high intraarticular transgenic expression of an IL–1 receptor antagonist. This has had promising antiarthritic effects in animal models. Soon a human trial using this principle is expected to appear.[261] A similar study in dogs has demonstrated that the injection of transduced synovial cells (with human IL–1a gene) can slow the progression of experimentally induced osteoarthritis.[263]

By gene transfer, it is also anticipated that therapeutic growth factors to be expressed *in vivo* at high concentrations for an extended period of time in order to enhance tissue repair. The feasibility of this concept has been tested by Kang et al.[264] in an articular cartilage defect model. They found that rabbit chondrocytes were susceptible to *in vitro* retrovirally mediated gene transfer, and that transgene expression persisted for four weeks following allotransplantation into full-thickness articular defects. This is a new, potentially very effective, approach to orthopaedic tissue repair. Further investigation is needed.

In the area of tendon and ligament repair, a number of growth factors have the potential to enhance the healing process, but they are extremely difficult to deliver clinically. Several studies have demonstrated that a possible solution to this problem is the use of gene transfer. Direct injection of an adenoviral vector carrying the lacZ gene and allotransplantation of lacZ containing tendon fibroblasts has resulted in lacZ gene expression throughout the body of the tendon itself. The effects lasted for six weeks, which may be long enough for clinical usefulness.[265,266]

REFERENCES

1. Smith, A. J., "The treatment of pain and suffering in laboratory animals," in *Handbook of Laboratory Animal Science*, Svendsen, P. and Hau, J., Eds., CRC Press, Boca Raton, 1994, Chapter 20.
2. González del Pino, J., Knapp, K., Gómez Castresana, F., and Benito, M., "Revascularization of femoral head ischemic necrosis with vascularized bone graft: a CT scan experimental study," *Skeletal Radiol.*, 19, 197, 1990.
3. Tiedeman, J. J., Lippiello, L., Connolly, J. F., and Strates, B. S., "Quantitative roentgenographic densitometry for assessing fracture healing," *Clin. Orthop.*, 253, 279, 1990.
4. Young, F. A., Spector, M., and Kresch, C. H., "Porous titanium endosseous dental implants in rhesus monkeys: microradiography and histological evaluation," *J. Biomed. Mater. Res.*, 13, 843, 1979.
5. Kälebo, P. and Jacobsson, M., "Recurrent bone regeneration in titanium implants. Experimental model for determining the healing capacity of bone using quantitative microradiography," *Biomaterials*, 9, 295, 1988.
6. Yan, W.-Q., Nakamura, T., Kobayashi, M., Kim, H. M., Miyaji, F., and Kokubo, T., "Bonding of chemically treated titanium implants to bone," *J. Biomed. Mater. Res.*, 37, 267, 1997.
7. Bachus, K. N., Bloebaum, R. D., Rubman, M. H., Bachus, K. N., and Plaster, R. L., "Microscopic analysis of autograft bone applied at the interface of porous-coated devices in human cancellous bone," *Cells Mater.*, 2, 3, 1992.
8. Brueton, R. N., Brookes, M., and Heatley, F. W., "The vascular repair of an experimental osteotomy held in an external fixator," *Clin. Orthop.*, 257, 286, 1990.
9. Schliephake, H., Neukam, F. W., Hutmacher, D., and Wustenfeld, H., "Experimental transplantation of hydroxylapatite-bone composite grafts," *J. Oral Maxillofac. Surg.*, 53, 46, 1995.

10. Seitz, H., Hausner, T., Schlenz, I., Lang, S., and Eschberger, J., "Vascular anatomy of the ovine anterior cruciate ligament. A macroscopic, histological and radiographic study," *Arch. Orthop. Traum Surg.*, 116, 19, 1997.
11. Schliephake, H., Reiss, G., Urban, R., Neukam, F. W., and Guckel, S., "Metal release from titanium fixtures during placement in the mandible: an experimental study," *J. Oral Maxillofac. Implants*, 8, 502, 1993.
12. Horsman, A. and Leach, A. E., "The estimation of the cross-sectional area of ulna and radius," *Am. J. Phys. Anthrop.*, 40, 173, 1963.
13. An, Y. H., Kang, Q., and Friedman, R. J., "The mechanical symmetry of rabbit long bones studied by bending and indentation test," *Am. J. Vet. Res.*, 57, 1786, 1996.
14. Ross, W. D., Rempel, R. D., Quibell, R. W., Smith, D., and Bird, L., "Technical note: an electronic caliper designed for measuring bone breadths in living subjects," *Am. J. Phys. Anthropol.*, 90, 373, 1993.
15. Horsman, A. and Simpson, M., "The measurement of sequential changes in cortical bone geometry," *Br. J. Radiol.*, 48, 471, 1979.
16. Chumlea, W. C., Mukherjee, D., and Roche, A. F., "A comparison of methods for measuring cortical bone thickness," *Am. J. Phys. Anthropol.*, 65, 83, 1984.
17. Rico, H. and Hernandez, E. R., "Bone radiogrametry: caliper versus magnifying glass," *Calcif. Tissue Int.*, 45, 285, 1989.
18. Tabensky, A. D., Williams, J., DeLuca, V., Briganti, E., and Seeman, E., "Bone mass, areal, and volumetric bone density are equally accurate, sensitive, and specific surrogates of the breaking strength of the vertebral body: an *in vitro* study," *J. Bone Miner. Res.*, 11, 1981, 1996.
19. McPherson, E. J., Friedman, R. J., An, Y. H., Chokesi, R., and Dooley, R. L., "Anthropometric study of normal glenohumeral relationship," *J. Shoulder Elbow Surg.*, 6, 105, 1997.
20. Pfeifer, T., Mahlo, R., Franzreb, M., Heiss, U., Lutz, P., et al., "Computed tomography in the determination of leg geometry," *In Vivo*, 9, 257, 1995.
21. Gualdi-Russo, E. and Russo, P., "A new technique for measurements on long bones: development of a new instrument and techniques comparison," *Anthropol. Anz.*, 53, 153, 1995.
22. Rumph, P. F. and Hathcock, J. T., "A symmetric axis-based method for measuring the projected femoral angle of inclination in dogs," *Vet. Surg.*, 19, 328, 1990.
23. Hermanussen, M., Bugiel, S., Aronson, S., and Moell, C., "A non-invasive technique for the accurate measurement of leg length in animals," *Growth Dev. Aging*, 56, 129, 1992.
24. Yang, R. S., "A new goniometer," *Orthop. Rev.*, 21, 877, 1992.
25. Hanlock, F. P., Kent, W. R., Loyd, J. L., Harrist, R. B., Deter, R. L., and Park, S. K., "An evaluation of two methods for measuring fetal head and body circumferences," *J. Ultrasound Med.*, 1, 359, 1982.
26. Sommerlath, K. and Gillqist. J., "The effect of a meniscal prosthesis on knee biomechanics and cartilage. An experimental study in rabbits," *Am. J. Sports Med.*, 20, 73, 1992.
27. Skinner, R. A., Hickmon, S. G., Lumpkin, C. K., Aronson, J., and Nicholas, R. W., "Decalcified bone: twenty years of successful specimen management," *J. Histotechnol.*, 20, 267, 1997.
28. Sanderson, C., "Entering the realm of mineralized bone processing: a review of the literature and techniques," *J. Histotechnol.*, 20, 259, 1997.
29. Pazzaglia, U. E., Bernini, F., Zatti, G., and Di Nucci, A., "Histology of the metal-bone interface: interpretation of plastic embedded slides," *Biomaterials*, 15, 273, 1994.
30. Donath, K. and Brenner, G., "A method for the study of undecalcified bones and teeth with attached soft tissue," *J. Oral Pathol.*, 11, 318, 1982.
31. Van der Lubbe, H. B. M., Klein, C. P., and de Groot, K., "A simple method for preparing thin histological sections of undecalcified plastic embedded bone with implants," *Stain Technol.*, 63, 171, 1988.
32. Klein, C. P., Sauren, Y. H. M. F., Modderman, W. E., and van der Waerden, J. P. C. M., "A new saw technique improves preparation of bone sections for light and electron microscopy," *J. Appl. Biomater*, 5, 369, 1994.
33. Burr, D. B., Milgrom, C., Boyd, R. D., Higgins, W. L., Robin, G., and Radin, E. L., "Experimental stress fractures of the tibia. Biological and mechanical aetiology in rabbits," *J. Bone Joint Surg.*, 72B, 370, 1990.
34. Parr, J. A., Young T, Dunn-Jena, P., and Garetto, L. P., "Histomorphometrical analysis of the bone-implant interface: comparison of microradiography and brightfield microscopy," *Biomaterials*, 17, 1921, 1996.

35. Gruber H. E., "Adaptations of Goldner's Masson trichrome stain for the study of undecalcified plastic embedded bone," *Biotech. Histochem.,* 67, 30, 1992.
36. Recker, R. R., *Bone Histomorphometry: Techniques and Interpretation,* CRC Press, Boca Raton, FL, 1983.
37. Takeshita, F., Ayukawa, Y., Iyama, S., Murai, K., and Suetsugu, T., "Long-term evaluation of bone-titanium interface in rat tibiae using light microscopy, transmission electron microscopy, and image processing," *J. Biomed. Mater. Res.,* 37, 235, 1997.
38. Gruber, H. E., Marshall, G. J., Kirchen, M. E., Kang, J., and Massry, S. G., "Improvements in dehydration and cement line staining for methacrylate embedded human bone biopsies," *Stain Technol.,* 60, 337, 1985.
39. Kayser, M. V., Downes, S., and Ali, S. Y., "An electron microscopy study of intact interfaces between bone and biomaterials used in orthopaedics," *Cells Mater.,* 4, 353, 1991.
40. Green, M. R. and Pastewka, J. V., "Simultaneous differential staining by a cationic carbocyanine dye of nucleic acids, proteins and conjugated proteins. II. Carbohydrate and sulfated carbohydrate-containing proteins," *J. Histochem Cytochem,* 22, 774, 1974.
41. Gruber, H. E., "Application of Stains-all for demarcation of cement lines in methacrylate-embedded bone," *Biotech. Histochem.,* 66, 181, 1991.
42. Frost, H. M., "Tetracycline-based histological analysis of bone remodeling," *Calcif. Tissue Res.,* 3, 211, 1969.
43. Thompson, R. C., Jr., Oegema, T. R., Jr., Lewis, J. L., and Wallace, L., "Osteoarthrotic changes after acute transarticular load. An animal model," *J, Bone Joint Surg.,* 73A, 990, 1991.
44. Kiviranta, I., Tammi, M., Jurvelin, J., Saamanen, A. M., and Helminen, H. J., "Demonstration of chondroitin sulphate and glycoproteins in articular cartilage matrix using periodic acid-Schiff (PAS) method," *Histochemistry,* 83, 303, 1985.
45. Stanka, P., Bellack, U., and Lindner, A., "On the morphology of the terminal microvasculature during endochondral ossification in rats," *Bone Miner.,* 13, 93, 1991.
46. Wallace, C. D. and Amiel, D., "Vascular assessment of the periarticular ligaments of the rabbit knee," *J. Orthop. Res.,* 9, 787, 1991.
47. Gruber, H. E., Marshall, G. J., Nolasco, L. M., Kirchen, M. E., and Rimoin, D. L., "Alkaline and acid phosphatase demonstration in human bone and cartilage: effects of fixation intervals and methacrylate embedments," *Stain Technol.,* 63, 299, 1988.
48. Piattelli, A., Scarano, A., and Piattelli, M., "Detection of alkaline and acid phosphatases around titanium implants: a light microscopical and histochemical study in rabbits," *Biomaterials,* 16, 1333, 1995.
49. Lucena, S. B., Duarte, M. E. L., and Fonseca, E. C., "Plastic embedded undecalcified bone biopsies: an immunohistochemical method for routine study of bone marrow extracellular matrix," *J. Histotechnol.,* 20, 253, 1997.
50. Okimura, A., Okada, Y., Makihira, S., Pan, H., Yu, L., et al., "Enhancement of cartilage matrix protein synthesis in arthritic cartilage," *Arthritis Rheum.,* 40, 1029, 1997.
51. Nerlich, A. G., Wiest, I., and von der Mark, K., "Immunohistochemical analysis of interstitial collagens in cartilage of different stages of osteoarthrosis," *Virch. Arch. B. Cell Pathol.,* 63, 249, 1993.
52. Claassen, H. and Kirsch, T., "Temporal and spatial localization of type I and II collagens in human thyroid cartilage," *Anat. Embryol.,* 189, 237, 1994.
53. Morrison, E. H., Ferguson, M. W., Bayliss, M. T., and Archer, C. W., "The development of articular cartilage. I. The spatial and temporal patterns of collagen types," *J. Anat.,* 189, 9, 1996.
54. Visco, D. M., Johnstone, B., Hill, M. A., Jolly, G. A., and Caterson, B., "Immunohistochemical analysis of 3-B-(-) and 7-D-4 epitope expression in canine osteoarthritis," *Arthritis Rheum.,* 36, 1718, 1993.
55. Bland, Y. S. and Ashhurst, D. E., "Development and aging of the articular cartilage of the rabbit knee joint: distribution of the fibrillar collagens," *Anat. Embryol.,* 194, 607, 1996.
56. Gao, J., Messner, K., Ralphs, J. R., and Benjamin, M., "An immunohistochemical study of enthesis development in the medial collateral ligament of the rat knee joint," *Anat. Embryol.,* 194, 399, 1996.

57. Comer, J. S., Kincaid, S. A., Baird, A. N., Kammermann, J. R., Hanson, R. R., Jr., and Ogawa, Y., "Immunolocalization of stromelysin, tumor necrosis factor (TNF) alpha, and TNF receptors in atrophied canine articular cartilage treated with hyaluronic acid and transforming growth factor beta," *Am. J. Vet. Res.*, 57, 1488, 1996.
58. Lust, G., Burton-Wurster, N., and Leipold, H., "Fibronectin as a marker for osteoarthritis," *J. Rheumtol.*, 14 (Suppl), 28, 1987.
59. von Recum, A. F., Opitz, H., and Wu, E., "Collagen types I and III at the implant/tissue interface," *J. Biomed. Mater. Res.*, 27, 757, 1993.
60. Liu, S. H., Panossian, V., al-Shaikh, R., Tomin, E., Shepherd, E., et al., "Morphology and matrix composition during early tendon to bone healing," *Clin. Orthop.*, 339, 253, 1997.
61. Fromm, B., Schafer, B., Parsch, D., and Kummer, W., "Reconstruction of the anterior cruciate ligament with a cyropreserved ACL allograft. A microangiographic and immunohistochemical study in rabbits," *Int. Orthop.*, 20, 378, 1996.
62. Overgaard, S., Søballe, K., Josephsen, K., Hansen, E. S., and Bunger, C., "Role of different loading conditions on resorption of hydroxyapatite coating evaluated by histomorphometric and stereological methods," *J. Orthop. Res.*, 14, 888, 1996
63. Tanzer, M., Harvey, E., Kay, A., Morton, P., and Bobyn, J. D., "Effect of noninvasive low intensity ultrasound on bone growth into porous-coated implants," *J. Orthop. Res.*, 14, 901, 1996.
64. Parfitt, A. M., Drezner, M. K., Glorieux, F. H., et al., "Bone histomorphometry: standardization of nomenclature, symbols, and units," *J. Bone Miner. Res.*, 2, 595, 1987.
65. Kinney, J. H., Lane, N. E., and Haupt, D. L., "*In vivo*, three-dimensional microscopy of trabecular bone," *J. Bone Miner. Res.*, 10, 264, 1995.
66. Müller, R., Hildebrand, T., Hauselmann, H. J., and Ruegsegger, P., "*In vivo* reproducibility of three-dimensional structural properties of noninvasive bone biopsies using 3D-pQCT," *J. Bone Miner. Res.*, 11, 1745, 1996.
67. Odgaard, A., "Three-dimensional methods for quantification of cancellous bone architecture," *Bone*, 20, 315, 1997.
68. Kang, Q., An, Y. H., Butehorn, H. F., and Friedman, R. J., "Morphological and mechanical study of the effects of experimentally induced inflammatory knee arthritis on rabbit long bones," *J. Mater. Sci. Mater. Med.*, 9, 463, 1998.
69. West, P. G., Rowland, G. R., Budsberg, S. C., and Aron, D. N., "Histomorphometric and angiographic analysis of bone healing in the humerus of pigeons," *Am. J. Vet. Res.*, 57, 1010, 1996.
70. Wolff, D., Goldberg, V. M., and Stevenson, S., "Histomorphometric analysis of the repair of a segmental diaphyseal defect with ceramic and titanium fibermetal implants: effects of bone marrow," *J. Orthop. Res.*, 12, 439, 1994.
71. DeVries, W. J., Runyon, C. L., Martinez, S. A., and Ireland, W. P., "Effect of volume variations on osteogenic capabilities of autogenous cancellous bone graft in dogs," *Am. J. Vet. Res.*, 57, 1501, 1996.
72. Breinan, H. A., Minas, T., Hsu, H. P., Nehrer, S., Sledge, C. B., and Spector, M., "Effect of cultured autologous chondrocytes on repair of chondral defects in a canine model," *J. Bone Joint Surg.*, 79A, 1439, 1997.
73. Mohr, H. and Kragstrup, J., "Morphostereometry of heterotopic ossicles in the rat," *Acta Orthop. Scand.*, 62, 257, 1991.
74. Ishaug-Riley, S. L., Crane, G. M., Gurlek, A., Miller, M. J., Yasko, A. W., et al., "Ectopic bone formation by marrow stromal osteoblast transplantation using poly(DL-lactic-co-glycolic acid) foams implanted into the rat mesentery," *J. Biomed. Mater. Res.*, 36, 1, 1997.
75. Hacker, S. A., Healey, R. M., Yoshioka, M., and Coutts, R. D., "A methodology for quantitative assessment of articular cartilage histomorphometry," *Osteoarthr. Cartil.*, 5, 343, 1997.
76. Holm, I. E., Bunger, C., and Melsen, F., "A histomorphometric analysis of subchondral bone in juvenile arthropathy of the dog knee," *Acta Pathol. Microbiol. Immunol. Scand.*, 93, 299, 1985.
77. Bogoch, E., Gschwend, N., Bogoch, B., Rahn, B., and Perren, S., "Juxta-articular bone loss in experimental inflammatory arthritis," *J. Orthop. Res.*, 6, 648, 1988.
78. Yoshioka, M., Shimizu, C., Harwood, F. L., Coutts, R. D., and Amiel, D., "The effects of hyaluronan during the development of osteoarthritis," *Osteoarthritis Cartil.*, 5, 251, 1997.

79. An, Y. H., Friedman, R. J., Draughn, R. A., Jiang, M., LaBreck, J. C., Butehorn, H. F., and Bauer, T. W., "Bone ingrowth to implant surfaces in an inflammatory arthritis model." *J. Orthop. Res.,* 16, 576, 1998.
80. Lukoschek, M., Schaffler, M. B., Burr, D. B., Boyd, R. D., and Radin, E. L., "Synovial membrane and cartilage changes in experimental osteoarthrosis," *J. Orthop. Res.,* 6, 475, 1988.
81. Friedman, R. J., An, Y. H., Jiang, M., Draughn, R. A., and Bauer, T. W., "*In vivo* mechanical and histological evaluation of bone ingrowth and apposition to metal implants of different surface textures in the rabbit femur," *J. Orthop. Res.,* 14, 455, 1996.
82. Vigorita, V. J., Minkowitz, B., Dichiara, J. F., and Higham, P. A., "A histomorphometric and histologic analysis of the implant interface in five successful, autopsy-retrieved, noncemented porous-coated knee arthroplasties," *Clin. Orthop.,* 293, 211, 1993.
83. Moroni, A., Caja, V. L., Egger, E. L., Trinchese, L., and Chao, E. Y., "Histomorphometry of hydroxyapatite coated and uncoated porous titanium bone implants," *Biomaterials,* 15, 926, 1994.
84. Fini, M., Nicoli Aldini, N., Gandolfi, M. G., Mattioli Belmonte, M., et al., "Biomaterials for orthopaedic surgery in osteoporotic bone: a comparative study in osteopenic rats," *Int. J. Artif. Org.,* 5, 291, 1997.
85. Therin, M., Christel, P., and Meunier, A., "Analysis of the general features of the soft tissue response to some metals and ceramics using quantitative histomorphometry," *J. Biomed. Mater. Res.,* 28, 1267, 1994.
86. Paquay, Y. C., de Ruijter, A. E., van der Waerden, J. P., and Jansen, J. A., "A one stage versus two stage surgical technique. Tissue reaction to a percutaneous device provided with titanium fiber mesh applicable for peritoneal dialysis," *ASAIO J.,* 42, 961, 1996.
87. Shrive, N., Chimich, D., Marchuk, L., Wilson, J., Brant, R., and Frank, C., "Soft-tissue 'flaws' are associated with the material properties of the healing rabbit medial collateral ligament," *J. Orthop. Res.,* 13, 923, 1995.
88. Karim, M. A., Miller, D. D., Farrar, M. A., Eleftheriades, E., Reddy, B. H., et al., "Histomorphometric and biochemical correlates of arterial procollagen gene expression during vascular repair after experimental angioplasty," *Circulation,* 91, 2049, 1995.
89. Wright, C. D., Vedi, S., Garrahan, N. J., Stanton, M., Duffy, S. W., and Compston, J. E., "Combined inter-observer and inter-method variation in bone histomorphometry," *Bone,* 13, 205, 1992.
90. Whitehouse, W. J., Dyson, E. D., and Jackson, C. K., "The scanning electron microscope in studies of trabecular bone from a human vertebral body," *J. Anat.,* 108, 481, 1971.
91. Jayasinghe, J. A. P., Jones, S. J., and Boyde, A., "Scanning electron microscopy of human lumbar vertebral trabecular bone surfaces," *Virch. Arch. A Pathol. Anat.,* 422, 25, 1993.
92. Zhou, H., Cherncky, R., and Davies, J. E., "Scanning electron microscopy of the osteoclast-bone interface *in vivo*," *Cells Mater.,* 3, 141, 1993.
93. Kobayashi, S., Yonekubo, S., and Kurogouchi, Y., "Cryoscanning electron microscopic study of the surface amorphous layer of articular cartilage," *J. Anat.,* 187, 429, 1995.
94. Nakashima, Y., Hayyashi, K, Inadome, T., et al., "Hydroxyapatite coating on titanium arc sprayed titanium implants," *J. Biomed. Mater. Res.,* 35, 287, 1997.
95. Wang, A., Essner, A., Stark, C., and Dumbleton, J. H., "Comparison of the size and morphology of UHMWPE wear debris produced by a hip joint simulator under serum and water lubricated conditions," *Biomaterials,* 17, 865, 1996.
96. Takeshita, F., Murai, K., Ayukawa, Y., and Suetsugu, T., "Effects of aging on titanium implants inserted into the tibiae of female rats using light microscopy, SEM, and image processing," *J. Biomed. Mater. Res.,* 34, 1, 1997.
97. McNamara, A. and Williams, D. F., "Scanning electron microscopy of the metal-tissue interface. II. Observations with lead, copper, nickel, aluminium, and cobalt," *Biomaterials,* 3, 165, 1982.
98. Orr, R. D., de Bruijn, J. D., and Davies, J. E., "Scanning electron microscopy of the bone interface with titanium, titanium alloy and hydroxyapatite," *Cells Mater.,* 2, 241, 1992.
99. Nanci, A., McCarthy, G. F., Zalzal, S., et al., "Tissue response to titanium implamts in the rat tibia: ultrastructural, immunochemical and lectin-cytochemical characterization of the bone-titanium interface," *Cells Mater.,* 4, 1, 1994.
100. Pannarale, L., Morini, S., D'Ubaldo, E., Gaudio, E., and Marinozzi, G., "SEM corrosion-casts study of the microcirculation of the flat bones in the rat," *Anat. Rec.,* 247, 462, 1997.

101. Pannarale, L., Gaudio, E., and Marinozzi, G., "Microcorrosion casts in the microcirculation of skeletal muscle," *Scan. Electron Microsc.,* Pt 3, 1103, 1986.
102. He, S.-Z., Xiu, Z. H., Hansen, E. S., and Bunger, C., "Microvascular morphology of bone in arthrosis. Scanning electron microscopy in rabbits," *Acta Orthop. Scand.*, 61, 195, 1990.
103. Kraus, B. L., Kirker-Head, C. A., Kraus, K. H., Jakowski, R. M., and Steckel, R. R., "Vascular supply of the tendon of the equine deep digital flexor muscle within the digital sheath," *Vet. Surg.*, 24, 102, 1995.
104. Read, N. D. and Jeffree, C. E., "Low-temperature scanning electron microscopy in biology," *J. Microsc.*, 161, 59, 1991.
105. Ratcliffe, A., Fryer, P. R., and Hardingham, T. E., "The distribution of aggregating proteoglycans in articular cartilage: comparison of quantitative immunoelectron microscopy with radioimmunoassay and biochemical analysis," *J. Histochem. Cytochem.*, 32, 193, 1984.
106. Young, R. D., Lawrence, P. A., Duance, V. C., Aigner, T., and Monaghan, P., "Immunolocalization of type III collagen in human articular cartilage prepared by high-pressure cryofixation, freeze-substitution, and low-temperature embedding," *J. Histochem. Cytochem.*, 43, 421, 1995.
107. Clark, J. M., Norman, A., and Notzli, H., "Postnatal development of the collagen matrix in rabbit tibial plateau articular cartilage," *J. Anat.*, 191, 215, 1997.
108. Landis, W. J., Hodgens, K. J., Arena, J., Song, M. J., and McEwen, B. F., "Structural relations between collagen and mineral in bone as determined by high voltage electron microscopic tomography," *Microsc. Res. Tech.*, 33, 192, 1996.
109. Spurr, A. R., "A low-viscosity epoxy resin embedding medium for electron microscopy," *J. Ultrastruct. Res.*, 26, 31, 1969.
110. Chehroudi, B., Ratkay, J., and Brunette, D. M., "The role of implant surface geometry on mineralization *in vivo* and *in vitro*; A transmission and scanning electron microscopic study," *Cells Mater.*, 2, 89, 1992.
111. Clokie, C. M. and Warshawsky, H., "Morphologic and radioautographic studies of bone formation in relation to titanium implants using the rat tibia as a model," *Int. J. Oral Maxillofac. Implants*, 10, 155, 1995.
112. Holgers, K. M., Thomsen, P., Tjellstrom, A., and Ericson, L. E., "Electron microscopic observations on the soft tissue around clinical long-term percutaneous titanium implants," *Biomaterials*, 16, 83, 1995.
113. Linder, L., Albrektsson, T., Branemark, P. I., et al., "Electron microscopic analysis of the bone-titanium interface," *Acta Orthop. Scand.*, 54, 45, 1983.
114. Bjursten, L. M., Emanuelsson, L., Ericson, L. E., Thomsen, P., Lausmaa, J., et al., "Method for ultrastructural studies of the intact tissue-metal interface," *Biomaterials*, 11, 596, 1990.
115. Hemmerlé, J. and Voegel, J.-C., "Ultrastructural aspects of the intact titanium implant-bone interface from undecalcified ultrathin sections," *Biomaterials*, 17, 1913, 1996.
116. Piattelli, A., Trisi, P., Passi, P., Piattelli, M., and Cordioli, G. P., "Histochemical and confocal laser scanning microscopy study of the bone-titanium interface: an experimental study in rabbits," *Biomaterials*, 15, 194, 1994.
117. Takeshita, F., Iyama, S., Ayukawa, Y., Akedo, H., and Suetsugu, T., "Study of bone formation around dense hydroxyapatite implants using light microscopy, image processing and confocal laser scanning microscopy," *Biomaterials*, 18, 317, 1997.
118. Ark, J. W., Gelberman, R. H., Abrahamsson, S. O., Seiler, J. G., III., and Amiel, D., "Cellular survival and proliferation in autogenous flexor tendon grafts," *J. Hand Surg.*, 19A, 249, 1994.
119. Ohlendorf, C., Tomford, W. W., and Mankin. H. J., "Chondrocyte survival in cryopreserved osteochondral articular cartilage," *J. Orthop. Res.*, 14, 413, 1996.
120. Freed, L. E., Grande, D. A., Nohria, A., Emmanural, J., Mikos, A. G., and Langer, R., "Joint resurfacing using chondrocytes and synthetic biodegradable polymer scaffolds," *J. Biomed. Mater. Res.*, 28, 891, 1994.
121. Wotton, S. F., Jeacocke, R. E., Maciewicz, R. A., Wardale, R. J., and Duance, V. C., "The application of scanning confocal microscopy in cartilage research," *Histochem. J.*, 23, 328, 1991.
122. Niyibizi, C., Visconti, C. S., Kavalkovich, K., and Woo, S. L., "Collagens in an adult bovine medial collateral ligament: immunofluorescence localization by confocal microscopy reveals that type XIV collagen predominates at the ligament-bone junction," *Matrix Biol.*, 14, 743, 1995.

123. Niyibizi, C., Sagarrigo Visconti, C., Gibson, G., and Kavalkovich, K., "Identification and immunolocalization of type X collagen at the ligament-bone interface," *Biochem. Biophys. Res. Commun.*, 222, 584, 1996.
124. Lohmander, L. S., "Markers of cartilage metabolism in arthrosis. A review," *Acta Orthop. Scand.*, 62, 623, 1991.
125. Thonar, E. J., Shinmei, M., and Lohmander, L. S., "Body fluid markers of cartilage changes in osteoarthritis," *Rheum. Dis. Clin. North Am.*, 19, 635, 1993.
126. Malemud, C. J., "Markers of osteoarthtitis and cartilage research in animal models," *Curr. Opin. Rheumatol.*, 5, 494, 1993.
127. Laurent, T. C., Laurent, U. B., and Fraser, J. R., "Serum hyaluronan as a disease marker," *Ann. Med.*, 28, 241, 1996.
128. Garnero, P. and Delmas, P. D., "New developments in biochemical markers for osteoporosis," *Calcif. Tissue Int.*, 59 (Suppl 1), S2, 1996.
129. Kleerekoper, M., "Biochemical markers of bone remodeling," *Am. J. Med. Sci.*, 312, 270, 1996.
130. Russell, R. G. G., "The assessment of bone metabolism *in vivo* using biochemical approaches," *Horm. Metab. Res.*, 29, 138, 1997.
131. Arican, M., Carter, S. D., Bennett, D., and May, C., "Measurement of glycosaminoglycans and keratan sulphate in canine arthropathies," *Res. Vet. Sci.*, 56, 290, 1994.
132. Alwan, W. H., Carter, S. D., Bennett, D., May, S. A., and Edwards, G. B., "Cartilage breakdown in equine osteoarthritis: measurement of keratan sulphate by an ELISA system," *Res. Vet. Sci.*, 49, 56, 1990.
133. Todhunter, R. J., Fubini, S. L., Freeman, K. P., and Lust, G., "Concentrations of keratan sulfate in plasma and synovial fluid from clinically normal horses and horses with joint disease," *J. Am. Vet. Med. Assoc.*, 210, 369, 1997.
134. Ratcliffe, A., Shurety, W., and Caterson, B., "The quantitation of a native chondroitin sulfate epitope in synovial fluid lavages and articular cartilage from canine experimental osteoarthritis and disuse atrophy," *Arthritis Rheum.*, 36, 543, 1993.
135. Segawa, Y., Nakamura, T., Aota, S., Tanaka, Y., Yoshida, K., et al., "Changes in urinary deoxypyridinoline level and vertebral bone mass in the development of adjuvant-induced arthritis in rats," *Bone*, 17, 57, 1995.
136. Dessauer, A., "Analytical requirements for biochemical bone marker assays," *Scand. J. Clin. Lab. Invest.*, 57 (Suppl 227), 84, 1997.
137. Thonar, E. J., Lenz, M. E., and Klintworth, G. K., "Quantification of keratan sulfate in blood as a marker of cartilage catabolism," *Arthritis Rheum.*, 28, 1367, 1985.
138. Ratcliffe, A, Doherty, M., Maini, R. N., and Hardingham, T. E., "Increased concentrations of proteoglycan components in the synovial fluids of patients with acute but not chronic joint disease," *Ann. Rheum. Dis.*, 47, 826, 1988.
139. Pastoureau, P., Meunier, P. J., and Delmas, P. D., "Serum osteocalcin (bone Gla-protein), an index of bone growth in lambs. Comparison with age-related histomorphometric changes," *Bone*, 12, 143, 1991.
140. Price, J. S., Jackson, B., Eastell, R., et al., "Age related changes in biochemical markers of bone metabolism in horses," *Equine Vet. J.*, 27, 201, 1995.
141. Tulamo, R. M., Houttu, J., Tupamaki, A., and Salonen, M., "Hyaluronate and large molecular weight proteoglycans in synovial fluid from horses with various arthritides," *Am. J. Vet. Res.*, 57, 932, 1996.
142. Yoshida, K., Miyauchi, S., Kikuchi, H., Tawada, A., and Tokuyasu, K., "Analysis of unsaturated disaccharides from glycosaminoglycuronan by high-performance liquid chromatography," *Ann. Biochem.*, 177, 327, 1989.
143. Uebelhart, D., Gineyts, E., Chapuy, M. C., and Delmas, P. D., "Urinary excretion of pyridinium crosslinks: a new marker of bone resorption in metabolic bone disease," *Bone Miner.*, 8, 87, 1990.
144. Kent, G. N., "Standardization of marker assays — pyridinoline/deoxypyridinoline," *Scand. J. Clin. Lab. Invest. Suppl.*, 227, 73, 1997.
145. Cheung, H. S. and Ryan, L. M., "A method for determining DNA and chondrocyte content of articular cartilage," *Anal. Biochem.*, 116, 93, 1981.
146. Mankin, H. J. and Baron, P. A., "The effect of aging on protein synthesis in articular cartilage of rabbits," *Lab. Invest.*, 14, 658, 1965.

147. Howell, D. S., Muniz, O., Pita, J. C., and Enis, J. E., "Extrusion of pyrophosphate into extracellular media by osteoarthritic cartilage incubates," *J. Clin. Invest.*, 56, 1473, 1975.
148. Stegemann, H. and Stadler, K., "Determination of hydroxyproline," *Clin. Chim. Acta*, 18, 267, 1967.
149. Ratcliffe, A., Azzo, W., Saed-Nejad, F., Lane, N., Rosenwasser, M. P., and Mow, V. C., "*In vivo* effects of naproxen on composition, proteoglycan metabolism, and matrix metalloproteinase activities in canine articular cartilage," *J. Orthop. Res.*, 11, 163, 1993b.
150. Furukawa, T., Eyre, D. R., Koide, S., and Glimcher, M. J., "Biochemical studies on repair cartilage resurfacing experimental defects in the rabbit knee," *J. Bone Joint Surg.*, 62A, 79, 1980.
151. O'Driscoll, S. W., Salter, R. B., and Keeley, F. W., "A method for quantitative analysis of ratios of types I and II collagen in small samples of articular cartilage," *Anal. Biochem.*, 145, 277, 1985.
152. Amiel, D., Frank, C., Harwood, F., Fronek, J., and Akeson, W., "Tendons and ligaments: a morphological and biochemical comparison," *J. Orthop. Res.*, 1, 257, 1984.
153. Carney, S. L., Billingham, M. E., Caterson, B., Ratcliffe, A., Bayliss, M. T., et al., "Changes in proteoglycan turnover in experimental canine osteoarthritic cartilage," *Matrix*, 12, 137, 1992.
154. Farndale, R. W., Sayers, C. A., and Barrett, A. J., "A direct spectrophotometric microassay for sulfated glycosaminoglycans in cartilage cultures," *Connect. Tissue Res.*, 9, 247, 1982.
155. Sandy, J. D., Adams, M. E., Billingham, M. E., Plaas, A., and Muir, H., "*In vivo* and *in vitro* stimulation of chondrocyte biosynthetic activity in early experimental osteoarthritis," *Arthritis Rheum.*, 27, 388, 1984.
156. Ratcliffe, A., Fryer, P. R., and Hardingham, T. E., "The distribution of aggregating proteoglycans in articular cartilage: comparison of quantitative immunoelectron microscopy with radioimmunoassay and biochemical analysis," *J. Histochem. Cytochem.*, 32, 193, 1987.
157. Woessner, J. F., Jr., and Taplin, C. J., "Purification and properties of a small latent matrix metalloproteinase of the rat uterus," *J. Biol. Chem.*, 263, 16918, 1988.
158. Dean, D. D. and Woessner, J. F., Jr., "A sensitive, specific assay for tissue collagenase using telopeptide-free [3H]acetylated collagen," *Ann. Biochem.*, 148, 174, 1985.
159. Dean, D. D. and Woessner, J. F., Jr., "Extracts of human articular cartilage contain an inhibitor of tissue metalloproteinases," *Biochem. J.*, 218, 277, 1984.
160. Hardingham, T., "Immunochemical methods in cartilage research," in *Methods in Cartilage Research*, Maroudas, A. and Kuettner, K., Eds., Academic Press, London, 1990, Chapter 7.
161. Jingushi, S., Heydemann, A., Kana, S. K., Macey, L. R., and Bolander, M. E., "Acidic fibroblast growth factor (aFGF) injection stimulates cartilage enlargement and inhibits cartilage gene expression in rat fracture healing," *J. Orthop. Res.*, 8, 364, 1990.
162. Nishimoto, S. K., Chang, C. H., Gendler, E., Stryker, WF., and Nimni, M. E., "The effect of aging on bone formation in rats: biochemical and histological evidence for decreased bone formation capacity," *Calcif. Tissue Int.*, 37, 617, 1985.
163. Nimni, M. E., Bernick, S., Cheung, D. T., Ertl, D. C., and Nishimoto, S. K., "Biochemical differences between dystrophic calcification of cross-linked collagen implants and mineralization during bone induction," *Calcif. Tissue Int.*, 42, 313, 1988.
164. Ding, M., Dalstra, M., Danielsen, C. C., Kabel, J., Hvid, I., and Linde, F., "Age variations in the properties of human tibial trabecular bone," *J. Bone Joint Surg.*, 79B, 995, 1997.
165. Weser, U., Etspuler, H., and Kaup, Y., "Enzymatic and immunological activity of 4000 years aged bone alkaline phosphatase," *FEBS Lett.*, 375, 280, 1995.
166. Kain, C. C., Russell, J. E., Burri, R., Dunlap, J., McCarthy, J., and Manske, P. R., "The effect of vascularization on avian flexor tendon repair. A biochemical study," *Clin. Orthop.*, 233, 295, 1988.
167. Abrahamsson, S. O., Gelberman, R. H., Amiel, D., Winterton, P., and Harwood, F., "Autogenous flexor tendon grafts: fibroblast activity and matrix remodeling in dogs," *J. Orthop. Res.*, 13, 58, 1995.
168. Watanabe, M., Nojima, M., Shibata, T., and Hamada, M., "Maturation-related biochemical changes in swine anterior cruciate ligament and tibialis posterior tendon," *J. Orthop. Res.*, 12, 672, 1994.
169. Spindler, K. P., Andrish, J. T., Miller, R. R., Tsujimoto, K., and Diz, D. I., "Distribution of cellular repopulation and collagen synthesis in a canine anterior cruciate ligament autograft," *J. Orthop. Res.*, 14, 384, 1996.
170. Amiel, D., Billings, E., Jr., and Harwood, F. L., "Collagenase activity in anterior cruciate ligament: protective role of the synovial sheath," *J. Appl. Physiol.*, 69, 902, 1990.

171. Amiel, D., Foulk, R. A., Harwood, F. L., and Akeson, W. H., "Quantitative assessment by competitive ELISA of fibronectin (Fn) in tendons and ligaments," *Matrix*, 9, 421, 1989.
172. Silver, F. H., *Biological Materials: Structure, Mechanical Properties, and Modeling of Soft Tissues*, NYU Press, New York, 1987.
173. Fung, Y. C., *Biomechanics. Mechanical Properties of Living Tissues*, Springer-Verlag, New York, 1993.
174. Black, J., *Orthopaedic Biomaterials in Research and Practice*, Churchill Livingstone, New York, 1988.
175. Mow, V. C. and Ratcliffe, A., "Structure and function of articular cartilage and meniscus," in *Basic Orthopaedic Biomechanics*, 2nd ed., Mow, V. C. and Hayes, W. C., Eds., Lippincott-Raven, Philadelphia, 1997. Chapter 4.
176. Wright, T. M. and Burstein, A. H., *Musculoskeletal Biomechanics*, Churchill Livingstone, New York, 1990, 231.
177. Heimke, G., "Ceramics," in *Handbook of Biomaterials Evaluation*, von Recum, A. F., Ed., Macmillan, New York, 1986, Chapter 3.
178. DePalma, A. F., McKeever, C. D., and Subin, D. K., "Process of repair of articular cartilage demonstrated by histology and autoradiography with tritiated thymidine," *Clini. Orthop.*, 48, 229, 1966.
179. Greiff, J., "Autoradiographic studies of fracture healing using 99Tcm-Sn-polyphosphate," *Injury*, 9, 271, 1978.
180. Tilden, R. L., Jackson, J., Jr., Enneking, W. F., DeLand, F. H., and McVey, J. T., "99m Tc-polyphosphate: histological localization in human femurs by autoradiography," *J. Nucl. Med.*, 14, 576, 1973.
181. Hansen, E. S., Holm, I. E., Bunger, C., Noer, I., Christensen, S. B., and Knudsen, V., "99mTc-DPD uptake in juvenile arthritis. Scintimetry and autoradiography of the knee in dogs," *Acta Orthop. Scand.*, 57, 299, 1986.
182. Landry, P. S., Sadasivan, K. K., Marino, A. A., and Albright, J. A., "Electromagnetic fields can affect osteogenesis by increasing the rate of differentiation," *Clin. Orthp.*, 338, 262, 1997.
183. Subramanian, G. and McAffee, J. G., "A new complex of ^{99}Tc for skeletal imaging," *Radiology*, 99, 192, 1971.
184. Bushberg, J. T., Hoffer, P. B., Schreiber, G. J., Lawson, A. J., Lawson, J. P., and Lord, P., "Comparative uptake of ^{67}Ga and ^{99}mTc MDP in rabbits with a benign noninfected bone lesion (fracture)," *Invest. Radiol.*, 20, 498, 1985.
185. Ruland, L. J. III, Wang, G. J., Teates, C. D., Gay, S., and Rijke, A., "A comparison of magnetic resonance imaging to bone scintigraphy in early traumatic ischemia of the femoral head," *Clin. Orthop.*, 285, 30, 1992.
186. Oster, Z. H., Som, P., Srivastava, S. C., et al., "The development and *in vivo* behavior of tin containing radiopharmaceuticals — II. Autoradiographic and scintigraphic studies in normal animals and in animal models of bone disease," *Int. Nucl. Med. Biol.*, 12, 175, 1985.
187. Sela, J., Shani, J., Kohavi, D., Soskolne, W. A., Itzhak, K., et al., "Uptake and biodistribution of 99mtechnetium methylene-[32P] diphosphonate during endosteal healing around titanium, stainless steel and hydroxyapatite implants in rat tibial bone," *Biomaterials*, 16, 1373, 1995.
188. Markel, M. D., Wikenheiser, M. A, Morin, R. L, Lewallen, D. G., and Chao, E. Y., "The determination of bone fracture properties by dual-energy X ray absorptiometry and single-photon absorptiometry: a comparative study," *Calcif. Tissue Int.*, 48, 392, 1991.
189. Augat, P., Merk, J., Genant, H. K., and Claes, L., "Quantitative assessment of experimental fracture repair by peripheral computed tomography," *Calcif. Tissue Int.*, 60, 194, 1997.
190. Schumacher, B., Albrechtsen, J., Keller, J., Flyvbjerg, A., and Hvid, I., "Periosteal insulin-like growth factor I and bone formation. Changes during tibial lengthening in rabbits," *Acta Orthop. Scand.*, 67, 237, 1996.
191. Gasser, J. A., "Assessing bone quantity by pQCT," *Bone*, 17(4 Suppl), 145S, 1995.
192. Bonse, U., Busch, F., Gunnewig, O., Beckmann, F., Pahl, R., et al., "3D computed X ray tomography of human cancellous bone at 8 microns spatial and 10(–4) energy resolution," *Bone Miner.*, 25, 25, 1994.
193. Gluer, C. C., Wu, C. Y., Jergas, M., Goldstein, S. A., and Genant, H. K., "Three quantitative ultrasound parameters reflect bone structure," *Calcif. Tissue Int.*, 55, 46, 1994.
194. Feldkamp, L. A., Goldstein, S. A., Parfitt, A. M., Jesion, G., and Kleerekoper, M., "The direct examination of three-dimensional bone architecture *in vitro* by computed tomography," *J. Bone Miner. Res.*, 4, 3, 1989.

195. Ciarelli, M. J., Goldstein, S. A., Kuhn, J. L., Cody, D. D., and Brown, M. B., "Evaluation of orthogonal mechanical properties and density of human trabecular bone from the major metaphyseal regions with materials testing and computed tomography," *J. Orthop. Res.,* 9, 674, 1990.
196. Ferretti, J. L., Capozza, R. F., and Zanchetta, J. R., "Mechanical validation of a tomographic (pQCT) index for noninvasive estimation of rat femur bending strength," *Bone,* 18, 97, 1996.
197. Hodgson, R. J., Barry, M. A., Carpenter, T. A., Hall, L. D., Hazleman, B. L., and Tyler, J. A., "Magnetic resonance imaging protocol optimization for evaluation of hyaline cartilage in the distal interphalangeal joint of fingers," *Invest. Radiol.,* 30, 522, 1995.
198. An, H. S., Nguyen, C., Haughton, V. M., Ho, K. C., and Hasegawa, T., "Gadolinium-enhancement characteristics of magnetic resonance imaging in distinguishing herniated intervertebral disc versus scar in dogs," *Spine,* 19, 2089, 1994.
199. Ho, C., Cervilla, V., Kjellin, I., Haghigi, P., Amiel, D., et al., "Magnetic resonance imaging in assessing cartilage changes in experimental osteoarthrosis of the knee," *Invest. Radiol.,* 27, 84, 1992.
200. Tyler, J. A., Watson, P. J., Koh, H. L., Herrod, N. J., Robson M., and Hall, L. D., "Detection and monitoring of progressive degeneration of osteoarthritic cartilage by MRI," *Acta Orthop. Scand. Suppl.,* 266, 130, 1995.
201. Markel, M. D., Wikenheiser, M. A., Morin, R. L., Lewallen, D. G., and Chao, E. Y., "Quantification of bone healing. Comparison of QCT, SPA, MRI, and DEXA in dog osteotomies," *Acta Orthop. Scand.,* 61, 487, 1990.
202. Viljanen, J., Kinnunen, J., Bondestam, S., Majola, A., Rokkanen, P., and Tormala, P., "Bone changes after experimental osteotomies fixed with absorbable self-reinforced poly-L-lactide screws or metallic screws studied by plain radiographs, quantitative computed tomography and magnetic resonance imaging," *Biomaterials,* 16, 1353, 1995.
203. Schwartz, H. S., Shockley, T. E., Lennington, W. J., and Mackey, E. S., Jr., "The significance of skeletal magnetic resonance imaging after open bone biopsy," *J. Orthop. Res.,* 9, 120, 1991.
204. Arnoczky, S. P., Cooper, T. G., Stadelmaier, D. M., and Hannafin, J. A., "Magnetic resonance signals in healing menisci: an experimental study in dogs," *Arthroscopy,* 10, 552, 1994.
205. Kladny, B., Bail, H., Swoboda, B., Schiwy-Bochat, H., Beyer, W. F., and Weseloh, G., "Cartilage thickness measurement in magnetic resonance imaging," *Osteoarthr. Cartil.,* 4, 181, 1996.
206. Eckstein, F., Adam, C., Sittek, H., Becker, C., Milz, S., et al., "Non-invasive determination of cartilage thickness throughout joint surfaces using magnetic resonance imaging," *J. Biomech.,* 30, 285, 1997.
207. Nishino, M., Matsumoto, T., Nakamura, T., and Tomita, K., "Pathological and hemodynamic study in a new model of femoral head necrosis following traumatic dislocation," *Acta Orthop. Trauma Surg.,* 116, 259, 1997.
208. Seiler, J. G., III, Kregor, P. J., Conrad, E. U., III, and Swiontkowski, M. F., "Posttraumatic osteonecrosis in a swine model. Correlation of blood cell flux, MRI and histology," *Acta Orthop. Scand.,* 67, 249, 1996.
209. Volk, A., Cremieux, A. C., Belmatoug, N., Vallois, J. M., Pocidalo, J. J., and Carbon, C., "Evaluation of a rabbit model for osteomyelitis by high field, high resolution imaging using the chemical-shift-specific-slice-selection technique," *Magn. Reson. Imaging,* 12, 1039, 1994.
210. Jara, H., Wehrli, F. W., Chung, H., and Ford, J. C., "High-resolution variable flip angle 3D MR imaging of trabecular microstructure *in vivo,*" *Magn. Reson. Med.,* 29, 528, 1993.
211. Chung, H.-W., Wehrli, F. W., Williams, J. L., and Wehrli, S. L., "Three-dimensional nuclear magnetic resonance microimaging of trabecular bone," *J. Bone Miner. Res.,* 10, 1452, 1995.
212. Hipp, J. A., Jansujwicz, A., Simmons, C. A., and Snyder, B. D., "Trabecular bone morphology from micro-magnetic resonance imaging," *J. Bone Miner. Res.,* 11, 286, 1996.
213. Mehta, B. V., Rajani, S., and Sinha, G., "Comparison of image processing techniques (magnetic resonance imaging, computed tomography scan and ultrasound) for 3D modeling and analysis of the human bones," *J. Digit. Imaging,* 10 (3 Suppl 1), 203, 1997.
214. Ito, M., Hayashi, K., Uetani, M., et al., "Bone mineral and other bone components in vertebrae evaluated by QCT and MRI," *Skeletal Radiol.,* 22, 109, 1993.
215. Kroger, H., Vainio, P., Nieminen, J., and Kotaniemi, A., "Comparison of different models for interpreting bone mineral density measurements using DXA and MRI technology," *Bone,* 17, 157, 1995.
216. Bradbeer, J. N., Kapadia, R. D., Sarkar, S. K., et al., "Disease-modifying activity of SK&F 106615 in rat adjuvant-induced arthritis. Multiparameter analysis of disease magnetic resonance imaging and bone mineral density measurements," *Arthritis Rheum.,* 39, 504, 1996.

217. Jergas., M. D., Majumdar, S., Keyak, J. H., Lee, I. Y., Newitt., D. C., et al., "Relationships between Young's modulus of elasticity, ash density, and MRI derived effective transverse relaxation T2* in tibial specimens," *J. Comput. Assist. Tomogr.*, 19, 472, 1995.
218. Aro, H. T., Wippermann, B. W., Hodgson, S. F., Wahner, H. W., Lewallen, D. G., and Chao, E. Y. S., "Prediction of properties of fracture callus by measurement of mineral density using micro-bone densitometry," *J. Bone Joint Surg.*, 71A, 1020, 1989.
219. Nordsletten, L., Kaastad, T. S., Skjeldal, S., Reikeras, O., Nordal, K. P., et al., "Fracture strength prediction in rat femoral shaft and neck by single photon absorptiometry of the femoral shaft," *Bone Miner.*, 25, 39, 1994.
220. Borg, J., Mollgaard, A., and Riis, B. J., "Single X ray absorptiometry: performance characteristics and comparison with single photon absorptiometry," *Osteoporosis Int.*, 5, 377, 1995.
221. Keenan, M. J., Hegsted, M., Jones, K. L., et al., "Comparison of bone density measurement techniques: DXA and Archimedes' principle," *J. Bone Miner. Res.*, 12, 1903, 1997.
222. Markel, M. D., Sielman, E., and Bodganske, J. J., "Densitometric properties of long bones in dogs, as determined by use of dual-energy X ray absorptiometry," *Am. J. Vet. Res.*, 55, 1750, 1994.
223. Mosheiff, R., Klein, B. Y., Leichter, I., Chaimsky, G., Nyska, A., et al., "Use of dual-energy X ray absorptiometry (DEXA) to follow mineral content changes in small ceramic implants in rats," *Biomaterials*, 13, 462, 1992.
224. Lu, P. W., Briody, J. N., Howman-Giles, R., Trube, A., and Cowell, C. T., "DXA for bone density measurement in small rats weighing 150–250 grams," *Bone*, 15, 199, 1994.
225. Bagi, C. M., Ammann, P., Rizzoli, R., and Miller, S. C., "Effect of estrogen deficiency on cancellous and cortical bone structure and strength of the femoral neck in rats," *Calcif. Tissue Int.*, 61, 336, 1997.
226. Sievänen, H., Kannus, P., Nieminen, V., Heinonen, A., Oja, P., and Vuori I., "Estimation of various mechanical characteristics of human bones using dual energy X ray absorptiometry: methodology and precision," *Bone*, 18(Suppl 1), 17S, 1996.
227. Markel, M. D., Bogdanske, J. J., Xiang, Z., and Klohnen, A., "Atrophic nonunion can be predicted with dual energy X ray absorptiometry in a canine ostectomy model," *J. Orthop. Res.*, 13, 869, 1995.
228. Vanderschueren, D., Van Herck, E., Schot, P., Rush, E., Einhorn, T., et al., "The aged male rat as a model for human osteoporosis: evaluation by nondestructive measurements and biomechanical testing," *Calcif. Tissue Int.*, 53, 342, 1993.
229. Turner, A. S., Alvis, M., Myers, W., Stevens, M. L., and Lundy, M. W., "Changes in bone mineral density and bone-specific alkaline phosphatase in ovariectomized ewes," *Bone*, 17, 395S, 95.
230. Kramer, M., Gerwing, M., Hach, V., and Schimke, E., "Sonography of the musculoskeletal system in dogs and cats," *Vet. Radiol. Ultrasound*, 38, 139, 1997.
231. Kim, H. K., Babyn, P. S., Harasiewicz, K. A., Gahunia, H. K., Pritzker, K. P., and Foster, F. S., "Imaging of immature articular cartilage using ultrasound backscatter microscopy at 50 MHz," *J. Orthop. Res.*, 13, 963, 1995.
232. Martino, F., Ettorre, G. C., Patella, V., Macarini, L., Moretti, B., et al., "Articular cartilage echography as a criterion of the evolution of osteoarthritis of the knee," *Int. J. Clin. Pharmacol. Res.*, 13 (Suppl) 35, 1993.
233. Saied, A., Cherin, E., Gaucher, H., Laugier, P., Gillet, P., et al., "Assessment of articular cartilage and subchondral bone: subtle and progressive changes in experimental osteoarthritis using 50 MHz echography *in vitro*," *J. Bone Miner. Res.*, 12, 1378, 1997.
234. Mutalib, A., Holland, M., Barnes, H. J., and Boyle, C., "Ultrasound for detecting osteomyelitis in turkeys," *Avian Dis.*, 40, 321, 1996.
235. Amo, C., Revilla, M., Hernandez, E. R., Gonzalez-Riola, J., Villa, L. F., et al., "Correlation of ultrasound bone velocity with dual-energy X ray bone absorptiometry in rat bone specimens," *Invest. Radiol.*, 31, 114, 1996.
236. Njeh, C. F., Boivin, C. M., and Langton, C. M., "The role of ultrasound in the assessment of osteoporosis: a review," *Osteoporosis Int.*, 7, 7, 1997.
237. Gregg, E. W., Kriska, A. M., Salamone, L. M., et al., "The epidemiology of quantitative ultrasound: a review of the relationships with bone mass, osteoporosis and fracture risk," *Osteoporosis Int.*, 7, 89, 1997.
238. Ashman, R. B., "Experimental techniques," in *Bone Mechanics,* Cowin, S. C., Ed., CRC Press, Boca Raton, FL, 1989, 91.

239. Ashman, R. B., Rho, J. Y., and Turner, C. H., "Anatomical variation of orthotropic elastic moduli of the proximal human tibia," *J. Biomech.*, 22, 895, 1989.
240. van der Perre, G. and Lowet, G., "*In vivo* assessment of bone mechanical properties by vibration and ultrasonic wave propagation analysis," *Bone*, 18, 29S, 1996.
241. Altman, R. D, Kates, J., Chun, L. E., Dean, D. D., and Eyre, D., "Preliminary observations of chondral abrasion in a canine model," *Ann. Rheum. Dis.*, 51, 1056, 1992.
242. Bjornland, T., Rorvik, M., Haanaes, H. R., and Teige. J., "Degenerative changes in the temporomandibular joint after diagnostic arthroscopy. An experimental study in goats," *Int. J. Oral Maxillofac. Surg.*, 23, 41, 1994.
243. Marshall, K. W. and Chan, A. D., "Bilateral canine model of osteoarthritis," *J. Rheumatol.*, 23, 344, 1996.
244. Rutili, G. and Arfors, K.-E., "Measurement of blood flow in the tenuissimus muscle with tracer Sephadex," *Microvas. Res.*, 11, 269, 1976.
245. Sejrsen, P., "Atraumatic local labeling of skin by inert gas: epicutaneous application of xenon 133," *J. Appl. Physiol.*, 24, 570, 1968.
246. Holloway, G. A., Jr., and Watkins, D. W., "Laser-Doppler measurement of cutaneous blood flow," *J. Invest. Dermatol.*, 69, 306, 1977.
247. Lindén, M., Sirsjo, A., Lindbom, L., Nilsson, G., and Gidlof. A., "Laser-Doppler perfusion imaging of microvascular blood flow in rabbit tenuissimus muscle," *Am. J. Physiol.*, 269, H1496, 1995.
248. Jain, R., Podworny, N., Anderson, G,I., and Schemitsch, E. H., "Assessment of the relationship between standard probe and implantable fiber measurements of cortical bone blood flow: a canine study," *Calcif. Tissue Int.*, 59, 64, 1996.
249. Bridge, J. A., Nelson, M., McComb, E., McGuire, M. H., Rosenthal, H., et al., "Cytogenetic and molecular cytogenetic techniques in orthopaedic surgery," *J. Bone Joint Surg.*, 75A, 606, 1993.
250. Sandberg, M. M., Aro, H. T., and Vuorio, E. I., "Gene expression during bone repair," *Clin. Orthop.*, 289, 292, 1993.
251. Shore, E. M. and Kaplan, F. S., "Tutorial. Molecular biology for the clinician. Part I. General principles," *Clin. Orthop.*, 306, 264, 1994.
252. Shore, E. M. and Kaplan, F. S., "Tutorial. Molecular biology for the clinician. Part II. Tools of molecular biology," *Clin. Orthop.*, 320, 247, 1995.
253. Hicks, D. G., Stroyer, B. F., Teot, L. A., and O'Keefe, R. J., "*In situ* hybridization in skeletal tissues utilizing non-radioactive probes," *J. Histotech.*, 20, 215, 1997.
254. McNicol, A. M. and Farquharson, M. A., "*In situ* hybridization and its diagnostic applications in pathology," *J. Pathol.*, 182, 250, 1997.
255. Bridge, J. A., "Cytogenetic findings in 73 osteosarcoma specimens and a review of the literature," *Cancer Genet. Cytogenet.*, 95, 74, 1997.
256. Chen, D. S., Zhu, N. L., Hung, G., Skotzko, M. J., Hinton, D. R., Tolo, V., et al., "Retroviral vector-mediated transfer of an antisense cyclin G1 construct inhibits osteosarcoma tumor growth in nude mice," *Hum. Gene Ther.*, 20, 1667, 1997.
257. Kivirikko, K. I., "Collagens and their abnormalities in a wide spectrum of diseases," *Ann. Med.*, 25, 113, 1993.
258. Prockop, D. J. and Kivirikko, K. I., "Collagens: molecular biology, diseases, and potentials for therapy," *Ann. Rev. Biochem.*, 64, 403, 1995.
259. de Crombrugghe, B., Katzenstein, P., Mukhopadhyay, K., Lefebvre, V., Zhou, G., et al., "Transgenic mice with deficiencies in cartilage collagens: possible models for gene therapy," *J. Rheumatol. Suppl.*, 43, 140, 1995.
260. Ko, S. C., Cheon, J., Kao, C., Gotoh, A., Shirakawa, T., Sikes, R. A., et al., "Osteocalcin promoter-based toxic gene therapy for the treatment of osteosarcoma in experimental models," *Cancer Res.*, 56, 4614, 1996.
261. Evans, C. H. and Robbins., P. D., "Pathways to gene therapy in rheumatoid arthritis," *Curr. Opin. Rheumatol.*, 8, 230, 1996.
262. Le, C. H., Nicolson, A. G., Morales, A., and Sewell, K. L., "Suppression of collagen-induced arthritis through adenovirus-mediated transfer of a modified tumor necrosis factor alpha receptor gene," *Arthritis Rheum.*, 40, 1662, 1997.

263. Pelletier, J. P., Caron, J. P., Evans, C., Robbins, P. D., Georgescu, H. I., et al., "*In vivo* suppression of early experimental osteoarthritis by interleukin–1 receptor antagonist using gene therapy," *Arthritis Rheum.,* 40, 1012, 1997.
264. Kang, R., Marui, T., Ghivizzani, S. C., Nita, I. M., Georgescu, H. I., et al., "*Ex vivo* gene transfer to chondrocytes in full-thickness articular cartilage defects: a feasibility study," *Osteoarthr. Cartil.,* 5, 139, 1997.
265. Gerich, T. G., Kang, R., Fu, F. H., Robbins, P. D., and Evans, C. H., "Gene transfer to the rabbit patellar tendon: potential for genetic enhancement of tendon and ligament healing," *Gene Ther.,* 3, 1089, 1996.
266. Lou, J., Manske, P. R., Aoki, M., and Joyce, M. E., "Adenovirus-mediated gene transfer into tendon and tendon sheath," *J. Orthop. Res.,* 14, 513, 1996.

7 Histological Study in Orthopaedic Animal Research

Helen E. Gruber and Audrey A. Stasky

CONTENTS

I. Introduction 115
II. Experimental Design 116
    A. Desired Endpoints for Examination..116
    B. Decalcify or Not Decalcify 117
    C. Bone Labelling 117
III. Specimen Harvesting..118
IV. Fixation and Transportation 119
V. Specimen Proccessing and Sectioning...119
    A. Decalcified Specimens 119
    B. Undecalcified Specimens 120
        1. Embedding of Undecalcified Specimens 120
        2. Sectioning of Plastic Embedded Specimens..121
    C. Soft Tissues 121
VI. Staining Techniques 122
    A. Staining of Bone and Cartilage..122
        1. Decalcified Sections 122
        2. Undecalcified Sections 123
        3. Ground Sections 123
    B. Staining of Soft Tissues 123
    C. Histochemical Staining 123
VII. Histological Evaluation and Histomorphometry 125
VIII. Other Important Procedures...126
    A. Immunohistochemistry 126
    B. *In situ* Hybridization 127
    C. Tissues Containing Implants 128
References 129
Appendix 1   Selected Protocols for Embedding and Staining.....................................132
Appendix 2   Useful Equipment and Tools for the Orthopaedic Histology Laboratory 137
Appendix 3   On-Line Help and Other Sources of Information 138

I. INTRODUCTION

The focus of this chapter is to provide a practical selection of useful guidelines and technical procedures relevant to the histologic study of bone, cartilage and associated soft tissues in the orthopaedic research laboratory setting. Occasional key references are cited, but the limited scope of a single chapter does not permit inclusion of all the publications which have provided technological advances and useful methodologies in bone and cartilage histology throughout the years. Important reference sources for detailed discussions of bone histology and detailed methods for histologic procedures are found in a number of excellent publications which are also highly recommended to the interested reader and researcher.[1-5]

Histology continues to remain a valuable tool in orthopaedic research. Specialized histologic techniques for bone and cartilage have historically built upon the methods developed for examination of bones from clinical conditions such as bone tumors, skeletal dysplasias, and, since the mid 1960s evaluation of undecalcified bone biopsies, from patients with osteoporosis, osteomalacia, or other metabolic bone diseases. However, routine clinical laboratory histology methods used in the pathology laboratory are sometimes inappropriate for the specific needs of orthopaedic examination of experimental bone and cartilage specimens. In the sections below, guidelines have been developed to assist the researcher in choosing decalcified or undecalcified bone processing schemes, and to suggest protocols best suited for the needs of the investigator.

Interest in the histologic examination of bone continues to grow as evidenced by a recent issue of the *Journal of Histotechnology* totally devoted to topics of bone and cartilage studies.[6] Such interest bodes well for the continued development and utilization of specialized techniques for study of these challenging tissues. The authors wish to thank all the investigators who have contributed valuable technical and interpretative methods for the study of bone, cartilage and associated soft tissues; not all important work was able to be cited in the space of this chapter. We hope that the reader will find the focus of this chapter of practical value in the study of bone, cartilage and associated soft tissues.

II. EXPERIMENTAL DESIGN

The focus of this chapter is to provide selected helpful guidelines of practical use to the orthopaedic researcher with little or no previous histologic experience, and to provide specialized information for the investigator with more advanced needs. The importance of early planning to define the histologic endpoints of interest and choice of the histologic methods which will achieve the desired endpoints is emphasized.

A. Desired Endpoints for Examination

It is very important to have a specific histologic region of interest defined at the start of the study. The first step is to clearly define a sampling site. For orthopaedic studies, the sites of interest can range from cartilage (e.g., changes in the physis or growth plate during surgical procedures, administration of drugs or metabolic agents, or unweighting of a limb; changes in articular cartilage with aging or disease), to bone (trabecular [metaphyseal or epiphyseal], cortical [including the endosteal or periosteal envelopes]), or soft tissues surrounding the bone and cartilage. For each of these sampling sites, a well defined experimental plan is needed to ensure that the correctly harvested and trimmed specimen is submitted for the appropriate histology method.

Also critical is appropriate careful dissection of the site of interest at harvest prior to submission to histology. If a study requires the examination of multiple sites and the exact anatomic orientation and location of specific regions are critical, histologic marking dyes are very useful aids; these come in several colors and will not harm the histologic preparation, and are best applied during specimen harvest by the investigator (see Appendix 1).

Pilot studies during which the exact skeletal site of interest is harvested using the exact conditions which mimic the planned experiment are possibly the most often overlooked part of the early experimental plan. The ability of a histology lab to successfully deal with bone and cartilage specimens varies with the research setting and the technical expertise available. If the orthopaedic department has access to a lab experienced in bone and cartilage histology, the investigator's planning task is much easier. If, on the other hand, specimens are going to be submitted to a general histology laboratory, early planning and practice specimens are well worth the extra effort.

TABLE 1
Key Questions in the Early Experimental Design Involving the Histologic Study of Bone and Cartilage

Histologic Feature of Interest	If "Yes," then	If "No," then
Are you interested in osteoid seam width or area?	Utilize undecalcified processing and embedding in methyl methacrylate.	Decalcify and embed in paraffin or glycol methacrylate.
Are you interested in measurement of bone formation or mineralization?	Double labelling with tetracyclines prior to animal euthanasia and then use undecalcified processing and embedding in methacrylate. Record dates of label administration and date of euthanasia.	Decalcify and embed in paraffin or glycol methacrylate.
Are you interested in histologic studies of surgical sites?	Consider pre–op and post–op radiographs and photographs of site of interest.	
Be sure that histologic embedment uses appropriate orientation to locate sites of interest (use marking dyes).		
If you are decalcifying your specimens, will you need to localize acid or alkaline phosphatase?	Use a decalcification method which does not block histochemical localizations.	
If you are interested in cortical bone, are you interested in osteoid seam width or area, bone formation or bone mineralization? Cortical endosteum vs. periosteum?	Use methyl methacrylate preps of longitudinal sections of the bone for tetracycline labelling; for cortical cross-sections, use hand ground thin sections (without decalcification).	Decalcified paraffin-embedded cross-sections and longitudinal sections can be used.
Are you interested in the periosteum of the cortex?	Be sure that during harvest of the specimen the periosteum and a thin layer of adjacent muscle is left intact.	

It is also worth noting that confirmation of normal anatomy prior to surgical procedures is important if costly animal models and/or costly research studies are involved.

B. DECALCIFY OR NOT DECALCIFY

Table 1 summarizes frequently encountered questions which the researcher must answer before a new project is started. There are several reasons for choosing methods which do not decalcify specimens prior to embedment: (1) to differentiate osteoid seams from mature mineralized bone; (2) to qualitatively or quantitatively assess tetracycline incorporation to study the rates of bone apposition, mineralization and formation; (3) to localize certain enzymes such as alkaline and acid phosphatase or other substances which can be histochemically localized in methacrylate embedded specimens; and (4) to analyze implant-containing specimens. The latter include implant-bone specimens and implant-soft tissue specimens. When a study does not focus on tetracycline labelling of bone, does not reqire for localization of ezymes, or the specimen does not contain a metal implant, decalcification of bone specimens may be the method of choice.

C. BONE LABELLING

Bone is unique as a tissue in that we can use time-spaced tetracycline and other labels to determine dynamic indices of bone formation and mineralization (Figure 1). It is interesting to note that Klein et al have modified the tetracycline bone labelling scheme and developed a method to assay bone resorption *in vivo* with ^{3}H-tetracycline.[7]

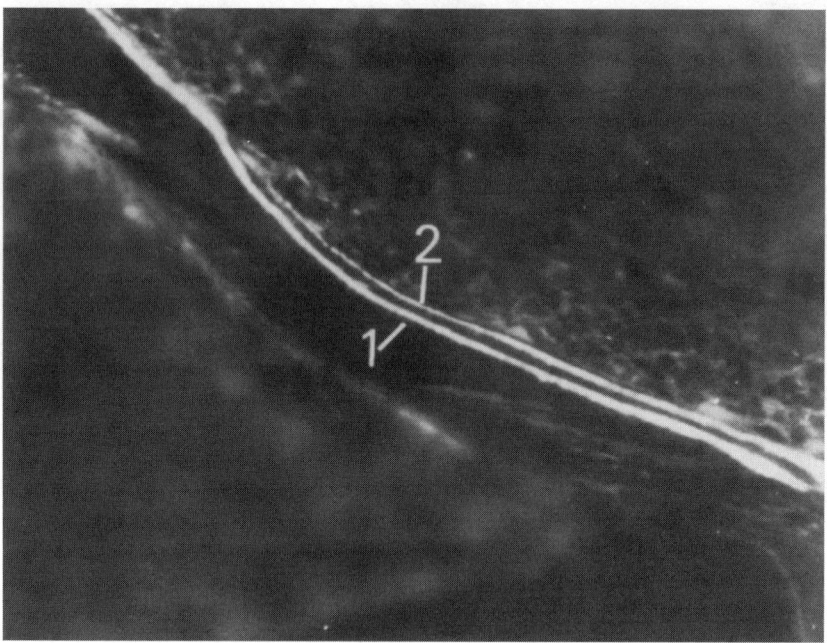

FIGURE 1. Photomicrograph of a methacrylate-embedded biopsy of the anterior iliac crest of a 56-year-old female who received two courses of tetracycline prior to biopsy. When viewed with fluorescence microscopy, double labels of tetracycline mark the bone sites where active bone formation and mineralization were present when the first (1) and second (2) courses of tetracycline were administered. (Unstained section under fluorescence, ×140.) (From Gruber, H. E., et al., Semin. Haematol., 18, 258, 1981. With permission.)

Many bone labelling agents are available.[8] The commonly used tetracyclines for humans (Achromycin, Declomycin, oxytetracycline) can readily be used for bone labelling in animal models. Other agents, such as alizarin red S, can also be used.[8–10] Most tetracycline antibiotics form stable tetracycline-calcium chelates which fluoresce intensely at wavelengths readily obtained with the standard fluorescent microscope. These chelates form only at sites of new bone deposition where the bone contains 20% or less of the maximum mineral content.[11] These chelates are locked into bone during further mineralization and remain in the bone until it is later resorbed by osteoclasts. Tetracyclines also mark the mineralization front in calcifying cartilage. By use of two different tetracyclines which fluoresce different colors, first and final labels can be distinguished in bone forming sites which show only one label incorporation. Such sites either stopped forming after incorporating the first label, or started forming just before administration of the final label.

Usually two injections are used, time-spaced to ensure that the labels are adequately separated for clear microscopic visualization. If your experiment uses an unusual animal model, or if you are studying very old animals with low bone turnover, a pilot study should be carried out testing the labelling intervals for the labelling agents. Labelling studies require that the dates of administration of the labels and the date of euthanasia are recorded. Critical is the processing of specimens in an undecalcified manner with embedment in methacrylate (see below).

III. SPECIMEN HARVESTING

Bone and cartilage are among the most challenging of tissues for histologic examination. Access to the appropriate tools and equipment is a critical aspect of successful orthopaedic histology. As with all tissue obtained for microscopic study, it is important to obtain the specimen quickly and place it in the appropriate fixative as soon as possible. During harvest of large bone specimens which require harvest with an autopsy saw or other saw, it is important to keep the specimen moist during harvest and sawing. Subdivision with small bench saw or other saw unit is sometimes necessary. If possible, this should be done under a fume hood with the specimen kept wet with fixative during the separation process; bone fragments caused by cutting should be gently removed. If ventilation is not available, buffers or saline can be used to wet the cut surface of the bone in place of formalin.

For studies of iliac crest bone and other trabecular bone sites which crush easily, the researcher should consider the advantages of using an electric drill for transiliac and vertical bone core specimens. If manual methods are to be used, Jamshidi core needle biopsies can be obtained. In both instances, it is important to develop good skill in minimizing drilling to prevent heat and crush artifacts in the specimens. Cores must also be of adequate size to permit reproducible histomorphometric measurements if quantitative analysis is needed.[1]

Once the gross specimen has been obtained, smaller saws and tools in the histology lab can be used to further trim the specimen. As mentioned above, colored dyes can be a useful aid for maintaining bone orientation and surgical landmarks. Cork sheets placed under specimens during sawing help to absorb vibration and make the specimens easier to handle. Small hand held saws and even dental drills can be adapted with diamond-coated circular saw blades for fine trimming of specimens. Such trimming should be done under a fume hood with care taken for safe operation of drills and protection against bone fragments dislodged during sawing. A diamond-coated wire saw (Histosaw, Delaware Diamond Knives, Wilmington, DE) is very useful for precise trimming of bones prior to and after methacrylate embedment.

IV. FIXATION AND TRANSPORTATION

Older studies utilized ethanol as a primary fixative because of concerns that aqueous based formalin fixatives might produce some leaching of tetracycline labels from the mineralizing front of bone specimens. Although still used by some investigators,[1] ethanol does not give good cellular preservation. Currently in most laboratories 10% buffered formalin is used as a general primary fixative. Studies using *in situ* hybridization may require other types of fixatives.[12] Such studies again require careful pilot trials to ensure appropriate processing procedures.

As mentioned above, it is important with large specimens to ensure appropriate penetration of the fixative solution. Vacuum should be used at the start of fixation. A vacuum pump attached to a desiccator with inlet and outlet ports provides a simple way to pull a vacuum on bone specimens to ensure adequate penetration of fixative into the tissue. Vacuums should be pulled until bubbles cease to emerge from the specimen. For large specimens, it may be advantageous to let specimens remain in fixative under vacuum for the whole period of fixation. For large specimens, trimming the muscle carefully from the bone aids in penetration. If only the diaphyseal bone is the site of interest, it is helpful to remove the proximal and distal metaphyses, which provides openings for access of fixative to the interior of the diaphysis. A general rule is that the volume of fixative should be at least 10 times that of the volume of the specimen to be fixed.

Fixation time is a critical issue for bone and cartilage as it is for histological procedures with other tissues. The two enzymes of major interest in bone histology, acid phosphatase and alkaline phosphatase, show different sensitivities to exposure to formalin.[13] Therefore, a total fixation time of 4–6 hours is recommended. If transportation is needed the specimen should be shipped in 70% ethanol. Specimens can be stored in 70% ethanol (higher concentrations tend to cause the specimen to become brittle).

V. SPECIMEN PROCESSING AND SECTIONING

A. DECALCIFIED SPECIMENS

Decalcification procedures in general use either acids which react with the calcium in bone to form soluble calcium salts, or chelating agents which complex the calcium ions. Acid decalcifying agents are known to have the drawback of altering the staining properties of the embedded bone.[14] Some decalcifying solutions also contain formalin, and thus increase the possibility of aldehyde groups increasing in the tissue and blocking reactions.

A large number of commercially available decalcification solutions offer a wide range of decalcification times. It is important to remember that many commercial products have been developed to be used for rapid decalcification of large specimens in the clinical pathology lab; these products should not automatically be assumed to be suitable for specialized orthopaedic research applications. In our laboratory, we routinely use a formic acid decalcification method (Appendix 1) which preserves the ability to localized tartrate-resistant acid phosphatase and alkaline phosphatase, and can be used for anti-Factor VIII immunohistochemical localization of vasculature.

If the experiment involves specialized histology, such as localization of tartrate-resistant acid phosphatase or alkaline phosphatase, of *in situ* hybridization studies, it is important that the decalcification method is compatible with the special techniques. Although rapid decalcification solutions are quick, there is the danger that they have destroyed the ability to localize key enzymes and to do other special procedures on the decalcified tissue. An alternative to paraffin embedding of decalcified tissue is embedment in glycol methacrylate; this affords a crisper resolution of cellular detail (see below) and is often overlooked as a method of choice. It is again advisable that a pilot study be used to ensure compatibility of the chosen decalcification/embedment method with the desired histologic outcome.

Although raising the temperature can hasten decalcification, it also may further deteriorate cell and tissue details. Placing the specimen in the decalcifying solution on a shaker and frequently changing the solution are important. Generally a 20:1 decalcifying solution:specimen volume ratio is recommended.[14]

The two most frustrating and disastrous outcomes from improper decalcification are (1) incompletely decalcified specimens which are embedded too soon and are impossible to section successfully, and (2) specimens which have been decalcified in a harsh manner so that cellular detail is lost and enzyme/immunohistochemical localizations cannot be performed.

Several methods are useful to determine whether a specimen has been completely decalcified and is ready for paraffin processing. For long bones of small mammals, a test can be made by testing the cutting of the cortex. If a scalpel cuts it easily, decalcification is complete. For larger bones, if one has access to an X ray unit, taking a radiograph of the specimen is a useful technique. The bone should have the same radiological contrast as surrounding muscle in the appropriately decalcified specimen. Chemical tests to determine if calcium is still being removed from the specimen (indicating incomplete decalcification) include the calcium oxalate test,[14] and determination of calcium content of the solution with atomic absorption spectrophotometry.[15]

For decalcified specimens, regular microtomes are efficient to cut 3–6 μm thick sections.

B. Undecalcified Specimens

1. Embedding of Undecalcified Specimens

In the past, various plastic embedding techniques have been successfully applied to bone and teeth.[13] Older methodologies relied upon UV polymerization with the specimens cooled in a water bath during polymerization. More recently, methods to remove commercially added inhibitors, and use of chemical initiators and catalysts, have greatly simplified plastic embedding. With careful variations on protocols, even large specimens, such as intact canine spine, can be embedded.[17]

Processing of undecalcified bone specimens is still a relatively long process. In Appendix I methods are summarized for our procedures for glycol and for methyl methacrylate embedding. Glycol methacrylate embedding results in good cartilage preservation and good preservation of osteoblast-associated alkaline phosphatase. Methyl methacrylate embedding results in a harder plastic providing superior support for cortical and trabecular bone. The goal is to match hardness of the embedding support medium to the hardness of the bone or cartilage in order to produce successful sections. The reader is also referred to previously cited sources for other useful embedding methodologies, such as Spurr's embedding method.

Large bone specimens are technically challenging both for decalcified and undecalcified preparations. The three most important steps are to ensure adequate fixation of the specimen (without masking desired enzymes or antibodies by exposure to aldehydes), and, for decalcified preparations, to achieve decalcification without loss of special staining/enzyme capabilities, and to obtain adequate infiltration of the specimen for either paraffin or methacrylate processing. As previously noted, a pilot study where the actual skeletal site of interest is used for laboratory practice is strongly recommended.

Our laboratory has recently reviewed large specimen methods in our study of the canine spine and has developed a procedure for infiltration and embedment of large bone specimens.[17] Pulling a vacuum on large specimens is very important to aid in infiltration of solutions and of the final polymerization solutions. If one is solely interested in the epiphysis and metaphysis, the diaphysis can be cut at mid-shaft; this provides an open face of bone and marrow into which solutions can more easily penetrate. Infiltration under vacuum can also be used. The presence of bubbles in the final polymerized plastic block indicates a rapid, uncontrolled polymerization which generated heat and damaged tissue. For methacrylate embedments, a heat sink can be incorporated into the procedures. Water is an excellent heat sink provided the surface is sealed with oil so that humidity does not build up in the polymerization chamber. Large specimens also sometimes require special methods to mount them for microtome sectioning; listed in Appendix 2 are useful molds and sealants which allow firm specimen anchorage during sectioning by large microtomes. Care should always be taken in trimming the initial submitted specimen if the cortical periosteum is of interest. Muscle should never be pulled from the cortical surface; gentle trimming of muscle layers with a scalpel ensures that a thin layer of muscle can be left overlying the periosteum.

2. Sectioning of Plastic Embedded Specimens

Based on the specific purpose of the study, the size of the specimen, or the existence of an implant, the following sectioning methods may be used for plastic embedded specimens.

Small undecalcified bone specimens embedded in glycol or methyl methacrylate can be cut using an appropriate motor-driven microtome and tungsten carbide knives (for methyl methacrylate or glycol methacrylate) or large glass knives (for glycol methacrylate).

For ground sectioning, traditionally, a slow speed diamond saw is used to cut approximately 1.0 mm thick sections off non-embedded or plastic embedded specimens. The sections are then glued onto plexiglass slides for grinding by hand on a grinding wheel to produce sections with average thicknesses of 25–100 µm. Subsequently, the sections are stained with a variety of procedures for microscopic evaluation or viewed directly with UV microscopy for identification of fluorescent labels.[18] In most places, this method remains the routine selection for study of plastic embedded specimens. It does not, however, provide good resolution of histologic detail at the cellular level because of the thickness of the specimen.

A diamond-coated wire saw, Histosaw (Delaware Diamond Knives, Wilmington, DE), is a relative new sectioning system. According to our experience and others', sections with a thickness down to 30 µm can be created using this method in a matter of minutes. The sections are then readily glued onto regular glass microslides, stained, and coverslipped.

C. Soft Tissues

Muscle, tendons and ligaments can usually be processed as other soft tissues either in paraffin or in glycol methacrylate. Orientation is critical for muscle, and in studies either longitudinal or tangential sections of muscle bundles may be desired. Tendons and ligaments can curl and twist

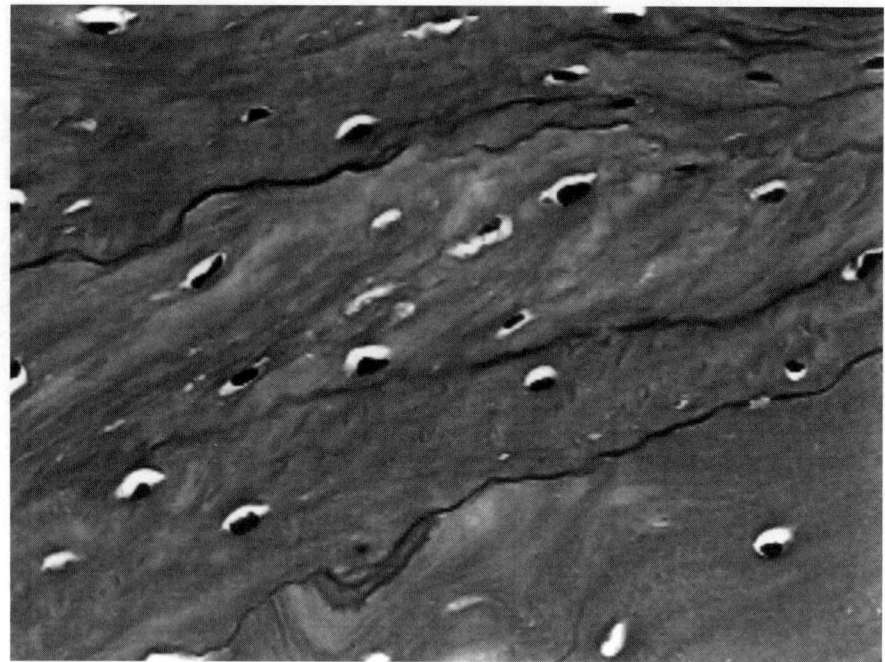

FIGURE 2. Photomicrograph illustrating good visualization of cement lines in a decalcified specimen of rat femur stained with hematoxylin and eosin (×420).

during processing; this can sometimes successfully be dealt with by anchoring the ends to a wooden applicator stick or tongue depressor to maintain specimen orientation during processing and sectioning.

If the insertion sites onto bone are the regions of interest, specimens must either be decalcified or processed through plastic undecalcified. Synthetic materials utilized to reconstruct tendons or ligaments may require longer infiltration processing intervals for paraffin or may alternatively be processed in methacrylates.

VI. STAINING TECHNIQUES

A. Staining of Bone and Cartilage

Many excellent reviews have previously presented selected methods for staining bone and cartilage.[4,14,19]

1. Decalcified Sections

Many of the routine trichrome, hematoxylin/eosin, toluidine blue and other stains have been adapted for study of bone and cartilage in paraffin sections (Figure 2). Methylene blue/basic fuchsin and other metachromatic stains are useful if one wishes to view entire long bones with epiphysis, physis and metaphysis present in one specimen. If the cartilage matrix components are of interest, the metachromatic stains, including azure A and the critical electrolyte staining series with Alcian blue, are very valuable (Figure 3).[11] Periodic acid-Schiff can be applied both to decalcified and plastic embedded specimens to investigate the mucopolysaccharide content of cartilage.[21]

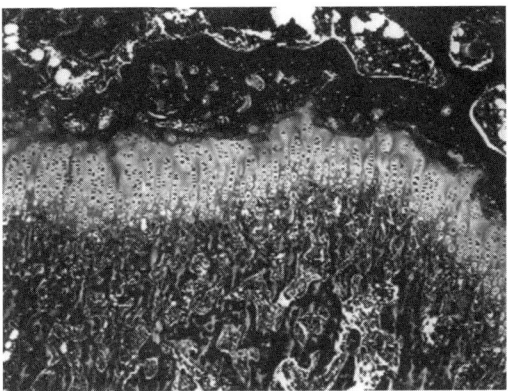 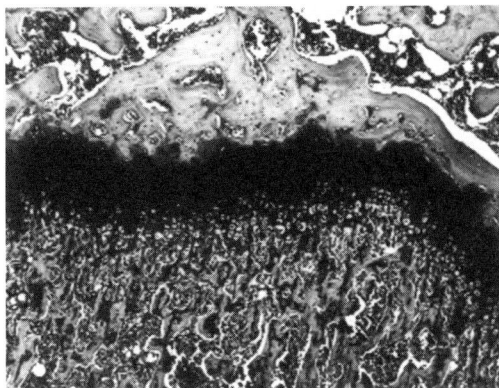

FIGURE 3. Photomicrographs illustrating differential staining of the growth plate of a decalcified rat femur obtained with Masson trichrome stain (left) (bone dark, growth plate pale) and with safranin-O (right) (bone pale, growth plate dark) ($\times 44$).

2. Undecalcified Sections

Staining regimes for plastics require modification because these procedures are usually performed without the removal of the plastic and thus different timing regimens are utilized. In Appendix 1 we list references for the staining procedures which we routinely carry out. Goldner's stain and von Kossa's stain both result in differentiation of osteoid from mature mineralized bone matrix; Goldner's stain (a modified Masson trichrome) has a hematoxylin component which provides excellent cellular staining.[21] Cement lines, which are specialized regions of the bone matrix which mark past sites of bone remodelling, can be visualized with a number of staining procedures, including Stains-All,[22] toluidine blue or methylene blue/basic fuchsin,[16] thionine,[23] a combined method using thionine, toluidine blue, methylene blue chloride or methylene violet,[16] or modifications of Bodian silver stain.[17]

Villanueva et al.[24] developed a method for block staining of the specimen followed by hand grinding. With other methods where specimens are not stained en bloc, only the surface of ground sections is stained. Bone embedded in plastic initially can also be cut, polished and stained. This may offer some advantage if small rodent long bones are being studied in cross section since the plastic margin makes the specimen easier to handle. Other useful suggestions in handling and staining ground sections have been summarized by Schenk et al.[18]

B. Staining of Soft Tissues

Soft tissue-bone interfaces can be examined using special staining methods to distinguish fibrocartilage, bone and soft tissues; Hurov has presented a battery of stains which help histologically define this specialized site (Verhoeff's elastic tissue stain, Alcian blue and periodic acid-Schiff for acid mucopolysaccharides, hematoxylin and eosin, and an aldehyde fuchsin counterstained with Van Gieson's picro-fuchsin to demarcate cartilage and bone).[25]

C. Histochemical Staining

Enzyme histochemistry has played an important role in the study of chondro-osseous tissue. The two phosphohydrolases with special relevance are alkaline phosphatase, an ectoenzyme present in the osteoblast and in matrix vesicle membranes, and tartrate-resistant acid phosphatase (TRAP), a lysosomal enzyme whose localization provides a sensitive method of osteoclast identification (Figure 4).

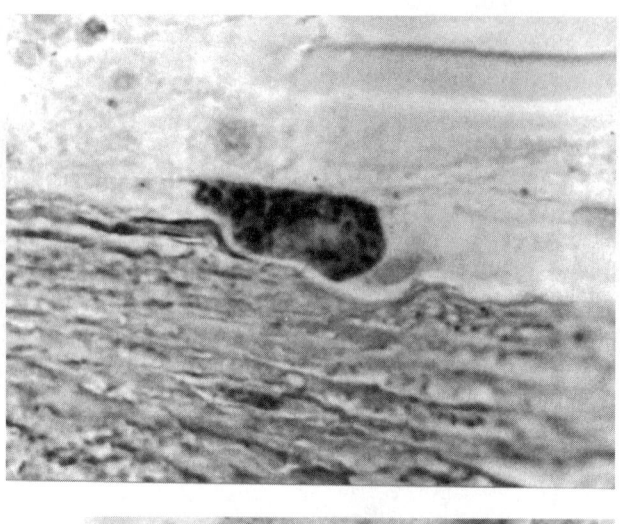
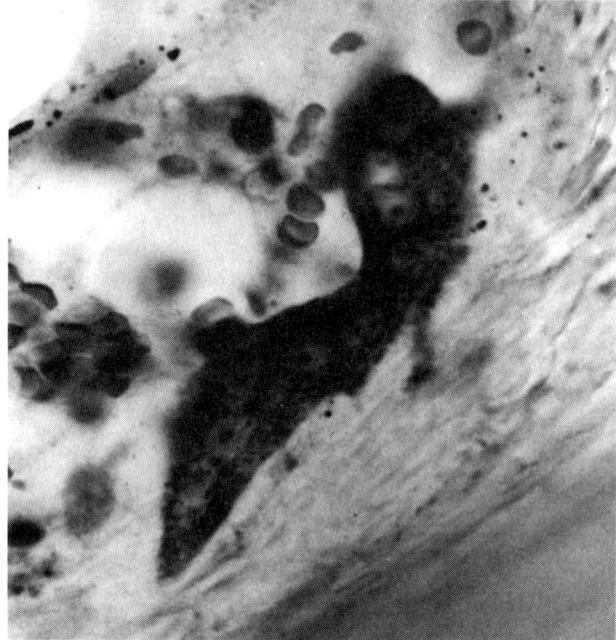

FIGURE 4. Localization of tartrate-resistant acid phosphatase (TRAP) is seen here in identification of both small (top) and large (bottom) osteoclasts. (methyl methacrylate undecalcified preparation, ×1080). (From Gruber, H. E., et al., *Semin. Haematol.*, 18, 258, 1981. With permission.)

Three factors influence one's success in histologically localizing enzymes: optimum procurement of the specimen, proper fixation, and proper embedding. For bone and cartilage, the latter condition can be satisfied by using methacrylate embedding and avoiding decalcification and paraffin processing.

A critical factor is appropriate fixation. Results from the authors' laboratory have shown that optimum localization of acid phosphatase (Figure 4) and alkaline (Figure 5) in fresh tissue is achieved after fixation not exceeding 15 hours in neutral buffered formalin followed by embedment in methyl methacrylate.[9] As a rule of thumb, 4–6 hours provides a safe fixation period. Occasionally, it is desirable to retrieve frozen tissue from −70°C collections. When frozen tissue is to be evaluated, good localization of both enzymes can be achieved after fixation for 30 min. and embedment in methyl methacrylate, or after five min. fixation followed by embedment in glycol methacrylate.

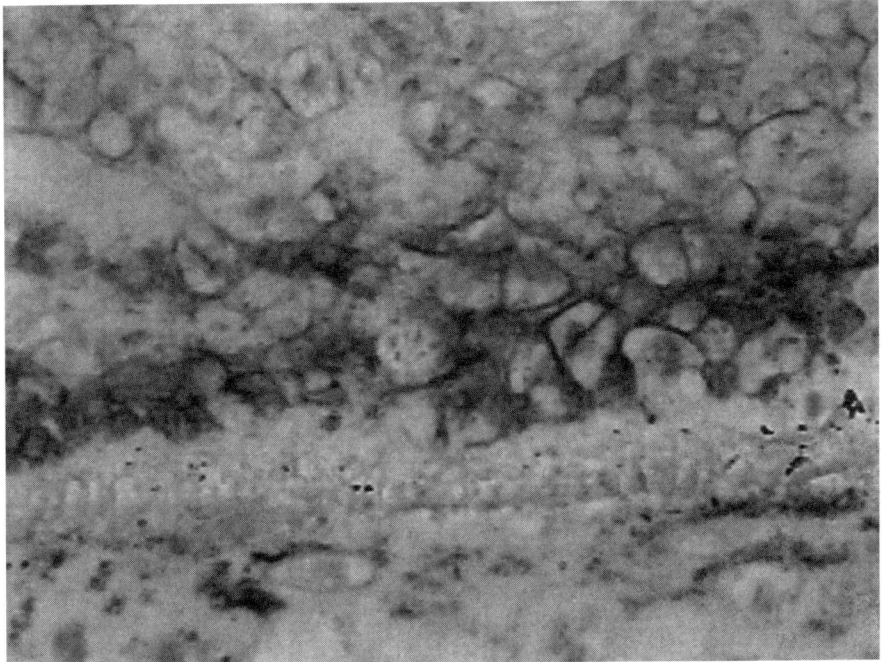

FIGURE 5. Localization of alkaline phosphatase in osteoblasts. (glycol methacrylate, ×420).

Often cytochemical studies of many enzymes are done on minimally fixed tissue, often sectioned on a cryostat. This, however, requires considerable technical skill with fragile growth plate or brittle large bone specimens, so it is well worth the effort to define the appropriate fixation period for each enzyme of interest.

It is important to have a positive control included in all runs with alkaline phosphatase and TRAP localizations.

VII. HISTOLOGICAL EVALUATION AND HISTOMORPHOMETRY

Qualitative evaluation of histologic changes in bone and cartilage is usually the first level of evaluation of specimens from an orthopaedic experimental study. Initial qualitative studies are of value because they help focus the investigator on specific sampling sites of interest under the microscope and help to further refine the investigator's notion of what type of quantitative information should be collected in the study. For large projects involving many investigators, this is also a useful first approach to the histologic study of the specimens of interest whereby teams of researchers discuss, evaluate and agree on the quantitative data to be collected. Qualitative studies are also of value should experimental treatment produce biochemical changes in matrices which can be revealed by special staining techniques. Histology in such research can play a valuable role prior to biochemical tissue analysis.

For most well defined studies of bone and cartilage, however, current research leads to quantitative evaluation of specific features of the tissue of interest. The most common types of data collected are those measuring tissue areas (areas of bone, osteoid, and fibrous tissue in the marrow cavity), widths (trabecular width, osteoid seam width, growth plate widths, widths between tetracycline labels), lengths (the fraction of endosteal surface lined by osteoid, or with an associated tetracycline label), or counts (osteoblast or osteoclast indices). The advent of PC-based, interactive image analysis systems designed for the specialized requirements of bone histomorphometry have

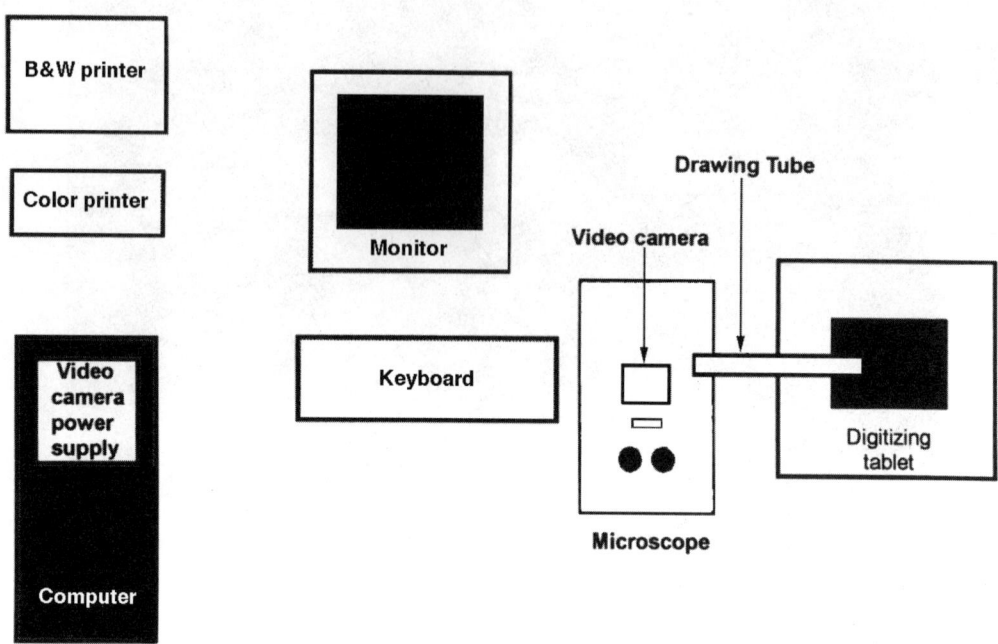

FIGURE 6. Basic instrumental components of a histomorphometry system. (From Gruber, H. E., *Biotechnol. Software J.,* 12, 30, 1995. With persission.)

not only made measurements easier and faster, but have increased the amount of histomorphometric data that can be acquired.

The basic instrument components for a PC-based interactive system are shown in Figure 6. Most systems now use video cameras interfaced between the microscope and the PC system; drawing tubes still have favor in some laboratories. When used with a drawing tube, the microscopic field shows both the microscopic image and red diode on the cursor. With the video system application, a cross-hair appears on the active measurement field on the monitor which is displaying a real-time image of the specimen viewed with the microscope. Older video systems sometimes had the drawback of loss of intensity of fluorescent label intensity viewed on the computer terminal, but newer generations of video cameras and imaging systems help to overcome this.

A standardized nomenclature now exists for reporting bone histomorphometric data and should be used for all presentations with appropriate choice of bone surface, tissue area or tissue volume referrents.[26]

VIII. OTHER IMPORTANT PROCEDURES

A. IMMUNOHISTOCHEMISTRY

Immunocytochemistry and immunohistochemistry are valuable techniques for specialized bone, cartilage, tendon and ligament studies. In immunohistochemistry, biologic substances of interest are localized by the precise attachment of a complex or label which subsequently can be visualized in the cell or tissue of interest. Visualization can be either by bright field microscopy (viewing of a chromogenic reaction product) or via UV microscopy using a fluorescent coupling agent. The visible product binds by the attraction between immunogen (antigen) and immunoglobulin (antibody). Several excellent reviews and instructional texts are available for further study of principles and newer immunologic and non-immunologic visualization methods.[27,28] For orthopaedic research purposes, the specimens are usually either fresh frozen and cut on a cryostat, or fixed appropriately as described below.

Since immunohistochemical methods are technically complex, most studies on bone have utilized decalcified preparations. There are, however, excellent studies using plastic embedded tissues which are best recommended to the more advanced histology laboratory.[29-31]

As noted above for enzyme histochemical studies of bone, an initial step in attempting to use a new antibody localization method is to determine the type of fixation, if any, which is required. As noted by Myers,[32] fixatives which maintain excellent morphologic detail in a tissue may not be at all successful in preserving immunoreactivity. Again, for the novice investigating an antibody for the first time, testing of fixation agents (1 or 10% neutral buffered formalin, Bouin's fixative, or Zamboni's fixative) is the first step. The reader is referred to detailed texts for further information about fixation types and regimes which may be of interest.[27,28] A methodological necessity is inclusion of known positive control tissue, and conscientious inclusion of negative controls, with each assay run.

Other common problems encountered with immunohistochemistry are poor visualization resulting from insufficient specific staining, high levels of nonspecific (background) staining, the inability to achieve localization due to masking by prolonged fixation or embedding methods, and lack of success with initial trials of new antibodies. Thorough rinsing between steps is important to control non-specific staining. Some investigators have reported success in retrieving antigenicity in specimens fixed for inappropriately long periods (or older archived fixed tissue) with use of antigen retrieval methodologies as recently reviewed elsewhere.[33-37] These methods include techniques such as enzyme pretreatment of sections and microwave irradiation.

There are two other common problems to be resolved when beginning new studies: the first is to identify reliable antibodies for your specific needs, and if the antibody is an anti-human one, to determine whether or not it can successfully be used to localize antigens in nonhuman tissue. Appendix 3 suggests some sources for information should routine literature searches and vendor product information be insufficient. It is important to remember that not all commercially available antibodies work equally well, and that there is no certification system for antibodies similar to that which ensures chemical histologic stain quality through the Biological Stain Commission and its issued certification numbers for stains.

B. *In Situ* Hybridization

This is a powerful technical procedure to localize either DNA or RNA nucleic acid sequences within cells, and as such gives the investigator the tool with which to visualize gene expression and other events at the cell level. This technological marriage between molecular biology techniques and techniques used for immunocytochemical imaging raises the complexity over each of the individual methods. Fixation and methods for pretreatment of sections must be optimized for each tissue and probe. It is also important to utilize a method wherein the reporter molecule in indirect procedures does not interfere with the hybridization reaction or the resultant hybrid stability.[39] The reader is referred to several excellent recent texts and laboratory manuals which discuss probe size, radioactive vs non-radioactive markers, and specialized strategies.[38-40]

In situ polymerase chain reaction (PCR) can be utilized to detect low-copy DNA and RNA; optimization is required for fixation, protease digestion, DNase digestion for RNA targets, amplification solution, and for DNA targets, the hot start procedure.[41]

An increasing number of *in situ* hybridization kits are becoming available for specific probes of interest,[40] and in our laboratory we routinely localize cells undergoing programmed cell death (apoptosis) with such a kit procedure (Figure 7).

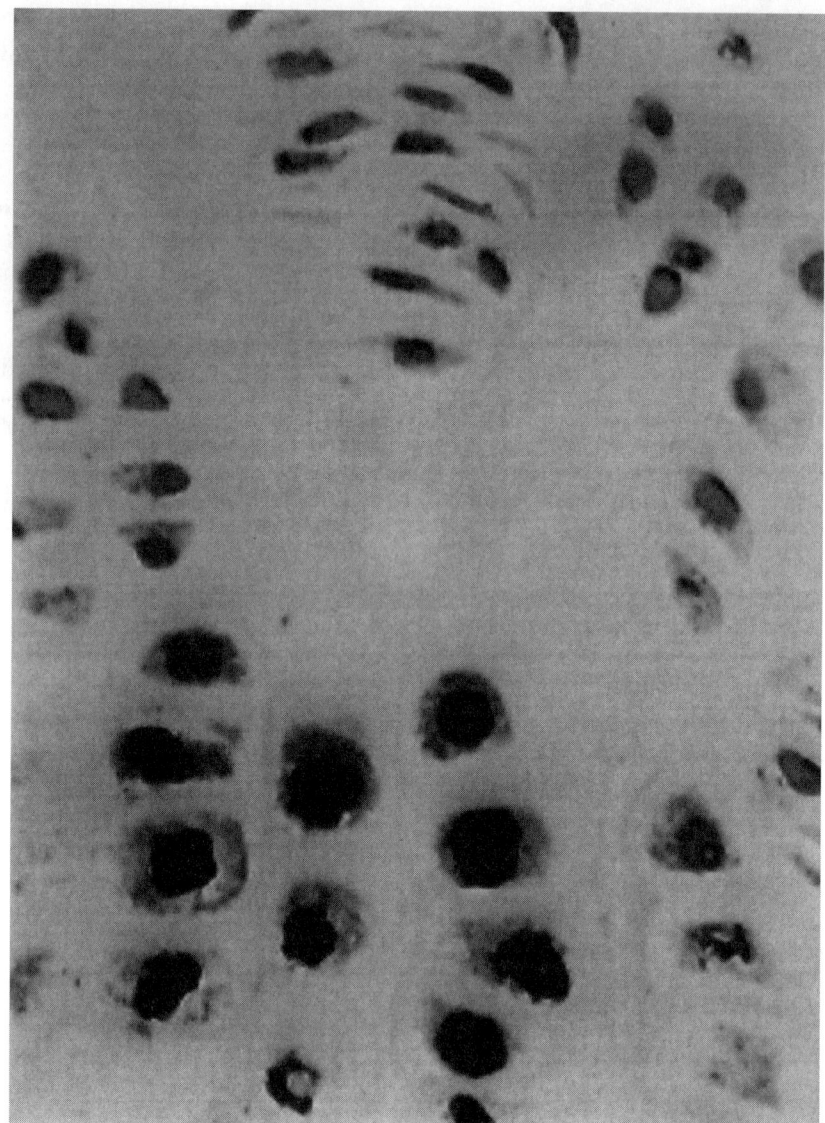

FIGURE 7. Photomicrograph illustrating *in situ* localization of apoptotic chondrocytes (darker cells) in the hypertrophic zone of the growth plate of a rat. (decalcified paraffin-embedded specimen, ×420).

C. Tissues Containing Implants

The diversity of biomaterials employed as implants is expanding rapidly. Implants are studied in orthopaedic research both for their application in architectural reconstruction (to restore, maintain or improve function), to repair defects, and as drug or bioactive agent delivery systems. Both permanent and biodegradable implants are being utilized. Implant types vary greatly; implant materials may be metal, may be ceramic, may be coralline, may be plastic, may be solid or porous, surfaces may be polished or may be roughened or may be coated with other materials in an attempt to achieve a solid bone-implant interface. Both natural and synthetic polymers in many configurations (sheets, sponges, gels) are playing a role in investigations to explore tissue engineered growth of bone, cartilage and other soft tissues.[42]

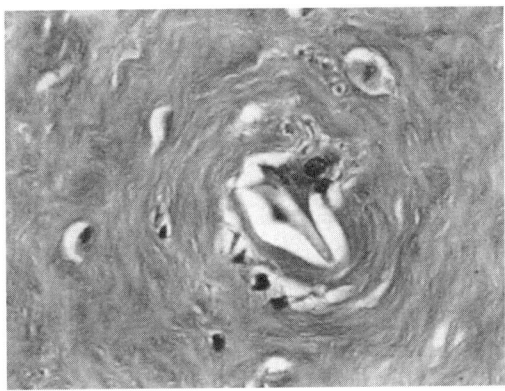 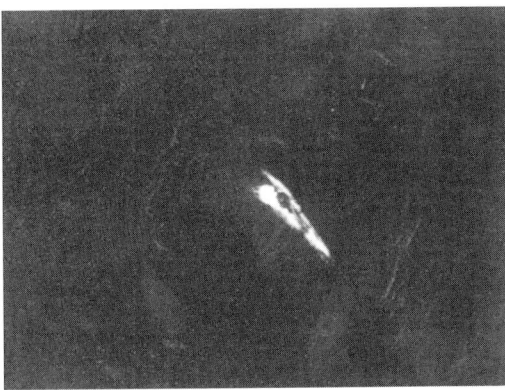

FIGURE 8. Photomicrographs illustrating wear particles in the fibrous capsule surrounding a deteriorating hip prosthesis in a patient with severe rheumatoid arthritis. Left, bright field microscopy shows a fibrous tissue region with a giant cell surrounding a possible wear particle; Right, the same microscopic field viewed with polarized light microscopy: the wear particle stands out well and is easily visualized because of its birefringent nature ($\times 420$).

The methods for general histologic examination of bone described above all have a role in the histologic examination of implant sites. Biodegradable or gel implant materials can be associated with mineralization *in situ* and require decalcification if paraffin embedding or plastic embedding as the method of choice. Porous implants of metal and metal alloys require diamond saw (or ceramic carborundum) blades for cutting and subsequent grinding and surface staining. Methacrylate infiltration and embedding help preserve the bone-implant interface. Mechanized grinding wheels may be required for large specimens with large metal implants.

As with all bone histologic studies, a sampling site must be identified clearly and the sections obtained in the appropriate plane with minimum artifact during gross specimen harvest. At tissue harvest, the gross features of nearby tissue reaction, encapsulation, or other tissue responses should be noted and photographically documented. Radiographic documentation at harvest is also often very important. Large specimens require the appropriate lengthened infiltration regimes for methacrylate embedding.

The histologic evaluation of implant specimens should include evaluation of any tissue reaction to the implant (tissue necrosis, inflammation, giant cells, neutrophils, macrophages, fibroblasts, fibrosis, encapsulation, scarring). For some implants, evaluations should include assessment of possible sarcoma formation around metal implants. The response of blood vessels near the implant, and ingrowth of vasculature into porous or biodegradable implants, is an important feature which can be monitored in microtome sections using anti-Factor VIII immunohistochemistry to denote blood vessels. For implant studies which need to assess new bone formation on or within the construct, undecalcified methods will allow assessment of the extent and quality of osteoid seams and tetracycline labelling evaluation if desired. Care should be taken to note the quality as well as quantity of new bone formed. Polarized light microscopy should be used to identify areas of poorly modelled, woven bone and well modelled lamellar bone. Polarized light microscopy is also valuable to detect wear debris in capsular material (Figure 8). With careful harvesting and preparative techniques, the interface between the implant and bone can be successfully preserved.

REFERENCES

1. Malluche, H. H., Faugere, M.-C., *Atlas of Mineralized Bone Histology*, Karger, Basel, 1986.
2. Erikson, E. F., Axelrod, D. W., and Melsen, F., *Bone Histomorphometry*, Raven Press, New York, 1994.
3. Recker, R. R., *Bone Histomorphometry: Techniques and Interpretation*, CRC Press, Boca Raton, FL, 1983.

4. Anderson, C., *Manual for the Examination of Bone*, CRC Press, Boca Raton, FL, 1982.
5. Dickson, G. R., *Methods of Calcified Tissue Preparation*, Elsevier, Amsterdam, 1984.
6. Elias, J. M., "Bones are diamonds in the rough," *J. Histotechnol.*, 20, 181, 1997.
7. Klein, L. and Jackman, K. V., "Assay of bone resorption *in vivo* with 3H-tetracycline," *Calcif. Tissue Res.*, 20, 275, 1976.
8. Frost, H. M., "Bone histomorphometry: choice of marking agent and labeling schedule," in *Bone Histomorphometry: Techniques and Interpretation*, Recker, R. R., Ed., CRC Press, Boca Raton, FL, 1983, 4.
9. Tonna, E. A., Singh, I. J., and Sandhu, H. S., "Non-radioactive tracer techniques for calcified tissues," in *Methods of Calcified Tissue Preparation*, Dickson, G. R., Ed., Elsevier, Amsterdam, 1984, 333.
10. Harris, W. H., Travis, D. F., Friberg, U., and Radin, E., "The *in vivo* inhibition of bone formation by alizarin red S," *J. Bone Joint Surg.*, 46A, 493, 1964.
11. Frost, H. M., Villanueva, A. R., Roth, H., and Stanisavljevi, C. S., "Tetracycline bone labelling," *J. New Drugs*, 1, 206, 1962.
12. Hicks, D. B., Stroyer, B. F., Teot, L. A., and O'Keefe, R. J., "*In situ* hybridization in skeletal tissues utilizing non-radioactive probes," *J. Histotechnol.*, 20, 215, 1997.
13. Gruber, H. E., Marshall, G. J., Nolasco, L. M., Kirchen, M. E., and Rimoin, D. L., "Alkaline and acid phosphatase demonstration in human bone and cartilage: effects of fixation intervals and methacrylate embedments," *Stain Technol.*, 63, 299, 1988.
14. Page, K. M., "Bone and the preparation of bone sections," in *Theory and Practice of Histological Techniques*, Bancroft, J. D. and Stevens, A., Eds., Churchill Livingstone, London, 1982; 297.
15. Kiviranta, I., Tammi, M., Lappalainen R., Kuusela, T., and Helminen, H. J., "The rate of calcium extraction during EDTA decalcification from thin bone slices as assessed with atomic absorption spectrophotometry," *Histochemistry*, 68, 119, 1980.
16. Gruber, H. E., Marshall, G. J., Kirchen, M. E., Kang, J., and Massry, S. G., "Improvements in dehydration and cement line staining for methacrylate embedded human bone biopsies," *Stain Technol.*, 60, 337, 1985.
17. Gruber, H. E. and Stasky, A. A., "Large specimen bone embedment and cement line staining," *Biotech. Histochem.*, 72, 198, 1997.
18. Schenk, R. K., Olah, A. J., and Herrmann, W., "Preparation of calcified tissues for light microscopy," in *Methods of Calcified Tissue Preparation*, Dickson, G. R., Ed., Elsevier, Amsterdam, 1984, Chapter 1.
18. Baron, R., Vignery, A., Neff, L., Silverglate, A., and Maria, A. S., "Processing of undecalcified bone specimens for bone histomorphometry," in *Bone Histomorphometry: Techniques and Interpretation*, Recker, R. R., Ed., CRC Press, Boca Raton, FL, 1983, Chapter 3.
20. Gruber, H. E., Massry, S. G., and Brautbar, N., "Effect of relatively long-term hypomagnesemia on the chondro-osseous features of the rat vertebrae," *Miner. Electrolyte Metabol.*, 20, 282, 1994.
21. Gruber H. E., "Adaptations of Goldner's Masson trichrome stain for the study of undecalcified plastic embedded bone," *Biotech. Histochem.*, 67, 30, 1992.
22. Gruber, H. E., "Application of Stains-All for demarcation of cement lines in methacrylate-embedded bone," *Biotech. Histochem.*, 66, 181, 1991.
23. Derkx, P. and Birkenhäger-Frenkel, D. H., "A thionin stain for visualizing bone cells, mineralizing fronts and cement lines in undecalcified bone sections," *Biotech. Histochem.*, 70, 70, 1995.
24. Villaneuva, A. R., Hattner, R. S. and Frost, H. M., "A tetrachrome stain for fresh mineralized bone sections," *Stain Technol.*, 39, 87, 1964.
25. Hurov, J. R., "Soft-tissue bone interface: How do attachments of muscles, tendons, and ligaments change during growth? A light microscopic study," *J. Morphol.*, 189, 313, 1986.
27. Parfitt, A. M., "Bone histomorphometry: standardization of nomenclautre, symbols and units. Summary of proposed system," *Bone Mineral*, 4, 1, 1988.
27. Larsson, L.-I., *Immunocytochemistry: Theory and Practice*, CRC Press, Boca Raton, FL, 1988.
28. Javois, L. C., "Direct immunofluorescent labeling of cells," *Methods Mol. Biol.*, 34, 117, 1994.
29. Lucena, S. B., Duarte, M. E. L., and Fonseca, E. C., "Plastic embedded undecalcified bone biopsies: an immunohistochemical method for routine study of bone marrow extracellular matrix," *J. Histotechnol.*, 20, 253, 1997.

30. Tacha, D. E., Bowman, P. D., and McKinney, L., "High resolution light microscopy and immunocytochemistry with glycol methacrylate embedded sections and immunogold-silver staining," *J. Histotechnol.*, 16, 13, 1993.
31. Hermanns, W., Colbatzky, F., Gunther, A., and Steiniger, B., "Ia antigens in plastic-embedded tissues: a post-embedding immunohistochemical study," *J. Histochem. Cytochem.*, 34, 827, 1986.
32. Myers, J. D., "Development and application of immunocytochemical staining techniques: a review," *Diagnost. Cytopathol.*, 5, 318, 1989.
33. Tsuji, Y., Kusuzak, K., Hirasawa, Y., Serra, M., and Baldini, N., "Ki-67 antigen retrieval in formalin- or ethanol-fixed, paraffin-embedded tissues: an enhancement method for immunohistochemical staining with autoclave treatment," *Acta Histochem. Cytochem.*, 30, 251, 1997.
34. Wakamatsu, K., Ghazizadeh, M., Ishizaki, M., Fukuda, Y., and Yamanaka, N., "Optimizing collagen antigen unmasking in paraffin-embedded tissues," *Histochem. J.*, 29, 65, 1997.
35. Shi, S.-R, Cote, R. J., Chen, T., and Taylor, C. R., "Antigen retrieval technique: an important approach to standardization of immunohistochemistry," *Cell Vision*, 3, 235 1996.
36. Shi, S. R., Cote, R. J., and Taylor, C. R., "Antigen retrieval immunohistochemistry: past, present, and future," *J. Histochem. Cytochem.*, 45, 327, 1997.
37. Taylor, C. R., Shi, S. R., Chen, C., Young, L., Yang, C., and Cote, R. J., "Comparative study of antigen retrieval heating methods: microwave, microwave and pressure cooker, autoclave, and steamer," *Biotech. Histochem.*, 71, 263, 1996.
38. *Nonradioactive in situ Hybridization Application Manual*, Mannheim: Boehringer Mannheim GmbH, Biochemica, 1996.
39. Choo, K. H. A., *In Situ Hybridization Protocols*, Humana Press, Totowa, NJ, 1994.
40. Leitch, A. R., Schwarzacher, T., Jackson, D., and Leitch, I. J., *In Situ Hybridization: A Practical Guide*, BIOS Scientific, Oxford, 1994.
41. Nuovo, G. J., "Keys to successful *in situ* PCR," *Cell Vision*, 3, 197, 1996.
42. Holder, W. D., Gruber, H. E., Roland, W. D., et al., "Increased vascularization and heterogeneity of vascular structures occurring in polyglycolide matrices containing aortic endothelial cells implanted in the rat," *Tissue Eng.*, 3, 149, 1997.

APPENDIX 1
SELECTED PROTOCOLS FOR EMBEDDING AND STAINING

A. Paraffin Embedding

1. Fixation

If enzyme histochemistry is desired, routine fixation should be in 10% buffered formalin for 4–6 hours with a vacuum pull until bubbles cease to emerge from the specimen. Immediately after placing the specimen in fixative, make a small identification tag with plastic paper and a lead pencil. This tag follows the specimen through all solution changes and into embedment, thus assuring correct specimen identification. The specimen can be transferred to, and held in, 70% ethanol.

2. Decalcification Methods

a. EDTA Method

EDTA solution
40 mg EDTA in 300 ml distilled H_2O (dH_2O).
Add 12.6 g NaOH pellets; add slowly, four pellets at a time.
The pH should be 7.3 when solution is prepared.

Place the specimen in a large volume of EDTA decalcifying solution (at least 4–5 times the volume of the specimen). Keep the specimen rotating or being stirred for several days. Rinse the specimen for 2–3 hours after decalcification. The specimen can be held in 70% ethanol for further processing. For small bones, decalcification may take 3–4 days. Change the solution at least once during this period. This method is useful for most types of immunohistochemistry.

b. Formic-Citrate Method

Solution A   50 g sodium citrate in 250 ml dH_2O.
Solution B   50 ml (90%) formic acid in 50 ml dH_2O.
Working solution   Mix equal portions of Solution A and Solution B.

Decalcify in 4–5× the volume of the specimen. Place the specimen in a container on a shaker or rotator at room temperature. The size of the specimen will influence the length of time required for decalcification. A rat tibia requires approximately 4 hours. Rinse the specimen for 2–3 hours after decalcification. The specimen can be held in 70% ethanol for further processing.

B. Methyl Methacrylate Embedding of Small to Medium Size Specimens

1. Preparation and Fixation of Small to Medium Size Specimens

Specimen should be sketched or photographed when obtained. A detailed description of the tissue should be kept when dissection begins.

Soft tissue that is not necessary for the analysis should be dissected off. Measurements of small bones can be taken prior to trimming by using a caliper. The trimmed piece should be placed in 10% buffered formalin for no more than 24 hours. An exception to this rule is the inclusion of acid phosphatase or alkaline phosphatase localization. In this case, specimens should be fixed for 4–6 hours. A paper label (plastic paper written in pencil) with the identification code should accompany the specimen throughout processing.

2. Dehydration and Infiltration Schedule for Small to Medium Size Specimens

After 24 hours, or the shorter fixation times, the tissue should be placed in 70% ethanol. A vacuum condition should be created until air bubbles cease to emerge from the specimen. Keep tissue in 70% ethanol until dehydration or for long term storage. Depending on the size of the specimen, use the guidelines listed in Appendix Table 1 for dehydration and infiltration:

APPENDIX TABLE 1
Dehydration and Infiltration of Small to Medium Size Specimens

Processing step*	Mouse bones	Rat tibia or fibula	Rat femur, vertebrae
95% ethanol	2–3 hours†	3–4 hours	4–6 hours
95% ethanol	Change, 2–3 hours	Change, 3–4 hours	Change, overnight
100% ethanol	Overnight	Overnight	Change, 4–6 hours
100% ethanol	Change in morning, 2–3 hours	Change in the morning, 6–7 hours	Change in the afternoon, cap, overnight
Working Soluton A (1st infiltration)	Cap 2–3 days	Place in methyl in the afternoon, cap, 2–3 days	Place in methyl in the morning, cap, 3–4 days
Working Solution A (2nd infiltration)	Change, cap, 2–3 days	Change, cap, 3–4 days	Change, cap, 3–4 days

* See section C below for preparation of methacrylate solutions
† Vacuum condition is needed for every step.

3. Methyl Methacrylate Infiltration and Embedding

a. Infiltration

After dehydrating the specimen according to the protocols listed in the Appendix Table 1, place the specimen into the Working Solution A described in Section C below and add 0.9 gm/100 ml benzoyl peroxide. This solution is the infiltration solution. All infiltrations are carried out in either a cold room or refrigerator. Vacuum applications must be performed under a vented fume hood.

b. Embedding

Embedding solution
   100 ml of Working Solution A
   0.9 gm of benzoyl peroxide
   1 ml of JB-4 Solution B (from the JB-4 kit supplied by Polysciences)

Choose a size of embedding mold appropriate for the size of the specimen. Orient the specimen so that the crucial side is on the bottom, then pour the embedding solution over the specimen. Insert a label (made with plastic paper and a No. 2 pencil). Place an aluminum embedding block carefully over the specimen. Try not to move the mold while doing this. Add more embedment solution to the specimen so the level of solution is surrounding the aluminum block holder.

Carefully transfer the molds to a vacuum desiccator to remove any excess air bubbles. Pull a vacuum for 5–10 minutes. Release the vacuum gently and transfer the molds to a GasPak container and seal it off. The oxygen has to be removed to allow polymerization to occur. Flush the GasPak with nitrogen gas for 5–10 minutes and close off the tubing without letting any oxygen in. The rate of flow from the nitrogen tank should be fairly slow so it does not dry out the specimens and is maintained continuously to allow the removal of the air. After flushing, seal the container and leave it undisturbed overnight at room temperature. The polymerization process should be complete in less than 16 hours.

Open the GasPak under a fume hood and remove the molds. Allow the specimens to air out under the fume hood for a few hours. The specimens are now ready to be sectioned on the Leica 2065 microtome or Polycut E microtome with a tungsten carbide knife.

Cautions   Avoid all skin contact with methyl methacrylate. Take extreme caution not to inhale fumes from methacrylate solutions. Gloves, lab coats, goggles, and a vented fume hood should be used when handling these solutions.

4. Sectioning the Methyl Methacrylate for Small to Medium Size Specimens

Remove excess methacrylate from the block with a hacksaw, bandsaw, hand-held small trimming saw (Dremel), or a Histosaw. Only a small amount of methyl methacrylate is needed to be left around the specimen.

Before sectioning, make sure the blade is securely set in the knife holder and the specimen is locked in place. Never wear any loose clothing or jewelry when sectioning. Keep fingers away from the knife. Leave guard in place when stepping away from microtome or changing the orientation of the block.

To set up the microtome, face off the block with a trimming knife. Face off the block at the area of interest and then change knives to take the good sections.

The sections are placed in 6, 12, 24, or 48 well tissue culture plates depending on the size of the section. Collect four sections per well and continue collecting adjacent sections to fill 6–12 wells. This is dependent upon which staining protocols will be performed and how many sections are kept in reserve. If the sample has been labelled with tetracycline, save any sections adjacent to the measuring sections for tetracycline analysis.

Methyl methacrylate sections can be stained free-floating, dried, and coverslipped using Permount (Fisher).

C. PREPARATION OF METHACRYLATE SOLUTIONS

1. Removal of Inhibitors from Methacrylate Solutions

Use basic alumina AG–10 for column filtration to remove commercially added inhibitors from solutions to be used in infiltration and embedding.

a. Methyl Methacrylate

Mount a column (use a narrow stem one with a funnel placed in it) on a ring stand in the fume hood. Fill the column 3/4 full with basic alumina. Fill the column with methyl methacrylate and allow it to filter through the alumina and be collected in a container. Do not allow the column to run dry. If it is stored, place a clamped rubber tube at the bottom of the column. Store filtered solution in a refrigerator for no longer than six months. Discard the alumina in the column when it turns grey blue. This means it can no longer act chemically to remove the inhibitors.

b. Glycol Methacrylate (2-hydroxyethyl methacrylate)

Shake the glycol methacrylate and basic alumina together and store tightly capped in the refrigerator. Shake frequently. Filter or centrifuge when needed for infiltration solution preparation. Store filtered solution in a refrigerator for no longer than six months. Discard the alumina when it turns grey blue; this means it can no longer act chemically to remove the inhibitors.

Cautions   Avoid all skin contact with methyl methacrylate. Take extreme caution not to inhale fumes from methacrylate solutions. Gloves, lab coats, goggles, and a vented fume hood should be used when handling these solutions.

2. Preparation of Methyl Methacrylate Solutions

Working Solution A
    85 ml methyl methacrylate
    10 ml glycol methacrylate
    5 ml dibutyl phthalate
    5 gm polyethylene glycol (PEG 600)

Heat mixture gently on a stirred hot plate until PEG is dissolved. Stir covered. Let solution cool to room temperature, place in glass container well covered and store in refrigerator. Solution is good for six months.

D. METHYL METHACRYLATE EMBEDDING OF THE LARGE SPECIMEN

For our large specimen methodology, see Gruber and Stasky[17] for a procedure which employs increased dehydration intervals and use of a heat sink for polymerization in cooler temperatures.

E. GLYCOL METHACRYLATE PROTOCOLS FOR SMALLER BONES AND CARTILAGE

1. Infiltration

This methodology utilizes the JB-4 kit provided by Polysciences. Note that this method does not use serial ethanol dehydration of tissues. After fixation, the infiltration schedule is carried out as the steps listed in Appendix Table 2. Be sure that specimens are well capped during processing.

APPENDIX TABLE 2
Glycol Methacrylate Infiltration Schedule*

Processing step	Solution	Time/condition
1	70:30 JB-4 solution A:dH$_2$O	1 hr/shaker in cold room
2	85:15 JB-4 solution A:dH$_2$O	2–3 hours/shaker in cold room
3	95:5 JB-4 solution A:dH$_2$O	2–3 hours/shaker in cold room
4	10ml JB-4 Solution A + 0.09g JB-4 catalyst	2–3 days/shaker in cold room

* See section C for preparation of methacrylate solutions.

2. Embedding

Materials needed   5 ml Solution A + 0.1 ml Solution B + 0.045 gm JB-4 catalyst. Mix the catalyst into solution A until completely dissolved. Add Solution B, mix for one minute and start the embedding procedure. Put the solution, specimen and ID label in a mold with care to properly orient the specimen. Place the block holder on the mold, add additional solution, and seal the edges with melted paraffin wax. Polymerization will be complete at room temperature in 4–5 hours; let sit overnight for best results. Successful embedding produces blocks with no bubbles formed during polymerization.

Cautions   Avoid all skin contact with methacrylates. Take extreme caution not to inhale fumes from methacrylate solutions. Gloves, lab coats, goggles and a vented fume hood should be used when handling these solutions.

3. Sectioning Guidelines for Glycol-embedded Specimens

Large amounts of excess methacrylate should be trimmed from the specimen but a sufficient margin of plastic left to support the specimen during sectioning. Trimming can be done with a hacksaw, Dremel, or Histosaw. Avoid contact of dust with skin.

Use an older trimming knife to face off the specimen down to the area of interest and then change to a good sectioning knife. Critical to good sections is a sharp knife edge; always use one site for only sectioning one specimen.

Glycol sections are collected and floated in a staining dish filled with room temperature dH_2O. Sections are collected on a slide and dried on a slide warmer. Toluidine blue provides a rapid screening stain to ensure that the region of interest has been reached in the specimen block. Glycol sections after staining are coverslipped using Pro-Texx Mounting Medium (American Scientific Products).

F. Staining and Enzyme Localizations

See references provided in the main body of this chapter for excellent surveys of staining procedures for bone and cartilage, such as Goldner's stain,[21] alkaline and acid phosphatase enzyme localization,[13] or cement line visualization.[16,17,22]

APPENDIX 2
USEFUL EQUIPMENT AND TOOLS FOR THE ORTHOPAEDIC HISTOLOGY LABORATORY

Adhesive for binding large specimens to molds and rings prior to sectioning   Technovit 3040 glue: Energy Beam Sciences, Agawam, MA; Heraeus Kulzer GmbH, Philipp-Reiss-Strasse 8/13, D-61273, Wehrheim, Germany.

Bone histomorphometry   OsteoMeasure: OsteoMetrics, Inc., 2103 N. Decatur Road, Suite 140, Atlanta, GA, Tel: 404-876-1004; Fax: 404-876-4004; email: support@osteometrics.com. Provides both software and hardware support.

Diamond circular saw   Isomet Low Speed Saw: Buehler Ltd., Lake Bluff, IL.

Diamond wire saws   Well type 3241 Precision Wire Saw: Delaware Diamond Knives, Wilmington, DE.

Dyes for marking tissue orientation   Marking dyes: Triangle Biomedical Sciences, Durham, NC.

Embedding rings for large specimens   Histoprep embedding ring: Fisher Scientific.

Hand-held small trimming saw   Dremel Model 732: Stoelting Instruments, Wooddale, IL. Also a source for diamond-coated cutting discs for the Dremel.

Megacassettes for embedding large specimens   Surgipath Medical Instruments, Richmond, IL.

Microtomes — Small to medium sized specimens   Jung RM2065 microtome: Vashaw Scientific, Norcross, GA; *Large specimens*: Leica Polycut E.

Microtome knives   Delaware Diamond Knives, Wilmington, DE. This vendor also resharpens used knives. For undecalcified sectioning, tungsten-carbide knives are required. Small to medium sized specimens: 30° angle; large specimens: 50° angle.

Vacuum pump   Gast Manufacturing Corp, Benton Harbor, MI, Model No. IHAB-25 MIOOX.

APPENDIX 3
ON-LINE HELP AND OTHER SOURCES OF INFORMATION

histonet@pathology.swmed.edu   An email based source for placing questions which other users can answer on line.

http://www.antibodies.probes.com/   A web page which gives instructions on how to subscribe to this service which is a source for antibody information.

boneworld@osteometrics.com   An email based source for asking questions about bone histology; free; interested parties can subscribe; questions are automatically broadcast to all members and can respond. Subscribe at: subscribe@osteometrics.com

Linscott's Directory of Immunological and Biological Reagents. Available both in soft bound and diskette formats. Address: 4877 Grange Road, Santa Rosa, CA 95404, Tel: (707) 544-9555; Fax: (415) 389-6025. Useful information/sources with cross-indices for mono- and poly-clonal antibodies, kits, and other data.

8 Mechanical Properties and Testing Methods of Bone

Yuehuei H. An and Robert A. Draughn

CONTENTS

I. Introduction 139
II. Composition and Structure of Bone 140
III. Mechanical Properties of Bone 140
    A. Mechanical Parameters 140
    B. Mechanical Properties of Cortical Bone 142
    C. Mechanical Properties of Cancellous Bone 143
        1. Structural Properties of Cancellous Bone 144
        2. Material Properties of Cancellous Bone 144
IV. General Considerations of Mechanical Testing of Bone 144
    A. Specimen Harvesting and Storage 145
    B. Sample Preparation 145
    C. Measurement of Bone Densities and Mineral Content 148
    D. Mechanical Testing and Data Collection 149
V. Mechanical Testing Techniques 151
    A. Bending Test 151
    B. Compression and Tensile Test 153
    C. Indentation Test 155
    D. Torsional Test 155
    E. Screw Pullout Test 155
    F. Strain Gauge 156
    G. Ultrasonic Methods 157
VI. Mechanical Testing of Bone-Implant Interface 157
    A. Pushout and Pullout Test 157
    B. Screw Pullout Test 158
    C. Removal Torque 159
    D. Other Tests 159
References 159

I. INTRODUCTION

Mechanical properties are basic parameters which reflect the structure and function of bone. The mechanical properties of bone obey Newton's and Hook's laws of mechanics.[1] The mechanical behavior of bone in normal physiological situations is similar to that of an elastic material. However, unlike inorganic materials, bone has the ability to repair itself and can alter its mechanical properties and morphology in response to increased or decreased function.

Mechanical testing of bone is involved in most bone-related animal studies. A survey of articles published in the *J. Orthop. Res.* in 1995 revealed that in 77% (23/30) of bone-related animal studies mechanical tests were utilized or biomechanical principles or theories were employed.

Because most orthopaedic surgeons and many basic science researchers are not well versed in biomechanics and help from an expert is not always available, this chapter emphasizes basic biomechanical theories and how to do common mechanical tests.

II. COMPOSITION AND STRUCTURE OF BONE

Bone is a composite material which consists of organic matrix (mainly collagen) and inorganic hydroxyapatite. Water accounts for about 20% of the wet weight of cortical bone, hydroxyapatite (HA) makes up approximately 45%, and the organic substances (90–95% collagen, 1% glycosaminoglycans, 5% other proteins) account for the remaining 35%. The presence of mineral is responsible for the strength and hardness of bone.

Bone structure can be observed at several levels:[2] (1) At the most fundamental level, HA crystals are embedded between the ends of adjoining collagen fibrils. This composite of rigid HA and flexible collagen provides a material that is superior in mechanical properties to either of them alone. Bone is more ductile than hydroxyapatite, allowing the absorption of more energy before failure, and more rigid than collagen, permitting greater load bearing. (2) At the second level, the collagen-HA fibrils are formed into sheets or lamellae with a preferred direction. The orientations of the fibers define directions of maximum and minimum strengths for a primary loading direction. (3) The third level consists of the arrangement of the lamellae. A circular concentric structure produces a tubular Haversian osteon with maximum strength along its long axis. Alternately, lamellae may be arranged in sheets, as found in plexiform bone, in which case the strength of the material transverse to the longitudinal axis is lower in the direction in which lamellae are being pulled apart at cement lines and higher in the plane of the sheets. (4) The fourth level of structure is the macroscopic materials, cortical or trabecular bone. The main factors determining strength at this level are the density of the bone and trabecular orientation.

Good articles on bone structure and composition and their relation to mechanical properties were presented by Carter and Spengler,[3] Katz,[4] Hoesler,[5] Cowin,[1] Tencer,[2] Hayes and Bouxsein.[6]

III. MECHANICAL PROPERTIES OF BONE

A. Mechanical Parameters

Bone has two main mechanical properties: stiffness and strength. Stiffness is expressed by the elastic modulus which is a measure of the stress required to elastically deform the bone. Strength defines the stress required to fracture the bone. Measures of stiffness and strength need to be recorded and calculated during a mechanical test. If the testing is used for comparing differences between groups, directly recorded values of stiffness and ultimate load are adequate for property evaluation. If the purpose of the study is comparing differences between values generated from different laboratories or from different experimental settings, elastic modulus and strength must be calculated to provide standardized values of stiffness and load.

The above-mentioned mechanical parameters are based on load-displacement curve which is recorded by a chart recorder or computer. The curve gives the yield load, ultimate load, deformation, and stiffness directly (Figure 1). Since the dimensions of specimens are readily measured, the values of elastic modulus and strength are easily calculated.

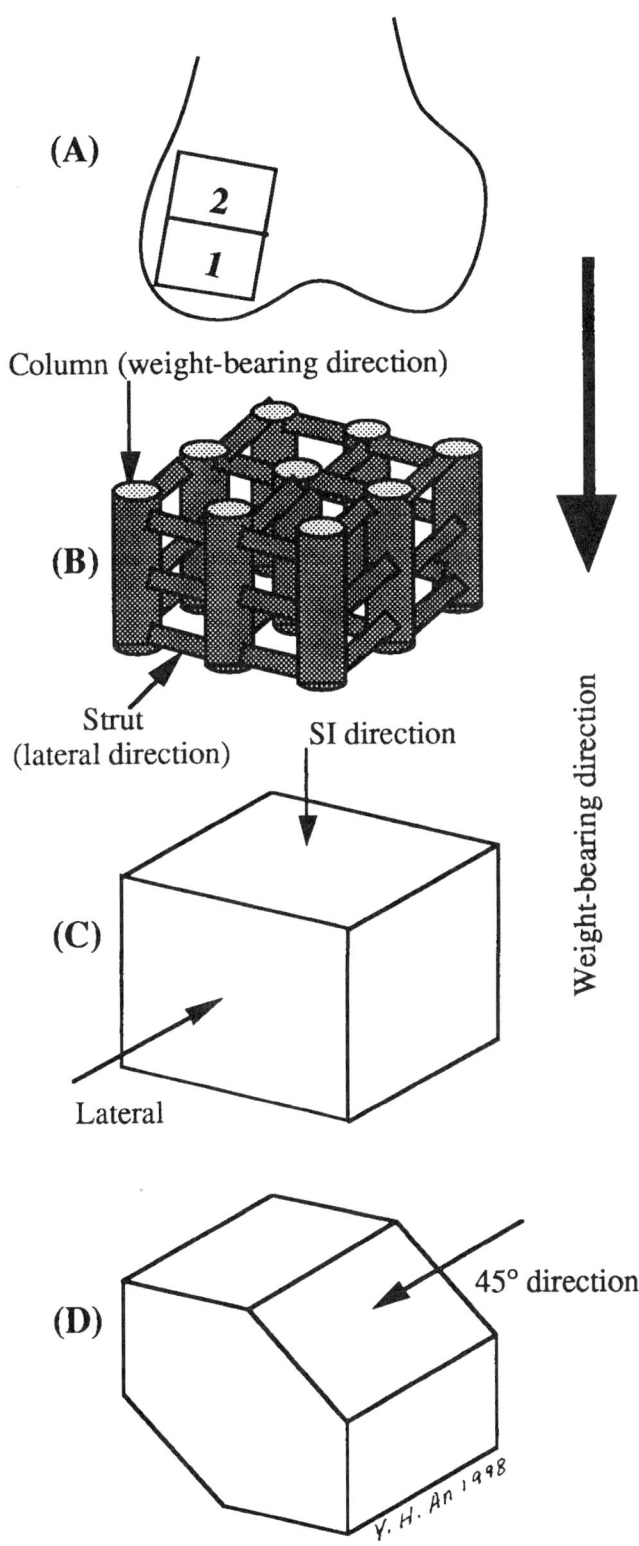

FIGURE 1. A test of screw-holding power of bovine cancellous bone from different directions. (A) represents sampling site and orientation. In (B) the structure of the bone block is idealized into a column-strut model. The columns indicate thicker trabeculae in the weight-bearing direction. The struts represent the trabecular connections between columns (column trabeculae). It is assumed that columns are thicker and stronger and struts are thinner and weaker. (C) and (D) show directions of screw insertions.

TABLE 1
Bending properties of human and animal cortical bones (selected data from the literature)

Species	Bone	Specimen	Mechanical testing	Strength (MPa)	Elastic modulus (GPa)	First author, year[Ref.]
Human	Femur	2×5×350 mm beam	3 pt. bending	181	15.5	Sedlin 1966[7]
		3×3×30 mm beam	4 pt. bending	103–238*	9.82–15.7*	Keller 1990[8]
		0.4×5×7 mm beam	3 pt. bending	225 ± 28	12.5 ± 2.1	Lotz 1991[9]
		2.0×3.4×40 mm	3 pt. bending	142–170*	9.1–14.4*	Currey 1997[10]
Monkey	Tibia	Whole bone	3 pt. bending	—	9.0 ± 1.3	Kasra 1994[11]
Cattle	Femur	2×3.5×30 mm beam	3 pt. bending	—	18.5 ± 2.8	Currey 1988[12]
		2×4×35 mm beam	3 pt. bending	228 ± 5	19.4 ± 0.7	Currey 1988[13]
		23×0.4 mm beam	3 pt. bending	209 ± 13	18.1 ± 0.5	Currey 1995[14]
	Tibia	4×4×35 mm beam	Bending	—	14.1	Simkin 1973[15]
		4×10×80 mm beam	3 pt. bending	230 ± 18	21.0 ± 1.9	Martin 1993[16]
Horse	Femur	2×2×40 mm beam	4 pt. bending	204–247*	17.1–19.9*	Schryver 1978[17]
		2×3.5×30 mm beam	3 pt. bending	—	21.2 ± 1.9	Currey 1988[12]
	Radius	2×2×40 mm beam	4 pt. bending	217–249*	16.2–20.2*	Schryver 1978[17]
	Metacarpus	2×2×40 mm beam	4 pt. bending	226–240*	17.0–18.4*	Schryver 1978[17]
	3MT, 3MC†	1.8×4.5×70 mm	4 pt. bending	195–226*	14–16*	Bigot 1996[18]
Sheep	Metacarpus	2×3.5×30 mm beam	3 pt. bending	—	18.9 ± 2.2	Currey 1988[12]
Donkey	Radius	2×3.5×30 mm beam	3 pt. bending	—	17.6 ± 2.0	Currey 1988[12]
Dog	Humerus	Whole bone	3 pt. bending	193 ± 35	2.7 ± 0.6‡	Kaneps 1997[19]
Pig	Femur	Whole bone	3 pt. bending	39.9	0.37‡	Crenshaw 1981[20]
	Rib	Whole bone	3 pt. bending	35.6	2.24‡	Crenshaw 1981[20]
	3MC†	Whole bone	3 pt. bending	37.2	0.22‡	Crenshaw 1981[20]
Goose	Femur	1.0×25 mm beam	3 pt. bending	232–283*	16.9–20.7*	McAlister 1983[21]
Cat	Femur	Whole bone	3 pt. bending	36 ± 9.47	7.1 ± 0.9	Ayers 1996[22]
	Tibia	Whole bone	3 pt. bending	60.5 ± 12	11.4 ± 3.2	Ayers 1996[22]
Rabbit	Femur	Whole bone	3 pt. bending	130 ± 5	13.6 ± 0.4	An 1996[23]
		Whole bone	3 pt. bending	88 ± 20	10.7 ± 2.5	Ayers 1996[22]
	Tibia	Whole bone	3 pt. bending	195 ± 6	21.3 ± 0.7	An 1996[23]
		Whole bone	3 pt. bending	192 ± 47	23.3 ± 7.0	Ayers 1996[22]
	Humerus	Whole bone	3 pt. bending	167 ± 5	13.3 ± 0.6	An 1996[23]
Rat	Femur	Whole bone	3 pt. bending	180 ± 6	6.9 ± 0.3	Jørgensen 1991[24]
		Whole bone	3 pt. bending	134 ± 4	8.0 ± 0.4	Barengolts 1993[25]
		Whole bone	3 pt. bending	153 ± 45	4.9 ± 4	Ejersted 1993[26]
Mouse	Femur	Whole bone	3 pt. bending	104–173*	8.8–11.4*	Simske 1992[27]
		Whole bone	3 pt. bending	40 ± 13	5.3 ± 1.8	Ayers 1996[22]
	Tibia	Whole bone	3 pt. bending	78 ± 12	8.9 ± 0.2	Ayers 1996[22]

* Range of average values; †Third metatarsus and third metacarpus; ‡Value is questionable.

B. Mechanical Properties of Cortical Bone

The dense nature of cortical bone determines its strong and stiff mechanical properties compared to cancellous bone. The mechanical properties of cortical bone depends on the type of mechanical testing. Although the tensile test is the standard method for testing mechanical properties of cortical bone, bending tests are used the most often (Table 1). The bending strength and elastic modulus of cortical bone ranges from 35 to 283 MPa and from 5 to 23 GPa (excluding the values marked with ‡) respectively. The strength and elastic modulus by tensile and compression tests ranges from 92 to 295 MPa and from 7 to 34 GPa respectively (Table 2). The tensile strength is about 2/3 of

TABLE 2
Mechanical properties of human and animal cortical bones tested by compression, tensile, and torsional testing (selected data from the literature)

Mechanical test	Species	Bone	Specimen dimensions	Strength (Mpa)	Elastic modulus (GPa)	First author, year[ref.]
Compression	Human	Femur	2×2×6 mm dumbbell	167–215*	14.7–19.7*	Reilly 1974[28]
			2×2×6 mm dumbbell	179–209*	15.4–18.6*	Burstein 1976[29]
			3 mm dia. cylindrical dumbbell	205–206*	—	Cezayirlioglu 1985[30]
		Tibia	2×2×6 mm dumbbell	183–213*	24.5–34.3*	Burstein 1976[29]
			3 mm dia. cylindrical dumbbell	192–213*	—	Cezayirlioglu 1985[30]
	Bovine	Femur	3.8×2.3×76 mm dumbbell	133	24.1–27.6*	McElhaney 1964[31]
			2×2×6 mm dumbbell	240–295*	21.9–31.4*	Reilly 1974[28]
		Tibia	4×5 mm rectangular	165	23.8 ± 2.2	Simkin 1973[15]
			2×2×6 mm dumbbell	228 ± 31	20.9 ± 3.26	Reilly 1974[28]
			3 mm dia. cylindrical dumbbell	217 ± 27	—	Cezayirlioglu 1985[30]
	Goose	Femur	0.8 mm dia./2.4 mm cylinder	164–203*	12.2–14.6*	McAlister 1983[21]
Tensile	Human	Femur	3.8×2.3×76 mm dumbbell	66–107*	10.9–20.6*	Evans 1951[32]
			2×2×6 mm dumbbell	107–140*	11.4–19.7*	Reilly 1974[28]
			2×2×6 mm dumbbell	120–140*	15.6–17.7*	Burstein 1976[29]
			3 mm dia. cylindrical dumbbell	133–136*	—	Cezayirlioglu 1985[30]
		Tibia	2×2×6 mm dumbbell	145–170*	18.9–29.2*	Burstein 1976[29]
			1.7×1.8×25 mm deam	162 ± 15	19.7 ± 2.4	Vincentelli 1985[33]
			3 mm dia. cylindrical dumbbell	154–158*	—	Cezayirlioglu 1985[30]
	Bovine	Femur	3.8×2.3×76 mm dumbbell	92	20.5	McElhaney 1964[31]
			2×2×6 mm dumbbell	129–182*	23.1–30.4*	Reilly 1974[28]
			3 mm dia. cylindrical dumbbell	162 ± 14*	—	Cezayirlioglu 1985[30]
		Tibia	4×5×30 mm dumbbell	136	7.1 ± 1.1	Simkin 1973[15]
			2×2×6 mm dumbbell	152 ± 17	21.6 ± 5.3	Reilly 1974[28]
			2×2×6 mm dumbbell	188 ± 9	28.2 ± 6.4	Burstein 1975[34]
Torsional	Human	Femur	?	53	—	Hazama 1964[35]
			?	54 ± 0.6	3.2	Yamada 1970[36]
			2×2×6 mm dumbbell	—	3.1–3.7*	Reilly 1974[28]
			3×3×6 mm dumbbell	65–71*	—	Reilly 1975[37]
			3 mm dia. cylindrical dumbbell	68–71*	—	Cezayirlioglu 1985[30]
		Tibia	3 mm dia. cylindrical dumbbell	66–71*	—	Cezayirlioglu 1985[30]
	Bovine	Femur	3×3×6 mm dumbbell	62–67*	—	Reilly 1975[37]
			3 mm dia. cylindrical dumbbell	76 ± 6	—	Cezayirlioglu 1985[30]

* Range of average values from different subjects.

compression strength. The torsional strength is about 60 MPa in average. The torsional (shear) strength is approximately 1/3 to 1/2 of the values of the longitudinal modulus (tested by bending, tensile or compressive tests) (Table 2). And the torsional (shear) modulus is about 1/5 of the longitudinal modulus. The mechanical properties of cortical bone also depend on loading directions of the testing method. The longitudinal (normally the weight bearing direction) elastic modulus is about two times that of the transverse (lateral directions) elastic modulus.[6,38]

C. MECHANICAL PROPERTIES OF CANCELLOUS BONE

The porous nature of cancellous bone, with bony trabecular columns and struts and marrow-filled pores or cavities (a two phase structure[39]), lends itself to a mechanical description by both structural and material properties.

1. Structural Properties of Cancellous Bone

The structural properties of cancellous bone are commonly measured by compression, tensile, or bending tests. The common phrase "mechanical properties of cancellous bone" means the structural properties. It is known that the strength and elastic modulus by tensile tests are smaller than that by compression tests. For example, the strength by tensile test is approximately 60% of the value by compression test reported by Kaplan et al.,[40] and the elastic modulus by tensile test is approximately 70% of the value by compression test reported by Keaveny et al.[41,42] The mechanical properties of cancellous bone depend on anatomic location and function. According to the data list summarized by Goldstein (21 sets of data generated using compression test), the average values of strength and elastic modulus of human cancellous bone from different locations (femur, tibia, humerus, radius, vertebrae, and iliac crest) are 6.6–36.2 MPa and 130–1080 MPa respectively.[43] According to the selected data from the literature (Table 3), the values of strength and elastic modulus of cancellous bone are 1.5–38 MPa and 10–1570 MPa respectively. The structural properties of cancellous bone are much smaller than those of cortical bone. The average value of elastic modulus is several hundred MPa for cancellous bone,[43] compared to 5–21 GPa for cortical bone.[12]

Several investigations have addressed the orthogonal mechanical properties of cancellous bone of both human and animals.[52,61,62] The strength and elastic modulus of cancellous bone depend on the direction of the load employed, as normally measured at SI (superior-interior), AP (anterior-posterior), or ML (medial-lateral) directions. Ciarelli et al.[62] found the highest overall mean of elastic moduli of human long bone metaphyseal locations to be in the SI direction, which is about 2.5 times the value at the AP direction. The AP direction is higher than that of the ML direction. An earlier study using vertebral cancellous bone specimens by Galante et al.[61] also showed a similar pattern. In a recent study in the authors' laboratory, it was found that the screw pullout strength of bovine cancellous bone also depends on the direction of the screw insertion (loading direction). The strength was strongest (55±5 MPa) at the SI direction (0 degree), weakest (37±5 MPa) at lateral direction (90 degree), and was intermediate (43±4 MPa) at a direction of 45 degrees.[63] This phenomenon may be explained by a column-strut model proposed in this study (Figure 1).

The strength and stiffness of cancellous bone also varies in different epiphyseal-metaphyseal locations both in human[62] and animals. In the authors' laboratory, the strength and elastic modulus of epiphyseal-metaphyseal bones of animals, such as rats,[60] rabbits,[23] dogs,[53,64] and goats[64] have been investigated using compression and indentation tests. Generally, for both humans and animals the cancellous bones of lower limbs (hind limbs) are stronger and stiffer than those of upper limbs (front limbs).

2. Material Properties of Cancellous Bone

The material properties of cancellous bone are defined by the intrinsic properties of individual trabeculae, which have been measured by mechanical testing of single trabeculae using methods such as buckling analysis,[65] compression test,[66] microtensile test,[67] cantilever test plus finite element modeling,[68] finite element modeling,[69] or ultrasound methods.[67] The elastic modulus of trabecular bone material (individual trabeculae) is less (10–30%) than that of cortical bone. For example, the elastic modulus is 14.8 GPa for trabeculae and 20.7 GPa for cortical bone measured by an ultrasonic technique and 10.4 GP and 18.6 GPa respectively using a microtensile test.[67]

Articles on the mechanical properties of animal cancellous bones include studies of bovine, canine, or goat distal femur, proximal tibia, and vertebrae determined by compression test (Table 3), or of canine, rabbit, or rat epiphysometaphyseal bones examined using indentation test.[53,60,70]

IV. GENERAL CONSIDERATIONS OF MECHANICAL TESTING OF BONE

Because of the limited size and inhomogeneity of bone, accurate measurement of its mechanical properties, especially elastic modulus (fortunately, it is not always needed for comparing properties between groups), is a challenging endeavor. The difficulty of standardizing technical procedures (sampling and testing) of mechanical testing is the main reason for the large differences among the reported data.[71] Also, many intrinsic and extrinsic factors have tremendous influence on the mechanical properties. The investigator should always be aware of factors which may be involved in the determination of mechanical properties of bone.[2,7,72]

A. SPECIMEN HARVESTING AND STORAGE

Bone specimens to be used for mechanical testing should be harvested with sufficient extra tissue around the area of interest. A hand saw or wire saw is efficient for cutting bone. Keeping surrounding soft tissue (muscle, fascia, or skin) intact is very helpful for protecting the bone from drying. Preparation of samples for testing always should be done immediately before mechanical testing.

The common method for storing bone specimens is freezing at –20°C. The bones should be wrapped in paper or cloth towels, immersed in saline and placed in an air-tight plastic bag. Thawing always should be done in saline. The effects of storage on mechanical properties of bone at –20°C for short periods of time are minor. The maximum effect reported is a 4.6% reduction of torsional strength of canine long bones.[73] However, after thawing, enzymes such as collagenases and proteases may become active and degrade the tissue. Also, enzymatic degradation is not completely arrested at –20°C.[74] With concerns about the effects of enzymes[54] and evaporation,[75] a question arises if there are significant effects of long term storage at –20°C. Panjabi et al.[76] found no significant effects of freezing for 7–8 months on the mechanical properties of human vertebrae bone. Roe et al.[77] found that bones frozen at –20°C for eight months did not become significantly weaker. Because time periods longer than eight months have not been reported for frozen storage at –20°C, storage at this temperature for more than eight months is not recommended. Alternatively, –70°C, –80°C, or even lower temperatures or liquid nitrogen are suggested for long term bone storage, since these temperatures may minimize evaporation[75] and markedly reduce enzyme activity.[54]

Owing to complexity of an experiment or unforeseen circumstances, sometimes a specimen must be thawed and frozen multiple times. The question arises whether multiple freezing and thawing is harmful to the mechanical properties. This question has been partially answered by Linde and Sørensen,[78] who found that freezing and thawing five times did not alter the compressive properties of cancellous bones. In our recent study, the proximal portion of the tibia of adult cows was sectioned to produce bone slices. They were then subjected to four freezing-thawing conditions: freezing with and without saline solution, then thawing in saline solution or exposed to the air. The mechanical properties of the bone before and after the treatments (5 cycles of freezing and thawing) were measured using an indentation test. It was found that there is no significant effect on the ultimate load and stiffness of the bone. Only a slight difference was noticed for the specimens frozen without saline soaking and thawed in air.[79] This work supports the practice to freezing and thawing bone specimens in saline solution.

B. SAMPLE PREPARATION

Rough cuts can be made with a regular bandsaw equipped with a 1/4-inch fine tooth saw blade. To prevent burning, low speed should be used with sufficient saline irrigation. This kind of cutting may only affect 1 mm depth of bone at the surface, which can be ground off using a polishing wheel. Finer cuts can be made using an ultra fine jigsaw. A diamond wafering or diamond wire saw (Histosaw, Delaware Diamond Knives, Wilmington, DE) is particularly good for making smooth, parallel cuts. For fabrication of cylindrical samples, a table top drill press is sufficient for relatively large samples (>7 mm diam.). For 4–5 mm diameter samples, a lathe or milling machine is recommended.

TABLE 3
Mechanical properties and densities of human and animal cancellous bones (selected data from the literature)

Species	Bone	Specimen	Ultimate strength (MPa)	Elastic modulus (MPa)	Apparent density (gm/cm3)	Ash density (gm/cm3)	First author, year[Ref.]
Human	Femoral head	8 mm diam. cylinder	9.3 ± 4.5	900 ± 710	—	—	Martens 1983[44]
	Proximal femur	8 mm diam. cylinder	6.6 ± 6.3	616 ± 707	—	—	Martens 1983[44]
	Distal femur	8 mm cube	5.6 ± 3.8	298 ± 224	0.43 ± 0.15	0.26 ± 0.08	Kuhn 1989[45]
		10.3mm dia., 5 mm cylinder	1.5–45†	10–500†	0.24 ± 0.09	—	Carter 1977[39]
		5 mm dia./7.5 mm cylinder	5.96	103–1058†	0.46	—	Odgaard 1989[46]
	Proximal tibia	7.5:7.5 mm cylinder	5.3 ± 2.9	445 ± 257	—	—	Linde 1989[47]
	Vetebral body	Cylinders	—	165 ± 110	0.14 ± 0.06	—	Keaveny 1997[48]
Monkey	Femoral head	5 mm dia./6 mm cylinder	23.1 ± 5.4	372 ± 54	—	—	Kasra 1994[11]
Cattle	Distal femur	5.5mm dia./8mm cylinder	8.5 ± 4.2	117 ± 61	—	—	Poumarat 1993[49]
	Proximal tibia	15 mm cube, Ultrasonic method	—	648 ± 430	0.41 ± 0.16	—	Rho 1997[50]
	Proximal humerus	Cylinders	—	1570 ± 628	0.71 ± 0.22	—	Keaveny 1997[48]
	Vetebral body	6 mm dia./7.5 mm cylinder	7.1 ± 3.0	173 ± 97	0.45 ± 0.09	0.19 ± 0.06	Swartz 1991[51]
Dog	Femoral head	5 mm cube	12 ± 5.8	435	—	—	Vahey 1987[52]
	Distal femur	8 mm cube	7.1 ± 4.6	209 ± 140	0.44 ± 0.16	0.26 ± 0.08	Kuhn 1989[45]
		4 mm dia./5 mm cylinder	13–28*	210–394*	0.69–0.98	0.40–0.56*	Kang 1998[53]
	Proximal tibia	4 mm dia./5 mm cylinder	5–24*	106–426*	0.41–0.83*	0.22–0.44*	Kang 1998[53]
		12.5mm dia./10mm cylinder	—	301–850	—	—	Sumner 1994[54]
		5 mm cube	—	344–1278	—	—	Sumner 1994[54]

Mechanical Properties and Testing Methods of Bone 147

Species	Location	Specimen					Reference
	Humeral head	4 mm dia./5 mm cylinder	18 ± 6	350 ± 171	0.84 ± 0.17	0.43 ± 0.06	Kang 1998[53]
	Distal humerus	6 mm dia./15 mm cylinder	13 ± 3	1490 ± 300	—	—	Kaneps 1997[19]
	Vertebral body	5 mm dia./8 mm cylinder	10.1 ± 2.6	530 ± 40	—	—	Acito 1994[55]
Goat	Femoral head	4 mm dia./5 mm cylinder	19.2 ± 6.9	502 ± 268	0.91 ± 0.04	0.48 ± 0.03	An (unpublished data)
	Distal femur	4 mm dia./5 mm cylinder	14.1–23.5*	399–429*	0.54–0.66*	0.32–0.40*	An (unpublished data)
	Proximal tibia	4 mm dia./5 mm cylinder	24.7–26.1*	532–566*	0.93–1.1*	0.50–0.56*	An (unpublished data)
	Humeral head	4 mm dia./5 mm cylinder	10.0 ± 1.0	247 ± 20	0.75 ± 0.03	0.36 ± 0.01	An (unpublished data)
Sheep	Femoral neck	8 mm dia./10 mm cylinder	3.2 ± 0.3	2.0 ± 0.2§	—	—	Geusens 1996[56]
	Vertebral body	7 mm dia./9 mm cylinder	23.6 ± 4.4	—	—	—	Deloffre 1995[57]
		7.5 mm dia./9 mm cylinder	22.3 ± 7.1	1510 ± 784	0.60 ± 0.16	0.37 ± 0.11	Mitton 1997[58]
Pig	Vertebral body	7 mm dia./5 mm cylinder	27.5 ± 3.4	1080 ± 470	—	0.46 ± 0.04	Mosekilde 1987[59]
Rabbit	Epiphyseal long bones	Indentation test	35–81	—	—	—	An 1996[23]
Rat	Epiphyseal long bones	Indentation test	38–71	—	—	—	An 1997[60]

† Range of values; *Range of average values from different parts; §Value is questionable (too low).

C. Measurement of Bone Densities and Mineral Content

The density of a material is its mass per unit volume. The material density of cortical bone is the wet weight divided by the specimen volume. Cortical bone has a density of approximately 1.9 gm/cm^3.[80,81] The common ways to measure the volume of a cortical bone specimen include the use of a gravity bottle based on Archimedes' principle or directly measuring the dimensions of the specimen. The latter requires the specimen having a regular shape such as a cylinder.

For cancellous bone there are different material characteristics arising from the two phase structure (trabeculae and marrow).[39] Therefore, two mechanical properties are generally considered, the structural and material properties, which are based on their structural (apparent) density and material density, respectively. The measurement of structural (apparent) density (ρ_a) is achieved by weighing the cancellous structure without free water in its marrow cavities (wet weight, w_b) and dividing the wet weight by the structural volume (including both of trabeculae and marrow cavities):

$$\rho_a = w_b/(\pi d^2 h/4) \qquad (1)$$

where d and h represent diameter and height of a cylindrical specimen. Other specimen shapes, such as cubic, can be used, but they are technically more demanding and have more sharp corners than cylinders, which may cause bone materials to fracture from the specimen during defatting or marrow removal. An accurate method for bone volume is using a gravity bottle based on Archimedes' principle (before marrow removal). The compressive strength (σ in MPa) of cancellous bone is related to its apparent density (ρ in g/cm^2) by a power law of the form:[6]

$$\sigma = 60\rho^2 \qquad (2)$$

Similarly, the compressive modulus (E, in MPa) of cancellous bone is related to the apparent density (ρ, in g/cm^2) by:

$$E = 2915\rho^2 \qquad (3)$$

Selected reports on apparent densities of human and animal bones are listed in Table 3. The apparent density of cancellous bone ranges from 0.14 to 1.10 gm/cm^3 (average: 0.62 gm/cm^3, n=16).

Material density of cancellous bone is measured using the weight of bone material (only trabeculae) divided by the volume of only trabeculae, which is a little smaller than that of cortical bone, being 1.6–1.9 gm/cm^3.[80] The principle is again that the marrow needs to be cleaned thoroughly before the measurements of weight and volume. Using a gravity bottle based on Archimedes' principle is the common way to measure both the weight and volume of the bone specimen. To make the measurement, the marrow has to be removed first and no air bubbles or water can be trapped inside the marrow cavities.

Many methods have been reported for removing bone marrow, including boiling in water with detergent, high pressure water jet, or chemical solvent. Depending on the size and shape of the specimen, an individualized combination of the above-mentioned methods is appropriate. In the authors' laboratory the following procedure has been used for small specimens (for example, 4 mm diam., 5 mm length cylinder): (1) Defatting in 50/50 acetone/ethanol mixture with agitation for 24 hours; (2) Removing marrow in low concentration bleach (1.0 to 1.5% sodium hypochlorite) with agitation for 12 hours; and (3) Removing marrow residues with a high pressure water jet (using a syringe).

For measuring bone mineral content, bone specimens are ignited in air in a 500°C furnace and the ash weighed. Instead of "ash weight" or "ash fraction," the authors prefer to use ash density, which is defined as ash weight per unit bone volume. It is suggested that the crucibles be dried at 500°C overnight, weighed, loaded with bone specimen, and heated at 500°C for 18 hours to remove the organic phase. Then, the crucible containing the ash is weighed to determine the weight of ash. No predrying is needed for this method, while for ash fraction the specimen has to be dried at least for one week to get a base weight for calculating ash fraction. Selected reports on ash densities of human and animal bones are listed in Table 3. Less frequently used methods for determining bone mineral content include the use of decalcifying solution or measuring the radiographic density of whole bone or bone sections. The latter is more suitable for *in vivo* conditions.

Selected data of ash densities of human and animal bones are listed in Table 3, ranging from 0.19 to 0.50 gm/cm^3 with an average of 0.37±0.10 gm/cm^3 (n=13), which is about 60% of the value of apparent density. The latter is calculated from the 12 data sets containing both values of apparent density and ash density.

D. MECHANICAL TESTING AND DATA COLLECTION

1. General Consideration

Mechanical testing of materials involves the application of measurable loads to specimens of uniform dimensions. The applied stress is calculated by dividing the applied force by the area over which the force acts. Change in specimen dimension divided by the original specimen dimension defines strain. Dependent upon the direction in which the force is applied, the test may be tensile, compression, or bending. A simple way to record the data is a load-displacement curve from which the ultimate load, stiffness, and displacement are obtained. A stress-strain curve is not always plotted. Ultimate strength and elastic modulus (if applicable) are often calculated using the recorded loads, ultimate load and displacements, and dimensions of the specimen.

The mechanical test machine is operated in displacement control for most tests. The machine linear variable displacement transducer (LVDT) should be calibrated periodically using an extensometer. Loading is commonly conducted at a constant slow rate (1 mm/min is the rate used in the author's lab.). Load at the peak point of the load-displacement curve is taken as the ultimate load. A stiffness measure is obtained by measuring the slope of the linear portion of the curve (Figure 2). If the test machine is controlled with a linear displacement rate and if the specimen fixture is very rigid, the time base of the recorder can be converted to specimen deformation. An extensometer is recommended when a tensile test is conducted or for other tests when a complicated or less rigid specimen fixture is employed. The deformation measured by a built-in LVDT includes the deformation of the specimen and potential displacements within the specimen fixture or at the specimen-fixture interface. An extensometer attached to the specimen provides a direct measurement of specimen strain without the complications of machine or fixture deformation.

2. Artifacts of Testing Procedures

a. Machine Compliance

If several fixture parts are used in the testing assembly, a significant machine compliance can exist. A routine testing for machine compliance is recommended by testing the fixture column without specimen in place.[23,58] If the machine and fixture deformation is found to be significant (more than 10% of the specimen value), it should be accounted for in the data analysis.[23] Because the testing gear should be the same for each specimen, only the mean machine compliance (or machine stiffness, S_m) is needed. The stiffness of the bone sample (S_b) is calculated using the following equations:

$$S_b = P/(d_{b+m} - d_m) \qquad (4)$$

$$= P/(P/S_{b+m} - P/S_m) \qquad (5)$$

$$= S_{b+m}S_m/(S_m - S_{b+m}) \qquad (6)$$

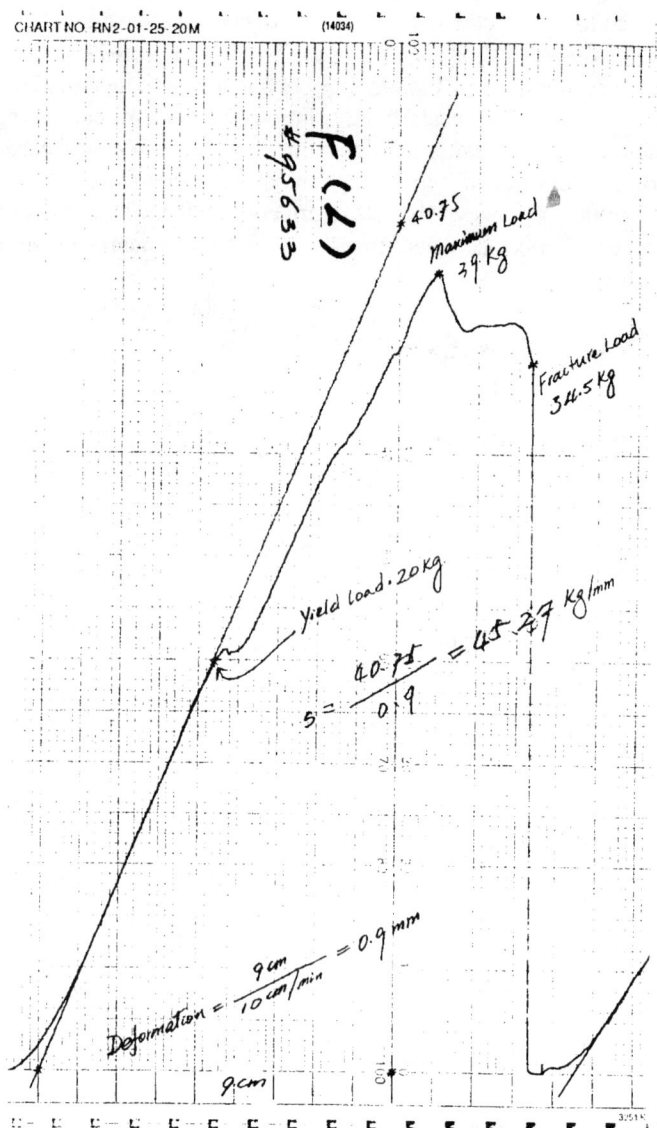

FIGURE 2. A typical load-displacement curve from a three point bending test on a rabbit femur.

where P is the load where deformation of the testing machine (d_m) or deformation of the machine plus bone specimen (d_{b+m}) are taken. S_{b+m} is the tested stiffness value (the stiffness of the machine plus bone specimen).

b. *Specimen-Fixture Interface*

When a tensile test is employed, the effect of specimen-fixture interface should be considered. Any loosening or low rigidity at the interface will lead to the underestimation of the specimen's true values. Therefore, a rigid connection between the specimen and the fixture is essential. Using a dumbbell-shaped specimen or PMMA end-coated specimen are two common strategies to achieve a good bonding between the specimen and fixture.[28,82] An external extensometer should be used in this kind of situation to accurately measure the specimen deformation.

When a compression test is used, friction at the specimen and platens should be considered.[84] An uneven specimen surface causes a triaxial stress field, leading to overestimation of the specimen stiffness. This effect can be limited by using a more accurate procedure for specimen fabrication to achieve a parallel end surfaces. An overestimation of specimen stiffness can be also caused by the horizontal friction between the surfaces of the specimen and the platen. It is known that both the axial and lateral deformations of a specimen in between the upper and lower platen are larger at the ends of the specimen than the central part of the specimen (end phenomenon or end effect). Any restrictions to the lateral expansion, such as a rough platen surface, will cause an overestimation of the true specimen stiffness. Common methods for reducing this kind of friction include the use of grease at the interface and using low friction stainless steel platen surfaces (polished "mirror" surfaces).

For compression and tensile tests and the most of other mechanical testing procedures, pre-loading with a small load is useful for "tightening" the specimen-fixture interface, to further limit the effect of the interface.

V. MECHANICAL TESTING TECHNIQUES

A. BENDING TEST

A bending test commonly is used for measuring mechanical properties of cortical bones.[85] Using bending tests, standard specimens cut from cortex or the whole intact bone can be used for testing. For the former, straight specimens (beams) with uniform cross-sectional shape and area are commonly prepared for testing. A three-point bending test is more often used than four-point bending, although theoretically it is not as sound as four-point bending. Elastic moduli determined from whole bone bending tests cannot be considered accurate because of the inconsistent cross-sectional shape and area, small length-to-diameter ratio (ideally, it should be 20:1, but for rabbit long bones only 6–8:1), and the inconsistent structures in the medullary canal (trabecular bone portion and marrow).[28,80] However, in orthopaedic research, more and more whole bone bending tests have been reported using intact bone, which is likely because for most research projects, the relative differences between treated and control bones are of more concern than the actual values of the mechanical properties of the bone.

Bending tests have been used for testing long bones of mouse, rat, rabbit, cat, dog, and sheep. In such experiments, it is usually assumed that the cross-sectional area of the shaft of the bone is a hollow ellipse with walls of uniform thickness and the portion of the bone to be tested (in between the two supports) is assumed to be consistent in cross-sectional size and shape.

For a three-point bending test, the loading point is chosen at the mid shaft of the tubular bone. The surface facing the loading striker is often at the position of the smaller diameter simply because in this way the bone could be positioned steadily on the two supports. The specimen supports are commonly two steel rods, 4–6 mm in diameter with a span determined by the length of the bone. After the bone is positioned on the supports, the striker (4–6 mm diam. rod) is driven into the bone surface at a constant rate (1 mm/min. in authors' lab.) until the bone fractures (Figure 3). The following equation is used to calculate the ultimate strength:

$$\sigma = PLa/8I \tag{7}$$

where P is the ultimate load, L is the distance between the supporting bars, a is the average value of the external diameters of the cross sections at the loading position of the bone being tested, and I is the area moment of inertia. The latter is calculated using the following equation:

$$I = \pi (a^3 b - a'^3 b')/64 \tag{8}$$

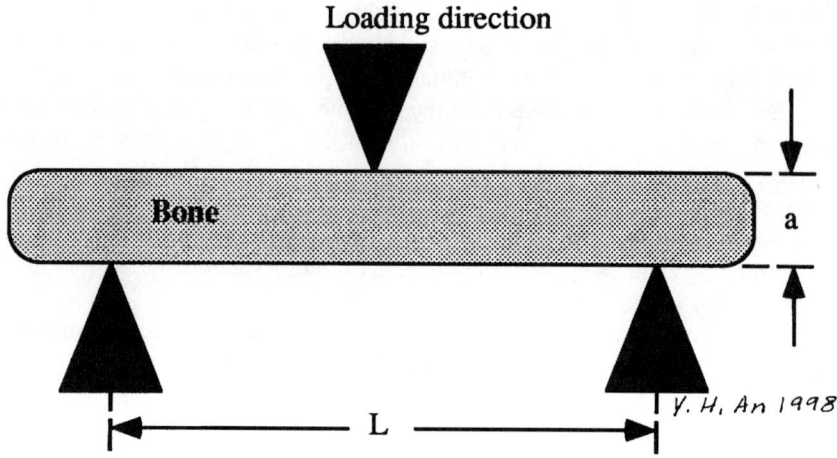

FIGURE 3. The setup of three-point bending test.

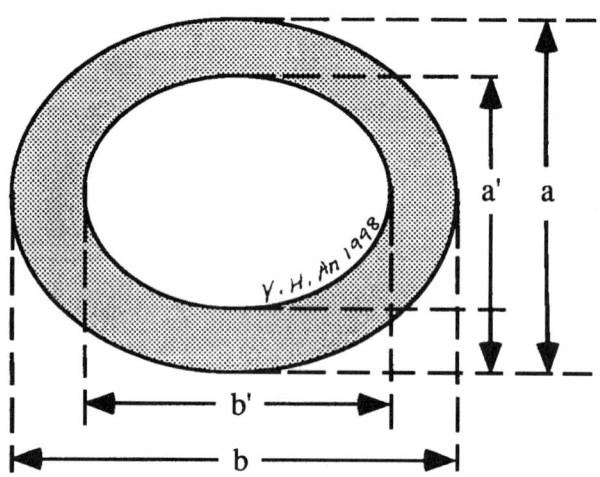

FIGURE 4. Illustration of the external and internal anteroposteral and side-to-side diameters for the cross-sections at the loading points of the bone (a, a', b, and b').

where a, a', b, and b' are the mean external and internal anteroposteral and side-to-side diameters for the cross-sections at the loading points of the bone (Figure 4). The external diameters (a and b) were measured before testing by use of a digital caliper. After testing, the pieces are glued together and cut transversely at the break point. The dimensions of the medullary canal (a' and b') are then measured. The following equation is used to calculate elastic modulus:

$$E = SL^3/48I \qquad (9)$$

where S is the stiffness and L is the distance between the two supporting bars.

FIGURE 5. Photograph of compression test.

B. COMPRESSION AND TENSILE TEST

A compression test is commonly used for testing mechanical properties of cancellous and cortical bones.[83] Theoretically, a compression test is not as solid as a tensile test due to the edge effect and the axial inaccuracies.[28,85] But, because of the easier sample preparation (commonly trephining), the compression test remains a popular and useful mechanical test.

For testing procedures, the testing machine should be operated under displacement control. The upper and lower platens should have polished stainless steel surfaces. Figure 5 shows a compression testing in action. Ultimate compression strength (σ) is calculated by the following equation if the bone specimen is cylindrical:

$$\sigma = 4 P/\pi d^2 \tag{10}$$

where P is the maximum load and d is the diameter of the bone cylinder. The elastic modulus by compression test (E) is calculated by:

$$E = SL/A \qquad (11)$$

where S is the compression stiffness, L is the length of the bone cylinder and A is the end-face area of the specimen.

Keaveny et al.[48] compared the results by an accurate nondestructive method and the platens compression test and found a significant influence of specimen aspect ratio. A specimen with an aspect ratio of 2:1 (5 mm diam., 10 mm length) was the geometry which had the least difference between the two methods. They recommend that a 2:1 cylinder be used as a standard specimen in studies of uniaxial elastic modulus and strength of cancellous bone.[86]

A tensile test is the standard engineering test for determining material properties. The specimen is machined into a beam with a smaller diameter in the central part of the beam (Figure 6). It needs a relatively complicated fixture with two universal joints above and below the specimen for specimen gripping and to line up the specimen. It has been successfully applied to both cortical[28,42,87] and cancellous bone[80,88] specimens. However, because of the difficulties of specimen fabrication, problems at the interface of specimen and grips of the machine, or simply because there is not enough volume of bone for making specimens, this test has not been utilized as commonly as compression testing for mechanical properties of bone, especially cancellous bones.

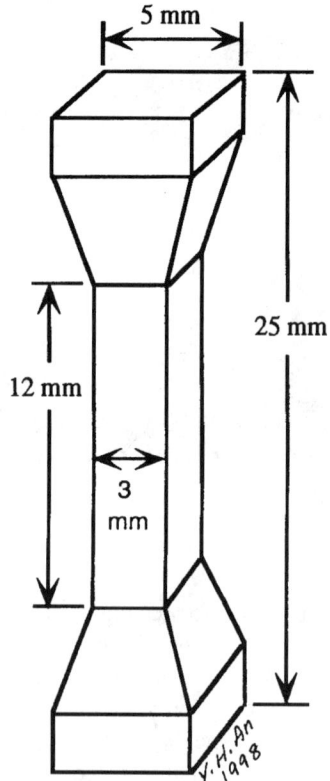

FIGURE 6. Illustration of a specimen fabricated with a gauge section reduced in size for tensile testing.

C. INDENTATION TEST

In the indentation test, an indentor is driven into a sectional surface of bone. Although the failure mechanisms are more complicated and less clear than the conventional compression test, it is useful for examining the mechanical properties of cancellous bones of different species.[23,54,63,79,89–92] Because of the ease of specimen fabrication, the use of the indentation test has increased in recent years.[23] The test is simpler than the compression test which uses cubed or cylindrical samples. Only a flat surface of a sample is needed for indentation tests. The test is particularly useful for testing small bones. Recent reports describe the use of the indentation test for measurement of the mechanical properties of rabbit and rat cancellous bones.[23,60] Indentation tests have also been used for testing the mechanical properties of fracture callus.[93]

The selected bone specimen is cut and ground to a proscribed level or depth in the cancellous bone to create a surface for testing. After the first surface is created, a parallel cut is made to create a second surface to be used for setting the specimen against the specimen-holding platform. Instead of performing the second cut, the specimen can be potted in dental stone or plaster of Paris for positioning on the platform. The latter is especially suitable for small specimens. A cylindrical steel indentor with a flat end surface (2–5 mm diam.) is driven into the surface at a constant slow rate (1 mm/min.). Ultimate indentation strength is calculated using the following equation:

$$\sigma = 4 \, P/\pi d^2 \qquad (12)$$

where P is the ultimate indentation load and d is the diameter of the indentor.

D. TORSIONAL TEST

Torsional tests are often used for testing mechanical properties of whole long bones such as the tibia or femur of dogs,[64,94,95] rabbits,[96,97] or rat.[98] The test is especially useful for testing larger bones. Bone ends can be embedded in plaster of Paris,[99] dental stone,[98] or epoxy resin[97] for mounting onto the testing machine. Commonly, maximum torque capacity (Nm), maximum angle of deformation (degree), torsional stiffness (N/m), torsional strength (MPa), shear modulus (MPa or N/m^2) are calculated. Readers may refer to the books on this subject edited by Evans[100] and Hayes and Bouxsein.[6]

E. SCREW PULLOUT TEST

Screw pullout tests are used for testing the holding power of different screw designs. When one screw is used on different bone specimens,[101,102] on bones treated with different preparation and storage methods,[77] or from different directions of one bone block,[63] variations of mechanical properties can be demonstrated. For the screw pullout test, a steel fixture is used, consisting of an upper screw gripping bar (connected to the load cell) and a lower specimen holding frame (Figure 7). It is always desirable to employ universal joints above and below the specimen for better specimen alignment. After the specimen is positioned on the fixture, the test machine is operated at a constant slow displacement rate (1 mm/min.) until the screw is pulled out. Ultimate strength (σ) is calculated using:

$$\sigma = P/\pi d h \qquad (13)$$

where P is the ultimate pullout load (Newton), d (mm) is the major diameter of the screw, and h is the length of effective threads in the bone.

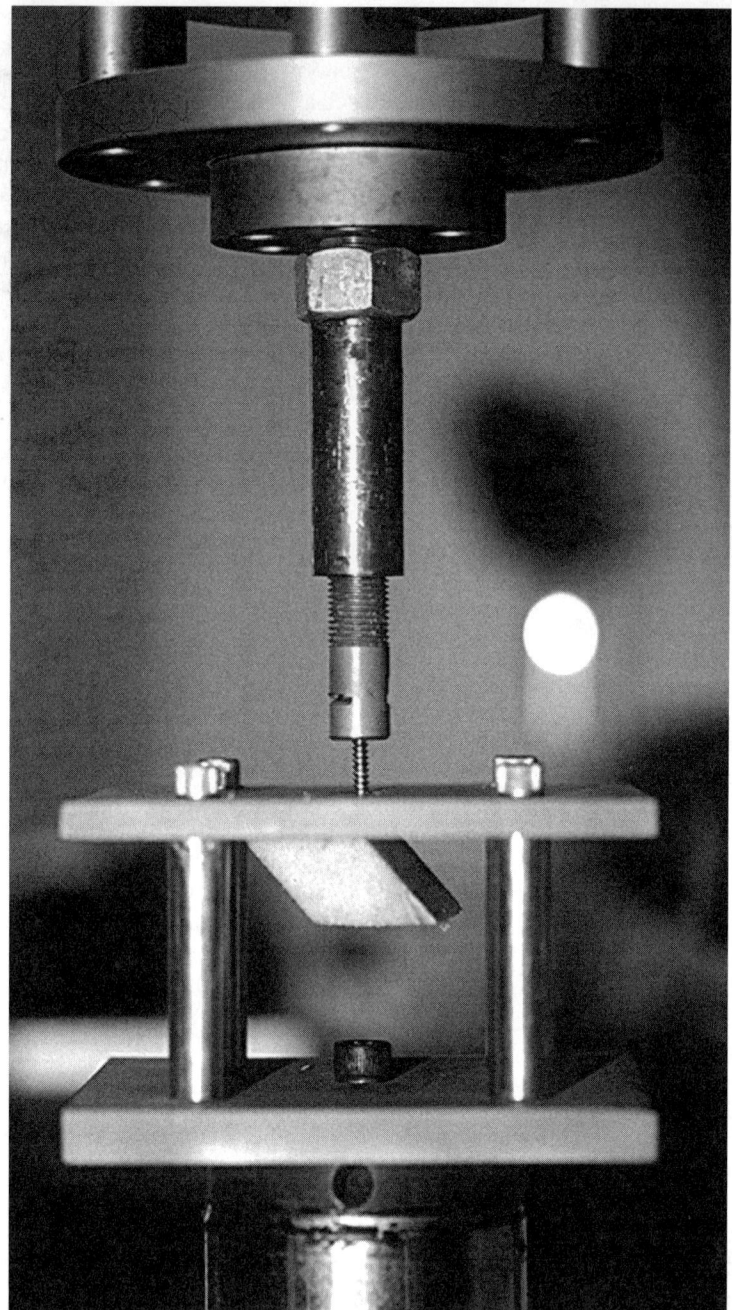

FIGURE 7. Photograph of screw pullout test.

F. Strain Gauge

A strain gauge is a strain transducer made of etched foil grid patterns with a polymer backing. It is bonded to the bone surface, usually with cyanoacrylate adhesives, to measure bone strain. Single axis strain gauges measure strain in a particular direction while strain gauge rosettes (three gauges oriented at different directions mounted on a single polymer backing) allow the measurement of both magnitude and direction of principal strain.[80,103,104] Strain gauges have been used in both *in vivo*[103,104] and *in vitro*[105] loading conditions. If the elastic modulus of the bone, location and orientation of strain gauges, and cross-sectional area and moment of inertia are known, then stresses, loads, and torque in the bone can be calculated from the measured strains.

G. ULTRASONIC METHODS

Ultrasonic techniques offer some advantages over mechanical testing for measuring elastic modulus of bone.[80] Specifically, the specimens can be smaller with less complicated shape (cylinder or cube) and several anisotropic properties can be tested using one (cubic) specimen.[106] The general relationship between elastic property (E) and the velocity of wave propagation (v) is as follows:[80]

$$E = \rho v^2 \tag{14}$$

where ρ is the apparent density of the bone specimen. The exact form of Equation 14 depends on the mode (or direction) of propagation, the wavelength (for cortical bone: 2–10 MHz;[107] for cancellous bone: 50–100 kHz[88]), and the cross-sectional dimensions of the material.[80,106,107] Details of ultrasonic methods can be found in the papers by Ashman's group.[80,82,88,106,107]

VI. TESTING OF BONE-IMPLANT INTERFACE

A. PUSHOUT AND PULLOUT TEST

The pushout test is the most popular test for evaluating the stiffness and strength of the bone-implant (cylindrical plug) interface.[108] Work done with a rabbit femoral condyle plug model is representative of the procedures of the pushout test.[109-111] In this work, specimens to be tested (5.5 mm diam.) are prepared by sectioning the distal femurs containing the implant and hand grinding the bone sections to expose both ends of the implant. The specimen containing the implant is placed on the test assembly so that the implant is concentric with the hole in the specimen support (Figure 8). The diameter of the hole in the specimen-support of the pushout test assembly is 7.55 mm. The diameter of the loading rod should be smaller than the implant and large enough to provide uniform loading to the implant. The test machine is operated under displacement control mode with a ramp function and a displacement rate of 1.0 mm/min. Two important parameters should be calculated, interfacial stiffness and interfacial shear strength. The latter is calculated by dividing the ultimate load by the bone-implant interface area:

$$\sigma = P/A = P/\pi d h \tag{15}$$

where P is the maximum load applied to the implant, A is the nominal surface area of the implant, and d and h are the diameter and height of the cylindrical implant. Interfacial stiffness is taken as the slope of the linear portion of the load/deflection curve and has the units of MPa/mm.

Like the pushout test, the pullout test also measures the implant-bone interfacial stiffness and strength.[112] Theoretically, if testing two identical specimens (identical implants and bones) using pushout and pullout test (one test for one specimen), the results should be the same. The only difference in the tests is the mode of loading, pushing or pulling. It is more complicated to design a fixture for pullout tests than for pushout tests. Pullout tests require complicated fixtures with two universal joints for specimen gripping and alignment of the specimen. Pushout tests require only a push rod and a block of metal with a hole drilled in the center. However, the pullout test has some advantages over the pushout test. The accuracy of the pushout test depends on how well the two bone surfaces are prepared. The surfaces should be parallel to each other and perpendicular to the axial line of the implant. Another advantage of the pullout test is the improvement in specimen

FIGURE 8. Photograph of pushout test.

alignment by the use of universal joints. For pullout tests, the implant-bone interfacial strength is calculated using Equation 15. Recently, a nondestructive pullout test has been proposed.[113]

The size of the bone is critical to the success of pullout/pushout tests. Rat bones are not appropriate for implantation using pushout or pullout testing because of their small size. Rabbit, canine or sheep femur or tibia are commonly used for metaphyseal or transcortical implant models. For cylindrical implants, the length/diameter ratio of the implant needs to be more than 1:1. Specimens with severe inflammatory changes or suspected infection should be excluded.

B. Screw Pullout Test

The holding power of different screw designs are tested using screw pullout tests.[114-116] The key for designing a screw pullout test is that the bone specimen to be used should have uniform structural and mechanical properties so that differences measured are caused by different screw designs. Uniform materials, such as Dacro porous foam (a polyurethane foam formed by mixing a resin and an isocyante),[117] also can be used as a model system for screw pullout tests. The testing procedure is the same as that described in section V. Ultimate strength (σ) is calculated using

Equation 15. In the authors' laboratory, a pullout test has been used to test the holding power of a threaded cylindrical ligament anchor.[118,119]

C. REMOVAL TORQUE

The bonding between bone and screw type implants can be evaluated by a removal torque method described by Carlsson at al.[120] The strength at the bone-implant interface is measured by using a torque wrench to determine the maximum torque necessary for manual removal of the implant.

D. OTHER TESTS

A variety of mechanical tests which combine and adapt aspects of the standard tests have been employed to study the properties of the bone-implant interface region. Of particular interest are two special tensile tests which have been reported for testing the bond strength at a bone-HA interface[121] and a bone-cement interface.[122]

REFERENCES

1. Cowin, S. C., "The mechanical properties of cancellous bone," in *Bone Biomechanics*, Cowin, S. C., Ed., CRC Press, Boca Raton, FL, 1989, 130.
2. Tencer, A. F. and Johnson, K. D., *Biomechanics in Orthopaedic Trauma. Bone Fracture and Fixation*, M. Dunitz, London, J. B. Lippincott, Philadelphia, 1994, 18.
3. Carter, D. R. and Spengler, D. M., "Mechanical properties and composition of cortical bone," *Clin. Orthop.*, 135, 192, 1978.
4. Katz, J. L., "The structure and biomechanics of bone," *Symp. Soc. Exp. Biol.*, 34, 137, 1980.
5. Hoesler, H., "The history of some fundamental concepts in bone biomechanics," *J. Biomech.*, 20, 1025, 1987.
6. Hayes, W. C. and Bouxsein, M. L., "Biomechanics of cortical and trabecular bone: implications for assessment of fracture risk," in *Basic Orthopaedic Biomechanics*, Mow, V. C., Hayes, W. C., Eds., 2nd ed., Lippincott-Raven, Philadelphia, 1997, Chapter 3.
7. Sedlin, E. D. and Hirsch, C., "Factors affecting the determination of the physical properties of femoral cortical bone," *Acta Orthop. Scand.*, 37, 29, 1966.
8. Keller, T. S., Mao, Z., and Spengler, D. M., "Young's modulus, bending strength, and tissue physical properties of human compact bone," *J. Orthop. Res.*, 8, 592, 1990.
9. Lotz, J. C., Gerhart, T. N., and Hayes, W. C., "Mechanical properties of metaphyseal bone in the proximal femur," *J. Biomech.*, 24, 317, 1991.
10. Currey, J. D., Foreman, J., Laketic, I., Mitchell, J., Pegg, D. E., and Reilly, G. C., "Effects of ionizing radiation on the mechanical properties of human bone," *J. Orthop. Res.*, 15, 111, 1997.
11. Kasra, M. and Grynpas, M. D., "Effect of long-term ovariectomy on bone mechanical properties in young female cynomolgus monkeys," *Bone*, 15, 557, 1994.
12. Currey, J. D., "The effect of porosity and mineral content on the Young's modulus of elasticity of compact bone," *J. Biomech.*, 21, 439, 1988.
13. Currey, J. D., "The effect of drying and re-wetting on some mechanical properties of cortical bone," *J. Biomech.* 21, 131, 1988.
14. Currey, J. D., Brear, K., Zioupos, P., and Reilly, G. C., "Effect of formaldehyde fixation on some mechanical properties of bovine bone," *Biomaterials*, 16, 1267, 1995.
15. Simkin, A. and Robin, G., "The mechanical testing of bone in bending," *J. Biomech.*, 6, 31, 1973.
16. Martin, R. B. and Boardman, D. L., "The effects of collagen fiber orientation, porosity, density, and mineralization on bovine cortical bone bending properties," *J. Biomech.*, 26, 1047, 1993
17. Schryver, H. F., "Bending properties of cortical bone of the horse," *Am. J. Vet. Res.*, 39, 25, 1978.
18. Bigot, G., Bouzidi, A., Rumelhart, C., and Martin-Rosset, W., "Evolution during growth of the mechanical properties of the cortical bone in equine cannon-bones," *Med. Eng. Phys.*, 18, 79, 1996.

19. Kaneps, A. J., Stover, S. M., and Lane, N. E., "Changes in canine cortical and cancellous bone mechanical properties following immobilization and remobilization with exercise," *Bone*, 21, 419, 1997.
20. Crenshaw, T. D., Peo, E. R., Jr., Lewis, A. J., Moser, B. D., and Olson, D., "Influence of age, sex and calcium and phosphorus levels on the mechanical properties of various bones in swine," *J. Anim. Sci.*, 52, 1321, 1981.
21. McAlister, G. B. and Moyle, D. D., "Some mechanical properties of goose femoral cortical bone," *J. Biomech.*, 16, 577, 1983.
22. Ayers, R. A., Miller, M. R., Simske, S. J., and Norrdin, R. W., "Correlation of flexural structural properties with bone physical properties: a four species survey," *Biomed. Sci. Instrum.*, 32, 251, 1996.
23. An, Y. H., Kang, Q., and Friedman, R. J., "The mechanical symmetry of rabbit long bones studied by bending and indentation test," *Am. J. Vet. Res.*, 57, 1786, 1996.
24. Jørgensen, P. H., Bak, B., and Andreassen, T. T., "Mechanical properties and biochemical composition of rat cortical femur and tibia after long-term treatment with biosynthetic human growth hormone," *Bone*, 12, 353, 1991.
25. Barengolts, E. I., Curry, D. J., Bapna, M. S., and Kukreja, S. C., "Effects of endurance exercise on bone mass and mechanical properties in intact and ovariectomized rats," *J. Bone Miner. Res.*, 8, 937, 1993.
26. Ejersted, C., Andreassen, T. T, Oxlund, H., et al., "Human parathyroid hormones (1–34) and (1–84) increase the mechanical strength and thickness of cortical bone in rats," *J. Bone Miner. Res.*, 8, 1097, 1993.
27. Simske, S. J., Guerra, K. M., Greenberg, A. R., and Luttges, M. W., "The physical and mechanical effects of suspension-induced osteopenia on mouse long bones," *J. Biomech.*, 25, 489, 1992.
28. Reilly, D. T., Burstein, A. H., and Frankel, V. H., "The elastic modulus for bone," *J. Biomech.*, 7, 271, 1974.
29. Burstein, A. H., Reilly, D. T., and Martens, M., "Aging of bone tissue: mechanical properties," *J. Bone Joint Surg.*, 58A, 82, 1976.
30. Cezayirlioglu, H., Bahniuk, E., Davy, D. T., and Heiple, K. G., "Anisotropic yield behavior of bone under combined axial force and torque," *J. Biomech.*, 18, 61, 1985.
31. McElhaney, J. H., Fogle, J., Byars, E., and Weaver, G., "Effect of embalming on the mechanical properties of beef bone," *J. Appl. Physiol.*, 19, 1234, 1964.
32. Evans, F. G. and Lebow, M., "Regional differences in some of the physical properties of the human femur," *J. Appl. Physiol.*, 3, 563, 1951.
33. Vincentelli, R. and Grigoroy, M., "The effect of Haversian remodeling on the tensile properties of human cortical bone," *J. Biomech.*, 18, 201, 1985.
34. Burstein, A. H., Zika, J. M., Heiple, K. G., and Klein L., "Contribution of collagen and mineral to the elastic-plastic properties of bone," *J. Bone Joint Surg.*, 57A, 956., 1975
35. Hazama, H., "Study on the torsional strength of the compact substance of human beings," *J. Kyoto Pref. Med. Univ.*, 60, 167, 1956.
36. Yamada, H., *Strength of Biological Materials*, Williams & Wilkins, Baltimore, MD, 1970.
37. Reilly, D. T. and Burstein, A. H., "The elastic and ultimate properties of compact bone tissue," *J. Biomech.*, 8, 393, 1975.
38. Hirsch, C. and da Silva, O., "The effect of orientation on some mechanical properties of femoral cortical specimens," *Acta Orthop. Scand.*, 38, 45, 1967.
39. Carter, D. R. and Hayes, W. C., "The compressive behavior of bone as a two-phase porous structure," *J. Bone Joint Surg.*, 59A, 954, 1977.
40. Kaplan, S. J., Hayes, W. C., Stone, J. L., and Beaupre, G. S., "Tensile strength of bovine trabecular bone," *J. Biomech.*, 18, 723, 1985.
41. Keaveny, T. M., Guo, X. E., Wachtel, E. F., McMahon, T. A., and Hayes, W. C., "Trabecular bone exhibits fully linear elastic behavior and yields at low strains," *J. Biomech.*, 27, 1127, 1994.
42. Keaveny, T. M., Wachtel, E. F., Ford, C. M., and Hayes, W. C., "Differences between the tensile and compressive strengths of bovine tibial trabecular bone depend on modulus," *J. Biomech.*, 27, 1137, 1994.
43. Goldstein, S. A., "The mechanical properties of trabecular bone: dependence on anatomical location and function," *J. Biomech.* 20, 1055, 1987.

44. Martens, M., Van Audekercke, R., Delport, P., De Meester, P., and Mulier, J. C., "The mechanical characteristics of cancellous bone at the upper femoral region," *J. Biomech.*, 16, 971, 1983.
45. Kuhn, J. L., Goldstein, S. A., Ciarelli, M. J., and Matthews, L. S., "The limitations of canine trabecular bone as a model for human: a biomechanical study," *J. Biomech.*, 22, 95, 1989.
46. Odgaard, A., Hvid, I., and Linde, F., "Compressive axial strain distributions in cancellous bone specimens," *J. Biomech.*, 22, 829, 1989.
47. Linde, F. and Hvid, I., "The effect of constraint on the mechanical behaviour of trabecular bone specimens," *J. Biomech.*, 22, 485, 1989.
48. Keaveny, T. M., Pinilla, T. P., Crawford, R. P., Kopperdahl, D. L., and Lou, A., "Systematic and random errors in compression testing of trabecular bone," *J. Orthop. Res.*, 15, 101, 1997.
49. Poumarat, G. and Squire, P., "Comparison of mechanical properties of human bone, bovine bone and a new processed bone xenograft," *Biomaterials,* 14, 337, 1993.
50. Rho, J. Y., Flaitz, D., Swarnakar, V., and Acharya, R. S., "The characterization of broadband ultrasound attenuation and fractal analysis by biomechanical properties," *Bone,* 20, 497, 1997.
51. Swartz, D. E., Wittenberg, R. H., Shea, M., White, A. A., III., and Hayes, W. C., "Physical and mechanical properties of calf lumbosacral trabecular bone," *J. Biomech.*, 24, 1059, 1991.
52. Vahey, J. W., Lewis, J. L., and Vanderby, R., Jr., "Elastic moduli, yield stress, and ultimate stress of cancellous bone in the canine proximal femur," *J. Biomech.*, 20, 29, 1987.
53. Kang, Q., An, Y. H., and Friedman, R. J., "The mechanical properties and bone densities of canine cancellous bones," *J. Mater. Sci. Mater. Med.*, 9, 463, 1998.
54. Sumner, D. R., Willke, T. L., Berzins, A., and Turner, T. M., "Distribution of Young's modulus in the cancellous bone of the proximal canine tibia," *J. Biomech.*, 27, 1095, 1994.
55. Alcito, A. J., Kasra, M., Lee, J. M., and Grynpas, M. D., Effects of intermittent administration of pamidronate on the mechanical properties of canine cortical and trabecular bone," *J. Orthop. Res.,* 12, 742, 1994.
56. Geusens, P., Boonen, S, Nijs, J., et al., "Effect of salmon calcitonin on femoral bone quality in adult ovariectomized ewes," *Calcif. Tissue Int.,* 59, 315, 1996.
57. Deloffre, P., Hans, D., Rumelhart, C., Mitton, D., Tsouderos, Y., and Meunier, P. J., "Comparison between bone density and bone strength in glucocorticoid-treated aged ewes," *Bone,* 17, 409S, 1995.
58. Mitton, D., Rumelhart, C., Hans, D., and Meunier. P. J., "The effects of density and test conditions on measured compression and shear strength of cancellous bone from the lumbar vertebrae of ewes," *Med. Eng. Phys.*, 19, 464, 1997.
59. Mosekilde, Li., Kragstrup, J., and Richards, A., "Compressive strength, ash weight, and volume of vertebral trabecular bone in experimental fluorosis in pigs," *Calcif. Tiss. Int.*, 40, 318, 1987.
60. An, Y. H., Zhang, J. H., Kang, Q., and Friedman, R. J., "Mechanical properties of rat epiphyseal cancellous bones studied by indentation test," *J. Mater. Sci. Mater. Med.*, 8, 493, 1997.
61. Galante, J., Rostoker, W., and Ray, R. D., "Physical properties of trabecular bone," *Calcif. Tissue Res.*, 5, 236, 1970.
62. Ciarelli, M. J., Goldstein, S. A., Kuhn, J. L., Cody, D. D., and Brown, M. B., "Evaluation of orthogonal mechanical properties and density of human trabecular bone from the major metaphyseal regions with materials testing and computed tomography," *J. Orthop. Res.*, 9, 674, 1990.
63. An, Y. H., Kang, Q., Friedman, R. J., and Young, F. A., "The effect of microstructure of cancellous bone on screw pull-out strength," *Trans. Soc. Biomater.*, 20, 385, 1997.
64. An, Y. H., Kang, Q., Friedman, R. J., Gharpuray, V. M., and Powers, D. L., "Do mechanical properties of epiphyseal cancellous bones vary?" *J. Invest. Surg.*, 10, 221, 1997c.
65. Runkle, J. C. and Pugh, J., "The micro-mechanics of cancellous bone," *Bull. Hosp. J. Dis.*, 36, 2, 1975.
66. Townsend, P. R., Rose, R. M., and Radin, E. L., "Buckling studies of single human trabeculae," *J. Biomech.,* 8, 199, 1975.
67. Rho, J. Y., Ashman, R. B., and Turner, C. H., "Young's modulus of trabecular and cortical bone material: ultrasonic and microtensile measurements," *J. Biomech.*, 26, 111, 1993.
68. Mente, P. L. and Lewis, J. L., "Experimental method for the measurement of the elastic modulus of trabecular bone tissue," *J. Orthop Res.*, 7, 456, 1989.
69. Van Rietbergen, B., Weinans, H., Huiskes, R., and Odgaard, A., "A new method to determine trabecular bone elastic properties and loading using micromechanical finite-element models," *J. Biomech.*, 28, 69, 1995.

70. Kang, Q., An, Y. H., Butehorn, H. F., and Friedman, R. J., "Morphological and mechanical study of the effects of experimentally induced inflammatory knee arthritis on rabbit long bones," *J. Mater. Sci. Mater. Med.*, in press, 1998.
71. Keaveny, T. M. and Hayes, W. C., "A 20-year perspective on the mechanical properties of trabecular bone," *J. Biomech. Eng.*, 115, 534, 1993.
72. Smith, J. W. and Walmsley, R., "Factors affecting the elasticity of bone," *J. Anat.*, 93, 503, 1959.
73. Strömberg, L. and Dalén, N., "The influence of freezing on the maximum torque capacity of long bones," *Acta Orthop. Scand.*, 37, 254, 1976.
74. Tomford, W. W., Doppelt, S. H., Mankin, H. J., and Friedlaender, G. E., "1983 bone banking procedures," *Clin. Orthop.*, 174, 15, 1983.
75. Malinin, T. I., Martinex, O. V., and Brown, M. D., "Banking of massive osteoarticular and intercalary bone grafts — 12 years' experience, *Clin. Orthop.*, 196, 44, 1985.
76. Panjabi, M. M., Krag, M., Summers, D., and Videman, T., "Biomechanical time-tolerance of fresh cadaveric human spine specimens," *J. Orthop. Res.*, 3, 292, 1985.
77. Roe, S. C., Pijanowski, G. J., and Johnson, A. L., "Biomechanical properties of canine cortical bone allografts: effects of preparation and storage," *Am. J. Vet. Res.*, 49, 873, 1988.
78. Linde, F. and Sørensen, H. C. F., "The effect of different storage methods on the mechanical properties of trabecular bone," *J. Biomech.*, 26, 1249, 1993.
79. Kang, Q., An, Y. H., and Friedman, R. J., "Effects of multiple freezing and thawing on the indentation strength of bovine cancellous bone," *Am. J. Vet. Res.*, 58: 1171, 1997.
80. Ashman, R. B., "Experimental techniques," in *Bone Mechanics*, Cowin, S. C., Ed., CRC Press, Boca Raton, 1989, 91.
81. Spatz, H.-Ch., O'Leary, E. J., and Vincent, J. F., "Young's moduli and shear moduli in cortical bone," *Proc. R. Soc. Lond. B. Biol. Sci.*, 263 (1368), 287, 1996.
82. Ashman, R. B. and Rho, J. Y., "Elastic modulus of trabecular bone material," *J. Biomech.*, 21, 177, 1988.
83. Linde, F., "Elastic and viscoelastic properties of trabecular bone by a compression testing approach," *Dan. Med. Bull.*, 41, 119, 1994.
84. Burstein, A. H. and Frankel, V. H., "A standard test for laboratory animal bone," *J. Biomech.*, 4, 155, 1971.
85. Wall, J. C., Chatterji, S., and Jeffery, J. W., "On the origin of scatter in results of human bone strength tests," *Med. Biol. Eng.*, 8, 171, 1970.
86. Keaveny, T. M., Borchers, R. E., Gibson, L. J., Hayes, W. C., "Trabecular bone modulus and strength can depend on specimen geometry," *J. Biomech.*, 26, 991, 1993.
87. Dickenson, R. P., Hutton, W. C., and Stott, J. R., "The mechanical properties of bone in osteoporosis," *J. Bone Joint Surg.*, 63B, 233, 1981.
88. Ashman, R. B., Corin, J. D., and Turner, C. H., "Elastic properties of cancellous bone: measurement by an ultrasonic technique," *J. Biomech.*, 20, 979, 1987.
89. Aitken, G. K., Bourne, R. B., Finlay, J. B., Rorabeck, C. H., and Andreae, P. R., "Indentation stiffness of the cancellous bone in the distal human tibia," *Clin. Orthop.*, 201, 264, 1985.
90. Josechak, R. G., Finlay, J. B., Bourne, R. B., and Rorabeck, C. H., "Cancellous bone support for patellar resurfacing," *Clin. Orthop.*, 220, 192, 1987.
91. Finlay, J. B., Bourne, R. B., Kraemer, W. J., Moroz, T. K., and Rorabeck, C. H., "Stiffness of bone underlying the tibial plateaus of osteoarthritic and normal knees," *Clin. Orthop.*, 247, 193, 1989.
92. Saitoh, S., Nakatsuchi, Y., Latta, L., and Milne, E., "An absence of structural changes in the proximal femur with osteoporosis," *Skeletal Radiol.*, 22, 425, 1993.
93. Markel, M. D., Wikenheiser, M. A., and Chao, E. Y., "A study of fracture callus material properties: relationship to the torsional strength of bone," *J. Orthop. Res.*, 8, 843, 1990.
94. Aro, H. T. and Chao, E. Y., "Bone-healing patterns affected by loading, fracture fragment stability, fracture type, and fracture site compression," *Clin. Orthop.*, 293, 8, 1993.
95. O'Sullivan, M. E., Bronk, J. T., Chao, E. Y., and Kelly. P. J., "Experimental study of the effect of weight bearing on fracture healing in the canine tibia," *Clin. Orthop.*, 302, 273, 1994.
96. White, A. A., III, Panjabi, M. M., and Southwick, W. O., "The four biomechanical stages of fracture repair," *J. Bone Joint Surg.*, 59A, 188, 1977.
97. Paavolainen, P., "Studies on mechanical strength of bone. I. Torsional strength of normal rabbit tibiofibular bone," *Acta Orthop. Scand.*, 49, 497, 1978.

98. Lepola, V., Vaananen, K., and Jalovaara P., "The effect of immobilization on the torsional strength of the rat tibia," *Clin. Orthop.,* 297, 55, 1993.
99. Davis, P. K., Mazur, J. M., and Coleman, G. N., "A torsional strength comparison of vascularized and nonvascularized bone grafts," *J. Biomech.,* 15, 875, 1982.
100. Evans, F. G., *Mechanical Properties of Bone,* Charles C. Thomas, Springfield, IL, 1973, Chapter 2.
101. Flahiff, C. M., Gober, G. A., and Nicholas, R. W., "Pullout strength of fixation screws from polymethylmethacrylate bone cement," *Biomaterials,* 16, 533, 1995.
102. Halvorson, T. L., Kelley, L. A., Thomas, K. A., Whitecloud, T. S., III., and Cook, S. D., "Effects of bone mineral density on pedicle screw fixation," *Spine,* 19, 2415, 1994.
103. Caler, W. E., Carter, D. R., and Harris, W. H., "Techniques for implementing an *in vivo* bone strain gage system," *J. Biomech.,* 14,503, 1981.
104. Biewener, A. A., "*In vivo* measurement of bone strain and tendon force," in *Biomechanics. Structure and System,* Biewener, A. A., Ed., IRL Press, Oxford, 1992, Chapter 6.
105. Battraw, G. A., Miera, V., Anderson, P. L., and Szivek, J. A., "Bilateral symmetry of biomechanical properties in rat femora," *J. Biomed. Mater. Res.,* 32, 285, 1996.
106. Ashman, R. B., Rho, J. Y., and Turner, C. H., "Anatomical variation of orthotropic elastic moduli of the proximal human tibia," *J. Biomech.,* 22, 895, 1989.
107. Ashman, R. B., Cowin, S. C., Van Buskirk, W. C., and Rice, J. C., "A continuous wave technique for the measurement of the elastic properties of cortical bone," *J. Biomech.,* 17, 349, 1984.
108. Dhert, W. J. A., Verheyen, C. C. P. M., Braak, L. H., et al., "A finite element analysis of the push-out test: Influence of test conditions," *J. Biomed. Mater. Res.,* 26, 119, 1992.
109. Friedman, R. J., Bauer, T. W., Garg, K., Jiang, M., An, Y. H., and Draughn, R. A., "Histological and mechanical comparison of hydroxyapatite coated cobalt-chrome and titanium implants," *J. Appl. Biomater.,* 6, 231, 1995.
110. Friedman, R. J., An, Y. H., Jiang, M., Draughn, R. A., and Bauer, T. W., "*In vivo* mechanical and histological evaluation of bone ingrowth and apposition to metal implants of different surface textures in the rabbit femur," *J. Orthop. Res.,* 14, 455, 1996.
111. Friedman, R. J., An, Y. H., Jiang, M., Butehorn, H. F., Draughn, R. A., and Bauer. T. W., "Bone ingrowth into implant surfaces in experimental inflammatory arthritis," *Trans. Orthop. Res. Soc.,* 22, 503, 1997.
112. Shirazi-Adl, A., Dammak, M., and Zukor, D. J., "Fixation pull-out response measurement of bone screws and porous-surfaced posts," *J. Biomech.,* 27, 1249, 1994.
113. Berzins, A., Shah, B., Weinans, H., and Sumner, D. R., "Nondestructive measurements of implant-bone interface shear modulus and effects of implant geometry in pull-out tests," *J. Biomed. Mater. Res.,* 34, 337, 1997.
114. Koranyi, E., Bowman, C. E., Knecht, C. D., and Janssen, M., "Holding power of orthopaedic screws in bone," *Clin. Orthop.,* 72, 283, 1970.
115. Skinner, R., Maybee, J., Transfeldt, E., Venter, R., and Chalmers, W., "Experimental pullout testing and comparison of variables in transpedicular screw fixation. A biomechanical study," *Spine,* 15, 195, 1990.
116. Wittenberg, R. H., Lee, K. S., Shea, M., White, A. A., III., and Hayes, W. C., "Effect of screw diameter, insertion technique, and bone cement augmentation of pedicular screw fixation strength," *Clin. Orthop.*, 296, 278, 1993.
117. Szivek, J. A. and Thomas, M., "Benjamin, J. B., "Characterization of a synthetic foam as a model for human cancellous bone," *J. Appl. Biomater.,* 4, 269, 1993.
118. Young, F. A. and An, Y. H., "A new artificial ACL anchor," *Mater. Res. Soc. Symp. Proc.,* 394, 31, 1995.
119. An, Y. H., Friedman, R. J., and Young, F. A., "A new artificial ACL anchor," *Trans. Soc. Biomater.,* 20, 386, 1997.
120. Carlsson, L., Rostlund, T., Albrektsson, B., and Albrektsson, T., "Removal torques for polished and rough titanium implants," *Int. J. Oral Maxillofac. Implants,* 3, 21, 1988.
121. Hong, L., Xu, H. C., and de Groot, K., "Tensile strength of the interface between hydroxyapatite and bone," *J. Biomed. Mater. Res.,* 26, 7, 1992.
122. Mann, K. A., Ayers, D. C., Werner, F. W., Nicoletta, R. J., and Fortino, M. D., "Tensile strength of the cement-bone interface depends on the amount of bone interdigitated with PMMA cement," *J. Biomech.*, 30, 339, 1997.

9 Mechanical Testing of Cartilage

Martine LaBerge

CONTENTS

I. Introduction 165
II. Indentation Testing of Articular Cartilage 166
   A. Indentation Theory for Compliant Layers 166
   B. Specifications of Articular Cartilage Indentation 167
   C. Lubrication of Articular Cartilage and Friction Measurement....................168
   D. Use of Indentation and Friction Measurements in an Orthopaedic Animal Model 169
   E. *In Vivo* Indentation Testing 172
III. Conclusion..173
References 173

I. INTRODUCTION

Due to a layered collagen network and the preferred orientation of the collagen fibers, articular cartilage is considered an inhomogeneous and anisotropic material. Both of these characteristics have been verified by tensile testing of cartilage sections obtained in different orientations.[1] This microscopic arrangement has been shown to influence strength and stiffness. Articular cartilage exhibits a nonlinear stress-strain behavior under large compressive strains[2] and also demonstrates a nonlinear stress-strain behavior in tension.[3] When cartilage is placed under physiological compressive stresses, the modulus of the tissue increases with increasing strain, demonstrating an ability to limit excessively large strains in the tissue.[4] The mechanical response of articular cartilage is time dependent, and exhibits phenomena such as stress relaxation, creep, and hysteresis. In addition, the material is sensitive to loading rate. Therefore, authors have used several mechanical tests to define its properties.

Commonly known experimental set-ups proposed for cartilage testing in compression include the confined and the unconfined compression test,[5] and the indentation test. Confined compression tests consist of using an apparatus designed to restrict radial expansion of the specimen[6,7] and allow uniaxial deformation.[8] A typical experimental set-up for confined tests involves placing small cylindrical cartilage specimens, which are attached onto subchondral bone, in a confining chamber. The cartilage surface is in contact with a porous filter which allows free fluid flow as it is exuded from the tissue during compression. Tests are conducted either at room temperature or physiological temperatures. A small tare load (0.2 N) is applied to ensure that the specimen fills the confining chamber, and is in contact with the porous filter. Loads providing physiological stresses are then applied with a plunger contacting the subchondral surface and the resulting creep deformation recorded over time, until equilibrium is reached. This equilibrium condition can take 40–60 min.

as reported in the literature and defined by Kwan et al.[9] Articular plugs must be excised for confined and unconfined compression tests.

Besides being the preferred method of characterization of the state of degeneration of articular cartilage in animal models, the biphasic indentation technique has also provided a means for the evaluation of Poisson's ratio,[10,11] the shear modulus, and permeability of articular cartilage.[12,13] An indentation test can be used to evaluate the response of excised cartilage or intact articular surfaces to compressive loading. It requires a testing apparatus that can control the applied load and monitor deformation.

The response of articular cartilage to compressive loading has been extensively studied through animal models in order to better understand mechanisms of degeneration due to aging,[14] joint diseases, and following joint hemiarthroplasty.[15] The mechanical testing of articular cartilage is either done *in vitro*, following the resection of the targeted tissue,[15] or *in vivo* through arthrotomy,[16] or arthroscopy.[17,18] The major goal of these experiments is to determine the response of the tissue when subjected to the action of an applied load. An indentation of the tissue is normally performed to quantify the hardness and the viscoelastic behavior of cartilage in different environments. The coefficient of friction of the cartilage surface, an indication of the breakdown of its surface properties and its boundary lubrication are also measured experimentally. The response of articular cartilage to these modes of loading provides the basis for an assessment of the integrity or structure of the tissue, and of the potential failure of the cartilage, and depends on its structure at the time of testing. This chapter focuses on the use of indentation in the evaluation of cartilage properties, and presents an overview of different methodologies used by authors to characterize the properties of articular cartilage as often used in the design of orthopedic animal models.

II. INDENTATION TESTING OF ARTICULAR CARTILAGE

A. Indentation Theory for Compliant Layers

Any material can be indented, whether its behavior is elastic or viscoelastic as well as linear or nonlinear. However, many mathematical models of indentation assume linear elastic behavior to simplify the analysis and most of the quantitative indentation methods have been developed to characterize materials in the linear elastic region.[19] The layer thickness affects the stresses in the material due to the influence of the substrate properties and interfacial boundary conditions. Many studies choose to relate the stresses in the layer to the stresses computed by an indentation of a half-space made of the same material which is given by Hertz theory for point contact.[19] The contact mechanics of layers require a numerical solution of integral equations in order to account for the additional length variable of the layer thickness and the boundary condition at the base of the layer. Such solutions are not as easy to apply as the standard Hertz equations.

For linear-elastic materials, data from an indenter test typically include the depth of penetration and the applied load. The data can then be used to compute the stresses if the material properties are known. Also, if the area of contact is experimentally measured, the local material properties can be determined. A general empirical equation (Eq.1) for a linear-elastic (rubber) thin layer indented by a rigid sphere and bonded on a rigid substrate was proposed by Finkin.[20]

$$E = \frac{9PR}{16H^3}\left[\left(\frac{Rd}{H^2}\right)^{0.5} + 0.252\left(\frac{Rd}{H^2}\right) + 0.1588\left(\frac{Rd}{H^2}\right)^{1.5} + 0.2245\left(\frac{Rd}{H^2}\right)^{2.0} + 0.3069\left(\frac{Rd}{H^2}\right)^{2.5} + 0.298\left(\frac{Rd}{H^2}\right)^{3.0}\right]^{-3.0} \quad (1)$$

where E is the elastic modulus, R the radius of the indenter, H the depth of penetration of the indenter, P the load, and d the thickness of the specimen. This model derives explicit equations for Young's modulus in terms of either penetration of the indenter or contact radius for a rubber layer of thickness greater than the radius of contact.

B. Specifications of Articular Cartilage Indentation

An indentation test is a sensitive way to quantify the viscoelastic response of cartilage by displaying both elastic and time-dependent properties under a compressive load.[4] While the tissue remains intact on the joint surface, a nominal static compressive load is applied with an indenter and the deformation of the cartilage is monitored. Upon unloading, an instantaneous recovery of the cartilage deformation is followed by a time-dependent recovery.[21] The monitored deformation of the test cartilage during loading is plotted against time and compared to that of normal cartilage, a contralateral joint, or different locations on the same joint surface. Modifications of the mechanical properties of articular cartilage induced by aging and pathological degeneration processes were investigated with indentation tests. Hirsch,[22] Sokoloff,[23] and Kempson et al.[24] determined that the degenerated cartilage on subchondral bone was more compliant than normal cartilage when loaded compressively with an indenter.

Hayes et al.[25] used a numerical analysis to derive a set of equations and constants that can be used to determine the Young's modulus and the Poisson's ratio of cartilage from indentation test measurements. In this case, the cartilage was modeled as a layer of linear-elastic material and an analysis was applied to the two extremes of an indentation curve: the initial response (immediately after loading) and the final deformation. Hayes et al.[25] derived two solutions for plane-ended and spherically ended indenters and their dependence on the contact radius, Poisson's ratio, and surface stresses were studied. Even though this model was not experimentally validated, researchers have used these equations to obtain material properties for articular cartilage. This model assumed the material to be homogeneous.

Even though an indentation test appears by definition to be a simple approach to quantify cartilage degeneration or evaluate its mechanical properties, there are many difficulties inherent to indentation measurements consequently leading to unreproducible and inaccurate data. Therefore, certain points must be addressed when designing an indentation test. These include: (1) design or selection of an apparatus that can accurately measure cartilage deformation accounting for the geometry of the indenter, (2) the selection of loading conditions such as the rate of loading and the load applied, (3) the environmental conditions during testing, and (4) the information to be extracted from the curves. The following conditions should be respected during indentation testing:

- The load must be applied perpendicularly to the cartilage surface in order to obtain an axially symmetric loading on the tissue. In this respect, tare loads used to facilitate the alignment of the indenter deform the cartilage surface and should be carefully monitored since they will modify the resulting indentation even though very small.
- Given that the cartilage surface is curved, alignment of a plane-ended indenter is very difficult, and leads to the formation of high stress concentrations resulting in cartilage damage. The use of spherically-ended indenters allows for a better alignment and for a constant contact with the surface and does not cause stress concentrations.
- The load applied on the surface should remain constant during the indentation to minimize variation in indentation depth due to load application. Load application techniques should be selected to minimize excessive loads due to inertia.[26] Swann and Seedhom[26] have recommended the use of linear bearings, and Athanasiou et al.,[13] air bearings. The rate of loading must also be controlled since cartilage is a viscoelastic material. The deformation therefore is dependent on its loading rate, so it will appear stiffer loaded at high strain rates than at low strain rates.
- A common limit of most commercial testing equipment is the precision of the equipment. As an example, a strain of 10% measured with a precision of 1% in a sample measuring 10 mm requires a displacement precision of ±0.01 mm, a common limit in most commercial testing equipment. Cartilage samples are normally in the range of 1 mm thick

requiring displacements of ±1 μm. Aspden et al.[4] proposed a computer-controlled mechanical testing machine for small samples of biological viscoelastic materials as articular cartilage. This apparatus allows displacement in steps of about 1 μm in a predetermined time.

- The simulation of physiological environmental conditions will allow for a closer representation of the behavior of the cartilage in its natural environment. Such systems should allow humidity and temperature to be controlled to simulate physiological conditions.[15,16] Elmore et al.[27] experimentally demonstrated that the deformation and recovery of cartilage depends on the fluid movement between the interstitial liquid and the environment. The accuracy of an indentation test is based on the presence of a liquid environment allowing for a recovery of the deformation after load removal. According to Tkaczuk et al.,[28,29] a change in properties might occur by alterations of the hydrochemical properties of the tissue after resection of the tissue. Therefore, immersion of the resected cartilage in physiological saline is the common practice during indentation tests.[15]
- The deformation of articular cartilage under compressive loads is influenced by its thickness. As discussed earlier, the cartilage lying on subchondral bone can be represented by an elastic layer connected rigidly to a rigid base. Therefore, the measurement of the cartilage thickness is an intrinsic part of an indentation protocol. Cartilage thickness can be measured on resected plugs.[9,30,31] However, sectioning results in changes in cartilage thickness as a result of dehydration or swelling depending if the tissue is in contact with air or with physiological saline. Rushfeldt and Mann[32] suggested a non-destructive method to measure the thickness of articular cartilage *in vitro* where a detailed thickness distribution of the articular cartilage in the human hip joint was obtained using an ultrasonic transducer. The cartilage thickness was measured with the pulse-echo "A-scan" ultrasonic technique, in which the ultrasonic transducer produced a pressure pulse that traveled through the immersion medium as a longitudinal wave. As the wave passes through interfaces between media of different acoustic impedances (such as cartilage surface, calcified cartilage, and bone) reflections return to the transducer and generate electrical signal in the transducer proportional to intensity. However, other authors have reported several difficulties when using ultrasound to measure cartilage thickness.[40] The use of a needle attached to a micrometer and inserted in the tissue was also proposed as a measurement technique.[33]

C. Lubrication of Articular Cartilage and Friction Measurement

The synovial joint is a highly efficient bearing with two major types of lubrication regimes: fluid lubrication and boundary lubrication. In conditions of low speed and high load, fluid lubrication mechanism is assisted by boundary lubrication achieved in a layer of molecules attached to the surface of cartilage also known as surfactant.[34] Low coefficients of friction (μ) have been measured in healthy joints (from cadavers) with values ranging from 0.0044 to 0.042, suggesting that fluid film would be the primary mode of lubrication.[40] Yet, the operating conditions seem unfavorable to fluid film lubrication due to impact loads and oscillating motion, conditions better supportive of boundary lubrication as observed by Barnett and Cobbold.[36] Additionally, the coefficient of friction decreases progressively as the load on the joint increases.[37] This situation is a classic indication of boundary lubrication. The impairment in the cartilage surfactant layer or its boundary lubrication properties would be an indication of a degeneration of the tissue.[38] Experimental set ups used to assess the tribological properties of cartilage tissue have mainly associated with normal cartilage. Animal or human articular surfaces are articulated against a mating cartilage surface or an artificial material such as metal or glass. Oscillatory (pin-on-disc system) or reciprocating motion is used. The tissue is normally kept under lubricated conditions in an environmental chamber simulating physiological temperature and atmosphere. Lubricants such as physiological saline, synovial fluid obtained from the joint at resection, and culture media typically are used. A compressive load is applied approximating physiological stresses, and the friction test is conducted at 1 Hz or other frequencies of physiological relevance. Test duration or number of cycles should be selected to provide the best assessment of the frictional behavior of the articular surface without leading to excessive post-mortem degeneration of the tissue.[39] Static and kinetic coefficients of friction are measured using different means, load cells and force transducers being commonly used.[40]

D. Use of Indentation and Friction Measurements in an Orthopaedic Animal Model

A study conducted by Hall[40] involved the use of indentation and frictional measurements to characterize the long term effect of an injectable non-steroid anti-inflammatory drug (NSAID), ketorolac tromethamine (Syntex Laboratories, Palo Alto, CA), used intra-articularly on the properties of lapine knee cartilage. In this model, the mechanical properties and tribological properties, focusing on the boundary lubrication, of articular cartilage were compared to its microstructure.

A total of six one month old NZW female rabbits were used for the study. Both knees were shaved and prepped with an alternating series of absolute alcohol and Betadine solution. The right knee was injected with a 0.5 cc solution of 30mg/ml NSAID and the left knee with 0.5 ml of sterile Ringer's solution. The rabbits were housed for 12 months, exercised weekly, and then euthanized in a CO^2 chamber following anesthesia and injections of xylazine (0.15 ml/kg body weight) and Ketaset (0.2 ml/kg body weight), according to the standards of the American Veterinary Medicine Association. Two additional 12 month old female rabbits were euthanized using the protocol described previously and used as controls. The guidelines of the PHS guide for care and use of laboratory animals were followed during the study. The knees were then resected. Each tibio-femoral joint was taken undisturbed and placed in a bag of sterile saline. The joints were then cleaned and all extraneous tissue removed, only leaving the bone, cartilage, and a few ligament attachments. The condyles were separated along the patellar groove. The testing procedures were conducted on fresh condyles.

An indentation test was conducted in order to evaluate the effect of NSAID treatment on the compressive behavior of articular cartilage. In this study, a desktop tensile testing machine was used for indentation testing (Vitrodyne V1000, Chatillon, NC). This benchtop testing system is designed to measure the mechanical properties of materials on a small scale as compared to standard testing specimens. This equipment consists of a precision linear actuator located on a testing stand, a control and interface module, interchangeable load cell, and custom computer software for machine control and data analysis. A precision linear actuator is mounted on a high-strength testing frame and used to convert rotary motion of a screw to slow linear motion of the main loading axis. The actuator is coupled with an optical position encoder to assure a displacement resolution of 1.0 µm for a total travel distance of 7.5 cm. The system can be used with nine interchangeable load cells, with maximum capacities from 30 grams to 10 kg with a force measuring accuracy of ±0.5% (2 to 100% of load cell capacity) and a load cell resolution of 0.05% of load cell maximum capacity. Testing can be conducted with a strain rate on the order of 10 µm/sec up to 200 mm/sec. Standard grips for test specimens can support testing in tension, compression, and bending. Specimens can be submersed in fluids or mounted in an environmental chamber. This system is fully computer controlled. The software specifically designed for this system features real-time graphic display of test results and supports standard tests such as load-displacement, stress-strain, stress relaxation, and continuous dynamic cycling at fixed or scanning frequency. Testing was performed in physiological saline at room temperature. Each test was completed on six different condyles for both the NSAID and saline injected knees, as well as four condyles for the control group.

An acrylic specimen holder mounted on a bearing system was designed to allow rotation of the condyle fixed with polymethylmethacrylate (Simplex, Howmedica) to any angle that would allow a perfect alignment of the indenter on the surface and to locking into place for testing. One indentation was performed on the selected condyles using a highly polished titanium alloy (Ti-6Al-4V) spherically ended cylindrical indenter with a radius of 1.5 mm. Load-displacement curves were obtained for a maximum indentation depth of 500 µm at a rate of 10 µm/sec. The system was activated and the indenter was positioned perpendicularly to the cartilage surface with a maximum preload of 0.5 g. After testing, the cartilage thickness was measured and the specimens were fixed in 10% buffered formalin solution for soft tissue histological preparation. The cartilage thickness was measured using techniques proposed by others.[33]

A pin-on-disc friction apparatus was specifically designed to assess the effect of intra-articular injection on the friction coefficient of the cartilage of remaining femoral condyles (Figure 1). This system allows for a normal force to be exerted along the vertical axis providing a point contact geometry between the specimen and bearing surface. The support does not hinder any measurement because the lubricant and bearings in the actuator housing allow for near frictionless movement and allow normal force to be applied directly to the specimen. The shoulder joint which is the hinge upon which the specimen rotates is also fitted with bearings and coated with lubricant, in order to reduce any frictional or external forces. A load cell (1000G, No. 360817, Omega Inc.) is fastened securely on the stationary plate in line with the axis of the actuator in order to provide a horizontal force, which can be regarded as the resistance of movement of the specimen, thus the frictional force. The voltage of the load cell is transmitted to a data acquisition package (Labview 3, National Instruments, AZ) and converted into gram force. A magnetically geared motor (Model 4Z728A, Dayton Electric Inc., IL) was used and tests were conducted at a frequency of 1 Hz. The frictional force was monitored for a total travel distance of 50 m. Highly cleaned flat glass plates (mean roughness average [Ra] = 0.72±0.09 nm, Topo-3D, Wyko Corp., 20X magnification head) were used as counterpart surfaces. A new cleaned plate was used for each test. Friction tests were conducted in a Plexiglass chamber at room temperature in a saturated nitrogen environment using normal saline as a lubricant to prevent cartilage dehydration. Tests were conducted at an entraining velocity of 30 mm/sec with a load of 600 g providing an average contact stress of 2.75 MPa. The calibration curve for the load cell was used to determine the frictional resistance based on the voltage output recorded during each test. The use of these values enabled the calculation of the coefficient of friction for each treatment. The signals were filtered through a low-pass resistor/capacitor filter in order to minimize the amount of low frequency noise.

A structural analysis was conducted to characterize the cellularity and the structural integrity of the test specimens. Histological sections of femoral condyles were stained with a safranin-O stain and hematoxylin-eosin stain. The image analysis of the histological sections was conducted by mounting a video camera (Sony, AVC-D7), powered by a camera adapter (Sony, MA-D2, Japan) to a large Zeiss Universal Research Microscope (Oberkochen, Germany) to capture the image observed under the microscope. The captured image was transferred to a monitor (Panasonic, WV-5410) and sent to a computer (Macintosh Quadra 900) to be digitized and saved. The actual cellularity determination was done by using the threshold mechanism in the imaging software package (Adobe Photoshop and NIH Image). The grayscale image could be thresholded into a density pattern and the nucleated cells would be differentiated from the rest of the tissues. The software could then tally the number of threshold areas which gives the number of cells per image. This analysis was conducted for 20 sections on all samples.

Results have shown that the control condyles required an average of 249.6±19.3 gm, an average of 81.3+23.1 gm for the saline group, and 123.4±54.3 gm for the NSAID treated group for 20% deformation. The NSAID treated specimens were statistically more compliant than the control specimens but not different from the saline treated surfaces (ANOVA, 95% confidence interval).

For friction data analysis, the static coefficient of friction was chosen to be an average value of the first measurement for each experiment in one group. The kinetic coefficient of friction was defined as the average value of the remaining linear points in the data set and was evaluated at

Mechanical Testing of Cartilage

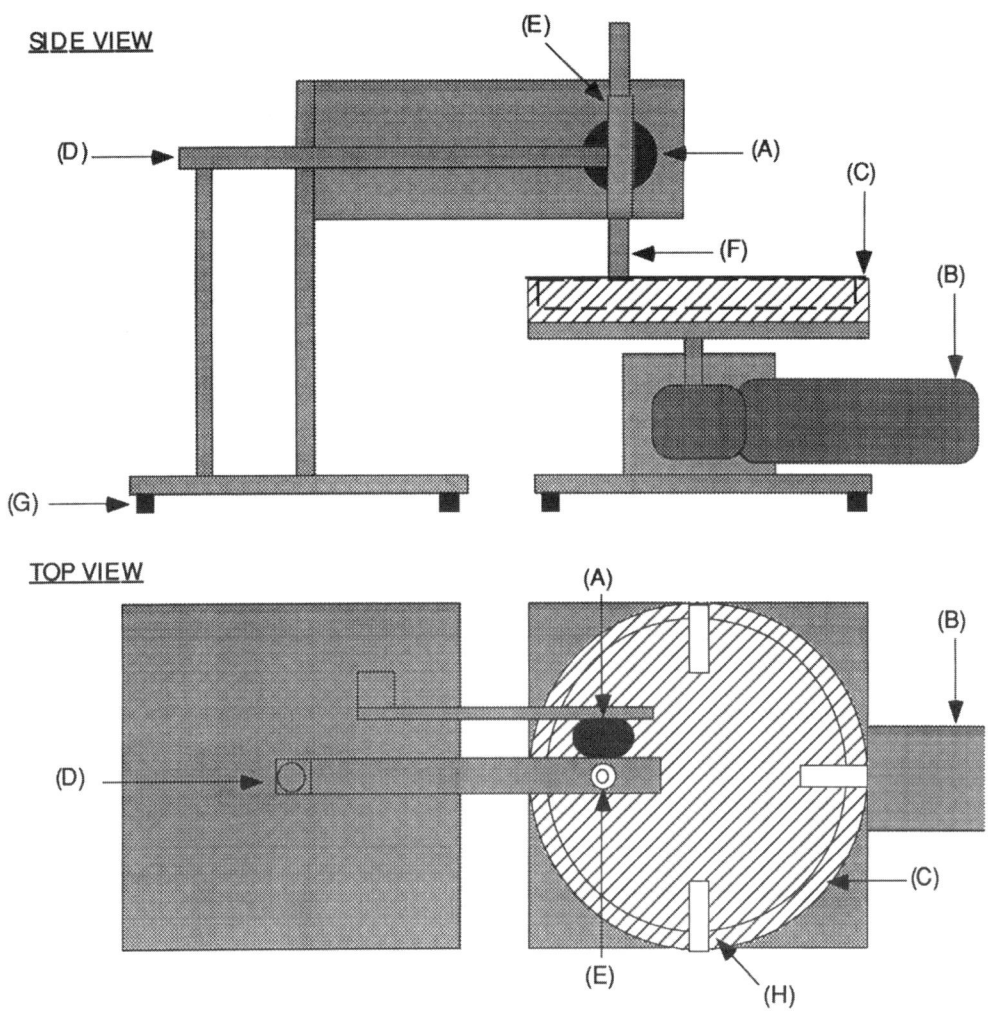

FIGURE 1. Schematic representation of the pin-on-disc friction apparatus specifically designed for articular cartilage testing: (A) 1000G load cell, (B) magnetic geared motor, (C) acrylic bowl and plate holder, (D) lubricated shoulder joint, (E) lubricated casing for actuator, (F) actuator and specimen holder (Dayton Electric Co.), (G) rubber feet, and (H) acrylic clamps for counterpart plate.

10 m sliding distance intervals. The statistical analysis (ANOVA, 95% confidence interval) revealed no significant differences between the test specimens, saline and NSAID, and the control specimens for the static coefficient of friction. However, during the kinetic phase the NSAID had a much higher average than the control and saline. This trend was followed throughout the linear region of the trial. A significant difference was determined for all of these cases. The difference remained significant as a function of time, and the NSAID samples remained higher than both the control and the saline samples. No statistical difference was observed between the saline and control specimens.

A statistically significant hypercellularity was observed for the NSAID specimens compared to the other groups. No statistical difference was observed for cellularity between the saline treated and control condyles. Overall, saline and NSAID groups had only 80% of the area positively stained for PGs as compared to controls. However, the superficial zone of NSAID specimens was abnormally positively stained for PGs. A decrease in stiffness was also correlated to a decrease in overall positive staining for PGs.

In conclusion, the properties of articular cartilage were successfully investigated using indentation and friction analysis. Results have shown that the intra-articular injection of ketorolactromethamine affects the structural properties of articular cartilage, associated with an alteration in the positive staining for proteoglycans in the matrix and cellularity, as well as its mechanical and tribological properties as compared to normal cartilage. The characterization of the tribological properties, mainly friction coefficient, allows for the evaluation of the response of articular cartilage to a specific experimental treatment.

E. *In Vivo* Indentation Testing

The indentation test remains a useful tool for assessing the compressive stiffness of cartilage, especially in survey work where the objective is to compare between cartilage stiffness rather than to obtain absolute values of intrinsic mechanical properties.[26] Indentation tests can be classified into two types: *in vitro* or *in vivo* indentation testing. Most of the mechanical apparatus and systems proposed by authors for the indentation of articular cartilage involve *in vitro* testing following the resection of the specimens as discussed previously. During *in vitro* testing, the post-mortem degeneration of cartilage should be minimized and mechanical tests performed in a few hours to resection or preserve in conditions that would not affect its properties.[41,42] Since indentation tests normally are performed on test and control specimens in orthopedic animal models, the preservation technique selected should be used for both groups of specimens.

Tkaczuk et al.[28] and Tkaczuk[29] pointed out the importance of performing mechanical indentation testing of cartilage in its physiological environment *in vivo* and proposed a cartilage elastometer for use on articular joints during orthopedic procedures. The apparatus essentially consisted of a device for attaching the elastometer to the bone, and force and displacement transducers. This procedure involved an arthrotomy of the joint to fix the elastometer to the bone. The deformation of cartilage as a response to compressive loading required breaking through the cartilage surface for measurement and, therefore, can be considered a destructive method. LaBerge et al[16] developed a non-destructive invasive method for *in situ* indentation testing to quantify cartilage deformation following an arthrotomy. The degeneration of articular cartilage in a canine closed chondromalacia patellae model was monitored quantitatively as a function of time. The portable autoclavable indentation system consisted of a load application device, a displacement measuring device, and a positioning device. The system subtracted out the deformation of the surrounding soft tissue and provided a more realistic measurement of the cartilage deformation. Depths of indentation and hardness of cartilage have been measured with an accuracy of 0.005 mm and a repeatability of 0.6%. A data acquisition system allowed to monitor cartilage deformation under a spring loaded hemispherically ended indenter.

A non-destructive approach was used by Dashefsky[17] to measure the chondromalacia of the medial patellar facet. Under arthroscopic control, a microminiature pressure transducer was used to measure the resistance of cartilage to a predetermined amount of deformation. This technique was used in a clinical trial and intended to provide standard data to measure patellar softening. The advantage of this innovative technique is the use of arthroscopy. However, the testing technique proposed by Dashefsky[17] did not provide a method to measure the angle between the probe and the cartilage, or the positioning of the device. Also, the use of a pressure transducer suffers intrinsic limitations. The required force (input) to apply a pressure should be kept constant and monitored, and the pressure (output) should be measured as a function of time since the cartilage is a viscoelastic material. The non-monitoring of these two factors can jeopardize the data interpretation and comparison. An indenter for use during arthroscopic procedures that allows testing of cartilage stiffness under minimally invasive procedures was also proposed by Lyrra et al.[18] This instrument imposes a constant deformation on the cartilage and the maximal indentation force by which the cartilage resists the deformation, is used as a measure for cartilage thickness. This instrument was also tested on cadaver knees and on elastomeric sheets and provided reproducible data.

III. CONCLUSION

Despite the fact that articular cartilage is a very complex material in terms of structure and mechanics, several mechanical tests can be performed on articular cartilage to assess its integrity or evaluate the effects of orthopedic procedures or implant materials on its behavior. Indentation testing of cartilage has been demonstrated by several authors as an effective method to observe the response of cartilage to compressive loading. The design of an indentation protocol should allow for reproducibility and accuracy, and should take into consideration the time-dependent response of the tissue, its biphasic nature, and its geometry (thickness and curvature).

REFERENCES

1. Mow, V. C., Ratcliffe, A., and Poole, R. A., "Cartilage and diarthrodial joints as paradigms for hierarchical materials and structures," *Biomaterials*, 13, 67, 1992.
2. Kwan, M. K., Lai, W. M., and Mow, V. C., "A finite deformation theory for cartilage and other soft hydrated connective tissues, I. Equilibrium results," *J. Biomech.*, 23, 145, 1990.
3. Woo, S. L.-Y., Lubock, P., Gomez, M. A., Jemmott, G. F., Juei, S. C., and Akeson, W. H., "Large deformation nonhomogeneous and directional properties of articular cartilage in uniaxial tension," *J. Biomech.*, 12, 437, 1979.
4. Aspden, R. M., Iardodon, T., Svensson, R., and Heinegård, D., "Computer-controlled mechanical testing machine for small samples of biological viscoelastic materials," *J. Biomed. Eng.*, 13, 521, 1991.
5. Suh, J.-K., "Dynamic unconfined compression of articular cartilage under a cyclic compressive load," *Biorheology*, 33, 289, 1996.
6. Mow, V. C., Kuei, S. C., Lai, W. M., and Armstrong, C. G., "Biphasic creep and stress relaxation of articular cartilage: theory and experiments," *J. Biomech. Eng.*, 102, 73, 1980.
7. Armstrong, C. G. and Mow, V. C., "Variations in the intrinsic mechanical properties of human cartilage with age, degeneration, and water content," *J. Bone Joint Surg.*, 64A, 88, 1982.
8. Kwan, M. K., Wayne, J. S., Woo, S. L.-Y., Field, F. P., Hoover, J., and Meyers, M., "Histological and biomechanical assessment of articular cartilage from stored osteochondral shell allografts," *J. Orthop. Res.*, 7, 637, 1989.
9. Kwan, M. K., Hacker, S. A., Woo, S. L.-Y., and Wayne, J. S., "The effect of storage on the biomechanical behavior of articular cartilage — A large strain study," *J. Biomech. Eng.*, 114, 149, 1992.
10. Mak, A. F., "The apparent viscoelastic behavior of articular cartilage: the contributions from the intrinsic matrix viscoelasticity and interstitial fluid flows," *J. Biomech. Eng.*, 108, 123, 1987.
11. Mow, V. C., Gibbs, M. C., Lai W. M., Zhu, W. H., and Athanasiou, K. A., "Biphasic indentation of articular cartilage II. A numerical algorithm and an experimental study," *J. Biomech.*, 22, 853, 1989.
12. Athanasiou, K. A., Rosenwasser, M. P., Buckwalter, J. A., Malinin, T. I., and Mow, V. C., "Interspecies comparisons of *in situ* intrinsic mechanical properties of distal femoral cartilage," *J. Orthop. Res.*, 9, 330, 1991.
13. Athanasiou, K. A., Agrawal, A., and Dzida, F. J., "Comparative study of the intrinsic mechanical properties of the human acetabular and femoral head condyle," *J. Orthop., Res.*, 12, 340, 1994.
14. Freeman, M. A. R. and Meachim, G., *Aging and Degeneration, in Adult Articular Cartilage*, Pitman Medical, New York, 1981, 487.
15. LaBerge, M., Bobyn, J. D., Drouin, G., and Rivard, C. H., "Evaluation of metallic personalized hemiarthroplasty: a canine patellofemoral model," *J. Biomed. Mater. Res.*, 26, 239, 1992.
16. LaBerge, M., Audet, J., and Drouin, G., "Structural and *in vivo* mechanical characterization of canine patellar cartilage: a closed chondromalacia patellae model," *J. Invest. Surg.*, 6, 105, 1993.
17. Dashefsky, J. H., "Arthroscopic measurement of chondromalacia of patellar cartilage using a microminiature pressure transducer," *Arthroscopy*, 3, 80, 1987.

18. Lyrra, T. Jurvelin, J., Pitkänen, P., Väätäinen, U., and Kiviranta, I., "Indentation instrument for the measurement of cartilage stiffness under arthroscopic control," *Med. Eng. Phys.*, 17, 395, 1995.
19. Johnson, K. L., *Contact Mechanics*, Cambridge University Press, New York, 1985.
20. Finkin, K. F., "The determination of Young's modulus from the indentation of rubber sheets by spherically tipped indentors," *Wear*, 19, 277, 1972.
21. Hori, R. Y. and Mockros, L. F., "Indentation tests of human articular cartilage," *J. Biomech.*, 9, 259, 1976.
22. Hirsch, C., "A contribution to the pathogenesis of chondromalacia of the patella," *Acta Chirurgica Scand.*, 90, 1, 1944.
23. Sokoloff, L., "Elasticity of aging cartilage," *Proc. Fed. Am. Soc. Exp. Biol.*, 25, 1089, 1966.
24. Kempson, G. E., Freeman, M. A. R., and Swanson, S. A. V., "The determination of a creep modulus for articular cartilage from indentation tests on the human femoral head," *J. Biomech.*, 4, 239, 1971.
25. Hayes, W. C., Keer, L. M., Herrman, G., and Mockros, L. F., "A mathematical analysis for indentation tests of articular cartilage," *J. Biomech.*, 5, 541, 1972.
26. Swann, A. C. and Seedhom, B. B., "Improved techniques for measuring the indentation and thickness of articular cartilage," *Proc. Inst. Mech. Eng.*, 203, 143, 1989.
27. Elmore, S. M., Sokoloff, L., Norris, G., and Carmeci, P., "Nature of imperfect elasticity of articular cartilage," *J. Appl. Physiol*, 18, 329, 1963.
28. Tkaczuk, H., Norrbom, H., and Werelind, H., "A cartilage elastometer for use in the living subject," *J. Med. Eng. Tech.*, 6, 104, 1982.
29. Tkaczuk, H., "Human cartilage stiffness," *Clin. Orthop.*, 206, 301, 1986.
30. Parsons, J. R. and Black, J., "The viscoelastic shear behavior of normal rabbit articular cartilage," *J. Biomech.*, 10, 21, 1977.
31. Jurvelin, J. S, Buschmann, M. D., and Hunziker, E. B., "Optical and mechanical determination of Poisson's ratio of adult bovine humeral articular cartilage," *J. Biomech.*, 30, 235, 1997.
32. Rushfeldt, P. D., Mann, R. W., and Harris, W. H., "Improved techniques for measuring in vitro the geometry and pressure distribution in the human acetabulum. Ultrasonic measurement of the acetabular surfaces, sphericity and cartilage thickness," *J. Biomech.*, 14, 253, 1981.
33. Hoch, D. H., Gridzinsky, A. J., Knob, T. J., Albert, M. L., and Eyre, D. R., "Early changes in material properties of rabbit articular cartilage after meniscectomy," *J. Orthop. Res.*, 1, 4, 1983.
34. Ogston, A. G. and Stanier, J. E., "The physiological function of hyaluronic acid in synovial fluid: viscous, elastic, and lubricant properties," *J. Physiol.*, 119, 244, 1953.
35. Roberts, B. J., Unsworth, A., and Mian, N., "Modes of lubrication in human hip joints," *Ann. Rheum.*, 41, 217, 1982.
36. Barnett, C. H. and Cobbold, A. F., "Lubrication within living joints," *J. Bone Joint Surg.*, 44B, 662, 1962.
37. Freeman M. A. R., Swanson S. A. V., and Manley P. T., "Stress lowering function of articular cartilage," *Med. Biolol. Eng.*, 13, 245, 1975.
38. Williams, P. F., Powell, G. L., and LaBerge, M., "Sliding friction analysis of phosphatidylcholine as a boundary lubricant for articular cartilage," *J. Eng. Med. (H)*, 207, 59, 1993.
39. LaBerge, M., Rogers, J. M., and Medley, J. B., "Friction and Wear," in *Handbook of Biomaterials Evaluation: Scientific, Technical, and Clinical Testing*, A. F. von Recum Ed., Taylor & Francis, Bristol, PA, 1998, Chapter 4.
40. Hall, H. T., IV, "Structural, mechanical, and tribological response of lapine articular cartilage to intra-articular keterolactromethamine," Master's Thesis, Clemson University, Clemson SC, 1995, 101.
41. Kawabe, N. and Yoshinao, M., "Cryopreservation of cartilage," *Int. Orthop.*, 14, 231, 1990.
42. Kiefer, G. N., Sundby, K., McAllister, D., Shrive, N. G., Frank, C. B., et al., "The effect of cryopreservation on the biomechanical behavior of bovine articular cartilage," *J. Orthop. Res.*, 7, 494, 1989.

10 Mechanical Testing of Ligaments and Tendons

Savio L-Y. Woo, Theodore T. Manson, and Tracy M. Vogrin

CONTENTS

- I. Introduction ... 175
- II. Biomechanical Properties of Ligaments and Tendons ... 176
- III. Experimental Factors ... 178
  - A. Stress Measurements: The Determination of Cross-Sectional Areas ... 178
  - B. Determination of Strain ... 178
  - C. Clamping of Testing Specimens ... 179
  - D. Effects of Specimen Orientation ... 180
  - E. Strain Rate ... 180
- IV. Biological Factors ... 180
  - A. Effects of Anatomical Location and Functional Role ... 180
  - B. Effects of Maturation and Age ... 181
  - C. Effects of Immobilization and Exercise ... 181
- V. Environmental Factors ... 183
  - A. Effects of Dehydration ... 183
  - B. Effects of Temperature ... 183
  - C. Effects of Freezing, Storing and Thawing ... 183
- VI. Viscoelastic Properties of Ligaments and Tendons ... 184
- VII. Determining the Functional Role of Ligaments/Tendons ... 184
  - A. The Robot/UFS Testing System ... 187
- VIII. Summary and Conclusion ... 188
- Acknowledgments ... 188
- References ... 189

I. INTRODUCTION

Ligaments and tendons are soft connective tissue structures which work in conjunction with each other to stabilize a single synovial joint. They are composed primarily of water and parallel oriented collagen fibers. The collagen is primarily Type I (approximately 70% dry weight), a small amount (3–10%) of Type III, and minute amounts of Types V, X, XII and XIV. Other non-collagenous proteins are present as well, including proteoglycans and elastin.

The longitudinal arrangement and crimped nature of the collagen fibrils serve to both guide joint motion and provide restraint at extremes of motion; in other words, at low loads the crimp pattern straightens easily so that joint motion is guided. At higher loads, the fibrils are straightened and loaded in tension, thus restraining excessive joint motion.

In this chapter, various methods to determine the biomechanical properties of ligaments and tendons will be discussed. With respect to animal models, the effects of experimental, biological

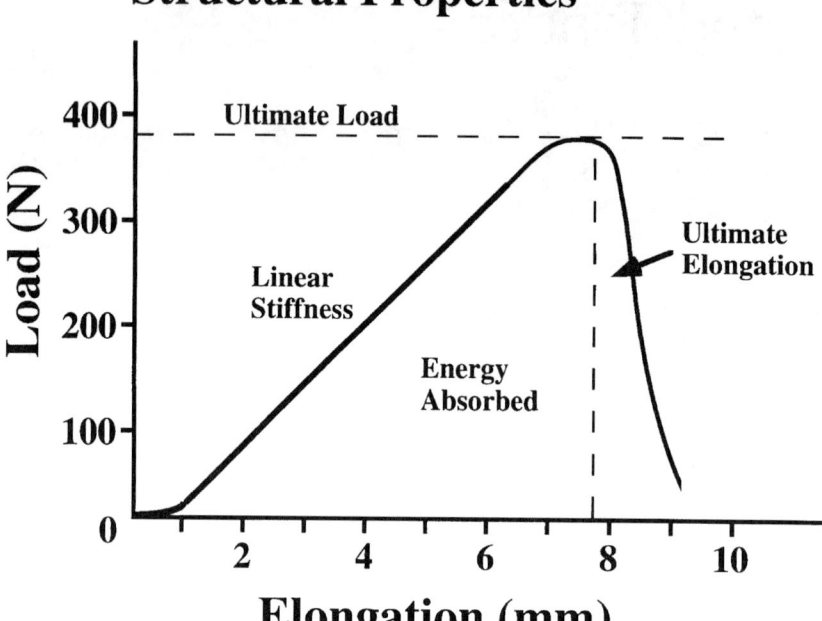

FIGURE 1. Typical load-elongation curve of a bone-ligament-bone complex.

and environmental factors on these properties will be reviewed. The flexor tendons of the hand and foot and the MCL of the knee are the two primary examples that will be our focus in this chapter because they are among the most studied; however, there is considerable information available on other ligaments and tendons as well, from shoulder to spinal ligaments to patellar and extensor tendons.

II. BIOMECHANICAL PROPERTIES OF LIGAMENTS AND TENDONS

Mechanical testing is used to elucidate the inherent mechanical properties of a material. Since ligaments and tendons are primarily loaded in tension, the result of a tensile test, the load-elongation curve is used to obtain ligament and tendon uniaxial properties. In the following discussion, the testing of ligaments will be used as examples. From the data curves, *structural* properties of a bone ligament bone complex and *mechanical* properties of the tissue substance can be calculated.

When we elongate a bone-ligament-bone complex in the materials testing machine, we simultaneously measure the load in this structure corresponding to that elongation from its resting length. Plotting the elongation as the independent variable and load as the dependent variable, a non-linear load-elongation curve is obtained (Figure 1). From this curve, important properties related to the entire bone-ligament-bone complex as a whole can be inferred. These properties include the linear stiffness (in Newtons/mm), the ultimate load (in Newtons), the ultimate deformation (in mm), and the energy absorbed at failure (in Newtons-mm).

The *mechanical* properties of the ligament substance can also be determined from the same uniaxial tensile test by normalizing both the load and elongation. When the load is normalized by the cross-sectional area to account for different widths and thicknesses, a quantity known as the

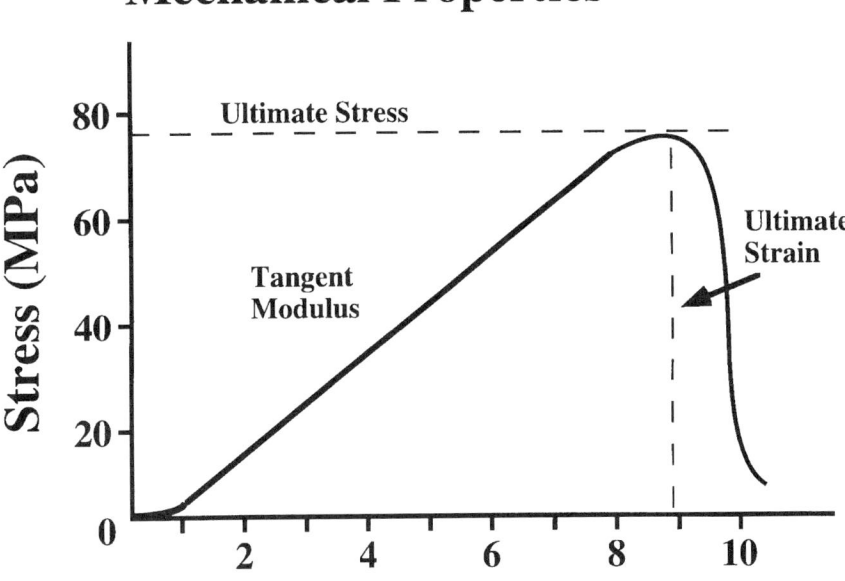

FIGURE 2. Typical stress-strain curve describing the mechanical properties of the ligament substance.

stress in the tissue is defined. Similarly, when the elongation of the specimen is normalized by the resting length, or "gage length" of the specimen, a quantity known as the strain in the tissue is defined. Plotting the strain as the independent variable and the stress as the dependent variable, a nonlinear stress-strain curve can be obtained for the ligament substance as shown in Figure 2. Between the range of strain 1–8% in this figure, the stress produced in the tissue is linearly proportional to the strain and the slope of the curve in this region is defined as the tangent (or Young's) modulus of the material. A ligament (or tendon) with more collagen cross linking or stouter collagen fibrils will be more difficult to elongate or strain and will show a *higher* Young's modulus.

The reason that the linear portion of the stress-strain curve is used to define Young's modulus is that any elongation or strain is reversible. Unloading the specimen will cause the specimen to contract back to near its initial length. There is no permanent change in the material. At around 8% strain, the stress/strain curve shows the stress/strain curve becoming nonlinear. If stressed to this high degree, the tissue will not return to its previous resting length after relaxing the load. This is known as the yield strength of the tissue. When the stress in a ligament exceeds this point, the effectiveness of the ligament is reduced because the ligament now has to be elongated more before it will start to resist the elongation. The point on the stress-strain curve where the material actually fails is known as the tensile strength of the material and represents the maximum stress that the material can sustain prior to failure.

Much work has been performed in the determination of these biomechanical properties. However, the task of measuring many of these properties is challenging. There are many experimental and biological factors that can affect the outcome. In the following sections, these factors will be discussed in detail.

III. EXPERIMENTAL FACTORS

A. Stress Measurements: The Determination of Cross-Sectional Areas

The cross sectional areas of ligaments and tendons are very difficult to measure because they are irregular in shape and also soft and deformable. The literature is divided on the approach to use, consisting of contact and non-contact methods. The contact methods include such methodologies as using calipers to measure the width and thickness of the specimen,[1] forming molds of the specimen,[2,3] and the area micrometer system. The area micrometer forces a ligament into a slot of known width. The micrometer mounted indenter is used to compress the specimen into a rectangular shape to measure its height. As with other contact methods, the measurements are dependent on the amount of pressure applied to the specimen.[4-6] Due to the fact that these methods alter the shape and cross section of the soft tissue during measurement, other investigators have preferred newer methods based on using non-contact technology. Examples of this technology include the shadow amplitude method[5] and the profile method.[4,7,8]

In our research center, a non-contact method using a laser micrometer system has been developed.[9,10] Utilizing a collimated laser beam field and a background screen detector, we can accurately map the surface profiles of a ligament or tendon. The data is reconstructed such that an accurate assessment of specimen cross sectional shape and area can be obtained. A study comparing this technology to digital caliper and area micrometer measurements determined that while the cross-sectional area measurements (of the relatively flat medial collateral ligament (MCL) using an area micrometer were 21% less than those using the laser micrometer while the digital calipers overestimated the cross sectional area by 2.3%.[9] Having proven to be accurate and reproducible, the laser micrometer system presents an excellent alternative to contact methods of cross-sectional area measurement. The one limitation of this system is its inability to measure the concavities present on the ligament surface such as those in the anterior cruciate ligament (ACL). While the number of ligaments exhibiting this geometry is minimal, additional methodologies utilizing laser reflectance transducers have also been developed.[11] The laser reflectance transducer uses an emitter to project a 1mm laser beam onto the surface of the specimen. A receiver collects the laser light reflected off of the specimen while the whole system is rotated 360 degrees around the specimen (Figure 3). From this data, accurate cross-sectional shape can be determined so that it can be integrated to determine the cross-sectional area. Since this system takes into account concavities present in the specimen, it does provide a realistic reconstruction of cross-sectional shape.

B. Determination of Strain

In order to effectively measure the strain present in soft tissues during a uniaxial tensile test, accurate measurements of the initial length of the specimen and the elongation at any point during the test must be made. Many different methodologies have been used to measure this elongation.[6,12-17] Literature use of percent elongation based on the clamp to clamp distance involved contributions of not only the ligament itself, but also its insertions. To measure the strain in the tissue, however, measurement of the strain in the ligament substance must be made. The technology available to measure tissue strain can again be divided into contact and non-contact designs.

Contact methods of measuring strain in ligament or tendon tissue include the use of strain gauges and Hall effect transducers.[12] Liquid mercury strain gauges were used by some researchers to measure the percent elongation in knee ligaments.[18] Both of these contact methodologies again involved direct interference with the tissue during testing, which may introduce errors into the measurements.

In our research center and others, non-contact methods of measuring strain have been used. Due to the fact that continuous measurements of tissue length must be made, two systems have enjoyed widespread use. The first is the video dimension analyzer (VDA) system.[19,20] In the use of this system, reference lines are marked perpendicular to the loading axis with Verhoff's elastin

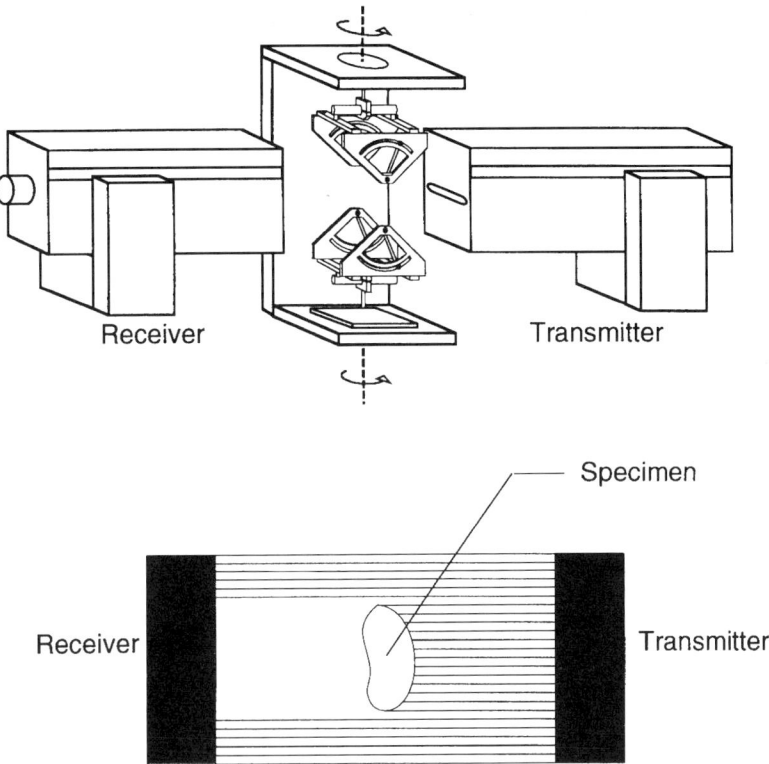

FIGURE 3. Schematic of the laser micrometer. (From Livesay, G. A., et al., Anatomy and biomechanics of the human posterior cruciate ligament, in *Clinical Biomechanics and Related Research*, Livesay, G. A., et al., Eds., Springer-Verlag, Tokyo, 1994, 200. With permission.)

stain. The specimen is then elongated in a materials testing machine while a video image of the specimen is captured. Using the VDA system, the videotape is played back and threshold "windows" placed over each reference line. The VDA hardware is able to track these thresholded lines and generate a continuous output voltage depending on the distance between the lines. Dividing the varying output voltages by the initial output voltage, the strain present in the tissue at any particular time can be calculated.[19]

A second system used is the Motion Analysis System (Motion Analysis, Santa Rosa, CA).[21-23] Consisting of a CCD camera and an image processing system, a video image file is captured from each of the three cameras over time. Marks made on the tissue using high contrast stain can be tracked over time by thresholding their outlines and then tracking these outlines over time. Ease of use as well as a reduction in data analysis make this an attractive system for non-contact strain measurement.[23]

Some additional advantages of a non-contact system are that midsubstance strains can be measured independently of those at the insertion sites. In addition, with careful stain marking, regional variation of strain can be measured.[23]

C. Clamping of Testing Specimens

The ligament or tendon to be tested has to be mounted in the tensile testing machine in such a way that the specimen is gripped without slippage at the clamps. In the past, many different methods have been devised for gripping directly to ligament and tendon tissue including sinusoidal[24] and cryo-clamp designs.[25] Even when this slipping is prevented, however, the direct clamping of ligamentous tissue causes stress concentrations which may introduce premature failure at the clamps. These problems can be avoided by leaving the ligament or tendon insertions to bone intact, thus allowing the specimen to be gripped on the bone substance and eliminating the aforementioned difficulties.[19]

On the other hand, when the bone-ligament-bone complex fails, there are three main modes of failure. The first is a frank ligament or tendon substance tear. The second is by bony avulsion, where the bone adjacent to the insertion site fails, the ligament remaining attached to bony debris. The third is by soft tissue pullout at the epiphyseal region with no bony involvement.

D. Effects of Specimen Orientation

The structural properties of this bone-ligament-bone complex are very dependent on the direction of the applied load during testing. For instance during testing of the canine and rabbit femur-anterior cruciate ligament-tibia complex (FATC), it was shown that the structural properties of the ACL changed with knee flexion angle being greatest at 0° of flexion and the least at 90° of flexion.[26,27] We felt the differences are related to the uniformity of load distribution across the specimen. As our data for the rabbit FATC further revealed, load was either applied along the axis of the ACL (even load distribution) or along the axis of the tibia (uneven load distribution). For the specimens loaded along the ligament axis, structural properties were not dependent on knee flexion angle and most failures occurred by bony avulsion. In the specimens loaded along the tibial axis, structural properties varied with flexion angle and most failures occurred in the ligament midsubstance.[28]

E. Strain Rate

In addition to specimen orientation, another consideration is the strain rate with which the specimen is tested. In a study of both rabbit ACLs and patellar tendons, the mechanical properties were shown to vary with strain rate, although these differences were relatively small compared to other factors.[29,30] For example, a strain rate increase from 0.15%/sec. to 222%/sec. showed only minor effects on the resulting load-elongation and stress-strain curves. In the case of a higher strain rate, some increase in the tensile strength of the specimens was noted.[29]

Many experimental factors contribute to the behavior shown by ligaments and tendons during testing, and careful attention must be paid to ensure elucidation of the correct mechanical properties.

IV. BIOLOGICAL FACTORS

It is well documented in the literature that the morphological, biomechanical and biochemical properties of soft tissue are sensitive to the tissue's environment. Factors such as maturation and age, immobilization and exercise, and especially the structure's anatomical location and functional role in the body will result in different properties in a ligament or tendon.

A. Effects of Anatomical Location and Functional Role

Although their biochemical compositions are almost identical and their morphologies similar, the reported mechanical properties for ligaments and tendons in the literature vary considerably. For example, the ultimate strain values for ligaments, which can range from 12% to over 50%, tend to be somewhat larger than those for tendons, which have been reported to range from 9 to 30%.[13,31–34]

A major factor contributing to this variability is the species and the anatomical location. For example, ultimate strain values on the order of 10–12% have been measured in the rabbit MCL and ACL,[34] the swine digital flexor tendon[35] and the tendons of the equine foreleg.[36] These values vary, however, as ultimate strains have been reported of 8.1% in equine superficial digital flexor tendon,[37] 6% in the swine digital extensor,[35] and just 1.6% in the ligamenta flava of pig lumbar spines.[38]

B. Effects of Maturation and Age

The effects of aging and maturation on soft tissues, including skin, ligament and tendon, are well known. In general, it appears that biomechanical properties of ligaments and tendons improve rapidly as the animals reach skeletal maturity. Vogel and Morein observed that ultimate load, as well as Young's modulus and tensile strength, increase during early maturity of the rat tail tendon.[39,40] Similarly, in the rabbit MCL, rapid increases were observed in cross sectional area, stiffness, and ultimate load.[20,41] An important finding to note is the mechanism of failure also changed with skeletal maturity; because the epiphyses of these young animals are not closed, failure of the ligament most often occurs by tibial avulsion. On the other hand, once the epiphyses are closed, the ligament is most likely to fail at its midsubstance.[41]

After skeletal maturity is reached, however, little change in the structural properties of the rabbit femur-MCL-tibia complex (FMTC) occur, even after senescence.[41] The structural properties, as well as the modulus, were almost constant after 12 months of age and decreased only slightly in the older animals. The human femur-ACL-tibia complex (FATC), on the other hand, does not follow this trend, as studies performed in our research center and others have demonstrated a significant decrease in structural properties of the FATC with age.[26,42]

C. Effects of Immobilization and Exercise

The effects of joint immobilization, as well as exercise, cause profound changes in ligament and tendon properties. It has been demonstrated that stress deprivation can result in pannus formation and cartilage necrosis in high-contact regions,[43] while causing erosion of cartilage in non-contact areas.[44]

The effects of immobilization followed by remobilization have been investigated in our research center.[45] Tensile testing of rabbit femur-MCL-tibia complexes was performed after nine and 12 weeks of immobilization as well as nine weeks of immobilization followed by nine weeks of remobilization. It was found that the nine and 12 weeks immobilized groups had ultimate loads of only 31% and 29%, respectively, of the contralateral controls ($p<0.01$), with all specimens failing by tibial avulsion. In the remobilized group, the mechanical properties of the MCL substance returned nearly to control values. However, structural properties of the FMTC remained inferior to the controls and the mode of failure was still by tibial avulsion.

A study by Newton et al. has reported that the cross-sectional area of the rabbit ACL is also significantly decreased after nine weeks of immobilization. He found no significant differences in mechanical properties, though he did note a 32–40% increase in strain in the immobilized joints.[46]

Interestingly, while stress deprivation causes profound detrimental effects in a relatively short amount of time, the positive contributions of exercise to the biomechanical properties of soft tissue are much less significant. Tipton and colleagues investigated this effect, observing a decrease in ligament water content, as well as a loss of waviness in collagen fibers, immediately post-exercise.[47] Subsequent studies have found minimal improvement of structural properties of bone-ligament-bone complexes after exercise.[48–53]

In our research center, we have found that the effects of exercise on ligaments and tendons may be dependent upon the soft tissue structure as well as its anatomical location. The effects of life-long exercise and concomitant aging on the mechanical properties of the MCL were evaluated in the beagle.[54] Nine animals were exercised at 3 km/hr, 75 min./day, five days/wk, while wearing an 11 kg backpack, for 9–12 yrs. Sedentary control groups were also used to evaluate the effects of aging. While aging was found to significantly reduce the tensile strength and strain at failure of the MCL substance, exercise was found to induce no significant differences in either the structural properties of the FMTC or the material properties of the ligament substance.

We also compared the effects of short term (3 months) and long term (12 months) exercise in the swine digital extensor and flexor tendons.[35,55] Animals were exercised by running at the speed of 6–8 km/hour for a total of 40 km/wk. In the extensor tendon of the forepaw, exercise had no significant effects on mechanical properties in the short term, but long term increases in cross-sectional area and tensile strength over those of age-matched, non-exercised controls were observed. On the other hand, in the flexor tendon, there were no significant effects on its mechanical properties nor cross-sectional area of the tissue substance, but there was a 19% increase in ultimate load which could be attributed to an increase in strength of the tendon-bone junction.

For the femur-MCL-tibia complex from these animals, when the specimens were subjected to tensile tests to failure, there was a 38% increase in ultimate load and 14% increase in stiffness, though these changes were not statistically significant. The modulus, tensile strength and ultimate strain of the MCL substance also increased slightly though not significantly.[56] In addition, no changes in collagen or elastin concentration were observed between the two groups.

In general, there is a highly nonlinear relationship between levels of stress and ligament properties (Figure 4).[57] Immobilization can significantly compromise both the structural properties of the bone-ligament-bone complex as well as the mechanical properties of the ligament substance. Exercise, on the other hand, only moderately enhances ligament properties, even over the course of many years.[54]

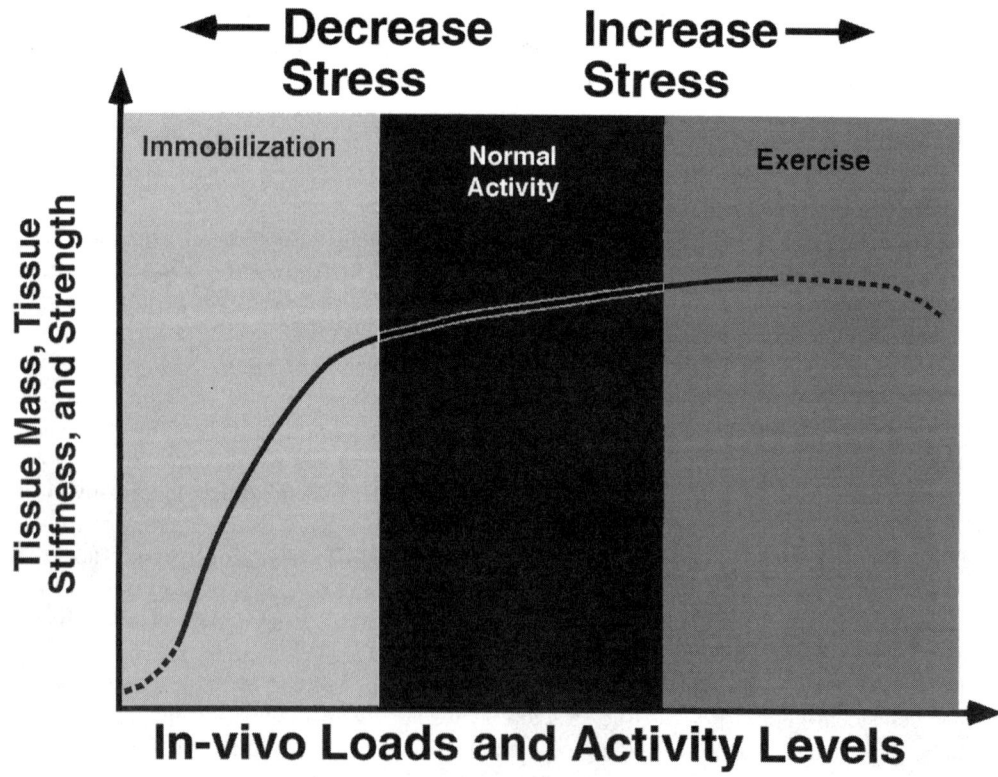

FIGURE 4. Hypothetical response of ligaments to levels of stress.[57]

V. ENVIRONMENTAL FACTORS

A. EFFECTS OF DEHYDRATION

Because 65–70% of the composition of ligaments and tendons is water, it can be expected that their mechanical properties will vary with the moisture content in the structure. Some investigators have tested soft tissues in air, both with and without a saline drip applied, while others have immersed the specimen in a temperature controlled saline bath. Chimich et al. observed that for the rabbit MCL, there was an increased stress relaxation in response to cyclic loading when more moisture is contained in the ligament.[58] It has also been observed that the human patellar tendon will exhibit a greater elastic modulus and tensile strength when tested in a saline bath than with a drip solution applied.[59] For these reasons, it is very important that soft tissue be kept hydrated and that conditions be noted when testing ligaments and tendons.

B. EFFECTS OF TEMPERATURE

Another factor which may contribute to variability in results between studies is the temperature at which the soft tissues are tested. Numerous investigations have presented the data on biomechanics of ligaments and tendons when the tests were done at room temperature, while other data have been based on a temperature controlled bath environment.

Investigators have studied the relationship between temperature and the tensile properties of ligaments and tendons, with varying results. Rigby et al. reported no significant differences in mechanical properties of ligaments when testing between 0° and 37°C,[60] while others have reported a decline in elastic modulus and stiffness with increasing temperature.[61,62] In our research center, the canine femur-MCL-tibia compex was tested in a saline bath at temperatures varying from 2–37°C under cyclic loading.[63] Each specimen was tested over the range of temperatures. This allowed the effects of the temperature change to be evaluated within each specimen, thus minimizing intra-specimen variability. It was found that an inverse relation exists between stiffness and temperature, and that the ligament relaxed to lower values under cyclic loading with higher temperatures. It should be noted that more than one hour of resting between tests was required for the ligament to return to its untested resting characteristics because of the viscoelastic nature of the tissue. These findings demonstrate the importance of both controlling and reporting temperatures when performing biomechanical testing.

C. EFFECTS OF FREEZING, STORING AND THAWING

Because it is often practically necessary to store specimens prior to testing or to use the specimens as allografts for reconstruction, the effects of freezing and storage on these tissues is of great interest. Several studies investigated the effect of post-mortem storage on tissue properties, but with conflicting results.[64–67] It has been reported that the rabbit ACL becomes "less extensible" just one hour after death,[65] while others have found no changes up to 96 hours later in the rabbit FATC.[66] Authors have reported no significant effects of freezing on the structural properties of monkey ACLs[42] while Dorlot et al. found an increase in the stiffness of the canine ACL.[13]

In our research center, the rabbit femur-MCL-tibia complex was used to study the effects of storage at −20°C for 1–3 months.[68] In order to protect the tissue from dehydration, the muscle and soft tissue were left intact, wrapped in saline-soaked gauze, and sealed in airtight plastic bags. Prior to testing, specimens were thawed overnight in the refrigerator (4°C). Previously frozen FMTCs were subjected to cyclic loading and then tensile loaded to failure. The same testing protocol was used on the controlateral fresh controls. No changes in cyclic stress relaxation, ligament cross-sectional area, ultimate load, ultimate deformation or energy absorbed to failure were observed after this method of freezing and storage. In addition, the mechanism of failure was the same in all specimens, suggesting that freezing had no significant effects on ligament properties, as well as its insertion sites. However, during the initial loading and unloading of the FMT complex, a significant decrease in the area of hysteresis was observed in the frozen specimens as compared to the fresh specimens. Thus, we recommend that care be exercised in the storage of ligament and tendons in order to preserve their biomechanical properties.

VI. VISCOELASTIC PROPERTIES OF LIGAMENTS AND TENDONS

Because of their complex collagen and protein ultrastructure, ligaments and tendons display both time- and history-dependent viscoelastic properties. Uniaxial tensile testing of a ligament will exhibit a hysteresis loop as the unloading portion will not follow that of the loading portion, an indication that there is energy dissipation (Figure 5). Ligaments and tendons also exhibit the phenomena of creep and stress relaxation. When a constant load is applied to the ligament, the deformation increases over time, known as creep (Figure 6A); when a constant deformation is applied, a decrease occurs over time which is known as stress relaxation (Figure 6B).[69] This viscoelastic response has important physical and clinical implications.[70,71] During cyclic loading and unloading, there is also a corresponding cyclic stress relaxation. Cyclic stress relaxation may help to prevent fatigue of ligaments when a large number of cyclic loads are applied, such as would occur when jogging. Conversely, stretching or prolonged exercising may enlist a gradual creep within a ligament or tendon. These effects manifest themselves clinically as temporary softening and increased laxity in joints after exercising. However, after a period of rest, these tissues can recover and return to their original lengths such that the joint returns to its normal stiffness.

VII. DETERMINING THE FUNCTIONAL ROLE OF LIGAMENTS/TENDONS

Characterizing the structural properties of the bone-ligament-bone complex and the mechanical properties of the ligament or tendon substance can aid in the understanding of the functional role of these tissues. By function, we focus on the contribution of ligaments and tendons to joint kinematics, as well as their forces *in situ*, when external loads are applied to the joint. Canine, rabbit, goat, and monkey models are among those which have been used to investigate the functional role of the ACL or MCL in the knee, as well as the effectiveness of various reconstructive techniques in restoring knee kinematics to as normal as possible.[72–81]

Studies are sometimes performed in which an external load is applied and the resulting joint kinematics are recorded. However, it is important to recognize that constraining the joint motion in one or more degrees of freedom will yield vastly different data on joint kinematics.[72,78,82]

In our research center, the anterior tibial translation in response to a 110 N anterior tibial load was determined in both 1 and 5 degrees of freedom (DOF) in the porcine knee.[82] In 5 DOF, the anterior tibial loading of the unconstrained knee was 1.4±0.2 times greater than of the constrained knee at 30° of flexion and 1.3±0.1 times greater at 60° and 90° of flexion. This effect was significant for all flexion angles.

In order to demonstrate the complex roles of the ACL and MCL in restraining varus/valgus rotation, our research center had applied a varus-valgus bending moment to the canine knee at 90 degrees of flexion in three and five DOF.[72] It was observed that with 3 DOF, varus-valgus knee laxity increased 171% after sectioning the MCL. However, when an identical test was performed in 5 DOF, it was found that VV laxity increased only 21%. This effect was attributed to the axial tibial rotation with is coupled with varus-valgus rotation and the results indicate that the ACL may compensate for the MCL under this loading condition when it is injured. The results of these studies indicate the importance of allowing multiple DOF motion in kinematic tests.

Also of interest when investigating the function of ligaments in a joint are the *in situ* forces within the ligament when an external load is applied. A number of methods have been used to

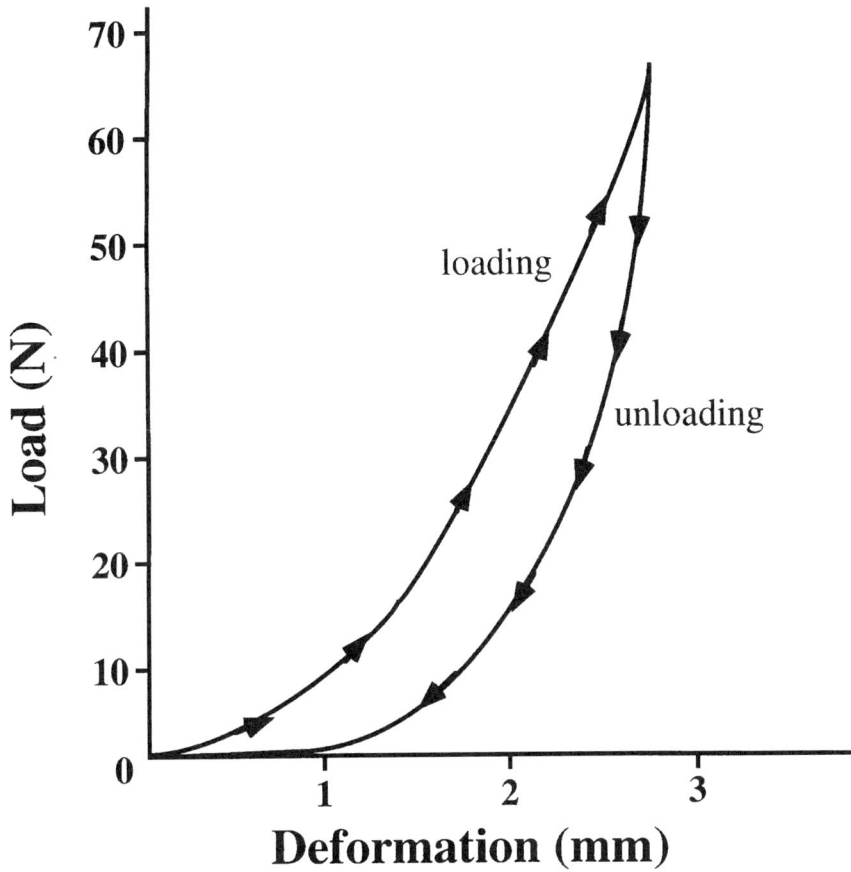

FIGURE 5. Typical hysteresis loop of a ligament subjected to tensile testing.

measure these forces in both animal and human models. Methods used have included the buckle transducer,[83–85] and implantable transducers in the ligament midsubstance.[86] In our research center, kinematic linkages[87] have been used to measure length changes in the ACL which are then used to calculate the *in situ* force by correlating the changes with those of length-tension data of the FATC.

Recently, a Universal Force Moment Sensor (UFS) was used in combination with an Instron material testing machine in order to measure and compare the *in situ* forces in the ACL of goats, pigs, sheep and humans.[88] The magnitude and direction of the *in situ* force in the anteromedial and posterolateral bundles of the ACL was determined with the knee at 90° of flexion in response to an A-P load in 1 DOF. The results indicate that both the magnitude and direction of the *in situ* force in the ACL of the sheep were significantly different from that of the human; the porcine model was the only specimen not significantly different from the human, not only in the magnitude and direction of the total ACL force, but also that in its two bundles.

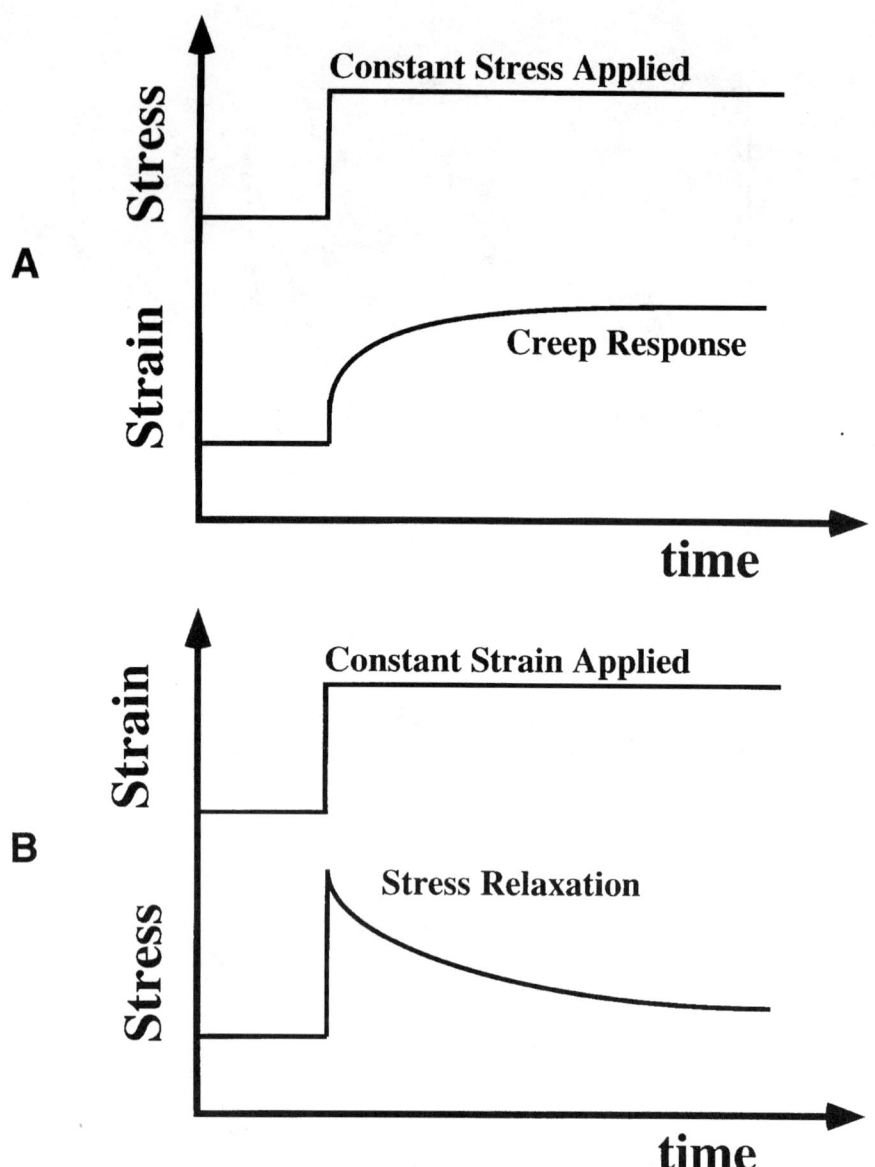

FIGURE 6. Schematic demonstrating (A) creep response (increasing deformation over time under a constant load) and (B) stress relaxation (decreasing stress over time under a constant deformation).

A. THE ROBOT/UFS TESTING SYSTEM

We have developed a new and unique testing system which combines a robotic manipulator with a universal force/moment sensor in order to measure the *in situ* forces in ligaments as well as determine joint kinematics (Figure 7). The robot (Unimate, Puma-762) is a six DOF, position-controlled device which can also perform in a force-controlled mode via force-feedback from

Mechanical Testing of Ligaments and Tendons

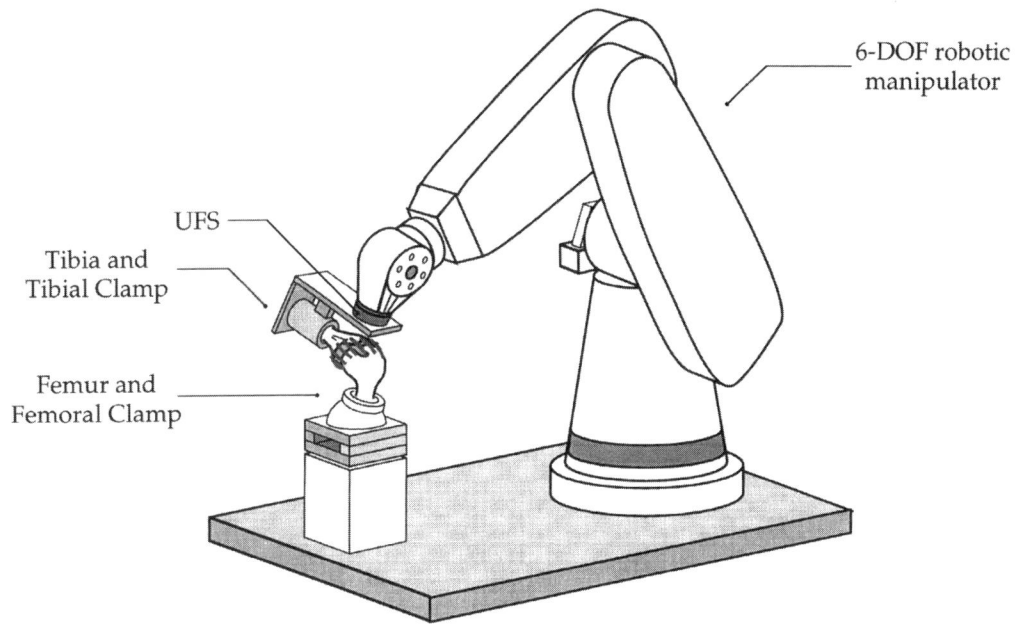

FIGURE 7. A schematic diagram of robot/UFS test system with a knee specimen in place.[82]

the UFS. The UFS (JR3, model 4015) is capable of measuring three forces and three moments along its Cartesian axes. The UFS guides the robot to "learn" an exact path of motion in a specimen under force control and the robot later repeats the learned positions (path of motion) under position control.[89] The system offers a highly accurate, non-contact method of measuring the *in situ* forces in ligaments. When a ligament is sectioned, under position control the previously determined kinematics are repeated. At this point, the UFS is now used for data acquisition, as it measures the resulting forces after sectioning the ligament. The vector decrease in forces measured by the UFS yields the *in situ* forces of the ligament using the principle of superposition. The direction and point of application of the force can also be determined.

In addition to measuring *in situ* forces, the system can also be used to evaluate the changes in kinematics which occur after a structure is sectioned. By using force control to apply an external load to the joint before and after sectioning, the resulting kinematic changes in 5 DOF (flexion is fixed) can be determined. A further advantage is that all effects, such as that of sectioning a ligament or of a reconstruction, can be tested in the same knee. This minimizes interspecimen variability, making statistical comparisons more powerful and reducing the number of specimens needed.

A number of animal studies have been performed using the robot/UFS testing system.[82,89-92] We have found the porcine knee to be an excellent model. The anatomy, size and geometry of the joint are comparable to those of the human knee, and the knees are easily and inexpensively obtained. Using this model, the effect of axial compression and anterior-posterior tibial loads on the *in situ* forces in the ACL and knee kinematics was evaluated. It was observed that the addition of a 200 N axial compressive load to a 100 N anterior-posterior load significantly increased anterior tibial translation of the knee, but decreased posterior tibial translation. It also caused a significant increase in the *in situ* forces in the ACL at 30°, 60° and 90° of flexion.[90]

The porcine model has also been used to test the effect of tunnel placement for ACL and PCL replacement grafts on graft forces and knee kinematics.[91,92] A 110 N anterior/posterior load was first applied to the intact knee at 30°, 60° and 90° of flexion and the intact knee kinematics and ACL or PCL *in situ* forces determined. The ligament under investigation was then reconstructed with a BPTB graft with tunnels drilled in several locations. It was found that the ACL replacement graft was most successful in restoring normal AP kinematics when the tibial tunnel is drilled proximally as compared to a more central or distal fixation. In the PCL recontructed knee, no significant difference in kinematics was observed with a proximal or anterior placement of the femoral tunnel; however, it was observed that the anteriorly-placed tunnel better replicated the trend of increasing *in situ* forces in the PCL occurring in the intact knee.

This system offers the potential of one day determining the *in vivo* forces in ligaments in animal models. If the *in vivo* kinematics of the joint are determined, the robot could later repeat these kinematics on the joint and thus determine the forces in the ligaments *in vivo*.

VIII. SUMMARY AND CONCLUSION

There have been many recent advances in the biomechanical testing of ligaments and tendons. The measurement of these properties is important to evaluate both reconstruction strategies and healing processes. However, there are many factors affecting the outcome of these measurements and readers are cautioned to evaluate the various experimental and biological factors in their interpretation of the published experimental data. Fortunately new bioengineering technologies are being developed such that it should be possible to standardize the mechanical testing of ligaments and tendons. By constantly building on past work, mechanical testing of soft tissue will enjoy further standardization as well as quality data management.

ACKNOWLEDGMENTS

The authors gratefully acknowledge the financial support of the RR&D grant AA188-3RA of the Veterans Administration, National Institutes of Health grants AR 14918, AR 33097, AR 34264 and AR 39683 and the Malcolm and Dorothy Coutts Institute for Joint Reconstruction and Research. Some of the work detailed here was performed in collaboration with the senior author during his tenure at the University of California, San Diego.

REFERENCES

1. Wright, D. G. and Rennels, D. C., "A study of the elastic properties of plantar fascia," *J. Bone Joint Surg.*, 46A, 482, 1964.
2. Race, A. and Amis, A. A., "A molding method to find cross-sections of soft tissue bundles with complex shapes," *Trans. Orthop. Res. Soc.*, 19, 783, 1994.
3. Race, A. and Amis, A. A., "The mechanical properties of the two bundles of the human posterior cruciate ligament," *J. Biomech.*, 27, 13, 1994.
4. Walker, L. B., Harris, E. H, and Benidict, J. V., "Stress-strain relationship in human cadaveric plantaris tendon — A preliminary study," *Med. Elect. Biol. Eng.*, 2, 31, 1964.
5. Ellis, D. G., "Cross-sectional area measurements for tendon specimens: A comparison of several methods," *J. Biomech.*, 2, 175, 1969.
6. Butler, D. L., Kay, M. D., and Stouffer, D. C., "Comparison of material properties in fascicle-bone units from human patellar tendon and knee ligaments," *J. Biomech.*, 19, 425, 1986.
7. Gupta, B. N., Subramanian, K. N., Brinker, W. O., and Gupta, A. N., "Tensile strength of canine cranial cruciate ligaments," *Am. J. Vet. Res.*, 32, 183, 1971.
8. Njus, G. O. and Njus, N. M., "A non-contact method for determining cross sectional area of soft tissues," *Trans. Orthop. Res. Soc.*, 32, 126, 1986.
9. Lee, T. Q. and Woo, S. L.-Y., "A new method for determining cross sectional shape and area of soft tissues," *J. Biomech. Eng.*, 110, 110, 1988.
10. Woo, S. L.-Y., Danto, M. I., Ohland, K. J., Lee, T. Q., and Newton, P. O., "The use of a laser micrometer system to determine the cross-sectional shape and area of ligaments: a comparative study with two existing methods," *J. Biomech. Eng.*, 112, 426, 1990.
11. Chan, S. S., Livesay, G. A., Morrow, D. A., and Woo, S. L.-Y., "The development of a low-cost laser reflectance system to determine the cross-sectional shape and area of soft tissues," *ASME Adv. Bioeng.*, BED-31, 123, 1995.
12. Arms, S. W., Pope, M., II, Boyle, J. B., Davignon, P. J., and Johnson, R. J., "Knee medial collateral ligament strain," *Trans. Orthop. Res. Soc.*, 7, 47, 1982.
13. Dorlot, J. M., Ait ba sidi, M., and Tremblay, G. M., "Load-elongation behavior of the canine anterior cruciate ligament," *J. Biomech. Eng.*, 102, 190, 1980.
14. Meglan, D., Zuelzer, W., and Buck, W. N. B., "The effect of quadriceps force upon strain in the anterior cruciate ligament," *Trans. Orthop. Res. Soc.*, 11, 55, 1986.
15. Monahan, J. J., Grigg, P., and Pappas, A. M., "*In vivo* strain patterns in the four major canine knee ligaments," *J. Orthop. Res.*, 2, 408, 1986.
16. Trent, P. S., Walker, P., and Wolf, B., "Ligament length patterns, strength and rotational axes of the knee joint," *Clin. Orthop.*, 117, 263, 1976.
17. Warren, L. F., Marshall, J. L., and Girgis, F., "The prime static stabilizers of the medial side of the knee," *J. Bone Joint Surg.*, 56A, 665, 1974.
18. Kennedy, J. C., Hawkins, R. J., and Willis, R. B., "Strain gauge analysis of knee ligaments," *Clin. Orthop.*, 129, 225, 1977.
19. Woo, S. L.-Y., Gomez, M. A., Seguchi, Y., Endo, C., and Akeson W. H., "Measurement of mechanical properties of the medial collateral ligament substance from a bone-ligament-bone preparation," *J. Orthop. Res.*, 1, 22, 1983.
20. Woo, S. L.-Y., Orlando, C. A., Gomez, M. A., Frank, C. B., and Akeson, W. H., "Tensile properties of medial collateral ligament as a function of age," *J. Orthop. Res.*, 4, 133, 1986.
21. Lee, T. Q. and Danto, M. I., "Application of a continuous video digitizing system for tensile testing on bone-soft tissue-bone complex," *ASME Adv. Bioeng.*, BED-22, 87, 1992.
22. Harner, C. D., Xerogeanes, J. W., Livesay, G. A., Carlin, G. A., and Woo, S. L.-Y., "The human posterior cruciate ligament: An interdisciplinary study," *Am. J. Sports Med.*, 23, 736, 1995.

23. Levine, R. E., Hildebrand, K. A., and Woo, S. L.-Y., "High surface strains occur at the failure site of rabbit medial collateral ligaments," *Ann. Biomed. Eng.,* 24, 455, 1996.
24. Debski, R. E., McMahon, P. J., Thompson, W. O., et al., "A new dynamic shoulder testing apparatus to study glenohumeral joint motion," *J. Biomech.,* 28, 869, 1995.
25. Des Jardins, J. D., MacWilliams, B. A., Wilson, D. R., and Chao, E. Y. S., "The *in vitro* assessment of knee joint kinematics under physiologic loading," *Trans. Orthop. Res. Soc.,* 44, 260, 1997.
26. Woo, S. L.-Y., Hollis. J. M., Adams, D. J., Lyon, R. M., and Takai, S., "Tensile properties of the human femur-anterior cruciate ligament-tibia complex: the effect of specimen age and orientation," *Am. J. Sports. Med.,* 19, 217, 1991.
27. Figgie, H. E., Bahniuk, E. H., Heiple, G. K., and Davy, D. T., "The effects of the tibial-femoral angle on the failure mechanics of the canine anterior cruciate ligament," *J. Biomech.,* 19, 89, 1986.
28. Woo, S. L.-Y., Hollis. J. M., Roux, R. D., et al., "Effects of knee flexion on the structural properties of the rabbit femur-anterior cruciate ligament-tibia complex (FATC)," *J. Biomech.,* 20, 557, 1987.
29. Peterson, R. H. and Woo, S. L.-Y., "A new methodology to determine the mechanical properties of ligaments at high strain rates," *J. Biomech. Eng.,* 108, 465, 1986.
30. Danto, M. I. and Woo, S. L.-Y., "The mechanical properties of skeletally mature rabbit anterior cruciate ligament and patellar tendon over a range of strain rates," *J. Orthop. Res.,* 11, 58, 1993.
31. Abrahams, M., "Mechanical behaviour of tendon *in vitro*. A preliminary report," *Med. Biol. Eng. Comput.,* 5, 433, 1967.
32. Kennedy, J. C., Hawkins, R. J., Willis, R. B., and Danylchuck, K. D., "Tension studies of human knee ligaments. Yield point, ultimate failure, and disruption of the cruciate and tibial collateral ligaments," *J. Bone Joint Surg.,* 58A, 350, 1976.
33. Noyes, F. R., "Functional properties of knee ligaments and alterations induced by immobilization: a correlative biomechanical and histological study in primates," *Clin. Ortho.,* 123, 210, 1977.
34. Woo, S. L.-Y., Newton, P. O., MacKenna, D. A., and Lyon, R. M., "Comparative evaluation of the mechanical properties of the rabbit medial collateral and anterior cruciate ligaments," *J. Biomech.,* 25, 377, 1992.
35. Woo, S. L.-Y., Gomez, M. A., Amiel. D., et al., "The effects of exercise on the biomechanical and biochemical properties of swine digital flexor tendons," *J. Biomech. Eng.,* 103, 51, 1981.
36. Herrick, W. C., Kingsbury, H. B., and Lou, D. Y., "A study of the normal range of strain, strain rate, and stiffness of tendon," *J. Biomed. Mater. Res.,* 12, 877, 1978.
37. Crevier, N., Pourcelot, P., Denoix, J. M., et al., "Segmental variations of *in vitro* mechanical properties in equine superficial digital flexor tendons," *Am. J. Vet. Res.,* 57, 1111, 1996.
38. Sikoryn, T. A., and Hukins, D. W., "Mechanism of failure of the ligamentum flavum of the spine during *in vitro* tensile tests," *J. Orthop. Res.,* 8, 586, 1990.
39. Vogel, H. G., "Age dependence of mechanical properties of rat tail tendons," *Aktuelle Gerontologie,* 13, 22, 1983.
40. Morein, G., Goldgefter, L., Kobyliansky, E., Goldschmidt-Nathan, M., and Nathan, H., "Change in mechanical properties of rat tail tendon during postnatal ontogenesis," *Anat. Embryol.,* 154, 121, 1978.
41. Woo, S. L.-Y., Ohland, K. J., and Weiss, J. A., "Aging and sex-related changes in the biomechanical properties of the rabbit medial collateral ligament," *Mech. Ageing Devel.,* 56, 129, 1990.
42. Noyes, F. R. and Grood, E. S., "The strength of the anterior cruciate ligament in humans and rhesus monkeys: age-related and species-related changes," *J. Bone Joint Surg.,* 56A, 1406, 1976.
43. Salter, R. B. and Field, P., "The effects of continuous compression on living articular cartilage: an experimental investigation," *J. Bone Joint Surg.,* 42A, 31, 1960.
44. Evans, E. B., Eggers, G. W. N., Butler, J. K., and Blumel, J., "Experimental immobilization and remobilization of rat knee joints," *J. Bone Joint Surg.,* 42A, 737, 1960.
45. Woo, S. L.-Y., Gomez, M. A., Sites, T. J., et al., "The biomechanical and morphological changes in the medial collateral ligament of the rabbit after immobilization and remobilization," *J. Bone Joint Surg.,* 69A, 1200, 1987.

46. Newton, P. O., Woo, S. L.-Y., MacKenna, D. A., and Akeson, W. H., "Immobilization of the knee joint alters the mechanical and ultrastructural properties of the rabbit anterior cruciate ligament," *J. Orthop. Res.,* 13, 191, 1995.
47. Tipton, C. M., Schild, R. J., and Tomanek, R. J., "Influence of physical activity on strength of knee ligaments in rats," *Am. J. Physiol.,* 221, 783, 1967.
48. Noyes, F. R., DeLucas, J. L., and Torvik, P. J., "Biomechanics of ligament failure. II. An analysis of immobilization, exercise and reconditioning in primates," *J. Bone Joint Surg.,* 56A, 1406, 1974.
49. Zuckerman, J. and Stull, G. A., "Effects of exercise on knee ligament separation force in rats," *J. Appl. Physiol.,* 26, 716, 1969.
50. Tipton, C. M., Matthes, R. D., Maynard, J. A., and Carey, R. A., "The influence of physical activity on ligaments and tendons," *Med. Sci. Sports Exer.,* 7, 165, 1975.
51. Tipton, C. M., Matthes, R. D., and Sandage, D. S., "*In situ* measurement of junction strength and ligament elongation in rats," *J. Appl. Physiol.,* 37, 758, 1974.
52. Tipton, C. M., James, S. L., Mergner, W., and Tcheng, T,K., "Influence of exercise on the strength of the medial collateral ligaments of dogs," *Am. J. Physiol.,* 218, 758, 1970.
53. Cabaud, H. E., "Exercise effects on the strength of the rat anterior cruciate ligaments," *Am. J. Sports Med.,* 8, 79, 1980.
54. Wang, C. W., Weiss, J. A., Albright, J., et al., "Life-long exercise and aging effects on the canine medial collateral ligament," *Trans. ORS,* 15, 518, 1990.
55. Woo, S. L.-Y., Ritter, M. M., Amiel, D., et al., "The biomechanical and biochemical properties of swine tendons. Long-term effects of exercise on the digital extensors," *Conn. Tiss. Res.,* 7, 177, 1980.
56. Woo, S. L.-Y., Kuei, S. C., Gomez, M. A., et al., "The effect of immobilization and exercise on the strength characteristics of bone-medial collateral ligament-bone complex," *ASME Biomech. Symp.,* 32, 67, 1979.
57. Woo, S. L.-Y., Chan, S. S., and Yamaji, T., "Biomechanics of knee ligament healing, repair, and reconstruction: ISB Keynote Lecture," *J. Biomech.,* 30, 431, 1997.
58. Chimich, D., Shrive, N., Frank, C., Marchuk, L., and Bray, R., "Water content alters viscoelastic behavior of the normal adolescent rabbit medial collateral ligament," *J. Biomech.,* 25, 831, 1995.
59. Haut, R. C. and Powlison, A. C., "The effects of test environment and cyclic stretching on the failure properties of human patellar tendons," *J. Orthop. Res.,* 8, 532, 1990.
60. Rigby, B., Hirai, N., Spikes, J., and Eyring, H., "The mechanical properties of rat tail tendon," *J. Gen. Physiol.,* 53, 265, 1958.
61. Apter, J., "Influence of composition on thermal properties of tissues," in *Biomechanics: Its Foundations and Objectives,* Fung, Y. C., Perrone, N., and Anliker, M., Eds., Prentice-Hall, Englewood Cliffs, NJ, 1972.
62. Hunter, J. and Williams, M. G., "A study of the effect of cold on joint temperature and mobility," *Can. J. Med. Sci.,* 29, 255, 1951.
63. Woo, S. L.-Y., Lee, T. Q., Gomez, M. A., Sato, S., and Field, F. P., "Temperature dependent behavior of the canine medial collateral ligament," *J. Biomech. Eng.,* 109, 68, 1987.
64. Matthews, L. S. and Ellis, D., "Viscoelastic properties of cat tendon: effects of time after death and preservation by freezing," *J. Biomech.,* 1, 65, 1968.
65. Smith, J. W., "The elastic properties of the anterior cruciate ligament of the rabbit," *J. Anat.,* 88, 369, 1954.
66. Viidik, A., Sanquist, L., and Magi, M., "Influence of postmortem storage on tensile strength characteristics and histology of rabbit ligaments," *Acta Orthop Scand (Suppl),* 79, 1, 1965.
67. Wertheim, M. G., "Memoirs sur l'elatsticite et la cohesion des principaux tissu du corps humain," *Ann. Chim. (Phys),* 21, 385, 1847.
68. Woo, S. L.-Y., Orlando, C. A., Camp, J. F., and Akeson, W. H., "Effects of postmortem storage by freezing on ligament tensile behavior," *J. Biomech.,* 19, 399, 1986.

69. Woo, S. L.-Y., Young, E. P., and Kwan, M. K., "Fundamental studies in knee ligament mechanics," in *Knee Ligaments: Structure, Function, Injury and Repair*, Daniel, D. M., Akeson, W. H., John J. and O'Connor, J. J., Eds., Raven Press, New York, 1990.
70. Woo, S. L.-Y., Gomez, M. A., Woo, Y. K., and Akeson, W. H., "Mechanical properties of tendons and ligaments. I. Quasi-static and nonlinear viscoelastic properties," *Biorheology*, 19, 385, 1982.
71. Woo, S. L.-Y., Gomez, M. A., and Akeson, W. H., "The time and history dependent viscoelastic properties of the canine medial collateral ligament," *J. Biomech. Eng.*, 103, 293, 1981.
72. Inoue, M., McGurk-Burleson, E., Hollis, J. M., and Woo, S. L.-Y., "Treatment of the medial collateral ligament injury. I. The importance of anterior cruciate ligament on the varus-valgus knee laxity," *Am. J. Sports Med.*, 15, 15, 1987.
73. Holden, J. P, Grood, E. S., and Cummings, J. F., "The effects of flexion angle and tibial rotation on measurement of anteromedial band force in the goat ACL," *Trans. Orthop. Res. Soc.*, 16, 588, 1991.
74. Arnoczky, S. P. and Marshall, J. L., "The cruciate ligaments of the canine stifle: an anatomical and functional analysis," *Am. J. Vet. Res.*, 38, 1807, 1977.
75. Clancy, W. G. J., Narechania, R. G., Rosenberg, T. D., et al., "Anterior and posterior cruciate ligament reconstruction in rhesus monkeys: a histological, microangiographic and biomechanical analysis," *J. Bone Joint Surg.*, 63A, 1270, 1981.
76. Berman, A. C., "The goat model for prosthetic anterior cruciate ligament reconstruction," *Trans. Soc. Biomater.*, 15, 1989.
77. Korvick, D. L., Pijanowski, G. J., and Schaeffer, D. J., "Three dimensional kinematics of the intact and cranial cruciate ligament-deficient dog stifle," *J. Biomech.*, 27, 77, 1993.
78. Woo, S. L.-Y., Young, E. P., Ohland, K. J., et al., "The effects of transection of the anterior cruciate ligament on healing of the medial collateral ligament: a biomechanical study of the knee in dogs," *J. Bone Joint Surg.*, 72A, 382, 1990.
79. Shino, K., Kawasaki, T., Hirose, H., et al., "Reconstruction of the anterior cruciate ligament by allogenic tendon graft: an experimental study in the dog," *J. Bone Joint Surg.*, 66B, 672, 1984.
80. O'Donoghue, D. H. and Rockwood, C. C., "Repair of the anterior cruciate ligament in dogs," *J. Bone Joint Surg.*, 48A, 503, 1966.
81. Oster, D. M., Grood, E. S., Feder, S. M., Butler, D. L., and Levy, M. S., "Primary and coupled motions in the intact and ACL deficient knee: an *in vitro* study in the goat model," *J. Orthop. Res.*, 10, 476, 1992.
82. Livesay, G. A., Rudy, T. W., Woo, S. L.-Y, et al., "Evaluation of the effect of joint constraints on the *in situ* force distribution in the anterior cruciate ligament," *J. Orthop. Res.*, 15, 278, 1997.
83. Lewis, J. L., Lew, W. D., and Schmidt, J., "Description and error evaluation of an *in vitro* knee joint testing system," *J. Biomech. Eng.*, 110, 238, 1988.
84. Lewis, J. L., Lew, W. D., and Schmidt, J., "A note on the application and evaluation of the buckle transducer for the knee ligament force measurement," *J. Biomech. Eng.*, 104, 125, 1982.
85. Barry, D. and Ahmed, A. M., "Design and performance of a modified buckle transducer for the measurement of ligament tension," *J. Biomech. Eng.*, 108, 149, 1986.
86. Holden, J. P., Grood, E. S., Korvick D. L., Cummings, J. F., and Butler, D. L., "*In vivo* forces in the anterior cruciate ligament: direct measurements during walking and trotting in a quadruped," *J. Biomech.*, 27, 517, 1994.
87. Hollis. M. J., Marcin, J. P., Horibe, S., and Woo, S. L.-Y., "Load determination in ACL fiber bundles under knee loading," *Trans. Orthop. Res. Soc.*, 13, 58, 1988.
88. Xerogeanes, J. W., Fox, R. J., Takeda, Y., et al., "A functional comparison of animal anterior cruciate ligament models to the human anterior cruciate ligament," *Ann. Biomed. Eng.*, 26, 345, 1998.
89. Rudy, T. W., Livesay, G. A., Woo, S. L.-Y., and Fu, F. H., "A combined robotic/universal force sensor approach to determine *in situ* forces of knee ligaments," *J. Biomech.*, 29, 1357, 1996.
90. Li, G., Rudy, T. W., Allen, C., Sakane, M., and Woo, S. L.-Y., "Effect of combined axial compressive and anterior tibial load on *in situ* forces in the anterior cruciate ligament: a porcine study," *J. Orthop. Res.*, 16, 122, 1998.

91. Ishibashi, Y., Rudy, T. W., Livesay, G. A., et al., "The effect of anterior cruciate ligament graft fixation site at the tibia on knee stability: evaluation using a robotic testing system." *Arthroscopy*, 13, 177, 1997.
92. Stone, J. D., Carlin, G. J., Ishibashi, Y., Harner, C. D., and Woo, S. L.-Y., "Assessment of posterior cruciate ligament graft performance using robotic technology," *Am. J. Sports. Med.*, 24, 824, 1996.

Part III

Animal Models of Bone Conditions

11 Animal Models of Bone Fracture or Osteotomy

Yuehuei H. An, Richard J. Friedman, and Robert A. Draughn

CONTENTS

I. Introduction 197
II. Fracture Healing Process 198
III. Animal Models of Diaphyseal Fractures 198
    A. Models of Diaphyseal Fractures 198
    B. Authors' Preferred Diaphyseal Models 202
        1. Rat Tibial Fracture (A Tibial Model Developed in Authors' Laboratory)...............202
        2. Other Rat Fracture Models 202
        3. Tibial Fractures in Rabbits, Dogs and Sheep 203
        4. Radial Fractures in Rabbits and Dogs 204
IV. Animal Models of Epiphysometaphyseal Models 205
V. Animal Models of Delayed Union and Nonunion 205
VI. Evaluation Methods..205
    A. Radiography 205
    B. Histology and Histomorphometry...208
    C. Mechanical Testing...208
    D. Other Evaluation Methods 209
VII. Methods for Enhancing Fracture Healing 209
    A. Bone Grafting..209
    B. Biomaterials..210
    C. Local Use of Growth Promoting Substances 210
    D. Systemic Use of Hormones or Drugs 210
    E. Biophysical Stimulation 210
VIII. Other Factors Affecting Fracture Healing 211
    A. Effects of Fixation Devices...211
    B. Effects of Drugs or Medication 211
References 212

I. INTRODUCTION

The processes of normal and abnormal fracture healing and the various factors affecting them have been widely studied since the days of Galen and Hippocrates. Today, with advanced technology and the desire of finding new solutions for fracture related problems, this field is still the object of numerous investigations. In the 1940s, Urist and Johnson[1] mentioned that there were more than 4,000 publications on this subject already in the literature, and this figure has been growing since then. One of the most important elements in studies of fracture healing or fracture fixation is establishing a standard method to make a reproducible fracture, which would allow results from different centers to be compared. Numerous models of experimental fractures have been published. This chapter attempts to summarize them and to pave an easier road for researchers to find valid models for new projects.

II. FRACTURE HEALING PROCESS

Normal fracture healing undergoes three phases: inflammation, reparation, and remodeling.[2-4] At the inflammatory stage, hemorrhage and cell death causes an inflammatory response. It evolves into granulation tissue with mesenchymal fibroblasts, macrophages, and lymphocytes. The granulation tissue absorbs dead tissue and provides chondrogenic and osteogenic precursor cells, forming the foundation for further healing phases. During the reparation phase (external callus formation), the mesenchymal cells in the granulation tissue undergo rapid chondrogenesis, followed by endochondral ossification supplemented by appositional bone formation. The fracture callus formed during this phase provides stability for osteogenesis. The remodeling phase includes gradual resorption of the periosteal bony callus, maturation of bone structure and the restoration of the cortical bone structure.

There are three main types or modes of fracture healing mechanisms depending on the rigidity of fracture fixation:[4] (1) direct early healing (primary osteonal healing); (2) direct healing with substantial callus formation (secondary osteonal healing); and (3) non-osteonal healing by periosteal and endosteal callus formation. The three phases of fracture healing are closely correlated to the primary and second osteonal healing mode. The rigidity of immobilization or fixation of the fracture site determines the amount of cartilaginous callus formation and, consequently, determines the healing mode. The greatest amount of callus is seen in nonfixed unstable fractures, the smallest amount of chondral tissue in rigid fixed fractures (such as compression plating), and an intermediate amount of callus formation occurs in most less rigid fixed fractures (such as intramedullary canal nailing, external fixation, and less rigid plating).

III. ANIMAL MODELS OF DIAPHYSEAL FRACTURES

A. Models of Diaphyseal Fractures

Major diaphyseal fracture models are listed in Table 1. Several more diaphyseal models are included in Chapter 12 and they were used for testing bioabsorbable fixation devices. Different animals have been used, ranging from small to large or from low to high vertebrate animals including the mouse, rat, rabbit, cat, dog, sheep, and goat. The most commonly used are rat, rabbit, dog, and sheep.

There have been four major fracture-producing methods for diaphyseal fracture models, including (1) manual fractures;[7,17-19,77] (2) three-point bending methods, such as those fracture devices described by Ekeland,[8,75] Bak and Andreassen,[14,79,80] Tepic et al.[72] and Greiff;[21] (3) a guillotine-like fracture apparatus, including those described by Jackson,[8] Sarmiento,[10] Northmore-Ball,[78] Bonnarens and Einhorn,[12] and An et al.;[28] and (4) osteotomies using a saw or scissors.[29,79] The first three methods are more natural, mimicking accidental fractures.

Most diaphyseal fractures can be fixed with internal fixation such as intramedullary rods or pins, plate and screws, screws only, thread or wires, alone or in combination. Intramedullary pins can be introduced percutaneously in rat models to create a closed fracture. Plate and screws are often used in larger animals. One or two screws can be used to fix an oblique diaphyseal fracture, but external supports are often needed for a satisfactory fixation. Fixation may not be used in small animals such as mice or rats to facilitate a natural healing process or in the cases of radial, ulnar, or fibular fracture models because of the existing support from the companion bones, ulna, radius, or tibia. External fixators are often used for larger animals such as dogs or sheep. Casts or splints are sometimes used alone or for extra protection after internal fixation.

TABLE 1
Diaphyseal Fracture Models in Animals

Animal	Bone	First author[ref.]	Fracturing method	Fracture type	Fixation method	Purpose or treatment
Mouse	Tibia	Hsu 1969[5]	Manual	T*	No fixation	Study of fracture healing
Rat		Hiltunen 1993[6]	Impact device	T/O†	Intramedullary pin	Study of fracture healing
	Femur	Eskelund 1950[7]	Manual	T	No fixation	Study of fracture healing
		Jackson 1970[8]	Jackson's punch	T	Intramedullary pin	Study of fracture healing
		Kernex 1973[9]	Manual	T	No fixation	Study of fracture healing
		Sarmiento 1977[10]	Jackson's punch	T	Intramedullary pin	Effect of weightbearing on fracture healing
		Ekeland 1981[11]	Fracturing forceps	T	No fixation	Study of fracture healing
		Bonnaren 1984[12]	Impact device	T	Intramedullary pin	Study of fracture healing
		Huo 1991[13]	Manual	T	Intramedullary pin	Effect of ibuprofen on fracture healing
		Grundnes 1993[14]	Saw, osteotomy	T	Intramedullary pin	Study of fracture healing
		Olmedo 1994[15]	Three point bending clamp	T	Metal catheter	Drug delivery to fracture site
		Hietaniemi 1996[16]	Saw, osteotomy	T	Unstable fixation	Creation of non-union
	Tibia	Urist 1941[17]	Manual	T	No fixation	Study of fracture healing
		Penttinen 1972[18]	Manual	T	No fixation	Study of fracture healing
		Hulth 1964[19]	Manual	T	No fixation	Effect of cortisone on fracture healing
		Greiff 1978[20]	Fracturing forceps	T	Intramedullary pin	Study of fracture healing
		Greiff 1978[21]	Fracturing forceps	T	Intramedullary pin	Autoradiographic study
		Molster 1982[22]	Saw, osteotomy	O	Intramedullary pin	Effect of instability on fracture healing
		Lowe 1983[23]	Guillotine device	T	Intramedullary pin	Study of fracture healing
		Bak 1988[24]	Fracturing forceps	T	Intramedullary pin	Study of fracture healing
		Aro 1989[25]	Manual	T	Intramedullary pin	Study of fracture healing
		Keller 1993[26]	Saw, osteotomy	T	Plate and screws	Effect of prostaglandin on fracture healing
		Nyman 1993[27]	Manual	T	Intramedullary pin	Effect of clodronate on fracture healing
		An 1994[28]	Guillotine device	T	Intramedullary pin	New fracture device
		Nielsen 1994[29]	Fracturing forceps	T	Intramedullary pin	Effect of TGF-β on fracture healing
		Otto 1995[30]	Special pliers	T	No fixation	New fracturing method
		David 1996[31]	Ring-cutter saw	T	Intramedullary pin	Effect of laser on fracture healing
	Fibulae	Herbsman 1966[32]	Scissors	T	No fixation	Study of fracture healing
		Kawaguchi 1994[33]	Bone cutter	T	No fixation	Effect of bFGF on fracture healing
	Radius	Volpin 1986[34]	Scissors	T	No fixation	Study of fracture healing

TABLE 1
Diaphyseal Fracture Models in Animals

Animal	Bone	First author[ref.]	Fracturing method	Fracture type	Fixation method	Purpose or treatment
Rabbit	Radius	Chai 1985[35]	Osteotomy	T	No fixation	Effect of a herb on fracture healing
		Bushberg 1985[36]	Bending with a metal bar	T	Splinting and tapping	Uptake of ^{67}Ga and ^{99m}Tc MDP
	Femur	Manninen 1993[37]	Osteotomy	T	SR-PLA rods	Use of bioabsorbable material on fracture fixation
		Albanese 1996[38]	Osteotomy	T	Plate and screws	Effect of osteotomy on bone growth
	Tibia	White 1977[39]	Osteotomy	T	External fixator	Effect of compression force on fracture healing
		Paavolainen 1979[40]	Osteotomy	T	Plate and screws	Effect of compression plate on fracture healing
		Ashhurst 1982[41]	C-clamp fracture device	T	Plate and screws	New fracture device
		Court-Brown 1985[42]	Osteotomy	T	Ext. fixator/cast	Effect of fixation on blood flow
		Terjesen 1986[43]	Osteotomy	T	External fixation	Effect of fixation stiffness on fracture healing
		Aalto 1987[44]	Osteotomy	T	External fixator	Effect of rigidity of fixation on fracture healing
		Laftman 1989[45]	Osteotomy	T	Rigid internal plate fixation	Studying stress shielding of the rigid fixation
		Burr 1990[46]	Impulsive loading	Micro	No fixation	New stress fracture model
		Carpenter 1992[47]	Dental burr osteotomy	T	External fixator	Growth hormone on fracture healing
		Kaatinen 1993[48]	Osteotomy	T	Intramedullary nail	Observation of healing patterns
		Jacob 1993[49]	Drill and bone biter	T	Plate and screws	Local use of cefazolin to prevent infection
		Worlock 1994[50]	C-clamp fracture device	T	DC plate or intramedul. nail	Effect of fracture stability on infection rate
		Nash 1994[51]	Osteotomy	V-shaped	Intramedullary PGA rod	Effect of PDGF on fracture healing
	Fibula	Pienkowski 1994[52]	Osteotomy	T	No fixation	Effect of electromagnetic stimulation on fracture healing
	Ribs	Brighton 1991[53]	Manual fracture	T	No fixation	Study of fracture healing
Dog	Femur	Rhinelander 1983[54]	Saw osteotomy	5-cm O	Pins and nylon straps	Effect of plain nylon straps on fracture fixation
		Miettinen 1992[55]	Osteotomy	T	SR-PLA or SR-PGA	Fracture fixation with absorbable rods
		An 1997[56]	Trephined	Circular	PGA/PLA screws	Fixation of cortical bone piece with absorbable screws

Animal	Bone	Author Year	Method	Type*	Fixation	Purpose
	Tibia	Skirving 1987[57]	"Bone breaker"	T	Plate and screws	Effect of different plates on fracture healing
		Gilbert 1989[58]	Osteotomy	T	Different ext. fixators	Effect of fixation stiffness on fracture healing
		Smith 1990[59]	Osteotomy	T	Plate and screws	Effect of fixation on blood flow
		Tiedeman 1990[60]	Osteotomy	T	External fixator	X ray assessing fracture healing
		Markel 1990[61]	Osteotomy	T	External fixator	Fracture healing assessed by QCT, SPA, DEXA, and MRI
		Aro 1993[62]	Osteotomy	T/60°O†	External fixator	Effect of fracture stability, fracture type, loading on fracture healing
		O'Sullivan 1994[63]	Saw osteotomy	T	External fixator	Effect of weight bearing on fracture healing
	Radius	Lenehan 1985[64]	Wire wheel puller	T	External splinting	Effect of EHDP‡ on fracture healing
		Chakkalakal 1990[65]	Saw osteotomy	T	Pins between radius and ulna	Study of fracture healing
		Peter 1996[66]	Three point bending by wire	T	Coaptation splinting	Effect of alendronate on fracture healing
Sheep	Tibia	Heitemeyer 1990[67]	Triple wedge osteotomy	C‖	Plate + screws, bridging plate intramedul. nail, or ext. fixator	Fixation methods for comminuted fracture
		Goodship 1993[68]	Osteotomy	T	External fixator	Fixator frame stiffness on fracture healing
		Schemitsch 1994[69]	Tepic's fracture device	Spiral	Intramedullary nail	Reamed and unreamed nailing on cortical blood flow
		Wallace 1995[70]	Osteotomy	T	External fixator	Serum angiogenic factor level after tibial fracture
		Schemitsch 1996[71]	Tepic's fracture device	Spiral	Intramedullary nail	Reamed or unreamed nailing on soft tissue blood flow
		Tepic 1997[72]	Tepic's fracture device	T	Point contact plate and screws	Testing effect of point contact fixator on bone healing
		Augat 1997[73]	Osteotomy	T	External	Assessment of fracture by peripheral CT
Goat	Tibia	Curtis 1994[74]	Osteotomy	T	External or internal	Effect of fixation method on infection
Cat	Femur	Hara 1994[75]	Saw osteotomy	T	PLA intramedullary rods	Fixation of femoral diaphyseal fracture
Calf	Metacarpal	Illi 1992[76]	Saw osteotomy	45°	PLA or AO metallic screws	Fixation of metacarpal diaphyseal fracture

* T = Transverse
† O = Oblique
‡ EHDP = Ethane-1-hydroxy-1,1-diphosphonate
C‖ = comminuted fracture.

B. AUTHORS' PREFERRED DIAPHYSEAL MODELS

1. Rat Tibial Fracture (A Tibia Model Developed in Authors' Laboratory)

When attempting to produce a standard closed fracture in the rat femur, we found it very difficult because the bulky soft tissues around the femoral bone made positioning of the rat thigh on the support anvil of the fracture apparatus very difficult. The rat tibia, on the other hand, is subcutaneously located and by its anatomic location better suited than the femur for the production of a closed fracture. So, modified from the method of Bonnarens and Einhorn,[12] a fracture apparatus has been built in the authors' laboratory (Figure 1). The support anvil was made with an adjustable foot rest, which ensures that all of the fractures are at the same level by positioning the rat leg on the anvil with the foot against the foot rest.[28]

To test the apparatus, 88 SD rats (300–350 gm) underwent percutaneous bilateral intramedullary pinning prior to the production of a fracture using this apparatus. Under general anesthesia, a hole was made 4 mm proximal and 2 mm medial to the tibial tuberosity percutaneously using a 20-gauge needle (Figure 2). The needle was driven directly into the medullary canal, and by rotating it, reamed the canal to within 2–5 mm of the ankle joint. A 0.9 mm Kirschner wire was then placed down the intramedullary canal, and the end of the wire was cut as short as possible so that the skin could roll over. No stitches were needed. Reaming was a necessary part of the procedure to allow the wire to seat and prevent it from perforating the cortex.

Thereafter, the rat was placed supine and the lower leg was positioned on the anvil of the fracture apparatus in an abducted and externally rotated position and the guillotine blade was lined up with the middle portion of the lower leg (about 3–5 mm proximal to the junction of the tibia and fibula). The travel distance of the guillotine ramming system was set at 2.5–3.0 mm. The 500-gm weight was dropped from a height of 40 centimeters, driving the guillotine blade to fracture the tibia. The legs were then radiographed to examine the fracture and fixation.

Animals were sacrificed three or five weeks after the surgery and the fracture healing was evaluated by X ray (Figure 2), histological and mechanical methods. Using this method, the fracture site is easy to control because of the adjustable foot rest built into the support anvil. Mechanical testing and histological studies showed that a standard fracture healing process was obtained. It is concluded that this modified method creates a standard, reproducible transverse closed fracture of rat tibia.

2. Other Rat Fracture Models

Manual techniques for fracture production are no longer acceptable, as they have no controls over the force produced and there can be wide variations in the fracture site and configuration. The lack of fixation leads to variable degrees of displacement and motion of the fragments during the healing period, resulting in a slow healing process. Three-point bending using specially designed forceps[11,24,78-80] and Greiff's fracture device[21] are easy to use and control the location of the fracture site. However, there is not enough control built in, and the force applied remains unknown. The method described by Jackson[8] can create a standard, closed, transverse, internally stabilized mid-shaft femur fracture in the rat with a minimal soft tissue injury. The fracture mechanism mimics the traumatic situation, but the fracture apparatus is complicated. The method described by Bonnarens and Einhorn[12] has the same advantages and the fracture device can be easily built. But the shortcoming of using the femur is that the fracturing procedure is difficult and sometimes frustrating, since the fracture site is not easy to control because of the bulky soft tissue around it.

To achieve rapid healing and avoid angulation and movement, open[8,12,24,77] or closed[77,79] intramedullary pinning, both before[9] or after fracturing,[76] have been used successfully. The closed pinning technique used in authors' laboratory has been proved to be simple and reliable.[28]

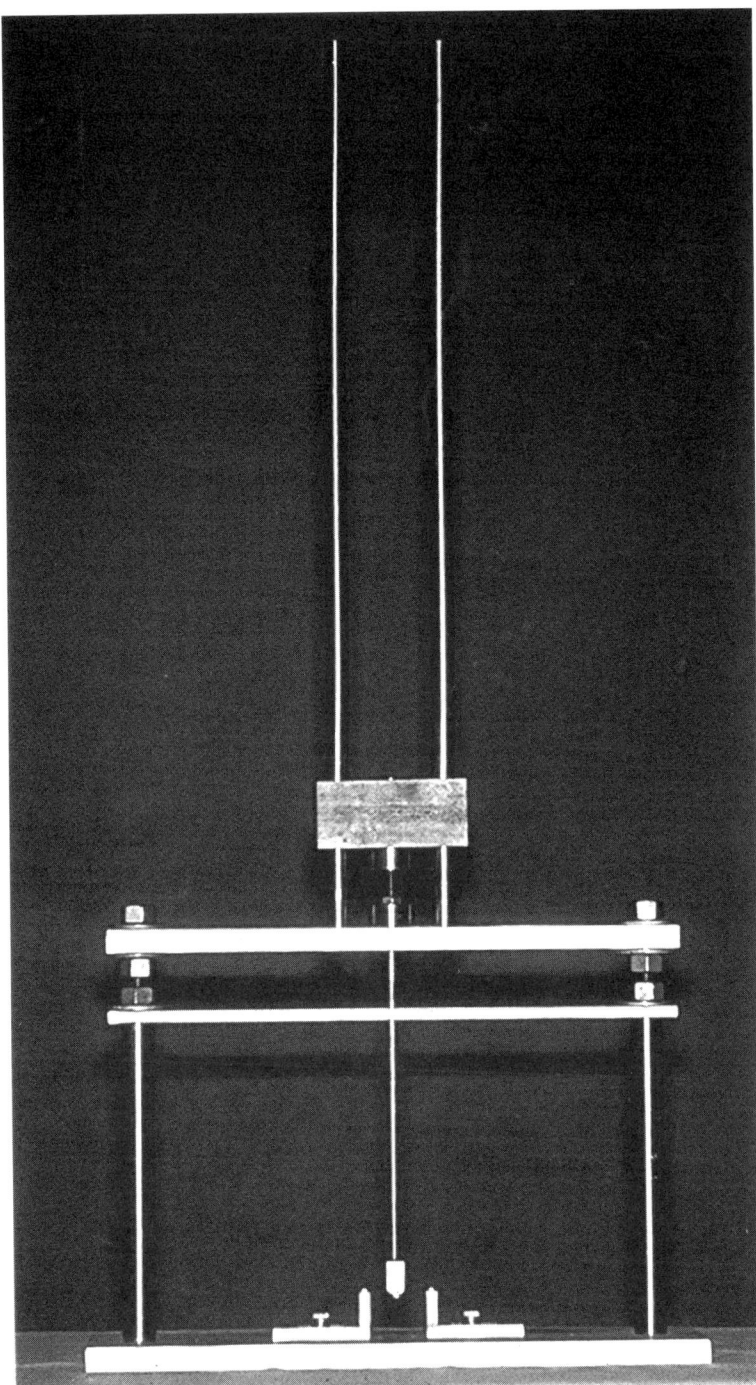

FIGURE 1. Photograph of the fracture device used in the authors' laboratory.

3. Tibial Fractures in Rabbits, Dogs and Sheep

Rabbit tibial fracture is another major fracture model (Table 1). The fracturing methods used in this model include saw osteotomy, fracture devices, and drill plus bone cutter. A high-speed dental burr also has been reported for creating a tibial osteotomy.[47] Compared to rat long bones,

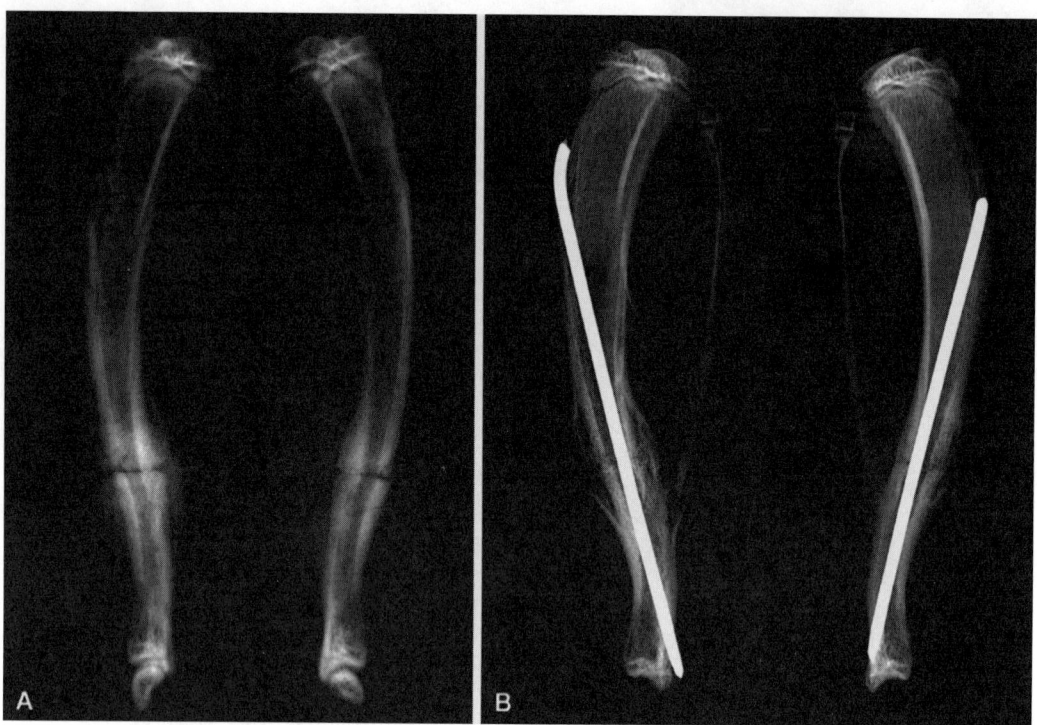

FIGURE 2. Lateral radiographic image of a rat tibia three weeks following intramedullary nailing and fracturing (A). Note the intramedullary nails were removed. The same type of fracture fully healed at five weeks (B).

rabbit tibia is suitable for different kinds of fixation methods, such as plate and screws, intramedullary nailing, external fixator, and even long-leg cast. This model has been used successfully for many purposes such as the evaluation of normal fracture healing, the effect of fixation devices, or effect of growth factors, hormones, or prostaglandin E_2 on fracture healing. A unique stress fracture model of rabbit tibia was created by using repeated application of nontraumatic impulsive loads for three to nine weeks.[46] Changes in the bone were monitored by radiograph and bone scan and the presence of stress fractures was confirmed in some cases at six weeks.

Aside from being higher level vertebrates, the unique side of dogs and sheep is that they have large size bones which allow the direct use of the fixation devices designed for human. Osteotomies have been created by using a saw, a "bone breaker,"[57] or Tepic's fracturing device.[72] The osteotomies have been fixed with plate and screws, intramedullary nails, or external fixators. These tibial fracture models have been used successfully for studying fracture healing, factors affecting bone healing, efficacy of fixation devices, and effects of fracture on bone blood flow (Table 1).

4. Radial Fractures in Rabbits and Dogs

Radial osteotomy in rabbits and dogs can be made using a powered saw. In the authors' laboratory, a small diamond circular saw blade driven by nitrogen gas through a dental handpiece has been proven to be accurate and caused limited trauma to the bone and surrounding tissues in a rabbit model (Figure 3). A three point bending fracture device could also be used to create a fracture of the radius.[64] The advantage of this model is that no fixation is needed, especially in the rabbit, so that bone healing can be observed without the influence of invasive fixation devices. However, because of the active nature of dogs, ulnar fracture may occur. The risk of ulnar fracture can be minimized by the use of an external support such as a coaptation splint for extra protection.[64,66] The fractured radius in dogs can also be transfixed internally to the ulna with Steinmann pins.[65]

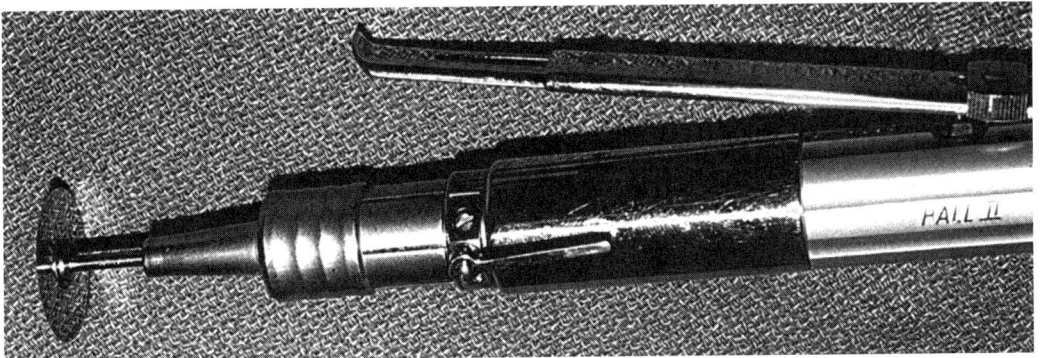

FIGURE 3. A small diamond circular saw used for creating radial osteotomies in the rabbit.

IV. ANIMAL MODELS OF EPIPHYSOMETAPHYSEAL MODELS

Many epiphysometaphyseal osteotomy models have been reported in the literature. Most of the models are used for testing bioabsorbable fixation devices (see Chapter 12). Only the models not used for testing absorbable materials and several representative ones used for testing absorbable materials are listed in Table 2, such as the femoral transcondylar osteotomy in the rabbit,[83] the osteotomy of medial tibial plateau in the rabbit,[87] Salter-Harris type IV fracture in the goat,[90] the subcapital femoral neck fracture in the sheep,[89] and the lateral femoral condyle osteotomy in the dog (see Figure 3 in Chapter 12).[28]

V. ANIMAL MODELS OF DELAYED UNION AND NONUNION

Delayed union and nonunion are special types of fractures models, representing abnormal fracture healing. They are used to study physiopathological conditions and treatment methods of delayed union and nonunion. Delayed unions and nonunions of animal radius, femur, tibia and fibula have been reported in rats, rabbits, and dogs (Table 3).

The basic principle of creating a delayed union and nonunion is the depletion of osteoconductive and osteoinductive factors at the fracture site. The common methods include a 3–10 mm resection of the diaphyseal or metaphyseal bone and the extensive stripping of surrounding periosteum. External fixators have been used for fixing osteotomies and keeping the bone ends apart. The latter can be also achieved by using a silicone rubber spacer.[95] To devitalize the bone ends, freezing at –20°C has been reported.[98] Loose intramedullary nailing (by excessive reaming) also facilitate the establishment of delayed union or nonunion.[16,91]

VI. EVALUATION METHODS

A. RADIOGRAPHY

Radiography is the basic method for evaluating fracture healing. Radiographs should be taken immediately after surgery to examine the location of the fracture and the quality of fixation. Periodic radiographies are essential for monitoring the process of fracture healing. After the animals are sacrificed, the bone specimens should be radiographed again using a high resolution X ray machine, such as Faxitron (Hewlett Packard, McMinnville, OR). The high resolution images may be used for a variety of measurements, such as bone density or bone dimensions. For comparison, it is very important that radiographs are taken before surgical procedures and also include the control limb.

For long bone fractures or osteotomies, healing parameters, such as periosteal reaction (callus formation), quality of union, and bone remodeling are quantitated based on a radiographic scoring

TABLE 2
Animal Models of Epiphysometaphyseal Osteotomies and Other Osteotomies

Animal	Bone	Osteotomy or fracture	Fixation treatment	Purpose	First author, year[ref.]
Rabbit	Femur	Transcondylar	PGA thread and rods	Testing absorbable fixation devices	Vainionpää 1986[83]
		Salter-Harris III and IV fracture	Fixed with screw or no fixation	Testing effect of rigid fixation	Gomes 1993[84]
		Medial condyle	No fixation	Observation of bone remodeling	Bogoch 1993[85]
	Tibia	Upper metaphyseal	No fixation	Simulating metaphyseal fracture	Aronson 1990[86]
		Proximal tibial	PLA or stainless steel screws	Testing absorbable fixation devices	Matsusue 1991[87]
Dog	Femur	Lateral femoral condyle	PGA/PLA screws	Testing absorbable fixation devices	An 1997[28]
Pig	Femur	Lateral femoral condyle	No fixation	Fracture produced by shear force	Tomatsu 1992[88]
Sheep	Femur	Subcapital osteotomy	SR-PLA lag-screws	Testing absorbable fixation devices	Vasenius 1993[89]
Goat	Femur	Salter-Harris IV fracture	PLA, polydioxanone pins	Testing absorbable fixation devices	Donigian 1993[90]

TABLE 3
Animal Models of Bone Nonunion

Animal	Bone	Method of fixation	Delayed union or Nonunion	Purpose or treatment	First author, year[ref.]
Rat	Femur	Reaming, manual fracture, nailing	Nonunion	New method to produce nonunion	Hietaniemi 1995[91]
		As above	Nonunion	Connective tissues in bone nonunion	Hietaniemi 1996[16]
	Fibula	8 mm periosteum stripped at the distal fibula	Nonunion	New method to produce nonunion	Aro 1985[92]
Rabbit	Radius	10 mm defect, periosteum resection	Delayed union	Treatment with coupled electric fields	Rijal 1994[93]
	Tibia	Osteotomy, marrow removing, silastic sheath wrapping of bone ends	Nonunion	A new nonunion model	Oni 1995[94]
	Fibula	Silicone rubber spacer for 48 days	Delayed union	Treatment with direct electric current stimulation	Petersson 1983[95]
Dog	Radius	3 mm defect, 10 mm periosteum stripping, no fixation	Nonunion	New method to produce nonunion	Santos Neto 1984[96]
	Radius	5 mm defect, 2 cm periosteum stripping, no fixation	Nonunion	New method to produce nonunion	Volpon 1994[34]
	Radius	6 mm defect, external fixator	Nonunion	Treatment with injectable demineralized bone matrix	Tiedman 1991[97]
	Tibia	5 mm osteotomy, external fixator bone ends frozen twice at −20°C	Nonunion	Prediction of nonunion with dual energy X ray absorptiometry	Markel 1995[98]

TABLE 4
Radiographic Scoring System for Fracture Healing

Categories	Scores
Periosteal reaction	
Full (across the defect)	3
Moderate	2
Mild	1
None	0
Bone union	
Union	3
Moderate bridge (>50%)	2
Mild bridge (<50%)	1
Nonunion	0
Remodeling	
Full remodeling cortex	2
Intramedullary canal	1
No remodeling	0
Maximum total score	8

system modified from the ones by Bos et al.,[99] Lane and Sandhu,[100] Yang et al.,[101] and Johnson et al.[102] (Table 4).

B. HISTOLOGY AND HISTOMORPHOMETRY

Histology is another basic method for evaluating fracture healing. Generally, longitudinal sections through the fracture callous and surrounding area are cut and stained with H&E or other stains. Common histological parameters include the following categories: callus formation, bone union, marrow changes, and cortex remodeling, which can be semi-quantitated based on a scoring system modified from the ones by Bos et al.,[99] Nilsson et al.,[103] Lane and Sandhu,[100] Heiple et al.,[104] and Suh et al.[105] (Table 5). Parameters for histomorphometry include original and new bone area, chondral tissue, fibrocartilage, fibrous vascular tissue (marrow).[106] See Chapters 6 and 7 for more information on bone histomorphometry.

C. MECHANICAL TESTING

To evaluate the mechnical properties of bone associated with fracture healing, bending and torsional tests are the most popular. For bending tests, the classic article is the one reported by Burstein et al.[107] The bending test has been used to measure the mechanical properties of mouse tibia,[6] rat femur,[10,11] rat tibia,[22,24,28,29] rabbit tibia,[26,44,45–51] canine radius,[64–66] canine tibia,[57,58,61] and sheep tibia.[67,72,73]

Torsional tests often are used for larger tubular bones such as the tibia or femur of rabbits[39,40] and dogs.[28,62,63] Based on torque-angle curves and radiographic findings, for mechanical stages of fracture healing were classified by White et al.:[39] I — fracture through the original fracture site (low stiffness); II — through the original fracture site (high stiffness); III — partially through the original fracture site and partially through intact bone (high stiffness); and IV — entirely through intact bone (high stiffness). These stages correlate with the quality of fracture healing and the healing time.

Indentation tests have been used to evaluate the mechanical properties of the fracture callus.[61] Occasionally, tensile tests have been used for testing bone mechanical properties as a function of fracture healing.[11,27]

TABLE 5
Histological Scoring System for Fracture Healing

Categories	Scores
Callus formation	
Full (across the defect)	3
Moderate	2
Mild	1
None	0
Bone union	
Full bone bridge (union)	3
Moderate bridge (>50%)	2
Mild bridge (<50%)	1
No new bone in the fracture line (nonunion)	0
Marrow changes	
Adult type fatty marrow	4
2/3 replaced by new tissue	3
1/3 replaced by new tissue	2
Fibrous tissue	1
Red	0
Cortex remodeling	
Full remodeling cortex	2
Intramedullary canal	1
No remodeling	0
Maximum total score	12

* For some osteogeneic materials such as collagen sponge/GFs, this category may not apply because of the nearly absent periosteal reaction.

D. OTHER EVALUATION METHODS

Other methods or special devices have been used for evaluating fracture healing, such as autoradiography,[20] quantitative roentgenographic densitometry (QRD),[60,61] single-photon absorptiometry (SPA),[25,61] dual energy X ray absorptiometry (DEXA),[61,98,108] bone scan (for osteotomies[36] or stress fractures[46]), quantitative CT (QCT),[61,73] or MRI.[61,109] Biological markers for bone healing and bone formation can be detected in serum (see Chapter 6).

VII. METHODS FOR ENHANCING FRACTURE HEALING

This topic is too big for the size of this chapter. Readers may want to refer to the comprehensive review on this subject by Einhorn,[110] who categorized the methods of enhancing fracture healing as biological enhancement and biophysical enhancement. The former includes local (osteogenic, osteoconductive, and osteoinductive methods) and systemic approach (prostaglandins and other circulating osteogenic substances). We want to add to the category of systemic approach the enhancing effect of certain hormones or drugs on fracture healing. The biophysical enhancements include local mechanical, electrical, and ultrasonic stimulation.

A. BONE GRAFTING

Bone grafting is an osteogenic approach including the use of autogenic or allogenic bone grafts, demineralized bone matrix (DBM) graft, autogenic bone marrow, or the combined use of autogenic marrow with the other grafts.[110] These grafts have the potential to facilitate direct osteogenesis, osteoinduction, and osteoconduction (not for marrow grafted alone) (See chapter 14 for more details).

B. Biomaterials

Biomaterials, such as porous hydroxyapatite, tricalcium phosphate, bioactive glasses, or some biodegradable polymers are osteoinductive, meaning that they support the ingrowth of blood vessels, perivascular tissues, and osteoprogenitor cells from the recipient bed into the graft (or scaffold).[110] Osteoconductive materials cannot induce bone formation at extraskeletal sites. Bone formation can be achieved by adding osteoinductive or osteogenic substances such as *in vitro* cultured osteoblasts (tissue engineering technique), autogenic marrow, GFs, or DBM (see Chapter 14).

C. Local Use of Growth Promoting Substances

Growth factors (GFs) (peptide-signaling molecules) such as BMP, TGF-β, FGF, and PDGF have been identified at the fracture site and have been verified to have osteoinductive function by stimulating the generation of osteoprogenitor cells from undifferentiated perivascular mesenchymal cells and enhancing osteoblast proliferation and migration.[110–112] The number of reports on animal models for *in vivo* effects of GFs on fracture healing has increased since several years ago, including the effect of local application of TGF-β on rat or rabbit tibial fracture,[29,113] PDGF on rabbit tibial osteotomies,[51] rhbFGF on rat fibular fracture,[33] and NGF-α (nerve growth factor) on rat rib fracture.[114] GFs for clinical use will become commercially available in the near future.[112]

Besides its systemic effect on fracture healing (see below), prostaglandin E^2 (PGE^2) has been demonstrated being released at the fracture site,[115] which has been believed to function as a local mediator for fracture healing.[110,116] Local infusion of PGE^2 in a rabbit fracture model has shown a stimulating effect on fracture healing (callus formation).[26]

D. Systemic Use of Hormones or Drugs

Animal studies have indicated that growth hormone has stimulatory effects on fracture healing.[79,80,117] Growth hormone has an initial stimulatory effect on external callus formation and also stimulates the haemopoietic system. Besides its local effect, PGE^2 also showed systemic stimulation of fracture healing in canine models.[118,119] There are other systemic hormones or factors having potential effects of stimulating fracture healing, such as estrone,[120] 1α-OH-D3,[121] Factor XIII,[122] or calcitonin.[123]

A few drugs or agents have been found to have enhancing effects on fracture healing, which are diphenylhydantoin (Dilantin),[124] clodronate,[27] and vitamin D3.[125] Chinese herbal medicine has been used in Asian countries for fracture healing for more than two thousand years. However, only a few articles reporting animal studies on herbs have been published in the English literature.[126,127]

E. Biophysical Stimulation

Mechanical stimulation, such as controlled micromotion, distraction, or weightbearing, has the ability of enhancing fracture healing,[110] while early full weightbearing delays fracture healing.[128] The commonly used animal models for testing the effects of mechanical stimulation or weightbearing include fractures or osteotomies of rat femur,[10] sheep tibia,[129] rabbit fibulae,[130] and canine tibia.[62] Clinical trials showed that mechanical stimulation appears to enhance fracture healing applied through an external fixator.[131] Encouraging clinical results on enhanced fracture healing by weight bearing have also been reported.[132]

Electrical stimulation was a hot topic in the late 1970s and 1980s. Although only a small number of articles have been published in recent years, the validity of electrical stimulation for the treatment of delayed union and nonunion has been established. Generally, three types of electrical stimulation devices have been used, including constant direct current stimulation, electromagnetic stimulation, and capacitive coupling. The commonly used animal models for testing the effects of electrical stimulation include delayed unions of rabbit fibulae[95] or radius,[93] nonunions of canine ulna,[133] fresh fracture of rabbit fibulae[52] or tibia,[134] and the long bone lengthening model of canine tibia.[135]

Duarte was the first to introduce the use of ultrasound for stimulating the growth of bone and fracture healing in fibular bone defects in the rabbit.[136] His results were further validated by others using the rabbit fibular model[137] and rat femoral model.[138] In a rabbit fibular model, it was shown that ultrasound stimulated fracture healing and the endogenous PGE^2 level was also significantly elevated, paralleling the enhanced bone healing,[139] leading to the suggestion that bone healing stimulated by ultrasound may be mediated via the production of PGE^2. Ultrasound also accelerates the clinical and overall healing of human fractures.[140]

The mechanisms of enhancing fracture healing by physical stimulation methods may be that the physical signals trigger the activation or release of biochemical mediators for osteogenesis, such as GFs and PGE^2.[116]

VIII. OTHER FACTORS AFFECTING FRACTURE HEALING

A. Effects of Fixation Devices

Stability of the fracture site is the most important factor in the successful treatment of fractures. Generally, a rigid, stable fixation allows early primary healing through the formation of internal callus, a stable but less rigid fixation leads to a slower secondary healing through the formation of external callus, and an unstable fixation tends to fail leading to delayed union or nonunion.[116,141–143]

Controversy remains as to what kind of rigidity is the best for fracture healing. Rigid plate fixation (compression plating) stabilizes the fracture, allowing primary healing and early bone remodeling. However, rigid fixation minimizes granulation tissue and external callus, which may be caused by retardation of the release of GFs and PGE^2 at the fracture surfaces. Rigid fixation provides load shielding through the metal plate, causing lower bone strength. Less rigid plating creates slower secondary fracture healing and remodeling, causing less load shielding and resulting in higher ultimate strength of the bone union. The rigidity of external fixators has effects similar to plating, allowing primary healing with rigid fixation or secondary healing with less rigid or flexible fixation.

Intramedullary nailing is a successful procedure which allows some motion and loading at the bone ends (a less rigid fixation) and is usually associated with external callus formation. Intramedullary reaming causes circulatory disturbances in the inner 2/3 of the cortex, but it does not impede the formation of external callus and the damaged parts will be revascularized.

Based on the existing knowledge and authors' opinion, the future direction of research is to define a carefully selected range of rigidity which is rigid enough for a stable fixation and primary healing and also has certain flexibility for strong bone union by partial secondary healing. For testing fixation devices, larger animals such as rabbits,[35,37,40,42–44] dogs,[55,57,58,62] or sheep[67,68] should be used (Table 1) due to the need for good size bones, better comparison to human conditions, and the convenience of implant manufacture.

B. Effects of Drugs or Medication

Nonsteroidal antiinflammatory drugs (NSAIDs), especially indomethacin, are notorious substances for their inhibition effect on bone growth and fracture healing.[144–146] Other NSAIDs having inhibiting effects on fracture healing include aspirin[145] and ibuprofen[13]

Several other agents or medications having inhibiting or discouraging effects on fracture healing have been demonstrated, such as estrogen,[147] cortisone,[19] and irradiation.[148] There are also other drugs or substances for systemic use, such as ethane-1-hydroxy-1,1-diphosphonate (EHDP)[64] or alendronate,[66] which have been indicated to have no significant inhibiting effect on experimental fracture healing. Rat fracture models are ideal for testing the effect of systemic drugs on fracture healing.

REFERENCES

1. Urist, M. R. and Johnson, R. W., Jr., "Calcification and ossification. IV. The healing of fractures in man under clinical conditions," *J. Bone. Joint Surg.*, 25, 375, 1943.
2. McKibbin, B., "The biology of fracture healing in long bones," *J. Bone Joint Surg.*, 60B, 150, 1978.
3. Sevitt, S., *Bone Repair and Fracture Healing in Man.* Churchill Livingstone, Edinburgh, 1981.
4. Sandberg, M. M., "Gene expression during bone repair," *Clin. Orthop.*, 289, 292, 1993.
5. Hsu, J. D. and Robinson, R. A., "Studies on the healing of long-bone fractures in hereditary pituitary insufficient mice," *J. Surg. Res.*, 9, 535, 1969.
6. Hiltunen, A., Vuorio, E., and Aro, H. T., "A standardized experimental fracture in the mouse tibia," *J. Orthop. Res.*, 11, 305, 1993.
7. Eskelund, V. and Plum, C. M., "Experimental investigations into the healing of fractures. I. Healing of fractures in the femoral diaphysis in rats," *Acta Orthop. Scand.*, 19, 433, 1949.
8. Jackson, R. W. and Reed, C. A., "Production of a standard experimental fracture," *Can. J. Surg.*, 13, 415, 1970.
9. Kernek, C. B. and Wray, J. B., "Cellular proliferation in the formation of fracture callus in the rat tibia," *Clin. Orthop.*, 91, 197, 1973.
10. Sarmiento, A., Schaeffer, J. F., Beckermen, L., Latta, L. L., and Enis, J. E., "Fracture healing in rat femora as affected by functional weight-bearing," *J. Bone Joint Surg.*, 59A, 369, 1977.
11. Ekeland, A., Engesaeter, L. B., and Langeland, N., "Mechanical properties of fractured and intact rat femora evaluated by bending, torsional and tensile tests," *Acta Orthop. Scand.*, 52, 605, 1981.
12. Bonnarens, F. and Einhorn A., "Production of a standard closed fracture in laboratory animal bone," *J. Orthop. Res.*, 2, 97, 1984.
13. Huo, M. H., Troiano, N. W., Pelker, R. R., Gundberg, C. M., and Friedlaender, G. E., "The influence of ibuprofen on fracture repair: biomechanical, biochemical, histologic, and histomorphometric parameters in rats," *J. Orthop. Res.*, 9, 383, 1991.
14. Grundnes, O. and Reikeras, O., "Effects of instability on bone healing. Femoral osteotomies studied in rats," *Acta Orthop. Scand.*, 64, 55, 1993.
15. Olmedo, M. L. and Weiss, A. P., "An experimental rat model allowing controlled delivery of substances to evaluate fracture healing," *J. Orthop. Trauma*, 8, 490, 1994.
16. Hietaniemi, K., Paavolainen, P., and Penttinen, R., "Connective tissue parameters in experimental nonunion," *J. Orthop. Trauma*, 10, 114, 1996.
17. Urist, M. R. and McLean, F. C., "Calcification and ossification. I. Calcification in the callus in healing fractures in normal rats," *J. Bone Joint Surg.*, 23, 1, 1941.
18. Penttinen, R., "Biochemical studies on fracture healing in the rat," *Acta Chir. Scand. Suppl.*, 432, 1, 1972.
19. Hulth, A. and Olerud, S., "Early fracture callus in normal and cortisone treated rats," *Acta Orthop. Scand.*, 34, 1, 1964.
20. Greiff, J., "Autoradiographic studies of fracture healing using 99Tcm-Sn-polyphosphate," *Injury*, 9, 271, 1978.
21. Greiff, J., "A method for the production of an undisplaced reproducible tibial fracture in the rat," *Injury*, 9, 278, 1978.
22. Mölster, A., Gjerdet, N. R., Raugstad, T. S., Hvidsten, K., Alho, A., and Bang, G., "Effect of instability of experimental fracture healing," *Acta Orthop. Scand.*, 53, 521, 1982.
23. Lowe, J., Bab, I., Stein, H., and Sela, J., "Primary calcification in remodeling haversian systems following tibial fracture in rats," *Clin. Orthop.*, 176, 291, 1983.
24. Bak, B. and Andreassen, T. T., "Reduced energy absorption of healed fracture in the rat," *Acta Orthop. Scand.*, 59, 548, 1988.
25. Aro, H. T., Wippermann, B. W., Hodgson, S. F., Wahner, H. W., Lewallen, D. G., and Chao, E. Y. S., "Prediction of properties of fracture callus by measurement of mineral density using micro-bone densitometry," *J. Bone Joint Surg.*, 71A, 1020, 1989.
26. Keller, J., Klamer, A., Bak, B., and Suder P., "Effect of local prostaglandin E_2 on fracture callus in rabbits," *Acta Orthop. Scand.*, 64, 59, 1993.
27. Nyman, M. T., Paavolainen, P., and Lindholm, T. S., "Clodronate increases the calcium content in fracture callus. An experimental study in rats," *Arch. Orthop. Trauma Surg.*, 112, 228, 1993.

28. An, Y. H., Friedman, R. J., Parent, T., and Draughn R. A. "Production of a standard closed fracture in the rat tibia," *J. Orthop. Trauma.*, 8, 111, 1994.
29. Nielsen, H. M., Andreassen, T. T., Ledet, T., and Oxlund, H., "Local injection of TGF-beta increases the strength of tibial fractures in the rat," *Acta Orthop. Scand.*, 65, 37, 1994.
30. Otto, T. E., Patka, P., and Haarman, H. J., "Closed fracture healing: a rat model," *Eur. Surg. Res.*, 27, 277, 1995.
31. David, R., Nissan, M., Cohen, I., and Soudry, M., "Effect of low-power He-Ne laser on fracture healing in rats," *Lasers Surg. Med.*, 19, 458, 1996.
32. Herbsman, H., Asrani, U. F., and Shaftan, G. W., "An improved method for the evaluation of experimental fracture healing," *Surg. Forum.*, 17, 447, 1966.
33. Kawaguchi, H., Kurokawa, T., Hanada, K., Hiyama, Y., Tamura, M., Ogata, E., and Matsumoto, T., "Stimulation of fracture repair by recombinant human basic fibroblast growth factor in normal and streptozotocin-diabetic rats," *Endocrinology*, 135, 774, 1994.
34. Volpon, J. B., "Nonunion using a canine model," *Arch. Orthop. Trauma. Surg.*, 113, 312, 1994.
35. Chai, B. F. and Tang, X. M., "Ultrastructural investigation of healing of experimental fracture treated according to the principle of activating circulation and alleviating stagnation," *Chin. Med. J.*, 98, 409, 1985.
36. Bushberg, J. T., Hoffer, P. B., Schreiber, G. J., Lawson, A. J., Lawson, J. P., and Lord, P., "Comparative uptake of 67Ga and 99mTc MDP in rabbits with a benign noninfected bone lesion (fracture)," *Invest. Radiol.*, 20, 498, 1985.
37. Manninen, M. J. and Pohjonen, T., "Intramedullary nailing of the cortical bone osteotomies in rabbits with self-reinforced poly-L-lactide rods manufactured by the fibrillation method," *Biomaterials*, 14, 305, 1993.
38. Albanese, S. A., Spadaro, J. A., Chase, S. E., and Geel, C. W., "Bone growth after osteotomy and internal fixation in young rabbits," *J. Orthop. Res.*, 14, 921, 1996.
39. White, A. A., III, Panjabi, M. M., and Southwick, W. O., "The four biomechanical stages of fracture repair," *J. Bone Joint Surg.*, 59A, 188, 1977.
40. Paavolainen, P., Slatis, P., Karaharju, E., and Holmstrom, T., "The healing of experimental fractures by compression osteosynthesis. I. Torsional strength," *Acta Orthop. Scand.*, 50, 369, 1979.
41. Ashhurst, D. E., Hogg, J., and Perren, S. M., "A method for making reproducible experimental fractures of the rabbit tibia," *Injury*, 14, 236, 1982.
42. Court-Brown, C. M., "The effect of external skeletal fixation on bone healing and bone blood supply. An experimental study," *Clin. Orthop.*, 201, 278, 1985.
43. Terjesen. T. and Johnson, E., "Effects of fixation stiffness on fracture healing. External fixation of tibial osteotomy in the rabbit," *Acta Orthop. Scand.*, 57, 146, 1986.
44. Aalto, K., Holmstrom, T., Karaharju, E., Joukainen, J., Paavolainen, P., and Slatis, P., "Fracture repair during external fixation. Torsion tests of rabbit osteotomies," *Acta Orthop. Scand.*, 58, 66, 1987.
45. Laftman, P., Nilsson, O. S., Brosjo, O., and Stromberg, L., "Stress shielding by rigid fixation studied in osteotomized rabbit tibiae," *Acta Orthop. Scand.*, 60, 718, 1989.
46. Burr, D. B., Milgrom, C., Boyd, R,D., Higgins, W. L., Robin, G., and Radin, E. L., "Experimental stress fractures of the tibia. Biological and mechanical aetiology in rabbits," *J. Bone Joint Surg.*, 72B, 370, 1990.
47. Carpenter, J. E., Hipp, J. A., Gerhart. T. N., Rudman, C. G., Hayes, W. C., and Trippel, S. B., "Failure of growth hormone to alter the biomechanics of fracture-healing in a rabbit model," *J. Bone Joint Surg.*, 74A, 359, 1992.
48. Kaartinen, E., Paavolainen, P., Holmstrom, T., Slatis, P., and Happonen, R. P., "Different healing patterns of experimental osteotomies treated by intramedullary nailing," *Arch. Orthop. Trauma Surg.*, 112, 171, 1993.
49. Jacob, E., Cierny, G. III, Fallon, M. T., McNeill, J. F., Jr., and Siderys, G. S., "Evaluation of biodegradable cefazolin sodium microspheres for the prevention of infection in rabbits with experimental open tibial fractures stabilized with internal fixation," *J. Orthop. Res.*, 11, 404, 1993.
50. Worlock, P., Slack, R., Harvey, L., and Mawhinney, R., "The prevention of infection in open fractures: an experimental study of the effect of fracture stability," *Injury*, 25, 31, 1994.
51. Nash, T. J., Howlett, C. R., Martin. C., Steele, J., Johnson, K. A., and Hicklin, D. J., "Effect of platelet-derived growth factor on tibial osteotomies in rabbits," *Bone*, 15, 203, 1994.

52. Pienkowski, D., Pollack, S. R., Brighton, C. T., and Griffith, N. J., "Low-power electromagnetic stimulation of osteotomized rabbit fibulae. A randomized, blinded study," *J. Bone Joint Surg.*, 76A, 489, 1994.
53. Brighton, C. T. and Hunt, R. M., "Early histological and ultrastructural changes in medullary fracture callus," *J. Bone Joint Surg.*, 73A, 832, 1991.
54. Rhinelander, F. W. and Stewart. C. L., "Experimental fixation of femoral osteotomies by cerclage with nylon straps," *Clin. Orthop.*, 179, 298, 1983.
55. Miettinen, H., Mäkelä, E. A., Rokkanen, P., and Törmälä, P., "Fixation of femoral shaft osteotomy with intramedullary metallic or absorbable rod: an experiment study on growing dogs," *J. Biomater. Sci. Polym. Ed.*, 4, 135, 1992.
56. An, Y. H., Friedman, R. J., Powers, D. L., Draughn, R. A., and Latour, R. A., "Fixation of osteotomies using bioabsorbable scews in the canine femur," *Clin. Orthop.*, 355, 300, 1998.
57. Skirving, A. P., Day, R., Macdonald, W., and McLaren, R., "Carbon fiber reinforced plastic (CFRP) plates versus stainless steel dynamic compression plates in the treatment of fractures of the tibiae in dogs," *Clin. Orthop.*, 224, 117, 1987.
58. Gilbert, J. A., Dahners, L. E., and Atkinson, M. A., "The effect of external fixation stiffness on early healing of transverse osteotomies," *J. Orthop. Res.*, 7, 389, 1989.
59. Smith, S. R., Bronk, J. T., and Kelly, P. J., "Effect of fracture fixation on cortical bone blood flow," *J. Orthop. Res.*, 8, 471, 1990.
60. Tiedeman, J. J., Lippiello, L., Connolly, J. F., and Strates, B. S., "Quantitative roentgenographic densitometry for assessing fracture healing," *Clin. Orthop.*, 253, 279, 1990.
61. Markel, M. D., Wikenheiser, M. A., Morin, R. L., Lewallen, D. G., and Chao, E. Y., "Quantification of bone healing. Comparison of QCT, SPA, MRI, and DEXA in dog osteotomies," *Acta Orthop. Scand.*, 61, 487, 1990.
62. Aro, H. T. and Chao, E. Y., "Bone-healing patterns affected by loading, fracture fragment stability, fracture type, and fracture site compression," *Clin. Orthop.*, 293, 8, 1993.
63. O'Sullivan, M. E., Bronk, J. T., Chao, E. Y., and Kelly. P. J., "Experimental study of the effect of weight bearing on fracture healing in the canine tibia," *Clin. Orthop.*, 302, 273, 1994.
64. Lenehan, T. M., Balligand, M., Nunamaker, D. M., and Wood, F. E. Jr., "Effect of EHDP on fracture healing in dogs," *J. Orthop. Res.*, 3, 499, 1985.
65. Chakkalakal, D. A., Lippiello, L., Wilson, R. F., Shindell, R., and Connolly, J. F., "Mineral and matrix contributions to rigidity in fracture healing," *J. Biomech.*, 23, 425, 1990.
66. Peter, C. P., Cook, W. O., Nunamaker, D. M., Provost, M. T., Seedor, J. G., and Rodan, G. A., "Effect of alendronate on fracture healing and bone remodeling in dogs," *J. Orthop. Res.*, 14, 74, 1996.
67. Heitemeyer, U., Claes, L., Hierholzer, G., and Korber, M., "Significance of postoperative stability for bony reparation of comminuted fractures. An experimental study," *Arch. Orthop. Trauma Surg.*, 109, 144, 1990.
68. Goodship, A. E., Watkins, P. E., Rigby, H. S., and Kenwright, J., "The role of fixator frame stiffness in the control of fracture healing. An experimental study," *J. Biomech.*, 26, 1027, 1993.
69. Schemitsch, E. H., Kowalski, M. J., Swiontkowski, M. F., and Senft, D., "Cortical bone blood flow in reamed and unreamed locked intramedullary nailing: a fractured tibia model in sheep," *J. Orthop. Trauma.*, 8, 373, 1994.
70. Wallace, A. L., Makki, R., Weiss, J. B., and Hughes, S. P., "Measurement of serum angiogenic factor in devascularized experimental tibial fractures," *J. Orthop. Trauma*, 9, 324, 1995.
71. Schemitsch, E. H., Kowalski, M. J., and Swiontkowski, M. F., "Soft-tissue blood flow following reamed versus unreamed locked intramedullary nailing: a fractured sheep tibia model," *Ann. Plast. Surg.*, 36,70, 1996.
72. Tepic, S., Remiger, A. R., Morikawa, K., Predieri, M., and Perren, S. M., "Strength recovery in fractured sheep tibia treated with a plate or an internal fixator: an experimental study with a two-year follow-up," *J. Orthop. Trauma*, 11, 14, 1997.
73. Augat, P., Merk, J., Genant, H. K., and Claes, L., "Quantitative assessment of experimental fracture repair by peripheral computed tomography," *Calcif. Tissue Int.*, 60, 194, 1997.
74. Curtis, M. J., Brown, P. R., Dick, J. D., and Jinnah, R. H., "Contaminated fractures of the tibia: a comparison of treatment modalities in an animal model," *J. Orthop. Res.*, 13, 286, 1995.

75. Hara, Y., Tagawa, M., Ejima, H., et al., "Clinical evaluation of uniaxially oriented poly-L-lactide rod for fixation of experimental femoral diaphyseal fracture in immature cats," *J. Vet. Med. Sci.*, 56, 1041, 1994.
76. Illi, O. E., Weigum, H., and Misteli, F., "Biodegradable implant materials in fracture fixation," *Clin. Mater.*, 10, 69, 1992.
77. Aro, H. T., Wippermann, B. W., Hodgson, S. F., Wahner, H. W., Lewallen, D. G., and Chao, E. Y. S., "Prediction of properties of fracture callus by measurement of mineral density using micro-bone densitometry," *J. Bone Joint Surg.*, 71A, 1020, 1989.
78. Ekeland, A., Engesaeter, L. B., and Langeland, N., "Influence of age on mechanical properties of healing fractures and intact bones in rats," *Acta Orthop. Scand.*, 53: 527, 1982.
79. Bak, B., Jorgensen, P. H., and Andreassen, T. T., "Dose response of growth hormone on fracture healing in the rat," *Acta Orthop. Scand.*, 61, 54, 1990.
80. Bak, B., Jorgensen, P. H., and Andreassen, T. T., "The stimulating effect of growth hormone on fracture healing is dependent on onset and duration of administration," *Clin. Orthop.*, 264, 295, 1991.
81. Northmore-Ball, M. D., Wood, W. M. R., and Meggitt, B. F., "A biomechanical study of the effects of growth hormone in experimental fracture healing," *J. Bone Joint Surg.*, 62B, 391, 1980.
82. Volpin, G., Rees, J. A., Ali, S. Y., and Bentley, G., "Distribution of alkaline phosphatase activity in experimentally produced callus in rats," *J. Bone Joint Surg.*, 68B, 629, 1986.
83. Vainionpää, S., Vihtonen, K., Mero, M., Patiala, H., Rokkanen, P., Kilpikari, J., and Törmälä, P., "Fixation of experimental osteotomies of the distal femur of rabbits with biodegradable material," *Arch. Orthop. Trauma Surg.*, 106, 1, 1986.
84. Gomes, L. S. and Volpon, J. B., "Experimental physeal fracture-separations treated with rigid internal fixation," *J. Bone Joint Surg.*, 75A, 1756, 1993.
85. Bogoch, E., Gschwend, N., Rahn, B., Moran, E., and Perren, S., "Healing of cancellous bone osteotomy in rabbits. Part I. Regulation of bone volume and the regional acceleratory phenomenon in normal bone," *J. Orthop. Res.*, 11, 285, 1993.
86. Aronson, D. D., Stewart, M. C., and Crissman, J. D., "Experimental tibial fractures in rabbits simulating proximal tibial metaphyseal fractures in children," *Clin. Orthop.*, 255, 61, 1990.
87. Matsusue, Y., Yamamuro, T., Yoshii, S., Oka, M., Ikada, Y., Hyon, S-H., and Shikinami, Y., "Biodegradable screw fixation of rabbit tibia proximal osteotomies," *J. Appl. Biomater.*, 2, 1, 1991.
88. Tomatsu, T., Imai, N., Takeuchi, N., Takahashi, K., and Kimura, H., "Experimentally produced fractures of articular cartilage and bone. The effects of shear forces on the pig knee," *J. Bone Joint Surg.*, 74B, 457, 1992.
89. Vasenius, J., Laitinen, O., Pohjonen, T., Vainionpää, S., Törmälä, P., and Rokkanen, P., "Fixation of subcapital femoral osteotomies by poly-L-lactic acid pins. An experimental study in sheep," *Int. Orthop.*, 17, 144, 1993.
90. Donigian, A. M., Plaga, B. R., and Caskey, P. M., "Biodegradable fixation of physeal fractures in goat distal femur," *J. Pediatr. Orthop.*, 13, 349, 1993.
91. Hietaniemi, K., Peltonen, J., and Paavolaine, P., "An experimental model for non-union in rats," *Injury*, 26, 681, 1995
92. Aro, H., Eerola, E., and Aho, A. J., "Development of nonunions in the rat fibula after removal of periosteal neural mechanoreceptors," *Clin. Orthop.*, 199, 292, 1985.
93. Rijal, K. P., Kashimoto, O., and Sakurai, M., "Effect of capacitively coupled electric fields on an experimental model of delayed union of fracture," *J. Orthop. Res.*, 12, 262, 1994.
94. Oni, O. O., "A non-union model of the rabbit tibial diaphysis," *Injury*, 26, 619, 1995.
95. Petersson, C. J. and Johnell, O., "Electrical stimulation of osteogenesis in delayed union of the rabbit fibula," *Arch. Orthop. Trauma. Surg.*, 101, 247, 1983.
96. Santos Neto, F. L. D., and Volpon, J. B., "Experimental nonunion in dogs," *Clin. Orthop.*, 187, 260, 1984.
97. Tiedeman, J. J., Connolly, J. F., Etrates, B. S., and Lippiello, L., "Treatment of nonunion by percutaneous injection of bone marrow and demineralized bone matrix. An experimental study in dogs," *Clin. Orthop.*, 268, 294, 1991.
98. Markel, M. D., Bogdanske, J. J., Xiang, Z., and Klohnen, A., "Atrophic nonunion can be predicted with dual energy X ray absorptiometry in a canine ostectomy model," *Orthop. Res.*, 13, 869, 1995.

99. Bos, G. D., Goldberg, V. M., Powell, A. E., Heiple, K. G., and Zika, J. M., "The effect of histocompatibility matching on canine frozen bone allografts," *J. Bone Joint Surg.*, 65A, 89, 1983.
100. Lane, J. M. and Sandhu, H. S., "Current approaches to experimental bone grafting," *Orthop. Clin. North Am.*, 18, 213, 1987.
101. Yang, C. Y. Simmons, D. J., and Lozano, R., "The healing of grafts combining freeze-dried and demineralized allogeneic bone in rabbits," *Clin. Orthop.*, 298, 286, 1994.
102. Johnson, K. D., Frierson, K. E., Keller, T. S., et al., "Porous ceramics as bone graft substitutes in long bone defects: a biomechanical, histological, and radiographic analysis," *J. Orthop. Res.*, 14, 351, 1996.
103. Nilsson, O. S., Urist, M. R., Dawson, E. G., Schmalzried, T. P., and Finerman, G. A., "Bone repair induced by bone morphogenetic protein in ulnar defects in dogs," *J. Bone Joint Surg.*, 68B, 635, 1986.
104. Heiple, K. G., Goldberg, V. M., Powell, A. E., Bos, G. D., and Zika, J. M., "Biology of cancellous bone grafts," *Orthop. Clin. North Am.*, 18, 179, 1987.
105. Suh, H. and Lee, C., "Biodegradable ceramic-collagen composite implanted in rabbit tibiae," *ASAIO Journal*, 41, M652, 1995.
106. West, P. G., Rowland, G. R., Budsberg, S. C., and Aron, D. N., "Histomorphometric and angiographic analysis of bone healing in the humerus of pigeons," *Am. J. Vet. Res.*, 57, 1010, 1996.
107. Burstein, A. H. and Frankel, V. H., "A standard test for laboratory animal bone," *J. Biomech.*, 4, 155, 1971.
108. Hamanishi, C., Yoshii, T., Totani, Y., and Tanaka, S., "Bone mineral density of lengthened rabbit tibia is enhanced by transplantation of fresh autologous bone marrow cells. An experimental study using dual X ray absorptiometry," *Clin. Orthop.*, 303, 250, 1994.
109. Viljanen, J., Kinnunen, J., Bondestam, S., Majola, A., Rokkanen, P., and Törmälä, P., "Bone changes after experimental osteotomies fixed with absorbable self-reinforced poly-L-lactide screws or metallic screws studied by plain radiographs, quantitative computed tomography and magnetic resonance imaging," *Biomaterials*, 16, 1353, 1995.
110. Einhorn, T. A., "Current concepts review: Enhancement of fracture-healing," *J. Bone Joint Surg.*, 77A, 940, 1995.
111. Einhorn, T. A. and Trippel, S. B., "Growth factor treatment of fractures," *Instr. Course Lect.*, 46, 483, 1997
112. Lind, M., "Growth factors: possible new clinical tools. A review," *Acta Orthop. Scand.*, 67, 407, 1996.
113. Lind, M., "Transforming growth factor-beta enhances fracture healing in rabbit tibiae," *Acta Orthop. Scand.*, 64, 553, 1993.
114. Grills, B. L., "Topical application of nerve growth factor improves fracture healing in rats," *J. Orthop. Res.*, 15, 235, 1997.
115. Dekel, S., "Release of prostaglandins from bone and muscle after tibial fracture. An experimental study in rabbits," *J. Bone Joint Surg.*, 63B, 185, 1981.
116. Hulth, A., "Current concepts of fracture healing," *Clin. Orthop.*, 249, 265, 1989.
117. Mosekilde, L. and Bak, B., "The effect of growth hormone on fracture healing in rats: a histological description," *Bone*, 14, 19, 1993.
118. High, W. B., "Effects of orally administered prostaglandin E-2 on cortical bone turnover in adult dogs: a histomorphometric study," *Bone*, 8, 363, 1987.
119. Norrdin, R. W. and Shih, M. S., "Systemic effects of prostaglandin E^2 on vertebral trabecular remodeling in beagles used in a healing study," *Calcif. Tiisue Int.*, 42, 363, 1988.
120. Negulesco, J. A., "Effects of increased earth gravity and estrone treatment on intact and healing avian radii," *Calcif. Tiisue Res.*, 23, 291, 1977.
121. Lindholm, T. S. and Sevastifoglou, J. A., "The effect of 1alpha-hydroxycholecalciferol on the healing of experimental fractures in adult rats," *Arch. Orthop. Trauma. Surg.*, 49, 485, 1978.
122. Hellerer, O., Bruckner, W. L., Frey, K. W., Westerburg, K. W., and Klessinger, U., "Fracture healing under factor XIII medication.," *Arch. Orthop. Trauma. Surg.*, 97, 157, 1980.
123. Karachalios, T., "Calcitonin effects on rabbit bone. Bending tests on ulnar osteotomies," *Acta Orthop. Scand.*, 63, 615, 1992.
124. Frymoyer, J. W., "Fracture healing in rats treated with diphenylhydantin (Dilantin)," *J. Trauma*, 16, 368, 1976.
125. Omeroglu, S., Erdogan, D., and Omeroglu, H., "Effects of single high-dose vitamin D3 on fracture healing. An ultrastructural study in healthy guinea pigs," *Arch Orthop. Traum Surg.*, 116, 37, 1997.

126. Yan, S. Q., Wang, G. J., and Shen, T. Y., "Effects of pollen from *Typha angustata* on the osteoinductive potential of demineralized bone matrix in rat calvarial defects," *Clin. Orthop.*, 306, 239, 1994.
127. Huang, H. F. and You, J. S., "The use of Chinese herbal medicine on experimental fracture healing," *Am. J. Chin. Med.*, 25, 351, 1997.
128. Augat, P., Merk, J., Ignatius, A., Margevicius, K., Bauer, G., et al., "Early, full weightbearing with flexible fixation delays fracture healing," *Clin. Orthop.*, 328, 194, 1996.
129. Goodship, A. E. and Kenwright, J., "The influence of induced micromovement upon the healing of experimental tibial fractures," *J. Bone Joint Surg.*, 65B, 650,1985.
130. Usui, Y., Zerwekh, J. E., Vanharanta, H., Ashman, R. B., and Mooney, V., "Different effects of mechanical vibration on bone ingrowth into porous hydroxyapatite and fracture healing in a rabbit model," *J. Orthop. Res.*, 7, 559, 1989.
131. Kenwright, J. and Goodship, A. E., "Controlled mechanical stimulation in the treatment of tibial fractures," *Clin. Orthop.*, 241, 36, 1989.
132. Sarmiento, A., "Functional bracing of tibial and femoral fractures," *Clin. Orthop.*, 82, 2, 1972
133. Jacobs, R. R., Luethi, U., Dueland, R. T., and Perren, S. M., "Electrical stimulation of experimental nonunions," *Clin. Orthop.*, 161, 146, 1981.
134. Goh, J. C. H., "Effects of electrical stimulation on the biomechanical properties of fracture healing in rabbits," *Clin. Orthop.*, 233, 268, 1988.
135. Pepper, J. R., Herber, M. A., Anderson, J. R., and Bobechko, W. P., "Effect of capacitive coupled electrical stimulation on regenerate bone," *J. Orthop. Res.*, 14, 296, 1996.
136. Duarte, L. R., "The stimulation of bone growth by ultrasound," *Arch. Orthop. Trauma Surg.*, 101, 153, 1983.
137. Pilla, A. A., Mont, M. A., Nasser, P. R., Khan, S. A., Figueiredo, M., Kaufman, J. J., and Siffert R. S., "Non-invasive low-intensity pulsed ultrasound accelerates bone healing in the rabbit," *J. Orthop. Trauma.*, 4, 246, 1990.
138. Wang, S.-J., Lewallen, D. G., Bolander, M. E., Chao, E. Y., Ilstrup, D. M., and Greenleaf, J. F., "Low intensity ultrasound treatment increases strength in a rat femoral fracture model," *J. Orthop. Res.*, 12, 40, 1994.
139. Tsai, C. L., Chang, W. H., Liu, T. K., and Song, G. M., "Ultrasonic effect on fracture repair and prostaglandin E2 production," *Chin. J. Physiol.*, 35, 27, 1992.
140. Heckman, J. D., Ryaby, J. P., McCabe, J., Frey, J. J., and Kilcoyne, R. F., "Acceleration of tibial fracture-healing by non-invasive, low-intensity pulsed ultrasound," *J. Bone Joint Surg.*, 76A, 26, 1994.
141. Chao, E. Y., Aro, H. T., Lewallen, D. G., and Kelly, P. J., "The effect of rigidity on fracture healing in external fixation," *Clin. Orthop.*, 241, 24, 1989.
142. Cornell, C. N. and Lane, J. M., "Newest factors in fracture healing," *Clin. Orthop.*, 277, 297, 1992.
143. Uhthoff, H. K., "Internal fixation of long bone fractures: concepts, controversies, debates," *Can. Med. Assoc. J.*, 149, 837, 1993.
144. Bo, J., Sudmann, E., and Marton, P. F., "Effect of indomethacin on fracture healing in rats," *Acta Orthop. Scand.*, 47, 588, 1976.
145. Allen, H. L., Wase, A., and Bear, W. T., "Indomethacin and aspirin: effect of nonsteroidal anti-inflammatory agents on the rate of fracture repair in the rat," *Acta Orthop. Scand.*, 51, 595, 1980.
146. Altman, R. D., Latta, L. L., Keer, R., Renfree, K., Hornicek, F. J., and Banovac, K., "Effect of nonsteroidal antiinflammatory drugs on fracture healing: a laboratory study in rats," *J. Orthop. Trauma*, 9, 392, 1995.
147. Engin, A. E., "Effects of oestrogen upon tensile properties of healing fractured avian bone," *J. Biomed. Eng.*, 5, 49, 1983.
148. Pelker, R. R. and Friedlaender, G. E., "The Nicolas Andry Award, 1995. Fracture healing. Radiation induced alterations," *Clin. Orthop.*, 341, 267, 1997.

12 Animal Models for Testing Bioabsorbable Materials

Yuehuei H. An and Richard J. Friedman

CONTENTS

- I. Introduction 219
- II. Basics of Bioabsorbable Materials 220
  - A. Common Bioabsorbable Materials 220
  - B. Biodegradation 220
  - C. Mechanical Properties of Bioabsorbable Materials 220
- III. Commonly Used Animal Models 220
  - A. Biocompatibility and Biodegradation Test in Soft Tissues 220
  - B. Biodegradation in Bone Tissue 222
  - C. Fixation of Fracture and Osteotomy 225
    - 1. Epiphysometaphyseal Osteotomy Models 225
    - 2. Diaphyseal Fracture Models 225
    - 3. Maxillofacial Bone Fracture Models 227
  - D. Bone Replacement 229
  - E. Drug Delivery 229
  - F. Repair of Cartilage Defect 230
  - G. Repair of Meniscus 230
  - H. Repair of Tendons and Ligaments 234
  - I. Small Blood Vessel and Nerve Regeneration 234
- IV. Evaluation Methods 234
- Acknowledgment 234
- References 235

I. INTRODUCTION

Bioabsorbable materials have been studied in many aspects of orthopaedic surgery, including fixation of fractures, bone replacement, cartilage repair, meniscal repair, fixation of ligament, and drug delivery. However, the only major clinical application is fixation of fractures or osteotomies. Absorbable materials have been used in the form of screws, pins, and plates for orthopaedic, oral, and craniofacial surgery. Their mechanical properties can be altered to provide sufficient rigidity to allow bone healing, retain their mechanical properties for a period of time and then begin to undergo degradation. The most commonly used absorbable materials include polyglycolic acid (PGA), polylactic acid (PLA) and their copolymers.[1] Since the beginning of the concept of bioabsorbable materials by Kulkarni et al. in the 1960s,[2,3] animal models have contributed tremendously to the success of the clinical use of the materials.

II. BASICS OF BIOABSORBABLE MATERIALS

A. COMMON BIOABSORBABLE MATERIALS

Polymers with the greatest potential for medical applications are polyhydroxyacids (polyesters). There are only few commercially available polyhydroxyacids, including polylactides, polyglycolide, poly(glycolide-co-lactide) (PGA-PLA), poly(glycolide-co-trimethylene carbonate), poly(p-dioxanone) (PDS), and polyhydroxybutyrate/valerate.

B. BIODEGRADATION

PGA, PLA, and their copolymers degrade in tissues by nonspecific hydrolytic scission of their ester bonds. By hydrolysis, PLA are changed to lactic acid which enters the tricarboxylic acid cycle and excreted in the forms of water and carbon dioxide. It is known that PGA also is broken down by enzymes like esterase.[4]

The biodegradation rate depends on many factors,[5] such as the size of the implant, the kind of materials, the molecular weight of the material, the material phase (crystalline or amorphous), the presence of additives or impurities, the implantation sites (SC tissue or bone), the mechanism of hydrolysis (enzymes vs. water), and even the age[6] of the animals.

Generally, pure PGA resorbs rapidly. As a result, either too much of a polymer load is produced or it is released too rapidly for it to be absorbed and excreted. Polyglycolic implants have produced fluid filled sterile sinuses with subsequent drainage.[1,7] Pure PLA is highly crystalline and resorbs slowly. The amorphous state resorbs over 2–3 years, leaving behind crystallites that can elicit an inflammatory response. A PGA/PLA copolymer resorbs differently depending on the copolymer ratios.[8,9]

C. MECHANICAL PROPERTIES OF BIOABSORBABLE MATERIALS

The requirement for polymer screws is that they have to be strong enough to hold the fracture fragments together until the fracture heals, which is normally 4–8 weeks. Most of the polymer screws or rods have been applied in areas of low stress.[10] Polymer screws were not recommended for use without external support in places of high mechanical stress.[10,11] Displacement of bone fragments has been reported, ranging from minor (several millimeters) to severe.[12-14] Self-reinforced PGA (SR-PGA) has been reported to function better.[15-18]

The tensile yield strength has been reported as 11 MPa to 72 MPa for PLA and 45 MPa for PGA.[19] Flexural strengths for PLA are reported between 45 and 145 MPa. The highest flexural strengths are reported for PGA-fiber-reinforced PGA-composites, between 195 and 375 MPa. PGA reinforced by PLA fibers reaches a strength of 250 MPa. Although fiber reinforcement can improve the maximum strength of the material, it does not always improve the Young's modulus of the material, which remains very low (flexural modulus: 2 to 27 GPa for polymers compared to 200 GPa for stainless steel).

III. COMMONLY USED ANIMAL MODELS

A. BIOCOMPATIBILITY AND BIODEGRADATION TEST IN SOFT TISSUES

Subcutaneous or intramuscular implantation is the first *in vivo* step of testing a bioabsorbable material for its biocompatibility and degradation (Table 1). The rat is suggested as the first choice for soft tissue degradation studies because of its low cost and the rich background data available. In the rat, bone elongation ceases by age 6–9 months and 12–15 months of life remain after that. If the observation period is more than 15 months, rabbits should be used instead.

TABLE 1
Degradation of Bioabsorbable Materials *In Vivo* — Subcutaneous or Intramuscular Implantation

Animal	First author, year[Ref.]	Implant site	Material tested	Period (Wks)	Mass/mol. weight (MW) degradation	Inflammation reaction	Mechanical strength after implantation
Mouse	Gogolewski 1993[20]	SC*	PLA, PHB†, PHB/VA§	24	56–99%/24 wks.	+ or ++	N/A
Rat	Pistner 1993[21]	IM£	PLA (crystalline, 429,000 Mvis)	116	No change	+ at first few weeks	N/A
			PLA (amorphous, 203,000 Mvis)	116	100%	+++	N/A
			PLA (amorphous, 120,000 Mvis)	116	100%	++	N/A
	Gerlach 1993[22]	IM	PLA rods	108	N/A	+	50% of original at 4 wks.
	Bos 1991[23]	SC	PLA	143	14%/80 wks.	+ or ++	N/A
	Schakenraad 1989[24]	SC	Glycine/DL-lactic acid discs	10	100%	+++	N/A
Rabbit	Nakamura 1989[25]	IM	PLA, purified	52	No change/40 wks.	N/A	50% decrease
	Richards 1991[26]	IM	4 Poly (phosphoesters)	70	80%/70 wks.	+	N/A
	Törmälä 1991[18]	SC	PGA	8	N/A	N/A	Lost mech. strength at 4–7 wks.
	Matsusue 1992[27]	SC	PLA rods	78	91% (MW)/12 wks.	+	100% decrease at 25 wks.
	Kumta 1992[28]	SC	PGA rods	21	N/A	N/A	64% decrease at 2 wks.
	Tschakaloff 1994[29]	SC	PLA plates	6	87% (MW)	No	N/A
	Bhatia 1994[6]	SC, IM, bone	Absorbable pins (Orthosorb)	5	N/A	N/A	Decreased differently at each site
	Matsusue 1995[30]	SC	PLA rods	276	100%/276 wks.	No	100% decrease at 276 wks.
	Ertel 1995[31]	IM, bone	Poly(DTH carbonate) and PDS pins	26	Started at 26 wks. PDS partially absorbed	No ++, bone resorption	N/A N/A

* SC = subcutaneous
£ IM = intramuscular
† PHB = poly (3-hydroxybutyrate)
§ PHB/VA = poly (3-hydroxybutyrate-co-3-hydroxyvalerate)
+ = slight
++ = moderate
+++ = significant.

FIGURE 1. Schematic diagram showing different soft tissue implantations. Up to six disk implants can be inserted subcutaneously or intramuscularly on the back of a rat.

Implants are inserted into pouches in SC tissue or in muscles. It makes no difference where in the SC tissue (the back, inner aspect of the thighs, or abdominal wall) or in which muscle (the back muscles, thigh muscles, or abdominal muscles) the implant is placed. However, most researchers implant into the back SC tissue or the back muscles, simply because of the convenience and the large area or volume of the tissues.

Up to six implants (discs, 10 mm diam. and 1 mm thickness) could be implanted on the back of each rat, while 8–10 implants could be inserted into the back of each rabbit (Figure 1). Implantations can be done through one large midline incision or multiple small ones. The key is that the sample is placed in the right layer, either in the loose SC tissue space or in the paravertebral muscles. For testing mechanical degradation, rod-shaped samples (3.2 mm diam. and 50 mm length) are often used. At least six rods could be implanted into the back of an adult rabbit. The time periods used for the observation of degradation range from one month to five yrs. For more than 1.5 yrs' observation, rabbits should be used because the adult life of rats is less than 1.5 yr.

Standardizing sample shape and size is very important for comparing data between research groups. Soft tissue implantation is suitable for standardized testing of mechanical and mass degradation because the implant is more retrievable compared to in-bone implantations. An absorbable material sample measuring 3.2 mm in diameter and 50 mm in length is suggested for studying the mechanical degradation of materials, which has been used by many investigators.[18,22,27,28,30] A disc measuring 10 mm in diameter and 1 mm in thickness may be used for mass degradation studies using histological method or molecular weight degradation studies using a viscosimeter.

B. Biodegradation in Bone Tissue

Bioabsorbable implant degradation in bone tissues are the further step after soft tissue implantation, when the material is to be used inside of or in contact with bone (Table 2). These models test the biocompatibility and degradation pattern of the material. Although different animals have been used, the rabbit is the most popular one.

Implants in the forms of rods, plugs, pins, or screws are most often implanted in the cancellous bone of the distal femur in the rabbit through drill holes (Figure 2A,B). Diaphyseal implantation has also been used in different animals (Figure 2C).

TABLE 2
Animal Models of Bioabsorbable Implant Degradation in Bone Tissues

Animal	Bone	First author, year[Ref.]	Bone defect	Material tested	Period (wks.)	Degradation	Inflammation reaction	Mechanical properties of materials
Rat	Femur	Cutright 1974[8]	Upper femur drill hole	PGA, PLA, or copolymers	31	100% in 14–31 wks.	No	N/A
Rabbit	Femur	Vainionpää 1986[32]	Intercondylar area	PGA rods	12	Started in 6 wks.	No	N/A
		Mäkelä 1989[33]	Intercondylar area	PDS rods	36	Nearly 100% in 36 wks.	N/A	N/A
		Matsusue 1992[27]	Intercondylar area	PLA rods	70	22%/52 wks., 70%/78 wks.	+	100% loss at 25 wks.
		Matsusue 1995[30]	Intercondylar area	PLA rods	248	100%/24 in 8 wks.	No	N/A
		Kumta 1992[28]	Medullary canal	PGA rods	3	Broken down in 3 wks.	N/A	73% Bending str.
		Böstman 1994[34]	Intercondylar area	PGA pins and screws	48	100% in 36 wks.	?	N/A
		Fini 1995[35]	Medullary canal	PLA (high MW) rods	64	0%	No	Flexional stiffness
		Knowles 1992[36]	Diaphyseal drill holes	Polyhydroxybutyrate plugs	19	—	N/A	N/A for the material
	Tibia	Vainionpää 1986[37]	Diaphyseal drill holes	PGA rods	12	Started in 6 wks.	No	N/A
		Ertel 1995[31]	Medial tibial plateau	Poly(DTH carbonate) and PDS pins	26	Started at 26 wks. PDS partially absorbed	No	N/A
	Peri-orbital	Kellman 1994[38]	1.5 mm drill holes	PLA, PLA/TMC*, PGA screws	32	N/A	++, osteolysis	No strength at 32 wks.
Dog	Femur	Miettinen 1992[39]	Intramedullary implant	SR-PGA implant	24	Started in 3–6 wks.	++	N/A
		Suganuma 1993[40]	Distal metaphysis	PLA in a bone chamber	24	20% in 24 wks.	++	N/A
Goat	Femur	Verheyen 1993[41]	Diaphysis drill holes	PLA, PLA/HA plugs	104	Obvious for pure PLA	Lymph nodes	N/A
Pig	Mandible	Schliephake 1993[42]	Hole below the mandibular canal	Polydioxanone pins	26	100% in 22–26 wks.	N/A	N/A

* PLA/TMC = polylactic acid/trimethylene carbonate copolymer
+ = slight; ++ = moderate.

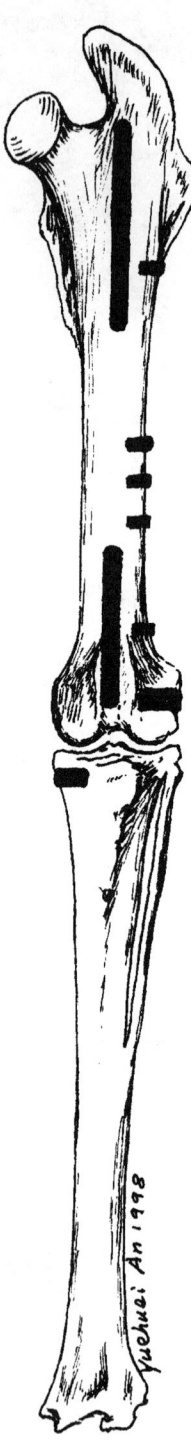

FIGURE 2. Schematic diagrams showing different in-bone implantations of bioabsorbable materials.

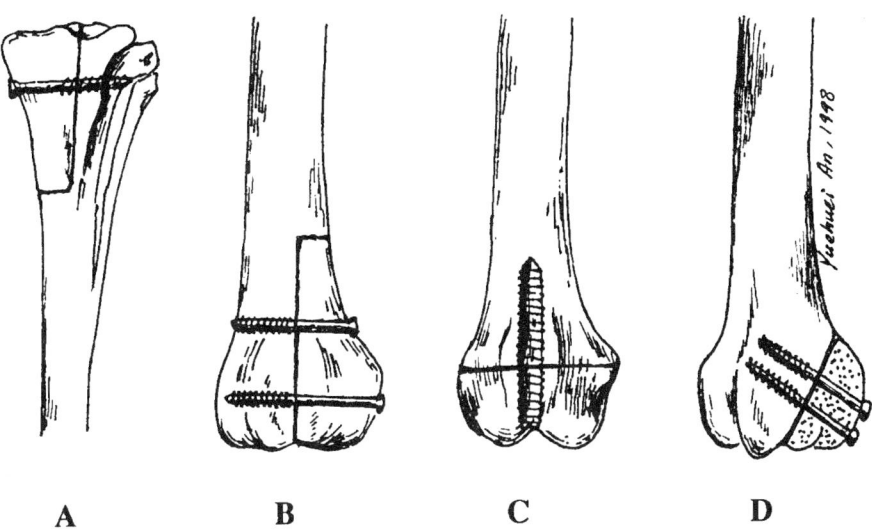

FIGURE 3. Schematic diagrams showing different epiphysometaphyseal implantation models: (A) osteotomy of medial tibial plateau in the rabbit; (B) Salter-Harris type IV fracture in the dog; (C) the femoral transcondylar osteotomy in the rabbit; and (D) the lateral femoral condyle osteotomy in the dog.

C. Fixation of Fracture and Osteotomy

1. Epiphysometaphyseal Osteotomy Models

The representative models of epiphysometaphyseal osteotomies (Table 3) include the osteotomy of medial tibial plateau in the rabbit (Figure 3A),[52] Salter-Harris type IV fracture in the dog (Figure 3B),[58] the femoral transcondylar osteotomy in the rabbit (Figure 3C),[37,50] and the lateral femoral condyle osteotomy in the dog (Figure 3D).[53]

Femoral transcondylar osteotomy in rabbits is a well-established model.[37,50] Salter-Harris type IV fracture in the dog or the similar osteotomy created at the medial tibial plateau by Matsusue et al[52] is another excellent model in the epiphysometaphyseal area. The forces that the absorbable screw is subjected to is simple, either on the lateral or the medial side of the knee joint. The proximal-to-distal shearing force is sustained by the transverse portion of the osteotomy. Another interesting model is the subcapital femoral neck osteotomy in sheep, which is more demanding for fixation because of weight-bearing.[55]

For the lateral femoral condyle osteotomy in dogs (from the authors' laboratory), an unilateral model is recommended because of the amount of surgical injury.[53] Through a lateral approach, an osteotomy from the superolateral aspect of the condyle to the intercondylar notch in the knee joint was made and the osteotomy was secured with two screws. The advantage of this model is the instability of the osteotomy (high shear force to the screws), which challenges the efficacy of the fixation devices. The effect of bioabsorbable screw on the cancellous bone healing can be evaluated with a trephined plug model in canine distal femur (Figure 4).[53]

2. Diaphyseal Fracture Models

Very few diaphyseal models are available (Table 4). The first diaphyseal model was reported by Vainionpää et al.,[59] in which a tibial osteotomy was fixed with a T-shaped PGA/PLA copolymer implant (Figure 5A). Oblique fractures of metacarpal bones in calves were also fixed with PLA

TABLE 3
Animal Models of Fixation of Osteotomies of Epiphysometaphyseal Area of Long Bones and Irregular Bones of the Limbs Using Bioabsorbable Materials

Animal	Bone	First author, year[Ref.]	Osteotomy or fracture	Fixation method	Period (Wks.)	Degradation	Inflammation reaction	Fracture healing
Rat	Femur	Majola 1991[43]	Transcondylar	SR-PLA, SR-PDLA/PLLA	48	Started in 12 wks.	No	95% healed at 48 wks.
Rabbit	Femur	Väinionpää 1986[37]	Transcondylar	PGA thread and rods	24	Started at six wks.	N/A	Healed at six wks.
		Vihtonen 1987[44]	Transcondylar	PGA thread (Dexon)	24	N/A	N/A	79% healed at six wks.
		Vihtonen 1988[45]	Transcondylar	Bone cement, PGA thread	48	N/A	N/A	65% healed
		Vasenius 1990[46]	Transcondylar	SR-PGA rods	48	100% in 24–36 wks.	No	99% healed
		Plaga 1992[47]	Oblique medial condyle	PDS pins	6	N/A	N/A	86% healed
		Böstman 1992[48]	Transcondylar	PGA screws	36	100%/36 wks.	++	Healed
		Päivärinta 1993[49]	Transcondylar	PGA screws	48	100%/36 wks.	+ or ++	Healed
			Transcondylar	PLA screws	48	0%/48 wks.	+ or ++	Healed
		Pihlajamäki 1994[50]	Transcondylar	PLA expansion plug	24	N/A	No	Healed
		Pihlajamäki 1994[51]	Transcondylar	SR-PLA expansion plug	24	N/A	No	75% healed at 24 wks.
	Tibia	Matsusue 1991[52]	Proximal tibial osteotomy	PLA or stainless steel screws	16	N/A	+	Healed
Dog	Femur	An 1997[53]	Lateral condyle osteotomy	PGA/PLA screws	68	>95% 68 wks.	No	Healed/8 wks.
	Total hip	Otsuka 1994[54]	Fixing acetabular component	PLA screws	14	N/A	N/A	Healed OK
Sheep	Femur	Jukkala-Partio 1997[55]	Subcapital osteotomy	SR-PLA lag-screws	12	N/A	?	86% healed
	Tibia	Weiler 1996[56]	Medial condyle	PGA rods (Biofix)	12	N/A	+++	Healed
	Ulnae	Manninen 1992[57]	Transverse olecranon	PLA screws	12	N/A	N/A	Healed
Goat	Femur	Donigian 1993[13]	Salter-Harris IV fracture	PLA, polydioxanone pins	8	N/A	No	Healed

*PHBA = poly-β-hydroxybutyric acid
+ = slight
++ = moderate
+++ = significant

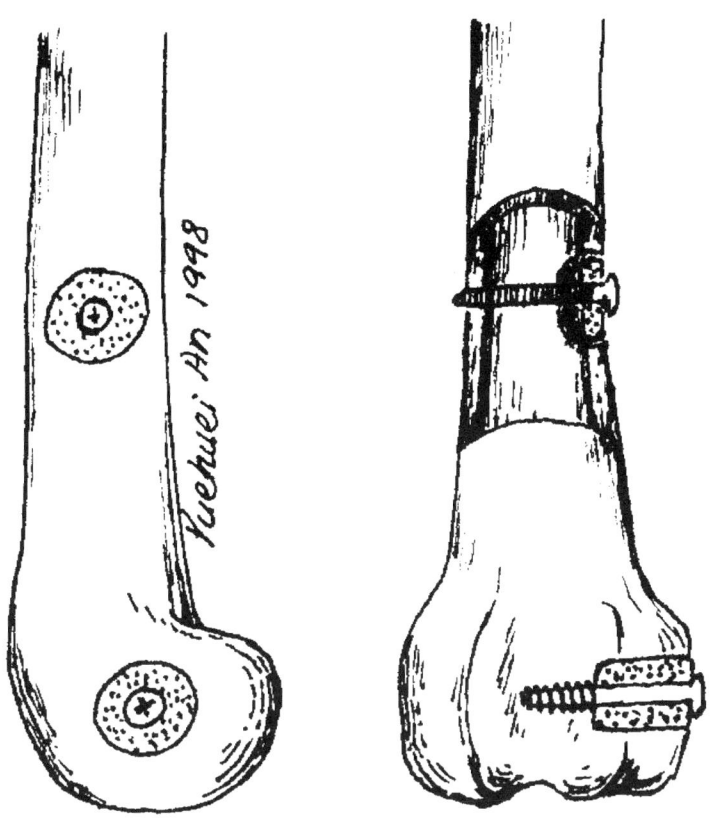

FIGURE 4. Schematic diagrams showing trephine osteotomies made at both diaphyseal and epiphysometaphyseal areas in the dog.

tension screws (Figure 5 B).[61] Models of femoral osteotomy fixed with PLA, SR-PLA, or SR-PGA intramedullary rods have been reported in the dog, rabbit, and cat (Figure 5C,D).[57,58,60] The intramedullary rod can be inserted through the greater trachanter or the intercondylar notch. The studies showed that the bioabsorbable rods were strong enough to be used in intramedullary nailing of femoral diaphyseal osteotomies in the three animal species. In the authors' laboratory, healing of a trephined femoral diaphyseal osteotomy in the dog has been used to observe the effect of fixation using a PGA/PLA (80/20 ratio) screw (Figure 4).[53] In most of the reports, bone consolidation normally occurred in 6–8 weeks.

3. Maxillofacial Bone Fracture Models

Models of mandibular osteotomies are the most popular and well established in dogs and sheep (Figure 6A). Rabbits are not suitable for mandibular models becasue they do not have a long mandible for fixation instrumentation. Very few models using zygomatic or nasal bones have been reported using rabbits (Figure 6B,C).[29,62] Based on anatomical features, dogs can also be used for these two models. The most common form of bioabsorbable fixation devices for these models are

TABLE 4
Animal Models of Fixation of Diaphyseal Osteotomy Using Bioabsorbable Materials

Animal	Bone	First author, year[ref.]	Osteotomy or fracture	Fixation method	Period (Wks.)	Degradation	Inflammation reaction	Fracture healing
Rabbit	Tibia	Vainionpää 1986[59]	T	Carbon fiber-reinforced PGA/PLA and PHBA* implant	24	Started at six wks.	N/A	Healed at six wks.
Dog	Femur	Manninen 1993[57]	T	SR-PLA intramedullary rods	48	N/A	No	Healed
Dog	Femur	Miettinen 1992[60]	T	SR-PLA or SR-PGA	48	Disappeared @ 24 wks.	No	Healed/6 wks.
		An 1997[53]	Trephined	PGA/PLA screws	68	>95% 68 wks.	No	Healed/8 wks.
Cat	Femur	Hara 1994[58]	T	PLA intramedullary rods	16	N/A	No	Healed/large callus
Calf	Metacarpal	Illi 1992[61]	45°O	PLA or AO metallic screws	6	0% or N/A	No	Healed

* PHBA = poly-β-hydroxybutyric acid

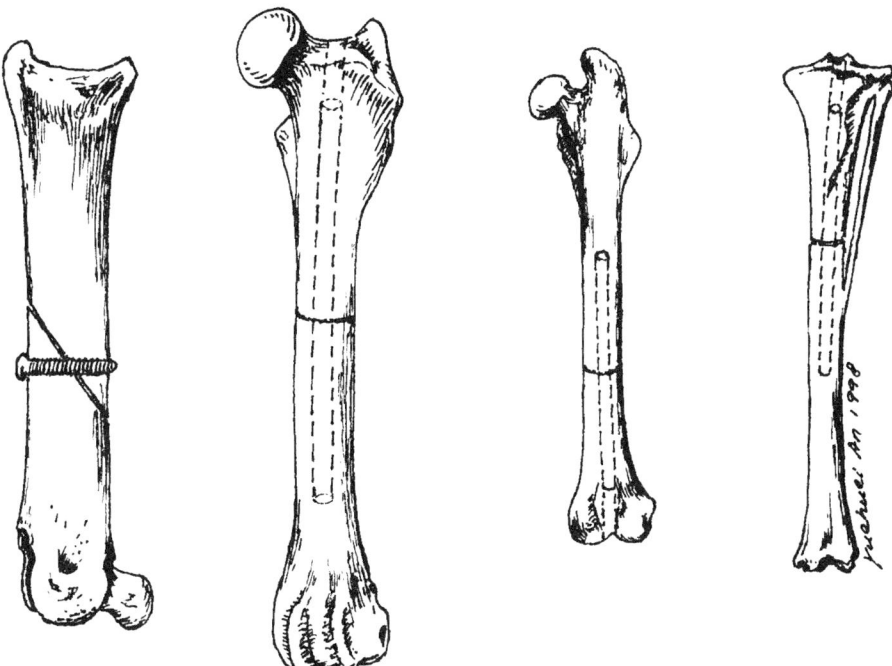

FIGURE 5. Schematic diagrams showing different diaphyseal fracture models: (A) an oblique fracture of metacarpal bone in calves fixed polymer screws (lateral view);[61] (B) a femoral osteotomy fixed with a polymer intramedullary rod inserted through greater trachanter in the dog;[60] (C) a femoral osteotomy fixed with a polymer intramedullary rod inserted through intercondylar notch in the rabbit;[57] and (D) a potential fracture fixation model using a intramedullary rod in the rabbit.

mini plate and screws although the use of absorbable suture and screw only have been reported (Table 5).

D. Bone Replacement

In the case of repairing a bone defect, bioabsorbable materials are often used independently for bone conduction,[69,70] or as a part of the substitute (with HA[71,72] or proteolipid[73]), or as a carrier for GFs,[74,75] BMP,[76–79] and other bone elements (Table 6).[84–87] The other function of absorbable materials may be no more than a scaffolding for bone conduction[69,70] or guided bone regeneration.[82]

E. Drug Delivery

Typical bioabsorbable drug delivery systems are designed to introduce antibiotics for local use (Table 6). PLA and oligomer dideoxykanamycin B were used as a delivery system to treat osteomyelitis in the rabbit femur.[101] Garvin et al.[103] reported the use of PLA/PGA gentamicin implant in the treatment of osteomyelitis in a canine model. Overbeck et al.[104] implanted a novel ciprofloxcin-containing PGA cylinder into the medullary canal of the proximal femur in the rabbit and showed that a significant greater concentration of ciprofloxcin into bone than has been reported for gentamicin cement beads. Langer reviewed drug delivery systems made from bioabsorbable materials.[111]

Degradable polymers can also be used for the controlled release of insulin, GFs,[74,78] indomethacin,[112] and angiogenesis inhibitors.

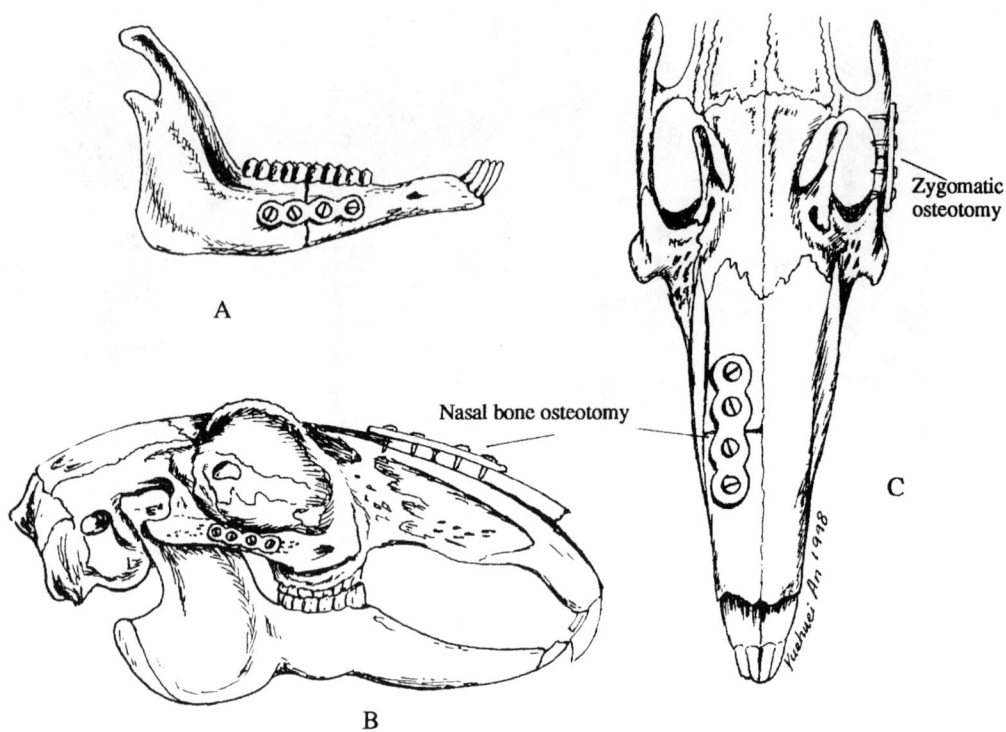

FIGURE 6. Schematic diagrams showing maxillofacial bone osteotomy and fixation with bioabsorbable plate and screws: (A) a mandibular osteotomy in the goat, (B) and (C) zygomatic bone and nasal bone osteotomies in the rabbit.

F. Repair of Cartilage Defect

With variable results, the typical studies of repairing cartilage defects using bioabsorbable materials include the use of PGA rods,[34] PGA or PLA combined with perichondrium in the rabbit[88,113] and PLA/PGA implants with impregnated DBM powder[114] or TGF-β[115] (Table 6).

Seeding cells onto porous PLA polymers is a new approach. Freed et al. studied neocartilage formation *in vitro* and *in vivo* using porous PGA and PLA scaffolds.[116] They also implanted chondrocyte-seeded absorbable implant to repair cartilage defects with promising results.[89] In recent years, more investigations have been reported on the repair of cartilage defect using cell-seeded absorbable implants.[90,91] There is no doubt that the use of tissue engineering principles for cartilage repair is the major direction of research.

G. Repair of Meniscus

A porous PLA scaffold was used by Klompmaker et al. to repair meniscal lesions.[92] It was found that the biodegradable implant with a porous structure (size 100–300 μm) could provide guidance for vascular ingrowth into the defect (Table 6). Recently, the same group reconstructed canine meniscus using a porous copoly(L-lactide/ε-caprolactone) material and found the gap formation between implant sides and meniscal tissue could be prevented, but the adherence of implant to the underlying meniscal tissue seemed to be a problem.[93] More recently, a new PGA/PLA staple has been tested for meniscal repair.[94] The results showed that the staple provided greater tensile strength than PDS sutures to the repaired meniscus.

TABLE 5
Animal Models of Fixation of Osteotomies of Maxillofacial Bones Using Bioabsorbable Materials

Animal	Bone	First author, year[ref].	Osteotomy or fracture	Fixation method	Period (Wks.)	Degradation reaction	Inflammation healing	Fracture
Rabbit	Zygomatic	Thaller 1993[62]	Saw osteotomy	PGA plate and screws	16	N/A	N/A	Healed
	Calvarial bone	Eppley 1995[63]	Circular defect	PLA/PGA plate and screws	52	100%/52 wks.	No	Healed/24 wks.
	Nasal bone	Tschakaloff 1994[29]	Transverse osteotomy	PLA plate and screws	6	75% mol. weight	No	N/A
Dog	Mandible	Getter 1972[64]	Transverse osteotomy	PLA plate and screws	32	100%/32 wks.	+	Healed
		Leenslag 1987[65]	Transverse osteotomy	PLA plate and screws	11	Some at 11 wks.	No	Healed
		Gerlach 1993[22]	Transverse osteotomy	PLA plate and screws	100	N/A	+	Healed/8 wks.
		Tams 1995[66]	Transverse osteotomy	PLA plate and screws	18	Not much/18 wks	No	Healed/12 wks.
		Hara 1994[67]	Salter-Harris IV fracture	PLA screws	24	Started/24 wks.	No	Healed/4–8 wks.
Sheep	Mandible	Leenslag 1987[65]	Transverse osteotomy	PLA plate and screws	11	Some at 11 wks.	No	Healed
		Bos 1989[68]	Transverse osteotomy	PLA plate and screws	11	Not much/11 wks.	No	Healed
		Suuronen 1991[17]	Neck of condylar process	SR-PLA screws	24	N/A	N/A	Healed

* PHBA = poly-β-hydroxybutyric acid; + = slight reaction.

TABLE 6
Applications of Bioabsorbable Materials in Animal Models and Human Subjects in Orthopaedic Research

Application	Subject	Location	Materials	First author, year[ref.]
Fixation of fracture or osteotomy	Animals	Various	Various	(For references see Tables 3, 4, 5)
Bone filler or substitute	Rat	Tibia and humerus drill holes	PLA/PGA	Hollinger 1983[69]
	Rat	Femur, drill hole	PLA/HA composite	Higashi 1986[71]
	Dog	Mandibular discontinuities	PLA/PGA + proteolipid	Hollinger 1987[73]
	Rabbit	Calvarial defect	PLA/PGA/HA	Antikainen 1992[72]
	Rat	Critical size defect (5 mm) in mandible	PLA/PGA membranes	Sanberg 1993[80]
	Rat	Femur/Around metaphyseal bone	SR-PGA membranes	Ashammakhi 1994[81]
	Rat	Femur/around bone surfaces	SR-PGA membranes	Ashammakhi 1995[82]
	Rabbit	Augmenting distal femoral drill holes	SR-PGA membranes	Ashammakhi 1995[70]
Carriers for osteogenic substances	Human	Femoral nonunion	h-BMP in PLA/PGA copolymer	Johnson 1988[83]
	Rabbit	Calvarial defect	DFDB* in PLA/PGA	Schmitz 1988[84]
	Rat	Osteogenesis in muscles	DBM† in polyorthoester	Pinholt 1991[85]
	Rabbit	Calvarial defect	Osteoinductive protein in PLA/PGA	Turk 1993[75]
	Dog	Osteogenesis in SC‡ tissue	PCBM§ in PLA mesh	Kinoshita 1993[86]
	Rat	Osteogenesis in SC tissue	DBM in PLA	Saitoh 1994[87]
	Rat	Calvarial defect	BMP in PLA disc	Miki 1994[76]
	Rat	Calvarial defect	BMP in PLA/PGA microparticles	Kenley 1994[77]
	Rat	Osteogenesis in calvarial defect	IGF-I in polyorthoester	Busch 1996[74]
	Rat	Osteogenesis in SC tissue	BMP in PLA/PGA/blood clot	Alpaslan 1996[79]
	Mouse	Osteogenesis in muscles	BMP in PLG/PEG reservoirs	Yamazaki 1996[78]

Application	Animal	Model	Material	Reference
Repair of cartilage defect	Rabbit	Defect in distal femoral joint	PLA/periosteum	von Schroeder 1991[88]
	Rabbit	Defect in distal femoral joint	PGA scaffold + chondrocytes	Freed 1994[89]
	Rabbit	Defect in distal femoral joint	PLA scaffold + chondroctes	Vasanti 1994[90]
	Rabbit	Defect on medial femoral condyle	Porous PLA + perichondrocytes	Chu 1995[91]
Repair of meniscus	Dog	Lateral menisci	Porous PLA implant	Klompmaker 1991[92]
	Dog	Lateral menisci	Copoly(L-lactide/ε-caprolactone)	de Groot 1997[93]
	Dog	Medial menisci peripheral third incison	PGA/PLA staples or 3-0 PDS suture	Koukoubis 1997[94]
Repair or replacement of ligament and tendon	Dog/Rabbit/Dog	Patellar/Achilles/med. collateral lig.	PLA-carbon scaffolding ribbons	Parsons 1983[95]
	Sheep	ACL reconstruction	Braided PLA augmentation	Laitinen 1993[96]
	Sheep	PDS-augmented patellar tendon	PDS cords	Holzmuller 1994[97]
	Rabbit	Achilles tendon laceration or defect	A PGA/Dacron device	Rodkey 1985[98]
Fixation of ligament and tendon	Human	Arthroscopic patellar tendon graft	PLA interference screws	Barber 1995[99]
	Human	Ulnar collateral ligament of the 1st metacarpophalangeal joint	SR-PLA mini tack fixation	Juutilainen 1996[100]
Drug (antibiotics) delivery	Rabbit	Implanted into distal femur	PLA/Kanamycin	Wei 1991[101]
	Rabbit	Tibial fracture, infection	PLA/cefazolin microspheres	Jacob 1993[102]
	Dog	Tibia osteomyelitis	PLA/PGA implant + gentamicin	Garvin 1994[103]
	Rabbit	Inserted into proximal femur	PGA/Ciprofloxacin cylinder	Overbeck 1995[104]
	Rabbit	Tibial osteomyelitis	PLA, PLA/PGA + vancomycin	Calhoun 1997[105]
Conduits for nerve repair	Human	Digital nerve defect	PGA tube	Mackinnon 1990[106]
	Primate	Median nerve defect	PGA tube	Hentz 1991[107]
	Rat	Ischiatic nerve defect	PLA/6-caprolactone conduit	Nicoli Aldini 1996[108]
Scaffold for vessel regeneration	Rat	Abdominal aorta graft	Microporous polyurethane/PLA	van der Lei 1987[109]
	Rabbit	Infrarenal aorta	Woven PDS graft	Greisler 1991[110]

* DFDB = demineralized freeze-dried bone
† DBM = demineralized bone matrix
‡ SC = subcutaneous tissue
§ PCBM = particulate cancellous bone and marrow
¶ PLG/PEG = poly(lactide-co-glycolide)/poly(ethylene glycol)

We believe that with the rapid development of tissue engineering, cell seeded and/or growth factor-impregnated implant is the main direction of research.

H. REPAIR OF TENDONS AND LIGAMENTS

The replacement of tendons and ligaments with bioabsorbable materials is still in the early stage of research (Table 6). The concept was given by Parson et al. in 1983.[95] They reported the use of a PLA-filamentous carbon fibers to replace tendons and ligaments. Rodkey et al.[98] found that a PGA/Dacron material had adequate strength and physical properties to be used both for primary tenorrhaphy and bridging tendon defect in a rabbit Achilles tendon model. Recently, braided PLA and PDS cords have been used to augment fascia lata and patellar tendon respectively in sheep for ACL reconstruction with promising results.[96,97]

Although no animal models have been reported, fixation of tendon and ligament to bone by an absorbable interference screw and a mini tack in human patients have been established recently.[99,100] Again, we believe that cell seeded and/or growth factor-impregnated porous implant made from bioabsorbable materials for the replacement of tendons and ligaments will be one of the future directions of the research.

I. SMALL BLOOD VESSEL AND NERVE REGENERATION

Absorbable materials have been used for small vessel and nerve regeneration (Table 6). The typical animal studies of vessel regeneration using absorbable materials were reported by van der Lei et al.[109] Microporous, compliant, vascular grafts made from a mixture of polyurethane (95%) and PLA (5%) can function as a scaffold for the regeneration of small-caliber arteries.[109] Common models for studying small vessel graft include rat abdominal aorta, rabbit carotid artery and infrarenal aorta, and canine femoral, carotid artery, and aorta.

For nerve regeneration, several materials, such as PGA tube,[106,107] PLA-co-6-caprolactone conduit,[108] and other forms of absorbable guides,[117] have been investigated with controversial results. The common model for experimental nerve regeneration is ischiatic nerve defect in the rat.

IV. EVALUATION METHODS

Animals should be inspected daily for evidence of foreign-body reactions (sterile sepsis). A radiograph is used for determining the location of an osteotomy, the quality of the fixation, the existence of osteolysis, and the progress of bone healing (see Chapters 11 and 13).

Histological analysis is used to evaluate bone healing (see Chapters 11 and 13), tissue response to bioabsorbable devices, and the mass degradation of the polymer. Ground sectioning can be chosen for its advantage that the residues of polymer may be preserved during specimen processing and suitability for specimens containing metal devices.

Bending and tensile tests are commonly used to testing the strength of absorbable devices before and after implantation. A screw pullout test also has been used. An indentation test can be used to test the strength of bone around a bioabsorbable screw or rod.[53] The strength of osteotomy sites of diaphysis can be tested to failure using a torsional test or bending test.

ACKNOWLEDGMENT

The work in this review was supported in part by Poly-Medics, Warsaw, IN, USA.

REFERENCES

1. Böstman, O. M., "Osteolytic changes accompanying degradation of absorbable fracture fixation implants," *J, Bone Joint Surg.*, 73B, 679, 1991.
2. Kulkarni, R. K., Pani, K. C., Neuman, C., and Leonard, F., "Polylactic acid for surgical implants," *Arch. Surg.*, 93, 839, 1966.
3. Kulkarni, R. K., Moore, E. G., Hegyeli, A. F., and Leonard, F., "Biodegradable poly(lactic acid) polymers," *J. Biomed. Mater. Res.*, 5, 169, 1971.
4. Willians, D. F. and Mort. E., "Enzyme-accelerated hydrolysis of polyglycolic acid," *J. Bioeng.*, 1, 231, 1977.
5. Vert, M., Li, S., and Garreau H., "New insights on the degradation of bioresorbable polymeric devices based on lactic and glycolic acids," *Clin. Mater.*, 10, 3 1992.
6. Bhatia, S., Shalaby, S. W., Powers, D. L., Lancaster, R. L., and Ferguson, R. L., "The effect of site of implantation and animal age on properties of polydioxanone pins," *J. Biomater. Sci. Polymer Ed.*, 6, 435, 1994.
7. Böstman, O., Hirvensalo, E., Mäkinen, J., and Rokkanen, P., "Foreign-body reactions to fracture fixation implants of biodegradable synthetic polymers," *J. Bone Joint Surg.*, 72B, 592, 1990.
8. Cutright, D. E., Perez, B., Beasley, J., Larson, W. J., and Posey, W. R., "Degradation rates of polymers and copolymers of polylactic and polyglycolic acids," *Oral Surg. Oral Med. Oral Pathol.*, 37, 142, 1974.
9. Miller, R. A., Brady, J. M., and Cutright, D. E., "Degradation rates of oral resorbable implants (polylactates and polyglycolates): rate modification with changes in PLA/PGA copolymer ratios," *J. Biomed. Mater. Res.*, 11, 711, 1977.
10. Hofmann, G. O., "Biodegradable implants in orthopaedic surgery — A review on the state of the art," *Clin. Mater.*, 10, 75, 1992.
11. Manninen, M. J., Päivärintä, U., Taurio, R., et al., "Polylactide screws in the fixation of olecranon osteotomies. A mechanical study in sheep," *Acta. Orthop. Scand.*, 63, 437, 1992.
12. Böstman, O., Mäkelä, E. A., Södergård, J., Hirvensalo, E., Törmälä, P., and Rokkanen, P., "Absorbable polyglycolic pins in internal fixation of fractures in children," *J. Pediatr. Orthop.*, 13, 242, 1993.
13. Donigian, A. M., Plaga, B. R., and Caskey, P. M., "Biodegradable fixation of physeal fractures in goat distal femur," *J. Pediatr. Orthop.*, 13, 349, 1993.
14. Hirvensalo, E., Böstman, O., and Rokkanen, P., "Absorbable polyglycolide pins in fixation of displaced fractures of the radial head," *Arch. Orthop. Trauma Surg.*, 109, 258, 1990.
15. Böstman, O., Hirvensalo, E., Vainiopää, S., Mäkelä, A., Vihtonen, K., et al., "Ankle fracture treated using biodegradable internal fixation," *Clin. Orthop.*, 238, 195, 1989.
16. Mäkelä, A., Böstman, O, Kekomäki, M., Södergård, J., Vainio, J., et al., "Biodegradable fixation of distal humeral physeal fracture," *Clin. Orthop.*, 283, 237, 1990.
17. Suuronen, R., "Comparison of absorbable self-reinforced poly-L-lactide screws and metallic screws in the fixation of mandibular condyle osteotomies: An experimental study in sheep," *J. Oral Maxillofac. Surg.*, 49, 989, 1991.
18. Törmälä, P., Vasenius, J., Böstman, O, Vainionpää, S., Laiho, J., et al., "Ultra-high-strength absorbable self-reinforced polyglycolide (SR-PGA) composite rods for internal fixation of bone fractures: *In vitro* and *in vivo* study," *J. Biomed. Mater. Res.*, 25, 1, 1991.
19. Claes, L. E., "Mechanical characterization of biodegradable implants," *Clin. Mater.*, 10, 41, 1992.
20. Gogolewski, S., Jovanovic, M., Perren, S. M., Dillon, J. G., and Hughes, M. K., "Tissue response and *in vivo* degradation of selected polyhydroxyacids: polylactides (PLA), poly (3-hydroxybutyrate) (PHB), and poly (3-hydroxybutyrate-co-3-hydroxyvalerate) (PHB/VA)," *J. Biomed. Mater. Res.*, 27, 1135, 1993.
21. Pistner, H., Bendix, D. R., Mühling, J., and Reuther, J. F., "Poly (L-lactide): a long-term degradation study *in vivo*. Part III. Analytical characterization," *Biomaterials*, 14, 291, 1993.
22. Gerlach, K. L., "*In vivo* and clinical evaluations of poly (L-lactide) plates and screws for use in maxillofacial traumatology," *Clin. Mater.*, 13: 21–28, 1993.
23. Bos, R. R. M., Rozema, F. R., Boering, G., et al., "Degradation of tissue reaction to biodegradable poly (L-lactide) for use as internal fixation of fractures: a study in rats," *Biomaterials*, 12, 32, 1991.

24. Schakenraad, J. M., Nieuwenhuis, P., Molenaar, I., Helder, J., Dijkstra, P. J., and Feijen, J., "*In vivo* and *in vitro* degradation of glycine/DL-lactic acid copolymers," *J. Biomed. Mater. Res.,* 23, 1271, 1989.
25. Nakamura, S., Ninomiya, S., Takatori, Y., Morimoto, S., Kusaba, I., and Kurokawa, T., "Polylactic screws in acetabular osteotomy: 28 dysplastic hips followed for one year," *Acta Orthop. Scand.,* 64, 301, 1993.
26. Richards, M., Dahiyat, B. I., Arm, D. M., Brown, P. R., and Leong, K. W., "Evaluation of polyphosphates and polyphosphonates as degradable biomaterials," *J. Biomed. Mater. Res.,* 25, 1151, 1991.
27. Matsusue, Y., Yamamuro, T., Oka, M., Shikinami, Y., Hyon, S. H., and Ikada, Y., "*In vitro* and *in vivo* studies on bioabsorbable ultra-high-strength poly(L-lactide) rods," *J. Biomed. Mater. Res.,* 26, 1553, 1992.
28. Kumta, S. M. Spinner, R., and Leung, P. C., "Absorbable intramedullary implants for hand fractures. Animal experiments and clinical trial," *J. Bone Joint Surg.,* 74B, 563, 1992.
29. Tschakaloff, A., Losken, H. W., von Oepen, R., et al., "Degradation kinetics of biodegradable DL-polylactic acid biodegradable implants depending on the site of implantation," *Int. J. Oral Maxillofac. Surg.,* 23, 443, 1994.
30. Matsusue, Y., Hanafusa, S., Yamamuro, T., Shikinami, Y., and Ikada, Y., "Tissue reaction of bioabsorbable ultra high strength poly (L-lactide) rod. A long-term study in rabbits," *Clin. Orthop.,* 317, 246, 1995.
31. Ertel, S. I., Kohn, J., Zimmerman, M. C., and Parsons, J. R., "Evaluation of poly(DTH carbonate), a tyrosine-derived degradable polymer, for orthopaedic applications," *J. Biomed. Mater. Res.,* 29, 1337, 1995.
32. Vainionpää, S., "Biodegradation of polyglycolic acid in bone tissue: an experimental study on rabbits," *Arch. Orthop. Trauma Surg.,* 104, 333, 1986.
33. Mäkelä, E. A., Vainionpää, S., Vihtonen, K., et al., "The effect of a penetrating biodegradable implant on the growth plate. An experimental study on growing rabbits with special reference to polydioxanone," *Clin. Orthop.,* 241, 300, 1989.
34. Böstman, O. and Päivärinta, U., "Restoration of tissue components after insertion of absorbable fracture fixation devices of polyglycolide through the articular surface: an experimental study in the distal rabbit femur," *J. Orthop. Res.,* 12, 403, 1994.
35. Fini, M., Giannini, S., Giardino, R., et al., "Resorbable device for fracture fixation: *in vivo* degradation and mechanical behaviour," *Int. J. Artif. Organs,* 18, 772, 1995.
36. Knowles, J. C., Hastings, G. W., Ohta, H., Niwa, S., and Boeree, N., "Development of a degradable composite for orthopaedic use: *in vivo* biomechanical and histological evaluation of two bioactive degradable composites based on the polyhydroxybutyrate polymer," *Biomaterials,* 13, 491, 1992.
37. Vainionpää, S., Vihtonen, K., Mero, M., Patiala, H., Rokkanen, P., et al., "Fixation of experimental osteotomies of the distal femur of rabbits with biodegradable material," *Arch. Orthop. Trauma Surg.,* 106, 1, 1986.
38. Kellman, R. M., Huckins, S. C., King, J., Humphrey, D., Marentette, L., and Osborn, D. C., "Bioresorbable screws for facial bone reconstruction: a pilot study in rabbits," *Laryngoscope,* 104, 556, 1994.
39. Miettinen, H., Mäkelä, E. A., Vainio, J., Rokkanen, P., and Törmälä P., "The effect of an intramedullary biodegradable self-reinforced polyglycolic acid implant on tubular bone. An experimental study on growing dogs," *J. Biomater. Sci. Polym. Ed.,* 3, 435, 1992.
40. Suganuma, J. and Alexander, H., "Biological response of intramedullary bone to poly-L-lactic acid," *J. Appl. Biomater.,* 4, 13, 1993.
41. Verheyen, C. C., de Wijn, J. R., van Blitterswijk, C. A., Rozing, P. M., and de Groot, K., "Examination of efferent lymph nodes after 2 years of transcortical implantation of poly(L-lactide) containing plugs: a case report," *J. Biomed. Mater. Res.,* 27, 1115, 1993.
42. Schliephake, H., Klosa, D., and Rahlff, M., "Determination of the 3-D morphology of degradable biopolymer implants undergoing *in vivo* resorption," *J. Biomed. Mater. Res.,* 27, 991, 1993.
43. Majola, A., Vainionpää, S., Vihtonen, K., Mero, M., Vasenius, J., et al., "Absorption, biocompatibility, and fixation properties of polylactic acid in bone tissue: an experimental study in rats," *Clin. Orthop.,* 268, 260, 1991.
44. Vihtonen, K., Vainionpää, S., Mero, M., Patiala, H., Rokkanen, P., et al., "Fixation of experimental osteotomy of the distal femur with biodegradable thread in rabbits," *Clin. Orthop.,* 221, 297, 1987.

45. Vihtonen, K., "Fixation of rabbit osteotomies with biodegradable polyglycolic acid thread," *Acta Orthop. Scand.*, 59, 279, 1988.
46. Vasenius, J., Vainionpää, S., Vihtonen, K., Mero, M., Mäkelä, A., et al., "A histomorphological study on self-reinforced polyglycolide (SR-PGA) osteosynthesis implants coated with slowly absorbable polymers," *J. Biomed. Mater. Res.*, 24, 1615, 1990.
47. Plaga, B. R., Royster, R. M., Donigian, A. M., Wright, G. B., and Caskey, P. M., "Fixation of osteochondral fractures in rabbit knees. A comparison of Kirschner wires, fibrin sealant, and polydioxanone pins," *J. Bone Joint Surg.*, 74B, 292, 1992.
48. Böstman, O., Päivärinta, U., Partio, E., Vasenius, J., Manninen, M., and Rokkanen, P., "Degradation and tissue replacement of an absorbable polyglycolide screw in the fixation of rabbit femoral osteotomies," *J. Bone Joint Surg.*, 74A, 1021, 1992.
49. Päivärinta, U., Böstman, O., Majola, A., Toivonen T., Törmälä, P., and Rokkanen, P., "Intraosseous cellular response to biodegradable fracture fixation screws made of polyglycolide or polylactide," *Arch. Orthop. Trauma Surg.*, 112, 71, 1993.
50. Pihlajamäki, H., Böstman, O., Manninen, M., et al., "Shear strength of a distal rabbit femur during consolidation of an osteotomy fixed with a polylactide expansion plug," *Biomaterials*, 15, 257, 1994.
51. Pihlajamäki, H., Böstman, O., Manninen, M., Päivärinta, U., Tormala, P., and Rokkanen, P., "Absorbable plugs of self-reinforced poly-L-lactic acid in the internal fixation of rabbit distal femoral osteotomies," *Clin. Orthop.*, 298, 277, 1994.
52. Matsusue, Y., Yamamuro, T., Yoshii, S., Oka, M., Ikada, Y., Hyon, S.-H., and Shikinami, Y., "Biodegradable screw fixation of rabbit tibia proximal osteotomies," *J. Appl. Biomater.*, 2, 1, 1991.
53. An, Y. H., Friedman, R. J., Powers, D.L., Draughn, R.A., and Latour, R. A., "Fixation of osteotomies using bioabsorbable screws in the canine femur," *Clin. Orthop.*, 355, 300, 1998.
54. Otsuka, N. Y., Binnington, A. G., Fornasier, V. L., and Davey, J. R., "Fixation with biodegradable devices of acetabular components in a canine model," *Clin. Orthop.*, 306, 250, 1994.
55. Jukkala-Partio, K., Laitinen, O., Partio, E. K., et al., "Comparison of the fixation of subcapital femoral neck osteotomies with absorbable self-reinforced poly-L-lactide lag-screws or metallic screws in sheep," *J. Orthop. Res.*, 15, 124, 1997.
56. Weiler, A., Helling, H. J., Kirch, U., Zirbes, T. K., and Rehm, K. E., "Foreign-body reaction and the course of osteolysis after polyglycolide implants for fracture fixation: experimental study in sheep," *J. Bone Joint Surg.*, 78B, 369, 1996.
57. Manninen, M. J. and Pohjonen, T., "Intramedullary nailing of the cortical bone osteotomies in rabbits with self-reinforced poly-L-lactide rods manufactured by the fibrillation method," *Biomaterials*, 14, 305, 1993.
58. Hara, Y., Tagawa, M., Ejima, H., et al., "Clinical evaluation of uniaxially oriented poly-L-lactide rod for fixation of experimental femoral diaphyseal fracture in immature cats," *J. Vet. Med. Sci.*, 56, 1041, 1994.
59. Vainionpää, S., Vihtonen, K., Mero, M., Patiala, H., Rokkanen, P., Kilpikari, J., and Törmälä, P., "Biodegradable fixation of rabbit osteotomies," *Acta Orthop. Scand.*, 57, 237, 1986.
60. Miettinen, H., Mäkelä, E. A., Rokkanen, P., and Törmälä, P., "Fixation of femoral shaft osteotomy with intramedullary metallic or absorbable rod: an experimental study on growing dogs," *J. Biomater. Sci. Polym. Ed.*, 4, 135, 1992.
61. Illi, O. E., Weigum, H., and Misteli, F., "Biodegradable implant materials in fracture fixation," *Clin. Mater.*, 10, 69, 1992.
62. Thaller, S. R., Hoyt, J., Borjeson, K., Dart, A., and Tesluk, H., "Polyglyconate plates and screws to stabilize zygomatic osteotomies in a rabbit model," *J. Craniofac. Surg.*, 4, 228, 1993.
63. Eppley, B. L. and Sadove, A. M., "A comparison of resorbable and metallic fixation in healing of calvarial bone grafts," *Plast. Reconstr. Surg.*, 96, 316, 1995.
64. Getter, L., Cutright, D. E., Bhaskar, S. N., and Augsburg, J. K., "A biodegradable intraosseous appliance in the treatment of mandibular fractures," *J. Oral Surg.*, 30, 344, 1972.
65. Leenslag, J. W., Pennings, A. J., Bos, R. R., Rozema, F. R., and Boering, G., "Resorbable materials of poly(L-lactide). VI. Plates and screws for internal fracture fixation," *Biomaterials*, 8, 70, 1987.
66. Tams, J., Joziasse, C. A., Bos, R. R., Rozema, F. R., Grijpma, D. W., and Pennings, A. J., "High-impact poly(L/D-lactide) for fracture fixation: *in vitro* degradation and animal pilot study," *Biomaterials*, 16, 1409, 1995.

67. Hara, Y., Tagawa, M., Ejima, H., et al., "Application of oriented poly-L-lactide screws for experimental Salter-Harris type 4 fracture in distal femoral condyle of the dog," *J. Vet. Med. Sci.*, 56, 817, 1994.
68. Bos, R. R., Rozema, F. R., Boering, G., Nijenhuis, A. J., Pennings, A. J., and Jansen, H. W., "Boneplates and screws of bioabsorbable poly(L-lactide) — an animal pilot study," *Br. J. Oral Maxillofac. Surg.*, 27, 467, 1989.
69. Hollinger, J. O., "Preliminary report on the osteogenic potential of a biodegradable copolymer of polylactide (PLA) and polyglycolide (PGA)," *J. Biomed. Mater. Res.*, 17, 71, 1983.
70. Ashammakhi, N., Mäkelä, E. A., Vihtonen, K., Rokkanen P., and Törmälä, P., "Repair of bone defects with absorbable membranes. A study on rabbits," *Ann. Chirurg. Gynaecol.*, 84, 309, 1995.
71. Higashi, S., Yamamuro, T., Nakamura, T., Ikada, Y., Hyon, S. H., and Jamshidi, K., "Polymerhydroxyapatite composites for biodegradable bone fillers," *Biomaterials*, 7, 183, 1986.
72. Antikainen, T., Ruuskanen, M., Taurio, R., Kallioinen, M., Serlo, W., et al., "Polylactide and polyglycolic acid-reinforced coralline hydroxy-apatite for the reconstruction of cranial bone defects in the rabbit," *Acta Neurochirurg.*, 117, 59, 1992.
73. Hollinger, J. O. and Schnitz, J. P., "Restoration of bone discontinuities in dogs using a biodegradable implant," *J. Oral Maxillofac. Surg.*, 45, 594, 1987.
74. Busch, O., Solheim, E., Bang, G., and Tornes, K., "Guided tissue regeneration and local delivery of insulinlike growth factor I by bioerodible polyorthoester membranes in rat calvarial defects," *Int. J. Oral Maxillofac. Implants*, 11, 498, 1996.
75. Turk, A. E., Ishida, K., Jensen, J. A., Wollman, J. S., and Miller, T. A., "Enhanced healing of large cranial defects by an osteoinductive protein in rabbits," *Plast. Reconstr. Surg.*, 92, 593, 1993.
76. Miki, T., Harada, K., Imai, Y., and Enomoto, S., "Effect of freeze-dried poly-L-lactic acid discs mixed with bone morphogenetic protein on the healing of rat skull defects," *J. Oral Maxillofac. Surg.*, 52, 387, 1994
77. Kenley, R., Marden, L., Turek, T., Jin, L., Ron, E., and Hollinger, J. O., "Osseous regeneration in the rat calvarium using novel delivery systems for recombinant human bone morphogenetic protein-2 (rhBMP-2)," *J. Biomed. Mater. Res.*, 28, 1139, 1994.
78. Yamazaki, Y., Oida, S., Ishihara, K., and Nakabayashi, N., "Ectopic induction of cartilage and bone by bovine bone morphogenetic protein using a biodegradable polymeric reservoir," *J. Biomed. Mater. Res.*, 30, 1, 1996 .
79. Alpaslan, C., Irie, K., Takahashi, K., Ohashi N., Sakai, H., et al., "Long-term evaluation of recombinant human bone morphogenetic protein-2 induced bone formation with a biologic and synthetic delivery system," *Br. J. Oral Maxillofac. Surg.*, 34, 414, 1996.
80. Sandberg, E., Dahlin, C., and Linde, A., "Bone regeneration by the osteopromotion technique using bioabsorbable membranes: an experimental study in rats," *J. Oral Maxillofac. Surg.*, 51, 1106, 1993.
81. Ashammakhi, N., Mäkelä, E. A., Vihtonen, K., Rokkanen, P., and Törmälä, P., "The effect of absorbable self-reinforced polyglycolide membrane on metaphyseal bone. An experimental study on rats," *Ann. Chirurg. Gynaecol.*, 83, 328, 1994.
82. Ashammakhi, N., Mäkelä, E. A., Vihtonen, K., Rokkanen, P., and Törmälä, P., "Effect of selfreinforced polyglycolide membranes on cortical bone: an experimental study on rats," *J. Biomed. Mater. Res.*, 29, 687, 1995.
83. Johnson, E. E., Urist, M. R., and Finerman, G. A., "Bone morphogenetic protein augmentation grafting of resistant femoral nonunions. A preliminary report," *Clin. Orthop.*, 230, 257, 1988.
84. Schmitz, J. P. and Hollinger, J. O., "A preliminary study of the osteogenic potential of a biodegradable alloplastic-osteoinductive alloimplant," *Clin. Orthop.*, 237, 245, 1988.
85. Pinholt, E. M., Solheim, E., Bang, G., and Sudmann, E., "Bone induction by composite of bioerodible polyorthoester and demineralized bone matrix in rats," *Acta Orthop. Scand.*, 62, 476, 1991.
86. Kinoshita, Y., Kirigakubo, M., Kobayashi, M., Tabata, T., Shimura, K., and Ikada, Y., "Study on the efficacy of biodegradable poly(L-lactide) mesh for supporting transplanted particulate cancellous bone and marrow: experiment involving subcutaneous implantation in dogs," *Biomaterials*, 14, 729, 1993.
87. Saitoh, H., Takata, T., Nikai, H., Shintani, H., Hyon, S. H., and Ikada, Y., "Effect of polylactic acid on osteoinduction of demineralized bone: preliminary study of the usefulness of polylactic acid as a carrier of bone morphogenetic protein," *J. Oral Rehabil.*, 21, 431, 1994.
88. von Schroeder, H. P., Kwan, M., Amiel, D., and Coutts, R. D., "The use of polylactic acid matrix and periosteal grafts for the reconstruction of rabbit knee articular defects," *J. Biomed. Mater. Res.*, 25, 329, 1991.

89. Freed, L. E., Grande, D. A., Nohria, A., Emmanural, J., Mikos, A. G., and Langer, R., "Joint resurfacing using chondrocytes and synthetic biodegradable polymers," *J. Biomed. Mater. Res.*, 28, 891, 1994.
90. Vacanti, C. A., Kim, W., Schloo, B., Upton, J., and Vacanti, J. P., "Joint resurfacing with cartilage grown *in situ* from cell-polymer structures," *Am. J. Sports Med.*, 22, 485, 1994.
91. Chu, C. R., Coutts, R. D., Yoshioka, M., et al., "Articular cartilage repair using allogeneic perichondrocyte-seeded biodegradable porous polylactic acid (PLA): a tissue-engineering study," *J. Biomed. Mater. Res.*, 29, 1147, 1995.
92. Klompmaker, J., Jansen, H. W., Veth, R. P., et al., "Porous polymer implant for repair of meniscal lesions: a preliminary study in dogs." *Biomaterials*, 12, 810, 1991.
93. de Groot, J. H., Zijlstra, F. M., Kuipers, H. W., et al., "Meniscal tissue regeneration in porous 50/50 copoly(L-lactide/epsilon-caprolactone) implants," *Biomaterials*, 18, 613, 1997.
94. Koukoubis, T. D., Glisson, R. R., Feagin, J. A. Jr., et al., "Meniscal fixation with an absorbable staple. An experimental study in dogs," *Knee Surg. Sports Traumatol. Arthrosc.*, 5, 22, 1997.
95. Parsons, J. R., Alexander, H., and Weiss, A. B., "Absorbable polymer-ligamentous carbon composites: A new concept in orthopaedic biomaterials," in *Biocompatible Polymers, Metals and Composites*, Szycher, M., Ed., Technomic, Lancaster, PA, 1983, 873.
96. Laitinen, O., Pohjonen, T., Törmälä, P., et al., "Mechanical properties of biodegradable poly-L-lactide ligament augmentation device in experimental anterior cruciate ligament reconstruction," *Arch. Orthop. Trauma Surg.*, 112, 270, 1993.
97. Holzmuller, W., Wehmeyer, M., Rehm, K. E., and Perren, S. M., "Histologic studies of replacement of the anterior cruciate ligament with PDS-augmented patellar tendon transplants," *Unfallchirurg*, 97, 144, 1994.
98. Rodkey, W. G., Cabaud, H. E., Feagin, J. A., and Perlik, P. C., "A partially biodegradable material device for repair and reconstruction of injured tendons. Experimental studies," *Am. J. Sports Med.*, 13, 242, 1985.
99. Barber, F. A., Elrod, B. F., McGuire, D. A., and Paulos, L. E., "Preliminary results of an absorbable interference screw," *Arthroscopy*, 11, 537, 1995.
100. Juutilainen, T., Vihtonen, K., Patiala, H., Rokkanen, P., and Törmälä, P., "Reinsertion of the ruptured ulnar collateral ligament of the metacarpophalangeal joint of the thumb with an absorbable self-reinforced polylactide mini tack," *Ann. Chirurgiae Gynaecol.*, 85, 364, 1996.
101. Wei, G., Kotoura, Y., Oka, M., et al., "A bioabsorbable delivery system for antibiotic treatment of osteomyelitis. The use of lactic acid oligomer as a carrier," *J. Bone Joint Surg.*, 73B, 246, 1991.
102. Jacob, E., Cierny, G., III, Fallon, M. T., McNeill, J. F., Jr., and Siderys, G. S., "Evaluation of biodegradable cefazolin sodium microspheres for the prevention of infection in rabbits with experimental open tibial fractures stabilized with internal fixation," *J. Orthop. Res.*, 11, 404, 1993.
103. Garvin, K. L., Miyano, J. A., Robinson, D., et al., "Polylactide/polyglycolide antibiotic implants in the treatment of osteomyelitis. A canine model," *J. Bone Joint Surg.*, 76A, 1500, 1994.
104. Overbeck, J. P., Winckler, S. T., Meffert, R., et al., "Penetration of ciprofloxacin into bone: a new bioabsorbable implant," *J. Invest. Surg.*, 8, 155, 1995.
105. Calhoun, J. H. and Mader, J. T., "Treatment of osteomyelitis with a biodegradable antibiotic implant," *Clin. Orthop.*, 341, 206, 1997.
106. Mackinnon, S. E. and Dellon, A. L., "Clinical nerve reconstruction with a bioabsorbable polyglycolic acid tube," *Plast. Reconstr. Surg.*, 85, 419, 1990.
107. Hentz, V. R., Rosen, J. M., Xiao, S. J., McGill, K. C., and Abraham. G., "A comparison of suture and tubulization nerve repair techniques in a primate," *J. Hand Surg.*, 16A, 251, 1991.
108. Nicoli Aldini, N., Perego, G., Cella, G. D., et al., "Effectiveness of a bioabsorbable conduit in the repair of peripheral nerves," *Biomaterials*, 17, 959, 1996.
109. van der Lei, B., Nieuwenhuis, P., Molenaar, I., and Wildevuur, C. R., "Long-term biologic fate of neoarteries regenerated in microporous, compliant, biodegradable, small-caliber vascular grafts in rats," *Surgery*, 101, 459, 1987.
110. Greisler, H. P., Cabusao, E. B., Lam, T. M., et al., "Kinetics of collagen deposition within bioresorbable and nonresorbable vascular prostheses," *ASAIO Transactions*, 37, M472, 1991.
111. Langer, R., "New methods of drug delivery," *Science*, 249, 1527, 1990.

112. Solheim, E., Pinholt, E. M., Andersen, R., Bang, G., and Sudmann, E., "Local delivery of indomethacin by a polyorthoester inhibits reossification of experimental bone defects," *J. Biomed. Mater. Res.,* 29, 1141, 1995.
113. Ruuskanen, M. M., Kallioinen, M. J., Kaarela, O. I., et al., "The role of polyglycolic acid rods in the regeneration of cartilage from perichondrium in rabbits," *Scand. J. Plast. Reconstr. Surg. Hand Surg.,* 25, 15, 1991.
114. Anthanasiou, K. A., Schenck, R. C., Constaintides, G., et al., "Biodegradable carriers of TGF-β in rabbit osteochondral defects," *Trans. Orthop. Res. Soc.,* 17, 172, 1992.
115. Anthanasiou, K. A., Schenck, R. C., Constaintides, G., et al., "The use of biodegradable implants for repairing large articular cartilage defects in the rabbit," *Trans. Orthop. Res. Soc.,* 18, 288, 1993.
116. Freed, L. E., Marquis, J. C., Nohria, A., et al., "Neocartilage formation *in vitro* and *in vivo* using cells cultured on synthetic biodegradable polymers," *J. Biomed. Mater. Res.,* 27, 11, 1993.
117. Brunelli, G. A., Vigasio, A., and Brunelli, G. R., "Different conduits in peripheral nerve surgery, *Microsurgery,*" 15, 176, 1994.

13 Animal Models of Bone Defect Repair

Yuehuei H. An and Richard J. Friedman

CONTENTS

I. Introduction 241
II. Animal Models for Bone Defect Repair 242
    A. Heterotopic Models of Osteogenesis 242
    B. Bone Defect Models and Animal Selection 243
III. Authors' Preferred Models 243
    A. Heterotopic Model (Rat Subcutaneous Model) 243
    B. Calvarial Defect Models 243
        1. Rabbit Calvarial Defect Model 243
        2. Rat Calvarial Defect Model 247
    C. Long Bone Defect Models 247
        1. Segmental Defects of Rabbit Forearm Bones 247
    D. The Second Choice 248
IV. Evaluation of Bone Defect Repair 248
    A. Radiography 248
    B. Histology and Histomorphometry 250
    C. Mechanical Testing 251
V. Bone Substitutes and Future Directions of Research 252
    A. Mechanisms of Bone Repair by Bone Grafting 252
    B. Autograft, Allograft, and Xenograft 252
    C. Demineralized Bone Matrix 252
    D. Biomaterials 252
    E. Bone Marrow 253
    F. Growth Factors 253
    G. Tissue Engineered Composite Graft 253
References 253

I. INTRODUCTION

The history of bone grafts for repairing defects can be traced back a few hundred years. Although the work was started at a time that lacked basic science and technology, the efforts led to the establishment of bone grafting techniques by the early twentieth century. Due to the complicated nature, bone defects caused by various conditions have been challenging orthopaedic surgeons and related biomedical scientists for some time, and are continuing to inspire them to seek better alternatives and new solutions. To achieve their goals, animal models of bone defects for bone grafting are essential.

TABLE 1
Selected Heterotopic Models of Osteogenesis

Animal	Implant site	Material tested	First author, year[Ref.]
Nude mouse	SC*	Human osteoblasts, diffusion chamber	Gotoh 1995[1]
	IP	Human osteoprogenitor cells in diffusion chamber	Gundle 1995[2]
	SC	Osteosarcoma cells	Hara 1996[3]
Syngenic mouse	IP‡	Rodent Achilles tendons in diffusion chamber	Gooney 1993[4]
Nude rat	IMO	Rat and human DBM	Aspenberg 1989[5]
Rat	SC	Inductive bovine DBM	Nathan 1988[6]
	SC, IM	Porous ceramics + marrow cells	Ohgushi 1989[7]
	IM	Fibroblast growth factor	Aspenberg 1989[8]
	SC	DBM powder	Mohr 1991[9]
	SC	Alkaline phosphatase	Beertsen 1992[10]
	IM	Rabbit BMP in diffusion chamber	Ono 1994[11]
	IM	Multiple bone substitutes (6 each)	Begley 1995[12]
	SC, IM	Porous HA/TCP	Yang 1996[13]
	MES§	Marrow stromal osteoblast in PLA/PGA foam	Ishaug-Riley[14]
Rabbit	IP	Rabbit marrow cells in diffusion chamber	Ashton 1980[15]
	IM	Bone marrow, DBM, collagen in diffusion chamber	Strates 1989[16]
	IM	Composite of HA and periosteum	Kurashina 1995[17]
	IM	Porous HA	Ripamonti 1996[18]
	SC, IM	Porous HA/TCP	Yang 1996[13]
Dog	IM	Porous HA	Ripamonti 1996[18]
	SC, IM	Porous HA/TCP	Yang 1996[13]
Pig	SC, IM	Porous HA/TCP	Yang 1996[13]
Goat	SC, IM	Porous HA/TCP	Yang 1996[13]
Primate	IM	Allogenic DBM and porous HA	el Deeb 1989[19]
	IM	Porous HA	Ripamonti 1996[18]

* SC = Subcutaneous tissue
† IM = Intramuscular
‡ IP = Intraperitoneal
§ MES = Mesentery

II. ANIMAL MODELS FOR BONE DEFECT REPAIR

A. HETEROTOPIC MODELS OF OSTEOGENESIS

There are several major heterotopic models for testing *in vivo* osteogenesis, the subcutaneous model, the intramuscular model, the intraperitoneal model, and the mesentery model, which are often used as the first step before a bone defect model (Table 1).[20] Substances, constructs, or diffusion chamber containing osteogenic materials were implanted in the three anatomical sites in mice, rats, rabbits, dogs, pigs, goats, or primates. After 3–24 weeks the implants were explanted and studied radiographically and histologically to identify the existence and extent of new bone. The effects of animal species on the amount of bone formation have been demonstrated.[13] There are also effects of implantation sites and local environment on the amount or extent of bone formation.[20,21]

The diffusion chamber is made from a plastic ring (2 mm-thick, 9 mm diam.) bounded by two porous cellulose acetate and nitrate membranes (100 μm thickness, 0.45 μm pore size) and a chamber volume about 130 μl (Millipore Corporation, Sedford, MA).[2,4,11,15] For example, a diffusion chamber containing rabbit BMP was implanted in the abdominal muscle of the rat. After implantation, cartilage differentiated around the chamber in 1–2 weeks and bone replaced the cartilage in 3–4 weeks.[11]

B. Bone Defect Models and Animal Selection

Selected bone defect models are listed in Table 2. There are mainly four types of defects including calvarial, long bone (or mandible) segmental, partial cortical (cortical window, wedge defect, or transcortical drill hole), and cancellous bone defect (drill holes). The calvarial defect and long bone segmental defect are used the most often. The commonly used animals are rabbits, rats, dogs and sheep. Rabbits and rats are first choice for calvarial models and rabbits and dogs for segmental defects. Dogs and sheep are often used for experimental conditions involving heavy internal or external fixations.

III. AUTHORS' PREFERRED MODELS

A. Heterotopic Model (Rat Subcutaneous Model)

Potential osteogenic materials can be implanted subcutaneously or intramuscularly in the rat. If the material is paste-like it can be injected using a syringe. The common place for implantation is the SC tissues on the back or upper abdominal area. Figure 1 shows radiographically and histologically a new bone nodule formed five weeks after the injection of a collagen gel impregnated with several osteogenic GFs. Osteogenic materials can also be implanted in muscle, peritoneal cavity, or mesentery.

B. Calvarial Defect Models

1. Rabbit Calvarial Defect Model

The rabbit calvarial defect model is very popular and appropriate for the following reasons: (1) the calvarial bone is a plate which allows creation of a uniform circular defect that enables convenient radiographic and histological analysis; (2) the calvarial bone has a good size for easier surgical procedures and specimen handling; (3) no fixation is required because of the supports by the dura and the overlying skin; (4) the model has been well studied and is reproducible, which permits precise comparison of a variety of graft substances; and (5) it is relatively economical compared to dogs.

When a calvarial model is selected, the critical size of the defect (CSD) for the animal species has to be considered. CSD is defined as the smallest size of a calvarial defect which does not heal spontaneously when left untreated for a certain period of time (often six months) (Table 3).[120] The CSD of adult NZW rabbit calvarial defect is 15 mm.[54]

For operative procedures, an anteroposterior midline skin incision (4–5 cm long) is made over the cranial vault. The periosteum is elevated and retracted to expose the cranial bone. A 15-mm diameter defect is created using a trephine (Figure 2A). The drill should be centered at a point on the midline 5 mm posterior to the transverse bone sutures between the frontal and parietal bones (Figure 2B). Copious saline irrigation is needed during the drilling. Care must be taken to avoid perforating the dura, which is usually achieved by time to time checking of the osteotomy depth. Usually, the bone disc needs to be taken out of the defect by breaking the left-over connection points in between the bone disc and the edge of the defect. A neural elevator is very helpful for checking the depth, breaking bone connections within the osteotomy, and separating the dura from the inner surface of the bone disc. The key is never cut the bone plate all way through. If this happens, it is very likely the dura has already been cut, which could cause an intracranial hematoma. After specific treatment, as often an implantation of a bone substitute (Figure 3), the periosteum is closed and strengthened by closing the adjacent muscles and SC tissue. Occasionally, a SC hemotoma may occur after surgery and it can be drained by cutting several skin stitches. Wound infection is rare.

TABLE 2
Animal Bone Defect Models Selected from the Literature

Animal	Bones	Types of defect	First author, year[Ref]
Rat	Cranial bone	Circular	Ray 1957,[22] Takagi 1982,[23] Noda 1989,[24] Hollinger 1991,[25] Kenly 1994,[26] Miki 1994,[27] Gombotz 1994,[28] Kobayashi 1995,[29] Sweeney 1995,[30] McKinney 1996,[31] el Montaser 1997[32]
	Radius	Segmental	Herold 1971,[33] Gepstein 1987,[34] Alper 1989,[35] Solheim 1992,[36] Nyman 1995[37]
	Femora	Segmental	Melcher 1962,[38] Einhorn 1984,[39] Pelker 1989,[40] Wolff 1994,[41] Feighan 1995,[42] Langenskiold 1996,[43] Hunt 1996,[44] Puelacher 1996,[45] Stevenson 1997[46]
	Tibia	Metaphyseal window	Taguchi 1990,[47] Uchida 1985[48]
	Fibula	Segmental	Narang 1971,[49] Chakkalakal 1994,[50] Bluhm 1995[51]
	Nasal bone	Circular	Dupoirieux 1994[52]
	Mandible	Circular	Zellin 1997[53]
Rabbit	Cranial bone	Circular	Frame 1980,[54] Schmitz 1988,[55] Damien 1990,[56] Kleinschmidt 1993,[57] Richardson 1993,[58] Arnaud 1994,[59] Meikle 1994,[60] Robinson 1995,[61] Ashby 1996,[62] Rabie 1996[63]
	Radius or ulna	Segmental	Herold 1971,[33] Tuli 1981,[64] Gupta 1982,[65] Wittbjer 1983,[66] Aspenberg 1986,[67] Bolander 1986,[68] Hopp 1989,[69] Iyoda 1991,[70] Yang 1994,[71] Ho 1995[72]
	Femoral greater trochanter	Drill hole	Heikkilä 1995[73]
	Femoral condyle	6 mm drill hole	Kühne 1994,[74] Roudier 1995[75]
	Upper femoral cortex	3.5 mm drill hole	McCormack 1993[76]
	Tibia	Circumscribed	Shimazaki 1985,[77] Suh 1987,[78] Uchida 1985,[48] Shimizu 1988[79]
	Fibulae	Segmental	Yang 1994,[71] Taguchi 1995[47]
	Scapular bone	Circular	Oikarinen 1978[80]
	Mandible	Rectangular defect	Eppley 1988[81]
Dog	Cranial bone	18–20 mm circular	Oklund 1986[82]
	Radius or ulna	Segmental	Holmes 1987,[83] Key 1934,[84] Nilsson 1986,[85] Moore 1987,[86] Delloye 1990,[87] Grundel 1991,[88] Cook 1994,[89] Johnson 1996,[90] Sciadini 1997[91]
	Radius	Cortical window	Bay 1993[92]
	Ulna	Segmental/bioreactor	Frayssinet 1991[93]
	Femur	Segmental	de Pablos 1994[94]
		Wedge defect	Black 1990,[95] St. John 1993[96]
	Tibia	Segmental	Tiedeman 1989,[97] Markel 1991[98]
	Fibula	Segmental	Enneking 1975,[99] Welter 1990[100]
	Mandible	Segmental	Holmes 1979,[101] Toriumi 1991[102]
Sheep	Cranial bone	18–20 mm circular	Lindholm 1988,[103] Viljanen 1996[104]
	Femur	Segmental	Ehrnberg 1993,[105] Brunner 1994[106]
	Tibia	6 mm drill hole	Hallfeldt 1995[107]
		Segmental	Gao 1995[108]
	Mandible	Holes	Gatti 1991[109]
		Segmental	Stoll 1992[110]
Goat	Femur	Transcortical hole	Radder 1996[111]
Primate	Cranial bone	15 mm circular	Hollinger 1989[112]
	Mandible	Grooves	Drury 1991[113]

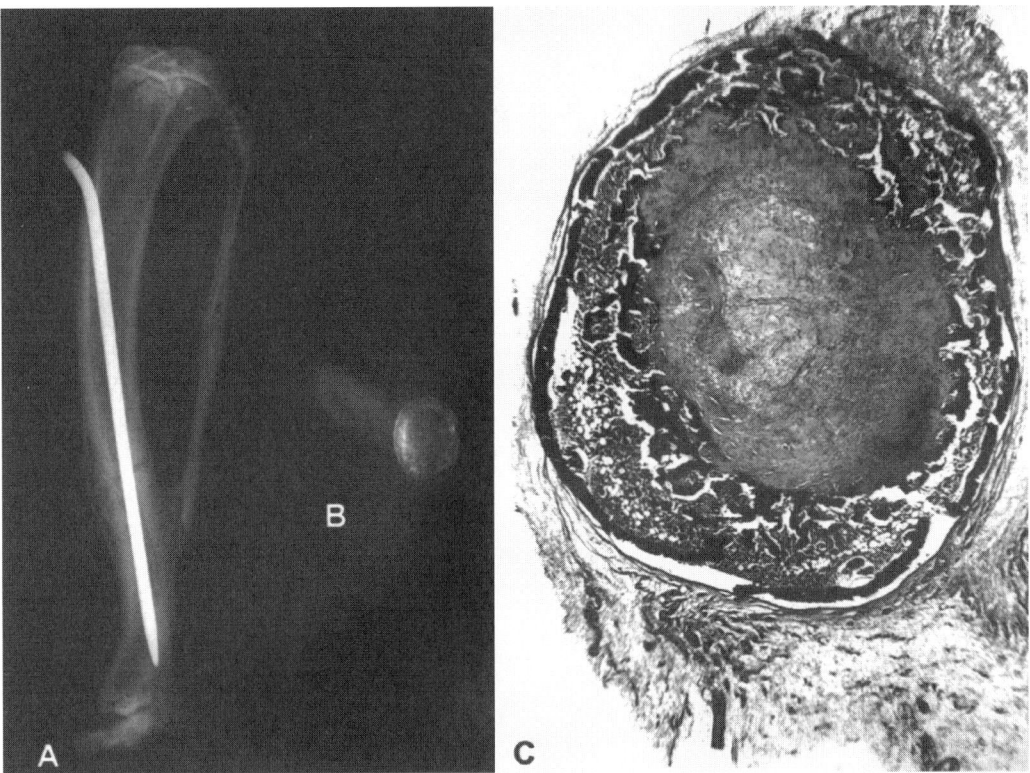

FIGURE 1. The radiographic (A,B) and histological (C) images of a new bone nodule five weeks after the injection of 100 µl collagen gel impregnated with osteogenic GFs (undecalcified paraffin section with H&E staining). The histological section shows that a bony shell, containing a central area of fibrous tissue, has two layers with vascularized adipose tissue in between resembling marrow. In areas this osseous tissue is lined with osteoblasts overlying a layer of osteoid. Also there is a transition zone of cartilage between the bone and fibrous tissue.

TABLE 3
The Critical Sizes of Common Animal Calvarial Defect Models

Animal	Strains	Defect size (Diameter)	Observation period (No healing, month)	First author year[Ref.]
Rat	Charles River	4 mm	6	Mulliken 1980[114]
	SD	8 mm (CSD)	3	Takagi 1982[23]
	—	8 mm	6 weeks	Schimitz 1990[115]
Rabbit	New Zealand White	8 mm	Healed in 4 weeks	Kramer 1968[116]
	NZreds-Half Lop*	15 mm (CSD)	6 and 9	Frame 1980[54]
Dog	Mongrel	17 mm (CSD)	5	Friedenberg 1962[117]
	Mongrel	20 mm	6	Prolo 1982[118]
Sheep	In Finland, Europe	18–20 mm (CSD)	3	Lindholm 1988[103]
Primate	Baboon	15 mm	2	Hollinger 1989[112]
		25 mm	3, 6, and 9	Ripamonti 1991[119]

* NZreds-Half Lop means the rabbits are crossbreeds of New Zealand reds and half lops.

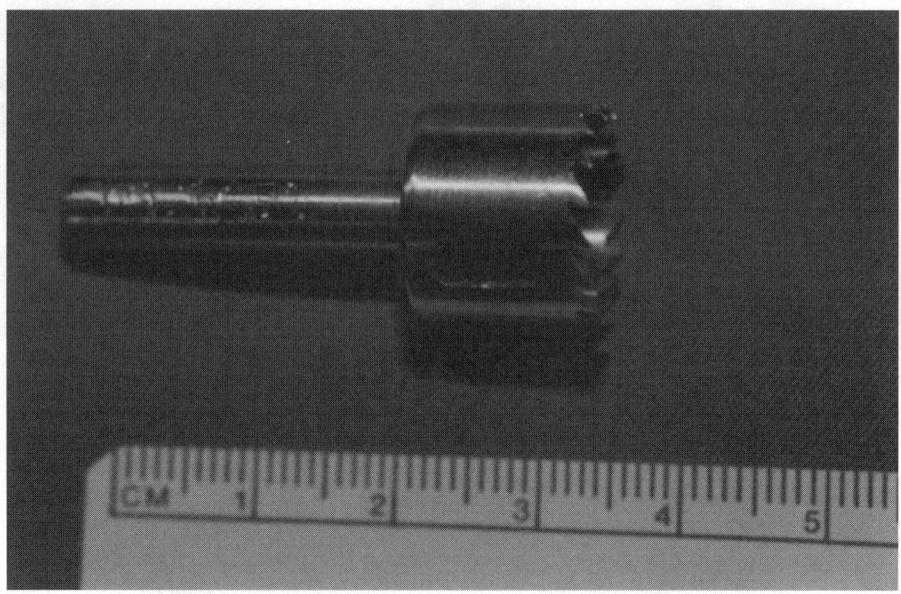

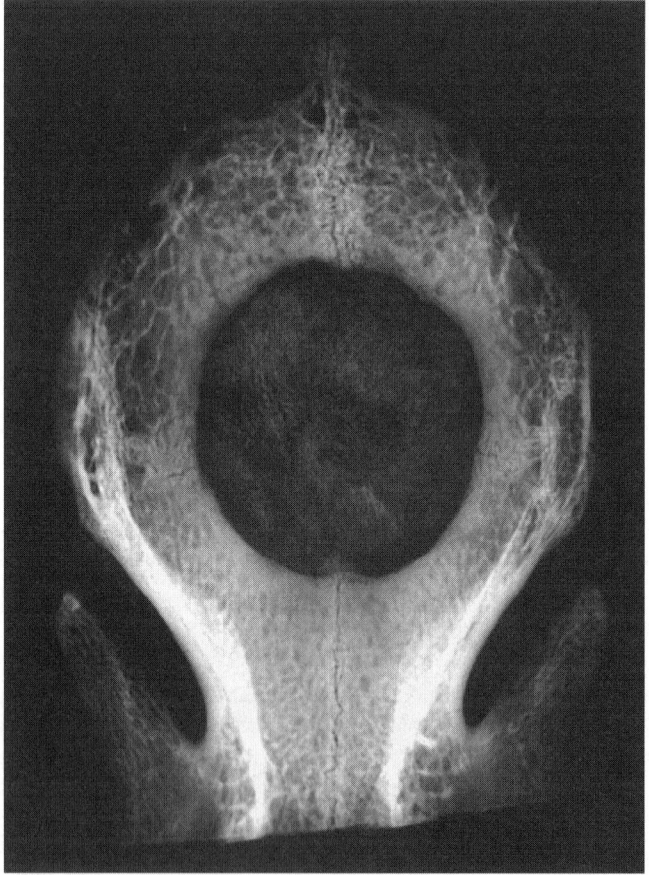

FIGURE 2. (A) Photograph of a trephine used for creating rabbit calvarial defect. (B) A defect three months after surgery showing no healing. It also shows where the trephine should be centered, which is the point on the midline 5 mm posterior to the transverse bone sutures between the frontal and parietal bones.

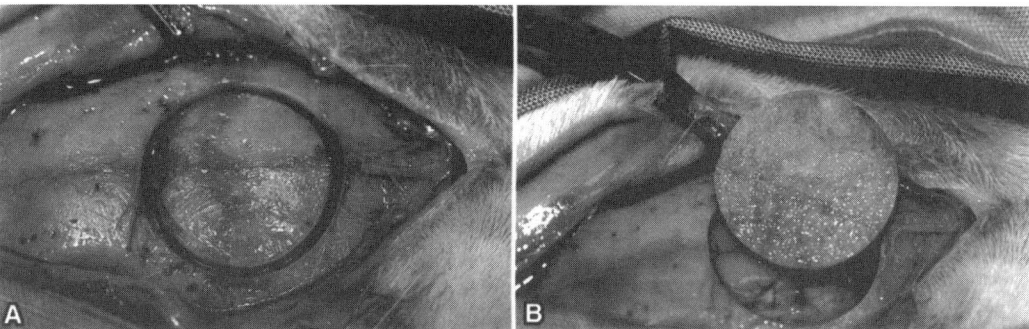

FIGURE 3. (A) A calvarial defect has been created with the bone disc still in place. (B) A disk implant (an HA-collagen complex) is ready to be placed in the defect.

2. Rat Calvarial Defect Model

The rat calvarial model is another popular one for the following reasons: (1) the calvarial bone is a plate which is large enough to allow creation of a circular defect that enables convenient radiographic and histological analysis; (2) the calvarial bone has a large enough size for easier surgical procedures and specimen handling; (3) no fixation is required; (4) the model has been well studied and is reproducible, which permits precise comparison of a variety of graft substances; and (5) it is much cheaper than rabbits. The CSD of a SD rat calvarial defect is considered as 8 mm in diameter (Table 3). There is one major concern about this model, which is the fast healing ability of the rat. The surgical procedure for the rat calvarial defect is about the same as that for rabbit. For the small size, a pair of operating glasses are essential.

C. LONG BONE DEFECT MODELS

1. Segmental Defects of Rabbit Forearm Bones

The rabbit radial model is popular and appropriate for the following reasons: (1) the radius bone is tubular, which allows creation of a segmental defect that enables convenient radiographic and histological evaluation; (2) the radius has a good size for easier surgical procedures and specimen handling; (3) no fixation is required because of the support of ulna; (4) the model has been well studied and is reproducible, which permits precise comparison of a variety of graft substances; and (5) it is relatively economical. The rabbit radial defect model was first described by Herold in 1971 to test the effect of growth hormone on the healing of bone defects.[33] It is well accepted that the CSD of long bones is two times the bone diameter. Because the diameter of the radius of adult NZW rabbits is about 5–6 mm, a radial defect should be no less than 12 mm long. In the author's laboratory, three out of twenty 12 mm defects healed spontaneously in 12 weeks. Therefore, it is the author's opinion that a 15 mm defect should be created.

Surgically, a longitudinal skin incision is made over the radial bone at the middle one third of the front leg. The periosteum is carefully separated from the surrounding muscles. A 15 mm defect located about 2.0 to 2.5 cm proximal to the radiocarpal joint (for the first cut) (Figure 4A) is created using a circular saw attached to a mini driver or dental handpiece (see Figure 3 in Chapter 11). The defect should be checked carefully and any periosteum left removed. A thorough wash with saline is needed before any treatment applied. The defect is then grafted with bone substitute or a spacer or left empty. No fixation is needed. Postoperatively, unrestricted weight bearing is normally allowed. Radiographs should be taken immediately after the surgery and also at any designated time periods. Figure 4B is a radiographic image showing a healed 15 mm radial defect by implantation of a porous form DBM in eight weeks. Adult rabbits with closed growth plate are preferred, which eliminated the possibility of epiphyseal slipping (Figure 5).

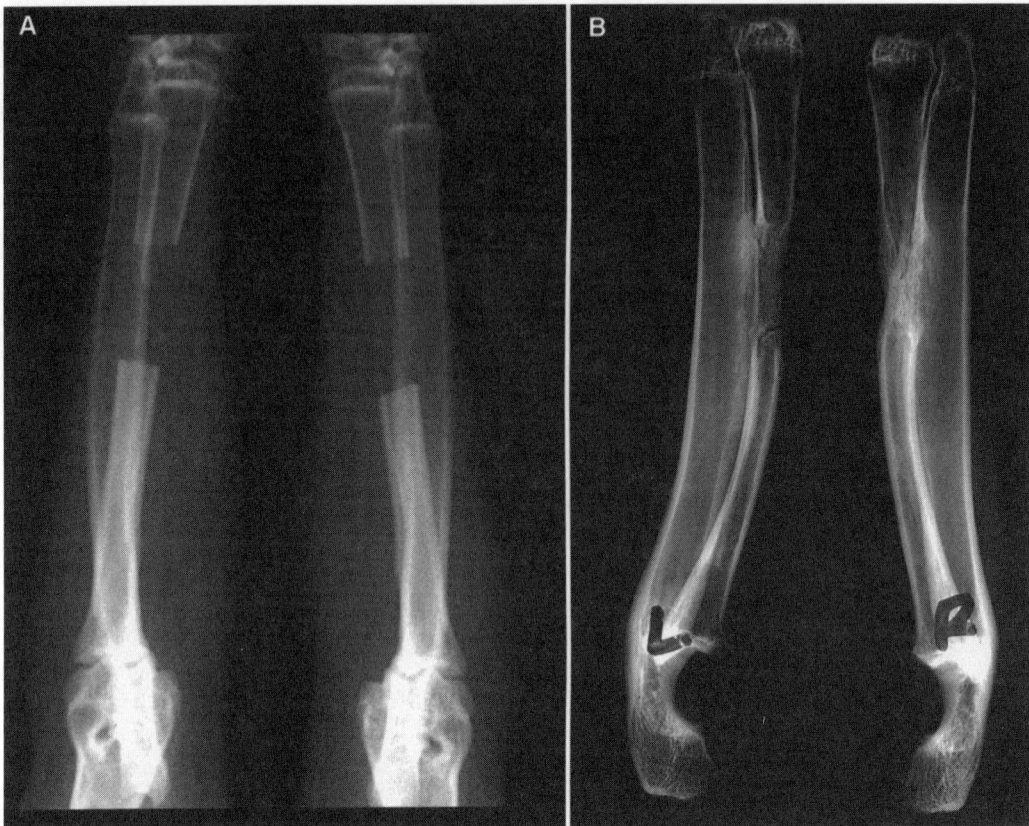

FIGURE 4. An X ray image showing a healed 15 mm radial defect by implantation of a porous DBM in eight weeks.

A rabbit ulnar defect is also popular, but the ulnae do not have the round shape as the radius, which creates difficulties and errors to the processes of implant preparation, implant positioning, and sample evaluation. Also because of its shape, doing mechanical testing, such as bending or torsional tests, on the ulnar bone is less favorable.

C. The Second Choice

If there are reasons for not using the three models mentioned above, the rat femoral defect model and the dog radial model could be selected. The former is a good model with only one shortcoming, the need for a secured internal fixation or external fixator. Due to the small size of the rat limb, it can be very frustrating to try to put an effective fixation device on. The dog radial defect model is popular and appropriate. Occasionally, because of the active nature of the dog, a fracture of the ulnae may occur, resulting in a tremendous loss compared to a rat or rabbit model.

IV. EVALUATION OF BONE DEFECT REPAIR

A. Radiography

Using radiographic analysis, information on the amount and quality of the new bone, such as bone density and structure, and continuity with the adjacent recipient bone, can be obtained. In recent years, computerized image analysis makes this more efficient. Radiographs taken at regular intervals

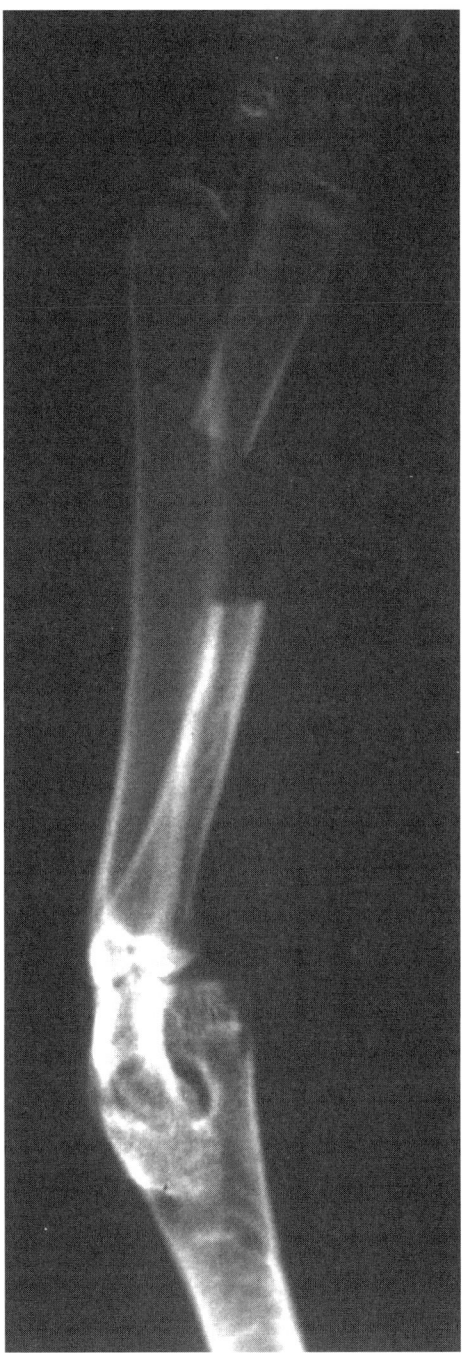

FIGURE 5. The radiograph shows a distal ulnar epiphyseal slipping when a young rabbit was used.

after surgery provide continuous information on the process of bone healing. For a long bone segmental defect, healing parameters, such as periosteal reaction (callus formation), appearance of the graft (bone formation), quality of union at both osteotomies, and bone remodeling are quantitated based on a radiographic scoring system modified from the ones by Bos et al.,[120] Lane and Sandhu,[121] Yang et al.,[71] and Johnson et al.[122](Table 4).

TABLE 4
Radiographic Scoring System

Categories	Scores
Periosteal reaction*	
Full (across the defect)	3
Moderate	2
Mild	1
None	0
Proximal osteotomy union	
Union	3
Moderate bridge (>50%)	2
Mild bridge (<50%)	1
Nonunion	0
Distal osteotomy union	
Union	3
Moderate bridge (>50%)	2
Mild bridge (<50%)	1
Nonunion	0
Appearance of graft	
Full replacement	3
Moderate replacement (>50%)	2
Mild replacement (<50%)	1
No change	0
Remodeling	
Full remodeling of cortex	2
Intramedullary canal	1
No remodeling	0
Maximum total score	14

* For some osteogeneic materials such as collagen sponge/GFs, this category may not apply because of the nearly absent periosteal reaction.

B. Histology and Histomorphometry

Histology is the most powerful method of examining the healing of bone defects. Common histological parameters include the following categories: bone union at the two osteotomies, callus formation, new bone formation in the defect, resorption of the bone graft, marrow changes, and cortex remodeling, which are semi-quantitated based on a scoring system modified from the ones by Bos et al.,[120] Nilsson et al.,[85] Lane and Sandhu,[121] Heiple et al.,[123] and Suh et al.[78](Table 5). Based on the nature of the graft to be used, the scoring system may be customized to suit the individual situation.

In recent years, computerized image analysis makes histomorphometry more efficient, especially for the calculation of the percentage of filling by the repair tissues in the defect and the fractions of different tissues in it. For histomorphometrical anaysis of ectopic bone formation, the following elements can be quantified: area and penetration depth of mineralized bonelike tissue, area and thickness of nonmineralized bonelike tissue, area of osteoblast-covered surfaces, thickness of trabeculae, area and thickness of cartilage tissue, area of fibrovascular tissue, and void space.[9,14] For analysis of repair tissues of bone defect the areas of mineralized bone, nonmineralized bone, lamellar bone, woven bone, chondral tissue, fibrocartilage, or fibrous vascular tissue can be used as paremeters for quantification.[41,124] See Chapter 6 for more information on bone histomorphometry.

TABLE 5
Histological Scoring System

Categories	Scores
Callus formation*	
Full (across the defect)	3
Moderate	2
Mild	1
None	0
Proximal osteotomy union	
Full bone bridge (union)	3
Moderate bridge (>50%)	2
Mild bridge (<50%)	1
No new bone in the osteotomy line (nonunion)	0
Distal osteotomy union	
Full bone bridge (Union)	3
Moderate bridge (>50%)	2
Mild bridge (<50%)	1
No new bone in the osteotomy line (nonunion)	0
Resorption of the bone graft	
Fully absorbed	3
Moderate adsorption (>50%)	2
Mild adsorption (<50%)	1
No change	0
New bone formation in the defect	
Full bone formation in the defect	3
Moderate bone formation (>50%)	2
Mild bone formation (<50%)	1
No new bone	0
Marrow changes	
Adult type fatty marrow	4
2/3 replaced by new tissue	3
1/3 replaced by new tissue	2
Fibrous tissue	1
Red	0
Cortex remodeling	
Full remodeling cortex	2
Intramedullary canal	1
No remodeling	0
Maximum total score	21

* For some osteogeneic materials such as collagen sponge/GFs, this category may not apply because of the nearly absent periosteal reaction.

C. MECHANICAL TESTING

Mechanical tests, such as torsional test,[40,69,71,87,89,95,122,125] bending test,[102] and tensile strength[126] of the repaired site, have been used effectively to evaluate the mechanical properties of the repair. For long bone segmental defect models, torsional testing is an appropriate and the most popular method.

V. BONE SUBSTITUTES AND FUTURE DIRECTIONS OF RESEARCH

A. Mechanisms of Bone Repair by Bone Grafting

The mechanisms of bone repair by bone grafting vary with the different types or composition of substitutes. Three basic mechanisms have been described for bone defect repair by grafting, osteoconduction, osteoinduction and direct osteogenesis. With osteoconduction, the grafted dead bone or bone substitute acts as a scaffolding for the ingrowth of blood vessels and new bone while itself being resorbed gradually by the host tissues. Osteoconduction is very important because without a scaffold defect, repair is less likely. Osteoinduction occurs when local mesenchymal cells, undifferentiated cells, or even muscle cells are transformed into bone-forming cells in the presence of certain stimulators, such as growth factors or hormones.[127] Direct osteogenesis is regulated by autogenic osteoblasts implanted with the graft or from the edges of the defect. Bone graft substitutes which can provide all three mechanisms for osteogenesis would be considered ideal bone grafts, since these mechanisms are often absent or insufficient at the sites of bone defect of nonunion and delayed union.

B. Autograft, Allograft, and Xenograft

In 1923, Albee reported 3000 cases of successful bone grafting.[128] In 1944, the use of iliac crest bone grafts in 75 cases were reported.[129] Since then, fresh autogenous bone grafts have been the most successful for skeletal reconstruction. The reasons for the high success rate include the osteoconductive ability for the ingrowth of blood vessels and new bone, the presence of pre-existing differentiated osteogeneic cells for immediate osteogenesis, the existing growth factors for osteoinduction, and the lack of immunologic rejection from the host. Unfortunately, autogenous bone graft cannot solve all osseous defects because of its limited quantity, donor site morbidity, and in some situations, the difficulty of fabricating the graft into a desired shape.

Despite the reported potential rejection from the host,[130] allogenic grafts are well tolerated, nontoxic, incorporated by host bone, and last a long period of time.[131] Due to its availability, allograft bone is one of the major methods for bone grafting in certain conditions, such as revision total joint arthroplasty, tumor surgery, or limb salvage.

Xenograft alone tends to fail (due to inflammatory reaction) in a rat model compared to allograft.[132] In another animal study, decalcified xenogenic bone impregnated with fresh autologous marrow was reported bridging a large cortical defect successfully.[133] Clinically, xenograft of hydrogen peroxide-macerated bone (Kiel bone) in combination with autologous marrow healed most of the bone defects or pseudoarthroses in two series.[134,135]

C. Demineralized Bone Matrix

By chemical or physical methods, allogenic, xenogenic, or even autogenous bone can be demineralized to make DBM which can be fabricated into powder or porous form or molded into certain solid shapes. DBM has been studied extensively since the early 1960s.[22,35,39,64,80,136] The osteogenic mechanisms of DBM include osteoconduction[137,138] and osteoinduction.[13,16] Because of its availability, diverse uses, and limited immunological response, DBM has a bright future in research and clinical practice for repairing bone and perhaps cartilage defect.[139]

D. Biomaterials

Biomaterials such as ceramics, polymers, or metals have been investigated extensively as bone substitutes with varying degrees of success. Hydroxyapatite,[35,92] tricalcium phosphate (TCP),[77] and composite forms such as HA/TCP ceramic[86] have been used for bone defects. The rate of bone regeneration into material pores varies depending upon the implant location, the availability of local osteogenic cells, and the stability of the implant. Clinical results reported by Bucholz et al.[140] showed that filling traumatic and tumor defects with HA and TCP healed in most of the cases. These materials act as osteoconductive scaffolds and have no osteoinductive ability. Attention has been paid to adding growth promoting factors such as GFs,[141] bone marrow,[88,142] or DBM[36] to the implants.

E. BONE MARROW

Adding autogenous bone marrow to bone substitutes (allograft, xenograft, DBM, or biomaterials) has been investigated since the early 1960s.[88,142-144] The purpose is to increase the osteogenic capability of the graft with osteogenic precursor cells which differentiate into bone forming cells in the recipient site.[15] Autogenous marrow is now being used clinically as an osteogenic material for repairing skeletal defects.[137,143]

F. GROWTH FACTORS

Bone tissue harbors many growth factors (GFs), such as bone morphogenic protein (BMP), transforming growth factor (TGF), platelet-derived growth factor (PDGF), epidermal growth factor (EGF), fibroblast growth factor (FGF), and insulin-like growth factor (IGF). It is known that bone structure is maintained through a balancing process of bone formation and resorption, and is mediated by cellular activities and regulated by systemic hormones and local GFs. GFs have the functions of stimulating osseous cellular proliferation, differentiation, DNA and protein synthesis, and extracellular matrix synthesis. They either act alone or in combinations. In the last two decades, a large number of *in vitro* and *in vivo* studies have been done and there has been significant evidence supporting the use of GFs for bone repair.[145,146]

G. TISSUE ENGINEERED COMPOSITE GRAFT

Bone grafting is aimed to provide the missing elements necessary for bone formation in a bone defect and thereby restore the bone integrity. Using cell-seeding to a substrate to make an implantable graft is not a new concept.[147,148] Recently, a few groups reported some preliminary and very important data on the use of cell-seeded implants for repairing osseous or chondral defects.[93,149-151] In 1991, Frayssinet et al. reported that bone cells from canine humeri were grown on HA granules and the cell-HA composite was placed in a bioreactor and implanted into a canine ulna defect. Osteogenesis was seen in active bioreactors three weeks after implantation.[93] Cultured chondrocytes bound to a HA block were implanted to repair a rabbit ulna defect.[70] Osteoblast-like cells (MC3T3-E1) were also used to study the potential of bioabsorbable polymers and ceramics to support osteoblastic growth for a bone-polymer composite in bone repair.[150]

The above-mentioned studies demonstrate that cell-seeded composite implants can induce bone tissue formation, leading to defect repair. Osteogenic cell-seeded composite implants act in a similar way as bone marrow transplantation does, and this technique has been referred to as tissue engineering.[152] Over other methods, it has certain advantages, such as the autogenic characteristics (if cells come from the same individual), minor morbidity of the donor site, no risk of disease transmission, and permitting potential production of a large quantity *in vitro*.

REFERENCES

1. Gotoh, Y., Fujisawa, K., Satomura, K., and Nagayama, M., "Osteogenesis by human osteoblastic cells in diffusion chamber *in vivo*," *Calcif. Tissue Int.,* 56, 246, 1995.
2. Gundle, R., Joyner, C. J., and Triffitt, J. T., "Human bone tissue formation in diffusion chamber culture *in vivo* by bone-derived cells and marrow stromal fibroblastic cells." *Bone,* 16, 597, 1995.

3. Hara, A., Ikeda, T., Nomura, S., Yagita, H., Okumura, K., and Yamauchi, Y., "*In vivo* implantation of human osteosarcoma cells in nude mice induces bones with human-derived osteoblasts and mouse-derived osteocytes," *Lab. Invest.,* 75, 707, 1996.
4. Rooney, P., Walker, D., Grant, M. E., and McClure, J., "Cartilage and bone formation in repairing Achilles tendons within diffusion chambers: evidence for tendon-cartilage and cartilage-bone conversion *in vivo*," *J. Pathol.,* 169, 375, 1993.
5. Aspenberg, P. and Andolf, E., "Bone induction by fetal and adult human bone matrix in athymic rats," *Acta Orthop. Scand.,* 60, 195, 1989.
6. Nathan, R. M., Bentz, H., Armstrong, R. M., Piez, K. A., Smestad, T. L., and Ellingsworth, L. R., "Osteogenesis in rats with an inductive bovine composite," *J. Orthop. Res.,* 6, 324, 1988.
7. Ohgushi, H., Goldberg, V. M., and Caplan, A. I., "Heterotopic osteogenesis in porous ceramics induced by marrow cells," *J. Orthop. Res.,* 7, 568, 1989.
8. Aspenberg, P. and Lohmander, L. S., "Fibroblast growth factor stimulates bone formation. Bone induction studied in rats," *Acta Orthop. Scand.,* 60, 473, 1989b.
9. Mohr, H. and Kragstrup, J., "Morphostereometry of heterotopic ossicles in the rat," *Acta Orthop. Scand.,* 62, 257, 1991.
10. Beertsen, W. and van den Bos, T., "Alkaline phosphatase induces the mineralization of sheets of collagen implanted subcutaneously in the rat," *J. Clin. Invest.,* 89, 1974, 1992.
11. Ono, Y., Kato, K., Oohira, A., Katoh, R., and Nogami, H., "Cell function during chondrogenesis and osteogenesis induced by bone morphogenetic protein enclosed in diffusion chamber," *Clin. Orthop.,* 298, 305, 1994.
12. Begley, C. T., Doherty, M. J., Mollan, R. A., and Wilson, D. J., "Comparative study of the osteoinductive properties of bioceramic, coral and processed bone graft substitutes," *Biomaterials,* 16, 1181, 1995.
13. Yang, Z., Yuan, H., Tong, W., Zou, P., Chen, W., and Zhang, X., "Osteogenesis in extraskeletally implanted porous calcium phosphate ceramics: variability among different kinds of animals," *Biomaterials,* 17, 2131, 1996.
14. Ishaug-Riley, S. L., Crane, G. M., Gurlek, A., Miller, M. J., Yasko, A. W., et al., "Ectopic bone formation by marrow stromal osteoblast transplantation using poly(DL-lactic-co-glycolic acid) foams implanted into the rat mesentery," *J. Biomed. Mater. Res.,* 36, 1, 1997.
15. Ashton, B. A., Allen, T. D., Howlet, C. R., Eaglesom, C. C., Hattori, A., and Owen M., "Formation of bone and cartilage by marrow stromal cells in diffusion chambers *in vivo*," *Clin. Orthop.,* 151, 294, 1980.
16. Strates, B. S. and Connolly, J. F., "Osteogenesis in cranial defects and diffusion chambers. Comparison in rabbits of bone matrix, marrow, and collagen implants," *Acta Orthop. Scand.,* 60, 200, 1989.
17. Kurashina, K., Kurita, H., Takeuchi, H., Hirano, M., Klein, C. P., and de Groot, K., "Osteogenesis in muscle with composite graft of hydroxyapatite and autogenous calvarial periosteum: a preliminary report." *Biomaterials,* 16, 119, 1995.
18. Ripamonti, U., "Osteoinduction in porous hydroxyapatite implanted in heterotopic sites of different animal models," *Biomaterials,* 17, 31, 1996.
19. el Deeb, M., Hosny, M., and Sharawy, M., "Osteogenesis in composite grafts of allogenic demineralized bone powder and porous hydroxylapatite," *J. Oral Maxillofac. Surg.,* 47, 50, 1989.
20. Ekelund, A., Brosjo, O., and Nilsson, O. S., "Experimental induction of heterotopic bone," *Clin. Orthop.,* 263, 102, 1991.
21. Urist, W. R., Hay, P. H., Dubuc, F., and Buring, K., "Osteogenetic competence," *Clin. Orthop.,* 64, 194, 1969.
22. Ray, B. and Holloway, J. A., "Bone implants," *J. Bone Joint Surg.,* 39A, 1119, 1957.
23. Takagi, K. and Urist, M. R., "The reaction of the dura to bone morphogenetic protein (BMP) in repair of skull defects," *Ann. Surg.,* 196, 100, 1982.
24. Noda, M. and Camilliere, J. J., "*In vivo* stimulation of bone formation by transforming growth factor-β," *Endocrinology,* 124, 2991, 1989.
25. Hollinger. J. O., Mark, D. E., Goco, P., Quigley, N., Desverreaux, R. W., and Bach, D. E., "A comparison of four particulate bone derivatives," *Clin. Orthop.,* 267, 255, 1991.
26. Kenley, R., Marden, L., Turek, T., Jin, L., Ron, E., and Hollinger, J. O., "Osseous regeneration in the rat calvarium using novel delivery systems for recombinant human bone morphogenetic protein-2 (rhBMP-2)," *J. Biomed. Mater. Res.,* 28, 1139, 1994.

27. Miki, T., Harada, K., Imai, Y., and Enomoto, S., "Effect of freeze-dried poly-L-lactic acid discs mixed with bone morphogenetic protein on the healing of rat skull defects," *J. Oral Maxillofac. Surg.*, 52, 387, 1994.
28. Gombotz, W. R., Pankey, S. C., Bouchard, L. S., Phan, D. H., and Puolakkainen, P. A., "Stimulation of bone healing by transforming growth factor-beta1 released from polymeric ceramic implants," *J. Appl. Biomer.*, 5, 141, 1994.
29. Kobayashi, K., Agrawal, K., Jackson, I. T., and Vega, J. B., "The effect of insulin-like growth factor 1 on craniofacial bone healing," *Plast. Reconstr. Surg.*, 97, 1129, 1996.
30. Sweeney, T. M., Opperman, L. A., Persing, J. A., and Ogle, R. C., "Repair of critical size rat calvarial defects using extracellular matrix protein gels," *J. Neurosurg.*, 83, 710, 1995.
31. McKinney, L. and Hollinger, J. O., "A bone regeneration study: transforming growth factor-β 1 and its delivery," *J. Craniofac. Surg.*, 7, 36, 1996.
32. el Montaser, M. A., Devlin, H., Sloan, P., and Dickinson, M. R., "Pattern of healing of calvarial bone in the rat following application of the erbium-YAG laser," *Laser Surg. Med.*, 21, 255, 1997.
33. Herold, H., Z. Hurvitz, and Tadmor, A., "The effect of growth hormone on the healing of experimental bone defects," *Acta Orthop. Scand.*, 42, 377, 1971.
34. Gepstein, R., Weiss, R. E., Saba, K., Hallel, T., and Israel, R. A., "Bridging large defects in bone by demineralized bone matrix in the form of a powder," *J. Bone Joint Surg.*, 69A, 984, 1987.
35. Alper, G., Bernick, S., Yazdi, M., and Nimni, M. E., "Osteogenesis in bone defects in rats: the effects of hydroxyapatite and demineralized bone matrix," *Am. J. Med. Sci.*, 298, 371, 1989.
36. Solheim, E., Pinholt, E. M., Andersen, R., Bang, G., Sudmann, B., and Sudmann, E., "The effect of a composite of polyorthoester and demineralized bone on the healing of large segmental defects of the radius in rats," *J. Bone Joint Surg.*, 74A, 1456, 1992.
37. Nyman, R., Magnusson, M., Sennerby, L., Nyman, S., and Lundgren, D., "Membrane-guided bone regeneration. Segmental radius defects studied in the rabbit," *Acta Orthop. Scand.*, 66, 169, 1995.
38. Melcher, A. H. and Irving, J. T., "The healing mechanism in artificially created circumscribed defects in the femora of albino rats," *J, Bone Joint Surg.*, 44B, 928, 1962.
39. Einhorn, T. A., Lane, J. M., Burstein, A. H., Kopman, C. R., and Vigorita, V. J., "The healing of segmental bone defects induced by demineralized bone matrix," *J. Bone Joint Surg.*, 66A, 274, 1984.
40. Pelker, R. R., Mckay, J., Jr., Troiano, N., Panjabi, M. M., and Friedlaender. G. E., "Allograft incorporation: a biomechanical evaluation in a rat model," *J. Orthop. Res.*, 7,585, 1989.
41. Wolff, D., Goldberg V. M., and Stevenson, S., "Histomorphometric analysis of the repair of a segmental diaphyseal defect with ceramic and titanium fiber-metal implants: effects of bone marrow." *J. Orthop. Res.*, 12, 439, 1994.
42. Feighan, J. E., Davy, D., Prewett, A. B., and Stevenson, S., "Induction of bone by a demineralized bone matrix gel: a study in a rat femoral defect model," *J. Orthop. Res.*, 13, 881, 1995.
43. Langenskiold, A., Hakkinen, S., and Ylinen, P., "Incorporation of cancellous bone into a diaphyseal defect of the radius in growing rabbits: tube-shaped versus homogeneous grafts," *J. Pediatr. Orthop.*, 16, 237, 1996.
44. Hunt, T. R., Schwappach, J. R., and Anderson, H. C., "Healing of a segmental defect in the rat femur with use of an extract from a cultured human osteosarcoma cell-line (Saos-2). A preliminary report," *J. Bone Joint Surg.*, 78A, 41, 1996.
45. Puelacher, W. C., Vacanti, J. P., Ferraro, N. F., Schloo, B., and Vacanti, C. A., "Femoral shaft reconstruction using tissue-engineered growth of bone," *Int. J. Oral Maxillofac. Surg.*, 253, 223, 1996.
46. Stevenson, S., Li, X. Q., Davy, D. T., Klein, L., and Goldberg, V. M., "Critical biological determinants of incorporation of non-vascularized cortical bone grafts. Quantification of a complex process and structure," *J. Bone. Joint. Surg.*, 79A, 1, 1997.
47. Taguchi, Y., Pereira, B. P., Kour, A. K., Pho, R. W., and Lee, Y. S., "Autoclaved autograft bone combined with vascularized bone and bone marrow," *Clin. Orthop.*, 320, 220, 1995.
48. Uchida, A., Nade, S., McCartney, and Ching, E. W., "Bone ingrowth into three different porous ceramics implanted into the tibia of rats and rabbits," *J. Orthop. Res.*, 3, 65, 1985.
49. Narang, R., Lloyd, W., and Wells, H., "Grafts of decalcified allogeneic bone matrix promote the healing of fibular fracture gaps in rats," *Clin. Orthop.*, 80, 174,1971.
50. Chakkalakal, D. A., Mashoof, A. A., Novak, J., Strates, B. S., and McGuire, M. H., "Mineralization and pH relationships in healing skeletal defects grafted with demineralized bone matrix," *J. Biomed. Mater. Res.*, 28, 1439, 1994.

51. Bluhm, A. E. and Laskin, D. M., "The effect of polytetrafluoroethylene cylinders on osteogenesis in rat fibular defects: a preliminary study," *J. Oral Maxillofac. Surg.*, 53, 163, 1995.
52. Dupoirieux, L., Costes, V., Jammet, P., and Souyris, F., "Experimental study on demineralized bone matrix (DBM) and coral as bone graft substitutes in maxillofacial surgery," *Int. J. Oral Maxillofac. Surg.*, 23, 395, 1994.
53. Zellin, G. and Linde, A., "Importance of delivery systems for growth-stimulatory factors in combination with osteopromotive membranes. An experimental study using rhBMP-2 in rat mandibular defects," *J. Biomed. Mater. Res.*, 35, 181, 1997.
54. Frame, J. W., "A convenient animal model for testing bone substitute materials," *J. Oral Surg.*, 38, 176, 1980.
55. Schmitz, J. P. and Hollinger, J. O., "A preliminary study of the osteogenic potential of a biodegradable alloplastic-osteoinductive alloimplant," *Clin. Orthop.*, 237, 245, 1988.
56. Damien, C. J., Parsons, J. R., Benedict, J. J., and Weisman, D. S., "Investigation of a hydroxyapatite and calcium sulfate composite supplemented with an osteoinductive factor," *J. Biomed. Mater. Res.*, 24, 639, 1990.
57. Kleinschmidt, J. C., Marden, L. J., Kent, D., Quigley, N., and Hollinger, J. O., "A multiphase system bone implant for regenerating the calvaria," *Plast. Reconstr. Surg.*, 91, 581, 1993.
58. Richardson, L., Zioncheck, T. F., Amento, E. P., Deguzman, L., Lee, W. P., Xu, Y., and Beck, L. S., "Characterization of radioiodinated recombinant human TGF-β1 binding to bone matrix within rabbit skull defects," *J. Bone Miner. Res.*, 8, 1407, 1993.
59. Arnaud, E., Morieux, C., Wybier, M., and de Vernejoul, M. C., "Potentiation of transforming growth factor (TGF-β1) by natural coral and fibrin in a rabbit cranioplasty model," *Calcif. Tissue Int.*, 54, 493, 1994.
60. Meikle, M. C., Papaioannou, S., Ratledge, T. J., et al., "Effect of poly DL-lactide-co-glycolide implants and xenogeneic bone matrix-derived growth factors on calvarial bone repair in the rabbit," *Biomaterials*, 15, 513, 1994.
61. Robinson, B. P., Hollinger, J. O., Szachowicz, E. H., and Brekke, J., "Calvarial bone repair with porous D,L-polylactide," *Otolaryngol. Head Neck Surg.*, 112, 707, 1995.
62. Ashby, E. R., Rudkin, G. H., Ishida, K., and Miller, T. A., "Evaluation of a novel osteogenic factor, bone cell stimulating substance, in a rabbit cranial defect model," *Plast. Reconstr. Surg.*, 98, 420, 1996.
63. Rabie, A. B., Deng, Y. M., Samman, N., and Hagg, U., "The effect of demineralized bone matrix on the healing of intramembranous bone grafts in rabbit skull defects," *J. Dent. Res.*, 75, 1045, 1996.
64. Tuli, S. M. and Gupta, K. B., "Bridging of large chronic osteoperiosteal gaps by allogeneic decalcified bone matrix implants in rabbits," *J. Trauma*, 21, 894, 1981.
65. Gupta, D. and Tuli, S. M., "Osteoinductivity of partially decalcified alloimplants in healing of large osteoperiosteal defects," *Acta Orthop. Scand.*, 53, 857, 1982.
66. Wittbjer, J., Palmer, B., Rohlin, M., and Thorngren, K-G., "Osteogenetic activity in composite grafts of demineralized compact bone and marrow," *Clin. Orthop.*, 173, 229, 1983.
67. Aspenberg, P., Wittbjer and J., Thorngren K-G., "Pulverized bone matrix as an injectable bone graft in rabbit radius defects." *Clin. Orthop.*, 206, 261, 1986.
68. Bolander, M. E. and Balian, G., "The use of demineralized bone matrix in the repair of segmental defects," *J. Bone Joint Surg.*, 68A, 1264, 1986.
69. Hopp, G., Dahners, L. E., and Gilbert, J. A., "A study of the mechanical strength of long bone defects treated with various bone autograft substitutes. An experimental investigation in the rabbit," *J. Orthop. Res.*, 7, 579, 1989.
70. Iyoda, K., Miura and T., Nogami, H., "Repair of bone defect with cultured chondrocytes bound to hydroxyapatite," *Clin. Orthop.*, 288, 287, 1993.
71. Yang, C. Y. Simmons, D. J., and Lozano, R., "The healing of grafts combining freeze-dried and demineralized allogeneic bone in rabbits," *Clin. Orthop.*, 298, 286, 1994.
72. Ho, M. L., Chang, J. K., and Wang, G. J., "Antiinflammatory drug effects on bone repair and remodeling in rabbits," *Clin. Orthop.*, 313, 270, 1995.
73. Heikkilä, J. T., Aho, H. J., Yli-Urpo, A., Happonen, R. P., and Aho, A. J., "Bone formation in rabbit cancellous bone defects filled with bioactive glass granules," *Acta Orthop. Scand.*, 66, 463, 1995.
74. Kühne, J. H., Bartl, R., Frisch, B., Hammer, C., Jansson, V., and Zimmer, M., "Bone formation in coralline hydroxyapatite. Effects of pore size studied in rabbits," *Acta Orthop. Scand.*, 65, 246, 1994.

75. Roudier. M., Bouchon, C., Rouvillain, J. L., et al., "The resorption of bone-implanted corals varies with porosity but also with the host reaction," *J. Biomed. Mater. Res.,* 29, 909, 1995.
76. McCormack, A. P., Anderson, P. A., and Tencer, A. F., "Effect of controlled local release of sodium fluoride on bone formation: filling a defect in the proximal femoral cortex," *J. Orthop. Res.,* 11, 548, 1993.
77. Shimazaki, K., and Mooney, V., "Comparative study of porous hydroxyapatite and tricalcium phosphate as bone substitute," *J. Orthop. Res.,* 3, 301, 1985.
78. Suh, H. and Lee, C., "Biodegradable ceramic-collagen composite implanted in rabbit tibiae," *ASAIO J.,* 41, M652, 1995.
79. Shimuzu, T., Zerwekh, J. E, Videman, T., Gill, K., Mooney, V., et al., "Bone ingrowth into porous calcium phosphate ceramics: influence of pulsing electromagnetic field," *J. Orthop. Res.,* 6, 248, 1988.
80. Oikarinen, J. and Korhonen, L. K., "The bone inductive capacity of various bone transplanting materials used for treatment of experimental bone defects," *Clin. Orthop.,* 140, 208, 1979.
81. Eppley, B. L., Doucet, M., Connolly, D. T., and Feder, J., "Enhancement of angiogenesis by bFGF in mandibular bone graft healing in the rabbit," *J. Oral Maxillofac. Surg.,* 46, 391, 1988.
82. Oklund, S. A., Prolo, D. J., Gutierrez, R. V., and King, S. E., "Quantitative comparisons of healing in cranial fresh autografts, frozen autografts and processed autografts, and allografts in canine skull defects," *Clin. Orthop.,* 205, 269, 1986.
83. Holmes, R. E., Bucholz, R. W., and Mooney, V., "Porous hydroxyapatite as a bone graft substitute in diaphyseal defects: a histometric study," *J. Orthop. Res.,* 5, 114, 1987.
84. Key, J. A., "The effect of a local calcium depot on osteogenesis and healing of fractures," *J. Bone Joint Surg.,* 16, 176, 1934.
85. Nilsson, O. S., Urist, M. R., Dawson, E. G., Schmalzried, T. P., and Finerman, G. A., "Bone repair induced by bone morphogenetic protein in ulnar defects in dogs," *J. Bone Joint Surg.,* 68B, 635, 1986.
86. Moore, D. C., Chapman, M. W., and Manske, D., "The evaluation of a biphasic calcium phosphate ceramic for use in grafting long-bone diaphyseal defects," *J. Orthop. Res.,* 5, 356, 1987.
87. Delloye, C., Verhelpen, M., d'Hemricourt, J., Govaerts, B., et al., "Morphometric and physical investigations of segmental cortical bone autografts and allografts in canine ulnar defects," *Clin. Orthop.,* 282, 273, 1990.
88. Grundel, R. E., Chapman, M. W., Yee, T., and Moore, D. C., "Autogeneic bone marrow and porous biphasic calcium phosphate ceramic for segmental bone defects in the canine ulna," *Clin. Orthop.,* 266, 244, 1991.
89. Cook, S. D., Baffes, G. C., Wolfe, M. W., Sampath, T. K., and Rueger, D. C., "Recombinant human bone morphogenetic protein-7 induces healing in a canine long-bone segmental defect model," *Clin. Orthop.,* 301, 302, 1994.
90. Johnson, K. D., August, A., Sciadini. M. F., and Smith, C., "Evaluation of ground cortical autograft as a bone graft material in a new canine bilateral segmental long bone defect model," *J. Orthop. Trauma,* 10, 28, 1996.
91. Sciadini, M. F., Dawson, J. M., and Johnson, K. D., "Bovine-derived bone protein as a bone graft substitute in a canine segmental defect model," *J. Orthop. Trauma,* 11, 496, 1997.
92. Bay, B. K., Martin, R. B., Sharkey, N. A., and Chapman, M. W., "Repair of large cortical defects with block coralline hydroxyapatite," *Bone,* 14, 225, 1993.
93. Frayssinet, P., Primout, I., Rouquet, N., Autefage, A., Guilhem, A., and Bonnevialle, P., "Bone cell grafts in bioreactor: a study of feasibility of bone cell autograft in large defects," *J. Mater. Sci. Mater. Med.* 2, 217, 1991.
94. de Pablos, J., Barrios, C., Alfaro, C., and Canadell, J., "Large experimental segmental bone defects treated by bone transportation with monolateral external distractors," *Clin. Orthop.,* 298, 259, 1994.
95. Black, R. J., Zardiackas, L. D., Teasdall, R., and Hughes, J. L., Jr., "The mechanical integrity of healed diaphyseal bone defects grafted with calcium hydroxyapatite/calcium triphosphate ceramic in a new animal model," *Clin. Mater.,* 6, 251, 1990.
96. St. John, K. R., Zardiackas, L. D., Terry, R. C., Teasdall, R. D., Cooke, S. E., and Mitias, H. M., "Histological and electron microscopic analysis of tissue response to synthetic composite bone graft in the canine," *J. Appl. Biomater.,* 6, 89, 1995.
97. Tiedeman, J. J., Connolly, J. F., Etrates, B. S., and Lippiello, L., "Treatment of nonunion by percutaneous injection of bone marrow and demineralized bone matrix. An experimental study in dogs," *Clin. Orthop.,* 268, 294, 1989.

98. Markel, M. D., Wikenheiser, M. A., and Chao, E. Y., "Formation of bone in tibial defects in a canine model. Histomorphometric and biomechanical studies," *J. Bone Joint Surg.,* 73A,914, 1991.
99. Enneking, W. F., Burchardt, H., Puhl, J. J., and Piotrowski, G., "Physical and biological aspects of repair in dog cortical-bone transplants," *J. Bone Joint Surg.,* 57A, 237, 1975.
100. Welter, J. F., Shaffer, J. W., Stevenson, S., et al., "Cyclosporin A and tissue antigen matching in bone transplantation," *Acta Orthop. Scand.,* 61, 517, 1990.
101. Holmes, R., "Bone regeneration within a coralline hydroxyapatite implant," *Plast. Reconstr. Surg.,* 63, 626, 1979.
102. Toriumi, D. M., Kotler, H. S., Luxenberg, D. P., Holtrop, M. E., and Wang, E. A., "Mandibular reconstruction with a recombinant bone-inducing factor," *Arch. Otolaryngol. Head Neck Surg.,* 117, 1101, 1991.
103. Lindholm, T. C., Lindholm, T. S., Alitalo, I., and Urist, M. R., "Bovine bone morphogenetic protein (bBMP) induced repair of skull trephine defects in sheep," *Clin. Orthop.,* 227, 265, 1988.
104. Viljanen, V. V., Gao, T. J., Lindholm, T. C., Lindholm, T. S., and Kommonen, B., "Xenogeneic moose (*Alces alces*) bone morphogenetic protein (mBMP)-induced repair of critical-size skull defects in sheep," *Int. J. Oral Maxillofac. Surg.,* 25, 217, 1996.
105. Ehrnberg, A., De Pablos, J., Martinez-Lotti, G., Kreicbergs, A., and Nilsson, O., "Comparison of demineralized allogeneic bone matrix grafting (the Urist procedure) and the Ilizarov procedure in large diaphyseal defects in sheep," *J. Orthop. Res.,* 113, 438, 1993.
106. Brunner, U. H., Cordey, J., Schweiberer, L., and Perren, S. M., "Force required for bone segment transport in the treatment of large bone defects using medullary nail fixation," *Clin. Orthop.,* 301, 147, 1994.
107. Hallfeldt K. K. Stutzle H. Puhlmann M. Kessler S. and Schweiberer L. "Sterilization of partially demineralized bone matrix: the effects of different sterilization techniques on osteogenetic properties," *J. Surg. Res.,* 59, 614, 1995.
108. Gao, T. J., Lindholm, T. S., Kommonen, B., Ragni, P., Paronzini, A., and Lindholm, T. C., "Microscopic evaluation of bone-implant contact between hydroxyapatite, bioactive glass and tricalcium phosphate implanted in sheep diaphyseal defects," *Biomaterials,* 16,1175, 1995.
109. Gatti, A. M. and Zaffe, D., "Short-term behaviour of two similar active glasses used as granules in the repair of bone defects," *Biomaterials,* 12, 497, 1991.
110. Stoll, P. and Wächter, R., "AO reconstruction plate systems for the repair of mandibular defects: 3-DBRP versus Thorp system," *J. Craniomaxillofac. Surg.,* 20, 40, 1992.
111. Radder, A. M., Leenders, H., and van Blitterswijk, C. A., "Application of porous PEO/PBT copolymers for bone replacement," *J. Biomed. Mater. Res.,* 30, 341, 1996.
112. Hollinger, J., Mark, D. E., Bach, D. E., Reddi, A. H., and Seyfer, A. E., "Calvarial bone regeneration using osteogenin," *J. Oral Maxillofac. Surg.,* 47, 1182, 1989.
113. Drury, G. I. and Yukna, R. A., "Histologic evaluation of combining tetracycline and allogeneic freeze-dried bone on bone regeneration in experimental defects in baboons," *J. Periodontal.,* 62, 652, 1991.
114. Mulliken, J. B. and Glowacki. J., "Induced osteogenesis for repair and construction in the craniofacial region," *Plast. Reconstr. Surg.,* 65, 553, 1980.
115. Schmitz, J. P., Schwartz, Z., Hollinger, J. O., and Boyan, B. D., "Characterization of rat calvarial nonunion defects," *Acta. Anat.,* 138, 185, 1990.
116. Kramer, I. R., Killey, H. C., and Wright, H. C., "A histological and radiological comparison of the healing of defects in the rabbit calvarium with and without implanted heterogeneous anorganic bone," *Arch. Oral. Biol.,* 13, 1095, 1968.
117. Friedenberg, Z. B. and Lawrence, R. R., "The regeneration of bone in defects of varying size," *Surg. Gynecol. Obstetr.,* 114, 721, 1962.
118. Prolo, D. J., Pedrotti, P. W., Burres, K. P., and Oklund, S., "Superior osteogenesis in transplanted allogeneic canine skull following chemical sterilization," *Clin. Orthop.,* 168, 230, 1982.
119. Ripamonti, U., "Bone induction in nonhuman primates. An experimental study on the baboon," *Clin. Orthop.,* 269, 284,1991.
120. Bos, G. D., Goldberg, V. M., Powell, A. E., Heiple, K. G., and Zika, J. M., "The effect of histocompatibility matching on canine frozen bone allografts," *J. Bone Joint Surg.,* 65A, 89, 1983.

121. Lane, J. M. and Sandhu, H. S., "Current approaches to experimental bone grafting," *Orthop. Clin. North Am.,* 18, 213, 1987.
122. Johnson, K. D., Frierson, K. E., Keller, T. S., et al., "Porous ceramics as bone graft substitutes in long bone defects: a biomechanical, histological, and radiographic analysis," *J. Orthop. Res.,* 14, 351, 1996
123. Heiple, K. G., Goldberg, V. M., Powell, A. E., Bos, G. D., and Zika, J. M., "Biology of cancellous bone grafts," *Orthop. Clin. North Am.,* 18, 179, 1987.
124. DeVries, W. J., Runyon, C. L., Martinez, S. A., and Ireland, W. P., "Effect of volume variations on osteogenic capabilities of autogenous cancellous bone graft in dogs," *Am. J. Vet. Res.,* 57, 1501, 1996.
125. Burstein, A. H. and Frankel, V. H., "A standard test for laboratory animal bone," *J. Biomech.,* 4, 155, 1971.
126. Paley, D., Young, M. C., Wiley, A. M., Fornasier, V. L., and Jackson, R. W., "Percutaneous bone marrow grafting of fractures and bony defects. An experimental study in rabbits," *Clin. Orthop.,* 208, 300, 1986.
127. Khouri, R., Koudsi, B., and Reddi, H., "Tissue transformation into bone *in vivo*," *J. Am. Med. Assoc.,* 266, 1953, 1991.
128. Albee, F. H., "Fundamentals in bone transplantation," *J. Am. Med. Assoc.,* 81, 1429, 1923.
129. Mowlem, R., "Cancellous chip bone-grafts. Report on 75 cases," *Lancet,* 1, 746, 1944.
130. Horowitz, M. C. and Friedlaender, G. E., "Immunologic aspects of bone transplantation," *Orthop. Clin. North Am.,* 18, 227, 1987.
131. Makin, H. J., Gebhardt, M. C., and Tomford, W. W., "The use of frozen cadaveric allografts in the management of patients with bone tumors of the extremities," *Orthop. Clin. North Am.,* 18, 275, 1987.
132. Thielemann, F. W., Spaeth, G., Veihelmann, D., and Schmidt, K., "Osteoinduction. Part I: Test model and comparative long term observation of allogenic and xenogenic matrix implants," *Arch. Orthop Trauma Surg.,* 99, 217, 1982.
133. Gupta, D., Khanna, S., and Tuli, S. M., "Bridging large bone defects with a xenograft composited with autologous bone marrow. An experimental study," *Int. Orthop.,* 6, 791, 982.
134. Salama, R., "Xenogeneic bone grafting in humans," *Clin. Orthop.,* 174, 113, 1983.
135. Horowitz, I. and Bodner, L., "Use of xenograft bone with aspirated bone marrow for treatment of cystic defect of the jaws," *Head Neck,* 11, 516, 1989.
136. Urist, M. R., "Bone: formation by autoinduction," *Science,* 150, 893, 1965.
137. Burwell, R. C., "Studies in the transplantation of bone. VII. The fresh composite homograft-autograft of cancellous bone: an analysis of factors leading to osteogenesis in marrow transplants and in marrow-containing bone grafts," *J. Bone Joint Surg.,* 46B, 110, 1964.
138. Nade, S. and Burwell, R. G., "Decalcified bone as a substrate for osteogenesis. An appraisal of the interrelation of bone and marrow in combined grafts," *J. Bone Joint Surg.,* 59B, 189, 1977.
139. Glowacki, J. and Mulliken, J. B., "Demineralized bone implants," *Clin. Plast. Surg.,* 12, 233, 1985.
140. Bucholz, R. W., Carlton, A., and Holmes, R. E., "Hydroxyapatite and tricalcium phosphate bone graft substitute," *Orthop. Clin. North Am.,* 18, 323, 1987.
141. Kawai, T., Mieki, A., Ohno, Y., et al., "Osteoinductive activity of composites of bone morphogenetic protein and pure titanium," *Clin. Orthop.,* 290, 296, 1993.
142. Goshima, J., Goldberg, V. M., and Caplan, A. I., "The origin of bone formed in composite grafts of porous calcium phosphate ceramic loaded with marrow cells," *Clin. Orthop.,* 269, 274, 1991.
143. Salama, R. and Weissman, S. L., "The clinical use of combined xenografts of bone and autologous red marrow," *J. Bone Joint Surg.,* 60B, 111, 1985.
144. Aspenberg, P., Wittbjer, J., and Thorngren, K.-G., "Bone matrix and marrow versus cancellous bone in rabbit radius defects," *Arch. Orthop. Trauma Surg.,* 106, 335, 1987.
145. Lind, M., "Growth factors: possible new clinical tools. A review," *Acta. Orthop. Scand.,* 67, 407, 1996.
146. Linkhart, T. A., Mohan, S., and Baylink, D. J., "Growth factors for bone growth and repair: IGF, TGF beta and BMP," *Bone,* 19, 1S, 1996.
147. Green, W. T., Jr., "Articular cartilage repair. Behavior of rabbit chondrocytes during tissue culture and subsequent allografting," *Clin. Orthop.,* 124, 237, 1977.
148. Herring, M. and Gardner, A., "A single-staged technique for seeding vascular grafts with autogenous endothelium," *Surgery,* 84, 498, 1978.

149. Vacanti, C. A., Kim, W., Upton, J., Mooney, D., Schloo, B., and Vacanti, J. P., "Tissue-engineered growth of bone and cartilage," *Transplant. Proc.,* 25, 1019, 1993.
150. Elgendy, H. M., Norman, M. E., Keaton, A. R., and Laurencin, C. T., "Osteoblast-like cell (MC3T3-E1) proliferation on bioerodable polymers: an approach towards the development of a bone-bioerodable polymer composite material," *Biomaterials,* 14, 263,1993.
151. Vacanti, C. A., Langer, R., Schloo, B., and Vacanti, J. P., "Synthetic polymers seeded with chondrocytes provide a template for new cartilage formation," *Plast. Reconstr. Surg.,* 88, 753, 1991.
152. Langer, R. and Vacanti, J. P., "Tissue engineering," *Science,* 260, 920, 1993.

14 Animal Models of Osteonecrosis

Kensaku Masuhara, Minoru Matui, Katsuya Nakata, and Keiro Ono

CONTENTS

I. Introduction 261
II. Nontraumatic Models of Osteonecrosis..262
    A. Spontaneously Hypertensive Rat (SHR) Model...............................262
    B. Stroke-Prone Spontaneously Hypertensive Rat (SHRSP) Model 264
    C. Steroid-Treated Rabbit Model..265
    D. Rabbits with Endotoxic (Shwartzman) Reactions 266
    E. Rabbits with Hypersensitivity Reactions (Study 1 by the Authors) ...267
    F. Rabbits with Hypersensitivity Reactions (Study 2 by the Authors) ...270
III. Traumatic Models of Osteonecrosis 271
    A. Goat Model..272
    B. Canine Model by Deep Freezing...273
    C. Canine Model by Dislocation and Vessel Ligation 274
    D. Swine Model 275
IV. Applications of Animal Models 276
References 277

I. INTRODUCTION

It has long been suggested that nontraumatic osteonecrosis (ON) of the femoral head may be associated with corticosteroid therapy, alcohol abuse, smoking, systemic lupus erythematosus, and renal transplantation. While much has been learned concerning the complex characteristics of nontraumatic ON, no means are yet available to prevent this disorder, and attempts to clarify pathogenesis of ON have been hampered by a number of problems. Incomplete early diagnosis of this disease, and the lack of useful experimental animal models of ON have prevented complete characterization of the pathological conditions present prior to the development of ON. With the advent of high performance magnetic resonance imaging (MRI) and optimized coil technology, an opportunity has arisen to detect subtle structural abnormalities in bone, enabling early diagnosis of bone diseases. Accordingly, more reproducible animal models of ON are needed in order to clarify the etiology and early pathogenesis of ON.

In the 1960s, 1970s, and 1980s, many attempts were made to induce ON traumatically in experimental animals such as intra-arterial infusion of oil (Lipiodol),[1] ligation of the blood vessels feeding the femoral epiphysis,[2,3] and intracapsular tamponade in the hip joint using wax or silicone.[4,5] However, the canine and rabbit models of ON produced by these methods cannot serve as animal models of nontraumatic ON in humans. In the 1990s, several animal models of nontraumatic

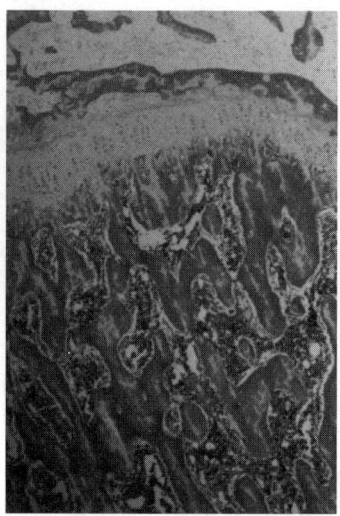

FIGURE 1. Femoral head necrosis in SHR (× 40).

ON have been successfully produced, from which a number of essential pathological findings have been obtained. The historical background, experimental protocol and principal findings obtained from studies using different animal models will be described below.

II. NONTRAUMATIC MODELS OF OSTEONECROSIS

A. SPONTANEOUSLY HYPERTENSIVE RAT (SHR) MODEL

1. Introduction

Skeletal disorders in SHR including osteoporosis and abnormal calcium metabolism have been reported, in addition to disorders of the endocrine and autonomic nervous systems.[6,7] In the late 1980s, it was incidentally observed that widespread ON frequently developed naturally in the epiphysis of the femoral head in growing SHR. In SHR with systemic skeletal growth retardation, ON naturally occurs in growing male rats and heals without remarkable deformity. These findings obtained in preliminary studies by Hirano et al. suggested that ON in SHR closely resembles Perthes' disease[8,9] The site and mechanism of the occlusion of the blood vessels feeding the femoral head were examined histologically and microangiographically by Iwasaki et al.[10]

2. Materials and Methods

One hundred fifteen growing male SHR (230 femurs) were used.[10] The femurs were fixed in formalin and decalcified with formic acid and hydrochloric acid before being embedded in paraffin. Thin coronal sections of the proximal femur were prepared and H&E stained. Histologic examination revealed three different sets of characteristic findings for the femoral head: (1) normal ossification (75/230); (2) abnormal (disturbed) ossification (78/230); and (3) ON (77/230). The group with ON was divided into subgroups based on stage of the healing process:

Stage 1 Fresh necrosis of trabeculae and bone marrow without repair tissue (Figures 1, 2)
Stage 2 Invasion of vascular granulation tissue into the intertrabecular space
Stage 3 Completely repaired ON with empty lacunae in the center to hypertrophic trabeculae

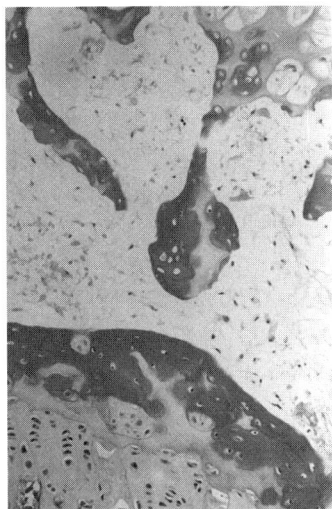 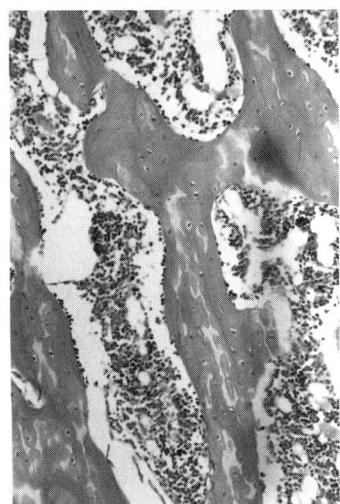

FIGURE 2. (A): Fresh necrosis of trabeculae and bone marrow without repair tissue in the epiphysis (×200). (B): Normal structure of trabeculae and bone marrow in the metaphysis (×200).

Microangiography using Micropaques solution[11] was performed on 30 male SHRs (10- to 40-week-old rats).

For assessment of the relationship between femoral head lesions and mechanical stress on the femoral head, 38 male SHR and 10 male WKY rats, each 6-weeks-old, were divided into five groups: A: no treatment, 10 WKYs (genetic control); B: no treatment, 10 SHRs; C: bilateral severance of the sciatic nerves, 8 SHRs; D: bilateral severance of the sciatic and femoral nerves, 10 SHRs; and E: amputation of the right lower hind limb, 10 SHRs.

After each treatment, the rats were killed at the age of 15 weeks and examined to determine the incidence of femoral head lesions such as ON, disturbed ossification, and abnormalities of the growth plate.

3. Essential Findings

In the femoral heads with fresh ON (Stage 1), the lateral epiphyseal vessels disappeared before entering the epiphysis. In the femoral heads with invasion of vascular granulation tissue (Stage 2), many slender vessels and capillaries entered the ossifying nucleus with the granulation tissue. In the femoral heads with completely repaired ON (Stage 3), blood vessels corresponding to the lateral epiphyseal vessels in normal femoral head entered the ossified nucleus.

Homogeneous distribution of blood vessels in the epiphysis was observed in the femoral heads with normal ossification and in heads with completely repaired ON (Stage 3), while absence of blood vessels throughout the epiphysis was demonstrated in the femoral heads with fresh ON without repaired tissue (Stage 1) and in the femoral heads in which no ossification had occurred at all. These findings suggest that ON and disturbed ossification may be the result of the same pathogenetic process.

The rats of Group C were able to walk using their hind limbs, in which quadriceps muscle function was intact. In the rats of Group D, active movement of the hind limbs was not observed. Locomotion was achieved using only the forelimbs, with dragging of the hind limbs. In the rats of Group E, the hip on the treated side was kept in flexion, and the amputated stump was suspended in midair. These observations suggested that the degree of stress on the femoral head was the highest in Groups A and B and the lowest in Groups D and E, Group C being intermediate. The incidence of ON, disturbed ossification, and growth plate abnormalities of the femoral head in each

TABLE 1
Incidence of ON, Disturbed Ossification and Growth Plate Abnormality

	Group A (n = 20)	Group B (n = 20)	Group C (n = 16)	Group D (n = 20)	Group E Amputation (n = 10)	Group E Nonamputation (n = 10)
Osteonecrosis	15%	60%	25%	0%	0%	40%
Disturbed ossification	0%	25%	12.5%	0%	0%	10%
Growth plate abnormality	15%	85–100%	50–69%	30%	30%	80–90%

(Adapted from Reference 10)

group are summarized in Table 1. Decrease in the incidence of femoral head lesions, such as ON, disturbed ossification, and abnormality of the growth plate accompanied a decrease in mechanical stress on the femoral head.

B. Stroke-Prone Spontaneously Hypertensive Rat (SHRSP) Model

1. Introduction

Marked hypertension and cerebral apoplexia are more common in SHRSPs than in SHRs. Histopathologically, SHRSPs often exhibit an age-dependent increase in ischemic lesions such as infarction in the brain, kidney, and heart, and some antihypertensive drugs completely prevent the hypertensive complications induced by angiospasm or arteriosclerosis in SHRSPs. Naito et al. reported that SHRSPs had high incidence of femoral head necrosis.[12]

2. Materials and Methods

A total of 135 male rats six to 36 weeks of age, including 40 WKYs (genetic control), 40 SHRs, and 55 SHRSPs, were used.[12] The length from the top of the major trochanter to the distal end of the femur was measured with a Vernier micrometer. The femur was placed in a small plastic phantom, and its bone mineral density (BMD) was measured using dual energy X ray absorptiometry (QDR-1000, Hologic). Portions of the femoral head, including the epiphyseal growth plates, of SHRSPs, SHRs, WKYs were resected at 12 weeks of age from the femur and subjected to a compressive load with the apparatus (Autography DCS-500, Shimazu, Japan). The hemispherical femoral head was placed between the plates, and a compressive load was administered vertically. The minimal compressive load at which the femoral head began to deform was recorded. The proximal femurs were fixed in 10% formalin solution and prepared for paraffin embedding after decalcification in EDTA. Thin sections through the teres ligament were stained with hematoxylin-eosin and toluidine-blue and examined by light microscopy.

3. Essential Findings

Among the 12-week-old rats, the mean bone length was least in SHRSPs and greatest in WKYs. This difference was statistically significant, but there were no significant differences in bone length among the groups after 12 weeks of age. The lowest mean femoral BMD was observed in SHRSPs and the highest in WKYs at all ages, while the BMD in SHRs was intermediate.

The mean compressive loads necessary to cause deformation of femoral heads in 12-wk-old SHRSPs, SHRs, and WKYs were 11.6±1.2 kg, 13.2±1.5 kg and 14.7±0.9 kg, respectively. The femoral heads in SHRSPs were the most easily deformed by loads applied during compression tests.

TABLE 2
Incidence of Fermoral Head Necrosis in SHRSPs as Compared to SHRs and WKYs

Strain	10–14 weeks	20–36 weeks
SHRSP	70%	45%
SHR	53.3%	20%
WKY	20%	13.3%

(Adapted from Reference 12)

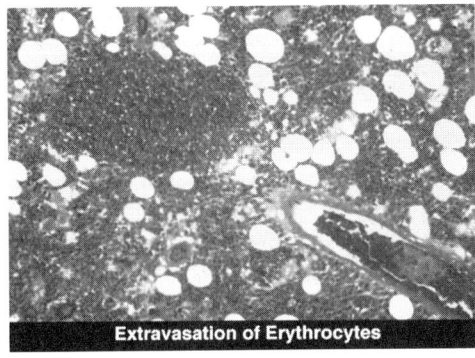

FIGURE 3. Extravasation of erythrocytes in bone marrow in the metaphysis in a rabbit with hypersensitivity. (×200)

Femoral head necrosis was observed in ossified regions with vascular invasion. In osteonecrotic regions in the young SHRSPs, dead trabeculae exhibited many empty lacunae without osteocytes and loss of marrow tissue. In osteonecrotic regions in the old SHRSPs, proliferation of the repairing tissue was observed. Empty lacunae were concentrated in the center of trabeculae, while the periphery of trabeculae was occupied by newborn osteocytes and osteoblasts. Infarctions were encountered on the lateral side of the epiphysis, but no thrombi were observed in any of the necrotic lesions. Femoral head necrosis was usually present in SHRSPs and SHRs from a young age (from about eight weeks). The incidence of femoral head necrosis in SHRSPs was 70% from 10 to 14 weeks of age and 45% from 24 to 36 weeks of age. The incidence of femoral head necrosis was highest in SHRSPs at all ages as shown in Table 2.

C. Steroid-Treated Rabbit Model

1. Introduction

Studies have been conducted in corticosteroid-treated rabbits in more than 10 different laboratories, and have suggested possible pathogeneses of corticosteroid-induced ON such as increase in the size of fat cells,[13] fatty degeneration of osteocytes,[14] subchondral fat embolism in the femoral head[15] and focal osteocytic death in subchondral bone.[16,17,18] However, these studies never successfully produced histologically definitive ON in animals by injection of corticosteroid alone. In a more recent study, Yamamoto et al. reported that a single high-dose injection of corticosteroid induced thrombocytopenia, hypofibrinogenemia, and hyperlipemia with multifocal ON in several bones.[19]

2. Materials and Methods

Twenty-six male adult (Japanese white) rabbits weighing 3.0–4.5 kg were injected once with 20 mg/kg body weight of methylprednisolone (MPSL) acetate into the right gluteus medius muscle.[19] Rabbits were sacrificed at four, six, eight, and 10 weeks after the injection of MPSL.

Bone samples were fixed with 10% phosphate buffered formalin, pH 7.4, for one week and decalcified in 25% formic acid for three days. The specimens were embedded in paraffin, sectioned and were routinely stained with H&E, elastica van Gieson and Masson trichrome.

Bone samples from the femur and humerus were histopathologically examined for the presence of hematopoietic cell necrosis (cytolysis, karyorrhexis, or karyolysis), fat cell necrosis (the loss of either nuclei or distinct cell borders) and ON. ON was blindly assessed by three pathologists, based on the diffuse presence of empty lacunae or pyknotic nuclei of osteocytes in the bone trabeculae, accompanied by surrounding bone marrow cell necrosis.

Hematologic and chemical examinations were performed to determine plasma levels of blood platelet, fibrinogen, free fatty acid, triglyceride, cholesterol, glutamine-oxaloacetic transaminase (GOT), and glutamic-pyruvic transaminase (GPT) in all animals before and after injection of MPSL.

3. Essential Findings

ON was observed in none of the rabbits in the control group, while multifocal osteonecrotic lesions were recognized in both femur and humerus at four weeks after steroid administration. ON was mainly found in the metaphysis and diaphysis, and was not in the epiphysis. In the femurs, the prevalence of ON was 43% at four weeks, 13% at 6, 25% at 8, and 25% at 10 weeks. The incidence of ON in the femoral condyle was 50% at four weeks. ON gradually accompanied the repair process after six weeks, and the necrotic bone marrow was sometimes almost entirely replaced by repair tissue by 10 weeks after injection.

Histopathologically, typical ON featured accumulation of bone marrow cell debris and bone trabeculae demonstrating empty lacunae occasionally containing some pyknotic nuclei of osteocytes. Repair tissue such as granulation tissue and appositional bone around ON varied based on the number of weeks after corticosteroid injection. At four weeks, little granulation tissue was present surrounding ON. At six weeks, fibrosis and vascular- or cellular-rich granulation tissue surrounded the necrotic area, while whether appositional bone formation was present was still unclear. At 10 weeks, necrotic bone tissue was surrounded by prominent appositional bone formation and dense fibrotic granulation tissue.

Platelet counts and fibrinogen levels were decreased significantly one week after the injection of MPSL, and then gradually recovered and reached or slightly surpassed normal levels after five weeks. On the other hand, plasma FFA, triglyceride, cholesterol, GOT and GPT levels were significantly increased at two weeks and then gradually returned to normal by eight weeks. At 10 weeks, no abnormal findings were obtained for any of the items examined.

D. Rabbits with Endotoxic (Shwartzman) Reactions

1. Introduction

There have been reports that bacterial endotoxic reactions may cause osteonecrosis in humans by inducing disseminated intravascular coagulation.[20–22] Yamamoto et al. have confirmed these clinical observations in rabbits in which corticosteroid was used to potentiate the Shwartzman reaction and increase the magnitude of ON.[23]

2. Materials and Methods

Male adult NZW rabbits weighing 3.0 to 4.5 kg were used in the following groups: A: 100 μg/kg of endotoxin (lipopolysaccharide [LPS] from *Escherichia coli*) intravenously twice at an interval of 24 h, 10 rabbits; B: 20 mg/kg of MPSL intramuscularly three times at intervals of 24

TABLE 3
Prevalence and Location of Femoral Bone Necrosis

Group	n	Epiphysis	Metaphysis	Diaphysis
A	10	0%	20%	0%
B	14	29%	86%	79%
C	12	0%	33%	0%
D	10	0%	0%	0%

(Adapted from Reference 23)

hours after LPS administration as in Group A, 10 rabbits; C: 20 mg/kg of MPSL intramuscularly three times at intervals of 24 h, six rabbits; and D: no treatment, five rabbits.

Bone samples were fixed with 10% formalin-0.1 M phosphate buffer, pH 7.4, decalcified in 25% formic acid and then neutralized with sodium sulfate buffer. The specimens were embedded in paraffin, sectioned and stained with hematoxylin and eosin, elastica van Gieson, and Masson trichrome.

All bone samples were examined histopathologically for the presence of hematopoietic cell necrosis (cytolysis, karyorrhexis, or karyolysis), fat cell necrosis (the loss of either nuclei or distinct cell borders), and ON. ON was assessed based on histopathologic features of bone necrosis (the presence of empty lacunae or pyknotic nuclei of osteocytes and bone marrow cell necrosis). Evidence of repair, such as the presence of granulation tissue, fibrosis, or appositional bone formation, was also examined.

3. Essential Findings

Histologically, necrotic regions exhibited an accumulation of cell debris, disappearance of hematopoietic and fat cells, and bone trabeculae either demonstrating empty lacunae or containing pyknotic nuclei. The formation of granulation tissue representing repair around ON varied among groups and depended on whether or not the rabbits received steroid. In Group A without steroid treatment, granulation tissue was well formed around regions of ON. In Group B with steroid treatment, the extent of formation of granulation tissue was less than in Group A. In Group C, with steroid treatment only, no formation of granulation tissue was observed, whereas exudative reaction (accumulation of serofibrinous exudate) was present around necrotic regions.

The incidence of ON was significantly higher in Group B than in the other groups (Table 3). In Group B, ON was frequently seen in the metaphysis (85.7%) and diaphysis (78.6%), and, occasionally, in the epiphysis (28.6%). No animal exhibited ON only in the epiphysis or diaphysis. In Groups A and C, ON was observed only in the metaphysis (20% and 33.3%, respectively). The distribution of the necrotic area in Group B was significantly wider than that in any other group.

E. RABBITS WITH HYPERSENSITIVITY REACTIONS (STUDY 1 BY THE AUTHORS)

1. Introduction

We successfully developed a rabbit model of ON.[25,26] This model is characterized by early microcirculatory injury (extravasation of erythrocytes and microthrombi in arterioles of the femoral metaphysis in the early stage) (Figures 4, 5) and immune complex deposition in the kidney. We determined serial changes in histological features of our rabbit model of inducible ON and examined whether immune complexes can be detected in femoral bone marrow, and whether immune complex deposition is present surrounding osteonecrotic regions and is related to early microcirculatory injury adjacent to ON.[27]

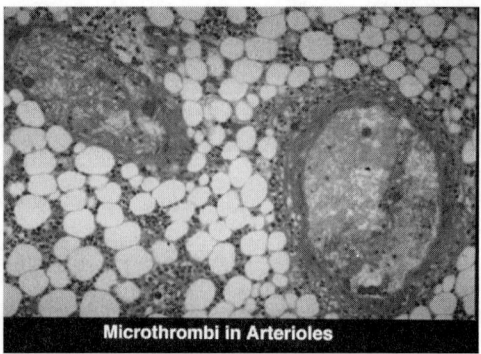

FIGURE 4. Microthrombi in arterioles in bone marrow of the femoral metaphysis in the same rabbit (×200).

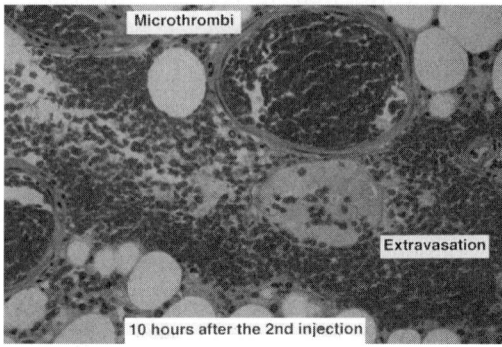

FIGURE 5. Major histological features including extravasation of erythrocytes and microthrombi in arterioles at 10 hours after the second injection of horse serum (×200).

2. Materials and Methods

One hundred and nine mature Japanese white rabbits, weighing 3.0–3.5 kg were used.[27] Whole horse serum was heat-inactivated at 56°C for 30 min. Ten ml/kg of heat-inactivated horse serum was given intravenously, and the same amount was injected again three weeks later.

Experimental Groups :

Group A Sacrificed prior to the second injection (at three weeks after the first injection), 17 rabbits.
Group B Sacrificed from one to four hours after the second injection of horse serum, 17 rabbits.
Group C Sacrificed from four to 12 hours after the second injection of horse serum, 17 rabbits.
Group D Sacrificed from 12 to 24 hours after the second injection of horse serum, 17 rabbits.
Group E Sacrificed from 24 to 72 hours after the second injection of horse serum, 21 rabbits.
Group F Sacrificed at one week after the second injection of horse serum, 20 rabbits.

The femurs and kidneys were excised bilaterally and fixed in buffered 4% paraformaldehyde saline (pH 7.4) at 4°C. The femurs were decalcified in EDTA (pH 7.4) at 37°C. These specimens were embedded in paraffin, sectioned, and stained with H&E stain. Phosphotungstic acid hematoxylin (PTAH) was used to demonstrate thrombi. In bone marrow, the presence of cytolysis, karyorrhexis, and karyolysis of marrow cells, and loss of nuclei and distinct cell borders of adipocytes was defined as bone marrow necrosis. We defined trabecular bone necrosis as being present when entirely empty lacunae of osteocytes were observed in the microscopic field at 100X magnification.

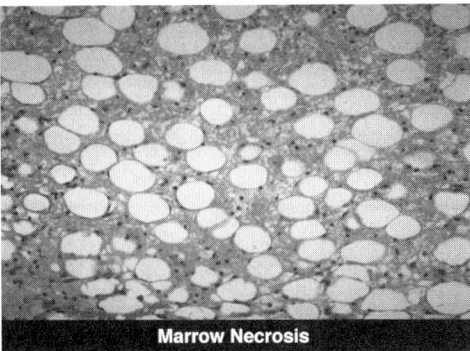

FIGURE 6. Bone marrow necrosis exhibiting necrotic debris and fibrosis at one week after the second injection of horse serum (×100).

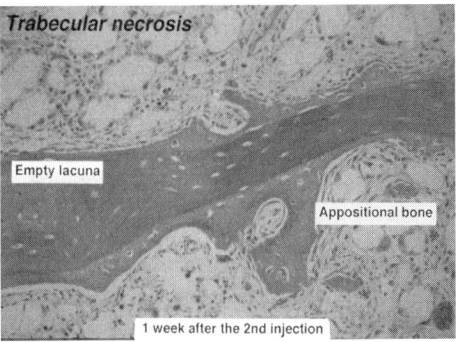

FIGURE 7. Trabecular bone necrosis surrounded by living appositional bone at one week after the second injection of horse serum (×100).

The femurs were embedded in paraffin, serially sectioned, and stained with the avidin-biotin-peroxidase complex method to detect immune complexes in bone marrow. An immunofluorescence method was performed to detect immune complexes in renal glomeruli.

3. Essential Findings

In Group A, neither extravasation of erythrocytes nor arteriolar microthrombi were observed in bone marrow of the femoral metaphysis. No necrosis of bone marrow and trabecular bone was observed immediately before the second injection of horse serum. Within 72 h after the second injection of horse serum (Group B, C, D, and E) major histological features included extravasation of erythrocytes (31%), arteriolar microthrombi (41%), and immune complex deposition in bone marrow of the femur. Extravasation of erythrocytes correlated well with the presence of arteriolar microthrombi ($p=0.0001$). In Group F, bone marrow necrosis and bone marrow replacement by fibrosis were observed in the femoral metaphysis in nine of 20 rabbits (45%) (Figure 6). Trabecular bone necrosis adjacent to bone marrow necrosis was observed in six of the nine rabbits (67%) with marrow necrosis, and necrotic trabeculae were surrounded by living appositional bone (Figure 7).

Immune complexes were demonstrated immunohistochemically in bone marrow (Figure 8) as well as in renal glomeruli (53%). Immune complex deposition both in the sinusoidal space of femoral bone marrow ($p=0.0385$) and in the renal glomeruli ($p=0.0209$) associated with extravasation of erythrocytes correlated well with the presence of arteriolar microthrombi in the early stage of this model. Early microcirculatory injury associated with immune complex deposition was present surrounding osteonecrotic regions.

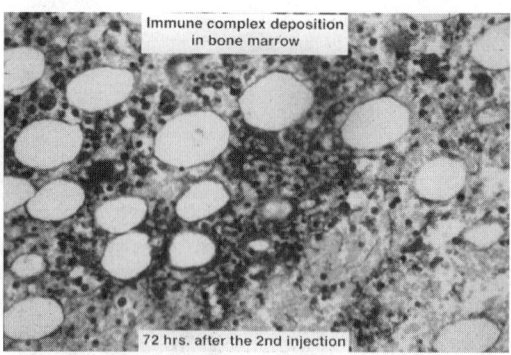

FIGURE 8. Immune complex deposition in the sinusoidal space of the femoral metaphysis (×100).

F. RABBITS WITH HYPERSENSITIVITY REACTIONS (STUDY 2 BY THE AUTHORS)

1. Introduction

We have developed a rabbit model of inducible ON.[25,26,27] In this model, we reproducibly observed many characteristic histological features quite similar to those of clinical ON. Intravascular coagulation of the intraosseous microcirculation, potentially activated by various factors, has been suggested to play an important role in the genesis of nontraumatic ON.[20] We examined whether any abnormalities related to coagulation occur early in ON.

2. Materials and Methods

Eighteen Japanese white rabbits weighing 3.0–3.5 kg were used. Whole horse serum was heat-inactivated at 56°C for 30 min. Ten ml/kg of sterile heat-inactivated horse serum was administered intravenously using the method of Rich and Gregory.[28] The horse serum (10 ml/kg) was intravenously administered to rabbits, and the same amount was given again three weeks later. The animals were sacrificed one week after the second injection of horse serum.

The femurs were excised bilaterally, fixed in buffered 4% paraformaldehyde saline (pH 7.4) at 4°C and decalcified in EDTA (pH 7.4) at 37°C. The femurs were cut along the sagittal plane for observation of trabeculae and bone marrow. The specimens were embedded in paraffin, sectioned and stained with H&E. The incidence of ON in histological sections of the femur was determined. We classified cases into two groups (ON group or Non-ON group) based on whether or not osteonecrosis or marrow necrosis in femoral bone was present.

Blood samples were serially obtained immediately before and at 0.5, 1, 2, 6, 24, 48 and 72 hours after the second injection of horse serum. Total platelet, leukocyte and erythrocyte counts, thromboxane-B2 (TXB2), blood viscosity, and blood coagulation and fibrinolysis [prothrombin time (PT) and activated partial thromboplastin time (APTT)] were examined.

Rabbit platelet-rich plasma (PRP) was prepared by centrifugation (1500 rpm at room temperature, 10 min) of blood collected from the carotid artery of rabbits, with 1/10 v/v 90 mM trisodium citrate used as an anticoagulant. Aggregation was measured with a dual channel aggregometer linked to a dual pen recorder at 37°C and a stirring rate of 1100 rpm. PRP was pre-incubated with stirring at 37°C for 2 min. before the addition of stimulator. Platelet activating factor (PAF-C16: 10^{-7} M) (Bachem, California, USA) was used as aggregation stimulator. The rate of change in light transmission caused by platelet aggregation was recorded, and the maximal rate was considered a measure of platelet aggregation.

3. Essential Findings

Histologically, ON was demonstrated in six of 18 (33%) rabbits. Total platelet counts decreased within 0.5 hour and remained low for 48 hours after the second injection. There were significant

TABLE 4
Total Platelet Count

Total Platelet Counts ($\times 10^5/\mu l$)	Time after the second injection (hours)						
	pre	0.5	1	2	6	24	48	168
ON group (n = 6)	4.4	1.8	1.9	1.8	1.7	1.8*	2.1	3.6
Non-ON group (n = 12)	3.7	2.4	2.5	2.7	3.2	3.5*	3.1	3.3

* Significant (unpaired t-test)

TABLE 5
Level of TXB2

TXB2 ($\times 10^3/\mu l$)	Time after the second injection (hours)						
	pre	0.5	1	2	6	24	168
ON group (n = 6)	0.8	6.7	11.4	19.1*	22.7*	2.3	1.0
Non-ON group (n = 12)	0.6	6.6	6.5	2.5*	1.2*	1.3	0.5

* Significant (unpaired t-test)

TABLE 6
Platelet-Aggregating Property

Platelet-aggregating property (%)	Time after the second injection (hours)						
	pre	0.5	1	2	6	24	168
ON group (n = 6)	41	35	36	55*	42	39	36
Non-ON group (n = 2)	45	37	37	36	38	41	33

* Significant (unpaired t-test)

differences in total platelet counts at 24 and 48 hours between ON group and Non-ON group (Table 4). TXB2 was maximal at six hours after the second injection of horse serum in the ON group (2.3×10^4 pg/ml), and was significantly higher than in the Non-ON group (Table 5). Platelet aggregation was transiently increased at six hours after the second injection of horse serum in the ON group (55%) (Table 6). There were no significant differences between the two groups in leukocyte count, erythrocyte count, PT, or APTT.

III. TRAUMATIC MODELS OF OSTEONECROSIS

There is a need for more reliable treatment modalities for both traumatic and atraumatic ON. The clinical picture and histology of traumatic ON of the femoral head (ONF) resemble those of nontraumatic ONF. For transcervical fracture of the femur, the incidence of ONF and the area of infarction become larger as the severity of injury (Garden stage) increases. Traumatic ONF occurs in 11–16% of patients with stage I or II fracture and in 20–28% of patients with stage III or IV fracture. MRI, as described earlier, is a more sensitive method for early diagnosis of ON. The reactive interface between necrotic and viable bone in traumatic ONF can be detected by MRI. We performed a prospective study in patients with transcervical fractures of the femur and found a high incidence of MRI abnormalities indicating ONF (8/17 patients) by one month after internal fixation of fracture.[29] Animal models of traumatically produced ON can also be used to identify useful diagnostic and therapeutic methods for ON if they are highly reproducible.

A. GOAT MODEL

In 1910, Legg, Calve, and Perthes individually described a disease (Legg-Calve-Perthes disease) manifesting as deformation of the developing hip in children. The exact etiology of this disease has not been determined, although extensive clinical, radiographic, and animal studies of it have been performed. None of the animal models developed has been able to reproduce the variety of lesions seen in children. The growth pattern resulting in the most severe femoral head deformity is one in which anterolateral physeal growth is arrested while perichondral ring growth and posteromedial epiphyseal growth continues or accelerates. An animal model that mimics this growth pattern and the resultant morphology of the affected femoral head is needed. Newton et al. reported a surgical method for creating morphologic changes similar to those observed in Legg-Calve-Perthes disease in an animal model.[30]

1. Materials and Methods

Two groups of animals were used in the experiment. Four mixed-breed goats between one and three months of age underwent surgery to develop the surgical procedure described below and were followed for approximately two months. The objective of the surgery was to ablate the physis anterolaterally while leaving as much of the perichondrial ring intact as possible. Another group of 10 pure-bred Alpine goats three months of age also underwent surgery using the same procedure and were followed radiographically for approximately 13 months. Under general anesthesia with halothane and oxygen, an anterolateral approach was used to expose the hip joint. To access the physis, a window approximately 2×5 mm in size was curetted from the articular cartilage over the physis. The electrocautery blade was introduced into the physis through this window to a depth of approximately 2 cm across the width of the growth plate. The blade was inserted into the physis three to four different times in a fan pattern and coagulated.

The first group of animals was sacrificed two months postoperatively, and the harvested femurs were fixed in neutral-buffered formalin. Measurement of the total length of the femur, the transverse and vertical diameters of the femoral head, and the distance across the femoral neck were made for the operated and control femurs. After proper fixation, the femoral head sections were decalcified, processed, and embedded in paraffin. The paraffin blocks were sectioned, and stained with hematoxylin and eosin. The extent and location of the growth plate by fibrous tissue, fibrocartilage, and bone bridges were evaluated quantitatively using planar morphometry. The anteroposterior radiographs were scanned using a X ray scanner. The radiographic images were enhanced using Adobe Photoshop (Adobe Systems), and a radiographic template was created using Aldus Freehand (Aldus, Seattle, WA, USA).

2. Essential Findings

In each case the test femur was shorter than the control femur ($p<0.05$), suggesting that longitudinal growth of the femur was partially arrested. Measurements of the remaining parameters for the test femurs were not significantly different from those for the control femurs. The gross observations indicated that the physis, and its potential for complete longitudinal growth, were affected by the surgical procedure. In test specimens, the physis exhibited irregular alignment of the growth plate chondrocytes compared with that in controls. Variable thickness of the growth plate, including the cartilage tongues from the physis in both specimens, may have resulted from localized avascularity of both the metaphysis and the epiphysis. Bone bridges, fibrous tissue, and fibrocartilage had replaced the physis. The histological findings following surgery were comparable to those reported for patients with Legg-Calve-Perthes disease. Quantitative histologic evaluation showed that the physis was affected in each femoral head, although the area of disturbance of the physis differed among the specimens. The overlays produced using graphic analysis showed the changes in growth of the proximal femur. The surgical procedure was effective in ablating the capital femoral physis, and induced changes in the growth of the femoral head. However, the resultant deformities did not mimic the changes identified in the graphic analysis study, perhaps because of inconsistencies in the surgical techniques used for ablation, which will require further modification.

B. Canine Model by Deep Freezing

1. Introduction

New treatment options for the management of ON have been proposed, such as electromagnetic stimulation and vascularized bone grafting procedures. However, the process of bone repair resulting from clinical use of these treatment modalities has not been clearly defined, since material for evaluation and histologic analysis usually is obtained after treatment has failed. An experimental model is thus needed to evaluate theses different therapeutic modalities. Malizos et al. investigated the efficacy of a surgically invasive technique, consisting of femoral head dislocation, soft-tissue stripping, and deep freezing, in the induction of ON in dogs, and also characterized the process of spontaneous bone repair within such lesions.[31]

2. Materials and Methods

Twenty-six mature beagles were used, four of which were evaluated in a pilot study. Dynamic bone remodeling in beagles is similar to that in humans. Under general anesthesia with halothane, the right hip was approached through a posterolateral incision. After dislocation of the femoral head, a surgical rubber tube was coiled around the femoral neck and liquid nitrogen was circulated through the tube for three min. Two animals died during the operation.

After sacrifice, the proximal third of both femurs was prepared for mineralized bone sections. Tissue fixation was performed in 70% alcohol for 48 hours. The samples were then embedded in methylmethacrylate and cut with a heavy duty microtome into complete sections in a frontal plane. Serial cuts from the central portion of the specimens were obtained for staining with Goldner trichrome and toluidine blue. Histomorphometric analysis was performed using a semiautomatic computerized system (OsteoMeasure, Osteometrics, Atlanta). To study the kinetics of bone healing, three different bone-labeling fluorochromes were administered intravenously for two consecutive days. Two dogs were killed after one week postoperatively, and two others, after two weeks. Four animals each were killed at 4, 8, 16, and 24 weeks postoperatively. At necropsy, both femurs were obtained and radiographed prior to the tissue preparation.

3. Essential Findings

A mixture of cellular and erythrocyte debris was observed in specimens examined two hours after the freezing procedure. By two weeks after the induction of ON, histologic changes in bone were more uniform. The osteocytic lacunae were empty or contained eccentrically located osteocytes. The marrow cavities contained amorphous debris, and mineralization was present on the surface of the trabeculae. One month postoperatively, soft-tissue debris was no longer present in the marrow cavities, and the repair process had already been established. Spontaneous healing originated mainly from the adjacent viable bone by migration of undifferentiated mesenchymal tissue into the necrotic bone, fibrosis, and, finally, formation of new bone. Planimeric measurement showed that at four weeks after the induction of necrosis the mesenchymal reparative tissue, occupied 13.1±4.1% of the total area of initial necrosis, and the zone of osteogenesis occupied 4.9±2% of this area. At eight weeks, the mesenchymal reparative tissue occupied 15.8±3.3% and neo-osteogenesis, 6.9±3.5% of the initially necrotic bone. At 16 weeks, the corresponding percentages were 55.3±7.7% and 20.1±8%, and at 24 weeks, 65.3±16.5% and 28.2±8.8%. Microscopic examination under fluorescent light revealed complete absence of fluorochrome fixation in the area of necrosis, confirming the lack of bone viability. In the 4-week specimens, labels were present in the form of patchy tallow fluorescent spots extending throughout the area of neo-osteogenesis. The rate of bone formation per unit area of surface was 64.1±16.1 $\mu m^3/\mu m^2/yr$, which exceeded the average formation (42.5 $\mu m^3/\mu m^2/yr$) in cancellous bone at distant sites in normal mature beagles. Radiographic changes became evident only four weeks after the induction of ON, and began as a band of radiodensity across the medulla at the margin of the necrotic area. More proximally, an extensive zone of marked osteopenia was present, including spots of complete trabecular loss. At 16 weeks, the areas with osteopenia and lysis in the trabecular bone had advanced further proximally near the femoral neck and at the greater trochanter. There was no loss of the contour of the articular surface at any of the time points of examination.

C. Canine Model by Dislocation and Vessel Ligation

1. Introduction

The etiology of femoral head necrosis following traumatic hip dislocation remains obscure, in large part because no suitable models have been established for the study of its pathophysiology. Microangiographic studies of the distribution and anastomosis of arteries supplying the femoral head may account for the difficulties in animal experiments to induce ON when the femoral head is not manipulated by unrealistic methods and the marrow cavity is preserved. Nishino et al. recently reported that they established a new model of ON by dislocating the hip joint and ligating the medial and lateral circumflex femoral arteries and veins.[32]

2. Materials and Methods

Mongrel adult dogs each weighing 8–12 kg were used. The dogs were divided into three groups, which were subjected to hip dislocation alone, ligation of blood vessels alone, or both hip dislocation and ligation of blood vessels. At two and four weeks after treatment, five animals in the hip dislocation only and ligation only groups, and 10 animals in the dislocation plus ligation group were killed, and pathologic studies were performed. In addition, MR images of the model were evaluated for the combined dislocation and ligation group. Femoral head blood flow was measured in 10 dogs to evaluate quantitatively to what extent dislocation and blood vessel ligation affected femoral head blood flow volume.

After the induction of anesthesia by ketamine, sodium pentobarbital (5mg/kg/h) was administered intraperitoneally with blood pressure monitoring to maintain general anesthesia. An approximately 1-cm-long incision was then made on the capsule through a lateral approach, the round ligament was served, and a cervical vertebra spreader used to create and maintain the dislocation. The hip was dislocated posteriorly, and the extent of the dislocation was maintained constant at a distance of one and a half femoral heads from the original position for nine hours. In the ligation group, the lateral and medial circumflex femoral arteries and veins leading to the femoral neck were ligated and cut.

The excised femoral heads were sliced parallel to the coronal plane and stained with hematoxylin and eosin for pathologic examination. In order to evaluate the effect of dislocation and blood vessel ligation on femoral head blood flow, femoral head blood flow volume was measured in 10 adult dogs at room temperature (20–22°C) using the electrochemically generated hydrogen clearance method. Measurements were made using the method of Stosseck et al.,[33] with the center of the femoral head selected as the site of measurement. MR images of excised femoral heads were examined within two hours after excision. MRI was performed using a Sigma 1.5-Tesla superconducting magnet. The coronal plane was used for all examinations.

3. Essential Findings

In the dislocation only group, no necrosis was observed in any of the 10 animals at either two or four weeks, although some cases exhibited congestive changes and a decrease in the number of marrow cells. In the ligation only group as well, no empty lacunae were found at either two or four weeks, and no case exhibited findings of ON. In the combined dislocation and ligation group, ON was observed at two weeks in eight of 10 (80%) and at four weeks in eight of 10 (80%) dogs. Also, in the majority of cases, appositional bone formation, which is thought to be a manifestation of the repair process of ON, was observed. The distribution of the necrotic area was similar to that for human femoral head necrosis, with necrosis more extensive on the joint side than on the neck side, and clearer manifestations of repair on the neck side. Although the extent of ON exhibited considerable individual variation, in most cases ON was found over a wide area centering on the weight-bearing region. The rate of femoral head blood flow in the control group prior to dislocation was 78.8±19.7 ml/min/100/ml. In the dislocation only group it was 32.1±15.4 ml/min/100ml, dropping to 40.7% of the predislocation value. In the dislocation plus ligation group, it was 11.6±10.3 ml/min/100/ml, representing a marked decrease to 14.7% of the predislocation value.

D. SWINE MODEL

1. Introduction

MRI is the diagnostic modality most useful for the early detection of femoral head necrosis. However, a negative MRI does not exclude the diagnosis of ON. In the setting of acute trauma, there have been few findings regarding the interval of time during which MRI reliably exhibit findings consistent with ON. Seiler et al. examined the correlation of the time sequence of MRI changes to changes with femoral head blood flow and the ultimate appearance of femoral heads in a miniature swine model of posttraumatic proximal femoral ON.[34]

2. Materials and Methods

Twelve 40–60 kg skeletally mature (age 14–20 months) Gottinger miniature swine were used. One animal died perioperatively from the anesthetic, leaving 11 swine available for study. Under general anesthesia with a mixture of halothane and oxygen, a posterolateral approach to the right hip was used for all animals. A 5.0 mm hollow aluminum screw with a Teflon coating and plastic sleeve assembly was inserted into the posterior wall of the acetabulum. The femoral head blood cell flux (BCF) was measured through the hollow screw and sleeve assembly with a 2.2 mm flexible laser Doppler flowmetry probe. A basilar femoral neck osteotomy (simulating a fracture) was then created with a 1.5 mm osteotome, and the BCF was again measured. The fracture was reduced under direct vision and internally fixed with two 1.6 mm commercially pure titanium Kirschner wires.

Repeated femoral head BCF values were obtained at one, two, four, and eight weeks postoperatively. At the time of the index operation, the end of the screw and sheath assembly were placed in subcutaneous tissue to permit subsequent access for percutaneous Doppler assessment. Doppler measurements were then obtained and recorded as previously described. All magnetic resonance images were obtained on a 1.5 Tesla imager. MRI intensity ratios were computed as the ratio of the intensity of the experimental hip divided by the intensity of the hip on the control side. Both femurs in each animal were explanted after sacrifice and immediately stored at −70°C. The femoral heads were sectioned and examined by light and fluoroscopic microscopy. The specimen was graded as having no (0), intermediate (1) or significant (2) involvement with each of the following seven histologic parameters associated with ON: (1) empty osteocyte lacunae; (2) marrow invasion with mesenchymal cells; (3) vascular ingrowth/ hypertrophy; (4) creeping substitution; (5) marrow fibrosis; (6) decrease in subchondral bone thickness; and (7) chondral collapse. Maximum total ON score was 14 points, with higher scores representing more extensive femoral head ON. Plain radiographs of all explanted femurs were obtained. Bone density data were calculated by a video image analyzer linked to image processing software via a Macintosh computer. Using this system, the trabecular bone surface area could be expressed as a percentage of a given field of view in the superolateral aspect of the femoral head.

3. Essential Findings

Significant inter-animal variation in femoral head blood cell flux, as assessed by Doppler, was observed. The mean baseline value was 1490±465 mV. After a femoral neck fracture, this value decreased to 869±211 mV. Significant decreases in femoral head blood cell flux were observed at week 1, 4, and 8. At week 8, BCF averaged 262±83 mV. Femoral head blood flow decreased immediately after fracture and continued to diminish with time. Seven animals had some degree of failure of internal fixation with loss of anatomic reduction by one week postoperatively, which was confirmed visually at the time of final exploration. In four animals, fixation remained intact throughout the period of observation and this group was analyzed separately. MRI signal intensities in the femoral head (MRI intensity ratio) at four and eight weeks were significantly less when fixation failed than when it was intact. ON histology grades averaged 7.6±0.8 on the experimental side and 1.6±0.4 on the control side ($p<0.01$). The range of histology grades was 0.0–4.5 for the control hip and 4.3–11.2 for the experimental hip. There was no significant difference between the intact fixation group (6.5±1.0) and failed fixation group (8.4±1.0) in histology grade. The mean bone density on the experimental side was 49%, while that on the control side was 56% ($p<0.01$). Histologic grade, bone density value and blood flow values were each unrelated to changes in MRI signal intensity.

IV. APPLICATIONS OF ANIMAL MODELS

Although many other experimental animal models of ON have been reported, only several recent, reproducible models are introduced in this chapter. The models described here can be used for various purposes. Nontraumatic models are more suitable for evaluation of the etiology of ON than traumatic models, the procedures in which are too radical. The pathology of ON could be clarified with different approaches to investigate the specimens much earlier than ON occurred.

Similarly, nontraumatic models may be more appropriate for use in identifying drugs or materials useful for preventing the genesis and/or progression of ON. Administration of those agents before the final procedure to induce nontraumatic ON is performed may reveal new therapeutic modalities for ON.

Nontraumatic and traumatic models are equally useful for development of new surgical approaches for ON. Clinical ON of the femoral head (ONF) is thought to follow a defined course once it has developed in the subchondral area of the femoral head. For both nontraumatic and traumatic ONF, the prognosis largely depends on the location and size of lesions. Regardless of the cause, continued weight-bearing on devitalized bone leads to collapse, joint deformity, degenerative change, and painful dysfunction in patients with relatively large areas of necrosis. In this regard, traumatic models may be more useful if the size and location of ON can be controlled well with specific surgical procedure.

REFERENCES

1. Jones, J. P., Jr., and Sakovich, L., "Fat embolism of bone. A roentgenographic and histological investigation, with use of intra-arterial Lipiodol in rabbits," *J. Bone Joint Surg.,* 48A, 149, 1966.
2. Sanchis, M., Zahir. A., and Freeman, M. A. R., "The experimental simulation of Perthes' disease by consecutive interruptions of blood supply to the capital femoral epiphysis in the puppy," *J. Bone Joint Surg.,* 55A, 335, 1973.
3. Inoue, A., Freeman, M. A. R., and Vernon-Roberts, B., "The pathogenesis of Perthes' disease," *J. Bone Joint Surg.,* 58B, 453, 1976.
4. Kemp, H. B. S., "Perthes' disease in rabbits and puppies," *Clin. Orthop.,* 209, 139, 1986.
5. Tachdjian, M. O. and Grana, L., "Response of the hip joint to increased intraarticular hydrostatic pressure," *Clin. Orthop.,* 61, 199, 1968.
6. Izawa, Y., Sagara, K., Kadota, T., and Makita, T., "Bone disorders in spontaneously hypertensive rat," *Calcif. Tissue. Int.,* 37, 605, 1985.
7. Lucas, P. A., Brown, R. C., Drueke, T., Lacour, B., Metz, J. A., and McCarron, D. A., "Abnormal vitamin D metabolism, intestinal calcium transport, and calcium status in the spontaneously hypertensive rat compared with its genetic control," *J. Clin. Invest.,* 78, 221, 1986.
8. Hirano, T., Iwasaki, K., and Yamane, Y., "Osteonecrosis of the femoral head of growing, spontaneously hypertensive rats," *Acta. Orthop. Scand.,* 59, 530, 1988.
9. Hirano, T., Iwasaki, K., Sagara, K., Nishimura, Y., and Kumashiro, T., "Necrosis of the femoral head in growing rats. Occlusion of lateral epiphyseal vessels," *Acta. Orthop. Scand.,* 60, 407, 1989.
10. Iwasaki, K., Sagara, K., Nishimura, Y., and Hirano, T., "Idiopathic necrosis of the femoral epiphyseal nucleus in rats," *Clin. Orthop.,* 277, 31, 1992.
11. Kujat, R., "The microangiographic pattern of the glenoid labrum of the dog." *Arch. Orthop. Trauma Surg.,* 105, 310, 1986.
12. Naito, S., Ito, M., Sekine, I., Ito, M., Hirano, T., et al., "Femoral head necrosis in stroke-prone spontaneously hypertensive rats (SHRSPs)," *Bone,* 14, 745, 1993.
13. Wang, G. J., Sweet, D. E., Reger, S. I., and Thompson, R. C., "Fat-cell changes as a mechanism of avascular necrosis of the femoral head in cortisone-treated rabbits," *J. Bone Joint Surg.,* 59A, 729, 1977
14. Kawai, K., Tamaki, A., and Horihata, K., "Steroid-induced accumulation of lipid in the osteocytes of the rabbit femoral head: a histochemical and electron microscope study," *J. Bone Joint Surg.,* 67A, 755, 1985.
15. Jaffe, W. F., Epstein, M., Heyman, N., and Mankin, H. J., "The effect of cortisone on femoral and humeral heads in rabbits," *Clin. Orthop.,* 82, 221, 1972.
16. Fisher, D. E., Bickel, W. H., Holley, K. E., and Ellefson, R. D., "Corticosteroid-induced aseptic necrosis. Experimental study," *Clin. Orthop.,* 84, 200, 1972.
17. Cruess, R. L., Ross, D., and Crawshaw, E., "The etiology of steroid-induced avascular necrosis of bone," *Clin. Orthop.,* 113, 178, 1975.
18. Warner, J. J. P., Philip, J. H., Brodsky, G. L., and Thornhill, T. S., "Studies of nontraumatic osteonecrosis. Manometric and histologic studies of the femoral head after chronic steroid treatment. An experimental study in rabbits," *Clin. Orthop.,* 225, 128, 1987.
19. Yamamoto, T., Irisa, T., Sugioka, Y., and Sueishi, K., "Effects of pulse methylprednisolone on bone and marrow tissue: corticosteroid-induced osteonecrosis in rabbits," *Arthritis Rheum.,* 40, 2055, 1997.
20. Jones, J. P., Jr., "Intravascular coagulation and osteonecrosis," *Clin. Orthop.,* 277, 41, 1992.
21. Robinow, M., Jonson, F., Nanagas, M. T., and Mesghali, H., "Skeletal lesions following meningococcemia and disseminated intravascular coagulation." *Am. J. Dis. Child.,* 137, 279, 1983.
22. Barre, P. S., Thompson, G. H., and Morrison, S. C., "Late skeletal deformities following meningococcal sepsis and disseminated intravascular coagulation," *J. Pediatr. Orthop.,* 5, 584, 1985.
23. Yamamoto, T., Hirano, K., Tsutsui, H., Sugioka, Y., and Sueishi, K., "Corticosteroid enhances the experimental induction of osteonecrosis in rabbits with Shwartzman reaction," *Clin. Orthop.,* 316, 235, 1995.
24. Siemsen, J. K., Brook, J., and Meister, L., "Lupus erythematosus and avascular bone necrosis. A clinical study of three cases and review of the literature," *Arthritis Rheum.,* 5, 492, 1962.
25. Matsui, M., Saito, S., Ohzono, K., Sugano, N., Saito, M., et al., "Experimental steroid-induced osteonecrosis in adult rabbits with hypersensitivity vasculitis," *Clin. Orthop.,* 277, 61, 1992.

26. Matsui, M., Ohzono, K., Nakamura, N., Sugano, N., Masuhara, K., et al., "The immune reaction to heterologous serum causes osteonecrosis in rabbits," *Virchows Arch.,* 427, 205, 1995.
27. Nakata, K., Masuhara, K., Nakamura, N., Shibuya, T., Sugano, N., et al., "Inducible osteonecrosis in a rabbit serum sickness model: deposition of immune complexes in bone marrow," *Bone,* 18, 609, 1996.
28. Rich, A. R. and Gregory, J. E., "The experimental demonstration that periarteritis nodosa is a manifestation of hypersensitivity," *Bull. Johns Hopkins Hosp.,* 72, 65, 1943.
29. Sugano N., Masuhara K., Nakamura N., Ochi T., Hirooka A., and Hayami Y., "MRI of early osteonecrosis of the femoral head after transcervical fracture," *J. Bone Joint Surg.,* 78B, 253, 1996.
30. Newton, A. S., Crawford, C. J., Powers, D. L., and Allen, B. L., Jr., "The immune goat as an animal model for Legg-Calve-Perthes disease," *J. Invest. Surg.,* 7, 417, 1994.
31. Malizos, K. N., Quarles, L. D., Seaber, A. V., Rizk, W. S., and Urbaniak, J. R., "An experimental canine model of osteonecrosis: characterization of the repair process," *J. Orthop. Res.,* 11, 350, 1993.
32. Nishino, M., Matsumoto, T., Nakamura, K., and Tomita, K., "Pathological and hemodynamic study in a new model of femoral head necrosis following traumatic dislocation," *Arch. Orthop. Trauma Surg.,* 116, 259, 1997.
33. Stosseck, K., Lubbers, D. W., and Cottin, N., "Determination of local blood flow (microflow) by electrochemically generated hydrogen. Construction and application of the measuring probe," *Pflugers Arch.,* 348, 225, 1974.
34. Seiler, J. G., III, Kregor, P. J., Conrad, E. U., III, and Swiontkowski, M. F., "Posttraumatic osteonecrosis in a swine model: Correlation of blood cell flux, MRI and histology," *Acta. Orthop. Scand.,* 67, 249, 1996.

15 Animal Models of Osteopenia or Osteoporosis

Donald B. Kimmel, Erica L. Moran, and Earl R. Bogoch

CONTENTS

I. Introduction 280
II. General Consideration 280
   A. The Criteria 280
   B. Bone Loss and Turnover Rate Rise after Estrogen Depletion 281
   C. Osteoporotic Fractures and Steady State Osteopenia 281
   D. Remodeling 281
   E. Timeframe Compression 281
III. Animal Models for Osteoporosis 282
   A. Mouse 282
      1. In General 282
      2. Practical Problems 282
      3. The SAM Mouse 282
   B. Rat 284
      1. In General 284
      2. Specific Recommendations for Rat Experiments 285
      3. Summary 285
   C. Avian, Guinea Pig, Rabbit, Ferret, and Cat 285
   D. Dog 285
   E. Pigs and Sheep 286
   F. Nonhuman Primate 286
   G. Summary 287
IV. Disuse Osteopenia 287
   A. In General 287
   B. Animals 288
   C. Common Procedures 289
      1. Neurectomy 289
      2. External Fixation 289
      3. Tenotomy 289
      4. Hindlimb Suspension 290
      5. Spaceflight 290
V. Glucocorticoid Osteopenia 290
   A. Small Animals 290
   B. Large Animals 291
   C. Summary 291

VI. Osteopenia Associated with Inflammation 291
    A. In General..291
    B. Animal Models..291
    C. Fracture Risk 292
References 293

I. INTRODUCTION

Much of the progress during the last one-sixth of the twentieth century in prevention and treatment of osteoporosis is attributable to today's excellent animal models. The purpose of this chapter is to describe the current status of animal models that address osteoporosis and osteopenia and then recommend strategies for applying them.

US Food and Drug Administration guidelines for using animals in preclinical tests of agents intended to treat osteoporosis[1] recommend animals either losing bone or have ovariectomy (OVX)-related osteopenia. Histologic, densitometric, biochemical, and biomechanical data from both the OVX rat and a larger species with Haversian remodeling are required.

Full parallelism of human symptoms with any single *in vivo* animal model does not exist. Thus, a strategy emphasizing small animals with limited use of a large animal is recommended. These criteria place the highest value on animal models that match the clinically apparent behaviors of osteoporosis and osteopenia. Models that give inconsistent results in the hands of numerous investigators, involve convoluted conditions, or require difficult species, win low marks. We find using relevant methods in an adult animal to produce consistent partial symptoms better than using less physiologic circumstances to develop a full set of symptoms.

II. GENERAL CONSIDERATION

A. THE CRITERIA

Animal models for the adult skeleton have been reviewed elsewhere.[2-8] This chapter not only updates previous points, but also raises new issues. Its approach is to: (1) highlight critical data that can be assessed *in vivo* in humans and (2) show how today's animal models match those data.

An animal model of osteoporosis should have lengthy growing and adult skeletal phases. Peak bone mass in women occurs between ages 25–50 yrs, and is important because of its likely role in the late-life development of osteoporosis.[9-10] A bone mass measurement at age 50 is the best predictor of future fracture in healthy persons.[11-12] Peak bone mass is precisely measurable and 80% heritable,[13-14] making it a good phenotype for quantitative trait linkage studies.[15]

Humans not only have a menarche and regular, frequent ovulatory cycles, but also experience bone loss at cessation of ovarian function. This is best shown by osteopenia in amenorrheic individuals,[16-18] the bone accumulation that occurs upon resumption of normal menses,[18-19] and the identification of oligomenorrhea/amenorrhea and late menarche as osteopenia risk factors.[20] Only mammals (and humans) with regular, frequent ovulatory cycles and high peaks of 17-estradiol (E2) may suffer estrogen-depletion bone loss. Regularly cycling female mammals accumulate an estrogen-related component of bone that integrates into the skeleton[21] and disappears at menopause. Animals with infrequent cycles and low E2 peaks might develop too small an estrogen-related bone compartment to allow estrogen depletion bone loss. Over two-thirds of women experience natural menopause. Of the rest, many have preserved ovarian function or prompt estrogen replacement therapy (ERT). Most animal skeletal models of estrogen-depletion employ surgical OVX.[22,23] No meaningful differences in bone behavior between surgical and natural menopause exist.[24]

B. BONE LOSS AND TURNOVER RATE RISE AFTER ESTROGEN DEPLETION

Following estrogen-depletion, bone loss accelerates in cancellous regions[25-26] and at endocortical surfaces[27-28] in multiple sites,[29] then decelerates into a semi-plateau phase.[30] Estrogen-depletion changes in Haversian remodeling are poorly documented. Cancellous and endocortical bone loss is accompanied by increased turnover[31-32] and a marked, transient negative calcium balance.[33] Histomorphometric signs of increased transmenopausal turnover are readily shown within individual humans.[34] Oophorectomized or menopausal women given estrogen replacement (ERT) show a smaller bone turnover rise,[31,35] less bone loss,[36-37] and fewer fractures than those given no ERT.[36,38] This ERT response is demonstrable by histomorphometric techniques in humans and animals.[35,39,40] All these behaviors should not only exist in an accurate animal model of the menopause, but be measurable by similar techniques.

C. OSTEOPOROTIC FRACTURES AND STEADY STATE OSTEOPENIA

Because of today's excellent ability to quantitate bone mass,[41] the World Health Organization endorses a numeric/symptomatic definition of osteoporosis, as all women with bone mass at one or more bone sites 2.5 standard deviations below the young adult normal, or with a history of low trauma fractures of the spine, hip, or wrist.[42] An animal model that develops fragility fractures after estrogen depletion would facilitate pre-clinical evaluation of agents' anti-fracture efficacy. However, today's excellent animal models of estrogen-depletion bone loss exhibit no low trauma fractures of critical bone sites.[43-44] The animals, unlike humans, may lack the contribution of low peak bone mass necessary to put them below their fracture threshold. One animal model with low peak bone mass, the SAM/P6 mouse,[45] has fragility fractures.[46]

D. REMODELING

Remodeling, the *in situ* removal and replacement of old bone tissue by new bone tissue, is the dominant process in the adult human skeleton.[47] Cancellous bone surfaces, including endocortical surfaces, experience the highest remodeling rates in the skeleton.[28] Mild, long-standing deficits in remodeling may cause cancellous osteopenia, a risk factor for fragility fractures.

Cortical bone plays a dominant role in skeletal strength. Adult humans have Haversian (intracortical) bone remodeling. Though the effect of estrogen depletion on Haversian remodeling is not well known,[48] post-menopausal osteoporosis is characterized by only minimal cortical porosity. Nonetheless, an accurate animal model should display Haversian remodeling because of its importance in the maintenance of cortical bone strength. Furthermore, first generation anti-osteoporosis agents tend to suppress remodeling. The inability to obtain cortical bone biopsy specimens from humans in Phase III/IV trials means that Haversian remodeling cannot be studied in humans. Therefore, adverse remodeling changes caused by anti-osteoporosis agents are best revealed in studies of animals with Haversian remodeling. An accurate animal model would thus display such activity in its skeleton.

E. TIMEFRAME COMPRESSION

The time from peak bone mass attainment until the development of fragility fractures is 30+ yrs. In women, accelerated estrogen-depletion bone loss lasts 5–8 yrs. An effective animal model known to experience peak bone mass and post-OVX bone loss should decrease both times by an order of magnitude. Convenience for animal models is denominated as cost of purchase, availability, housing, handling difficulties; and designing/implementing/validating new analysis procedures. Having validated small animal models is the best route to including the largest number of investigators in any research field. Using a highly accurate animal model occasionally can be so inconvenient that it may be more difficult than a human study.

III. ANIMAL MODELS FOR OSTEOPOROSIS

Animals that should receive major consideration in osteoporosis research are mouse, rat, dog, pig, sheep, and nonhuman primate. Because the above criteria are directed at clinical observations in humans, the animal models will be evaluated as to how they duplicate those outcomes (Table 1).

A. MOUSE

1. In General

The mouse is now rising as an *in vivo* model for osteoporosis research. It is the ideal model of osteopetrosis,[49–50] osteoclast and stromal cell ontogeny,[51] and cytokine and marrow studies.[52–54] It will become even more popular for the ease with which its genome can be manipulated.[55–58] Mice have been used to identify and characterize osteopetrosis genes,[50] both with linkage and transgenic animal studies. The disclosure of genes associated with osteopetrosis, a disease of osteoclast dysfunction, may lead to the discovery of agents that inactivate osteoclasts. Strains of mice with low and high peak bone mass[59] lend themselves to genetic investigations.[15] Considering the availability of SAM/P6, C57BL/6J, and C3H/HeJ,[59] proper breeding techniques with probing the whole genome for high density polymorphic markers should identify one or more genetic loci linked to bone mass in mice.

Estrogen has similar effects on the mouse and human skeletons. Cancellous,[58,60] but not cortical[61] bone loss occurs soon after OVX in the distal femur of strains like Swiss-Webster. Cancellous and cortical bone loss in the vertebrae and femur occurs after age one yr.[62–63] Hypogonadal female mice are osteopenic.[64] ERT at doses of ~10 mg/kg/d 2–3X/wk prevents estrogen-depletion bone loss.[60] Increased bone formation and woven bone deposition after E2 administration occurs at doses above 50 mg/kg/d,[60–61,65] a response never seen in humans. The dose-related E2 effects in the mouse do not stop the consideration of certain strains of mice as models of the estrogen-deplete human skeleton.

The ability to construct designer animal models (i.e., transgenic mice) is now driving mouse skeletal experiments. Validation of the mouse as an animal model for osteoporosis calls for targeted experimental work to document fully its skeletal response to E2. The time course and site specificity of estrogen-depletion osteopenia must be established in a strain specific fashion as it has been for the rat.[22,66] The bone response to OVX in strains that are the basis of transgenic models must be established.

2. Practical Problems

Specimens adequate for histomorphometric study of cancellous bone mass, structure, and turnover can be obtained from the distal femur. Mouse bones contain such small amounts of mineral that conventional dual energy X ray absorptiometry (DXA) cannot be used. Modified DXA (slower speed, smaller collimator) is the best alternative.[67] Equipment like pQCT (peripheral quantitative computed tomography) and peripheral DXA (pDXA) is useful,[59] but not yet widely applied.

3. The SAM Mouse

The SAM/P6 (senescence accelerated mouse) mouse has low peak bone mass and late-life fractures.[45–46,68] It is the only experimental animal with both low peak bone mass and fragility fractures of aging. The SAM mouse needs full genetic,[69] hormonal,[70] and biomechanical characterization, including site specificity for fractures. If it lacks collagen defects seen in osteogenesis imperfecta[55,71] it may, when combined with osteopenia prevention approaches like ERT or bisphosphonates, allow study of the role of low peak bone mass in late-life fractures. It will definitely provide chances to identify genes contributing to peak bone mass.

TABLE 1
Summary of *In Vivo* Animal Models for Osteoporosis

Attribute	Human	Avian	Mouse	Rat	Dog	Pig	Sheep	Primate
Growth and Adult Phases?	Yes	OK	OK	Yes	Yes	Yes	Yes	Yes
Menstrual/Estrus Cyclicity/Natural Menopause	28d/Yes	daily/No	inducible/Yes	4–5d/Yes	205d/No	21d/?	21d seasonal	21-28d/Yes
Bone Loss after Estrogen Depletion	Yes	?	Yes	Yes	Not Consistent	Weak	Not Consistent	Yes
Response to Estrogen	Turnover	Formation	Turnover (dose-related)	Turnover	Not Consistent	?	?	Turnover
Development of Osteoporotic Fractures	Yes	No	No	No+	No+	?	?	No+
Cancellous Remodeling	Yes	No	No	Yes	Yes	Yes	Yes	Yes
Haversian Remodeling	Yes (study site difficult)	No	No	Little; inducible	Yes	Yes	Yes	Yes
Timeframe Compression	No	?	Yes	Yes	No	Some	Some	Some
Convenience	OK	Yes	Yes	Yes	Weak	Yes	Yes	Yes
Cost Effectiveness	Yes	No	Yes	Yes	Weak	?	?	Yes

B. RAT

1. In General

The rat has long provided data about skeletal behavior. It gave the first evidence that osteoclasts ingest bone mineral[72] and early evidence about the hematogenous origin of osteoclasts.[73–74] The rat skeleton was once held unsuitable as a human skeletal model because many growth cartilages in male rats remain open past age 30 months.[75] Recent studies prove that growth cartilages at important sampling sites close by age 6–8 months in female rats, much earlier than males.[75–78] Those once holding reservations about the rat as an adult skeletal model because of its "continuous growth and lack of remodeling," have relented. They now only caution investigators to use female rats of age 6–9 months and avoid studying Haversian remodeling.[79] Periosteal expansion continues until about age 10 months, the age of peak bone mass in the female rat.[80] The mean healthy lifespan for rats is 18–21 months, but can be extended by OVX that prevents estrogen-dependent mammary tumors and restrictive feeding,[81] improving the possibility of doing FDA-mandated skeletal studies in rats.

Adult female rats have a regular estrus cycle with E2 levels spiking every four days.[82] After age one yr, the fraction of rats in constant diestrus rises gradually,[83] and cancellous bone loss frequently occurs. Spikes in E2 end as bone loss occurs, suggesting a similarity to human menopause. Following OVX, site-specific loss of cancellous bone mass and strength occurs, accompanied by increased bone turnover,[66,84–85] and decelerates into a plateau phase.[22,43–44,66] These features mimic well the bone changes following OVX or menopause in humans. Not all cancellous bone sites in the rat show such bone loss,[86] tightening the parallel of the rat and human skeletons. DXA, the current state-of-the-art for measuring bone mass in humans, is readily applied in rats.[87]

The OVX rat is a textbook example of a bench scientist introducing a highly relevant, widely-applied pre-clinical animal model.[22,40] OVX rats given prompt ERT show neither a turnover rise,[39–40,88] nor bone loss.[40,88–89] Bisphosphonates,[90–91] calcitonin,[92–93] and selective estrogen receptor modulators[94–95] also block the rise in turnover and bone loss in OVX rats, just like in humans.[96–97] Though one group suggests high dose E2 stimulates bone formation,[98–99] these data have been discounted by others using similar experimental designs who correctly consider the use of growing rats and a long fluorochrome labeling interval.[100]

The rat, like most osteoporosis animal models, lacks fragility fractures related to osteopenia. This problem has been overcome by mechanical testing of the vertebral body,[101–102] femoral shaft,[103] and proximal femur.[102,104] Using vertebral body strength as a surrogate for human vertebral body fracture likelihood may be a more direct test of pre-clinical anti-fracture efficacy of anti-osteoporotic agents than bone mass measurement. Such preclinical testing could avoid the problems observed with sodium fluoride as an osteoporosis treatment.[105–106] The recent development of *ex vivo* tests of bone fragility in small and large animals that now serve as surrogates for osteoporotic fracture is a similar example.[101–102,104,107]

Adult rats have adequate cancellous bone remodeling to permit useful experiments.[79,108–109] The relative amount of modeling and remodeling activity appears related to age. In the proximal tibial metaphysis, reversal lines at the base of cancellous osteons are mostly absent in four month old rats,[98] but present in cancellous osteons of 9–12 month old rats,[110] suggesting that remodeling activity increases with age in the rat. In most of its cortical bone, the rat has near zero levels of Haversian remodeling. However, processes resembling intracortical remodeling are induced by anabolic agents[111] or stressful metabolic conditions.[112–113] The rat has such low levels of Haversian remodeling that it cannot be used for studying Haversian remodeling, especially when evaluating agents that suppress remodeling. The female rat reaches peak bone mass by age 10 months, a thirty-fold timeframe compression when compared to the adult human. In three month old OVX rats, the phase of accelerated estrogen-depletion bone loss lasts three–four months in the proximal tibial metaphysis,[22] a 20-fold time saving over estrogen deplete women. The rat is among the most convenient of experiment animals to handle and house.

2. Specific Recommendations for Rat Experiments

Food intake in OVX rats should be restricted to block OVX-induced hyperphagia. It may involve matching intake to sham rats (~10% restriction), restricting intake to match weights of OVX rats to sham rats (~20% restriction), or caloric restriction (~30% restriction).[81] All types of food restriction accentuate bone loss in OVX rats,[114–115] creating more reliable cortical bone loss and quickening the onset of estrogen-depletion cancellous osteopenia.

Treatments that may prevent OVX-induced bone loss should start immediately after OVX. New techniques of bone structural study show that bone changes begin before 10 days post-OVX.[116,117] Delaying preventive treatment decreases the chance that a test agent can have its optimal effect, analogous to using an antibiotic in curative vs. preventive mode. Once the estrogen depletion bone loss process has begun, it may be difficult to stop.

3. Summary

The OVX rat duplicates the most important clinical features of the estrogen-deplete adult human skeleton. Its site-specific cancellous osteopenia is among the most prompt, certain physiologic responses in skeletal research. Ample time exists for experimental designs that either prevent estrogen-depletion bone loss or restore bone lost after estrogen depletion, particularly when using restrictive feeding. Its response to ERT parallels the human. The rat's low levels of Haversian remodeling present no problem for testing agents that may prevent loss of, or rebuild lost cancellous bone. Its lack of fragility fractures is overcome by biomechanical testing. Rats are convenient; unbred females aged 6–10 months are best for treatment phases of experiments. Densitometry, biochemistry, histomorphometry, and mechanical testing are used readily.

C. AVIAN, GUINEA PIG, RABBIT, FERRET, AND CAT

Adult female birds have a daily egg-laying cycle that corresponds to alternating deposition and resorption of medullary bone.[118] Bone accumulation occurs with rising serum E2; removal accompanies falling E2. E2 treatment of male birds causes medullary bone deposition.[119] While hypoestrogenemia in birds is associated with medullary bone loss, just as estrogen depletion in mammals is associated with osteopenia, the course of avian bone mass following OVX is unknown. Current data suggest that birds have little remodeling. The E2-related bone buildup may one day aid in understanding peak bone mass accumulation in pubertal females,[21,120–121] but will hinder experiments about osteoporosis, an adult human disease, because it has no counterpart in adult mammalian physiology. Birds are irrelevant for osteoporosis research because their skeletal behavior does not match known features of adult human osteoporosis.

Though guinea pigs, rabbits, ferrets, and cats have been used in osteoporosis research,[122–124] too few published experiments exist to assess model validity. OVX guinea pigs do not lose bone.[123] Adult (8 month old) rabbits might serve as a model for Haversian remodeling. The ferret has Haversian remodeling,[125] but its normal skeletal physiology, including the accumulation of estrogen dependent bone that accompanies normal cyclicity in other mammals, is light cycle dependent.[126]

D. DOG

The adult dog is generally a reliable model of the adult human skeleton. Haversian and cancellous osteons remodel as in humans, though more rapidly.[127] Skeletal findings in the adult dog parallel the adult human for corticosteroids,[128,129] uremia,[130] bisphosphonates,[131] and PTH excess.[132] In contrast to all other uses for the adult dog as a model of the adult human skeleton, the OVX dog is problematic. Many individual studies lack significance, but the data in bulk[133] suggest that 8–10% annual bone loss occurs in newly-OVX dogs. Most studies indicate a minor transient post-OVX rise in bone formation, but some[134] suggest that formation falls rapidly by 50%, without a transient increase. The latter suggests a dissimilarity to findings in transmenopausal humans and other animals, where the transient turnover rise[34] is well known.[31] ERT tends to suppress turnover, but with uncertain effects on bone mass.[135] E2 levels in the dog are usually very low, rising twice yearly.[136] E2 spikes in rats for 18 hours every four d[82] and monthly in women.[137] The estrus cycle in monkeys is similar to humans, but E2 peaks only about half as high.[138] Dogs' integrated E2 exposure, only marginally less than in rats, is only one-fourth that in humans and similar to primates, except for the peaks. These differences could cause the dog to develop only a small estrogen-dependent cancellous bone compartment.

The adult dog, with its Haversian remodeling, is an excellent model of the adult human skeleton except for its response to estrogen depletion. Though the OVX dog probably has estrogen-depletion osteopenia, the poor inter-laboratory reproducibility for this finding has caused most investigators to abandon it. The FDA guidelines advise not using the OVX dog.[1] The main problem is that most studies have insufficient power (N=6-9/group) to detect 8–10% bone loss.[133] Despite its inconsistent estrogen depletion bone loss, the dog is fine for testing the Haversian remodeling effects of agents with anabolic effects on cancellous bone. A strategy using one large animal for both Haversian remodeling and estrogen-depletion bone loss makes sense.

E. Pigs and Sheep

The OVX pig has been disappointing, showing only minor structural deterioration and bone loss.[139–141] Its regular estrus cycle is shorter than in the human. Pigs, with Haversian remodeling, have been used to study skeletal effects of bisphosphonates, fluoride, and exercise.[142–144] Peak bone mass occurs after age 3, confounding experimental efforts by introducing cost difficulties. More work is necessary for the pig to gain acceptance as a large animal model of estrogen depletion bone loss.

Ewes have a regular estrus cycle during fall and winter, but show anestrus during longer days.[145] Post-OVX skeletal behavior is less consistent than in post-menopausal women, rats, or monkeys.[146] Bone biomarkers indicate accelerated remodeling by three months post-OVX;[147] marginal relative osteopenia occurs at a few bone sites by 6–12 months post-OVX.[148] Fluoride and glucocorticoid effects in ewes parallel findings in other animals and humans.[122,129,149–151] Aged sheep pose little problem for handling. Existing data thus reveal inconsistent OVX-related bone loss as in pigs. New data that assess the age of peak bone mass and more data on post-OVX bone loss are needed to validate the adult ewe as a model of osteoporosis.

F. Nonhuman Primate

Peak bone mass occurs at age 9–11yrs in cynomolgous and rhesus monkeys and baboons.[152–154] Nonhuman primates have a 28 day menstrual cycle and experience natural menopause at age 18-20 yrs. Pedigree studies of nonhuman primates can establish genetic linkage to bone mass.[15,155] Combining genetic studies of the mouse and nonhuman primate is a multi-species approach to identifying genes that control peak bone mass in humans.

Nonhuman primates (NHP) show decreased bone mass and strength with increased turnover after OVX[156–159] or GnRH agonist treatment.[23] Absolute bone loss of ~5% occurs by three months post-OVX and plateaus at 8% by nine months post-OVX.[160] Though the ERT response has not been studied, bisphosphonates prevent post-OVX bone loss.[160] Adult NHPs show bone loss with age.[161–162] Post-OVX bone changes are frequently masked by using animals still gaining peak bone mass.[157,162] Histomorphometric studies of estrogen-depletion cancellous bone loss in NHPs and humans not only yield similar values,[163–165] but also allow studies of Haversian remodeling shortly after OVX.[48] Late life spinal pathology in baboons[166–167] and rhesus monkeys[168] is osteoarthritis, not osteoporosis. Baboons experience osteopenia[161] with an age-related decline in anterior vertebral height like that in osteoarthritis rather than osteoporosis;[166–168] no such decline is seen in rhesus.[154] Thus, NHPs may not be a model of fragility fractures, because osteoarthritis and osteoporosis tend to be mutually exclusive.[169] The interference of osteoarthritis in older primates with spine DXA should also be considered.[170]

NHPs have cancellous and Haversian remodeling comparable to humans. Animals aged 9–11 yrs, the age of peak bone mass, should be used. Bone loss after OVX lasts nine months, a six–eight-fold timeframe improvement over humans. NHPs are today's large animal model of choice for adult skeletal research when information about Haversian remodeling is required.

G. Summary

The 10 month old female rat is now the animal of choice for studies of every feature of human osteoporosis except Haversian remodeling. It has reached peak bone mass and can be manipulated through OVX to simulate clinical osteoporosis in adult women. Routine methods for humans, like histomorphometry, DXA, and, serum biochemistry work well. Though it develops no fragility fractures, mechanical testing is a good surrogate. Estrogen-deplete nonhuman primates are the large animal of choice when studies of Haversian remodeling are required.

Data about bone behavior in estrogen-deplete mice are now convincing as to similarity to post-menopausal women. Knockout and transgenic mice with interesting skeletal phenotypes will necessitate full characterization of the adult mouse skeleton in the next few years. Female mice are likely to have skeletal growth and maturation phases that can make them useful for peak bone mass experiments. Mice are likely to be uniquely useful in revealing genes that control peak bone mass, showing skeletal behavior after specific gene alterations, and providing animals with "designer bone disease."

IV. DISUSE OSTEOPENIA

A. In General

The prevalence of human disease associated with permanent or transient skeletal disuse is rising as more persons survive partially debilitating events like stroke, myocardial infarction, and central nervous system trauma. While there are reasonable animal models of marked declines in bone use, there are no models of the imperceptible decline in mechanical usage that impacts the aging skeleton. Experimental and cohort data from humans with disuse osteopenia is first reviewed to place the data from animal models in proper perspective.

The principal conditions of skeletal disuse are paraplegia or quadriplegia after spinal cord injury, spaceflight, enforced bedrest, or stroke with varying degrees of paralysis. Information is limited by the availability of participants for studies and ethical/operational considerations in experimental design. Disuse osteopenia was first described at the same time as post-menopausal osteoporosis.[171] Long-term disuse is characterized by localized bone mass decreases of ~30%, implied bone formation rate decreases of ~80%, and the replacement of red by fatty marrow.[172–173] Bone mass several years after spinal cord injury is reduced 40–60%.[174] In limbs immobilized one year after stroke, bone mineral density is 4–8% lower than in the opposite limb.[175–176] Transient disuse osteopenia has been documented following fracture,[177] but its etiology is intertwined with the RAP.[178]

The main disuse experiment in humans is bedrest in healthy fourth decade males.[179–187] All other data are from cohort studies.[172–173,175,188–191] Since few opportunities occur to observe the development of steady state disuse osteopenia in humans, its pathophysiology is poorly characterized. Four month bedrest experiments in healthy men imply transient elevation of resorption.[179] However, this elevation and a likely decline in formation[180] causes only a 1–10% loss of bone,[181] that is completely regained within seven months. Thirty to fifty percent bone loss is reported following bedrest for treatment of back pain.[189–190] Urinary calcium rises transiently in humans

TABLE 2
Summary of Various Immobilization Methods in Rats

Method	Advantages	Disadvantages
Sciatic Neurectomy and Others	Profound immobilization; straightforward; widely-used; long-term studies possible	Irreversible; proper sham surgery difficult (RAP); involvement of nerve changes; no monitoring of food consumption; foot chewing; mostly in growing rats; sham-op control and study of both limbs not done
Tail Suspension	Reversibility; non-surgical; consistency of findings; spaceflight relevance; ease of drug intervention	Unique equipment and confinement; only growing animals possible; short term, labor intensive; food intake never monitored; muscle motion possible
Hindlimb Taping	Reversibility; non-surgical; consistency of findings; simple equipment; suitable for growing and adult rats ease of drug intervention; long-term studies possible	High maintenance; leg motion possible; food intake never monitored
External Fixation	Reversibility; non-surgical; operator specific; suitable for growing and adult rats; ease of drug intervention; long-term studies possible	High maintenance; leg motion possible; food intake never monitored
Tenotomy	Partially reversible; straightforward; suitable for growing and adult rats; ease of drug intervention	Sham-surgery; only short-term studies possible w/o intervention to prevent healing
Spaceflight	Reversibility; non-surgical; true weightlessness	Muscle motion permitted; inconvenient; food intake effect not reported

during spaceflight.[191] The human data suggest that a few months' immobilization causes mild, reversible bone loss that is unlike that seen with severe immobilization following permanent injury. Thus, the descriptive pathophysiology of the development of acute disuse osteopenia in humans is incomplete, making the development of *in vivo* animal models somewhat difficult.

Though a few disuse osteopenia studies have been done in dogs, goats,[192] turkeys, and monkeys,[193] the principal animal is the rat. Permanent and transient methods exist. Cross model findings are concordant, but each has problems (Table 2). Many have surgical components that cause a RAP.[178] The transient ones require continuous attention to maintain substantial, though incomplete immobilization, creating less severe bone loss that may be relevant to milder disuse. Permanent ones give profound immobilization and are low maintenance, but preclude studying recovery, being most relevant to disuse osteopenia associated with CNS injury.

B. ANIMALS

Reversible forelimb immobilization in dogs by casting has been reported.[194–197] Progressive cortical and cancellous osteopenia in the immobilized limb of ~20–25% develops during 10–12 weeks, followed by a slower decline to a steady state of ~30–35% by 32 weeks. Young dogs develop more rapid and severe osteopenia, but experience more complete recovery than old dogs. Full recovery takes about 1.5–2 times as long as the immobilization period in animals remobilized after 16 weeks or less.[198] Immobilization longer than 12 weeks prevents complete recovery in older dogs. Others using similar methods in adult dogs concur.[199] Others report 30–40% osteopenia in growing dogs after four weeks immobilization that is partially prevented by bisphosphonates, tamoxifen, or NSAIDs.[200–201] There is thus good consensus that the immobilized dog forelimb responds to reversible immobilization with relative cancellous and cortical osteopenia. It is the main model for studying disuse effects on Haversian remodeling.

Permanent, surgical immobilization of one forelimb has been done in turkeys. Cortical osteopenia of ~12% develops during 4–6 weeks, and is preventable by four loading cycles per day.[202-203] This model has not been adequately controlled by sham-operation. It also requires unique skills and equipment. Its results have not been validated in multiple independent laboratories.

Permanent and temporary immobilization models in rats are ubiquitous. Permanent models include sciatic or caudal neurectomy, hemicordotomy, and amputation.[204] Temporary models include hindlimb taping; external fixation, including casting or tail encasement; tenotomy; hindlimb suspension; and spaceflight.

C. Common Procedures

1. Neurectomy

Neurectomy is the most frequently-used model for inducing disuse osteopenia. A 5 mm section of the sciatic nerve is resected, causing permanent distal paralysis of the operated limb. Immobilization times from 1–8 weeks have been studied.[205-214] It is suitable for both growing and adult rats, but is most frequently used in growing rats. Cancellous bone loss of 10–15% occurs after 10 days[207] that stabilizes at 40–75%[205-206] by four weeks. Loss is more rapid and prominent in cancellous than cortical bone.[205,209] The most prominent finding is decreased periosteal bone formation.[205] Bone formation is decreased while bone resorption is increased in cancellous bone.[205,206] While bone formation is universally reduced, bone resorption endpoints occasionally are found unchanged.[207] Sciatic neurectomy in older rats may cause both limbs to lose bone with more marked loss in the operated limb.[208] This systemic bone loss may also be an effect of reduced food intake or lack of movement associated with hindquarters injury. Various interventions prevent the bone loss induced by sciatic neurectomy.[205-206,208] At three weeks, caudal neurectomy causes depressed bone formation, but only minor bone loss.[213] Hemicordotomy causes more rapid and severe loss of bone than does sciatic neurectomy.[210]

2. External Fixation

Several reversible methods are available. Recovery from steady state immobilization is an excellent method to study cellular events during natural bone formation stimulation. These immobilization methods tend to be laboratory specific. The hindlimb taping model is most frequently applied.[215-222] One hindlimb is secured against the ventral body wall by adhesive tape. The tape is checked daily and replaced weekly as the immobilized limb is stretched. Studies of up to 26 weeks exist. Older rats are most frequently used, but it is suitable for growing rats. Cancellous bone loss of 10–12% is detected by two weeks that progresses to 20–25% at six weeks,[220] 45% by 10 weeks, and stabilizes at ~60% by 18 weeks.[216] For bone mineral density, loss can be detected at 10 weeks (12%) that stabilizes at ~18% by 18 weeks.[216] These changes are accompanied by a transient 50% rise in cancellous bone resorption during the bone mass decline, with an accompanying chronic 20% depression in bone formation.[216] Remobilization studies work well.[219,222]

Another reversible method applies padded tape to one limb.[223-225] Loss of 5–12% in BMD occurs by three weeks. Another method is by plaster cast.[226] Yet another reversible method is with filament tape,[227] that induces an 8–25% decline within four days in four-week-old rats, with a 40–50% decline in bone formation rate.[227] Tail encasement for 10 days causes 6% reduction in caudal vertebra BMD.[228] Another reversible method is suturing the knee in a partially flexed position, resulting in 10–15% loss of BMD in three month old rats within one week.[229]

3. Tenotomy

Tenotomy is a partially reversible surgical method of immobilization.[230-232] Reversal occurs as healing takes place and the rat adapts. After two weeks in four-week-old rats, 15% loss in cancellous bone volume occurs, with elevation in resorption surfaces and decline in formation surfaces.[231] Another study suggests 35% loss after three weeks, with a rise in resorption surfaces and no change in formation.[232] Comparative work suggests that tenotomy and sciatic neurectomy are similar.

4. Hindlimb Suspension

This model matches spaceflight duration and conditions.[233–238] Thus it is generally applied with rats aged 4–6 wks and a duration of 4–14 days. Support of the body weight by the tail with heavier rats is troublesome. Special cages are required. The rats are suspended by the tail from a rotating arm in the cage roof that permits the rat to walk only on its forelimbs at 30° head down tilt to match spaceflight fluid shifts. Hindlimbs of suspended rats have 20–30% lower bone density and cancellous bone volume than controls after 4–14 days. Bone mass in these growing rats recovers quickly upon remobilization.[236–237]

5. Spaceflight

Growing rats flown in space for 14 days lose 50–60% cancellous bone volume in regions accustomed to loading.[239] This loss is primarily accompanied by decreased formation.[240] Other shorter studies suggest that bone loss accompanied by decreased formation and occasional increased resorption also occurs.[239–244] Bone mass in these growing rats recovers in 1.5–2 times the duration of spaceflight.[243] Recent data suggest that spaceflight bone loss occurs in singly-housed, but not in multiply housed rats.

V. GLUCOCORTICOID OSTEOPENIA

Though human glucocorticoid (GC) osteoporosis has been recognized as both an endocrinologic and iatrogenic entity for over half a century,[246,247] its pathogenesis remains controversial.[248–252] It is marked by vertebral crush fractures, osteopenia with normal numbers of thinned trabeculae, and low bone formation rate.[129,253–255] Secondary hyperparathyroidism linked to reduced intestinal calcium absorption is also found, but the major long-term lesion is low bone formation.

A. SMALL ANIMALS

Most mouse and rat experiments on GC skeletal effects have used growing animals.[256–266] They yield the confusing result of increased metaphyseal cancellous bone mass or density with thinned epiphyseal growth cartilages, decreased bone elongation rate, and decreased rates of bone formation and resorption. The paradoxical finding of increased cancellous bone mass is due to the GC influence on the endochondral ossification process with decreased rate of disappearance of mineralized metaphyseal tissue, coupled with decreased bone elongation, leaving more mineralized tissue in a smaller bone. Low cortical bone mass and strength is also found in treated groups.[260–262] This is due to general GC inhibition of bone formation that decreases the rapid periosteal formation in cortical bone of growing animals. While decreased bone formation is a pertinent finding, its cause is not the same as in adult human GC osteopenia. The cortical osteopenia and low bone strength of GC treated growing rats is due to reduced periosteal expansion, not the same process producing GC osteopenia in adult humans.

As in all osteoporosis research, growing animals are poor models. For GC-induced osteopenia, growing animal data has led to the erroneous conclusion that all mouse and rat experiments are inappropriate. In fact, experiments in older GC-treated rats show either a trend to, or a significant decrease in metaphyseal bone mass.[267–274] The usual dose and treatment time is 0.5–2mg/kg/d prednisolone or equivalent for 6–12 weeks. Decreased formation rate is a universal finding. Bone density by Archimedes principle or defatted bone weight is ~6% lower[269–271] and lumbar spine bone mineral density declines about 10% after eight weeks.[274]

Skeletal studies in rats of any age show body weight declines with GC administration. A few suggest that the food intake reduction associated with the weight decline plays some role in creating osteopenia that might be mistakenly attributed to GC actions.[264,272] Pair-feeding of non-GC-treated rats to match the food intake of GC-treated rats is thus needed to separate the effects of GCs from those of hypophagia.

B. Large Animals

As in post-menopausal osteoporosis studies, the rat cannot disclose all details of the GC osteopenia disease process because of its low levels of Haversian remodeling. Existing data suggest that glucocorticoids affect Haversian remodeling. Large animal studies are thus necessary to disclose the whole picture of GC skeletal effects.[122,128,275–277] There is a phase of accelerated remodeling targeted to the endocortical one-third of the cortical bone of long bones and rib. Since a primary effect of GC treatment is suppression of bone formation, the formation phase that customarily follows the resorption phase of remodeling, proceeds extremely slowly if at all, leaving large holes, accounting for considerable bone loss.[122,128] Other studies in rabbits suggest that bones are weaker and bone density is lower after GC treatment.[277,278] Skeletal findings in GC-treated dogs are similar, showing a 13% loss of lumbar spine BMD over 12 months.[128,275] Studies of glucocorticoids in sheep suggest that bone formation is suppressed significantly, but that changes in remodeling due to the photoperiod of sheep are marked.[151,278–279]

C. Summary

Past experiments with growing rats have been misleading. Older rats are likely to be an appropriate model of human GC osteopenia, but experiments that document the influence of decreased food intake during GC treatment are needed. Large animal studies have proven that Haversian remodeling is influenced by GC treatment, causing bone loss in the endocortical one-third of cortical bone. Just as for post-menopausal osteoporosis, a combination of small and large animal studies is necessary to understand its overall behavior.

VI. OSTEOPENIA ASSOCIATED WITH INFLAMMATION

A. In General

Rheumatoid arthritis, the commonest inflammatory arthritis, is associated with loss of bone mass, alterations in bone structure, and clinically important loss of strength of the skeleton. Considerable morbidity experienced by rheumatoid arthritis (RA) patients results from the increased fracture risk associated with this disease.[280–282] The patients are also at risk of articular surface collapse, weaker fixation of orthopaedic implants and prostheses, and periprosthetic fractures.[283–285] Other chronic inflammatory conditions, including non-arthritic conditions, may be associated with osteoporosis and increased fracture risk.

B. Animal Models

Arthritis and non-articular inflammation can be induced in laboratory animals to create osteopenia. These models are relevant to the morphologic and biomechanical study of bone loss, including bone remodeling, bone cellular activity, microstructure, mineralization and various aspects of bone strength. Further, the study of fixation of implants to abnormal bone requires suitable models, as does the evaluation of interventions designed to prevent or reverse bone loss and defects. Because bone quality is important in the surgical management of musculoskeletal disease, the problem of the osteopenic skeleton in RA and other inflammatory conditions that frequently require orthopaedic reconstruction is an appropriate subject for orthopaedic research.

Minne et al.[286] described osteopenia in the rat that develops after subcutaneous injection of nonspecific irritants such as talcum (magnesium silicate) and cotton wool (cellulose), that initiate a chronic inflammation. Eight injections of 400 mg of magnesium silicate suspended in 0.5 ml saline at different subcutaneous sites on the back of female rats of the Chbb Thom strain created systemic inflammatory responses including loss of body weight, increase in spleen weight, and an increase of cells of myelopoietic origin in bone marrow. A transient decrease in circulating neutrophils, lymphocytes and monocytes was followed by an increase in peripheral blood leukocytes for the duration of the inflammation. In this model of inflammation-induced osteopenia not characterized by arthritis, there is a transient generalized depression of bone formation, measurable as early as three days after injection, due to a decrease in osteoblast function; bone resorption is not increased.[287,288] Vitamin D[289] and insulin,[290] but not salmon calcitonin[291] appear osteoprotective. The model has been applied to studies of diabetes-associated osteoporosis.[292]

Extensive investigation into the osteopenia of inflammatory arthritis has been carried out using intra-articular injection of a sterile 1% solution of carrageenan, an extract from species of seaweed. This model was first described by Gardner[293] and well characterized by others[294–297] who applied it in studies of articular cartilage damage in inflammatory arthritis. Subsequently, the model was found suitable for studies of osteoporosis because of the consistent occurrence of erosive, rapid juxta-articular remodeling and diaphyseal bone loss, resembling the changes of RA.[298–300] The model is described in detail in Chapter 19. Carrageenan-induced arthritis is described in mature[300] and immature[301] dogs and rabbits.[298–299]

C. Fracture Risk

A series of experiments performed in the carrageenan-injection model has elucidated mechanisms of fracture risk that may have relevance to clinical problems due to osteoporosis in RA patients. In cancellous bone, increased remodeling with bone formation increased by a factor of four, and resorption calculated to be increased by a greater ratio, results in net loss of ~20% of cancellous bone volume in 49 days.[298] Cortical thinning was observed as well as evidence of increased remodeling in the metaphysis and diaphysis of the femur, proximal to the involved tibiofemoral joint.[299] Since the bone was composed of a higher proportion of newly-formed material, bone mineralization was diminished compared to normal specimens.[302] A comparative study of bone mass and remodeling in immobilization and arthritis suggested that, in the carrageenan model, the abnormality resulted not from immobilization, but from inflammatory effects.[303] The rapid remodeling osteopenia is principally observed in the bones of the ipsilateral limb,[305] but there is inconclusive evidence that monoarticular inflammation may cause increased bone remodeling elsewhere in the skeleton.[303]

A further series of experiments focused on the mechanism of increased fracture risk in cortical bone. Inflammatory arthritis-induced loss of strength was demonstrated in rabbit femoral diaphyses from the limb affected by experimental tibiofemoral arthritis, showed a loss of ultimate strength of ~39%.[304] Kang et al. documented osteopenia and decreased mechanical strength of the entire ipsilateral femur and tibia.[305] Strips of femoral cortex from the carrageenan injection arthritis model, tested within the elastic range, demonstrated no significant alteration in the elastic and flexural modulus.[306] Large defects in the femoral cortex were described[307] and defined as disordered, giant resorption sites crossing osteonal borders and up to 1.2 mm in cross-sectional diameter.[308] Cooke and Takashima identified an analogous porous lesion in the femoral neck of RA patients who underwent total hip replacement. Enlarged Haversian systems contained osteoclasts and vascular changes with tall endothelial cells.[309] A simplified femoral shaft mathematical model predicted the loss of strength in the femoral cortex that would result from defects of the size observed in the carrageenan model. The calculated result, ~29% was reasonably concordant with the observed loss of strength (~39%).[304,307]

These data suggest that cortical porosity resulting from focal bone resorption could be a major cause of the loss of strength of long bones in inflammatory arthritis. Since a large increase in osteoclast numbers was observed in the arthritis model,[310] a pharmacologic approach to resorption suppression was proposed as a protection against long bone fracture. Pamidronate completely prevented loss of strength associated with carrageenan arthritis.[304] Zoledronate also prevented the formation of the giant osteonal resorption areas, supporting the hypothesis that cortical porosity due to focal osteoclastic resorptions is an important cause of increased risk of long bone fracture in inflammatory arthritis.[311] These experiments also provide a basis for trials of anti-resorptives in the management of the osteoporosis of inflammatory arthritis.

REFERENCES

1. Guidelines for preclinical and clinical evaluation of agents used in the prevention or treatment of postmenopausal osteoporosis, Food and Drug Administration Division of Metabolism and Endocrine Drug Products, Washington, DC, 1994.
2. Cesnjaj, M., Stavljenic, A., and Vukicevic, S., "*In vivo* models in the study of osteopenias," *Eur. J. Clin. Chem. Clin. Biochem.*, 29, 211, 1991.
3. Rodgers, J. B., Monier-Faugere, M. C., and Malluche, H. H., "Animal models for the study of bone loss after cessation of ovarian function," *Bone*, 14, 369, 1993.
4. Barlet, J. P., Coxam, V., Davicco, M. J., and Gaumet, N., "Animal models of post-menopausal osteoporosis," *Reprod. Nutr. Dev.*, 34, 221, 1994.
5. Miller, S. C., Bowman, B. M., and Jee, W. S. S., "Available animal models of osteopenia — small and large," *Bone*, 17 (Supp. 4), 117S, 1995.
6. Thompson, D. D., Simmons, H. A., Pirie, C. M., and Ke, H. Z., "FDA guidelines and animal models for osteoporosis," *Bone*, 17 (Supp. 4), 125S, 1995.
7. Kimmel, D. B., "A current assessment of *in vivo* animal models of osteoporosis," in *Osteoporosis*, Marcus, R., Feldman, D., and Kelsey, J., Eds., Academic Press, San Diego, 1996, 671.
8. Geddes, A., "Animal models of bone disease," in *Principles of Bone Biology*, Bilizekian, J. P., Raisz, L. G., and Rodan, G. A., Eds., Academic Press, San Diego, 1996, 1343.
9. Hansen, M. A., Kirsten, O., Riis, B. J., and Christiansen, C., "Role of peak bone mass and bone loss in postmenopausal osteoporosis: 12 year study," *Br. Med. J.*, 303, 961, 1991.
10. Matkovic, V., "Calcium intake and peak bone mass," *N. Eng. J. Med.*, 327, 119, 1992.
11. Hui, S. L., Slemenda, C. W., and Johnston, C. C., Jr., "Baseline measurement of bone mass predicts fracture in white women," *Ann. Int. Med.*, 111, 355, 1989.
12. Ross, P. D., Wasnich, R. D., and Vogel, J. M., "Detection of prefracture spinal osteoporosis using bone mineral absorptiometry," *J. Bone Miner. Res.*, 3, 1, 1988.
13. Smith, D. M., Nance, W. E., Ke, W. K., et al., "Genetic factors in determining bone mass," *J. Clin. Invest.*, 52, 2800, 1973.
14. Evans, R. A., Marel, G. M., Lancaster, E. K., et al., "Bone mass is low in relatives of osteoporotic patients," *Ann. Int. Med.*, 109, 870, 1988.
15. Lander, E. S. and Schork, N. J., "Genetic dissection of complex traits," *Science*, 263, 2037, 1994.
16. Drinkwater, B. L., Nilson, K., Chesnut, C. H., et al., "Bone mineral content of amenorrheic and eumenorrheic athletes," *N. Eng. J. Med.*, 311, 277, 1984.
17. Marcus, R., Cann, C., Madvig, P., et al., "Menstrual function and bone mass in elite women distance runners," *Ann. Intern. Med.*, 102, 158, 1985.
18. Klibanski, A. and Greenspan, S. L., "Increase in bone mass after treatment of hyperprolactinemic amenorrhea," *N. Eng. J. Med.*, 315, 542, 1986.
19. Drinkwater, B. L., Nilson, K., and Ott, S., "Bone mineral density after resumption of menses in amenorrheic athletes," *JAMA*, 256, 380, 1986.
20. Hansen, M. A., "Assessment of age and risk factors on bone density and bone turnover in healthy premenopausal women," *Osteo. Int.*, 4, 123, 1994.
21. Garn, S. M., *The Early Gain and Later Loss of Cortical Bone*, Thomas, Springfield, IL, 1970.

22. Wronski, T. J., Dann, L. M., Scott, K. S., and Cintron, M., "Long-term effects of ovariectomy and aging on the rat skeleton," *Calcif. Tissue Int.*, 45, 360, 1989.
23. Mann, D. R., Gould, K. G., and Collins, D. C., "A potential primate model for bone loss resulting from medical oophorectomy or menopause," *J. Clin. Endocr. Metab.*, 71, 105, 1990.
24. Hartwell, D., Riis, B. J., and Christiansen, C., "Changes in vitamin D metabolism during natural and medical menopause," *J. Endocrinol. Metab.*, 71, 127, 1990.
25. Nordin, B. E., Horsman, A., Brook, R., et al., "The relationship between oestrogen status and bone loss in post-menopausal women," *Clin. Endocrinol.*, 5, 353S, 1976.
26. Hedlund, L. R. and Gallagher, J. C., "The effect of age and menopause on bone mineral density of the proximal femur," *J. Bone Miner. Res.*, 4, 639, 1989 .
27. Arlot, M. E., Delmas, P. D., Chappard, D., and Meunier, P. J., "Trabecular and endocortical bone remodeling in postmenopausal osteoporosis: comparison with normal postmenopausal women," *Osteo. Int.*, 1, 41, 1990.
28. Keshawarz, N. M. and Recker, R. R., "Expansion of the medullary cavity at the expense of cortex in postmenopausal osteoporosis," *Metab. Bone Dis. Rel. Res.*, 5, 223, 1984.
29. Wasnich, R., Yano, K., and Vogel, J., "Postmenopausal bone loss at multiple skeletal sites: relationship to estrogen use," *J. Chron. Dis.*, 36, 781, 1983.
30. Horsman, A., Simpson, M., Kirby, P. A., and Nordin, B. E. C., "Non-linear bone loss in oophorectomized women," *Br. J. Radiol.*, 50, 504, 1977.
31. Recker, R. R., Heaney, R. P., and Saville, P. D., "Menopausal changes in remodeling," *J. Lab. Clin. Med.*, 92, 964, 1978.
32. Ebeling, P. R., Atley, L. M., Guthrie, J. R., et al., "Bone turnover markers and bone density across the menopausal transition," *J. Endocrinol. Metab.*, 81, 3366, 1996.
33. Heaney, R. P., Recker, R. R., and Saville, P. D., "Menopausal changes in calcium balance performance," *J. Lab. Clin. Med.*, 92, 953, 1978.
34. Coble, T., Kimmel, D. B., Lappe, J. M., and Recker, R. R., "Association of bone markers with cancellous bone behavior in normal premenopausal women," *J. Bone Miner. Res.*, 9 (Supp. 1), S260, 1994.
35. Steiniche, T., Hasling, C., Charles, P., et al., "A randomized study on the effects of estrogen/gestagen or high dose oral calcium on trabecular bone remodeling in postmenopausal osteoporosis," *Bone*, 10, 313, 1989.
36. Jensen, G. F., Christiansen, C., and Transbol, I., "Fracture frequency and bone preservation in postmenopausal women treated with estrogen," *Obstetr. Gyn.*, 69, 493, 1982.
37. Christiansen, C., Christensen, M. S., McNair, P., et al., "Prevention of early postmenopausal bone loss: controlled 2-year study in 315 normal females," *Eur. J. Clin. Invest.*, 10, 273, 1980.
38. Maxim, P., Ettinger, B., and Spitalny, G. M., "Fracture protecting provided by long-term estrogen treatment," *Osteo. Int.*, 5, 23, 1995.
39. Turner, R. T., Vandersteenhoven, J. J., and Bell, N. H., "The effects of OVX and estradiol on cortical bone in growing rats," *J. Bone Miner. Res.*, 2, 115, 1987.
40. Wronski, T. J., Cintron, M., Doherty, A. L., and Dann, L. M., "Estrogen treatment prevents osteopenia and depresses bone turnover in OVX rats," *Endocrinology*, 123, 681, 1988.
41. Johnston, C. C., Slemenda, C. W., and Melton, L. J., "Clinical use of bone densitometry," *N. Eng. J. Med.*, 324, 1105, 1991.
42. Kanis, J. A., Melton, L. J., Christiansen, C., et al., "The diagnosis of osteoporosis," *J. Bone Miner. Res.*, 9, 1137, 1994.
43. Kalu, D. N., "The ovariectomized rat as a model of postmenopausal osteopenia," *Bone Miner.*, 15, 175, 1991.
44. Wronski, T. J., "The ovariectomized rat as an animal model for postmenopausal bone loss," *Cells Mater.*, 1 (Supp. 1), 69, 1992.
45. Tsuboyama, T., Takahashi, K., Matsushita, M., et al., "Decreased endosteal formation during cortical bone modeling in SAM-P/6 mice with a low peak bone mass," *Bone Miner.*, 7, 1, 1989.
46. Matsushita, M., Tsuboyama, T., Kasai, R., et al., "Age-related changes in bone mass in the senescence-accelerated mouse (SAM)," *Am. J. Pathol.*, 125, 276, 1986.

47. Frost, H. M., "The skeletal intermediary organization," *Metab. Bone Dis. Rel. Res.*, 4, 281, 1983.
48. O'Rourke, C., Bare, S., Kimmel, D., et al., "Zoledronate (CGP 42,446) suppresses turnover in cortical bone of the ovariectomized nonhuman primate," *J. Bone Miner. Res.*, 12 (Supp. 1), S482, 1997.
49. Walker, D. G., "Congenital osteopetrosis in mice cured by parabiotic union with normal siblings," *Endocrinology*, 91, 916, 1972.
50. Marks, S. C. and Lane, P. W., "Osteopetrosis, a new recessive skeletal mutation on chromosome 12 of the mouse," *J. Heredity*, 67, 11, 1976.
51. Lennon, J. E. and Micklem, H. S., "Stromal cells in long-term murine bone marrow culture: FACS studies and origin of stromal cells in radiation chimeras," *Exp. Hematol.*, 14, 287, 1986.
52. Kyoizumi, S., Baum, C. M., Kaneshima, H., et al., "Implantation and maintenance of functional human bone marrow in SCI-hu mice," *Blood*, 79, 1704, 1992.
53. Passeri, G., Girasole, G., Jilka, R. L., et al., "Increased interleukin–6 production by murine bone marrow and bone cells after estrogen withdrawal," *Endocrinology*, 133, 822, 1993.
54. Miyaura, C., Onoe, Y., Inada, M., et al., "Increased B lymphopoiesis by interleukin 7 induces bone loss in mice with intact ovarian function: similarity to estrogen deficiency," *Proc. Natl. Acad. Sci. USA*, 94, 9360, 1997.
55. Cassella, J. P., Pereira, R., Khillan, J. S., et al., "An ultrastructural, microanalytical, and spectroscopic study of bone from a transgenic mouse with a COLA1.A1 pro-Alpha-1 mutation," *Bone*, 15, 611, 1994.
56. Ducy, P., Desbois, C., Boyce, B., et al., "Increased bone formation in osteocalcin-deficient mice," *Nature*, 382, 448, 1996.
57. Korach, K. S., Couse, J. F., Curtis, S. W., et al., "Estrogen receptor gene disruption: molecular characterization and experimental and clinical phenotypes," *Recent Prog. Horm. Res.*, 51, 159, 1996.
58. Poli V, Balena R, Fattori E, et al., "IL–6 deficient mice are protected from bone loss caused by estrogen depletion," *EMBO J.*, 13, 1189, 1994.
59. Beamer, W. G., Donahue, L. R., Rosen, C. J., and Baylink, D. J., "Genetic variability in adult bone density among inbred strains of mice," *Bone*, 18, 397, 1996.
60. Suzuki, H. K., "Effects of estradiol-17-β-n-valerate on endosteal ossification and linear growth in the mouse femur," *Endocrinology*, 60, 743, 1958.
61. Edwards, M. W., Bain, S. D., Bailey, M. C., et al., "17-β-estradiol stimulation of endosteal bone formation in the ovariectomized mouse: an animal model for the evaluation of bone-targeted estrogens," *Bone*, 13, 29, 1992.
62. Bar-Shira-Maymon, B., Coleman, R., Cohen, A., et al., "Age-related bone loss in lumbar vertebrae of CW-1 female mice: a histomorphometric study," *Calcif. Tissue Int.*, 44, 36, 1989.
63. Weiss, A., Arbell, I., Steinhagen-Thiessen, E., and Silbermann, M., "Structural changes in aging bone: osteopenia in the proximal femurs of female mice," *Bone*, 12, 165, 1991.
64. Smithson, G., Beamer, W. G., Shultz, K. L., et al., "Increased B lymphopoiesis in genetically sex steroid-deficient hypogonadal (hpg) mice," *J. Exp. Med.*, 180, 717, 1994.
65. Urist, M. R., Budy, A. M., and McLean, F. C., "Endosteal bone formation in estrogen-treated mice," *J. Bone Joint Surg.*, 32A, 143, 1950.
66. Wronski, T. J., Cintron, M., and Dann, L. M., "Temporal relationship between bone loss and increased bone turnover in OVX rats," *Calcif. Tissue Int.*, 42, 179, 1988.
67. Jilka, R. L., Weinstein, R. S., Takahashi, K., et al., "Linkage of decreased bone mass with impaired osteoblastogenesis in a murine model of accelerated senescence," *J. Clin. Invest.*, 97, 1732, 1996.
68. Okamoto, Y., Takahashi, K., Toriyama, K., et al., "Femoral peak bone mass and osteoclast number in an animal model of age-related spontaneous osteopenia," *Anat. Rec.*, 242, 21, 1995.
69. Tsuboyama, T., Takahashi, K., Tamamuro, T., et al., "Cross-mating study on bone mass in the spontaneously osteoporotic mouse (SAM-P/6)," *Bone Miner.*, 23, 57, 1993.
70. Takahashi, K., Tsuboyama, T., Matsushita, M., et al., "Modification of strain-specific femoral bone density by bone marrow-derived factors administered neonatally: a study on the spontaneously osteoporotic mouse, SAMP6," *Bone Miner.*, 24, 245, 1994.
71. Spotila, L. D., Constantinou, C. D., Sereda, L., et al., "Mutation in a gene for type I procollagen (COL1A2) in a woman with postmenopausal osteoporosis: evidence for phenotypic and genotypic overlap with mild osteogenesis imperfecta," *Proc. Natl. Acad. Sci. USA*, 88, 5423, 1991.
72. Arnold, J. S. and Jee, W. S., "Bone growth and osteoclastic activity as indicated by radioautographic distribution of 239Pu," *Am. J. Anat.*, 101, 367, 1957.

73. Gothlin, G. and Ericsson, J. L., "On the histogenesis of the cells in fracture callus," *Virch. Arch. Abt. B. Zellpath.*, 12, 318, 1973.
74. Marks, S. C. and Schneider, G. B., "Evidence for a relationship between lymphoid cells and osteoclasts: bone resorption restored in ia (osteopetrotic) rats by lymphocytes, monocytes and macrophages from a normal littermate," *Am. J. Anat.*, 152, 331, 1978.
75. Dawson, A. B., "The age order of epiphyseal union in the long bones of the albino rat," *Anat. Rec.*, 31, 1, 1925.
76. Swanson, H. E. and van der Werff Ten Bosch, J. J., "Sex differences in growth of rats, and their modification by a single injection of testosterone propionate shortly after birth," *J. Endocrinol.*, 26, 197, 1963.
77. Kimmel, D. B., "Quantitative histologic changes in the proximal tibial epiphyseal growth cartilage of aged female rats," *Cells Mater.*, 1 (Supp.), 11, 1992.
78. Turner, R. T., Hannon, K. S., Demers, L. M., et al., "Differential effects of gonadal function on bone histomorphometry in male and female rats," *J. Bone Miner. Res.*, 4, 557, 1989.
79. Frost, H. M. and Jee, W. S. S., "On the rat model of human osteopenias and osteoporoses," *Bone Miner.*, 18, 227, 1992.
80. Li, X. J., Jee, W. S. S., Ke, H. Z., et al., "Age related changes of cancellous and cortical bone histomorphometry in female Sprague-Dawley rats," *Cells Mater.*, (Supp.) 1, 25, 1992.
81. Weindruch, R., "Dietary restriction, tumors, and aging in rodents," *J. Gerontol.*, 44, 67, 1989.
82. Butcher, R. L., Collins, W. E., and Fugo, N. W., "Plasma concentration of LH, FSH, prolactin, progesterone, and estradiol-17-β throughout the four day cycle of the rat," *J. Endocrinol.*, 94,1704, 1974.
83. Lu, K. H., Hopper, B. R., Vargo, T. M., and Yen, S. S. C., "Chronological changes in sex steroid, gonadotropin, and prolactin secretion in aging female rats displaying different reproductive states," *Biol. Reprod.*, 21, 193, 1979.
84. Wronski, T. J., Walsh, C. C., and Ignaszewski, L. A., "Histologic evidence for osteopenia and increased bone turnover in ovariectomized rats," *Bone*, 7, 119, 1986.
85. Black, D., Farquharson, C., and Robins, S. P., "Excretion of pyridinium cross-links of collagen in OVX rats as urinary markers for increased bone resorption," *Calcif. Tissue Int.*, 44, 343, 1989.
86. Ma, Y. F., Ke, H. Z., and Jee, W. S. S., "PGE2 adds bone to a cancellous bone site with a closed growth plate and low bone turnover in OVX rats," *Bone*, 15, 137, 1994.
87. Amman, P., Rizzoli, R., Slosman, D., et al., "Sequential, precise in vivo measurement of bone mineral density in rats using DXA," *J. Bone Miner. Res.*, 7, 311, 1992.
88. Kalu, D. N., Liu, C. C., Salerno, E., et al., "Skeletal response of OVX rats to low and high doses of estradiol," *Bone Miner.*, 14, 175, 1991.
89. Wronski, T. J., Yen, C. F., and Scott, K. S., "Estrogen and diphosphonate protect against osteopenia in OVX rats," *J. Bone Miner. Res.*, 6, 387, 1991.
90. Wronski, T. J., Dann, L. M., Scott, K. S., and Crooke, L. R., "Endocrine and pharmacological suppressors of bone turnover protect against osteopenia in ovariectomized rats," *Endocrinology*, 125, 810, 1989.
91. Cheng, P. T., Chan, C., and Muller, K., "Cyclical treatment of osteopenic OVX adult rats with PTH (1–34) and pamidronate," *J. Bone Miner. Res.*, 10, 119, 1995.
92. Shen, Y., Li, M., and Wronski, T. J., "Calcitonin protects against cancellous bone loss in the femoral neck of OVX rats," *Calcif. Tissue Int.*, 60, 457, 1997.
93. Mazzuoli, G. F., Tabolli, S., Bigi, F., et al., "Effects of salmon calcitonin on the bone loss induced by OVX," *Calcif. Tissue Int.*, 47, 209, 1990.
94. Black, L. J., Sato, M., Rowley, E. R., et al., "Raloxifene (LY139481 HCI) prevents bone loss and reduces serum cholesterol without causing uterine hypertrophy in OVX rats," *J. Clin. Invest.*, 93, 63, 1994.
95. Ke, H. Z., Chen, H. K., Qi, H., et al., "Effects of droloxifene on prevention of cancellous bone loss and bone turnover in the axial skeleton of aged OVX rats," *Bone*, 17, 491, 1995.
96. Harris, S. T., Gertz, B. J., Genant, H. K., et al., "The effect of short term alendronate treatment on vertebral density and biochemical markers of bone remodeling in early postmenopausal women," *J. Clin. Endocrinol. Metabol.*, 76, 1399, 1993.
97. Szucs, J., Horvath, C., Kollin, E., et al., "Three-year calcitonin combination therapy for postmenopausal osteoporosis with crush fractures of the spine," *Calcif. Tissue Int.*, 50, 7, 1992.

98. Lean, J. M., Chow, J. W. M., and Chambers, T. J., "Estrogen induces bone formation on non-resorptive surfaces in the rat," *Bone*, 14, 297, 1993.
99. Chow, J. W. M., Lean, J. M., and Chambers, T. J., "17-β-estradiol stimulates cancellous bone formation in female rats," *Endocrinology*, 130, 3025, 1992.
100. Turner, R. T., Riggs, B. L., and Spelsberg, T. C., "Skeletal effects of estrogen," *Endocrinol. Rev.*, 15, 275, 1994.
101. Mosekilde, Li., Sogaard, C. H., Danielsen, C. C., et al., "The anabolic effects of human parathyroid hormone (hPTH) on rat vertebral body mass are also reflected in the quality of bone, assessed by biomechanical testing: a comparison study between hPTH-(1–34) and hPTH-(1–84)," *Endorinology*, 129, 421, 1991.
102. Toolan, B. C., Shea, M., Myers, E. R., et al., "Effects of 4-amino-1-hydroxybutylidene bisphosphonate on bone biomechanics in rats," *J. Bone Miner. Res.*, 7, 1399, 1992.
103. Ferretti, J. L., Tessaro, R. D., Delgado, C. J., et al., "Biomechanical performance of diaphyseal shafts and bone tissue of femurs from protein-restricted rats," *Bone Miner.*, 4, 329, 1988.
104. Sogaard, C. H., Danielsen, C. C., Thorling, E. B., and Mosekilde, Li., "Long-term exercise of young and adult female rats: effect on femoral neck biomechanical competence and bone structure," *J. Bone Miner. Res.*, 9, 409, 1994.
105. Kleerekoper, M., Peterson, E. L., Nelson, D. A., et al., "A randomized trial of sodium fluoride as a treatment for postmenopausal osteoporosis," *Calcif. Tissue Int.*, 49, 155, 1991.
106. Riggs, B. L., Hodgson, S. F., O'Fallon, W. M., et al., "Effect of fluoride treatment on the fracture rate in postmenopausal women with osteoporosis," *N. Eng. J. Med.*, 322, 802, 1990.
107. Kasra, M. and Grynpas, M. D., "Effect of long-term ovariectomy on bone mechanical properties in young female cynomolgus monkeys," *Bone*, 15, 557, 1994.
108. Vignery, A. and Baron, R., "Dynamic histomorphometry of alveolar bone remodeling in the adult rat," *Anat. Rec.*, 196, 191, 1980.
109. Baron, R., Tross, R., and Vignery, A., "Evidence of sequential modeling in rat trabecular bone: morphology, dynamic histomorphometry, and changes during skeletal maturation," *Anat. Rec.*, 208, 137, 1984.
110. Erben, R. G., "Trabecular and endocortical bone surfaces in the rat: modeling or remodeling?" *Anat. Rec.*, 246, 39, 1996.
111. Jee, W. S. S., Mori, S., Li, X. J., and Chan, S., "Prostaglandin E2 enhances cortical bone mass and activates intracortical bone remodeling in intact and ovariectomized female rats," *Bone*, 11, 253, 1990.
112. Ruth, E., "An experimental study of the Haversian-type vascular channels," *Anat. Rec.*, 112, 429, 1953.
113. deWinter, F. R. and Steendijk, R., "The effect of a low-calcium diet in lactating rats; observations on the rapid development and repair of osteoporosis," *Calcif. Tissue Int.*, 17, 303, 1975.
114. Wronski, T. J., Schenck, P. A., Cintron, M., and Walsh, C. C., "Effect of body weight on osteopenia in ovariectomized rats," *Calcif. Tissue Int.*, 40, 155, 1987.
115. Roudebush, R. E., Magee, D. E., Benslay, D. N., et al., "Effect of weight manipulation on bone loss due to ovariectomy and the protective effects of estrogen in the rat," *Calcif. Tissue Int.*, 53, 61, 1993.
116. Kinney, J. H., Lane, N. E., and Haupt, D., "Three dimensional in vivo microscopy of sequential changes in trabecular architecture following estrogen depletion in female rats," *J. Bone Miner. Res.*, 10, 264, 1995.
117. Shen, V., Birchman, R., Xu, R., et al., "Short term changes in histomorphometric and biochemical turnover markers and bone mineral density in estrogen and or dietary calcium-deficient rats," *Bone*, 16, 149, 1995.
118. Bloom, W., Bloom, M. A., and McLean, F. C., "Calcification and ossification: medullary bone changes in the reproductive cycle of female pigeons," *Anat. Rec.*, 79, 443, 1941.
119. Miller, S. C. and Bowman, B. M., "Medullary bone osteogenesis following estrogen administration to mature male Japanese quail," *Dev. Biol.*, 87, 52, 1981.
120. Gilsanz, V., Gibbens, D. T., Roe, T. F., et al., "Vertebral bone density in children: effect of puberty," *Radiology*, 166, 847, 1988.
121. Bonjour, J. P., Theintz, G., Buchs, B., et al., "Critical years and stages of puberty for spinal and femoral bone mass accumulation during adolescence," *J. Clin. Endocrinol. Metab.*, 73, 555, 1991.
122. Duncan, H., Hanson, C. A., and Curtiss, A., "The different effects of soluble and crystalline hydrocortisone on bone," *Calcif. Tissue Int.*, 12, 159, 1973.

123. Vanderschueren, D., Van Herck, E., Suiker, A. M., et al., "Bone and mineral metabolism in the adult guinea pig: long-term effects of estrogen and androgen deficiency," *J. Bone Miner. Res.*, 7, 1407, 1992.
124. Jowsey, J. and Gershon-Cohen, J., "Effect of dietary calcium levels on production and reversal of experimental osteoporosis in cats," *Proc. Soc. Exp. Biol. Med.*, 116, 437, 1964.
125. Mackey, M. S., Stevens, M. L., Li, X. J., et al., "The ferret: a small animal model with BMU-based remodeling for skeletal research," *J. Bone Miner. Res.*, 9 (Supp. 1), S319, 1994.
126. Li, X. J., Mackey, M. S., Stevens, M. L., et al., "The ferret: skeletal response to estrogen deficiency," *J. Bone Miner. Res.*, 9 (Supp. 1), S397, 1994.
127. Kimmel, D. B. and Jee, W. S. S., "A quantitative histologic study of bone turnover in young adult beagles," *Anat. Rec.*, 203, 35, 1982.
128. Jett, S., Wu, K., Duncan, H., and Frost, H. M., "Adrenalcorticosteroid and salicylate actions on human and canine Haversian bone formation and resorption," *Clin. Orthop.*, 68, 310, 1070.
129. Bressot, C., Meunier, P. J., Chapuy, M. C., et al., "Histomorphometric profile, pathophysiology and reversibility of corticosteroid-induced osteoporosis," *Metab. Bone Dis. Rel. Res.*, 1, 303, 1979.
130. Ritz, E, Krempien, B., Mehls, O., and Malluche, H. H., "Skeletal abnormalities in chronic renal insufficiency before and during maintenance hemodialysis," *Kidney Int.*, 4, 116, 1973.
131. Grynpas, M. D., Kasra, M., Dumitriu, M., et al., "Recovery from pamidronate (APD): a two year study in the dog," *Calcif. Tissue Int.*, 55, 288, 1994.
132. Podbesek, R. D., Edouard, C., Meunier, P. J., et al., "Effect of two treatment regimes with synthetic human parathyroid hormone fragment on bone formation and the tissue balance of trabecular bone in greyhounds," *Endocrinology*, 112, 1000, 1983.
133. Kimmel, D. B., "The oophorectomized beagle as an experimental model for estrogen-depletion bone loss in the adult human," *Cells Mater.*, (Supp.) 1, 75, 1992.
134. Malluche, H. H., Faugere, M. C., Rush, M., and Friedler, R. M., "Osteoblastic insufficiency is responsible for maintenance of osteopenia after loss of ovarian function in experimental beagle dogs," *Endocrinology*, 119, 2649, 1986.
135. Snow, G. R. and Anderson, C., "The effects of 17-β-estradiol and progestagen on trabecular bone remodeling in oophorectomized dogs," *Calcif. Tissue Int.*, 39, 198, 1986.
136. Concannon, P. W., Hansel, W., and Visek, W. J., "The ovarian cycle of the bitch: plasma estrogen, LH and progesterone," *Biol. Reproduct.*, 13, 112, 1975.
137. Baird, D. T. and Guevara, A., "Concentration of unconjugated estrone and estradiol in peripheral plasma in nonpregnant women throughout the menstrual cycle, castrate and postmenopausal women and in men," *J. Clin. Endocrinol.*, 29, 149, 1969.
138. Longcope, C., Hoberg, L., Steuterman, S., and Baran, D., "The effect of ovariectomy on spine bone mineral density in rhesus monkeys," *Bone*, 10, 341, 1989.
139. Mosekilde, Li., Weisbrode, S. E., Safron, J. A., et al., "Calcium-restricted ovariectomized Sinclair S-1 minipigs: an animal model of osteopenia and trabecular plate perforation," *Bone*, 14, 379, 1993.
140. Stevens, M. L., Sacco-Gibson, N., Combs, K. S., et al., "Evaluation of skeletal parameters in the Sinclair S-1 minipig: histomorphometric assessment of skeletal changes at 3 months post-OVX," *J. Bone Miner. Res.*, 9 (Supp. 1), S258, 1994.
141. Franks, A. F., Wyder, W. E., Li, X. J., et al., "Evaluation of skeletal parameters in the Sinclair S-1 minipig: longitudinal analysis of OVX-induced skeletal changes," *J. Bone Miner. Res.*, 9, S191, 1994.
142. Lafage, H. M., Balena, R., Battle., M. A., et al., "Comparison of alendronate and sodium fluoride effects on cancellous and cortical bone in minipigs," *J. Clin. Invest.*, 95, 2127, 1995.
143. Kragstrup, J., Richards, A., and Fejerskov, O., "Effects of fluoride on cortical bone remodeling in the growing domestic pig," *Bone*, 10, 421, 1989.
144. Raab, D. M., Crenshaw, T., Kimmel, D. B., and Smith, E. L., "Cortical bone response in adult swine after exercise," *J. Bone Miner. Res.*, 6, 741, 1991.
145. Malpaux, B. and Karsch, F. J., "A role for short days in sustaining seasonal reproductive activity in the ewe," *J. Reproduct. Fertil.*, 90, 555, 1990.
146. Jerome, C. P., Lees, C. J., and Weaver, D. S., "Development of osteopenia in OVX cynomolgus monkeys (*Macaca fascicularis*)," *Bone*, 17 (Supp 4), 403S, 1995.
147. Turner, A. S., Alvis, M., Myers, W., et al., "Changes in bone mineral density and bone-specific alkaline phosphatase in OVX ewes," *Bone*, 17 (Supp. 4), 395S, 1995.

148. Turner, A. S., Mallinckrodt, C. H., Alvis, M. R., and Bryant, H. U., "Dose-response effects of estradiol implants on bone mineral density in OVX ewes," *Bone*, 17 (Supp. 4), 421S, 1995.
149. Eriksen, E. F., Mosekilde, L., and Melsen, F., "Effect of sodium fluoride, calcium, phosphate, and vitamin D2 on trabecular bone balance and remodeling in osteoporotics," *Bone*, 6, 381, 1985.
150. Chavassieux, P., Pastoureau, P., Boivin, G., et al., "Fluoride-induced bone changes in lambs during and after exposure to sodium fluoride," *Osteoporos. Int.*, 2, 26, 1991.
151. Chavassieux P, Pastoureau P, Chapuy MC, et al., "Glucocorticoid-induced inhibition of osteoblastic bone formation in ewes: a biochemical and histomorphometric study," *Osteoporos. Int.*, 3, 97, 1993.
152. Pope, N. S., Gould, K. G., Anderson, D. C., and Mann, D. R., "Effects of age and sex on bone density in the rhesus monkey," *Bone*, 10, 109, 1989.
153. Jayo, M. J., Jerome, C. P., Lees, C. J., et al., "Bone mass in female cynomolgus macaques: a cross-sectional and longitudinal study by age," *Calcif. Tissue Int.*, 54, 231, 1994.
154. Champ, J. E., Binkley, N., Havighurst, T., et al., "The effect of advancing age on bone mineral content of female rhesus monkeys," *Bone*, 19, 485, 1996.
155. Rogers, J. and Kidd, K. K., "Nuclear DNA polymorphisms in a wild population of yellow baboons (Papio hamadryas cynocephalus) from Mikumi National Park Tanzania," *Am. J. Phys. Anthropol.*, 90, 477, 1993.
156. Jerome, C., Kimmel, D. B., McAlister, J. A., and Weaver, D. S., "Effects of OVX on iliac trabecular bone in baboons (*Papio anubis*)," *Calcif. Tissue Int.*, 39, 206, 1986.
157. Jerome, C. P., Turner, C. H., and Lees, C. J., "Decreased bone mass and strength in OVX cynomolgus monkeys (*Macaca fascicularis*)," *Calcif. Tissue Int.*, 60, 265, 1997.
158. Lundon, K. and Grynpas, M., "The long term effect of OVX on the quality and quantity of cortical bone in the young cynomolgus monkey: a comparison of density fractionation and histomorphometric techniques," *Bone*, 14, 389, 1993.
159. Lundon, K., Dumitriu, M., and Grynpas, M., "The long-term effect of OVX on the quality and quantity of cancellous bone in young macaques," *Bone Miner.*, 24, 135, 1994.
160. Binkley, N., Champ, J., Davidowitz, B., Kimmel, D., Schaffer, V., and Green, J., "Zoledronate prevents bone loss in ovariectomized rhesus monkeys," *J. Bone Miner. Res.*, 11 (Supp. 1), S339, 1996.
161. Aufdemorte, T. B., Fox, W. C., Miller D., et al., "A nonhuman primate model for the study of osteoporosis and oral bone loss," *Bone*, 14, 581, 1993.
162. Balena, R., Toolan, B. C., Shea, M., et al., "The effects of 2-year treatment with alendronate on bone metabolism, bone histomorphometry, and bone strength in OVX nonhuman primates." *J. Clin. Invest.*, 92:2577-2586.
163. Schnitzler, C. M., Ripamonti, U., and Mesquita, J. M., "Histomorphometry of iliac crest trabecular bone in adult male baboons in captivity," *Calcif. Tissue Int.*, 52, 447, 1993.
164. Bare, S., Kimmel, D., Binkley, N., et al., "Zoledronate (CGP 42,446) suppresses turnover without affecting mineralization in cancellous bone of the OVX nonhuman primate," *J. Bone Miner. Res.*, 12 (Supp. 1), No. S482, 1997.
165. Recker, R. R., Kimmel, D. B., Parfitt, A. M., et al., "Static and tetracycline-based bone histomorphometric data from 34 normal post-menopausal females," *J. Bone Miner. Res.*, 3, 133, 1988.
166. Kimmel, D. B., Lane, N. E., Kammerer, C. M., et al., "Spinal pathology in adult baboons," *J. Bone Miner. Res.*, 8 (Supp. 1), S279, 1993.
167. Hughes, K. P., Kimmel, D. B., Kammerer, C. M., et al., "Vertebral morphometry in adult female baboons," *J. Bone Miner. Res.*, 9 (Supp. 1), S209, 1994.
168. Carlson, C. S., Loeser, R. F., Jayo, M. J., et al., "Osteoarthritis in cynomolgus macaques: a primate model of naturally occurring disease," *J. Orthop. Res.*, 12, 331, 1994.
169. Dequeker, J., "The relationship between osteoporosis and osteoarthritis," *Clin. Rheum. Dis.*, 11, 271, 1985.
170. Orwoll, E. S., Oviatt, S. K., and Mann, T., "The impact of osteophytic and vascular calcifications on vertebral mineral density measurements in men," *J. Clin. Endocrinol. Metab.*, 70, 1202, 1990.
171. Albright, F., Burnett, C. H., Cope, O., and Parson, W., "Acute atrophy of bone simulating hyperparathyroidism," *J. Clin. Endocrinol. Metab.*, 1, 711, 1941.
172. Minaire, P., Meunier, P. J., Edouard, C., et al., "Quantitative histologic data on disuse osteoporosis," *Calcif. Tissue Res.*, 17, 57, 1974.

173. Minaire, P., Edouard, C., Arlot, M., and Meunier, P. J., "Marrow changes in paraplegic patients," *Calcif. Tissue Int.*, 36, 338, 1984.
174. Biering-Sorenson, F., Bohr, H. H., and Schaadt, O. P., "Longitudinal study of bone mineral content in the lumbar spine, the forearm and the lower extremities after spinal cord injury," *Eur. J. Clin. Invest.*, 20, 330, 1990.
175. del Puente, A., Pappone, N., Mandes, M. G., et al., "Determinants of bone mineral density in immobilization: a study on hemiplegic patients," *Osteoporos. Int.*, 6, 50, 1996.
176. Takamoto S, Masuyama T, Nakajima M, et al., "Alterations of bone mineral density of the femurs in hemiplegia," *Calcif. Tissue Int.*, 56, 259, 1995.
177. Houde, J. P., Schulz, L. A., Morgan, W. J., et al., "Bone mineral density changes in the forearm after immobilization," *Clin. Orthop.*, 317, 199, 1995.
178. Frost, H. M., "The regional acceleratory phenomenon," *Orthop. Clin. North Am.*, 12, 725, 1981.
179. Chappard, D., Alexandre, C., Palle, S., et al., "Effects of 1-hydoxy ethylidene-1,1 bisphosphonic acid on osteoclast number during prolonged bed rest in healthy humans," *Metabolism*, 38, 822, 1989.
180. Palle, S., Vico, L., Bourrin, S., and Alexandre, C., "Bone tissue response to four-month antiorthostatic bedrest: a bone histomorphometric study," *Calcif. Tissue Int.*, 51, 189, 1992.
181. Leblanc, A. D., Schneider, V. S., Evans, H. J., et al., "Bone mineral loss and recovery after 17 weeks of bed rest," *J. Bone Miner. Res.*, 5, 843, 1990.
182. LeBlanc, A., Schneider, V., Krebs, J., et al., "Spinal bone mineral after 5 weeks of bed rest," *Calcif. Tissue Int.*, 41:259, 1987.
183. Donaldson, C. L., Hulley, S. B., Vogel, J. M., et al., "Effect of prolonged bedrest on bone mineral," *Metabolism*, 19, 1071, 1970.
184. Chappard, D., Vico, L., Alexandre, C., et al., "Effects of a 120 day period of bedrest on bone mass and bone cell activities in man: attempts at countermeasures," *Bone Miner.*, 2, 383, 1987.
185. Mack, P. B. and LaChance, J. E., "Effects of recumbency and spaceflight on bone density," *Am. J. Clin. Nutr.*, 20, 1194, 1967.
186. Chappard, D., Minaire, P., Privat, C., et al., "Effects of tiludronate on bone loss in paraplegic patients," *J. Bone Miner. Res.*, 10, 112, 1995.
187. Bloomfield, S. A., "Changes in musculoskeletal structure and function with prolonged bed rest," *Med. Sci. Sports Exerc.*, 29, 197, 1997.
188. Sato, Y., Maruoka, H., Oizumi, K., and Kikuyama, M., "Vitamin D deficiency and osteopenia in the hemiplegic limbs of stroke patients," *Stroke*, 27, 2183, 1996.
189. Hansson, T. H., Roos, B. O., and Nachemson. A., "Development of osteopenia in fourth lumbar vertebra during prolonged bed rest after operation for scoliosis," *Acta Orthop. Scand.*, 46, 621, 1975.
190. Krolner, B. and Toft, B., "Vertebral bone loss: an unheeded side effect of therapeutic bed rest," *Clin. Sci.*, 64, 537, 1983.
191. Mazess, R. B. and Whedon, G. D., "Immobilization and bone," *Calcif. Tissue Int.*, 35, 265, 1983.
192. Welch, R. D., Ashman, R. B., Baker, K. J., and Browne, R. H., "Intraosseous infusion of PGE^2 prevents disuse-induced bone loss in the tibia," *J. Orthop. Res.*, 14, 303, 1996.
193. Young, D. R., Niklowitz, W. J., and Steele, C. R., "Tibial changes in experimental disuse osteoporosis in the monkey," *Calcif. Tissue Int.*, 35, 304, 1983.
194. Uhthoff, H. K., Sekaly, G., and Jaworski, Z. F. G., "Effect of long term nontraumatic immobilization on metaphyseal spongiosa in young adult and old beagle dogs," *Clin. Orthop.*, 192, 278, 1985.
195. Uhthoff, H. K. and Jaworski, Z. F. G., "Bone loss in response to long term immobilisation," *J. Bone Joint Surg.*, 60B, 420, 1978.
196. Jaworski, Z. F. G., Liskova Kiar, M., and Uhthoff, H. K., "Effect of long term immobilisation on the pattern of bone loss in older dogs," *J. Bone Joint Surg.*, 62B, 104, 1980.
197. Jaworski, Z. F. G. and Uhthoff, H. K., "Reversibility of nontraumatic disuse osteoporosis during its active phase," *Bone*, 7, 431, 1986.
198. Lane, N. E., Kaneps, A. J., Stover, S. M., et al., "Bone mineral density and turnover following forelimb immobilization and recovery in young adult dogs," *Calcif. Tissue Int.*, 59, 401, 1996.
199. Grynpas, M. D., Kasra, M., Renlund, R., and Pritzker, K. P., "The effect of pamidronate in a new model of immobilization in the dog," *Bone*, 17(Supp. 4), 225S, 1995
200. Caywood, D. D., Wallace, L. J., Olson, W. G., and Stevens, J. B., "Effects of 1-α-dihydroxycholecalciferol on disuse osteoporosis in the dog: a histomorphometric study," *Am. J. Vet. Res.*, 40, 89, 1979.

201. Waters, D. J., Caywood, D. D., Trachte, G. J., et al., "Immobilization increases bone prostaglandin E. Effect of acetylsalicylic acid on disuse osteoporosis studied in dogs," *Acta Orthop. Scand.*, 62, 238, 1991.
202. Rubin, C. T. and Lanyon, L. E., "Regulation of bone formation by applied dynamic loads," *J. Bone Joint Surg.*, 66A, 397, 1984.
203. Lanyon, L. E., Rubin, C. T., and Baust, G., "Modulation of bone loss during calcium insufficiency by controlled dynamic loading," *Calcif. Tissue Int.*, 38, 209, 1986.
204. Svesatikoglou, J. A. and Larsson, S. E., "Changes in composition and metabolic activity of the skeletal parts of the extremity of the adult rat following below knee amputation," *Acta Chir. Scand. Suppl.*, 476, 9, 1976.
205. Wakley, G. K., Baum, B. L., Hannon, K. S., and Turner, R. T., "The effects of tamoxifen on the osteopenia induced by sciatic neurotomy in the rat: a histomorphometric study," *Calcif. Tissue Int.*, 43, 383, 1988.
206. Murakami, H., Nakamura, T., Tsurukami, H., et al., "Effects of tiludronate on bone mass, structure, and turnover at the epiphyseal, primary, and secondary spongiosa in the proximal tibia of growing rats after sciatic neurectomy," *J. Bone Miner. Res.*, 9, 1355, 1994.
207. Weinreb, M., Rodan, G. A., and Thompson, D. D., "Osteopenia in the immobilized rat hind limb is associated with increased bone resorption and decreased bone formation," *Bone*, 10, 187, 1989.
208. Tarvainen, R., Arnala, I., Olkkonen, H., et al., "Clodronate prevents immobilization osteopenia in rats," *Acta Orthop. Scand.*, 65, 643, 1994.
209. Zeng, Q. Q., Jee, W. S. S., Bigornia, A. E., et al., "Time responses of cancellous and cortical bones to sciatic neurectomy in growing female rats," *Bone*, 19, 13, 1996.
210. Yoshida, S., Yamamuro, T., Okumura, H., and Takahashi, H., "Microstructural changes of osteopenic trabeculae in the rat," *Bone*, 12, 185, 1991.
211. Madsen, J. E., Aune, A. K., Falch, J. A., et al., "Neural involvement in post-traumatic osteopenia: an experimental study in the rat," *Bone*, 18, 411, 1996.
212. Shen, V., Liang, X. G., Birchman, R., et al., "Short term immobilization induced cancellous bone loss is limited to regions undergoing high turnover and/or modeling in mature rats," *Bone*, 21, 71, 1997.
213. Chow, J. W. M., Jagger, C. J., and Chambers, T. J., "Reduction in dynamic indices of cancellous bone formation in rat tail vertebrae after caudal neurectomy," *Calcif. Tissue Int.*, 59, 117, 1996.
214. Svesatikoglou, J. A. and Mattson, S., "Changes in composition and metabolic activity of the skeletal parts of the extremity of the adult rat following resection of the sciatic nerve," *Acta Chir. Scand. Suppl.*, 476, 16, 1976.
215. Lin BY, Jee WSS, Chen MM, et al., "Mechanical loading modifies ovariectomy-induced cancellous bone loss," *Bone Miner.*, 25, 199, 1994.
216. Li, X. J., Jee, W. S. S., Chow, S. Y., and Woodbury, D. M., "Adaptation of cancellous bone to aging and immobilization in the adult rat: a single photon absorptiometry and histomorphometry study," *Anat. Rec.*, 227, 12, 1990.
217. Bagi, C. M., Mecham, M., Weiss, J., and Miller, S. C., "Comparative morphometric changes in rat cortical bone following OVX and/or immobilization," *Bone*, 14, 877, 1993.
218. Lindgren, J. U. and Mattson, S., "The reversibility of disuse osteoporosis," *Calcif. Tissue Res.*, 23, 179, 1977.
219. Maeda, H., Kimmel, D. B., Lane, N., and Raab, D., "The musculoskeletal response to immobilization and recovery," *Bone*, 14, 153, 1993.
220. Lane, N., Maeda, H., Cullen, D. M., and Kimmel, D. B., "Cancellous bone behavior in hindlimb immobilized rats during and after naproxen treatment," *Bone Miner.*, 26, 43, 1994.
221. Mattson, S., "The reversibility of disuse osteoporosis," *Acta Orthop. Scand. Supp.*, 144, 59, 1972.
222. Ijiri, K., Jee, W. S. S., Ma, Y. F., and Yuan, Z., "Remobilization partially restored the bone mass in a non-growing cancellous bone site following long term immobilization," *Bone*, 17 (Supp. 4), 213S, 1995.
223. Kannus, P., Jarvinen, T. L. N., Sievanen, H., et al., "Effects of immobilization, three forms of remobilization, and subsequent deconditioning on bone mineral content and density in rat femora," *J. Bone Miner. Res.*, 11, 1339, 1996.
224. Tuukkanen, J., Peng, Z., and Vaananen, H. K., "The effect of training on the recovery from immobilization-induced bone loss in rats," *Acta Physiol. Scand.*, 145, 407, 1992.

225. Kannus, P., Sievanen, H., Jarvinen, T. L., et al., "Effects of free mobilization and low- to high-intensity treadmill running on the immobilization-induced bone loss in rats," *J. Bone Miner. Res.*, 9, 1613, 1994.
226. Thomaidis, V. T. and Lindholm, T. S., "The effect of remobilization on the extremity of the adult rat after short term immobilization in a plaster cast," *Acta Chir. Scand. Supp.*, 476, 36, 1976.
227. Yamaguchi, M. and Kishi, S., "Differential effects of insulin and IGF-I in the femoral tissues of rats with skeletal unloading," *Calcif. Tissue Int.*, 55, 363, 1994.
228. Fiorentino, S., Melillo, G., Fedele, G., et al., "Ketoprofen lysine salt inhibits disuse induced osteopenia in a new non traumatic immobilization model in the rat," *Pharm. Res.*, 33, 277, 1996.
229. Akai, M., Shirasaki, Y., Tateishi, T., and Yasuoka, S., "Localized osteoarticular change due to joint immobilization: biomechanical test and bone densitometry in rats hind limb model," *Arch. Orthop. Trauma Surg.*, 116, 129, 1997.
230. Shaker, J. L., Fallon, M. D., Goldfarb, S., et al., "WR-2721 reduces bone loss after hindlimb tenotomy in rats," *J. Bone Miner. Res.*, 4, 885, 1989.
231. Weinreb, M., Rodan, G. A., and Thompson, D. D., "Immobilization-related bone loss in the rat is increased by calcium deficiency," *Calcif. Tissue Int.*, 48, 93, 1991.
232. Thompson, D. D. and Rodan, G. A., "Indomethacin inhibition of tenotomy induced bone resorption in rats," *J. Bone Miner. Res.*, 3, 409, 1988.
233. Wronski, T. J. and Morey-Holton, E. R, "Skeletal response to simulated weightlessness: a comparison of suspension techniques," *Aviation, Space, Env. Med.*, 58, 63, 1987.
234. LeBlanc, A., Marsh, C., Evans, H., et al., "Bone and muscle atrophy with suspension of the rat," *J. Appl. Physiol.*, 58, 1669, 1985.
235. Globus, R. K., Bikle, D. D., and Morey,-Holton, E., "Effects of simulated weightlessness on bone mineral metabolism," *Endocrinology*, 114, 2264, 1984.
236. Bourrin, S., Palle, S., Genty, C., and Alexandre, C., "Physical exercise during remobilization restores a normal bone trabecular network after tail suspension-induced osteopenia in young rats," *J. Bone Miner. Res.*, 10, 820, 1995.
237. Sessions, N. D., Halloran, B. P., Bikle, D. D., et al., "Bone response to normal weight bearing after a period of skeletal unloading," *Am. J. Physiol*, 257, E606, 1989
238. Shaw, S. R., Zernicke, R. F., Vailas, A. C., et al., "Mechanical, morphological, and biochemical adaptations of bone and muscle to hindlimb suspension and exercise," *J. Biomech.*, 20, 225, 1987.
239. Wronski, T. J., Morey-Holton, E., and Jee, W. S. S., "COSMOS 1129: Spaceflight and bone changes," *Physiologist*, 23, S79, 1979.
240. Cavolina, J. M., Evans, G. L., Harris, S. A., et al., "The effects of orbital spaceflight on bone histomorphometry and messenger ribonucleic acid levels for bone matrix proteins and skeletal signaling peptides in ovariectomized growing rats," *Endocrinology*, 138, 1567, 1997.
241. Morey, E. R. and Baylink, D. J., "Inhibition of bone formation during spaceflight," *Science*, 201, 1138, 1978.
242. France, E. P., Oloff, C. M., and Kazarian, L. E., "Bone mineral analysis of rat vertebra following spaceflight: COSMOS 1129," *Physiologist*, 25, S147, 1982.
243. Wronski, T. J. and Morey, E. R., "Recovery of the rat skeleton from the adverse effects of simulated weightlessness," *Metab. Bone Dis. Rel. Res.*, 4, 347, 1983.
244. Wronski, T. J. and Morey-Holton, E. R., "Effect of spaceflight on periosteal bone formation in rats," *Am. J. Physiol*, 244, R305, 1983.
245. Jee, W. S. S., Wronski, T. J., Morey, E. R., and Kimmel, D. B., "Effects of spaceflight on trabecular bone in rats," *Am. J. Physiol.*, 244, R310, 1983.
246. Cushing, H., "The basophil adenomas of the pituitary body and their clinical manifestations (pituitary basophilism)," *Bull. Johns Hopkins Hosp.*, 50, 137, 1932.
247. Nelson, A. M. and Conn, D. L., "Glucocorticoids in rheumatic disease," *Mayo Clin. Proc.*, 55, 1980.
248. Need, A. G., "Corticosteroids and osteoporosis," *Aust. N.Z. J. Med.*, 17, 267, 1987.
249. Avioli, L. V., Peck, W., Gennari, C., et al., "Corticosteroids and bone," *Calcif. Tissue Int.*, 36, 4, 1984.
250. Canalis, E., "Mechanisms of glucocorticoid action in bone: implications to glucocorticoid-induced osteoporosis," *J. Clin. Endocrinol. Metabol.*, 81, 3441, 1996.
251. Reid, I. R., "Steroid osteoporosis," *Calcif. Tissue Int.*, 45, 63, 1989.
252. Hahn, T. J., "Steroid and drug-induced osteopenia," *Primer Metabol. Bone Dis. Disord. Miner. Metabol.*, 158, 1989.

253. Aaron, J. E., Francis, R. M., Peacock, M., and Makins, N. B., "Contrasting microanatomy of idiopathic and corticosteroid-induced osteoporosis," *Clin. Orthop.*, 243, 294, 1989.
254. Chappard, D., Legrand, E., Basle, M. F., et al., "Altered trabecular architecture induced by corticosteorids," *J. Bone Miner. Res.*, 11, 676, 1996.
255. Dempster, D. W., "Bone histomorphometry in glucocorticoid-induced osteoporosis," *J. Bone Miner. Res.*, 4, 137, 1989.
256. Young, R. H. and Crane, W. A. J., "Effect of hydrocortisone on the utilization of tritiated thymidine for skeletal growth in the rat," *Ann. Rheum. Dis.*, 23, 163, 1964.
257. Simmons, D. J. and Kunin, A. S., "Autoradiographic and biochemical investigations of the effect of cortisone on the bones of the rat," *Clin. Orthop.*, 55, 201, 1967.
258. King, C. S., Weir, E. C., Gundberg, C. W., et al., "Effects of continuous glucocorticoid infusion on bone metabolism in the rat," *Calcif. Tissue Int.*, 59, 184, 1996.
259. Geusens, P., Dequeker, J., Nijs, J., and Bramm, E., "Effect of ovariectomy and prednisolone on bone mineral content in rats: evaluation by single photon absorptiometry and radiogrammetry," *Calcif. Tissue Int.*, 47, 243, 1990.
260. Ortoft, G. and Occlude, H., "Reduced strength of rat cortical bone after glucocorticoid treatment," *Calcif. Tissue Int.*, 43, 376, 1988.
261. Ferretti, J. L., Capozza, R. F., and Zanchetta, J. R., "Mechanical validation of a tomographic (pQCT) index for noninvasive estimation of rat femur bending strength," *Bone*, 18, 97, 1996.
262. Ferretti, J. L., Delgado, C. J., Capozza, R. F., et al., "Protective effects of disodium etidronate and pamidronate against the biomechanical repercussion of betamethasone-induced osteopenia in growing rat femurs," *Bone Miner.*, 20, 265, 1993.
263. Aerssens, J., Van Audekercke, R., Talalaj, M., et al., "Effect of 1-α-vitamin D3 on bone strength and composition in growing rats with and without corticosteroid treatment," *Calcif. Tissue Int.*, 55, 443, 1994.
264. Ortoft, G., Oxlund, H., Jorgensen, P. H., and Andreassen, T. T., "Glucocorticoid treatment or food deprivation counteract the stimulating effect of growth hormone on rat cortical bone strength," *Acta. Paediatr.*, 81, 912, 1992.
265. Advani, S., LaFrancis, D., Bogdanovic, E., et al., "Dexamethasone suppresses *in vivo* levels of bone collagen synthesis in neonatal mice," *Bone*, 20, 41, 1997.
266. Altman, A., Hochberg, Z., and Silbermann, M., "Interactions between growth hormone and dexamethasone in skeletal growth and bone structure of the young mouse," *Calcif. Tissue Int.*, 51, 298, 1992.
267. Nakamuta, H., Nitta, T., Hoshino, T., and Koida, S., "Glucocorticoid-induced osteopenia in rats: histomorphometric and microarchitectural characterization and calcitonin effect," *Biol. Pharm. Bull.*, 19, 217, 1996.
268. Sjoden, G., Johnell, O., DeLuca, H. F., and Lindgren, J. U., "Effects of 1-α-OHD2 and 1-α-OHD3 in rats treated with prednisolone," *Acta Endocrinol.*, 106, 564, 1984.
269. Lindgren, J. U., Merchant, C. R., and DeLuca, H. F., "Effect of 1,25(OH)2D3 on osteopenia induced by prednisolone in adult rats," *Calcif. Tissue Int.*, 34, 253, 1982.
270. Lindgren, J. U., Johnell, O., and DeLuca, H. F., "Studies of bone tissue in rats treated by prednisolone and 1,25(OH)2D3," *Clin. Orthop.*, 181, 264, 1983.
271. Lindgren, J. U. and DeLuca, H. F., "Oral 1,25(OH)2D3: an effective prophylactic treatment for glucocorticoid osteopenia in rats," *Calcif. Tissue Int.*, 35, 107, 1983.
272. Ortoft, G., Bruel, A., Andreassen, T. T., and Oxlund. H., "Growth hormone is not able to counteract osteopenia of rat cortical bone induced by glucocorticoid with protracted effect," *Bone*, 17, 543, 1995.
273. Wimalawansa, S. J., Chapa, M. T., Yallampalli, C., et al., "Prevention of corticosteroid-induced bone loss with nitric oxide donor nitroglycerin in male rats," *Bone*, 21, 275, 1997.
274. Wimalawansa, S. J. and Simmons, D. J., "Prevention of corticosteroid-induced bone loss with alendronate," *Proc. Exp. Biol. Med.*, in press, 1998.
275. Lyles, K. W., Jackson, T. W., Nesbitt, T., and Quarles, L. D., "Salmon calcitonin reduces vertebral bone loss in glucocorticoid-treated beagles," *Am. J. Phys.*, 264, E938, 1993.
276. Lindgren, J. U., DeLuca, H. F., and Mazess, R. B., "Effects of 1,25(OH)2D3 on bone tissue in the rabbit: studies on fracture healing, disuse osteoporosis, and prednisone osteoporosis," *Calcif. Tissue Int.*, 36, 591, 1984.
277. Grardel, B., Sutter, B., Flautre, B., et al., "Effects of glucocorticoids on skeletal growth in rabbits evaluated by DPA, microscopic connectivity and vertebral compressive strength," *Osteoporos. Int.* 4, 204, 1994.

278. Chavassieux, P., Buffet, A., Vergnaud, P., et al., "Short-term effects of corticosteroids on trabecular bone remodeling in old ewes," *Bone*, 20, 451, 1997.
279. O'Connell, S. L., Tresham, J., Fortune, C. L., et al., "Effects of prednisolone and deflazacort on osteocalcin metabolism in sheep," *Calcif. Tissue Int.*, 53, 117, 1993.
280. Cooper, C. and Wickham, C., "Rheumatoid arthritis, corticosteroid therapy, and hip fracture," in *Osteoporosis*, Christiansen, C., and Overgaard, K., Eds., Osteopress, Copenhagen, 1990, 1578.
281. Hooyman, J. R., Melton, L. J., III, Nelson, A. M., et al., "Fractures after rheumatoid arthritis: a population based study," *Arthritis Rheum.*, 27, 1353, 1984.
282. Spector, T. K., Hall, G. M., McCloskey, E. V., and Kanis, J. A., "Risk of vertebral fracture in women with rheumatoid arthritis," *Br. Med. J.*, 306, 1993.
283. Bogoch, E., Hastings, D., Gschwend, N., et al., "Supracondylar fractures of the femur after total knee arthroplasty in patients with rheumatoid arthritis," *Clin. Orthop.*, 229, 223, 1988.
284. Bogoch, E. R., Ouellette, G., and Hastings, D., "Failure of internal fixation of displaced femoral neck fractures in patients with rheumatoid arthritis," *J. Bone Joint Surg.*, 73B, 7, 1991.
285. Bogoch, E. R., Ouellette, G., and Hastings, D., "Intertrochanteric fractures of the femur in rheumatoid arthritis patients," *Clin. Orthop.*, 294, 181, 1993.
286. Minne, H. W., Pfeilschifter, J., Scharla, S., et al., "Inflammation-mediated osteopenia in the rat; A new animal model for pathological loss of bone mass," *Endocrinology*, 115, 50, 1984.
287. Lempert, U. G., Minne, H. W., Fleisch, H., et al., "Inflammation-mediated osteopenia (IMO): no change in bone resorption during its development," *Calcif. Tissue Int.*, 48, 291, 1991.
288. Pfeilschifter, J., Wuster, C., Vogel, M., et al., "Inflammation-mediated osteopenia (IMO) during acute inflammation in rats is due to a transient inhibition of bone formation," *Calcif. Tissue Int.*, 41, 321, 1987.
289. Lempert, U. G., Minne, H. W., Albrecht, B., et al., "1,25-Dihydroxyvitamin D3 prevents the decrease of bone mineral appositional rate in rats with inflammation-mediated osteopenia (IMO)," *Bone Miner.*, 7, 149, 1989.
290. Orbai, P., Gozariu, L., and Saramet-Comsa, T., "Insulin secretion in magnesium silicate-induced osteopenia in rats," *Endocrinologie*, 29, 43, 1991.
291. Wuster, C., Ding, G., Minne, H. W., and Ziegler, R., "Effects of endogenous and exogenous calcitonin on inflammation-mediated osteopenia in the rat," *Horm. Res.*, 37, 119, 1992.
292. Hadjidakis, D., Lempert, U. G., Minne, H. W., and Ziegler, R., "Bone loss in experimental diabetes. Comparison with the model of inflammation mediated osteopenia," *Horm. Metabol. Res.*, 25, 77, 1993.
293. Gardner, D. L., "Production of arthritis in the rabbit by the local injection of the mucopolysaccharide carageenin," *Ann. Rheum. Dis.*, 19, 369, 1960.
294. Carmichael, D. J., Gillard, G. C., Lowther, D. A., et al., "Carrageenin-induced arthritis. IV. Rate changes in cartilage matrix proteoglycan synthesis," *Arthritis Rheum.*, 20, 834, 1977.
295. Gillard, G. C. and Lowther, D. A., "Carrageenin-induced arthritis. II. Effect of intraarticular injection of carrageenin on the synthesis of proteoglycan in articular cartilage," *Arthritis Rheum.*, 19, 918, 1976.
296. Lowther, D. A. and Gillard, G. C., "Carrageenin-induced arthritis. 1. The effect of intraarticular carrageenin on the chemical composition of articular cartilage," *Arthritis Rheum.*, 19, 769, 1976.
297. Santer, V., Sriratana, A., and Lowther, D. A., "Carrageenin-Induced arthritis: V. A morphologic study of the development of inflammation in acute arthritis," *Sem. Arthritis Rheum.*, 13, 160, 1983.
298. Bogoch, E., Gschwend, N., Bogoch, B., et al., "Juxta-articular bone loss in experimental inflammatory arthritis," *J. Orthop. Res.*, 6, 648, 1988.
299. Bogoch, E., Gschwend, N., Bogoch, B., et al., "Metaphyseal and diaphyseal changes in the femur proximal to experimental inflammatory arthritis of the rabbit knee," *Arthritis Rheum.*, 32, 617, 1989.
300. Søballe, K., Pedersen, C. M., Odgaard, A., et al., "Physical bone changes in carrageenin-induced arthritis evaluated by quantitative computed tomography," *Skel. Radiol.*, 20, 345, 1991.
301. Bunger, C., Bunger, E. H., Harving, S., et al., "Growth disturbances in experimental juvenile arthritis of the dog knee," *Clin. Rheum.*, 3, 181, 1984.
302. Lucas, S., Bogoch, E. R., Nespeca, R., and Grynpas, M. D., "Bone changes induced in a rabbit model of experimental arthritis," *Eur. J. Exp. Musc. Res.*, 1, 121, 1993.
303. Bogoch, E. R., Moran, E., Crowe, S., et al., "Arthritis not immobilization causes bone loss in the carrageenan injection model of inflammatory arthritis," *J. Orthop. Res.*, 13, 777, 1995.

304. Bellingham, C. M., Lee, J. M., Moran, E. L,, and Bogoch, E. R., "Torsional properties of the rabbit femoral diaphysis: bisphosphonate prevents arthritis-induced loss of fracture toughness," *J. Orthop. Res.*, 13, 876, 1995.
305. Kang, Q., An, Y. H., Butehorn, H. F., III, and Friedman, R. J., "Morpholological and mechanical study on the effects of experimentally induced inflammatory knee arthritis in rabbit long bones," *J. Mater. Sci. Mat. Med.*, 9, 463, 1998.
306. Moran, E. L. B., Lee, J. M., Reicheld, S., and Bogoch, E. R., "Material and geometric properties of bone affected by experimental inflammatory arthritis of the rabbit knee," *Orthop. Trans.*, 17, 364, 1992.
307. Pysklywec, M. W., Bogoch, E. R., Moran, E. L., and Fornasier, V. L., "Altered pore characteristics and geometry in diaphyseal cortical bone affected by inflammatory arthritis," *J. Orthop. Rheum.*, 9, 150, 1996.
308. Hajcsar, E. E., Roberts, E. M., Moran, E. L., et al., "Osteonal parameters in experimental inflammatory arthritis and treatment with zoledronate in rabbits," *Proc. Ann. Meet. Canad. Orthop. Assoc.*, 1997.
309. Takashima, T., Kawai, K., Hirohata, K., et al., "Inflammatory cell changes in Haversian canals," *J. Bone Joint Surg.*, 71B, 671, 1989.
310. Bogoch, E. R., Roberts, E., Moran, E. L., and Fornasier, V., "Pamidronate (APD) prevents rapid bone remodeling and trabecular bone loss in experimental inflammatory arthritis," *Trans. Orthop. Res. Soc.*, 226, 1995.
311. Pysklywec, M. W., Moran, E. L., and Bogoch, E. R., "Zoledronate (CGP 42'446), a biophosphonate protects against metaphyseal intracortical defects in experimental inflammatory arthritis," *J. Orthop. Res.*, 15, 858, 1997.

Part IV

Animal Models of Articular Cartilage and Joint Conditions

16 Animal Models of Articular Cartilage Defect

Yuehuei H. An and Richard J. Friedman

CONTENTS

   I. Introduction 309
  II. Normal Articular Cartilage and Repair Process 310
      A. Normal Articular Cartilage...310
      B. Repair Process of Cartilage Defect..311
 III. Animal Models for Cartilage Repair 311
      A. Heterotopic Models of Chondrogenesis 311
      B. Cartilage Defect Models 312
 IV. Authors' Preferred Animal Models..312
      A. Heterotopic Models (Nude Mice Model) 312
      B. Rabbit Distal Femoral Joint Defect 312
      C. The Second Choice 314
  V. Evaluation of Cartilage Defect Repair 314
      A. Macro Findings at Necropsy..314
      B. Histology and Histomorphometry..315
      C. Mechanical testing..317
      D. Other Methods..317
 VI. Grafts Investigated for Cartilage Repair...318
      A. Autograft, Allograft, and Xenograft 318
      B. Biomaterials..318
      C. Stimulus for Chondrogenesis 318
      D. Chondrocytes or Chondrogenic Cell Grafting 319
      E. Tissue Engineering Technique 319
References 319

I. INTRODUCTION

The regeneration potential of damaged articular cartilage is extremely limited. Although many methods have been investigated, none of them have given satisfactory long-term clinical results. Over the last few decades, artificial joint replacement has developed very rapidly and many arthritic conditions have been successfully treated. However, total joint arthroplasty does not last a lifetime and, therefore, is contraindicated for most children and young active adults. A need still exists for a method that biologically regenerates full thickness defects in articular cartilage defects. With the development of tissue engineering technique, there seems to be a bright future for biological repair of cartilage injuries. Without adequate *in vitro* alternatives so far, animal models of cartilage defect are still playing a very important role at this frontier.

TABLE 1
The Total Thickness of Articular Cartilage of Femoral Condyles of Common Subjects

Species	Age	Body weight (kg)	Cartilage thickness (mm)	First author, year[Ref.]
Human	Adult	—	2.2–2.4	Martino 1993[3]
Cow	1 year	273	3.17	Simon 1970[4]
Sheep	2–5 years	68	1.68	Simon 1970[4]
Dog	2–3 years	25	1.3	Simon 1970[4]
Rabbit	12–14 months	4.5±0.25	0.25–0.75	Athanasiou 1995,[5] An 1998[6]
Rat	12 months	300 (gm)	170 μm	Simon 1970[4]
Mouse	12 months	22 (gm)	30 μm	Simon 1970[4]

II. NORMAL ARTICULAR CARTILAGE AND REPAIR PROCESS

A. Normal Articular Cartilage

Morphologically, articular cartilage is composed of four layers.[1] (1) The superficial layer, about 10% of the cartilage thickness, is composed of mainly thin collagen fibers oriented along the outer surface of the joint. The lower part of the layer contains flattened chondrocytes aligned parallel to the surface. The load-bearing ability of the cartilage depends largely upon the integrity of this layer. (2) The intermediate layer is composed of vertically oriented collagen fibers and small spherical chondrocytes. The columnar orientation is less marked than in the deep layer. (3) The third or deep layer composed of large collagen fibers and larger chondrocytes. The collagen fibers are oriented mainly perpendicularly to the joint surface. The large chondrocytes form chondrons, which are composed of several chondrocytes surrounded by matrix and an outer layer of collagen/proteoglycan capsule.[2] The chondrons lie in columns oriented in vertical direction. (4) The fourth or deepest layer, about 10% of the cartilage thickness, is the calcified cartilage, which joins together the cartilage and subchondral bone and provides growth and remodeling of underlying bone tissue. Collagen fibers are arranged in vertical bundles with bridging fibrils in between. There is a basophilic line called "tidemark," aggregates of mineral associated with matrix vesicles, separating the uncalcified cartilage from the calcified layer. The total thickness of articular cartilage of femoral condyles of human and common animal subjects are listed in Table 1.

Biochemically, articular cartilage is composed of hydrated extracellular matrix (ECM) (collagen, proteoglycans [PGs], and noncollagenous proteins) in which chondrocytes are embedded.[7,8] Chondrocytes occupy only less than 1% to 10% of the tissue volume. A particular feature of these chondrocytes is that they lack contact between cells; thus, communication between cells are through the ECM. There are no blood and lymphatic vessels for the delivery of nutrients to and the removal of wastes from the cells, so these functions are performed via diffusion through the ECM. There are no nerve fibers in cartilage. Chondrocytes live in an anoxic environment and seem to carry out their metabolism mostly through anaerobic pathways. The collagen, 90% type II collagen and 10% type IX, X, XI, and VI collagen, represents more than 50% of the organic dry weight or 10–20% of the wet weight. The network of type II collagen fibrils provides the tensile strength of cartilage and is essential for maintaining tissue volume and shape. PGs, including aggregan, biglycan, and decorin, are a diverse group of heterogeneous macromolecules. Each type of PG is composed of a protein core and a number of glycosaminoglycan (GAG) chains. The most important GAGs are chondroitin sulfate, keratan sulfate, dermatan sulfate, and heparan sulfate. As many as 100 aggregan and link proteins can bind to a strand of hyaluronan to form a large aggregate. PGs function as a space filler, provide swelling pressure by attracting water, and assist the mechanical properties of the collagen network. Articular cartilage also contains a number of extracellular noncollageneous matrix proteins, such as chondrocalcin, anchorin, fibronectin, or thrombospondin. These proteins may be involved in the process of cartilage calcification, interaction between chondrocytes, tissue repair and remodeling.

Biomechanically, cartilage is considered to be a multilayered, heterogeneous, anisotropic, physically non-linear, viscoelastic, 2-phase porous structure. The average equilibrium compression modulus for the lateral condyle and femoral groove cartilages of normal human, canine, monkey, and rabbit and bovine meniscus are 0.41–0.89 MPa.[9,10] The average equilibrium tensile modulus by tensile test of human, bovine and canine cartilage is 1.0–15 MPa.[9,11] The aggregate modulus by indentation test is 0.81–1.82 MPa for cartilage of human femoral head[12] and 0.59±0.18 MPa for rabbit femoral condyles.[5] The average equilibrium shear modulus by torsional test of human patellar cartilage is 0.23 MPa.[13] The hydraulic permeability of cartilage is 0.63±0.28 × 10–15 m^4/N.sec (rabbit femoral condyle) by confined compression test[10] and 0.71–1.10 m^4/N.sec (human femoral head).[12]

B. Repair Process of Cartilage Defect

Unlike the repair process of other tissues, articular cartilage does not heal satisfactorily by itself although some studies have demonstrated close to complete repair with hyaline cartilage in immature animals or in small defect.[14,15] DePalma et al.[16] reported that partial-thickness defects show no significant repair up to more than one year, but that full-thickness wounds were completely filled with immature cartilage in four months. Campbell found that injuries to hyaline cartilage do not heal with normal hyaline cartilage, but mainly with fibrous tissues or fibrocartilage.[17]

Silver and Glasgold[8] pointed out that three factors are important in determining whether cartilage repair occurs: (1) the depth of the defect, (2) the maturity of the cartilage (better results in young subjects), and (3) the position of the defect on the surface. It is obvious that the size of the defect is also a determining factor of cartilage repair.[15,18] Significant better repair of full-thickness defects than partial-thickness defects indicated that the repair process appeared to be mediated by the proliferation and differentiation of mesenchymal cells of the marrow, and not by chondrocytes of the defect wall.[16,19] The repair of full thickness defects follows the following sequence: fibrin; granulation tissue; connective tissue; cartilage cells in connective tissue; fibrocartilage; and hyaline cartilage.[16,20] The ultrastructural changes of cells and matrix in the defect during the repair process have been reported by Ghadially et al.[21] In conclusion, cartilage only has an incomplete capacity for self-repair.

III. ANIMAL MODELS FOR CARTILAGE REPAIR

A. Heterotopic Models of Chondrogenesis

One of the major heterotopic models for *in vivo* chondrogenesis to test potential substances includes the subcutaneous, intramuscular, and intraperitoneal model (Table 2). After promising *in vitro* studies such as cell culture or cell seeding, subcutaneous implantation is often the initial step of *in vivo* studies. The animals used for heterotopic models include nude mice, syngenic mice, rats, and rabbits. Substances or constructs having potential effect of chondrogenesis were implanted or injected at the above-mentioned sites. After 2–4 months the implants were explanted and studied histologically to identify new cartilage tissues. Although Green[35] found similar cartilage formation at both of the intramuscular and subcutaneous model, the effects of inplantation sites and local environment are still not clear. Older rabbits (3.2–4.0 kg) were grafted with chondrocytes from younger rabbits (2.2–3.2 kg) into the anterior tibial compartment.[35] The result showed that freshly isolated articular chondrocytes reformed cartilage tissue in 10 days.

Another heterotopic defect model of chondrogenesis is a diffusion chamber implanted intramuscularly in rat or nude mice (Table 1). The diffusion chamber is made from a plastic ring (2 mm thick, 9 mm diam.) bounded by two microporous cellulose acetate and nitrate membranes of

TABLE 2
Selected Heterotopic Models of Chondrogenesis

Animal	Procedure	Material tested	First author, year[Ref.]
Nude mice	SC*	Human periosteal cells–porous ceramic	Nakahara 1991[22]
		Chondrocytes of rabbits and dogs	Lipman 1993[23]
		Chondrocyte–PGA/PLA construct	Vacanti 1991[24]
		Chondrocyte–PGA/PLA construct	Freed 1993[25]
		Chondrocyte–PGA/PLA construct	Puelacher 1994[26]
		Chondrocyte–calcium alginate construct	Paige 1996[27]
		Chondrocyte–collagen composite	Fujisato 1996[28]
	IP†	Chicken periosteal cells in diffusion chamber	Nakahara 1990[29]
	IP	Human periosteal cells in diffusion chamber	Nakahara 1991[22]
Syngenic mice	IP	Rodent Achilles tendons in diffusion chamber	Rooney 1993[30]
Rats	IM‡	Syngenic rat chondrocytes	Moskalewski 1993[31]
	IM	Rabbit BMP in diffusion chamber	Ono 1994[32]
	SC	BMP carriers	Kuboki 1995[33]
Rabbits	SC	Collagen sponge plus perichondrium	Matsuda 1995[34]
	IM	Chondrocytes from younger rabbits	Green 1977[35]

* SC = Subcutaneous tissue
† IP = Intraperitoneally
‡ IM = Intramuscularly

100 μm thickness and 0.45 μm pore size and a chamber volume about 130 μl (Millipore Corporation, MA).[22,29,30,32] A diffusion chamber containing rabbit bone morphogenetic protein (BMP) was implanted in the abdominal muscle of the rat. Outside of the chamber, cartilage differentiated 1–2 weeks after implantation, and bone replaced the cartilage after 3–4 weeks.[32]

B. Cartilage Defect Models

Selected cartilage defect models from the literature are listed in Table 3. There are mainly two types of defects, focal full thickness defect and partial-thickness defect. The commonly used animals are rabbits and dogs which are the first choices for the studies on cartilage repair.

IV. AUTHORS' PREFERRED ANIMAL MODELS

A. Heterotopic Models (Nude Mice Model)

The nude mice model for testing *in vivo* chondrogenesis of potential substance or construct is the most popular heterotopic model for chondrogeneis.[27,28] Athymic nude mice with an average age of 5±1 weeks could be selected. A total of 2–4 sample disks (9 mm diam., 2 mm thickness) can be implanted into noncommunicating subcutaneous pockets on the back of the mouse. The substance can be mixed with collagen gel or other gelly carrier and injected into SC tissue. Care needs to be taken where the material is inserted or injected, which should be in the loose SC tissue and not underneath the back muscles. Histological evaluation is efficient to examine the existence of new cartilage tissues.

B. Rabbit Distal Femoral Joint Defect

The basic steps in the testing of the cartilage repairing effect of an implant or construct include the manufacturing of the implant, the creation of a cartilage defect, the placement and securing of

TABLE 3
Selected Animal Cartilage Defect Models

Animal	Bone	Types of defect	First author, year[Ref.]
Rabbit	Distal femur	Round drill hole (s)	Ghadially 1971,[21] Green 1977,[35] Speer 1979,[36] Furukawa 1980,[37] Salter 1980,[38] Shimizu 1987,[39] Heatley 1985,[40] Aston 1986,[41] O'Driscoll 1986,[42] Amiel 1988,[43] Wakitani 1989,[44] Lipiello 1990,[45] Dahlberg 1991,[46] von Schroeder 1991,[47] Hogervorst 1992,[48] Shapiro 1993,[19] Klompmaker 1992,[49] Messner 1993,[50] Messner 1994,[51] Freed 1994,[52] Kreder 1994,[53] Chu 1995,[54] Kandel 1995,[55] Sumen 1995,[56] Specchia 1996,[57] Wei 1997,[58] Frankel 1997,[59] Sellers 1997[60]
		Rectangular or oval	Kawabe 1991,[61] Wakitani 1994,[62] Brittberg 1997,[63] Chu 1997[64]
		Large full-thickness	Menche 1996[65]
		Scalpel, superficial	Mochizuki 1993[66]
		Small partial-thickness	Hunziker 1996[67]
	Proximal tibia	Round hole	Bentley 1978,[68] Kon 1981[69]
	Patella	Round hole	Upton 1981,[70] Grande 1989,[71] Muckle 1989,[72] Brittberg 1996,[73] Specchia 1996[57]
	Humerus	Wedged osteotomy	Chesterman 1968[74]
Dog	Distal femur	Round or oval hole	Calandruccio 1962,[14] Klompmaker 1992,[49] Hale 1993,[75] Shortkroff 1996,[76] Breinan 1997[77]
	Distal femur	Condyle osteotomy	Campbell 1963[78]
	Distal femur	Superficial	Calandruccio 1962[14]
	Patella	Chondral excision	Engkvist 1979[79]
	Distal radius	Osteotomy	Campbell 1963,[78] Steveson 1989[80]
Rat	Distal femur	1.0–1.5 mm hole	Noguchi 1994,[81] Grundnes 1995,[82] Chang 1996[83]
		1.2 × 5 mm defect	Göransson 1995[84]
Goat	Tibial plateau	Wedged defect	Jackson 1993[85]
	Fem. condyle, patella	3.0 mm holes	Shahgaldi 1991[86]
	Femoral condyle	4.0 mm round hole	Butnariu-Ephrat 1996[87]
Horse	Distal femur	Circular hole	Hendrickson 1994,[88] Sams 1995[89]
	Radial carpal bone	Circular hole	Vachon 1992,[90] Howard 1994[91]
		Rectangular	Todhunter 1993[92]
	3rd carpal bone	Partial-thickness defect	Shamis 1989[93]
Primate	Mandibular condyle	Round hole	Robinson 1993[94]
	Femoral condyle	1.5 mm holes	Girdller 1993[95]
Chick	Tibiotarsal joint	3–4 mm round hole	Itay 1987,[96] Robinson 1990[97]
Sheep	Femoral condyle	5 × 10 mm defect	Homminga 1991[98]

the implant, specimen harvesting, and evaluation. The authors prefer the rabbit knee model with a defect on the distal femoral joint surface since (1) it has been widely used and well studied, (2) has a good size for easier surgical procedures and specimen handling, (3) is consistently reproducible, and (4) is relatively economical. By comparison, if there are no specific reasons, rats are perhaps too small for manipulation and specimen handling and dogs are more difficult to be justified ethically and economically.

Theoretically, there would be four major cartilage defect models including localized or extensive full-thickness defect with a depth beyond the subchondral bone plate, and localized or extensive partial-thickness defect within the cartilage layer. The localized models are suitable for investigating

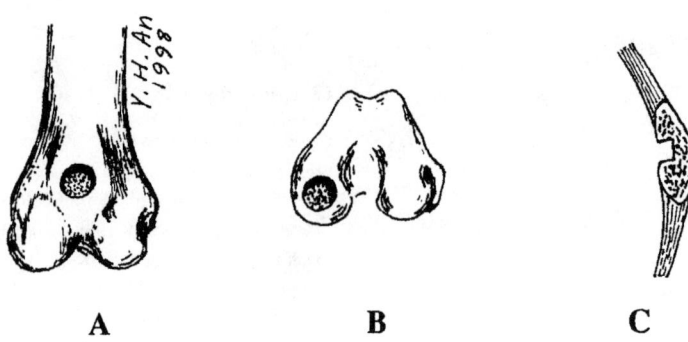

FIGURE 1. Illustration of commonly used articular defects in the rabbit: (A) intercondylar groove (partial weight bearing), (B) femoral condyle (weight bearing), and (C) patellar articular defect.

the effect of an implant or construct and the extensive ones are often used for examining other methods such as abrasion burr arthroplasty, enzymatic treatment,[66] or joint surface drilling. The rabbit knee models can be classified into weight-bearing and "partial-weight bearing" models. The former include the models with the defect created on the distal femoral joint surface, especially those located inferioposteriorly (rabbit knees stay in a flexed position most of the time). The latter are those with the defects created in the intercondylar groove.

Typically, the rabbit is put in a supine position and the knee joint is opened through a lateral parapatellar incision. The patella is dislocated medially to expose the articular cartilage of the patellar groove and the femoral condyles. A 3.5 mm diameter and 1.5–3.0 mm deep defect is created using a drill in either the intercondylar groove or the central portion of the lateral or medial femoral condyles (Figure 1). At least a 1.25 mm depth is needed if a full thickness defect is to be made since the total thickness of the cartilage and the subchondral bone plate is 1.0 mm in average for femoral condyle joint surfaces. It is a good idea that the deeper depth (3.5–5.0 mm) is created to facilitate the anchoring of a cylindrical implant (Figure 2A).[52] Implants or constructs can also be secured with sutures (Figure 2B). If gel materials, such as collagen gel containing chondrocytes, are to be used, the gel can be maintained in place with a piece of periosteum sutured to the defect rim (Figure 2C).

C. The Second Choice

If there are reasons of inappropriateness of using the rabbit model, a dog distal femur defect model may be selected. The dog knee joint is much larger than that of rabbits, allowing a larger defect. Also, because of the larger volume of cartilage it is easier to harvest enough cartilage tissue for chondrocytes isolation used in tissue engineering approach for autogenic cell seeding. In our laboratory, we experienced the difficulty of culturing chondrocytes based on a small piece of cartilage tissue taken from a 3.5 mm diam. cartilage defect in a rabbit model (at least 50% failure).

V. EVALUATION OF CARTILAGE DEFECT REPAIR

A. Macro Findings at Necropsy

The surface morphology of the repair tissue and adjacent cartilage should be recorded descriptively. Any limitation of joint movement, joint swelling, or degenerative changes (erosion of cartilage surface or formation of osteophytes) should be also documented. For most articles, macro photographs of the

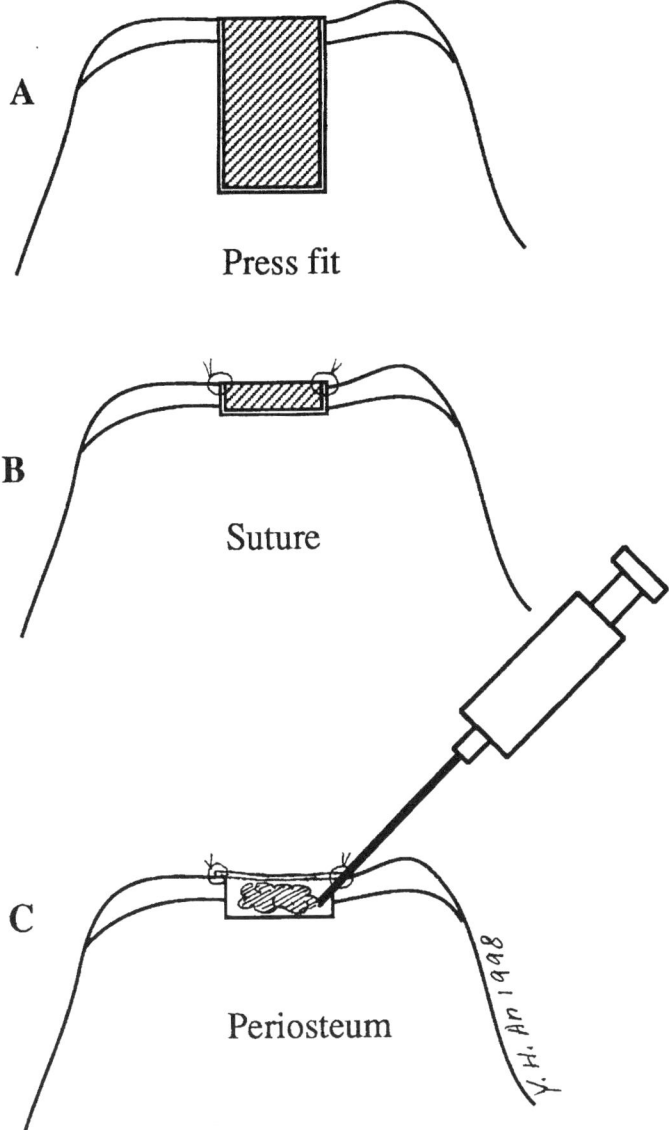

FIGURE 2. Illustration of three commonly used implant anchoring methods for repairing cartilage defects: (A) press fit, (B) suturing, and (C) periosteum (for gel materials).

specimen are basic for the results section, which give a general image of the quality of the repair. Photography of the repaired cartilage surface taken under dissecting microscope is also a useful documentation of the surface morphology. Although descriptive macro findings are important, no grading system has been reported.

B. Histology and Histomorphometry

Histology and histomorphometry may be the most powerful methods for examining the quality and quantity of the repair. Traditional histology is descriptive and semiquantitative. Two original grading systems can be found in the literature, the system described by Pineda et al.[99] and the one by O'Driscoll

TABLE 4
Modified Histological Grading System from the Original Systems in the Literature*

Category	Scores
Nature of the repair tissue	
Hyaline cartilage	4
Mostly hyaline cartilage	3
Mostly fibrocartilage	2
Mostly non-cartilage	1
Non-cartilage only	0
Matrix staining (Safranin-O)	
Normal staining	3
Moderate	2
Slight	1
None	0
Structural integrity	
Normal structure	2
Slight disruption	1
Severe disruption	0
Surface regularity	
Smooth and intact	2
Slight disruption	1
Severe disruption	0
Filling of the defect	
100%	2
>50%<100%, or >100%	1
<50%	0
Bonding to host tissue	
Bonded	2
Partially bonded	1
Not bonded	0
Degenerative changes of the repair tissue	
Normal cellularity and cell morphology	3
Mild hypocellularity and cell cluster	2
Moderate hypocellularity and cell degeneration	1
Severe hypocellularity and cell degeneration	0
Degenerative changes of the adjacent cartilage	
Normal cellularity, cell morphology, and matrix staining	3
Mild hypocellularity and cell clustering, moderate matrix staining	2
Moderate hypocellularity and cell degeneration, decreased matrix staining	1
Severe hypocellularity and cell degeneration, poor or no matrix staining	0
Maximum Possible Score (The most abnormal condition)	21

* Sources: O'Driscoll et al.,[48,100] Pineda et al.,[99] and Freed et al.[52]

et al.[42,100] They have been used by other investigators with modifications.[52,62] The two systems have common categories and each has its own advantages. A combined system may be more complete, which is proposed here for investigators who may be interested (Table 4). The mean and standard deviation are calculated for the individual categories and the total score for each of the graded specimens. The Fisher exact test, chi-square test, or Kruskal-Wallis test (a one-way non-parametric analysis of variance) could be used for analyzing the differences between the scores of different groups.[42,101]

In recent years, computerized image analysis makes histomorphometry more efficient, especially for the calculation of the percentage of filling by the repair tissues in the defect and the fractions of different tissues in it. Normally, the parameters for quantification include the percentage of total area of the defect that become filled with repair tissue, different types of tissues (hyaline cartilage, fibrocartilage, fibrous tissue), the integration of repair tissue with adjacent cartilage and the calcified cartilage at the base of the defect.[77] Yoshioka et al.[102] defined three parameters for quantifying cartilage defect repair: repair height, percent repair, and root mean square roughness. The later is a quantitative measure of the deviation of repair site surface from an idealized surface. See Chapter 6 for more information on cartilage histomorphometry.

C. MECHANICAL TESTING

A common method for mechanical testing of cartilage tissue is the confined compression test.[64,103] A cylindrical sample is tested by being placed in a PBS bath within a confined compression apparatus with the articular surface against a stainless steel filter and the subchondral surface supported on a rigid flat disc. The mechanical property of the cartilage is described by equilibrium confined compression modulus, which is about 0.41–0.89 MPa.[9,10]

Another important test is the indentation test which has been widely used to examine the mechanical properties of the repaired tissue.[62,104,105] Briefly, a needle-penetration technique is used to measure the thickness of the cartilage at the area to be tested for indentation depth. The force on the needle and its displacement are measured simultaneously so the thickness of the cartilage can be documented. For the indentation test, a 1.5-mm porous indenter is used to first apply a 2 gm preload to ensure uniform contact between the indenter and the cartilage surface. Then, an additional 2 gm load is applied to record the displacement of the indenter. The mechanical parameter to be measured is the compliance value, which is calculated using the following equation: Compliance = 1/stiffness = k × indentation depth/cartilage thickness (k is an instrument constant that was the same for all tests). Normal cartilage has small values (usually single digits), implying low compliance or high stiffness, while softer or unnormal tissues have a larger number (usually two digits), indicating high compliance or low stiffness.

Elastic modulus or compliance values should be measured for the repair tissue, the adjacent cartilage, and the same areas in the contralateral knee. The means of the values of the repair tissue or the adjacent cartilage of different groups can be compared using Student's t-test or ANOVA test.

D. OTHER METHODS

SEM (scanning electron microscopy) is an important method for examining morphology and surface structure on joint surface and cross-sectional surfaces of cartilage. The shortcoming of SEM is that the specimen has to be dried before observation, which deforms the real spatial structure and morphology.[107] This problem seems to have been solved by low-temperature or cryo-SEM systems.[107,108] TEM is the best method for examining ultrastructure and components of cartilage including matrix, chondrocytes, organelles, membranes, and large molecules (such as proteoglycan or collagen fibrils).[109–111] See Chapter 6 for more information on TEM techniques.

The relative amounts of type I and II collagen comprising the repair tissue can be determined by gel filtration high performance liquid chromatography (HPLC).[64,112] Total GAG content in the repair tissue is measured with a hexosamine method.[113] DNA content and synthesis can be measured with a fluorometric day assay[114] and ^{3}H-thymidine method.[115] See Chapter 6 for more biochemical assays for cartilage tissue.

VI. GRAFTS INVESTIGATED FOR CARTILAGE REPAIR

A. Autograft, Allograft, and Xenograft

Although there were promising results from animal studies,[14,78] the use of autogenous tissue grafts to repair localized cartilage defects has been associated with the difficulty of requiring another surgical procedure and the risk of damaging the donor site. Obviously, there is a very limited amount of autologous cartilage available for grafting in humans. Also, the overall preservation of cartilage before transplantation seems to have an adverse effect on its mechanical and morphological properties, and on a long term basis has proven to be less than satisfactory.[78,116,117]

The major problem associated with allogenic cartilage grafts is the formation of loose vascular fibrous tissue surrounding the grafts.[41,68,74] Osteochondral allografts are easy to anchor into the recipient site, but tend to cause a severe immunologic rejection from the host.[85,118] For some post-traumatic joint injuries or limb salvage procedures in tumor cases, allografts still play an important role.[119,121]

Periosteal grafts,[42,122] perichondrial grafts,[70,79] bone callus, and cortical bone[84] have been reported as alternative methods to repair cartilage defects by introducing chondrogenic or progenitor cells to the site of repair. Cartilage components such as hyaline cartilage or fibrocartilage have been generated in the defects of experimental animals and human patients.[123] Long-term results, however, have not been reported.

One major advantage of the xenograft in the studies of cartilage repair is that a large amount of cartilage tissue could be obtained inexpensively from a slaughterhouse. Limited number of reports on animal models of repairing cartilage defects using xenograft materials have been found in the literature with controversial results.[86,98]

B. Biomaterials

Various synthetic materials have been investigated as replacement materials for a cartilage defect. Elastomeric materials such as silicone rubber have replaced articular cartilage in prosthetic finger joints, but have limited fatigue resistance which limits their use in weight-bearing conditions.[124] Other elastomers such as hydrophilic polyurethanes,[49,124] polytetrafluoroethylene (PTFE), and polyester,[50] have been tested as cartilage replacements to provide increased compliance and greater mechanical compatibility. However, these materials did not display sufficient fatigue resistance for long-term use, and also induce degeneration of the bearing articular cartilage. Poly (HEMA) hydrogel,[69] PLA matrix,[47] TCP-collagen composite,[48] and carbon fiber pads[72] have also been tried. Although there have been many attempts on the development of artificial cartilage, currently there is no acceptable synthetic material for use in this type of application, i.e., a high strain to failure elastomeric biomaterial with high fatigue resistance, low friction, low wear characteristics, and more significantly, acceptable long-term results.

C. Stimulus for Chondrogenesis

Growth factors (GFs) play a very important role in the process of chondrogenesis. The major GFs with chondrogenic ability include TGF-β, FGF, EGF, IGF, and BMPs.[125,126] They stimulate DNA synthesis, chondrocyte proliferation and differentiation, and matrix production.[127,128] Several carriers or substrates have been reported to deliver GFs for *in vivo* chondrogenesis studies such as DBM,[129] and collagen.[28]

Continuous passive motion (CPM) has been used alone or in combination with osteoperiosteal grafts and demonstrated its positive effects on the formation of chondral tissue and the boundary between graft and defect.[38,42] Kim et al. found that full-thickness defects created by subchondral abrasion can heal by regeneration of hyaline-like cartilage and such healing is enhanced by CPM for two weeks postoperatively.[130]

Also, some evidence indicates that chondrocytes and progenitor cells respond to electrical or electrical field stimulation.[131,132] A pulsing direct current stimulation was also studied for *in vivo* chondrogenesis and the results showed an enhanced quality of repair.[45]

D. CHONDROCYTES OR CHONDROGENIC CELL GRAFTING

Chondrocytes or chondrogenic cells (or mesenchymal progenitor cells) have been isolated or culture-expanded from articular cartilage[44,71,74] growth plates,[61,68,96] perichondrium,[64] bone marrow,[62] or periosteum.[62] These cells were then transplanted into articular defects, secured by a piece of periosteum,[133] or embedded in collagen gel. These attempts have had mixed results. However, the earlier work by Grande et al.[71] in 1989 had already shed some light on the development of tissue engineering technique during the later years.

E. TISSUE ENGINEERING TECHNIQUE

Cell seeding to a substrate or scaffold for making a composite graft is not a new concept,[34,134–136] but it has not been well developed until the early 1990s, especially in the field of bone and cartilage repair. Green was the first to use this concept for repairing articular cartilage defects.[13] In the last five years, several groups have reported very important data on the use of cell-seeded implants for repairing cartilage defects.[24,25,52,59,64,88,137] All these studies have demonstrated that the cell-seeded composite implants induce cartilage formation leading to the final defect repair. This technique has been referred as "tissue engineering" by Langer and Vacanti.[138,139] It has certain advantages over the previous methods, such as autogeneic characteristics if the cells are taken from the same individual, minor morbidity of the donor site, and the unique properties produced in large quantities *in vitro*. Also, there is no risk of disease transmission.

Different scaffolds for cell-seeding have been used in cartilage defect repair, including porous collagen,[59] PGA mat,[25] porous PLA,[25,64] or PLA-PGA copolymer mesh.[24] Among them, porous collagen sponge has been used widely as a substrate or scaffold for cultured cell ingrowth to make implantable composite grafts, which is not only for repairing cartilage,[24,59,88,137] but also for other purposes such as bone defect repair, skin grafting, or esophageal grafting.

In summary, an ideal chondral graft substance should have the following properties, which most of the current methods lack: (1) the chondroconductive ability for the ingrowth of new chondral tissue; (2) the presence of differentiated autogenic chondrogenic cells for immediate chondrogenesis; (3) the presence of pre-existing GFs for chondroinduction, which convert mesenchymal or undifferentiated cells into cartilage-forming cells; (4) no donor site morbidity and no risk of disease transmission; and (5) the ability to be produced in large quantities *in vitro* and easily fabricated into any desired shape. Using tissue engineering techniques, a composite graft made *in vitro* according to the above specifications could be an optimal substance to heal a cartilage defect *in vivo*.

REFERENCES

1. Weiss, C., Rosenberg, L., and Helfet, A. J., "An ultrastructural study of normal young adult human articular cartilage," *J. Bone Joint Surg.*, 50A, 663, 1968.
2. Poole, C. A., Flint, M. H., and Beaumont, B. W., "Chondrons in cartilage: ultrastructural analysis of the pericellular microenvironment in adult human articular cartilages," *J. Orthop. Res.*, 5, 509, 1987.
3. Martino, F., Ettorre, G. C., Patella, V., Macarini, L., Moretti, B., Pesce, V., and Resta, L., "Articular cartilage echography as a criterion of the evolution of osteoarthritis of the knee," *Int. J. Clin. Pharmacol. Res.*, 13 (Suppl 35, 1993.
4. Simon, W. H., "Scale effects in animal joints. I. Articular thickness and compressive stress," *Arthritis Rheum.*, 13, 244, 1970.

5. Athanasiou, K. A., Fischer, R., Niederauer, G. G., and Puhl, W., "Effects of excimer laser on healing of articular cartilage in rabbits," *J. Orthop. Res.,* 13, 483, 1995.
6. An, Y. H., Friedman, R. J., Draughn, R. A., Jiang, M., LaBreck, J. C., et al., "Bone ingrowth to implant surfaces in an inflammatory arthritis model," *J. Orthop. Res.,* 16, 576, 1998.
7. Kuettner, K. E., "Biochemistry of articular cartilage in health and disease," *Clin. Biochem.,* 25, 155, 1992.
8. Silver, F. H. and Glasgold, A. I., "Cartilage wound healing," *Otolaryngol. Clin. North Am.,* 5, 847, 1995.
9. Mow, V. C. and Ratcliffe, A., "Structure and function of articular cartilage and meniscus," in *Basic Orthopaedic Biomechanics,* 2nd ed., Mow, V. C. and Hayes, W. C., Eds., Lippincott-Raven, Philadelphia, 1997, Chapter 4.
10. Sah, R. L., Yang, A. S., Chen, A. C., Hant. J. J., Halili, R. B., et al., "Physical properties of rabbit articular cartilage after transection of the anterior cruciate ligament," *J. Orthop. Res.,* 15, 197, 1997.
11. Akizuki, S., Mow, V. C., Muller, F., Pita, J. C., Howell, D. S., and Manicourt, D. H., "Tensile properties of human knee joint cartilage: I. Influence of ionic conditions, weight bearing, and fibrillation on the tensile modulus," *J. Orthop. Res.,* 4, 379, 1986.
12. Athanasiou, K. A., Agarwal, A., and Dzida, F. J., "Comparative study of the intrinsic mechanical properties of the human acetabular and femoral head cartilage," *J. Orthop. Res.,* 12, 340, 1994.
13. Zhu, W. B., Mow, V. C., Koob, T. J., and Eyre, D. R., "Viscoelastic shear properties of articular cartilage and the effects of glycosidase treatments," *J. Orthop. Res.,* 11, 771, 1993.
14. Calandruccio, R. A. and Gilmer, W. S., Jr., "Proliferation, regeneration, and repair of articular cartilage of immature animals," *J. Bone Joint Surg.,* 44A, 431, 1962.
15. Convery, F. R., Akeson, W. H., and Keown, G. H., "The repair of large osteochondral defects. An experimental study in horses," *Clin. Orthop.,* 82, 253, 1972.
16. DePalma, A. F., McKeever, C. D., and Subin, D. K., "Process of repair of articular cartilage demonstrated by histology and autoradiography with tritiated thymidine," *Clin. Orthop.,* 48, 229, 1966.
17. Campbell, D. J., "The healing of cartilage defect," *Clin. Orthop.,* 64, 45, 1969.
18. Vachon, A., Bramlage, L. R., Gabel, A. A., and Weisbrode, S., "Evaluation of the repair process of cartilage defects of the equine third carpal bone with and without subchondral bone perforation," *Am. J. Vet. Res.,* 47, 2637, 1986.
19. Shapiro, F., Koide, S. and Glimcher, M. J., "Cell origin and differentiation in the repair of full-thickness defects of articular cartilage," *J. Bone Joint Surg.,* 75A, 532, 1993.
20. Shands, A. R., Jr., "The regeneration of hyaline cartilage in joints: an experimental study," *Arch. Surg.,* 22, 137, 1931.
21. Ghadially, F. N., Fuller, J. A., and Kirkaldy-Willis, W. H., "Ultrastructure of full-thickness defects in articular cartilage," *Arch. Pathol.,* 92, 356, 1971.
22. Nakahara, H., Bruder, S. P., Haynesworth, S. E., et al., "Bone and cartilage formation in diffusion chambers by subcultured cells derived from the periosteum," *Bone,* 11, 181, 1990.
23. Lipman, J. M., McDevitt, C. A., and Sokoloff, L., "Xenografts of articular chondrocytes in the nude mouse," *Calcif. Tissue Int.,* 35, 767, 1983.
24. Vacanti, C. A., Langer, R., Schloo, B., and Vacanti. J. P., "Synthetic polymers seeded with chondrocytes provide a template for new cartilage formation," *Plast. Reconstr. Surg.,* 88, 753, 1991.
25. Freed, L. E., Marquis, J. C., Nohria, A., Emmanural, J., Mikos, A. G., and Langer, R., "Neocartilage formation *in vitro* and *in vivo* using cells cultured on synthetic biodegradable polymers," *J. Biomed. Mater. Res.,* 27, 11, 1993.
26. Puelacher, W. C., Mooney, D., Langer, R., Upton, J., Vacanti, J. P., and Vacanti, C. A., "Design of nasoseptal cartilage replacements synthesized from biodegradable polymers and chondrocytes," *Biomaterials,* 15, 774, 1994.
27. Paige, K. T., Cima, L. G., Yaremchuk, M. J., Schloo, B. L., Vacanti, J. P., and Vacanti, C. A. "De novo cartilage generation using calcium alginate-chondrocyte constructs," *Plast. Reconstr. Surg.,* 97, 168, 1996.
28. Fujisato, T., Sajiki, T., Liu, Q., and Ikada, Y., "Effect of basic fibroblast growth factor on cartilage regeneration in chondrocyte-seeded collagen sponge scaffold," *Biomaterials,* 17, 155, 1996.
29. Nakahara, H., Goldberg, V. M., and Caplan, A. I., "Culture-expanded human periosteal-derived cells exhibit osteochondral potential *in vivo,*" *J. Orthop. Res.,* 9, 465, 1991.

30. Rooney, P., Walker, D., Grant, M. E., and McClure, J., "Cartilage and bone formation in repairing Achilles tendons within diffusion chambers: evidence for tendon-cartilage and cartilage-bone conversion *in vivo*," *J. Pathol.*, 169, 375, 1993.
31. Moskalewski, S., Hyc, A., Grzela, T., and Malejczyk, J., "Differences in cartilage formed intramuscularly or in joint surface defects by syngeneic rat chondrocytes isolated from the articular-epiphyseal cartilage complex," *Cell Transplant.*, 2, 467, 1993.
32. Ono, Y., Kato, K., Oohira, A., Katoh, R., and Nogami, H., "Cell function during chondrogenesis and osteogenesis induced by bone morphogenetic protein enclosed in diffusion chamber," *Clin. Orthop.*, 298, 305, 1994.
33. Kuboki, Y., Saito, T., Murata, M., Takita, H., Mizuno, M., Inoue, M., Nagai, N., and Poole, A. R., "Two distinctive BMP-carriers induce zonal chondrogenesis and membranous ossification, respectively; geometrical factors of matrices for cell-differentiation," *Connect. Tissue Res.*, 32, 219, 1995.
34. Matsuda, K., Nagasawa, N., Suzuki, S., Isshiki, N., and Ikada, Y., "*In vivo* chondrogenesis in collagen sponge sandwiched by perichondrium," *J. Biomater. Sci. Polym. Ed.*, 7, 221, 1995
35. Green, W. T., Jr., "Articular cartilage repair. Behavior of rabbit chondrocytes during tissue culture and subsequent allografting," *Clin. Orthop.*, 124, 237, 1977.
36. Speer, D. P., Chvapil, M., Volz, R. G., and Holmes, M. D., "Enhancement of healing in osteochondral defects by collagen sponge implants," *Clin. Orthop.*, 144, 326, 1979.
37. Furukawa, T., Eyre, D. R., Koide, S., and Glimcher, M. J., "Biochemical studies on repair cartilage resurfacing experimental defects in the rabbit knee," *J. Bone Joint Surg.*, 62A, 79, 1980.
38. Salter, R. B., Simonds, D. F., Malcolm, B. W., Rumble, E. J., Macmicheal, D., and Clements, N. D., "The biological effect of continuous passive motion on the healing of full-thickness defects in articular cartilage," *J. Bone Joint Surg.*, 62A, 1232, 1980.
39. Shimizu, T., Videman, T., Shimazaki, K., and Mooney, V., "Experimental study on the repair of full thickness articular cartilage defects: effects of varying periods of continuous passive motion, cage activity, and immobilization," *J. Orthop. Res.*, 5, 187, 1987.
40. Heatley, F. W. and Revell, W. J., "The use of meniscal fibrocartilage as a surface arthroplasty to effect the repair of osteochondral defects: An experimental study," *Biomaterials*, 6, 161, 1985.
41. Compston, J. E. and Bentley, G., "Repair of articular surfaces by allografts of articular and growth-plate cartilage," *J. Bone Joint Surg.*, 68B, 29, 1986.
42. O'Driscoll, S. W. and Salter, R. B., "The repair of major osteochondral defects in joint surfaces by neochondrogenesis with autogenous osteoperiosteal grafts stimulated by continuous passive motion," *Clin. Orthop.*, 208, 131, 1986.
43. Amiel, D., Coutts, R. D., Harwood, F. L., Ishizue, K. K., and Kleiner, J. B., "The chondrogenesis of rib perichondrial grafts for repair of full thickness articular cartilage defects in a rabbit model: a one year postoperative assessment," *Connect. Tissue Res.*, 18, 27, 1988.
44. Wakitani, S., Kimura, T., Hirooka, A., Ochi, T., Yoneda, M., Yasui, N., Owaki, H., and Ono, K., "Repair of rabbit articular surfaces with allograft chondrocytes embedded in collagen gel," *J. Bone Joint Surg.*, 71B, 74, 1989.
45. Lippiello, L., Chakkalakal, D., and Connolly, J. F., "Pulsing direct current-induce repair of articular cartilage in rabbit osteochondral defects," *J. Orthop. Res.*, 8, 266, 1990.
46. Dahlberg, L. and Kreichergs, A., "Demineralized allogeneic matrix for cartilage repair," *J. Orthop. Res.*, 9, 11, 1991.
47. von Schroeder, H. P., Kwan, M., Amiel, D., and Coutts, R. D., "The use of polylactic acid matrix and periosteal grafts for the reconstruction of rabbit knee articular defects," *J. Biomed. Mater. Res.*, 25, 329, 1991.
48. Hogervorst, T., Meijer, D. W., and Klopper P. J., "The effect of a TCP-collagen implant on the healing of articular cartilage defects in the rabbit knee joint," *J. Appl. Biomater.*, 3, 251, 1992.
49. Klompmaker. J., Jansen, H. W. B., Veth, R. P. H., et al., "Porous polymer implants for repair of full-thickness defects of articular cartilage: an experimental study in rabbit and dog," *Biomaterials*, 13, 625, 1992.
50. Messner, K. and Gillquist J., "Synthetic implants for the repair of osteochondral defects of the medial femoral condyle: a biomechanical and histological evaluation in the rabbit knee," *Biomaterials*, 14, 513, 1993.

51. Messner, K., "Durability of artificial implants for repair of osteochondral defects of the medial femoral condyle in rabbits," *Biomaterials,* 15, 657, 1994.
52. Freed, L. E., Grande, D. A., Nohria, A., Emmanural, J., Mikos, A. G., and Langer, R., "Joint resurfacing using chondrocytes and synthetic biodegradable polymer scaffolds," *J. Biomed. Mater. Res.,* 28, 891, 1994.
53. Kreder, H. J., Moran, M., Keeley, F. W., and Salter, R. B., "Biologic resurfacing of a major joint defect with cryopreserved allogeneic periosteum under the influence of continuous passive motion in a rabbit model," *Clin. Orthop.,* 300, 288, 1994.
54. Chu, C. R., Coutts, R. D., Yoshioka, M., Harwood, F. L., Monosov, A. Z., and Amiel, D., "Articular cartilage repair using allogeneic perichondrocyte-seeded biodegradable porous polylactic acid (PLA): a tissue-engineering study," *J. Biomed. Mater. Res.,* 29, 1147, 1995.
55. Kandel, R. A., Chen, H., Clark, J., and Renlund, R., "Transplantation of cartilagenous tissue generated *in vitro* into articular joint defects," *Artif. Cells Blood Substit. Immobil. Biotechnol.,* 23, 565, 1995.
56. Sumen, Y., Ochi, M., and Ikuta, Y., "Treatment of articular defects with meniscal allografts in a rabbit knee model," *Arthroscopy,* 11, 185, 1995.
57. Specchia, N., Gigante, A., Falciglia, F., and Greco, F., "Fetal chondral homografts in the repair of articular cartilage defects," *Bull. Hosp. Joint Dis.,* 1996.
58. Wei, X., Gao, J., and Messner, K., "Maturation-dependent repair of untreated osteochondral defects in the rabbit knee joint," *J. Biomed. Mater. Res.,* 34, 63, 1997.
59. Frankel, S. R., Toolan, B., Menche, D., Pitman, M. I., and Pachence, J. M., "Chondrocyte transplantation using a collagen bilayer matrix for cartilage repair," *J. Bone Joint Surg.,* 79B, 831, 1997.
60. Sellers, R. S., Peluso, D., and Morris, E. A., "The effect of recombinant human bone morphogenetic protein-2 (rhBMP-2) on the healing of full-thickness defects of articular cartilage," *J. Bone Joint Surg.,* 79A, 1452, 1997.
61. Kawabe, N. and Yoshinao, M., "The repair of full-thickness articular cartilage defects. Immune response to reparative tissue formed by allogeneic growth plate chondrocyte implants," *Clin. Orthop.,* 268, 279, 1991.
62. Wakitani, S., Goto, T., Pineda, S. J., Young, R. G., Mansour, J. M., Caplan, A. I., and Goldberg, V. M., "Mesenchymal cell-based repair of large, full-thickness defects of articular cartilage," *J. Bone Joint Surg.,* 74A, 579, 1994.
63. Brittberg, M., Sjogren-Jansson, E., Lindahl, A., and Peterson, L., "Influence of fibrin sealant (Tisseel) on osteochondral defect repair in the rabbit knee," *Biomaterials,* 18, 235, 1997.
64. Chu, C. R., Dounchis, J. S., Yoshioka, M., Sah, R. L., Coutts, R. D., and Amiel, D., "Osteochondral repair using perichondrial cells. A 1-year study in rabbits," *Clin. Orthop.,* 340, 220, 1997.
65. Menche, D. S., Frenkel, S. R., Blair, B., Watnik NF., Toolan, B. C., Yaghoubian, R. S., and Pitman, M. I., "A comparison of abrasion burr arthroplasty and subchondral drilling in the treatment of full-thickness cartilage lesions in the rabbit," *Arthroscopy,* 12, 280, 1996.
66. Mochizuki, Y., Goldberg, V. M., and Caplan, A. I., "Enzymatical digestion for the repair of superficial articular cartilage lesions," *Trans. Orthop. Res. Soc.,* 18, 728,1993.
67. Hunziker, E. B. and Rosenberg, L. C., "Repair of partial-thickness defects in articular cartilage: cell recruitment from the synovial membrane," *J. Bone Joint Surg.,* 78A, 721, 1996.
68. Bentley, G., Smith, A. U., and Mukerjhee, R., "Isolated epiphyseal chondrocyte allografts into joint surfaces," *Ann. Rheum. Dis.,* 37, 449, 1978.
69. Kon, M. and de Visser, A. C., "A poly (HEMA) sponge for restoration of articular cartilage defects," *Plast. Reconstr. Surg.,* 67:289-293,1981.
70. Upton, J., Sohn, S. A., and Glowacki, J., "Neocartilage derived from transplanted perichondrium: What is it?" *Plast. Reconstr. Surg.,* 68, 166, 1981
71. Grande, D. A., Pitman, M. I., Peterson, L., Menche, D., and Klein, M., "The repair of experimentally produced defects in rabbit articular cartilage by autologous chondrocytes transplantation," *J. Orthop. Res.,* 7, 208, 1989.
72. Muckle, D. S. and Minns, R. J., "Biological response to woven carbon fibre pads in the knee," *J. Bone Joint Surg.,* 71B, 60, 1989.
73. Brittberg, M., Nilsson, A., Lindahl, A., Ohlsson, C., and Peterson, L., "Rabbit articular cartilage defects treated with autologous cultured chondrocytes," *Clin. Orthop.,* 326, 270, 1996.

74. Chesterman, P. J. and Smith, A. U., "Homotransplantation of articular cartilage and isolated chondrocytes," *J. Bone Joint Surg.,* 50B, 184, 1968.
75. Hale, J. E., Rudert, M. J., and Brown, T. D., "Indentation assessment of biphasic mechanical property deficits in size-dependent osteochondral defect repair," *J. Biomech.,* 26, 1319, 1993.
76. Shortkroff, S., Barone, L., Hsu, H. P., et al., "Healing of chondral and osteochondral defects in a canine model: the role of cultured chondrocytes in regeneration of articular cartilage," *Biomaterials,* 17, 147, 1996.
77. Breinan, H. A., Minas, T., Hsu, H. P., Nehrer, S., Sledge, C. B., and Spector M., "Effect of cultured autologous chondrocytes on repair of chondral defects in a canine model," *J. Bone Joint Surg.,* 79A, 1439, 1997.
78. Campbell, C. J., Ishida, H., Takahashi, H., and Kelly, F., "The transplantation of articular cartilage: An experimental study in dogs," *J. Bone Joint Surg.,* 45A, 1579, 1963.
79. Engkvist, O., "Reconstruction of patellar articular cartilage with free autologous perichondral grafts," *Scand, J. Plast. Reconstr. Surg.,* 13, 361, 1979.
80. Stevenson, S., Dannucci, G. A., Sharkey, N. A., and Pool, R. R., "The fate of articular cartilage after transplantation of fresh and cryopreserved tissue-antigen-matched and mismatched osteochondral allografts in dogs," *J. Bone Joint Surg.,* 71A, 1297, 1989.
81. Noguchi, T., Oka, M., Fujino, M., Neo, M., and Yamamuro. T., "Repair of osteochondral defects with grafts of cultured chondrocytes. Comparison of allografts and isografts," *Clin. Orthop.,* 302, 251, 1994.
82. Grundnes, O. and Reikeras, O., "Effects of function and weight-bearing on the healing of full-thickness cartilage defects in rats," *Scand. J. Med. Sci. Sports,* 5, 297, 1995.
83. Chang, J. K., Ho, M. L., and Lin, S. Y., "Effects of compressive loading on articular cartilage repair of knee joint in rats," *Kao-Hsiung I Hsueh Ko Hsueh Tsa Chih,* 12, 453, 1996.
84. Göransson, H., Lehtosalo, J., Vuola, J., Patiala, H., and Rokkanen, P., "Regeneration of defects in articular cartilage with callus and cortical bone grafts. An experimental study," *Scand. J. Plast. Reconstr. Surg. Hand Surg.,* 29, 281, 1995.
85. Jackson, D. W., Halbrecht, J., Proctor, C., van Sickle, D., and Simon, T. M., "Assessment of donor cell and matrix survival in fresh articular cartilage allografts in a goat model," *J. Orthop. Res.,* 14, 255, 1996.
86. Shahgaldi, B. F., Amis, A. A., Heatley, F. W., McDowell, J., and Bentley, G., "Repair of cartilage lesions using biological implants. A comparative histological and biomechanical study in goats," *J. Bone Joint Surg.,* 73B, 57, 1991.
87. Butnariu-Ephrat, M., Robinson, D., Mendes, D. G., Halperin, N., and Nevo Z., "Resurfacing of goat articular cartilage by chondrocytes derived from bone marrow," *Clin. Orthop.,* 330, 234, 1996.
88. Hendrickson, D. A., Nixon, A. J., Grande, D. A., et al., "Chondrocyte-fibrin matrix transplants for resurfacing extensive articular cartilage defects," *J. Orthop. Res.,* 12, 485, 1994.
89. Sams, A. E. and Nixon, A. J., "Chondrocyte-laden collagen scaffolds for resurfacing extensive articular cartilage defects," *Osteoarthr. Cartil.,* 3, 47, 1995.
90. Vachon, A. M., McIlwraith, C. W., Powers, B. E., McFadden, P. R., and Amiel, D., "Morphologic and biochemical study of sternal cartilage autografts for resurfacing induced osteochondral defects in horses," *Am. J. Vet. Res.,* 53, 1038, 1992.
91. Howard, R. D., McIlwraith, C. W., Trotter, G. W., et al., "Long-term fate and effects of exercise on sternal cartilage autografts used for repair of large osteochondral defects in horses," *Am. J. Vet. Res.,* 55, 1158, 1994.
92. Todhunter, R. J., Minor, R. R., Wootton, J. A., Krook, L., Burton-Wurster, N., and Lust. G., "Effects of exercise and polysulfated glycosaminoglycan on repair of articular cartilage defects in the equine carpus," *J. Orthop. Res.,* 11, 782, 1993.
93. Shamis, L. D., Bramlage, L. R., Gabel, A. A., and Weisbrode, S., "Effect of subchondral drilling on repair of partial-thickness cartilage defects of third carpal bones in horses," *Am. J. Vet. Res.,* 50, 290, 1989.
94. Robinson, P. D., "Histologic study of articular cartilage repair in the marmoset condyle," *J. Oral Maxillofac. Surg.,* 51, 1088, 1993.
95. Girdler, N. M., "Repair of articular defects with autologous mandibular condylar cartilage," *J. Bone Joint Surg.,* 75B, 710, 1993.

96. Itay, S., Abramovici, A., and Nevo, Z., "Use of cultured embryonal chick epiphyseal chondrocytes as grafts for defects in chick articular cartilage," *Clin. Orthop.*, 220, 284, 1987.
97. Robinson, D., Halperin, N., and Nevo, Z., "Regenerating hyaline cartilage in articular defects of old chickens using implants of embryonic chick chondrocytes embedded in a new natural delivery substance," *Calcif. Tissue Int.*, 46, 246, 1990.
98. Homminga, G. N., Bulstra, S. K., Kuijer, R., and van der Linden, A. J., "Repair of sheep articular cartilage defects with a rabbit costal perichondrial graft," *Acta. Orthop. Scand.*, 62, 415, 1991.
99. Pineda, S., Pollack, A., Stevenson, S., Goldberg, V. and Caplan, A., "A semiquantitative scale for histologic grading of articular cartilage repair," *Acta Anat.*, 143, 335, 1992.
100. O'Driscoll, S. W., Keeley, F. W., and Salter, R. B., "Durability of regenerated articular cartilage produced by free autogenous periosteal grafts in major full-thickness defects in joint surfaces under the influence of continuous passive motion. A follow-up report at one year," *J. Bone Joint Surg.*, 70A, 595, 1988.
101. Fienberg, S. E., *The Analysis of Cross-Classified Category Data*, MIT Press, Cambridge, 1980, 172.
102. Yoshioka, M., M., Coutts, R. D., Amiel, D., and Hacker, S. A., "Characterization of a model of osteoarthritis in the rabbit knee," *Osteoarthr. Cartil.*, 4, 87, 1996.
103. Mow, V. C., Kuei, S. C., Lai, W. M., and Armstrong, C. G., "Biphasic creep and stress relaxation of articular cartilage in compression? Theory and experiments," *J. Biomech. Eng.*, 102, 73, 1980.
104. Mow, V. C., Gibbs, M. C., Lai, W. M., Zhu, W. B., and Athanasiou, K. A., "Biphasic indentation of articular cartilage — II. A numerical algorithm and an experimental study," *J. Biomech.*, 22, 853, 1989.
105. Matsuura, T., Mansour, J. M., and Goldberg, V. M., "Indentation testing of rabbit distal femoral cartilage," *Proc. Biomech. Symp., Am. Soc. Mech. Eng,. Appl. Mech. Div.*, 120, 157, 1991.
107. Kobayashi, S., Yonekubo, S., and Kurogouchi, Y., "Cryoscanning electron microscopic study of the surface amorphous layer of articular cartilage," *J. Anat.*, 187, 429, 1995.
108. Read, N. D. and Jeffree, C. E., "Low-temperature scanning electron microscopy in biology," *J. Microsc.*, 161, 59, 1991.
109. Tavakol, K., Miller, R. G., Bazett-Jones, D. P., Hwang, W. S., McGann, L. E, and Schachar, N. S., "Ultrastructural changes of articular cartilage chondrocytes associated with freeze-thawing," *J. Orthop. Res.*, 11, 1, 1993.
110. Broom, N. D. and Silyn-Roberts, H., "The three-dimensional 'knit' of collagen fibrils in articular cartilage," *Connect. Tissue Res.*, 23, 261, 1989
111. Clark, J. M., Norman, A., and Notzli, H., "Postnatal development of the collagen matrix in rabbit tibial plateau articular cartilage," *J. Anat.*, 191, 215, 1997.
112. Amiel, D., Harwood, F. L., Hoover, J. A., and Meyers, M., "A histological and biochemical assessment of the cartilage matrix obtained from *in vitro* storage of osteochondral allografts," *Connect. Tissue Res.*, 18, 27, 1989.
113. Amiel, D., "Tendons and ligaments: a morphological and biochemical comparison," *J. Orthop. Res.*, 1, 257, 1984.
114. Kim, Y. J., Sah, R. L., Doong, J. Y., and Grodzinsky, A. J., "Fluorometric assay of DNA in cartilage explants using Hoechst 33258," *Anal. Biochem.*, 174, 168, 1988.
115. Toolan, B. C., Frenkel, S. R., Pachence, J. M., Yalowitz, L., and Alexander, H., "Effects of growth-factor-enhanced culture on a chondrocyte-collagen implant for cartilage repair," *J. Biomed. Mater. Res.*, 31, 273, 1996.
116. Oakeshott, R. D., Farine, I., Pritzker, K. P., Langer, F., and Gross, A. E., "A clinical and histologic analysis of failed fresh osteochondral allografts," *Clin. Orthop.*, 233, 283, 1988.
117. Kawabe, N. and Yoshinao, M., "Cryopreservation of cartilage," *Int. Orthop.*, 14, 231, 1990.
118. Friedlander, G. E., "Immune responses to osteochondral allografts," *Clin. Orthop.*, 174, 58,1983.
119. McDermott, A. G. P., Langer, F., Pritzker, K. P. H., and Gross, A. E., "Fresh small-fragment osteo-chondral allografts. Long-term follow-up study on first 100 cases," *Clin. Orthop.*, 197, 96, 1985.
120. Czitron, A. A., Langer, F., McKee, N., and Gross, A. E., "Bone and cartilage allotransplantation. A review of 14 years of research and clinical study," *Clin. Orthop.*, 208, 141, 1986.
121. Mankin, H. J., Gebhardt, M. C., and Tomford, W. W., "The use of frozen cadaveric allografts in the management of patients with bone tumors of the extremities," *Orthop. Clin. North Am.*, 18, 27, 1987.
122. Rubak, J. M., "Reconstruction of articular cartilage defects with free periosteal grafts," *Acta Orthop. Scand.*, 53, 175, 1982.

123. Niedermann, B., Boe, S., Lauritzen, J., and Rubak, J. M., "Glued periosteal grafts in the knee," *Acta Orthop. Scand.,* 56, 457, 1985.
124. Medley, J. B., Medley, J. B., Dilliar, R. M., Wong, E. W., and Strong, A. B., "Hydrophilic polyurethane elastomers for hemiarthroplasty: A preliminary *in vitro* wear study," *Eng. Med.,* 9, 59, 1980.
125. Mankin, H. J., Jennings, L. C., Treadwell, B. V., and Trippel, S. B., "Growth factors and articular cartilage," *J. Rheumatol. Suppl.,* 27, 66, 1991.
126. Coutts, R. D., Sah, R. L., and Amiel, D., "Effects of growth factors on cartilage repair," *Instr. Course. Lect.,* 46, 487, 1997.
127. Hill, D. J. and Logan, A., "Peptide growth factors and their interactions during chondrogenesis," *Prog. Growth Factor Res.,* 4, 45, 1992.
128. Trippel, S. B., "Growth factor actions on articular cartilage," *J. Rheumatol. Suppl.,* 43, 129, 1995.
129. Aspenberg, P. and Lohmander, L. S., "Fibroblast growth factor stimulates bone formation. Bone induction studied in rats," *Acta Orthop. Scand.,* 60, 473, 1989.
130. Kim, H. K., Moran M. E., and Salter, R. B., "The potential for regeneration of articular cartilage in defects created by chondral shaving and subchondral abrasion. An experimental investigation in rabbits," *J. Bone Joint Surg.,* 73A, 1301, 1991.
131. Baker, B., Spadaro J. A., and Becker, R. O., "Electrical stimulation of articular cartilage," *Ann. N.Y. Acad. Sci.,* 238, 491, 1974.
132. Brighton, C. T., Unger, A. S., and Stambough, J. L., "*In vitro* growth of bovine articular cartilage chondrocytes in various capacitively coupled electrical fields," *J. Orthop. Res.,* 2, 15, 1984.
133. Brittberg, M., Lindahl, A., Nilsson, A., Ohlsson, C., Isaksson, O., and Peterson, L., "Treatment of deep cartilage defects in the knee with autologous chondrocyte transplantation," *N. Eng. J. Med.,* 331, 889, 1994.
134. Russell, P. S., "Selective transplantation. An emerging concept," *Ann. Surg.,* 201, 255, 1985.
135. Weinberg, C. B. and Bell, E., "A blood vessel model constructed from collagen and cultured vascular cells," *Science,* 231, 397, 1986.
136. Webber, R. J., York, J. L., Vanderschilden, J. L., and Hough, A. J. Jr., "An organ culture model for assaying wound repair of the fibrocartilagious knee joint meniscus," *Am. J. Sports Med.,* 17, 393, 1989.
137. Solursh, M., "Formation of cartilage tissue *in vitro*," *J. Cell Biochem.,* 45, 258, 1991.
138. Langer, R. and Vacanti, J. P., "Tissue engineering," *Science,* 260, 920, 1993.
139. Vacanti, C. A., Kim, W., Upton, J., Mooney. D., Schloo, B., and Vacanti, J. P., "Tissue-engineered growth of bone and cartilage," *Transplant. Proc.,* 25, 1019, 1993.

17 Animal Models of Meniscal Repair

Jan Klompmaker and René P. H. Veth

CONTENTS

I. Introduction 327
II. Animal Selection 330
III. Commonly Used Models 332
    A. Meniscal Repair 332
        1. Humans 332
        2. Animals 332
    B. Meniscal Replacement 333
        1. Humans 333
        2. Animals 333
        3. Tissue Culture 334
IV. Evaluation Methods 334
    A. Gross Morphology 334
    B. Histology 334
    C. Mechanical Testing 335
    D. Others 335
V. Applications of the Models and Future Directions of Research 336
    A. Evaluation of the Pathophysiology and Natural Healing Process of Meniscal Injury 336
    B. Evaluation of Artificial Substitute of Meniscus 337
        1. In General 337
        2. Nondegradable Implants 338
        3. Degradable Implants 340
    C. Evaluation of Biological Graft 341
        1. Autografts 341
        2. Allografts 341
        3. Xenografts 343
        4. Cell-Seeding 343
VI. Summary and Conclusions 344
References 344

I. INTRODUCTION

Once menisci were described as the functionless remains of leg muscle[1] and it was generally believed that they could be removed without consequences. The immediate results of meniscectomy were, and still are, very satisfactory and, therefore, removal of a meniscus, when causing symptoms, was standard practice. Fairbank[2] was the first to call attention to the fact that removal of a meniscus is followed by roentgenographic signs of joint degeneration. He discussed the possibility that menisci have a significant role in weight-bearing and suggested that meniscectomy results in specific degenerative changes of the knee joint. The functions attributed to the meniscus are primarily load bearing, stress distribution and attenuation of shock waves within the knee joint, thus protecting the underlying cartilage from concentrations of stress and overloading.[3,4,5,6,7] Also, menisci are reported to increase joint congruency and facilitate the rotation of the opposing articular surfaces of the knee joint.[8] Providing stability is a secondary meniscal function when ligamentous injuries exist.[5,9,10] It has been proposed, although not proven, that the menisci also play a role in the distribution of lubricating fluid between femoral and tibial articular surfaces thus facilitating the lubrication of the joint.[11] Meniscectomy results in abnormal high stresses on articular cartilage in the meniscectomized compartment, which, over time, may provoke degenerative changes.[4,12] Since Fairbank's observations, it has been demonstrated many times by long-term surveys of meniscectomized patients, as well as by experimental studies using laboratory animals, that removal of a meniscus results in degeneration of articular cartilage in a high percentage of cases.[13,14,15,16] The increasing awareness of the consequences of meniscectomy has led to a more conservative approach in the treatment of meniscal lesions. Since it has been shown that the degree of degenerative changes is directly proportional to the amount of meniscal tissue removed, and is inversely related to the amount of fibrocartilage remaining[17,18] it is good clinical practice to preserve as much functional meniscal tissue as possible while addressing the clinical symptoms caused by tears. This can either be achieved by performing a partial meniscectomy or by repairing meniscal lesions. Total meniscectomy should be considered only in the relatively rare instances in which the extent of meniscal damage is so great that partial meniscectomy or meniscal repair is not appropriate.

Repair of meniscal lesions is the other alternative and may be the best option since all meniscal tissue can be preserved. King's experiments in 1936 set the biological limitations of meniscal healing. He showed that for meniscal lesions to heal, they must communicate with the peripheral vascular area of the meniscus.[18] Later it was established that only the outer 10–25% of the human's and dog's meniscus, throughout its attachment to the joint capsule, is vascularized (Figure 1) and this explains why there is no tendency for healing when tears occur in the avascular central part of the meniscus.[19,20] As a result only lesions limited to the vascular periphery can be repaired adequately by simple suturing. For lesions situated in the avascular central part of the meniscus no reliable methods exist, although new experimental techniques are in development. Successful repair of peripheral meniscal lesions by suturing or abrasion of parasynovial tissue has been widely applied with good results in animal experiments and in clinical trials on humans.[21,22] For repair of lesions in the avascular part of the meniscus the basic principle is to improve the vascularity of the defect by stimulating ingrowth of vascular tissue. Creation of a radial access channel from the vascular periphery towards a lesion is not very successful due to occlusion of this conduit, causing an insignificant increase in vascularity.[23] Repair by interposition of a synovial flap between lesion and periphery does stimulate healing but repair takes place by ingrowth of fibrous repair tissue not resembling meniscal fibrocartilage.[25,25] The peripheral rim can be resected back to a bleeding bed after which the meniscus can be reattached. However, when using this method, the cross-sectional area of the meniscus is diminished and load transmission is impaired, similar to partial meniscectomy.[23,26] Repair with fibrocartilage, resembling normal meniscal tissue, can be obtained after application of a fibrin clot.[27] A combination of fibrin glue and endothelial cell growth factor appeared to provide similar results.[28] However, it remains to be determined whether these methods are applicable for repair of large and highly stressed lesions.

After it had been shown that an access channel connecting a lesion in the avascular part of the meniscus to the periphery can result in healing of the tear, Veth et al. laid the basis for the present studies. They showed that repair of large meniscal lesions by fibrocartilage can be achieved by implantation of a porous polymer in the connecting defect.[29,30] Although several procedures discussed earlier result in improved healing of the tear, healing in general takes place by ingrowth of fibrous tissue not resembling normal meniscal fibrocartilage, which can also be observed in regenerated menisci. This fibrovascular scar tissue will have deviating biomechanical properties and

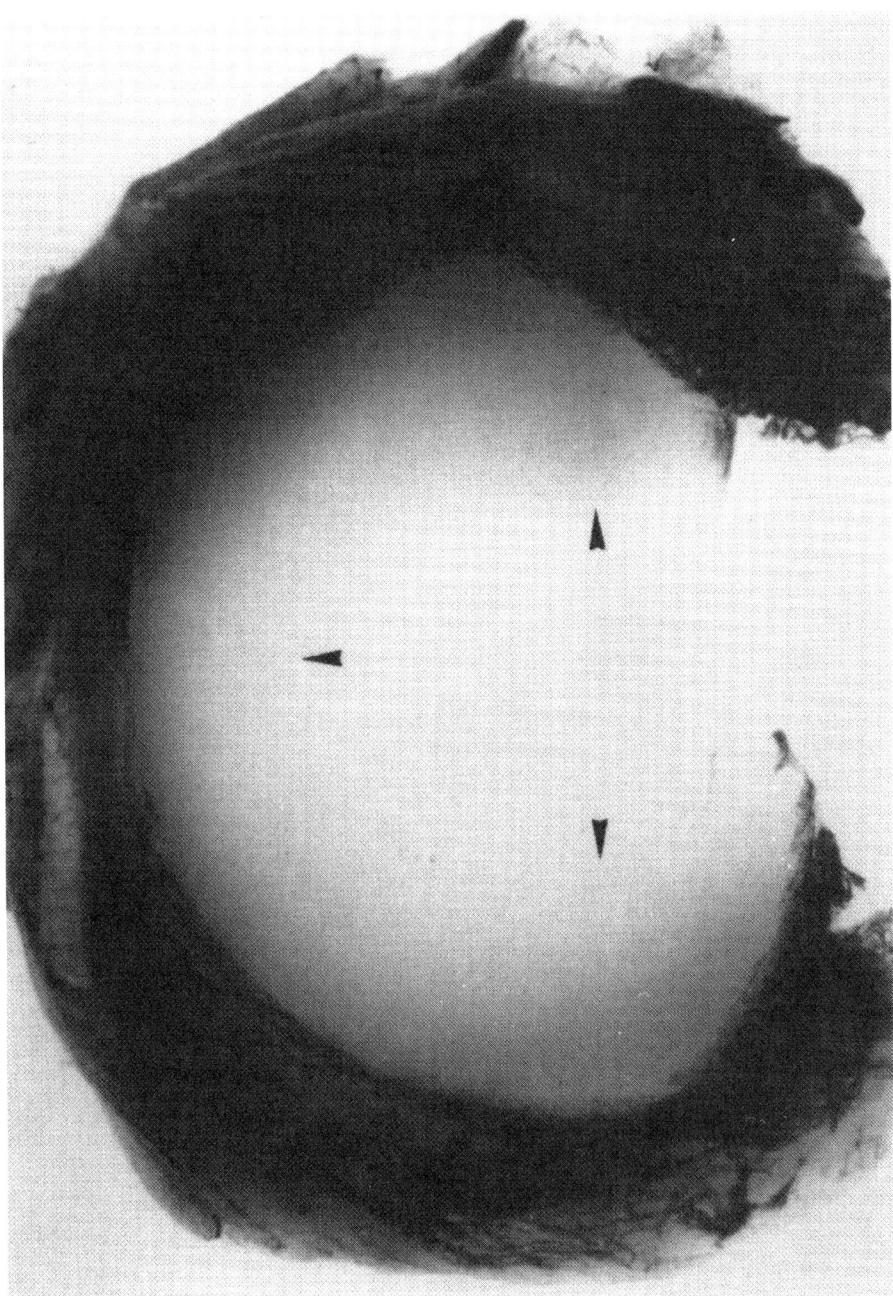

FIGURE 1. Vascularity of a normal meniscus. Only the periphery of the meniscus is vascularized. Arrows indicate the central rim ($\times 5$).

cannot be expected to function adequately in the long term.[24,31] Repair of meniscal lesions by fibrocartilage is not frequently encountered but has been observed in spontaneously healed meniscal lesions in the rabbit,[26] after application of a fibrin clot or glue[27,28] and implantation of a porous polymer.[29,30] Sometimes a meniscus is too severely damaged to be treated by a partial meniscectomy or meniscal repair. In this case, a total meniscectomy or replacement of the meniscus by a prosthesis or allograft are the two alternatives. The results of total meniscectomy, as discussed earlier, are poor, making replacement an attractive alternative. Until now, several methods have been applied,

TABLE 1
Species Used for *In Vivo* Studies of Meniscal Repair or Replacement

Subject	Meniscal replacement	Meniscal repair
Human	Rare	Numerous
Primate	Unknown	Unknown
Canine	Frequently	Frequently
Goat	Frequently	Rare
Sheep	Frequently	Unknown
Pig	Rare	Unknown
Rabbit	Frequently	Frequently
Rat	Unknown	Unknown
Mouse	Unknown	Unknown

but none is associated with long-term uniform success. The conclusion may be that, although allografts and meniscal prostheses may protect the articular cartilage, the results are highly variable and more research is needed.

Compared to articular cartilage, little is known about the cell biology and biochemistry of the meniscus. Studies carried out using a variety of animal species, show considerable similarities. The meniscal cells in the rabbit and human meniscus consist of fusiform cells in the superficial zone and rounded cells in the deeper zones, called fibrochondrocytes. The collagen fibers are organized into three layers: circumferentially oriented fibers, radial tie fibers and fibers woven in a meshlike fashion (Figure 2).[22,32,33,34,35,36] These fibers can be found in all animals studied. Five types of collagen have been found in both human and animal menisci: Predominantly type I (approximately 90%) and lesser amounts of type II, III, V and VI.[22,37] Furthermore, neural elements, fibronectin and thrombospondin are present.[38] All these studies suggest a large similarity between human and animal menisci but comparative studies between the species have not been carried out. To our knowledge no biochemical analysis of the primate meniscus has been carried out. One should expect a large similarity with the human meniscus. However, Le Minor[39] studied the morphology of the lateral meniscus in primates and found large differences between different primate species. In humans it has been shown that the meniscus size may be different for the right and left legs in the same individual.[40] Therefore, one should be careful in extrapolating results obtained in one animal species to another and maybe even from one leg to the other.

II. ANIMAL SELECTION

The subjects used for *in vivo* studies of meniscal repair or graft are listed in Table 1. In general one can say that the use of immature animals should be avoided because of the regeneration potential of the meniscus and surrounding tissues. Stone et al.[41,42] performed studies on a collagen meniscal template. In immature pigs, resection of 80% of the meniscus resulted in spontaneous healing. Therefore, the authors concluded that in this model they were not able to determine whether or not a collagen scaffold actually stimulated regeneration of the meniscus. Canine menisci probably are the best choice. It has been shown that their regeneration potential, like humans, is unpredictable and incomplete. Also, a structure is being formed which does not resemble normal meniscal tissue and does not protect the underlying articular cartilage. The human and canine meniscus have a strong similarity concerning vascular supply, anatomy and biochemistry whereas swine, goat, sheep and rabbit menisci have anatomical features that differ from dogs and humans.[15,16,19,20,25,30,42–52] The rabbit has often been used for meniscus experiments, but this model may be less suitable for studying cartilage changes after meniscus surgery because the rabbit knee is very prone to synovial

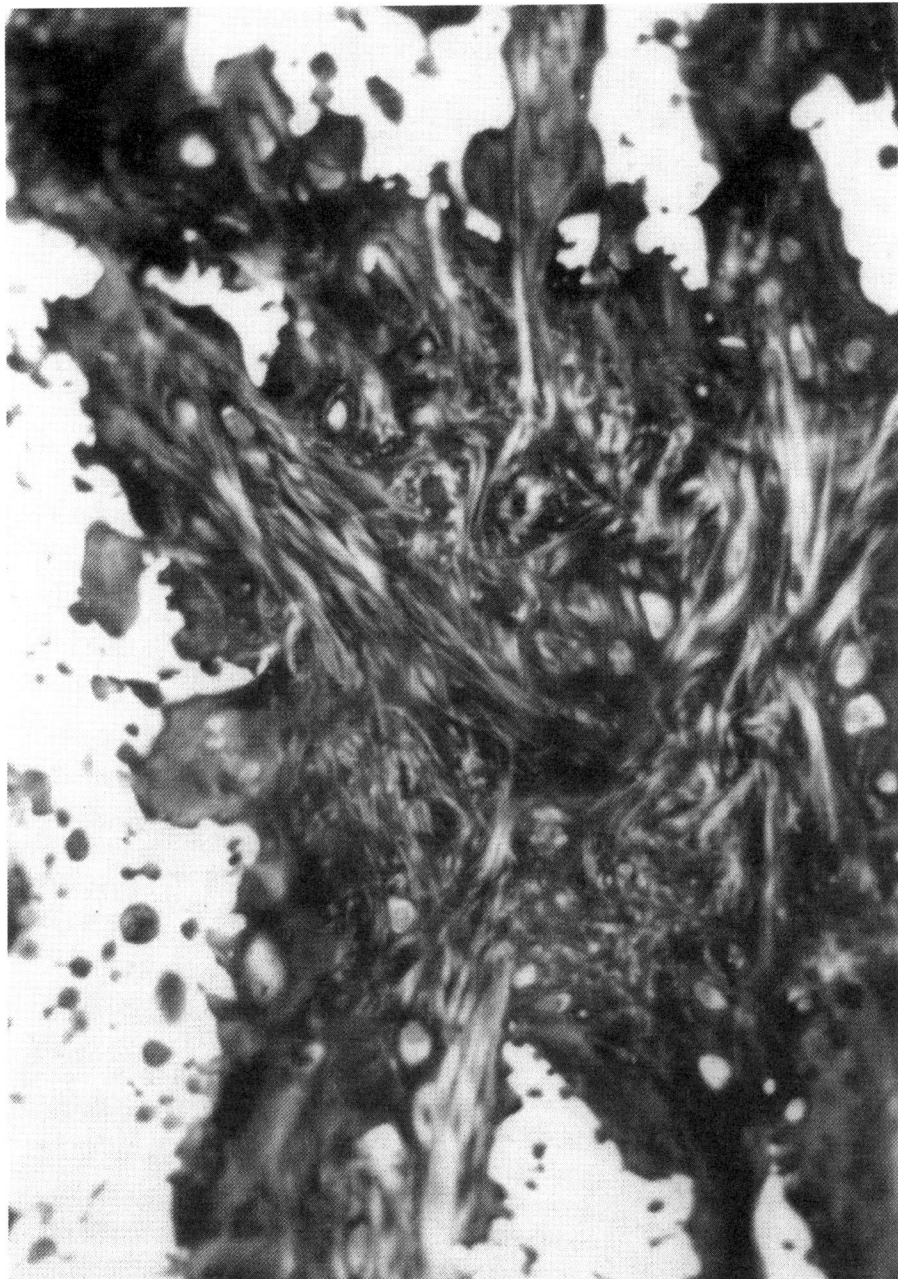

FIGURE 2. Fibrocartilage formation after implanting a porous polymer implant. Fibrochondrocytes are surrounded by an extracellular matrix of collagen fibrils.

irritation and osteophyte formation.[53] Knee instability after ACL resection in the rabbit causes more cartilage changes than in the dog. This may be due to the fact that the rabbit knee is more dependent on a functional ACL than, for example, a dog or human. In the rabbit the ACL is ten times stronger per unit of cross sectional area than a human ACL. Rodents like rat and mice are not suitable for meniscal repair and replacement because of their size and because of the fact that centers of ossification (ossicles) are found in their menisci. The occurrence of ossicles in human or canine menisci has been reported but they are extremely rare. Whether members of the cat family may be suitable is questionable because there have been no reports of their use in meniscus research, and the fact that ossicles may be frequently found, as they are in the tiger.[54,55]

III. COMMONLY USED MODELS

In fact there are three basic *in vivo* or *in vitro* models. All three have been used for meniscal repair and meniscal grafting: (1) Humans (many authors): Meniscal repair with sutures or arrows and transplantation of allografts; (2) Animals (many authors): Meniscal repair, allografts and autografts of meniscal or non-meniscal tissue, biological and synthetic scaffolds); (3) Tissue culture:[35,36] Basic research on fibrocartilage and culture of meniscal transplants.

A. MENISCAL REPAIR

1. Humans

Many studies on meniscal repair in humans exist. Success however, can only be achieved with meniscal lesions located in the peripheral, vascular part of the meniscus. With the use of sutures (ligature, arrows and other methods) success rates of 90% and more have been achieved. Repair of the central avascular part of the meniscus has not been successful. Suggested methods are the same as those used in animals and include the creation of an access channel extending from the periphery towards the lesion, synovial abrasion or the application of a blood clot.[24,27] Although some minor improvement in healing has been claimed, these methods are not generally accepted. Concerns and complications are the same as for animals and will be discussed below.

2. Animals

Many animal models have been employed in meniscal repair. The dog has been used for basic research,[5,17,20,43,56] for studying repair methods,[27,28,45–47,50,51] for implantation of a meniscal prosthesis[46,57,58] and for meniscal transplantation.[59] The rabbit also has been a much used animal, mainly for basic research on meniscal structure,[60,60] for studying meniscal healing and the effects of meniscectomy.[52] Reports on meniscal repair[26,52] and prosthetic replacement in the rabbit are frequent.[49,53,61–69] For transplantation purposes, rabbits have rarely been used.[70] As far as we know, goats[71] and sheep[72–76] have not been used for studies on meniscal repair but only for transplantation purposes. Successful repair of peripheral meniscal lesions by suturing or abrasion of parasynovial tissue has been widely applied with good results in animal experiments using rabbits, dogs and in clinical trials on humans.[21–23,26] This can be done by suturing using either the original suture wire or newer techniques using arrows. The basic principle is to fix the two opposing sides of the tear together, often after roughening of the surfaces.

For repair of lesions in the avascular part of the meniscus, the basic principle is to improve the vascularity of the defect by stimulating ingrowth of vascular tissue.

In the dog, this principle was realized by the creation of a radial access channel from the vascular periphery towards a lesion. Using a punch, a round channel is drilled starting at the periphery and ending in the meniscal lesion. This procedure was not very successful due to occlusion of this conduit, causing an insignificant increase in vascularity.[23]

Interposition of a synovial flap between lesion and periphery has been attempted both in the dog as in the rabbit. In this technique, a synovial flap is prepared free from the capsule. Its base is left untouched and contains the blood vessels, its top is sutured in the tear. This technique does stimulate healing but repair takes place by ingrowth of fibrous repair tissue, not resembling meniscal fibrocartilage.[24,25]

In the rabbit the longitudinal tear can be repaired by removing part of the meniscus located between tear and capsule. The central part of the meniscus can then be reattached to the joint capsule. This method does result in healing but the cross-sectional area of the meniscus is diminished and load transmission is impaired, as with partial meniscectomy.[23,26] Repair with fibrocartilage, resembling normal meniscal tissue, can be obtained after application of a fibrin clot in the dog.[27] A lesion was created in the avascular part of the meniscus and filled with an autologous fibrin clot. A combination of fibrin glue and endothelial cell growth factor appeared to provide similar results in dogs.[28] However, it remains to be determined whether these methods are applicable for repair of large and highly stressed lesions. Another way of promoting vascularity is by connecting the tear in the avascular part to the vascular periphery by implantation of a porous implant.[29,30,45-51] Composites consisting of polyurethane, polylactide and carbon fibers were used for successful reconstruction of large meniscal lesions in the dogs and rabbits and repair by fibrocartilaginous tissue was observed. The key question is not only whether a technique results in healing of a tear but also if healing takes place by fibrocartilage. Healing by fibrous tissue is seen in most of the above mentioned techniques. This tissue can also be observed in regenerated menisci. It has been proven that fibrous tissue is biomechanically inferior to normal meniscal fibrocartilage and does not protect articular cartilage in the long term.[24,25,31] Repair of meniscal lesions by fibrocartilage is not frequently seen but has been observed in spontaneously healed meniscal lesions in the rabbit,[26] after application of a fibrin clot or glue in the dog[27,28] and implantation of a porous polymer.[29,30,45-51]

B. MENISCAL REPLACEMENT

1. Humans

Replacement of the meniscus by an allograft is a new development in humans.[73-77] The ideal patient is a person with disabilitating osteoarthritis of the knee who cannot be helped by a corrective osteotomy and who is too young for an arthroplasty. Large series do not exist and the long-term results are not known. Therefore, it can be considered an experimental method in humans. Concerns and complications are the same as for animals and will be discussed below.

2. Animals

As mentioned above, meniscal replacement has been carried out in a variety of animals, including humans. It is not easy to say which animal is preferable for meniscal repair or replacement. Of course, the human being is the ideal model. However, only at repeat arthroscopy can one determine the effect of repair and grafting techniques. Repeat arthroscopy cannot be done as a standard procedure in patients who do not have complaints because of the ethical aspects. Therefore, follow-up is limited to patients who develop new complaints. When an arthroscopy can be carried out, it is of limited value. The outward aspect of a graft does not provide information about tissue structure and function. For example, is has been shown that after repair of articular cartilage defects the outward appearance of such defects may be that of normal healthy cartilage whereas microscopic analysis shows fibrocartilage of inferior quality. A biopsy can be taken during arthroscopy[77] but the amount of tissue thus obtained is minimal.

Animal models seem to be essential to obtain more information. It is not known which animal is preferable for meniscal repair or replacement. When repair is performed, one needs a certain meniscus size for surgical technical reasons. The meniscus of the dog, goat and sheep appear to meet this demand. The dog's meniscus may be preferable for reasons of comparing one's own results with those provided by literature since the most cited papers are done using dogs. Using dogs, in our country, is at least twice as expensive as using goats and sheep. Lodging of dogs is more complicated and the capacity is more limited than for goats and sheep. Getting permission to use dogs for terminal experiments from ethical committees may be more difficult than it is for other animals. Other animals might be chosen as an alternative for these reasons. When meniscal transplantation or replacement is carried out, the size of the meniscus is less a deciding factor in the choice of an animal. Technically it is possible to use the rabbit for these experiments (own experience). Our own approach in this matter is to use rabbits and goats for pilot and screening experiments and use a limited number of dogs for the key experiments.

3. Tissue Culture

Tissue culture of meniscal tissue and fibrochondrocytes was initiated by Webber et al.[34-36] The advantages are, obviously, the possibility of close monitoring of cellular processes and the need for less animal experiments. In explants a small portion of the meniscus is kept in culture. Webber et al. found that explants are able to produce proteoglycans identical to those of the normal meniscal matrix. The disadvantages of these explants are the small number of cells and the limited manipulation of the cells by the extracellular matrix. Also it could not be assessed whether the cells could proliferate and synthesize a matrix. Therefore, successful attempts have been made to culture cells *in vitro* after they have been released from the extracellular matrix. This is done by enzymes such as collagenase and trypsin. In this way they could demonstrate that meniscal fibrochondrocytes are able to proliferate and synthesize a matrix. One concern of culturing meniscal cells is their changing behavior when grown in different growth media. Another problem, which may even be more important, is the process of dedifferentiation. In many studies concerning the culturing of chondrocytes taken from articular cartilage it has been shown that chondrocytes change their morphology and biochemical behavior after several days in culture and become fibroblast-like cells. This process can, to some extent, be prevented by culturing them in a 3-dimensional matrix like a collagen gel but the question remains if the cells cultured really are similar to cells in the native tissue. Our own experience with cell culture of meniscal cells is limited but we found it to be difficult, time-consuming and we had little success using this method. Certainly this method will have great value in screening toxicity of biomaterials but maybe other cell lines than fibrochondrocytes can be used. A very special application for the use of *in vitro* techniques is culturing meniscal allograft transplants.[97] They can be kept in a nutrient medium for about 2–3 weeks without loss of viability, during which period the appropriate patient can be selected and prepared. Transplant risks like disease transmission can be avoided by testing the transplants during this period. It has been suggested that culturing can reduce the antigenic potential of meniscal tissue but rejection is not a problem in reality.

IV. EVALUATION METHODS

A. Gross Morphology

The morphology of the implant or graft is the first method for evaluating its success. Clearly, visible destruction is an indicator for failure. Apart from implant and graft, the appearance of the articular cartilage is important. Visible fibrillation, a non-glossy appearance or even gross destruction is indicative of failure. When more subtle degenerative changes of the articular cartilage have to be detected, Meachim's test can be used.[79] In this test the articular cartilage is pencilled with India ink which fills any irregularities in the articular surface, thus making them visible. One can detect minimal changes with the use of a magnifying-glass or stereomicroscope. Measurement of the size of meniscal transplants is important because shrinkage of grafts is one of the main reasons for failure.[73,76]

B. Histology

The next step is to prepare the meniscus and articular cartilage for histologic examination in order to investigate the results on a cellular level. After fixation of the tissues, they are embedded in paraffin or in plastic and thin slices are cut which can be stained. Several stains are available and are used depending on personal favor or on the specific feature one wants to see. The tissues and cells can be studied using ordinary stains like Giemsa or hematoxylin and eosin, but many other stains are suitable. For studying the cartilage-containing part of the tissue, a stain like toluidine-blue may be useful. It provides the proteoglycans in cartilage and fibrocartilage with a red color thus distinguishing these tissues from fibrous tissue. Microscopic sections taken from the articular cartilage are useful to detect minimal degenerative changes as indicated by a decrease of proteoglycan content, cloning of articular cartilage cells, fibrillation and other changes. Immunohistochemistry is a special technique using antibodies. These antibodies can detect specific cells or tissues and many are available. We have used them to show the collagen types present in meniscal repair tissue.[50] This technique is technically demanding, time consuming and relatively expensive. Polarized-light microscopy is useful in highlighting collagen fibrils.

C. Mechanical Testing

Mechanical testing should be performed whenever possible. Tensile testing and deformational testing can be performed. Mow and others[53,66,68,80] have used this technique which will answer one of the most important questions: Is the implant or graft biomechanically comparable to a normal meniscus? Numerous studies have shown that the cartilage-protecting effect of a meniscus is dependent on its biomechanical behavior. Using biomechanical testing methods, it has been found that the meniscus is a very complex structure. Its tensile and deformation behavior not only vary with location but also vary with direction. A strong correlation between the tensile properties and the collagen architecture exists. A greater degree of collagen fiber bundle orientation causes superior tensile strength and stiffness. Under compression meniscal tissue behaves as a viscoelastic tissue and its behavior makes it functionally a highly efficient shock absorber. The extracellular matrix (proteoglycans and glycosaminoglycans) of the meniscus varies continuously across its width in a manner consistent with increased compressive loading.[80] Shear tests have shown that the meniscus is anisotropic in shear, which can be explained on the ultrastructural basis. Not only the material properties of the meniscal tissue are important. The mechanical behavior of a meniscus is also highly dependent on the size, the attachments, the alignment of the knee and these factors must be taken into acount.[77,81,82] Because of the high complexity of measuring the impact of all these factors on the knee joint, most authors use computer models for finite element analysis.

D. Others

Arthroscopy can be a useful tool for the gross evaluation of an allograft or a graft.[47,76,77] It can provide data like synovial irritation, shrinkage, tearing, fraying and gross degenerative changes within the knee joint. In humans repeat arthroscopy after an operation is only done when new problems arise, whereas in animals it can be done at regular intervals. The big advantage of this procedure is that more follow-up data can be collected without the need to sacrifice the animal. However, minor changes and especially the processes going on inside the meniscus remain obscured. We have tried arthroscopy in the dog and think that technically, the procedure is easy to perform. Most likely, arthroscopy can be performed in other large animals too.

A biopsy at the time of arthroscopy can be useful to get insight in the processes inside of the tissue without killing the animal. The amount of tissue that can be removed without destroying the allograft or implant is limited. Therefore, no conclusions can be made concerning the structure as a whole.

Enzyme histochemistry can be used to test cell viability.[77] Mitochondrial enzymes (NADH-tetrazolium-reductase, alpha-glycerolphosphate oxidase), lysosomal enzymes (acid phosphatase, alphanaphtylacetate eaterase) give a biochemical impression of cell viability and degeneration of cells. Uptake of radioactive sulfur by the cells has been used as an alternative method to test cell viability in the meniscus.[83,84] Antigenic characteristics of cells and tissues and the amount of cell proliferation can be detected using monoclonal antibodies.[77]

Walking analysis can be useful to assess quality of gait and joint stability.[85] Also it can be used to study if the animals favor a joint with a certain implant compared to a joint with another implant, thus indirectly comparing several implants. This method provides information in a non-invasive manner and can be repeated as many times as desired. The disadvantage lies in the fact that this method provides only indirect information about the implant or graft.

V. APPLICATIONS OF THE MODELS AND FUTURE DIRECTIONS OF RESEARCH

A. Evaluation of the Pathophysiology and Natural Healing Process of Meniscal Injury

The pathophysiology of meniscal injury has been described by several authors.[18,31,86,87] Several types of tears exist. Tears in the medial meniscus are far more common than tears in the lateral meniscus, but the same type of tears can be encountered in either compartment. Most classifications used for meniscal tears are based on the direction of the tear. Tears in one direction are classified as: (1) Horizontal: These tears are also called horizontal cleavage lesions. This type of tear may be the result of abnormal compression forces as can be encountered in a primary varus deformity. It occurs at a later age thus causing less osteoarthritis at an early age. The lesion is caused by a degenerative process within the fibrocartilage, possibly because of a breakdown in nutrition. A decrease in the collagen/chondroitin sulfate ratio is seen at the site of the lesion, caused by an accumulation of mucopolysaccharides.[87] The femoral condyle sinks into the meniscus and extrudes the periphery causing pain, swelling and tenderness over the joint line. The degenerative changes inflicted on the femur are more pronounced than on the tibia. (2) Longitudinal: These tears are directed along the length of the meniscus and are traumatic in origin. It is more harmful because it is a lesion sustained in young patients thus causing early osteoarthritis. In case of a displaced longitudinal tear the articular cartilage is particularly susceptible to destruction. (3) Radial: These lesions are directed outward from the meniscal periphery starting at the central rim of the meniscus. In contrast to horizontal and longitudinal lesions, radial tears disrupt the circumferential collagen fiber structure. For this reason, meniscal function is more affected by radial tears.

Smillie[87] mentioned the main features of a meniscal lesion. Most importantly the symptoms are intermittent. The patient is older than the age of 30 and younger than 57. A history of trauma may be absent in case of a cleavage tear but is often present in case of a longitudinal tear. Local pain is present in the joint line. There is a feeling of giving-way and instability. Lack of extension (locking) of the knee is present when a flap is interposed between the femur and tibia. Effusion is not a specific symptom.

There's no doubt that the meniscus has a limited healing potential. King's experiments in 1936 set the biological limitations of meniscal healing. He showed that for meniscal lesions to heal, they must communicate with the peripheral vascular area of the meniscus.[18] Later it was established that only the outer 10–25% of the human and dog's meniscus, throughout its attachment to the joint capsule, is vascularized. This explains why there is no tendency for healing when tears occur in the avascular central part of the meniscus.[19,20] As a result only lesions limited to the vascular periphery can be repaired adequately by simple suturing. For lesions situated in the avascular central part of the meniscus no healing is observed. The microscopic appearance of meniscal tears is similar regardless of the type of tear. A repair reaction including hemorrhage, vascular invasion and formation of meniscal tissue is seen in the vascular periphery of the meniscus. Whether a perfectly normal microscopic structure is formed is not known. It may be that, especially in case of a healed radial lesion, the circumferential collagen fibers are never fully restored to normal.[47] Although tears located in the central avascular part of the meniscus do not heal spontaneously, healing can be achieved by several methods. The basic principle is to improve the vascularity of the defect by stimulating ingrowth of vascular tissue. Although several procedures do result in improved healing of the tear, healing in general takes place by ingrowth of fibrous tissue not resembling normal meniscal fibrocartilage, which can also be observed in regenerated menisci. This fibrovascular scar tissue will have deviating biomechanical properties and cannot be expected to function adequately in the long term.[24,25,31] Repair of meniscal lesions by fibrocartilage is not frequently encountered but has been observed in spontaneously healed meniscal lesions in the rabbit[26] after application of a fibrin clot or glue[27,28] and implantation of a porous polymer.[29,30] (Figure 3) After partial or total meniscectomy remodeling and regeneration has been observed both in animals as in humans.[35,36,44,85,88] Remodeling and regeneration are thought to be caused by migrating cells from the synovium or a proliferation of surrounding meniscal fibrochondrocytes, possibly stimulated by the presence of a fibrin clot and growth factors. However, these phenomena are not predictable and usually a smaller meniscus is being formed. This meniscus consists of tissue not resembling normal meniscal fibrocartilage. This tissue is biomechanically inferior to normal fibrocartilage and osteoarthritis is not prevented.

B. Evaluation of Artificial Substitute of Meniscus

1. In General

Perhaps the most attractive option for the repair and replacement of a torn meniscus is the use of artificial materials. In contrast to biological grafts they do not transmit diseases like hepatitis, AIDS and others. Sterilization of the implant will be easy, whereas in biological grafts sterilization will affect its properties. Also the limited availability of biological grafts is overcome since artificial substitutes can be made in unlimited numbers. The problems of storage do not exist, and many meniscal shapes and sizes could be obtained. Work on the use of artificial substitutes is not new but the older concepts of implanting a solid material have proven to be unsuccessful. Therefore, newer designs are based on porous materials which allow ingrowth of tissue.

Two groups of materials can be distinguished, nondegradable materials and degradable (or resorbable) materials. Theoretically, it seems unlikely that nondegradable materials will ever approach, both structural and biomechanically, a normal meniscus because of its enormous complexity. A degradable material ultimately forms a structure composed of normal body tissue. Therefore, it seems likely, although not proven, that a degradable material will result in a meniscal replica which most strongly resembles a normal meniscus. The rationale for the use of resorbable scaffolds is often based on the work of Yannas and Burke who were the first to use artificial skin.[89] The scaffolds must fulfill the following demands and more demands may be added in the future (several of these demands also apply for nondegradable implants):[42,45–51,89] (1) They must be biocompatible and should not evoke a foreign-body reaction. One should realize that the biocompatibility of any material used not only depends on its composition but also on its application. For example, the use of carbon fibers elicits a synovial inflammation in the knee whereas the same material can be used without problem in another part of the body. (2) They must be biodegradable to avoid any long-term complications. (3) Their degradation rate must be slow enough to permit ingrowth of pro-meniscal tissue and transformation into fibrocartilage. We have seen that quickly degradable implants do not give rise to the formation of fibrocartilage. (4) The pore size of the biomaterial must be optimized because it controls the speed of ingrowth into the implant and the amount of fibrocartilage formed. (5) They should have the same size and morphology as the normal meniscus, or acquire these features as soon as possible since incongruous grafts will lead to degeneration of the meniscus or the articular cartilage.[40] (6) The scaffolds should provide a tissue having identical biomechanical properties as a normal meniscus as soon as possible thus avoiding or retarding degenerative changes of cartilage. (7) The attachments must be anatomical. It has been shown that a non-anatomic position of the attachments leads to osteoarthritis. For this reason, allografts are transplanted together with the bone blocks at the attachment sites.

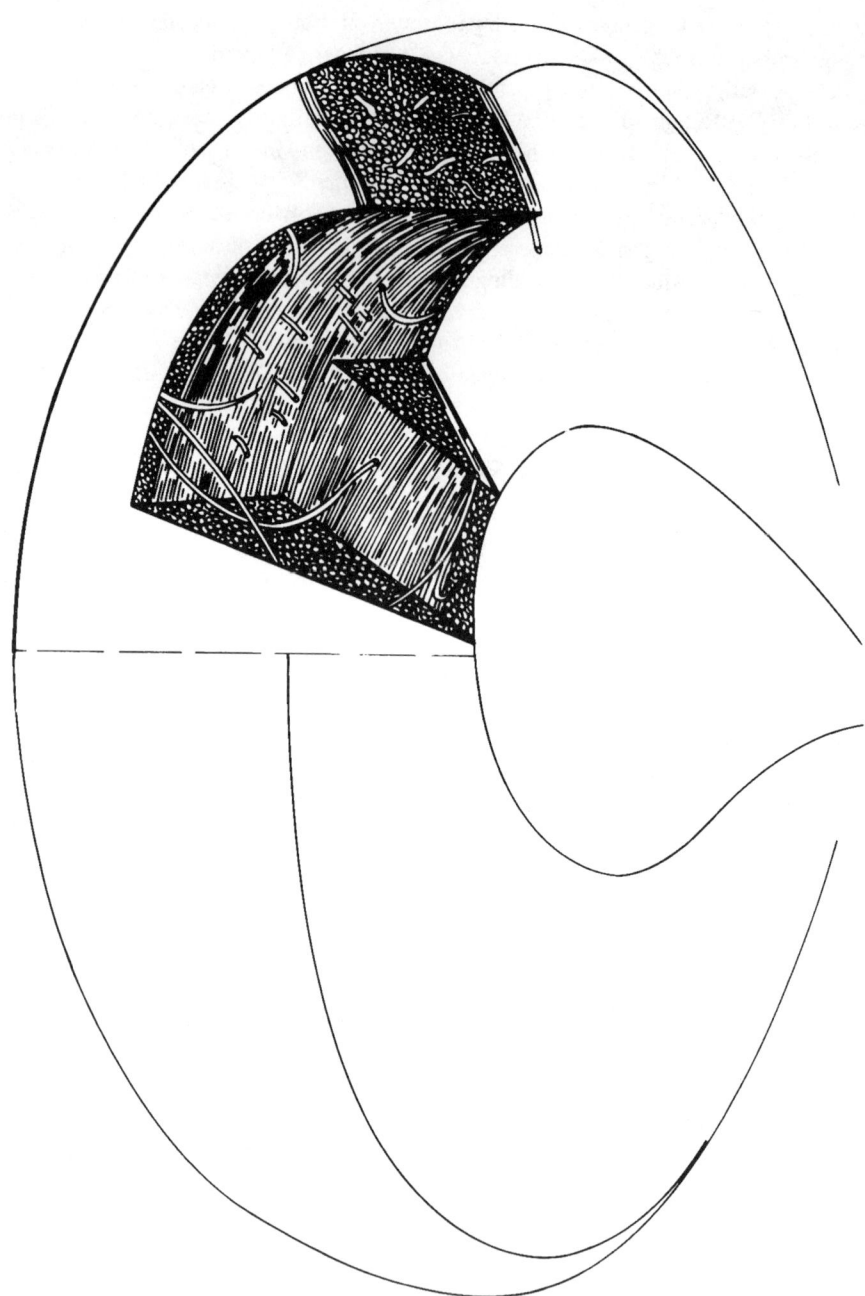

FIGURE 3. The complex meniscal structure. The main collagen fibrils are oriented in a concentric configuration interwoven by radial tie fibers.

2. Nondegradable Implants

A Teflon net prosthesis in the dog provided one of the first encouraging results in using an artificial meniscus.[57] A Teflon net was folded into a cylinder and sutured to the joint capsule. Although cartilage degeneration was not prevented, and was observed as soon as three months post-operatively, it was less severe compared to meniscectomy. Histologically, Teflon allowed ingrowth of tissue into its interstices; it also elicited a clear inflammatory response. After nine months cartilage cells could be observed. Clearly, the induction of cartilage-like tissue was a positive finding. It is speculated that joint instability, improper fixation and adhesion of the graft to the popliteal tendon thus impending knee motion are responsible for the joint degeneration and that Teflon as a material is superior to previously used materials. However, the early degenerative changes make Teflon net used in this form unsuitable for meniscal replacement.

Wood et al. used a prosthesis of concentrically carbon fibers ensheathed by woven polyester fibers to replace the rabbit's meniscus with disastrous results.[69] The prosthesis was secured using a transosseous tibial tunnel and it was sutured to the capsule. The carbon fibers fragmented staining the synovium black. Osteophytes were present in all knees and the menisci showed lateral displacement. There was no invasion of fibrous tissue and the implants were encapsulated instead of being incorporated into the surrounding tissue. A marked inflammatory response was shown around the polyester fibers. It was concluded that this prosthesis in its present form did not work. It appeared that this prosthesis showed the following negative features: (1) Carbon fibers may be compatible for certain applications, but they were not in the knee joint. (2) The prosthesis was not porous and its fixation not rigid enough. A knee with intact menisci can store more energy than a knee with the meniscus removed. Joint stiffness increases after meniscectomy. The biomechanical effects of a Dacron prosthesis with a polyurethane coating was tested in rabbits.[53,66–68] Knees with a prosthesis showed lower energy storage than sham operated knees and energy storage was similar to the meniscectomy group. After three months cartilage changes were found in 70% of the prosthesis group, compared to 100% in the meniscectomy group.

The cartilage changes on the femur were comparable for both groups; the tibia was less affected in the prosthesis group. Synovial changes with fibrosis were noted in 90% of the prosthetic knees but no foreign body reaction or loose Dacron particles were found. It was concluded that the prosthesis had a cartilage-protecting effect but the biomechanical behavior of the knee joint was similar to meniscectomy.

Partial ingrowth was seen in 50% of the prosthesis and fibrocartilage had not formed. The authors think that an improper size and inferior biomechanics of the prostheses may be responsible for the failures. Therefore, a new series of prosthesis was designed. These prostheses had better *in vitro* biomechanics than the previously tested prosthesis. A polyester (Dacron), a PTFE (polytetrafluorethylene) implant with a polyurethane coating on the upper surface and an uncoated PTFE implant were implanted in rabbits.[64] The PTFE prosthesis had compression values that matched the normal meniscus in contrast to polyester. After three months cartilage softening and osteophyte formation was found in all groups. Ingrowth of tissue was not complete and fibrocartilage was not found. The uncoated PTFE prosthesis, which had mechanical properties most closely resembling a normal meniscus, lost its shape and was prone to wear causing debris. Its effects on the cartilage was similar to meniscectomy. The polyurethane coating helped to conserve the shape and material of the implant. The polyester prosthesis showed little ingrowth, probably because of too little porosity. The coated PTFE prosthesis gave the best overall results. All prostheses elicited a synovial reaction and osteoarthritis was not prevented. The load-relaxation characteristics of joints with implanted prostheses were better than those of meniscectomized knees, in contrast to previous studies. This is explained by a better anatomic position and better sizing of the prosthesis.

The same authors also studied the effect of a meniscal prosthesis in an unstable knee after resection of the anterior cruciate ligament.[53] It appeared that no protective effect of a prosthesis is present in an unstable knee. Therefore, just as applies for meniscal repair, a stable knee joint seems to be a prerequisite for success. A combination of an artificial prosthesis of Teflon combined with a biological periosteal implant in rabbits did not improve results.[61] All grafts had changed in shape and were extruded towards the periphery of the joint. The cartilage no longer was covered and protected. Fibrocartilage was not formed and the meniscus failed to carry out meniscal function.

3. Degradable implants

A collagen-based scaffold has been reported to give an excellent meniscal replica in swine and canine.[41,42,85,90-94] Tissue from bovine Achilles tendon is formed into meniscal disks. These disks are infused with glycosaminoglycans in amount approximating those found in human menisci. These scaffolds are reported to be non-toxic because they allow the ingrowth of fibrochondrocytes. When implanted in the knee joint of immature pigs excellent regeneration of the meniscus was reported, which is not surprising because meniscal regeneration in immature animals is present anyway. After implantation in mature dogs, 63% of competent meniscal regeneration was noted compared to 25% for control menisci. After a maximum follow-up of 12 months no significant difference in the gross appearance was seen between joints that had received collagen implants and those that had not. Synthesis of proteoglycan by the templates had normalized by nine months. Histologically, an excellent resembling meniscal replica was formed. Implant material remained visible in the regenerated menisci at 12 months.

One restriction of using this technique is that the scaffolds are sutured to a remaining peripheral rim of meniscal tissue. In reality, a remaining peripheral rim may not be present. Therefore, at least in the animal, this technique has proved to be suitable for the partial replacement of a meniscus rather than for total meniscal replacement. Our own experience in artificial meniscal substitute is done with resorbable materials in the dog, rabbit and goat (Figure 4).[30,45-52] The first attempt was to repair meniscal lesions using a combination of carbon fibers and polyurethane-polylactide. These grafts appeared to be unsuitable because of carbon particle induced synovitis. Later, mixtures of polyurethane and polylactide were used to form several porous implants. Implants initially became filled with vascular fibrous tissue and displayed a mild foreign body reaction consisting of polynuclear giant cells, some macrophages, and lymphocytes. After two months the fibrous tissue filling the implants became transformed into metachromatic avascular fibrocartilage strongly resembling normal meniscal tissue. The repair tissue initially consisted of fibrous tissue containing type I collagen. Later, this vascular fibrous tissue was transformed into avascular fibrocartilage. Both type I and type II collagen, the major collagen types of normal meniscal fibrocartilage, could be detected in this newly formed fibrocartilage. In control defects, which were filled with vascular fibrous tissue without fibrocartilage, only type I collagen could be detected. Type II collagen was never found. This reaction took place in all implants used, based on porous physical mixtures of polyurethane, polylactide, and caprolacton. There were, however, considerable differences in ingrowth and fibrocartilage formation among the implants used. In contrast to the transformation of fibrous tissue into fibrocartilage, as can be observed in implants, empty control defects and synovial flap defects were filled with vascular connective tissue which did not transform into fibrocartilage. Roughly, two-thirds of the longitudinal tears could be repaired using this method.

It is well known that the implant's physical structure can be of great influence on its biological behavior. It has been shown for several tissues that it not only alters the rate of tissue ingrowth but also the degree and type of differentiation of ingrowing tissue, thus determining the ultimate type of tissue formed. Therefore, implants with varying pore structure were tested for their biological behavior in rabbit menisci. It appeared that tissue ingrowth was optimal in two large pore implants (macropores of 150–250 and 250–500 μm) whereas small pore implants (macropores of 50–90 and 90–150 μm) remained partially empty up to one year postoperative. Capsule formation and the foreign body reaction was severe for the small pore implants whereas this occurred to a lesser extent in the two large pore implants. Fibrocartilage formation, as assessed by morphology and antibody labeling for type I and type II collagen, was observed in a similar way in all implant types. It was concluded that for optimal ingrowth and incorporation of partial or total meniscal prostheses, macropore sizes should be in the range of 150–500 μm.

After it was shown that small defects could be repaired by porous polymers, whole menisci were made of the same porous polymer and implanted in the dog. Sixteen knees received a prosthesis. The tissue reaction was essentially identical to the one seen in small implants used for meniscal repair. In the first six knees the prostheses were secured using single sutures which were only pulled through the anterior and posterior prosthesis horns. At sacrifice four of these prostheses appeared to be dislocated due to tearing-out of the sutures. Therefore, the following prostheses were secured using two sutures running longitudinally through the entire prosthesis. Of the remaining 10 prostheses only one dislocated. The short-term implants had a connective tissue appearance, similar to dislocated implants. They were light-brown of color and had a soft consistency. After three months the prosthesis had a yellowish glistening appearance, and had a firm consistency. A rim of hyaline-like neocartilage had formed at the prosthesis' inner margin.

After an initial ingrowth of vascular fibrous tissue containing type I collagen only, the prostheses became filled with fibrocartilage strongly resembling normal meniscal fibrocartilage, containing both type I and II collagen. Degenerative changes of articular cartilage were present in all meniscectomized control knees. Moderate fibrillation of cartilage seen at eight weeks progressed to severe destruction exposing the subchondral bone after 20 weeks. Degeneration in association with a dislocated prosthesis was comparable to meniscectomy and was more severe when follow-up periods were longer. Degeneration associated with well incorporated prostheses was frequent, although less severe than seen after meniscectomy or dislocation. Intact cartilage was seen in five knees. In the remaining six knees, varying degrees of cartilage destruction were seen, although exposure of the subchondral bone did not occur. In two of these six knees it appeared that the drill holes were located in the central part of the tibial cartilage instead of in the eminentia. This surgical error may have contributed to the cartilage damage. Cartilage degeneration was not related to the length of follow-up or formation of fibrocartilage inside of the prosthesis. Tibial plateaus were more frequently and more severely affected than femoral condyles. Although less severe than seen after total meniscectomy, cartilage degeneration was frequent, possibly because tissue ingrowth in the prostheses occurred too slowly, because the size of the prosthesis was incorrect and due to surgical error. It was concluded that porous polymers can be useful for replacement of the meniscus, provided that chemical and physical properties are optimized.

C. EVALUATION OF BIOLOGICAL GRAFT

1. Autografts

Autograft tissue is harvested elsewhere in the body and is used for a meniscal substitute. The advantage of autograft is its availability and the absence of immunological reactions. It is thought that non-meniscal tissue can transform into meniscal tissue when implanted in the knee. In both humans and sheep the meniscus was replaced by the infrapatellar adipose body.[73–76] Although there were no adverse effects, all grafts resulted in a soft tissue not resembling a normal meniscus. Also the size of the grafts decreased in time. An alternative method is to use a strip of quadriceps tendon which is sutured in place.[40,72,82,90] This method is in development and clinical results have to be awaited.

2. Allografts

Allograft menisci comies from a different animal of the same species. Allografts are the state of art to replace a complete meniscus. Both human and animal trials have proven that transplantation of a meniscus is technically possible.[59,70,77,78,83,92] Theoretically, this method provides for a replacement meniscus that is as much identical to the original one as it possibly can be. Its disadvantages have been mentioned before and include the danger of disease transmission, the limited availability and storage time and problems of sterilization. In theory, allograft menisci could give rise to immunological reactions. Although the meniscus certainly has immunologic potential, rejection is not seen in practice. However, when the attachment sites are transplanted using bone blocks, the bone can elicit an immunological response in the recipient. In practice this does not result in

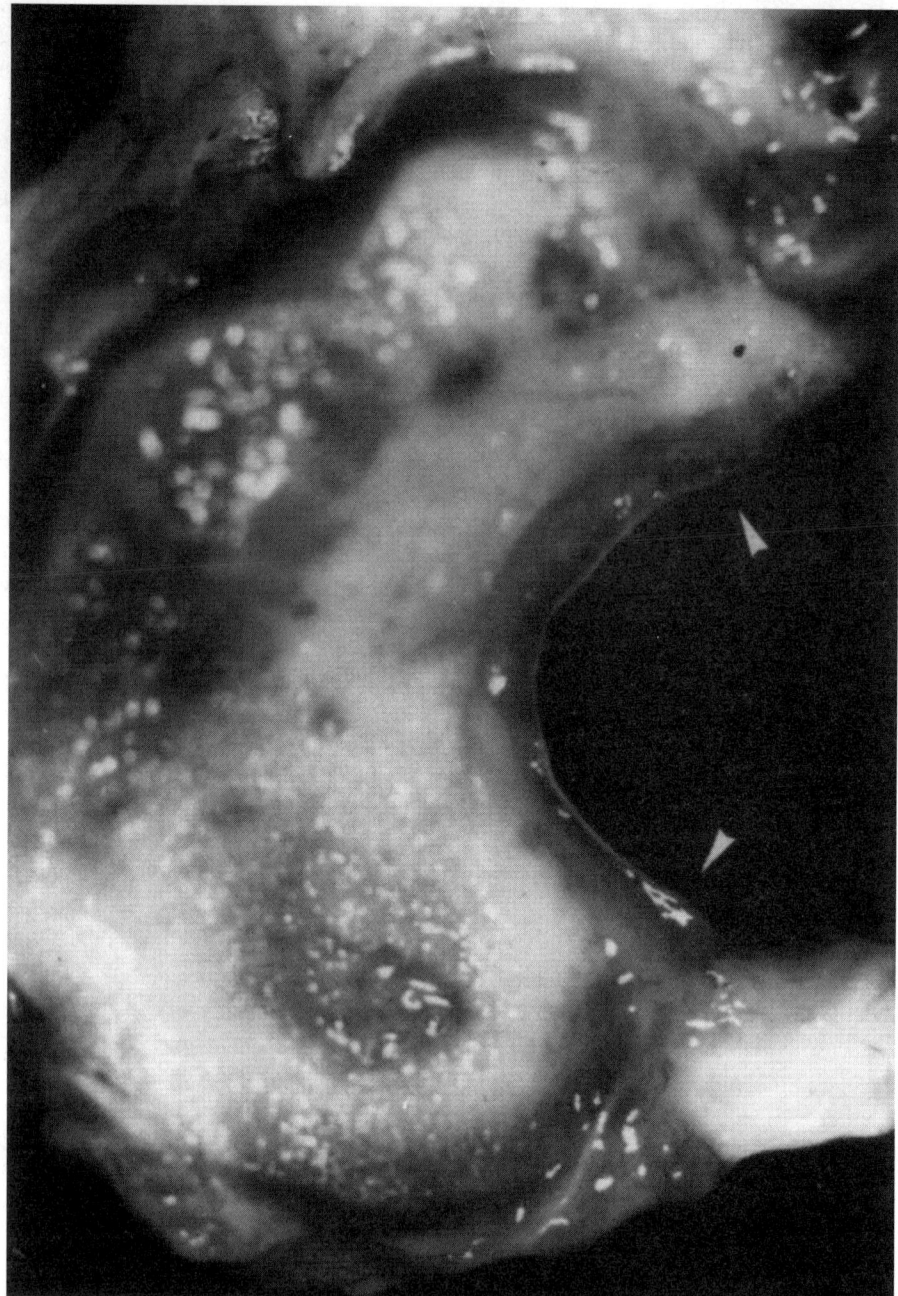

FIGURE 4. Degradable polyurethane implant after 12 weeks' implantation in the dog. The implant is covered by fibrocartilage. Arrows indicate the central rim.

rejection of the graft.[73,82] It is not exactly known whether the transplant must contain living cells or if it should only serve as a template which is repopulated by cells from the patient.

When living cells are to be transplanted, the meniscus cannot simply be frozen as this will kill the cells. Two options are available. First, fresh tissue can be used. This method has considerable logistical problems. The tissue can only be stored for a limited time. The main problem of fresh grafts is the danger of transmitting disease. The donor may be infected without yet being seropositive.

The risk of HIV transmission from soft tissue allografts in appropriately screened donors is about one in one million.[93] The second method is to use cryopreserved menisci. These menisci are frozen in a slow and controlled way using cryoprotectants. A small percentage (10–30) of the original cells will survive and preserve the mechanical properties of the transplant immediately after transplantation. Lyophilized menisci only consist of a collagen structure. The cells and proteoglycans are removed. A decrease in size was noted over time and the results are worse than for fresh and fresh-frozen grafts.[76,82] In contrast to lyophilized grafts, deep frozen menisci do show healing to the periphery but showed little revascularization or remodeling when implanted in sheep.[74] Cryopreserved menisci in dogs all healed to the periphery. Cellularity and metabolic activity returned to normal after six months. Damage to the tibial cartilage was present although it was less than after meniscectomy.[92] Menisci treated with glutaraldehyde have been used in animals but were not successful.[59] All knees showed an effusion and the menisci did not demonstrate healing to the periphery. Cellular repopulation of allografts was studied by Arnoczky.[83] He found that the deep frozen meniscus was repopulated by synovial cells from the recipient. However, the lack of repopulation of the meniscal center and the loss of the normal collagen orientation was a matter of concern. In humans fresh[73,76] or cryopreserved[77] allografts have been used with good results. This procedure can be done in conjunction with a reconstruction of the anterior cruciate ligament since meniscal transplantation in an ACL-deficient knee will result in failure. Allograft selection is a big problem. Especially, determination of the size is difficult. There is a large variation in size among patients and there is considerable difference in size between contralateral lateral and medial menisci. MRI and CT scans underestimate or overestimate the meniscal size an average of 2.8–4 mm and it is well known that improper sizing will deteriorate the results. If a meniscus is too large, the periphery may be trimmed but the circumferential collagen fibers must be kept intact. If the graft is too small it should not be used. The indications for transplantation are not well defined. Theoretically, the best time for transplantation would be immediately after meniscectomy because cartilage degeneration will be minimal and the reduction of progression of the wear would be maximal. However, clinically the authors consider patient age, knee stability, the degree of osteoarthritis and pain. A painless patient who does show cartilage degeneration on the X ray is not a candidate for transplantation. Also patient age is important. The majority of authors operate on patients who are in their twenties to forties and who are too young for a total knee arthroplasty. This indicates that meniscal transplantation still is in an experimental phase. Clinically fresh and cryopreserved allografts can heal to the periphery but their cartilage protective effect in the long term yet has to be determined.

3. Xenografts

Xenograft meniscal tissue (menisci from one animal species implanted into another species) does not seem to be a realistic option. Although rejection probably will not be a problem, there are large differences in meniscal sizes and composition among different animals. Since size and anatomy are crucial, this will result in failure.[81,96]

4. Cell-Seeding

Cell seeding of artificial scaffolds may accelerate ingrowth of fibrocartilage. This experimental technique has been used successfully for the repair of defects in articular cartilage in animals. Cells are harvested, released by enzymes and kept in appropriate culture media until they are implanted. Webber[35,36] has proven that meniscal cells can survive in tissue culture and are able to migrate. One of the problems could be the dedifferentiation process of the cells. From articular cartilage cells it is known that they lose their phenotype and become fibroblast-like cells when kept in culture. This process can be prevented when the cells are grown in a three-dimensional matrix such as a collagen gel. It is unknown if fibrocartilage cells show the same process of dedifferentiation.

Coating of scaffolds with adhesion molecules such as fibronectin or chondronectin, chemotactic factors or with growth factors may increase and accelerate cellular ingrowth and fibrocartilage formation. In tissue culture, fibrochondrocytes do respond to these factors, thereby enhancing ingrowth into a matrix.

VI. SUMMARY AND CONCLUSIONS

The meniscus is an important structure of the knee joint and is essential in preventing osteoarthritis. Every effort to preserve this structure should be made. When this is not possible, meniscal replacement is a good alternative. Allografts can provide good results in the short term but have important disadvantages. Synthetic grafts have, at least in theory, fewer disadvantages but until now none has proven to be able to match the results of allografts. Future research will have to improve their qualities in term of biomechanics, induction of fibrocartilage and others. Several animal species can be used for meniscal research but the dog may be the preferable animal.

REFERENCES

1. Sutton, J. B., *Ligaments: Their Nature and Morphology,* 2nd ed., H. K. Lewis, London, 1987.
2. Fairbank, T. J., "Knee joint changes after meniscectomy," *J. Bone Joint Surg.,* 30B, 664, 1948.
3. Fukubayashi, T. and Kurosawa, H., "The contact area and pressure distribution pattern of the knee. A study of normal and osteoarthritic joints," *Acta Orthop. Scand.,* 51, 871, 1980.
4. Kurosawa, H., Fukubayashi, T., and Nakajima, H., "Load-bearing of the knee joint," *Clin. Orthop.,* 149, 283, 1980.
5. Oretorp, N., Alm, A., Ekstrom, H., and Gillquist, J., "Immediate effects of meniscectomy on the knee joint. The effects of tensile load on knee joint ligaments in dogs," *Acta Orthop. Scand.,* 49, 407, 1978.
6. Shrive, N. G., O'Connor, J. J., and Goodfellow, J. W., "Load-bearing in the knee joint," *Clin. Orthop.,* 131, 279, 1978.
7. Voloshin, A. S. and Wosk, J., "Shock absorption of meniscectomized and painful knees: a comparative *in vivo* study," *J. Biomed. Eng.,* 5, 157, 1983.
8. Helfet, A. J., "Mechanism of derangements of the medial semilunar cartilage and their management," *J. Bone Joint Surg.,* 41B, 319, 1959.
9. Levy, I. M., Torzilli, P. A., and Warren, R. F., "The effect of lateral meniscectomy on motion of the knee," *J. Bone Joint Surg.,* 71A, 401, 1989.
10. Shoemaker, S. C. and Marlkolf, K. L., "The role of the meniscus in the anterior-posterior stability of the loaded anterior cruciate-deficient knee. Effects of partial versus total excision," *J. Bone Joint Surg.,* 68A, 71, 1986.
11. McConaill, M. A., "The movement of bones and joints. 3. The synovial fluid and its assistants," *J. Bone Joint Surg.,* 30B, 244, 1950.
12. Ahmed, A. M. and Burke, D. L., "*In vitro* measurement of static pressure distribution in synovial joints. Part I: Tibial surface of the knee," *J. Biomech. Eng.,* 105, 216, 1983.
13. Huckell, J. R., "Is meniscectomy a benign procedure? A long-term follow-up study," *Can. J. Surg.,* 8, 252, 1965.
14. Johnson, R. J., Kettelkamp, D. B., Clark, W., and Leaverton, P., "Factors affecting late results after meniscectomy," *J. Bone Joint Surg.,* 56A, 719, 1974.
15. Veth, R. P. H., "Clinical significance of knee joint changes after meniscectomy," *Clin. Orthop.,* 198, 56, 1985.
16. Veth, R. P. H., *Thesis: Over de resultaten van de mensiscectomie van de knie,* VRB Press, Groningen, The Netherlands, 1978.
17. Cox, J. S., Nye, C. E., Schaefer, W. W., and Woodstein, I. J., "The degenerative effects of partial and total resection of the medial meniscus in dogs' knees," *Clin. Orthop.,* 109, 178, 1975.
18. King, D., "The healing of the semilunar cartilages," *J. Bone Joint Surg.,* 18, 333, 1936.
19. Arnoczky, S. P. and Warren, R. F., "Microvasculature of the human meniscus," *Am. J. Sports Med.,* 10, 90, 1982.

20. Arnoczky, S. P. and Warren, R. F., "The microvasculature of the meniscus and its response to injury. An experimental study in the dog," *Am. J. Sports Med.*, 11, 131, 1983.
21. Cassidy, R. E. and Shaffer, A. J., "Repair of peripheral meniscus tears: a preliminary report," *Am. J. Sports Med.*, 9, 209, 1981.
22. DeHaven, K. E., "The role of the meniscus," in *Articular Cartilage and Knee Joint Function: Basic Science and Arthroscopy*, Ewing, J. W., Ed., Raven Press, New York, 1990, 103.
23. Henning, C. E. and Lynch, M. A., "Vascularity for healing of meniscus repairs," *Arthroscopy*, 3, 13, 1987.
24. Gershuni, D. H., Skyhar, M. J., Danzig, L. A., et al., "Experimental models to promote healing of tears in the avascular segment of canine knee menisci," *J. Bone Joint Surg.*, 71A, 1363, 1989.
25. Veth, R. P. H., den Heeten, G. J., Jansen, H. W. B., and Nielsen, H. K. L., "An experimental study of reconstructive procedures in lesions of the meniscus," *Clin. Orthop.*, 181, 250, 1983.
26. Heatley, F. W., "The meniscus. Can it be repaired? An experimental investigation in rabbits," *J. Bone Joint Surg*, 62B, 397, 1980.
27. Arnoczky, S. P., Warren, R. F., and Spivak, J. M., "Meniscal repair using an exogenous fibrin clot: an experimental study in dogs," *J. Bone Joint Surg.*, 70A, 1209, 1988.
28. Hashimoto, J., Kurosaka, M., Yoshiya, S., et al., "Meniscal repair using fibrin glue and endothelial cell growth factor (ECGF) — an experimental study in dogs." *Trans. Orthop. Res. Soc.*, 15, 562, 1990.
29. Leenslag, J. W., Gogolewski, S., and Pennings, A. J., "Resorbable materials of poly(L-lactide). V. Influence of secondary structure on the mechanical properties and hydrolyzability of poly(L-lactide) fibres produced by a dry-spinning method," *J. Appl. Polym. Sci.*, 29, 2829, 1984.
30. Veth, R. P. H., Jansen, H. W. B., Leenslag, J. W., et al., "Experimental meniscal lesions reconstructed with a carbon fibre polyurethane-poly(L-lactide) graft," *Clin. Orthop.*, 202, 286, 1986.
31. King, D., "Regeneration of the semilunar cartilage," *Surg. Gynecol. Obstetr.*, 36, 167, 1936.
32. Benjamin, M. and Evans, E. J., "Fibrocartilage," *Acta Anat.*, 171, 1, 1990.
33. Ghadially, F. N., Lalonde, J. M. A., and Wedge, J. H., "Ultrastructure of normal and torn menisci of the human knee joint," *J. Anat.*, 136, 773, 1983.
34. Mc Devitt, C. A., Miller, R. R., and Spindler, K. P., "The cells and cell matrix interactions of the meniscus," in *Knee Meniscus: Basic and Clinical Foundations*, Mow, V. C., Ed., Raven Press, New York, 1992, 29.
35. Webber, R. J., Harris, M. G., and Hough, A. J., "Cell culture of rabbit meniscal fibrochondrocytes. Proliferative and synthetic response to growth factors and ascorbate," *J. Orthop. Res.*, 3, 36, 1985.
36. Webber, R. J., York, L., Vander Schilden, J. L., and Hough, A. J., Jr., "Fibrin clot invasion of rabbit meniscal fibrochondrocytes in organ culture," *Trans. Orthop. Res. Soc.*, 12, 470, 1987.
37. Evans, E. J., Benjamin, M., and Pemberton, D. J., "Fibrocartilage in the attachment zones of the quadriceps tendon and patellar ligament of man," *J. Anat.*, 171, 155, 1990.
38. McDevitt, C. A. and Webber, R. J., "The ultrastructure and biochemistry of meniscal cartilage," *Clin. Orthop.*, 252, 8, 1990.
39. Le Minor, J. M., "Comparative morphology of the lateral meniscus of the knee in primates," *J. Anat.*, 170, 161, 1990.
40. Kohn, D. and Moreno, B., "Meniscus insertion," *Orthopade*, 23, 98, 1994.
41. Stone, K. R., Rodkey, W. G., Webber, R. J., et al., "Future directions. Collagen-based prostheses for meniscal regeneration," *Clin. Orthop.*, 252, 129, 1990.
42. Stone, K. R., Rodkey, W. G., Webber, R. J., et al., "Development of a prosthetic meniscal replacement," in *Knee Meniscus: Basic and Clinical Foundations*, Mow, V. C., Ed., Raven Press, New York, 1992, 165.
43. Arnoczky, S. P., Warren, R. F., and Kaplan, N., "Meniscal remodeling following partial meniscectomy — An experimental study in the dog," *Arthroscopy*, 1, 247, 1985.
44. Clark, C. R. and Ogden, J. A., "Development of the menisci in the human knee joint," *J. Bone Joint Surg.*, 65A, 538, 1983.
45. Klompmaker, J., Jansen, H. W. B., Veth, R. P. H., et al., "Meniscal repair by fibrocartilage? An experimental study in the dog," *J. Orthop. Res.*, 10, 359, 1992.
46. Klompmaker, J., Veth, R. P. H., Jansen, H. W. B., et al., "Meniscal replacement using a porous polymer prosthesis. A preliminary study in the dog," *Biomaterials*, 17, 1169, 1996.

47. Klompmaker, J., *Thesis: Porous polymers for repair and replacement of the knee joint meniscus and articular cartilage*, De Regenboog, Groningen, The Netherlands, 1992.
48. Klompmaker, J., Veth, R. P. H., Jansen, H. W. B., et al., "A porous polymer implant for repair of meniscal lesions: A preliminary study in dogs," *Biomaterials*, 12, 810, 1991.
49. Klompmaker, J., Jansen, H. W. B., Veth, R. P. H., et al., "Porous implants for knee joint meniscus reconstruction: A preliminary study on the role of pore sizes in ingrowth and differentiation of fibrocartilage," *Clin. Mater.*, 14, 1, 1994.
50. Klompmaker, J., Jansen, H. W. B., Veth, R. P. H., et al., "Meniscal repair by fibrocartilage in the dog: characterization of the repair tissue and the role of vascularity," *Biomaterials*, 17, 1685, 1996.
51. Klompmaker, J., Jansen, H. W. B., Veth, R. P. H., et al., "Porous polymer implants for repair of full-thickness defects of articular cartilage. An experimental study in the dog," *Biomaterials*, 13, 625, 1992.
52. Veth, R. P. H., den Heeten, G. J., Jansen, H. W. B., and Nielsen, H. K. L., "Repair of the meniscus. An experimental investigation in rabbits," *Clin. Orthop.*, 175, 258, 1983.
53. Sommerlath, K., "The effect of anterior cruciate ligament resection and immediate or delayed implantation of a meniscus prosthesis on knee joint biomechanics and cartilage," *Clin. Orthop.*, 289, 276, 1993.
54. Glass, R. S., Barnes, W. M., Kells, D. U., et al., "Ossicles of knee menisci. Report of seven cases," *Clin. Orthop.*, 111, 163, 1975.
55. Symeonides, P. P. and Ioannides, G., "Ossicles in the knee menisci. Report of three cases," *J. Bone Joint Surg.*, 54A, 1288, 1972.
56. O'Connor, B. L., "The histological structure of dog knee menisci with comments on its possible significance," *Am. J. Anat.*, 147, 407, 1976.
57. Toyonaga, T., Uezaki, N., and Chikama, H., "Substitute meniscus of Teflon-net for the knee joint of dogs," *Clin. Orthop.*, 179, 291, 1983.
58. White, R. A., Hirose, F. M., Sproat, R. W., Lawrence, R. S., and Nelson, R. J., "Histopathologic observations after short-term implantation of two porous elastomers in dogs," *Biomaterials*, 2, 171, 1981.
59. Canham, W. and Stanish, W., "A study of the biological behavior of the meniscus as a transplant in the medial compartment of a dog's knee," *Am. J. Sports Med.*, 14, 376, 1986.
60. Gigante, A., Specchia, N., and Greco, F., "Age-related distribution of elastic fibers in the rabbit knee," *Clin. Orthop.*, 308, 33, 1994.
61. Messner, K., "Meniscal substitution with a Teflon-periosteal composite graft: a rabbit experiment," *Biomaterials*, 15, 223, 1994.
62. Messner, K., "Review. The concept of a permanent synthetic meniscus prosthesis: a critical discussion after 5 years of experimental investigations using Dacron and Teflon implants," *Biomaterials*, 15, 243, 1994.
63. Messner, K., "Cartilage mechanics and morphology, synovitis and proteoglycan fragments in rabbit joint fluid after prosthetic meniscal substitution," *Biomaterials*, 14, 163, 1993.
64. Messner, K., "Prosthetic replacement of the rabbit medial meniscus," *J. Biomed. Mater. Res.*, 27, 1165, 1993.
65. Sommerlath, K., "The effects of an artificial meniscus substitute in a knee with a resected anterior cruciate ligament," *Clin. Orthop*, 289, 276, 1993.
66. Sommerlath, K., "The Effect of a meniscal prosthesis on knee biomechanics and cartage: an experimental study in rabbits," *Am. J. Sports Med.*, 20, 73, 1992.
67. Sommerlath, K. and Gillquist, J., "Late artificial meniscus implantation in an osteoarthritic knee: an experimental study in rabbits," *Trans. Eur. Soc. Biomech.*, 7, 25, 1990.
68. Sommerlath, K., "Biomechanical characteristics of different artificial substitutes for rabbit medial meniscus and effect of pore size on knee cartilage," *Clin. Biomech.*, 7, 97, 1992.
69. Wood, D. J., Minns, R. J., and Strover, A., "Replacement of the rabbit medial meniscus with a polyester-carbon fibre bioprosthesis," *Biomaterials*, 11, 13, 1990.
70. Zukor, D. J., Rubins, I. M., Daigle, M. R., et al., "Allotransplantation of frozen irradiated menisci in rabbits," *Trans. Orthop. Res. Soc.*, 15, 219, 1990.
71. Fabbriciani, C., Lucania, L., Milano, G., et al., "Meniscal allografts: cryopreservation vs deep-frozen technique. An experimental study in goats," *Knee Surg. Sports Traumatol. Arthrosc.*, 5, 124, 1997.

72. Kohn, D., "Medial meniscus replacement by a tendon autograft. Experiments in sheep," *J. Bone Joint Surg.*, 74B, 910, 1992.
73. Milachowski, K. A., Kohn, D., and Wirth, C. J., "Transplantation of allogeneic menisci," *Orthopade*, 23, 160, 1994.
74. Milachowski, K. A., Weismeier, K., and Wirth, C. J., "Homologous meniscus transplantation. Experimental and clinical results," *Int. Orthop.*, 13, 1, 1989.
75. Milachowski, K. A., Kohn, D., and Wirth, C. J., "The infrapatellar adipose body in meniscus replacement," *Unfallchirurg*, 16, 190, 1990.
76. Milachowski, K. A., Kohn, D., and Wirth, C. J., "Arthroscopische Befunde nach Meniscustransplantation und Meniskusersatz," *Arthroscopie*, 3, 57, 1990.
77. Boer de, H. H. and Koudstaal, J., "Failed meniscus transplantation. A report of three cases," *Clin. Orthop.*, 306, 155, 1994.
78. Verdonk, R., Van Daele, P., Claus, B., et al., "Viable meniscus transplantation," *Orthopade*, 23, 153, 1994.
79. Meachim, G., "Light microscopy of India ink preparations of fibrillated cartilage," *Ann. Rheum. Dis.*, 31, 457, 1972.
80. Mow, V. C., Ratcliffe, A., Chern, K. Y., and Kelly, M. A., "Structure and function relationships of the menisci of the knee," in *Knee Meniscus: Basic and Clinical Foundations*, Mow, V. C., Ed., Raven Press, New York, 37, 1992.
81. Chen, M. I., Branch, T. P., and Hutton, W. C., "Is it important to secure the horns during lateral meniscal transplantation? A cadaveric study," *Arthroscopy*, 12, 174, 1996.
82. Kohn, D., "Meniscus replacement," *Orthopade*, 23, 164, 1994.
83. Arnoczky, S. P., O'Brien, S. J., DiCarlo, E. F., et al., "Cellular repopulation of deep-frozen meniscal autografts. An experimental study in the dog," *Trans. Orthop. Res. Soc.*, 13, 145, 1988.
84. Arnoczky, S. P., "Material properties of the normal medial bovine meniscus," *J. Orthop. Res.*, 7, 771, 1989.
85. Rodkey, W. G., Stone, K. R., and Steadman, J. R., "Replacement of the irreparably injured meniscus," *Sports Med. Arthrosc. Rev.*, 1, 168, 1993.
86. King, D., "The function of semilunar cartilages," *J. Bone Joint Surg.*, 18, 1069, 1936.
87. Smillie, I. S., *Injuries of the Knee Joint*, 4th ed., Livingstone, Edinburgh, 1970, 98.
88. Smillie, I. S., "Regeneration of the semilunar cartilages in man," *Br. J. Surg.*, 31, 398, 1943.
89. Yannas, I. V. and Burke, J. F., "Design of an artificial skin I. Design principles," *J. Biomed. Mater. Res.*, 14, 65, 1980.
90. Kohn, D., "Autograft meniscus replacement: experimental and clinical results," *Knee Surg. Sports Traumatol. Arthrosc.*, 1, 123, 1993.
91. Stone, K. R., Rodkey, W. G., Webber, R. J., et al., "Meniscal regeneration with copolymeric collagen scaffolds. *In vitro* and *in vivo* studies evaluated clinically, histologically, and biochemically," *Am. J. Sports Med.*, 20, 104, 1992.
92. Arnoczky, S. P., Warren, R. F., and McDevitt, C. A., "Meniscal replacement using a cryopreserved allograft," *Clin. Orthop.*, 252, 121, 1990.
93. Stone, K. R., Rodkey, W. G., Webber, R. J., et al., "Future directions. Collagen-based prosthesis for meniscal regeneration," *Clin. Orthop.*, 252, 129, 1990.
94. Stone, K. R., Stoller, D. W., Irving, S. G., et al., "3D MRI volume sizing of knee meniscus cartilage," *Arthroscopy*, 10, 641, 1994.
95. Veltry, D. M., Warren, R. F., Wickiewicz, T. L., O'Brien, S. J., Current status of allograft meniscal transplantation, *Clin. Orthop.*, 303, 44, 1994.
96. Lazovic, D., Wirth, C. J., Knosel, T., et al., "Meniscus replacement using incongruent transplants — an experimental study," *Z. Orthop. Ihre Grenzgeb.*, 135, 131, 1997.

18 Animal Models of Osteoarthritis

Theodore R. Oegema, Jr. and Denise Visco

CONTENTS

I. Introduction 349
II. Animal Models of Osteoarthritis 350
    A. Spontaneous Models 350
    B. Animal Models of Mechanically/Surgically-Induced Disease 352
    C. Chemically-Induced Models 352
    D. Author-Preferred Model 352
    F. General Precautions 355
III. Evaluation Methods 355
    A. Clinical Evaluation 355
        1. Physical Exam 355
        2. Gait Analysis 355
        3. Kinematics 356
        4. Arthroscopy 356
        5. Radiography 356
        6. MRI 356
        7. Ultrasound 356
        8. Scintigraphy 357
        9. Fluid Analyses 357
    B. Macroscopic Analysis 357
    C. Microscopic Analysis 357
    D. Molecular Biology Methodology 358
    E. Biomechanical Properties 358
IV. Applications of the Models 359
    A. Evaluation of the Pathophysiology of Osteoarthritis 359
    B. Intervention and Osteoarthritis Models 359
    C. Identification of Markers of Osteoarthritis 359
    D. Bone Ingrowth to Implants under Osteoarthritic Conditions 360
Acknowledgments 360
References 361

I. INTRODUCTION

The development of osteoarthritis involves a complex interaction between environmental, biological (including sex and genetic contributions), and mechanical components and, at least in the latter stages of disease, involvement of all the tissues surrounding the joint.[1,2,3] Human tissues usually represent end-stage disease, therefore, animal models are critical to understanding the disease, and validating interventions.[1,4] This is a practical overview of the osteoarthritis models, their advantages and disadvantages, and is not comprehensive. Readers are referred to earlier reviews that outline relevance of models to human disease.[1,4,5]

All models have limitations, so first the questions being asked need to be clearly defined. For comparison to human disease, histologic characteristics of osteoarthritis that develop naturally in three commonly used species are illustrated in Figure 1. Each species demonstrates gross similarities and show subtle, or not so subtle, differences that should be kept in mind. There are the usual concerns, such as, are the molecular targets the same as in humans or even present, and especially for a drug study, is the metabolism of the cartilage and of the compound appropriate? However, some questions are unique to osteoarthritis models. These are: Will joint geometry make a difference; Does treatment evaluation require progression to the same end-stage disease as the human; What is the balance between how fast the model develops and the expected treatment effects. Human osteoarthritis develops over decades, thus a sustained, modest effect might be sufficient to delay progression, but may be undetectable if it is tested in an aggressive model. In general, it is recommended that an intervention be tried in three models involving two species with at least one being a surgical model.

After the model is selected, the choice of read-outs is the next most important parameter and should be determined in advance. The method of evaluation should be appropriate to questions that are being addressed, both in terms of sensitivity and specificity, and reflective of the aspect of joint physiology that will be followed. Additional information that will be collected must also be decided at the beginning. Some currently used methods are addressed later in the chapter.

Pritzker[4] defined availability criteria for animal models of osteoarthritis and he included the following: cost; previous research in the model (background, knowledge, and peer acceptance); ease of experimental manipulation; ease of handling; supply of animals of the appropriate age, size and sex; availability of experimental techniques for model evaluation and manipulation; anatomical and physiological attributes including sufficient joint tissue and fluids; quantitative resemblance to human tissue; weight-bearing or nonweight-bearing nature of joint; availability of matched controls; life span of animal relative to onset of disease and time-frame of disease progression; availability of nutritional and exercise history; and genetic background. To this list might be added: what phase of the process is being modeled (early-, middle- or end-stage), does bone quality and bone biology make a difference, relevance of use of immature, mature, or aged animals, and especially in rodents, what are the consequences of the presence of an open growth plate?

In broad terms, there are three types of models. They are spontaneous or naturally-occurring models, a group that includes genetically-manipulated mouse models, chemically-induced models, and mechanically-induced models. Mixed models such as combined mechanical and chemical alterations have seen only limited development. For additional discussion of early literature for many of these models, the readers are referred to older reviews.[1,4,5–7]

II. ANIMAL MODELS OF OSTEOARTHRITIS

A. Spontaneous Models

All vertebrate animal species develop osteoarthritis, but at different rates and sometimes with different joint distributions (Table 1).[18] Spontaneous models have the advantage in that they occur naturally, but their major disadvantage is the time they take to develop. These are being expanded with transgenic models where matrix or cellular proteins that rapidly cause osteoarthritis (such as types IX and XI collagen mutations)[20,21] are targeted, but where the phenotype is not so severe that it prevents development and growth. Mouse models will become very powerful tools for exploring the contribution of specific proteins to the osteoarthritic process and may provide accelerated models for evaluating interventions.

Animal Models of Osteoarthritis

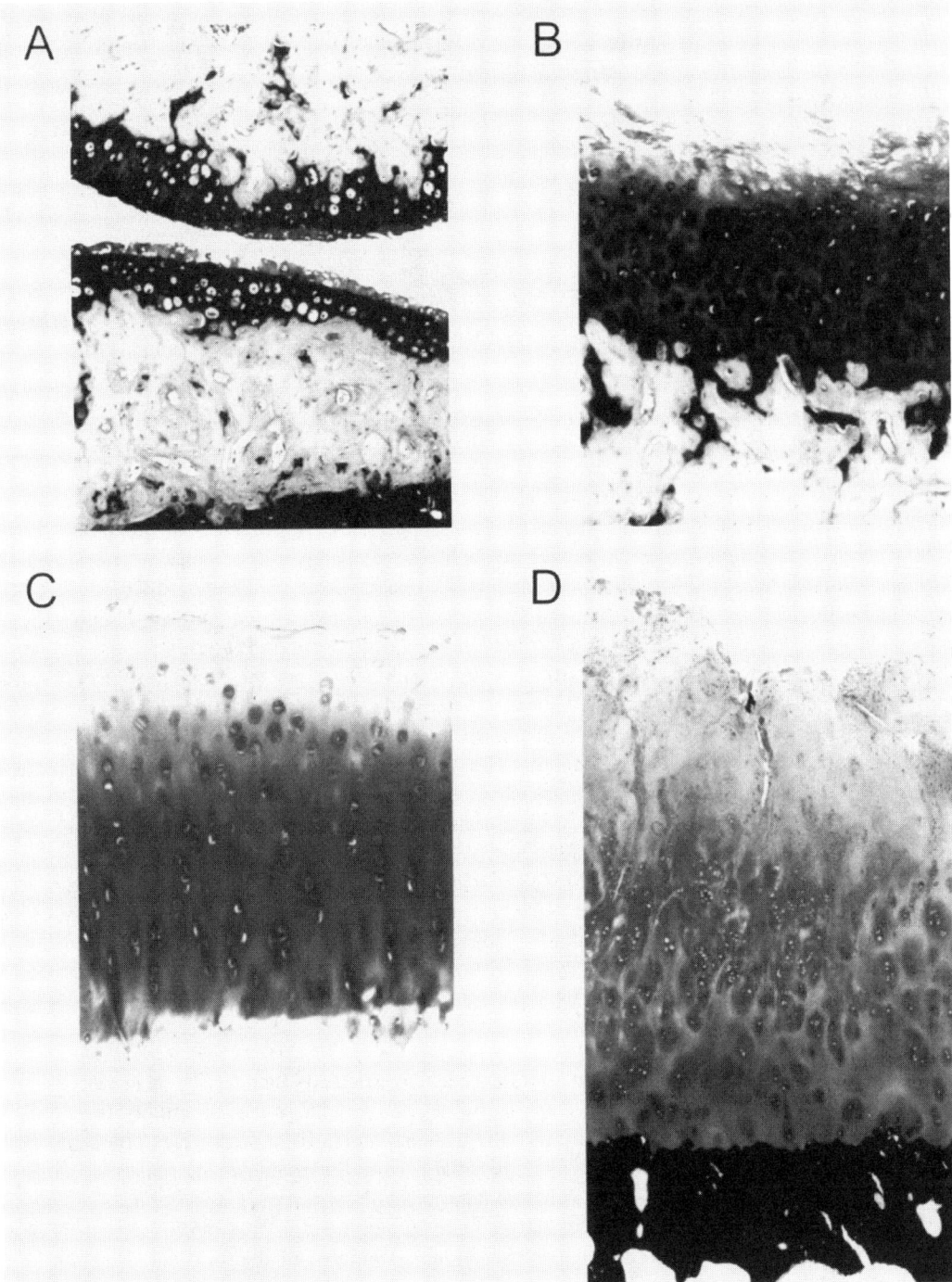

FIGURE 1. Photomicrographs from a sagittally sectioned medial portion of the knee joint of a partial medial meniscectomized mouse six weeks post surgery (A), sagittal section from the medial tibial condyle of a partial medial meniscectomized guinea pig nine weeks post surgery (B), sagittal section through the distal aspect of the medial trochlear ridge of the left stifle joint of a dog 12 weeks post transection of the anterior cruciate ligament (C), and osteoarthritic cartilage from the right tibial condyle of a 72 year old female who underwent arthroplasty surgery. Toluidine blue, original magnification 12.8×.

TABLE 1
Spontaneous Models

Animal	Strain	References
Mouse	C57BL	8,9
	STR/IN	10
	STR/ORT	11
Guinea pig	Hartley	12–15
Canine (hip)	Labrador retriever	16
Monkey	Macaca malatta	17
Rat		18
Hamster		19

B. Animal Models of Mechanically/Surgically-Induced Disease

Mechanical models were developed to provide an accelerated, but controlled model of osteoarthritis or, in some cases, to demonstrate the contribution of particular mechanical components or structural elements to the development of osteoarthritis (Table 2). Attempts to repair the initial alteration and potentially make the model reversible have not been done. The anterior (cranial) cruciate ligament transection model in the canine is considered the gold standard where the early, mid- and end-stage aspects of the disease have been carefully documented and eburnation to bone takes 3–4 years.[46,47] Meniscectomy is performed in several versions, and progression and extent of late changes depend on the model. Partial medial meniscectomy in the rabbit produces a mild, progressive disease.[1] The complete meniscectomy in the sheep produces rapid progression.[42,63] The partial meniscectomy, in combination with anterior cruciate ligament resection or anterior cruciate ligament and posterior cruciate transection, also rapidly produces extensive disease in the rabbit.[1,41,64] Extraarticular surgeries, such as varus and valgus osteotomies and muscle resection, produce disease more slowly (Table 2). Osteotomies need additional instrumentation and more operating time, and are technically challenging.

Especially in mechanical/surgical models, the animal often alters use of the opposite leg, which can cause persistent changes.[59] These changes do not progress, so the opposite limb may be an appropriate gross morphologic or histologic grading control for mid- and end-stage disease studies, but a poor control for detailed studies.[59]

C. Chemically-Induced Models

Direct injection of active agents into joints has been investigated (Table 3). These models usually show dose-dependent damage and at least at low levels of damage are potentially reversible.[78] Characteristically, for the enzymatic models, the damage begins from the surface down and may be varied throughout the joint because of differential access of reagent or differences in cartilage composition. Damaged areas are then vulnerable to normal joint forces. Progression can be accelerated with exercise or, in some cases, delayed with immobilization,[68] or improved with passive motion.[76] The reagent may also damage the secondary structures of the joint and alter the joint mechanics.[63] Mechanical and chemical models of other joints, besides the knee, have been reported (Table 4).

D. Author-Preferred Model

If the objective is to study the etiopathogenesis of osteoarthritis, then each model has its merits with many of the models elucidating changes in chondrocyte and bone metabolism. So the model may best be decided by the requirement of the methods that are going to be applied, such as the

TABLE 2
Mechanical/Surgical Models

Models	Animal[Ref.]	Comments
Focal defects	canine,[22] rabbit[23]	see Chapter 16
Scarification	rabbit,[24] canine[25]	canine: nc at one year
Freezing	rabbit[26]	nc in 6 months; sb in 12 months
Repetitive impulse loading	rabbit,[27] guinea pig[28]	sf, sb
Closed transarticular impact	rabbit,[29,30] canine[61,92]	sf, eb, cl, pl
Open transarticular impact	canine[32]	sf, eb, cl, pl
Tibial varus osteotomy	rabbit[33]	os, sf, cl in 34 weeks
Tibial valgus osteotomy	rabbit[33,34]	os, sf, cl in 34 weeks
Gluteal resection	guinea pig[35]	
Gluteal resection (intrapatellar ligament release)	guinea pig[35]	sf, cl, sb in 24 weeks
Denervation	rabbit,[36] canine[37]	rabbit: cd, canine: nc, 64 weeks
Denervation/ACL resection	canine[37]	accelerated change over ACL
Partial meniscectomy (anterior aspect medial 1/3)	rabbit,[1] mouse,[38] guinea pig[39]	rabbit, ~70% develop sf, cl, os in 4, 8 weeks, 20–25 extensive disease
Partial lateral meniscectomy (colateral sesmoid ligaments)	mouse[40]	
Partial meniscectomy + PCL + ACL	rabbit[41]	
Meniscectomy	sheep,[42] rabbit,[43] monkey[44]	
Anterior cruciate resection	rat,[45] canine,[46–48] rabbit[49]	
Posterior cruciate resection	rabbit[50]	mild focal pitting
Patellar dislocation	canine[22]	
Patellalectomy	rabbit,[51] sheep[52]	cd, sf, eb, os, pl
Polyethylene insert	rabbit[53]	rapid, eb, cl, os, pl
Step-off fractures	rabbit[54]	pl, cl, os, eb
Resection of femoral condyle	rat,[55] rabbit[56]	excess pannus
Immobilization	rat,[57] rabbit,[7,58] canine[59,60]	
Immobilization with compress	rabbit,[61] rat[62]	

cd = cell death
cl = cloning
eb = bone eburnation
nc = no changes
os = osteophyte
sb = subchondral bone
pl = proteoglycan loss
sf = surface fibrillation

specificity of antibodies or molecular biology probes. However, if the question is which model or models are best for the identification of disease modifying agents then the answer is extremely difficult. Naturally occurring arthritis occurs in many species including certain inbred stains of mice (C57bl, STR/1N, STR/ORT) and guinea pigs (Table 1). The difficulties with utilizing naturally occurring models are if they occur due to a genetic abnormality and the time of onset and severity.

Surgical models of mechanical instability which are mimics of chronic traumatic osteoarthritis are the most commonly utilized models. There are four prime advantages: incidence is 100%, time of onset is known, site where osteoarthritis develops is known and the duration can be short. But there are also disadvantages: variability in severity, size of animals (>20 kg dogs, 25 gm mice), age of animals (guinea pigs and mice immature while most dog ages are unknown), preexisting or naturally occurring osteoarthritis (guinea pig, dog, C57bl mice). The choice of animal again is

TABLE 3
Chemical Models with Joint Injection

Cell Perturbation	Species[Ref.]	Comments
Iodoacetic acid	rabbit,[65] guinea pig,[66] chicken,[67] canine[68]	cd, sf, eb, cl, os, pl
IL–1b	rat[69]	pl
Vitamin A	rabbit[70]	erosion, os
Fibronectin fragments	rabbit[71]	pl
H_2O_2 and exercise	rabbit[72]	

Direct Matrix Depletion	Species	Comments
Trypsin	rabbit[73]	
Papain	rabbit,[74] guinea pig[75]	cl, sf, pl
Chymopapain	rabbit[17,76]	pl, sl, eb, cl
Chondroitinase ABC	rabbit[77]	pl
Collagenase, bacterial	mouse[63]	

cd = cell death
cl = clones
eb = bone eburnation
os = osteophyte
sb = subchondral bone
pl = proteoglycan loss
sf = surface fibrillation

TABLE 4
Osteoarthritis Model for Joints Other Than the Knee

Model	Animal[Ref.]	Comments
Lumbar facet joint (intradiscal chymopapain)	canine[79]	transient, early changes
Cervical facets-bipedal	rat[80]	
Hip — papain injection	rabbit[81]	similar to knee
Hip — osteotomy	canine[82]	rapid aggressive change
Temporomandibular joint (natural)	mouse[83]	
Temporomandibular joint (surgical)	rabbit[84]	
Temporomandibular joint	rabbit[85]	
Wrist, rat, spontaneous	rat[18]	

For keys see table 3.

dependent upon the question to be answered. Surgical models are available for use in mice,[86] guinea pigs,[12] rabbits,[1,41] goats, sheep and dogs (Table 2). Mice are small, easy to handle and administer compounds, large numbers can be used, and transgenic animals can be studied; however, the anatomy of the joints differs significantly from that of humans in that mice do not go to full skeletal maturity (growth plates fail to close), the articular cartilage is only 3–5 cell layers thick (Figure 1), and the horns of menisci are ossified. Rabbits are more difficult to handle and to dose; the joint anatomy and gait (hop) differs significantly from that of humans. Goat and sheep are ruminants which are of concern when compounds are to be studied, are large (>50 kg), and stifle joint differs from humans anatomically in that it contains three patellar ligaments and synovial fossae develop.

Goats are also prone to develop caprine arthritis encephalitis, an inflammatory arthritis, which could complicate the study and analysis of joints.

Of all the surgical models, the canine anterior cruciate ligament transection model is the best characterized. The clinical, radiographic, morphologic, and biochemical changes that occur following surgery are similar to those found in naturally occurring osteoarthritis in dogs[48,87] and humans.[88] The instability can be created either open through a medial[38,89] or lateral[88] arthrotomy, or closed via a stab incision.[38,87] However, the major limitation for analysis of pharmacologic agents is length of time (approximately 3–4 years),[46] that the dog takes to progress to clinical, not early, osteoarthritis.

E. General Precautions

Good animal husbandry practices are important. The animal models, and especially the spontaneous models, may depend on the animal source and the cleanliness of the colony. If age is important, using animals of known birth date has advantages over weight, especially as they near maturity. If closed growth plates are required, X rays are recommended if the ages are unknown. Controlling stress levels is important and the environment (temperature, noise and lighting) and animal handling should be stress-free and controlled as much as possible. A controlled diet is also important since osteoarthritis progression is highly dependent on weight,[90] in the mouse,[8] guinea pig,[13] and dogs.[91] Modest exercise or conversely deconditioning may have dramatic effects.[19] This may be even more important as new larger runs are required for dogs and the animals may now be able to more frequently jump on their hind limbs and socialize. Enrichment for other animals may also change activity levels.

III. EVALUATION METHODS

A. Clinical Evaluation

In vivo evaluation methods vary dramatically with the size of the animal. The methods can include physical exam, gait analysis, kinematics, arthroscopy, conventional radiography, magnetic resonance imaging (MRI), diagnostic ultrasound, computerized tomography (CT), microCT,[35] scintigraphy, and fluid analyses.

1. Physical Exam

A good physical exam is a necessity as part of the records. This should include comments on general health, weight, and temperature, noting food and water intake and hair and eye quality. The animals should be examined for altered gait, signs of joint swelling, redness of skin, temperature over the joint, active and passive range of motion of the control and experimental limb, joint stiffness, and the presence of instability. After surgery, wound healing should also be monitored.

2. Gait Analysis

Gait analysis using a strain gauge force plate can be done on animals as small as a mouse (single channel) to as large as a horse (6 channel). Determination of the ground reaction forces is a simple procedure which describes the forces and moments of the foot to ground contact that can be used as an indicator of limb (not joint) use under varying conditions. The forces can be described by orthagonal vector components including the vertical (F^z), craniocaudal (F^y; braking and propulsion) and mediolateral (F^x). Gait analysis has been used in dogs to describe both the normal,[39] as well as that of dogs with either acute or chronic arthritis.[46,89,92] While initial costs can be as high as $30,000 to equip a gait analysis lab, the day to day costs are low. Equipment needs include a room, computer, force plate (AMTI or Kistler) and program.

3. Kinematics

Kinematics, the study of movements of the limb, has been used by investigators to study the normal gait and the changes that occur in dogs with anterior cruciate ligament deficiency,[93-95] and/or massive hind limb deafferentation.[95] The change in kinematic parameters after transection of the anterior cruciate ligament was shown not to be limited to just the destabilized joint, but also affected the other joints in that limb, joints in the contralateral limb, and the vertical movements of the rump.[94]

4. Arthroscopy

Arthroscopy can be utilized to visualize the surface of most components of the joint (articular cartilage, intra-articular ligaments, menisci and synovium), but it cannot be used to examine the deep structures (i.e., subchondral bone), joint effusion or extrasynovial structures. Furthermore, tactile inspection can also be made by probing and/or applying compression. Currently, due to the size of the scopes, arthroscopy is limited to larger animals. It is invasive which can have effects on the joint, can be painful thus impacting on the model if the animal does not use the limb, requires anesthesia and thus post-procedure care, significant operator training is necessary, and it is expensive.

5. Radiography

Imaging techniques have also been used to follow the development of arthritis. Conventional radiographs allows 2D evaluation of secondary changes of osteoarthritis, however, joint space (cartilage space) cannot be addressed unless views are made during weight-bearing which is difficult to impossible to accomplish in most animals. Radiographic grading schemes using standard views of synovial joints have been developed and utilized to describe the soft and hard tissue changes in both experimental and naturally-occurring osteoarthritis in most species.[96] In dogs, Widmer *et al.*[96] were able to describe the development of osteophytes as early as two weeks following anterior cruciate ligament transection in dogs. They attributed the early identification to their use of five different views. Arthrography can be used to delineate intra-articular surfaces of cartilage, menisci, ligaments and synovium. If the equipment is present on site, radiographs are inexpensive and easy to take; however, analysis can be difficult and requires training and rigorous standardization.

6. MRI

Magnetic resonance imaging has been used to describe the progression of osteoarthritis in mice,[11] rat,[97] guinea pigs,[98] rabbits,[99] and dogs.[46,96] MRI provides exceptional soft tissue contrast not afforded by other imaging methods. Therefore, internal components of synovial joints, i.e., menisci, synovium, ligaments, etc. can be seen. While MRI can be useful in following the development of osteoarthritis over time, it is limited in the size of animal that can be examined, the expense of the equipment and the training necessary to evaluate the images. MRI of small animals requires the availability of instruments with strong magnetic fields in order to obtain good resolution.[11,97,98]

7. Ultrasound

Diagnostic ultrasound also allows evaluation of internal components of synovial joints, but is limited to medium to large joints and is extremely operator dependent. The normal ultrasonic appearance of the canine stifle joint has been described,[47] however, the ultrasonic appearance of the arthritic joint has not been previously described.

8. Scintigraphy

Scintigraphy with bone seeking radiopharmaceuticals provides excellent physiologic and limited anatomic information regarding bone turnover and early soft tissue changes associated with osteoarthritis. These radiopharmaceuticals are used to study both hard and soft tissue changes associated with osteoarthritis. Bone phase scintigraphic evaluation is sensitive for identifying initiation of osteophytosis and subchondral remodeling. Blood pooling and soft tissue phase imaging will detect the early inflammatory changes of synovitis. In rabbits the radiopharmaceutical 99mTc methylene diphosphonate was increased in operated knee joints in the area of developing osteophytes as early as one week post destablization.[100] Later uptake occurred in the subchondral bone beneath areas of damaged cartilage. Tc-labeled macrophage methodology is also used as a way of detecting low levels of synovitis in rabbit osteoarthritis models.[101]

9. Fluid Analyses

Serum, urine, and synovial fluid can be evaluated biochemically for its content of cartilage breakdown products (i.e., keratan sulfate)[102,103] and/or production of mediators (i.e., TNF, IL–1, matrix metalloproteinases) as well as parent compound and metabolite levels.[104]

B. MACROSCOPIC ANALYSIS

Following euthanasia, the joints of interest should be inspected for changes in muscle mass, thickening of the joint capsule (specific anatomical areas noted), and the synovial fluid collected (color, turbidity, and amount noted) and centrifuged to remove cells, aliquoted and stored at –80°C. If inflammation is suspected, cytospins of the cells can be made, and cell counts and cell identification performed. The joint should be opened and the synovial membrane (all aspects), ligaments, articular cartilage (extent and depth), bone changes (osteophytes, eburnation, etc.) and if present menisci (striations and or tears) all inspected and scored.[105] To highlight early changes to the articular cartilage India ink can be used.[64] Menisci can be scored for texture (smooth, linear streaking, irregularities, or tears), thickness (normal, thinned and translucent, or enlarged inner circumference) or tears (location, size, number).[105] Samples from synovium, cartilage and bone should be taken either into a fixative or frozen for further analysis. Photographs taken at necropsy can provide useful records for review if needed.

C. MICROSCOPIC ANALYSIS

In setting up samples for evaluation for the first time with general histological, immunohistochemical or molecular biology protocols, specific methods and references should be consulted.[106] There are a few alterations that may be needed because of the unusual nature of cartilage with its low cellularity and presence of high concentrations of polyanionic proteoglycans. The presence of bone further complicates the situation with slow fixation, and the added need for decalcification.[107]

Troyer[7] suggests that preparation of samples for general histology by paraffin embedding of decalcified tissue should include fixation with a buffered fixative, such as formaldehyde, and with the smallest size sample that is practical. Decalcification in citrate and EDTA in the cold is the gentlest method, but time consuming.[7] For many purposes, 10% formic acid or 10% formic acid/3.7% formaldehyde at room temperature has given acceptable results.[31] For samples embedded in paraffin to prevent specimen hardening and being more difficult to cut, the paraffin temperature should be kept only a few degrees above the melting temperature and chloroform should replace xylene in the dehydration series. Glycol methacrylate resins offer an attractive alternative mounting media.[7]

Samples for immunohistochemistry may have to be fresh-frozen and a frozen section cut if the antisera only detects the native protein. In some cases, renaturation of fixed tissue in a microwave may be possible. For good antisera penetration and epitope unmasking, proteoglycans frequently must be removed by digestion with hyaluronidase or chondroitinase ABC or AC.[108]

Most investigators use histopathology as an end point in their studies. However, one major limitation is an adequate validated grading scheme. Many grading schemes have been developed and used to describe the changes that occur in the articular cartilage observed in osteoarthritis,[38,108,109] and some have even been semi-automated.[110] The most widely adopted grading method for evaluating microscopically hyaline cartilage changes was first described by Mankin et al. in 1971.[109] However, the Mankin scheme has been described by one group to be reproducible,[111] and by others to be inadequate.[38,112] Furthermore, the use of the histochemical stain Safranin O has also been questioned by some researchers for its lack of reproducibility.[113] Thus the need for a reproducible, standardized and validated grading scheme still exists and may involve scoring of area and severity of the surface fibrillation. For small animals, if no other analyses are planned, it may be possible to do this by scanning electron microscopy.[14,19]

Grading schemes have also been developed to assess the changes that occur in the synovial membrane.[38,114] In general the amount of mononuclear cell infiltrate, deposition of hemosiderin and recently the presence of formalin-resistant mast cells have been evaluated.[38]

Cartilage preparation for transmission electron microscopy (TEM) with good cell and matrix preservation requires additional care. The best example of cartilage preservation has been with high pressure, low temperature freeze substitution methods,[115] but this requires expensive equipment and is technically demanding. Routine fixation for TEM is usually done with a ruthenium salt present in the fixative if proteoglycan preservation is required, although this may slightly compromise the cell membrane details.[116] Scanning electron microscopy for surface detail requires careful regulation processing to prevent cracking artifacts.[14,19,117]

D. MOLECULAR BIOLOGY METHODOLOGY

In situ hybridization offers many of the advantages of immunolocalization since the cellular response in the different regions of cartilage can be easily determined and even quantitated.[118] The major advantage of *in* situ hybridization over immunohistochemistry is that by looking at message RNA, short-term cellular responses can be evaluated. The major caveat being that not all mRNA expression leads to protein production. For low abundance messages, the signal can be further enhanced using *in situ* RT-PCR.[119]

Extraction and purification of mRNA for cartilage and other dense connective tissues requires extra care to release the mRNA from the relatively acellular tissue and remove the contaminating proteoglycans and can be used on as little as 10 mg of tissue.[120,121] However, in the small animals, even this small amount of tissue represents pooled samples and may still include involved and noninvolved tissue. Although direct analysis of mRNA levels is something possible,[122] RT-PCR has been frequently employed in both semi-quantitative and quantitative competitive RT-PCR applications.[123,124]

E. BIOMECHANICAL PROPERTIES

Biochemical components and structure of articular cartilage and subchondral bone dertermine their mechanecal properties. Intrinsic (such as age effect) or external (impulsive loading or fracture) factors cause biochemical and morphological changes of articular cartilage and subchondral bone. Then, mechanical properties of these tissues follow. Biomechanical properties can provide important information about the process, especially early in the process before gross histologic changes occur. These very specialized measurements can extract intrinsic material properties of juxtaarticular bone.[58,60,125] See Chapter 9 and 16 for methods of cartilage mechanical testing.

IV. APPLICATIONS OF THE MODELS

A. Evaluation of the Pathophysiology of Osteoarthritis

Early-stage human samples are difficult to identify or obtain. The animal models have been used extensively for defining early and mid-stage disease and providing comparisons between disease progression in different species and different models. Because of the larger amounts of tissue available, most of the biochemical and biomechanical studies have largely focused on the canine, rabbit and sheep models.[13,72–74] Extensive histologic studies and some immunohistochemical studies have been done on all the models. Molecular biology methods, especially RT-PCR[122,124] and *in situ* hybridization methods have seen limited,[118] but expanding use.

The important questions currently being addressed with animal models are: what is the nature of early stages of the disease, what causes early matrix swelling and the prolonged hypertrophic phase, what are the enzymes involved in matrix turnover and what is their relative contribution to disease progression, how does initial trauma lead to subsequent degeneration, how much of an active role do chondrocytes play in progression, how do joint biomechanics contribute to disease progression, what role do other tissues, especially bone, play in osteoarthritis, what is the interplay of genetics, environment and mechanics in osteoarthritis, and at what stage is the process still reversible?

B. Intervention and Osteoarthritis Models

The frequency of osteoarthritis models used in interventions appears to be strongly driven by cost, ease of animal handling, especially of intra-articular injections, rapidity of disease development, and to a lesser extent, availability of the appropriate reagent. This means that, to date, rabbit models were used several times more often than the canine ACL model with a smattering of other species. Most of these interventions have been aimed at slowing early stages of progression with group sizes of 4–20 animals for 4–12 weeks, and either gross or histologic measures have been used to assess the extent of cartilage destruction, which are usually measured by fibrillation and loss of cartilage height as the read-outs. The majority of intervention studies have only been able to test a single level of compound and/or dosing, a single dosing schedule, so previous *in vitro* proof of efficacy and possible effective levels of drug are important. Table 5 compares the response of humans with osteoarthritic cartilage and animals to various drug treatments or pharmacologic interventions.

C. Identification of Markers of Osteoarthritis

The investigators in the field have an intense interest in identifying molecular markers that could be used to identify patients at risk, to follow disease progression, and be used to demonstrate efficacy of interventions. The possible complications of following development of the disease in a single joint by assaying urine, blood or synovial fluid have been extensively discussed.[103,146] Some of the problems include the signal to noise ratio from normal metabolism by the noninvolved joints or other tissues, the specificity of the marker for the tissue and disease process, or altered marker turnover either by changes in metabolism or flux (especially in synovial fluid) caused by disease.[147,148] Synovial fluid has the advantage of being directly from the involved joint. The major drawbacks are in obtaining enough sample volume and the relationship of the fluid levels to actual tissue levels and turnover in the tissue. Serum levels can reflect the summation of all possible sources, while the urine products can also be influenced by metabolism in the kidney.

Since animal models can be followed from the time of initiation of the process, models are being used to determine if and how a given marker changes with disease progression and to follow up biochemical leads from culture experiments. Samples from human osteoarthritic patients are being used to follow mid- and end-stage disease. The field has focused on cartilage-specific or cartilage-enriched markers, which change in the osteoarthritic process. There is renewed interest

TABLE 5
Comparison of Human Osteoarthritis with Animal Models of Osteoarthritis — Pharmacology and Therapeutics

	Human[Ref.]	Dog[Ref.]	Rabbit[Ref.]	Guinea Pig[Ref.]	Mouse[Ref.]
Modulation of Inflammation/pain					
Corticosteroids	↓126	↓127			
NSAIDS	↓128,129				
Tenidap					
Hyaluronic Acid	↓130				
Polysulfated GAGs	↓129				
MMP Inhibitors					
Doxycycline					
Modulation of Chondrodestruction/osteophytosis					
Corticosteroids	↑131	↓132	↓40	↓133	No134
NSAIDS	No/↑127	No132	No/↓135		No /↓135
Tenidap	?136	↓137			↓138
Hyaluronic Acid		↓139,140			
Polysulfated GAGs	↓129	↓127	↓127		No/↓127
MMP Inhibitors			↓141	↓142	
Doxycycline		No ↓143	↓144	↓145	

NSAIDS = nonsteroidal anti-inflammatory drugs
GAG = glycosaminoglycan
MMP = matrix metalloproteinase

in finding fragments of matrix molecules that could be used either as markers of catabolism or anabolism, or tissue remodeling.[104]

In animal models, a few examples include: the persistent upregulation of serum hyaluronic acid,[149] change in keratan sulfate levels,[102,103] specific sulfation epitopes in chondroitin sulfate,[150] specific degradation products of aggrecan, bone sialoprotein (a marker for bone remodeling), metalloproteases, cartilage oligomeric matrix protein, and type II collagen fragments including amino and carboxyl crosslink fragment(s) and link protein fragments.[104]

D. Bone Ingrowth to Implants under Osteoarthritic Conditions

Since there are many studies that demonstrate changes of the subchondral bone in osteoarthritis,[107,151] it is possible that as implants are developed that allow replacement of only part of the osteoarthritic joint, there may be problems with bone ingrowth. This has been reported for rheumatoid arthritis, diabetes, or revised joint replacements.[107] End-stage, large animal models of osteoarthritis would be useful for these studies. Ideally, the model would be allowed to develop to at least mid-stage where bone changes are extensive and then the mechanical defect corrected by surgery before the implant study is started.

ACKNOWLEDGMENTS

Supported by NIH AR41975. Thanks to Laurel Deloria and Andrea Chatfield in preparation of manuscript; Merck Visual communications: John Shockey, Sharon O'Brien, John Remminger, and Chad Orevillo; and Scott Hofsess and Amy Christen for their input and support.

REFERENCES

1. Moskowitz, R. W., "Experimental models of osteoarthritis," in *Osteoarthritis: Diagnosis and Medical/Surgical Management*, 2nd ed., Moskowitz, R. W., Howell, D. S., Goldberg, V. M., and Mankin, H. J., Eds., W. B. Saunders, Philadelphia, 1992, 213.
2. Bullough, P. G., "The pathology of osteoarthritis," in *Osteoarthritis: Diagnosis and Medical/Surgical Management*, 2nd ed., Moskowitz, R. W., Howell, D. S., Goldberg, V. M., and Mankin, H. J., Eds., W. B. Saunders, Philadelphia, 1992, 39.
3. Schiller, A. L., "Pathology of osteoarthritis," in *Osteoarthritic Disorders*, Kuettner, K. E. and Goldberg, V. M., Eds., American Academy of Orthopaedic Surgeons, Chicago, 1995, 95.
4. Pritzker, K. P., "Animal models for osteoarthris: processes, problems and prospects," *Ann. Rheum. Dis.*, 53, 406, 1994.
5. Malemud, C., "The biology of cartilage and synovium in animal models of osteoarthritis," in *CRC Handbook of Animal Models for the Rheumatic Diseases*, Vol. II, Greenwald, R. A. and Diamond, H. S., Eds., CRC Press, Boca Raton, 1988, 3.
6. Altman, R. D. and Dean, D. D., "Osteoarthritis research: animal models," *Semin. Arth. Rheum.*, 19, 21, 1990.
7. Troyer, H., "Experimental models of osteoarthritis: a review," *Semin. Arthritis Rheum.*, 11, 362, 1982.
8. Silberberg, M. and Silberberg, R., "Osteoarthritis in mice diets enriched with animal or vegetable fat," *Arch. Pathol.*, 70, 385, 1960.
9. Pataki, A., Reife, R., Witzemann, E., Graf, H. P., and Schweizer, A., "Quantitative radiographic diagnosis of osteoarthritis of the knee joint in the C57BL mouse," *Agents Actions*, 29, 301, 1990.
10. Schunke, M., Tillmann, B., Bruck, M., and Muller-Ruchholtz, W., "Morphologic characteristics of developing osteoarthritic lesions in the knee cartilage of STR/1N mice," *Arthritis Rheum.*, 31, 898, 1988.
11. Munasinghe, J. P., Tyler, J. A,. Hodgson, R. J., et al., "Magnetic resonance imaging, histology, and X ray of three stages of damage to the knees of STR/ORT mice," *Invest. Radiol.*, 31, 630, 1996.
12. Bendele, A. M. and White, S. L., "Early histopathologic and ultrastructural alterations in femorotibial joints of partial medial meniscectomized guinea pigs," *Vet. Pathol.*, 24, 436, 1987.
13. Bendele, A. M. and Hulman, J. F., "Effects of body weight restriction on the development and progression of spontaneous osteoarthritis in guinea pigs," *Arthritis Rheum.*, 34, 1180, 1991.
14. Bendele, A. M., White, S. L., and Hulman, J. F., "Osteoarthrosis in guinea pigs: histopathologic and scanning electron microscopic features," *Lab. Anim. Sci.*, 39, 115, 1989.
15. de Bri, E., Jönsson, K., Reinholt, F. P., and Svensson, O., "Focal destruction and remodeling in guinea pig arthrosis," *Acta Orthop. Scand.* 67, 498, 1996.
16. Wurster, N. B. and Lust, G., "Fibronectin in osteoarthritic canine articular cartilage," *Biochem. Biophys. Res. Comm.*, 109, 1094, 1982.
17. Chateauvert, J. M. D., Grynpas, M. D., Kessler, M. J., and Pritzker, K. P. H., "Spontaneous osteoarthritis in rhesus macaques. II. Characterization of disease and morphometric studies," *J. Rheumatol.*, 17, 73, 1990.
18. Sokoloff, L. and Jay, G. E., Jr., "Natural history of degenerative joint disease in small laboratory animals. IV. Degenerative joint disease in the laboratory rat," *Arch. Pathol.*, 62, 140, 1956.
19. Otterness, I. G., Eskra, J. D., Bliven, M. L., et al., "Exercise protects against articular cartilage degeneration in the hamster," *Arthritis Rheum.*, 41, 2068, 1998.
20. Jacenko, O. and Olsen, B. R., "Transgenic mouse models in studies of skeletal disorders," *J. Rheumatol.*, 22(S43), 39, 1995.
21. Decrombrugghe, B., Katzenstein, P., Mukhopadhyay, K., et al., "Transgenic mice with deficiencies in cartilage collagens — Possible models for gene therapy," *J. Rheumatol.*, 22, 140, 1995.
22. Bennett, G. A. and Bauer,W., "A study on the repair of the articular cartilage and the reaction of normal joints of adult dogs to surgically created defects of articular cartilage, 'joint mice', and patellar displacement," *Am. J. Pathol.*, 8, 499, 1932.
23. Lefkoe, T. P., Trafton, P. G., Ehrlich, M. G., et al., "An experimental model of femoral condylar defect leading to osteoarthrosis," *J. Orthop. Trauma*, 7, 458, 1993.
24. Meachim, G., "The effect of scarification on articular cartilage in the rabbit," *J. Bone Joint Surg.*, 45B, 150, 1963.

25. Thompson, R. C., Jr., "An experimental study of surface injury to articular cartilage and enzyme responses within the joint," *Clin. Orthop.*, 107, 239, 1975.
26. Simon, W. H., Richardson, S., Herman, W., Parsons, J. R., and Lane, J., "Long-term effects of chondrocyte death on rabbit articular cartilage *in vivo*," *J. Bone Joint Surg.*, 58A, 517, 1976.
27. Lukoschek, M., Boyd, R. D., Schaffler, M. B., and Radin, E. L., "Comparison of joint degeneration models. Surgical instability and repetitive impulsive loading," *Acta Orthop. Scand.*, 57, 349, 1986.
28. Simon, W. H., Radin, E. L., Paul, I. L., and Rose, R. M., "The response of joints to impact loading. II. *In Vivo* behavior of subchondral bone," *J. Biomech.*, 5, 267, 1972.
29. Mazieres, B., Berdah, L., Thiechart, M., and Viguier, G., "Diacetylrhein on a postcontusion model of experimental osteoarthritis in the rabbit," *Revue du Rhumatisme*, 60, 77S, 1993.
30. Haut, R. C., Ide, T. M., and De Camp, C. E., "Mechanical responses of the rabbit patello-femoral joint to blunt impact," *J. Biomech. Eng.*, 117, 402, 1995.
31. Donohue, M., Buss, D., Oegema, T. R., and Thompson, R. C., "The effects of indirect trauma on adult canine articular cartilage," *J. Bone Joint Surg.*, 65A, 948, 1983.
32. Oegema, T. R., Jr., Lewis, J. L., and Thompson, R. C., Jr., "Role of acute trauma in development of osteoarthritis," *Agents Actions*, 40, 220, 1993.
33. Wu, D. D., Boyd, R. D., Burr, D. B., and Radin, E. L., "Comparative bone and cartilage changes in two different models of OA," *Trans. Orthop. Res. Soc.*, 15, 210, 1990.
34. Goodman, S. B., Lee, J., Smith, R. L., Csongradi, J. C., and Fornasier, V. L., "Mechanical overload of a single compartment induces early degenerative changes in the rabbit knee: a preliminary study," *J. Invest. Surg.*, 4, 161, 1991.
35. Layton, M. W., Goldstein, S. A., Goulet, R. W., Feldkamp, L. A., Kubinski, D. J., and Bole, G. G., "Examination of subchondral bone architecture in experimental osteoarthritis by microscopic computed axial tomography," *Arthritis Rheum.*, 31, 1400, 1988.
36. Finsterbush, A. and Friedman, B., "The effect of sensory denervation on rabbits' knee joints: a light and electron microscopic study," *J. Bone Joint Surg.*, 57A, 949, 1975.
37. O'Connor, B. L., Visco, D. M., and Brandt, K. D., "The development of experimental osteoarthritis (OA) in dogs with extensively deafferented knee joints," *Arthritis Rheum.*, 32, S106, 1989.
38. Visco, D. M., Hill, M. A., Widmer, W. R., Johnstone, B., O'Connor, B. L., and Myers, S. L., "Experimental osteoarthritis in dogs: a comparison of the Pond-Nuki and medial arthrotomy methods," *Osteoarthritic Cartilage*, 4, 9, 1996.
39. Rumph, P. F., Lander, J. E., Kincaid, S. A., Baird, D. K., Kammermann, J. R., and Visco, D. M., "Ground reaction force profiles from force platform gait analyses of clinically normal mesomorphic dogs at the trot," *Am. J. Vet. Res.*, 55, 756, 1994.
40. Colombo, C., Butler, M., Hickman, L., Selwyn, M., Chart, J., and Steinetz, B., "A new model of osteoarthritis in rabbits. II. Evaluation of anti-osteoarthritic effects of selected antirheumatic drugs administered systemically," *Arthritis Rheum.*, 26, 1132, 1983.
41. Hulth, A., Lindberg, L., and Telhag, H., "Experimental osteoarthritis in rabbits," *Acta Orthop. Scand.*, 41, 522, 1971.
42. Smith, M. M., Little, C. B., Rodgers, K., and Ghosh, P., "Animal models used to evaluate anti-osteoarthritis drugs," *Pathol. Biol.*, 45, 313, 1997.
43. Korkala, O., Karaharju, E., Gronblad, M., and Aalto, K., "Articular cartilage after meniscectomy. Rabbit knees studied with the scanning electron microscope," *Acta Orthop. Scand.*, 55, 273, 1984.
44. Lutfi, A. M., "Morphological changes in the articular cartilage after meniscectomy. An experimental study in the monkey," *J. Bone Joint Surg.*, 57B, 525, 1975.
45. Williams, J. M., Felton, D. L., Peterson, R. G., and O'Connor, B. L., "Effects of surgically induced instability on rat knee articular cartilage," *J. Anat.*, 134, 103, 1982.
46. Brandt, K. D., Braunstein, E. M., Visco, D. M., O'Connor, B., Heck, D., and Albrecht, M., "Anterior (cranial) cruciate ligament transection in the dog: a bona fide model of osteoarthritis, not merely of cartilage injury and repair," *J. Rheumatol.*, 18, 436, 1991.
47. Brandt, K. D., "Insights into the natural history of osteoarthritis provided by the cruciate-deficient dog. An animal model of osteoarthritis," *Ann. N.Y. Acad. Sci.*, 732, 199, 1994.
48. McDevitt, C. A. and Muir, H., "Biochemical changes in the cartilage of the knee in experimental and natural osteoarthritis in the dog," *J. Bone Joint Surg.*, 58B, 94, 1976.

49. Vignon, E., Bejui, J., Mathieu, P., Hartmann, J. D., Ville, G., Evreux, J. C., and Descotes, J., "Histological cartilage changes in a rabbit model of osteoarthritis," *J. Rheumatol.*, 14, S104, 1987.
50. Davis, W. and Moskowitz, R. W., "Degenerative joint changes following posterior cruciate ligament section in the rabbit," *Clin. Orthop.*, 93, 1307, 1973.
51. Garr, E. L., Moskowitz, R. W., and Davis, W., "Degenerative changes following experimental patellectomy in the rabbit," *Clin. Orthop.*, 92, 296, 1973.
52. De Palma, A. F. and Flynn, J. J., "Joint changes following experimental partial and total patellectomy," *J. Bone Joint Surg.*, 40A, 395, 1958.
53. Mitrovic, D. R., Garcia, F., Front, P., and Guillermet, V., "Histochemical and cellular changes induced in the rabbit knee joint by an intraarticular implantation of a sheet of polyethylene," *Lab. Invest.*, 53, 228, 1985.
54. Lefkoe, T. P., Walsh, W. R., Anastasatos, J., Ehrlich, M. G., and Barrach, H. J., "Remodeling of articular step-offs. Is osteoarthrosis dependent on defect size?" *Clin. Orthop.*, 314, 253, 1995.
55. Hall, M. C., "Cartilage changes after experimental relief of contact in knee joint of the mature rat," *Clin. Orthop.*, 64, 64, 1969.
56. Engh, G. A. and Chrisman, O. D., "Experimental arthritis in rabbit knees: a study of relief of pressure on one tibial plateau in immature and mature rabbits," *Clin. Orthop.*, 125, 221, 1977.
57. Hall, M. C., "Cartilage changes after experimental immobilization of the knee joint of the young rat," *J. Bone Joint Surg.*, 45A, 36, 1963.
58. Sah, R. L., Yang, A. S., Chen, A. C., et al., "Physical properties of rabbit articular cartilage after transection of the anterior cruciate ligament," *J. Orthop. Res.*, 15, 197, 1997.
59. Jortikka, M. O., Inkinen, R. I., Tammi, M. I., et al., "Immobilization causes longlasting matrix changes both in the immobilized and contralateral joint cartilage," *Ann. Rheum. Dis.*, 56, 255, 1997.
60. Setton, L. A., Vow, V. C., Müller, F. J., Pita, J. C., and Howell, D. S., "Mechanical behavior and biochemical composition of canine knee cartilage following periods of joint disuse and disuse with remobilization," *Osteoarthritic Cartilage*, 5, 1, 1997.
61. Crelin, E. S. and Southwick, W. O., "Changes induced by sustained pressure in the knee joint articular cartilage of adult rabbits," *Anat. Rec.*, 149, 113, 1964.
62. Hall, M. C., "Articular changes in the knee of the adult rat after prolonged immobilization in extension," *Clin. Orthop.*, 34, 184, 1964.
63. van Osch, G. J., van der Kraan, P. M., Blankevoort, L., Huiskes, R., and van den Berg, W. B., "Relation of ligament damage with site specific cartilage loss and osteophyte formation in collagenase induced osteoarthritis in mice," *J. Rheumatol.*, 23, 1227, 1996.
64. Ehrlich, M. G., Mankin, H. J., Jones, H., Grossman, A., Crispen, C., and Ancona, D., "Biochemical confirmation of an experimental osteoarthritis model," *J. Bone Joint Surg*, 57A, 392, 1975.
65. Meachim, G., "Light microscopy of India ink preparation of fibrillated cartilage," *Ann. Rheum. Dis.*, 31, 457, 1972.
66. Williams, J. M. and Brandt, K. D., "Iodo-acetate (IA) causes osteoarthritis in guinea pigs," *Anat. Rec.*, 202, 201A, 1982.
67. Kalbhen, D. A., "Chemical model of osteoarthritis — a pharmacological evaluation," *J. Rheumatol.*, 14, S130, 1987.
68. Williams, J. M. and Brandt, K. D., "Immobilization ameliorates chemically-induced articular cartilage damage," *Arthritis Rheum.*, 27, 208, 1984.
69. Borella, L., Eng, C. P., DiJoseph, J., et al., "Rapid induction of early osteoarthritic-like lesions in the rabbit knee by continuous intra-articular infusion of mammalian collagenase or interleukin–1," *Agents Actions*, 34, 220, 1991.
70. Lapadula, G., Nico, B., Cantatore, F. P., La Canna, R., Roncali, L., and Pipitone, V., "Early ultrastructural changes of articular cartilage and synovial membrane in experimental vitamin A-induced osteoarthritis," *J. Rheumatol.*, 22, 1913, 1995.
71. Homandberg, G. A., Meyers, R., and Williams, J. M., "Intraarticular injection of fibronectin fragments causes severe depletion of cartilage proteoglycans *in vivo*," *J. Rheumatol.*, 20, 1378, 1993.
72. Kaiki, G., Tsuji, H., Yonezawa, T., et al., "Osteoarthrosis induced by intra-articular hydrogen peroxide injection and running load," *J. Orthop. Res.*, 8, 731, 1990.
73. Havdrup, T., "Trypsin induced mitosis in the articular cartilage of rabbits," *Acta Orthop. Scand.*, 50, 15, 1979.

74. Havdrup, T., Henricson, A., and Telhag, H., "Papain-induced mitosis of chondrocytes in adult joint cartilage; an experimental study in full-grown rabbits," *Acta Orthop. Scand.*, 53, 119, 1982.
75. Kopp, S., Mejersjo, C., and Clemensson, E., "Induction of osteoarthrosis in the guinea pig knee by papain," *Oral Surg. Oral Med. Oral Pathol.*, 55, 259, 1983.
76. Williams, J. M., Moran, M., Thonar, E. J., and Salter, R. B., "Continuous passive motion stimulates repair of rabbit knee articular cartilage after matrix proteoglycan loss," *Clin. Orthop.*, 304, 252, 1994.
77. Nahir, A. M., Shomrat, D., and Awad, M., "Chondroitinase ABC affects the activity of intracellular enzymes in rabbit articular cartilage chondrocytes," *J. Rheumatol.*, 22, 702, 1995.
78. Williams, J. M., Ongchi, D. R., and Thonar, E. J., "Repair of articular cartilage injury following intra-articular chymopapain-induced matrix proteoglycan loss," *J. Orthop. Res.*, 11, 705, 1993.
79. Godfried, Y., Bradford, D. S., and Oegema, T. R., Jr., "Facet joint changes after chemonucleolysis-induced disc space narrowing," *Spine*, 11, 944, 1986.
80. Gloobe, H. and Nathan, H., "Osteophyte formation in experimental bipedal rats," *J. Comp. Pathol.*, 83, 133, 1973.
81. Bently, G., "Papain-induced degenerative arthritis of the hip in rabbits," *J. Bone Joint Surg.*, 53B, 324, 1971.
82. Inerot, S., Heinegård, D., Olsson, S. E., Telhag, H., and Audell, L., "Proteoglycan alterations during developing experimental osteoarthritis in a novel hip joint model," *J. Orthop. Res.*, 9, 658, 1991.
83. Dreessen, D. and Halata, Z., "Age-related osteoarthrotic degeneration of the temporomandibular joint in the mouse," *Acta Anat.*, 139, 91, 1990.
84. Axelsson, S., Bjornsson, S., Holmlund, A., and Hjerpe, A., "Metabolic turnover of sulfated glycosaminoglycans and proteoglycans in rabbit temporomandibular joint cartilages with experimentally induced osteoarthritis," *Acta Odontol. Scand.*, 52, 65, 1994.
85. Mejersjo, C., "Long-term development after treatment of mandibular dysfunction and osteoarthrosis. A clinical-radiographic follow-up and an animal experimental study," *Swed. Dent. J.*, (Supp), 22, 1, 1984.
86. Visco, D. M., Orevillo, C. J., Kammermann, J., Kincaid, S. A., Widmer, W. R., and Christen, A. J., "Progressive chronic osteoarthritis in a surgically induced model in mice," *Trans. Orthop. Res. Soc.*, 21, 241, 1996.
87. Johnson, J. M. and Johnson, A. L., "Cranial cruciate ligament rupture: pathogenesis, diagnosis, and postoperative rehabilitation," *Vet. Clin. N. Am.*, 23, 717, 1993.
88. Adams, M. E. and Pelletier, J. P., "Canine anterior cruciate ligament transection model of osteoarthritis," in *CRC Handbook of Animal Models of Rheumatic Diseases* Vol II, Greenwald, R. A., and Diamond, H. S., Eds., CRC Press, Boca Raton, 1988, 57.
89. O'Connor, B. L., Visco, D. M., Heck, D. A., Myers, S. L., and Brandt, K. D., "Gait alterations in dogs after transection of the anterior cruciate ligament," *Arthritis Rheum.*, 32, 1142, 1989.
90. Wilhelmi, G., "Potential effects of nutrition including additives on healthy and arthrotic joints, I. Basic dietary constituents," *Zeitschrift Rheumatol.*, 52, 174, 1993.
91. Kealy, R. D., Lawler, D. F., Ballam, J. M., Lust, G., Smith, G. K., Biery, D. N., and Olsson, S. E., "Five-year longitudinal study on limited food consumption and development of osteoarthritis in coxofemoral joints of dogs," *J. Am. Vet. Med. Assoc.*, 210, 222, 1997.
92. Rumph, P. F., Kincaid, S. A., Visco, D. M., Baird, D. K., and Kammermann, J., "Redistribution of vertical ground reaction force in dogs with experimentally induced chronic hind limb lameness," *Vet. Surg.*, 24, 384, 1995.
93. Korvick, D. L., Pijanowski, G. J., and Schaffer, D. J., "Three-dimensional kinematics of the intact and cranial cruciate ligament deficient stifle of dogs," *J. Biomech.*, 27, 77, 1994.
94. Vilensky, J. A., O'Connor, B. L., Brandt, K. D., Dunn, E. A., and Rogers, P. I., "Serial kinematic analysis of the trunk and limb joints after anterior cruciate ligament transection. Temporal, spatial, and angular changes in a canine model of osteoarthritis," *J. Electromyogr. Kinesiol.*, 4, 181, 1994.
95. Vilensky, J. A., O'Connor, B. L., Brandt, K. D., Dunn, E. A., and Rogers, P. I., "Serial kinematic analysis of the canine hindlimb joints after deafferentation and anterior cruciate ligament transection," *Osteoarthritic Cartilage*, 5, 173, 1997.
96. Widmer, W. R., Buckwalter, K. A., Braunstein, E. M., Hill, M. A., O'Connor, B. L., and Visco, D. M., "Radiographic and magnetic resonance imaging of the stifle joint in experimental osteoarthritis of dogs," *Vet. Radiol. Ultrasound*, 35, 371, 1994.

97. Loeuille, D., Gonord, P., Guingamp, C., Gillet, P., Blum, A., Sauzade, M., and Netter, P., "*In vitro* magnetic resonance microimaging of experimental osteoarthritis in the rat knee joint," *J. Rheumatol.*, 24, 133, 1997.
98. Watson, P. J., Carpenter, T. A., Hall, L. D., and Tyler, J. A., "Cartilage swelling and loss in a spontaneous model of osteoarthritis visualized by magnetic resonance imaging," *Osteoarthritic Cartilage*, 4, 181, 1996.
99. Paul, P. K., O'Byrne, E., Blancuzzi, V., et al., "Magnetic resonance imaging reflects cartilage proteoglycan degradation in the rabbit knee," *Skeletal Radiol.*, 20, 31, 1991.
100. Christensen, S. B., "Localization of bone-seeking agents in developing, experimentally induced osteoarthritis in the knee joint of the rabbit," *Scand. J. Rheumatol.*, 12, 343, 1983.
101. Goupille, P., Chevalier, X., Valat, J. P., Garaud, P., Perin, F., and Le Pape, A., "Macrophage targeting with 99 mTc-labeled J001 for scintigraphic assessment of experimental osteoarthritis in the rabbit," *J. Rheumatol.*, 36, 758, 1997.
102. Thonar, E. J., Masuda, K., Lenz, M. E., Hauselmann, H. J., Kuettner, K. E., and Manicourt, D. H., "Serum markers of systemic disease processes in osteoarthritis," *J. Rheumatol.*, 43, S68, 1995.
103. Williams, J. M., Downey, C., and Thonar, E. J., "Increase in levels of serum keratan sulfate following cartilage proteoglycan degradation in the rabbit knee joints," *Arthritis Rheum.*, 31, 557, 1988.
104. Saxne, T. and Heinegard, D., "Matrix proteins: Potentials as body fluid markers of changes in the metabolism of cartilage and bone in arthritis," *J. Rheumatol.* (Supp.), 43, 71, 1995.
105. Baird, D. K., *Low-field magnetic resonance imaging: Application in the study of osteoarthritis and the canine stifle joint*, Ph.D. dissertation, Auburn University, pg. 87, June 13, 1997.
106. Gardner, D. L., Salter, D. M., and Oates, K., "Advances in the microscopy of osteoarthritis (Review)," *Microsc. Res. Tech.*, 37, 245, 1997.
107. Burr, D. B. and Schaffler, M. B., "The involvement of subchondral mineralized tissues in osteoarthrosis: quantitative microscopic evidence (Review)," *Microscopy Res. Tech.*, 37, 343, 1997.
108. Kammermann, J. R., Kincaid, S. A., Rumph, P. F., Baird, D. K., and Visco, D. M., "Immuno-localization of TNF-a, stromelysin, and TNF receptors in canine osteoarthritic cartilage," *Osteoarthritic Cartilage*, 4, 23, 1996.
109. Mankin, H. J., Dorfman, H., Lippiello, L., and Zarins, A., "Biochemical and metabolic abnormalities in articular cartilage from osteoarthritis human hips. II Correlation of morphology with biochemical and metabolic data," *J. Bone Joint Surg*, 53A, 523, 1971.
110. Yoshioka, M., Coutts, R. D., Amiel, D., and Hacker, S. A., "Characterization of a model of osteoarthritis in the rabbit knee,"*Osteoarthritic Cartilage*, 4, 87, 1996.
111. Van der Sluijs, J. A., Geesnik, R. G., van der Linden, A. J., Bulstra, S. K., Kujer, R., and Drukker, J., "The reliability of the Mankin score for osteoarthritis," *J. Orthop. Res.*, 10, 58, 1992.
112. Ostergaard, K., Peterson, J., Andersen, C. B., Bendtzen, K., and Salter, D. M., "Histologic/ histochemical grading system for osteoarthritic articular cartilage: reproducibility and validity," *Arthritis Rheum.*, 40, 1766, 1997.
113. Getzy, L. L., Malemud, C. J., Goldberg V. M., and Moskowitz, R. W., "Factors influencing metachromatic staining in paraffin-embedded sections of rabbit and human articular cartilage: a comparison of the saffranin-O and toluidine blue techniques," *J. Histotechnology*, 5, 111, 1982.
114. Myers, S. L., Flusser, D., Brandt, K. D., and Heck, D. A., "Prevalence of cartilage shards in synovium and their association with synovitis in patients with early and endstage osteoarthritis," *J. Rheum.*, 19, 1247, 1992.
115. Studer, D., Michel, M., Wohlwend, M., Hunziker, E. B., and Buschmann, M. D., "Vitrification of articular cartilage by high-pressure freezing," *J. Microscopy*, 179, 321, 1995.
116. Hunziker, E. B., Ludi, A., and Hermann, W., "Preservation of cartilage matrix proteoglycans using cationic dyes chemically related to ruthenium hexaamine trichloride," *J. Histochem. Cytochem.*, 40, 90, 1992.
117. Thompson, R. C., Jr., Vener, M., Griffiths, H., Lewis, J. L., Oegema, T. R., Jr., and Wallace, L., "Scanning electron microscopic and magnetic resonance imaging studies of injuries to the patellofemoral joint after acute transarticular loading," *J. Bone Joint Surg.*, 75A, 704, 1993.
118. Lefkoe, T. P., Nalin, A. M., Clark, J. M., Reife, R. A., Sugai, J., and Sandell, L. J., "Gene expression of collagen types IIA and IX correlates with ultrastructural events in early osteoarthrosis: new applications of the rabbit meniscectomy model," *J. Rheumatol.*, 24, 1155, 1997.

119. Mee, A. P., Denton, J., Hoyland, J. A., Davies, M., and Mawer, E. B., "Quantification of vitamin D receptor mRNA in tissue sections demonstrates the relative limitations of *in situ* reverse transcriptase-polymerase chain reaction," *J. Pathol.*, 182, 22, 1997.
120. Adams, M. E., Huang, D. Q., Yao, L. Y., and Sandell, L. J., "Extraction and isolation of mRNA from adult articular cartilage," *Anal. Biochem.*, 202, 89, 1992.
121. Reno, C., Marchuk, L., Sciore, P., Frank, C. B., and Hart, D. A., "Rapid isolation of total RNA from small samples of hypocellular, dense connective tissues," *Biotechniques*, 22, 1082, 1997.
122. Matyas, J. R., Adams, M. E., Huang, D., and Sandell, L. J., "Major role of collagen IIB in the elevation of total type II procollagen messenger RNA in the hypertrophic phase of experimental osteoarthritis," *Arthritis Rheum.*, 40, 1046, 1997.
123. Re, P., Valhmu, W. B., Vostrejs, M., Howell, D. S., Fischer, S. G., and Ratcliffe, A., "Quantitative polymerase chain reaction assay for aggrecan and link protein gene expression in cartilage," *Anal. Biochem.*, 225, 356, 1995.
124. Melching, L. I., Cs-Szabo, G., and Roughley, P. J., "Analysis of proteoglycan messages in human articular cartilage by a competitive PCR technique," *Matrix Biol.*, 16, 1, 1997.
125. Setton, L. A., Mow, V. C., Muller, F. J., Pita, J. C., and Howell, D. S., "Mechanical properties of canine articular cartilage are significantly altered following transection of the anterior cruciate ligament," *J. Orthop. Res.*, 12, 451, 1994.
126. Towheed, T. E. and Hochberg, M. C., "A systematic review of randomized controlled trials of pharmacological therapy in osteoarthritis of the knee, with an emphasis on trial methodology," *Sem. Arthritis Rheum.*, 26, 755, 1997.
127. Burkhardt, D. and Ghosh, P., "Laboratory evaluation of antiarthritic drugs as potential chondroprotective agents," *Semin. Arthritis Rheum.*, 17(S1), 3, 1987.
128. Pinals, R. S., "Pharmacologic treatment of osteoarthritis," *Clin. Therap.*, 14, 336, 1992.
129. Rejholec, V., "Long-term studies of anti-osteoarthritic drugs: an assessment," *Semin. Arthritis Rheum. Supp. 1*, 17, 35, 1987.
130. Namiki, O., Toyoshima, H., and Morisake, N., "Therapeutic effect of intra-articular injection of high molecular weight hyaluronic acid on osteoarthritis of the knee," *Int. J. Clin. Pharm. Ther. Toxicol.*, 20, 501, 1982.
131. Chandler, G. N. and Wright, B., "Deleterious effects of intra-articular hydrocortisone," *Lancet* 2, 661, 1958.
132. Pelletier, J-P. and Martel-Pelletier, J., "Protective effects of prophylactic treatment with tiaprofenic acid or intraarticular corticosteroids on osteoarthritic lesions in the experimental dog model," *J. Rheumatol.*, 27, S127, 1991.
133. Williams, J. M. and Brandt, K. D., "Triamcinolone hexacetonide protects against fibrillation and osteophyte formation following chemically induced articular cartilage damage," *Arthritis Rheum.*, 28, 1267, 1985.
134. Silberberg, M., Silberberg, R., and Hasler, M., "Fine structure of articular cartilage in mice receiving cortisone acetate," *Arch. Pathol.*, 82, 569, 1966.
135. Maier, R. and Wilhelmi, G., "Osteoarthrotosis-like disease in mice: effects of antiarthrotic and antirheumatic drugs," in *Studies in Osteoarthrotis:Pathogenesis, Intervention, and Assessment*, Lott, D. J., Jasani, M. K., and Birdwood, G. F. B., Eds., John Wiley, Chichester, 1987, 75.
136. Blackburn, W. D., "Management of osteoarthritis and rheumatoid arthritis: prospects and possibilities," *Am. J. Med.*, 100, 24S, 1996.
137. Fernandes, J. C., Caron, J. P., Martel-Pelletier, J., et al., "Effects of Tenidap on the progression of osteoarthritic lesions in a canine experimental model. Suppression of metalloprotease and interleukin–1 activity," *Arthritis Rheum.*, 40, 284, 1997.
138. Revell, P. A., Pirie, C. J., Osei, D., and Mackillop, N. G., "Effect of Tenidap on osteoarthritis in the STR/ORT mouse," *Arthritis Rheum S* 58, 1995.
139. Abatangelo, G., Botti, P., Del Bue, M., et al., "Intraarticular sodium hyaluronate injections in the Pond-Nuki experimental model of osteoarthritis in dogs. I. Biochemical results," *Clin. Orthop.*, 241, 278, 1989.
140. Kikuchi, T., Yamada, H., and Shimmei, M., "Effect of high molecular weight hyaluronan on cartilage degeneration in a rabbit model of osteoarthritis," *Osteoarthritic Cartilage*, 4, 99, 1996.

141. O'Byrne, E. M., Parker, D. T., Roberts, E. D., et al., "Oral administration of a matrix metalloproteinase inhibitor, CGS 27023A, protects the cartilage proteoglycan matrix in a partial meniscectomy model of osteoarthritis in rabbits," *Inflamm. Res.*, 44, S117, 1995.
142. O'Byrne, E. M., *Personal Communication*, 1997.
143. Yu, L. P., Jr., Smith, G. N., Jr., Brandt K. D., et al., "Reduction of the severity of canine osteoarthritis by prophylactic treatment with oral doxycycline," *Arthritis Rheum.*, 35, 1150, 1992.
144. Greenwald, R. A., "Treatment of destructive arthritis disorders with MMP inhibitors. Potential role of tetracyclines," *Ann. N.Y. Acad. Sci.*, 732, 181, 1994.
145. Golub, L. M., Ramamurthy, N. S., McNamara, T. F., et al., "Method to reduce connective tissue destruction," *United States Patent* 5,258,37, Nov., 1993.
146. Dieppe, P., "Osteoarthritis and molecular markers. A rheumatologist's perspective," *Acta Orthop. Scand. Suppl.*, 266, 1, 1995.
147. Myers, S. L., O'Connor, B. L., and Brandt, K. D., "Accelerated clearance of albumin from the osteoarthritic knee: implications for interpretation of concentrations of 'cartilage markers' in synovial fluid," *J. Rheumatol.*, 23, 1744, 1996.
148. Lindenhayn, K., Heilmann, H. H., Niederhausen, T., Walther, H. U., and Pohlenz, K., "Elimination of tritium-labelled hyaluronic acid from normal and osteoarthritic rabbit knee joints," *Eur. J. Clin. Chem. Clin. Biochem.*, 35, 355, 1997.
149. Manicourt, D. H., Cornu, O., Lenz, M. E., Druetz-van Egeren, A., and Thonar, E. J., "Rapid and sustained rise in the serum level of hyaluronan after anterior cruciate ligament transection in the dog knee joint," *J. Rheumatol.*, 22, 262, 1995.
150. Ratcliffe, A., Shurety, W., Caterson, B., "The quantitation of a native chondroitin sulfate epitope in synovial fluid lavages and articular cartilage from canine experimental osteoarthritis and disuse atrophy," *Arthritis Rheum.*, 36, 543, 1993.
151. Dequeker, J., Mokassa, L., Aerssens, J., and Boonen, S., "Bone density and local growth factors in generalized osteoarthritis," *Microsc. Res. Tech.*, 37, 358, 1997.

19 Animal Models of Rheumatoid Arthritis

Erica L. Moran and Earl R. Bogoch

CONTENTS

I. Introduction 370
II. Selection of an Animal Model of Rheumatoid Arthritis...371
    A. Adjuvant Arthritis...372
        1. Induction of Arthritis..372
        2. Clinical Observations 372
        3. Gross Pathology and Histology 372
        4. Immunology 373
    B. Collagen Induced Arthritis 373
        1. Induction of Arthritis..374
        2. Clinical Observations 375
        3. Gross Pathology and Histology 375
        4. Gender Effects..375
        5. Immunology 375
        6. Significance of the Collagen Induced Arthritis Model....................................376
    C. Proteoglycan Induced Arthritis 377
    D. Antigen Induced Arthritis..377
    E. Steptococcal Cell Wall Arthritis..377
        1. Pathology..378
        2. Immunology 378
    F. Carrageenan Induced Arthritis 378
        1. General 378
        2. Background..379
        3. Induction of Arthritis..379
        4. Species/Strain Differences 379
        5. Comparison with Other Models...380
III. Mechanistic Approaches to Models of Arthritis..380
    A. Cytokine-Induced Arthritis and Inhibition with Anti-Cytokine Antibodies...................380
    B. Mutation-Based Approaches 380
IV. Evaluation Methods..380
    A. Biochemical Assessment of Cartilage and Bone Turnover 381
    B. Biochemical Assessment of Inflammatory Disease 381
    C. Clinical Severity..381
    D. Roentgenographic Methods..381
    E. Necropsy: Gross Examination 382
    F. Histomorphometry: Assessment of Bone Turnover by *in vivo* Labelling....................382

V. Applications of Animal Models 382
   A. Evaluation of the Pathophysiology of Rheumatoid Arthritis 382
   B. Observations of Bone Morphology and Metabolism — Effects in Animal
      Models of Arthritis 383
   C. Fixation of Prosthetic Implants 383
   D. Synovectomy 385
   E. Soft Tissue Joint Constraints in Rheumatoid Arthritis 385
   F. Bone Structure and Strength 386
References 386

I. INTRODUCTION

Animal models of rheumatoid arthritis have been sought and employed in research for decades, but, as in other complex human conditions, no animal model is identical to the human disease. The disease states which are created as animal models of rheumatoid arthritis are thus inflammatory arthritides but cannot be considered to be "rheumatoid" arthritis.

Nevertheless, these models have afforded many opportunities to better understand the pathology, etiology and pathogenesis of rheumatoid arthritis (RA), as well as genetic factors in the disease. There is a growing number of reports on the use of animal models for experiments relevant to the surgical management of RA. Until the appearance of new methods of genetic manipulation permitting the development of animal strains which more closely mimic the natural course of disease, no specific species of animal was identified which naturally developed inflammatory arthritis similar to clinically observed human rheumatoid arthritis.

The earliest studies seeking an appropriate model of rheumatoid arthritis in the animal were conducted in the monkey, horse, sheep, goat, pig, dog rabbit, guinea pig, rat, mouse and chick embryo.[1] Arthritis was induced by injection of various antigens, either directly into the joint, or into subcutaneous or serous tissues for systemic effect. Many antigens have been injected in an effort to create inflammatory joint disease, and the arthritogenicity of various antigens was determined on an empirical basis.[1] The nomenclature for models of rheumatoid arthritis continues to reflect the specific antigen used to induce arthritis. Models of arthritis are sometimes categorized into monoarthritic and polyarthritic types. All animal models of arthritis are based on the injection of an irritant or antigen which results in an inflammatory or immunological response which is either local or generalized. Although sharing this common factor and often a similar general pathology, each model is characterized by a particular pathogenesis, course of disease, and pattern of joint and extraarticular involvement. The current rapid progress in understanding the genetics, cell biology, and cytokine chemistry of various models is increasing our fundamental knowledge of models which have, in some cases, been in use for decades.

The use of several different types of animal models, with known differences in etiology and pathogenesis, permits differential approaches to an understanding of mechanisms of arthritis. As interventions or treatment modalities are introduced into the model, hypotheses regarding the expected mechanism of action or pathological process which is affected can be tested against one or more of the models.[2] A considerable variety of animal models of rheumatic disease have been developed reflecting the wide range of signs and symptoms which characterize the various human diseases. Animal systems developed to model the spondyloarthropathies, systemic lupus erythematosus, systemic sclerosis, Sjögren's syndrome, polymyositis, and dermatomyositis have been comprehensively reviewed.[3] Models of inflammatory joint disease which are induced by viable infectious agents are of particular interest in the study of the "trigger" event which is presumed to initiate a pathological immune response in those individuals who are genetically susceptible to the development of autoimmune diseases such as rheumatoid arthritis. Mycoplasmas may be the most common cause of naturally occurring arthritis.

TABLE 1
Animals Employed for Models of Rheumatoid Arthritis

Animal	AA	CIA	SCW	AIA	CarA	Ref.
Mouse	DA	DBA/1		CBA		3, 5, 50
Rat	Wistar	Wistar, SD	SD, Lewis			8, 9, 10, 24, 39, 40
Rabbit				NZW, Old English	NZW	5, 42, 44, 68
Dog					mongrel	78
Pig					SPF	49
Monkey		squirrel, macaque				50

AA = adjuvant arthritis
CIA = collagen induced arthritis
SCW = streptococcal cell wall induced arthritis
AIA = antigen induced arthritis
CarA = carrageenan induced arthritis
SPF = specific pathogen free

Rheumatoid arthritis creates a specific spectrum of musculoskeletal pathology, different in important ways from other categories of disease, such as osteoarthrosis. Rheumatoid involvement of articular surfaces, ligaments and joint capsules, juxtaarticular and diaphyseal bone, and tendons and tendon sheaths creates characteristic symptoms and impairments requiring specific orthopaedic management. For this reason, an evaluation of the response of a bone, joint or soft tissue structure in rheumatoid disease to a manipulation or surgical intervention should be evaluated not in a normal animal, nor in a model of arthrosis, but, when possible, in an analogous animal model of inflammatory arthritis. The focus of this chapter has been placed upon those animal models of rheumatoid arthritis which are most relevant for experiments on the surgical management of the musculoskeletal system, and upon practical applications of these methods to surgical questions. A narrow definition of the scope of orthopaedic research has been avoided, and studies on implant fixation to abnormal bone, synovectomy, bone anatomy and biomechanics are reviewed, as well as studies of the mechanism of joint destruction in inflammatory disease. A detailed review of the genetics and biochemistry of models of inflammatory arthritis is beyond the scope of this chapter: Excellent reviews are available.[4–6]

II. SELECTION OF AN ANIMAL MODEL OF RHEUMATOID ARTHRITIS

Table 1 lists the most commonly used animal models and major references which provide detailed information on animal models of rheumatoid arthritis (RA). The murine species, the rat and the mouse, are the most commonly used because they are inexpensive and easy to handle and breed. The use of canine and porcine models of arthritis, and to a lesser degree lapine models, is limited by the cost of handling and maintenance of the animals. It is widely considered that results obtained in experiments in large, long-lived mammals are most relevant to human disease, especially where the anatomy or physiology of the smaller animal is known to differ from the human. One example of an important species-related difference is that the skeletons of small rodents do not mature, but continue to grow throughout life. This growth pattern is termed a modeling skeletal system, which differs from the pattern in humans and other large mammals, in which the skeleton is considered to be a remodeling system after the cessation of longitudinal growth. Notwithstanding this principle, where larger species have been employed in studies of arthritis, the findings have proved in general to be comparable to those for either the rat or the mouse. It is advisable to consider the specific anatomical or physiological factors of importance to the specific experiments to be carried out in order to determine if the proposed species is an appropriate model. In some cases, the use of larger species, particularly the rabbit or the dog, is based on a large existing literature describing the histology, pathology or physiology of the animal and its relevance to human disease. In studies in which biomechanical testing is performed, or where a surgical intervention or implantation is planned, larger specimens are usually employed because of the requirements of larger samples.

A. Adjuvant Arthritis

Adjuvant arthritis was introduced and developed by Pearson[7] and more extensively studied by Pearson,[8] Taurog,[9] van Eden, their coworkers and many others. The original and more recent literature on the topic of adjuvant arthritis is extensive, and the interested reader is directed to several reviews.[6,10–12] Early experiments studied the effect of various chemical irritants and antigens in inducing arthritis; ultimately a water-in-oil adjuvant was found to be effective.[7] Adjuvants are water-in-oil emulsions or simple mixtures in oil, which are more effective in inducing arthritis than are oil-in-water emulsions. Freund's adjuvant is termed "complete" when it contains killed mycobacteria of any of several species.

1. Induction of Arthritis

Adjuvant arthritis is induced in either the laboratory mouse or in the rat. The success of induction of arthritis is species-dependent and also strain-dependent.[13] In-depth reviews of the types of adjuvants and their use are given by Whitehouse et al.,[14] Herbert,[15] Whitehouse,[16] and Taurog.[6] In more recent research, the most commonly used adjuvant material is a complete Freund's adjuvant, prepared without water. Lyophilized *M. tuberculosis* or *M. butyricum* is prepared as a suspension in purified mineral oil or paraffin oil using a tissue grinder, and can be stored at 4°C for a period of up to six months. In commercially available complete Freund's adjuvant the concentration of active antigenic material may be lower than the required 0.5–1.0 mg per dose. Immediately before intradermal injection at the preferred site at the base of the tail, the material is mixed again using a vortex.

2. Clinical Observations

Depending on the species, size and strain of animal, the success of intradermal injection (providing a depot for slow release of the antigen), antigenic preparation, vehicle, and concentration and dose of the adjuvant employed, arthritis will become evident in the hind limbs within 11–21 days after the injection. Elevated temperature, redness and swelling of the hind feet, and occasional paresis of the hind limbs may also be observed. Scoring systems have been developed for the grading of these clinical signs. Necrosis may appear at the site of injection and extend distally along the length of the tail.

3. Gross Pathology and Histology

Histological examination of the affected hindpaw joints shows an acute synovitis, joint effusion and proliferation of synovium, usually without a purulent effusion; a subsynovial infiltration consisting of histiocytes, lymphocytes and plasma cells with a few polymorphonuclear leukocytes; peritendonitis and bursitis; invasion of subchondral bone by pannus; and new bone formation near the affected joints.

Further changes which may be observed from days 20–300 after the onset of arthritis include: fibrous thickening of the joint capsule, fibrous adhesions between articular surfaces, hypertrophy of synovial villi, a low grade inflammatory reaction with lymphocytic infiltrates near some joints, and periarticular ossification with occasional bone ankylosis, especially in the tail vertebrae.

TABLE 2
Comparison of Selected Animal Models of Arthritis to Rheumatoid Arthritis*

	RA**	AA	CIA	SCW	AIA	CarA
Onset/Course		100%	40–60%	100%	90%	100%
acute		10–12 days	14–60 days	3 days	21 days	28–49 days
chronic				14 days	40–300 days	
flare/remission	+			+	inducible flare	
Joints Affected						
peripheral	+	hind feet	hind feet	feet	tibio-femoral	tibio-femoral
axial		tail				
Extra-Articular Features	+	+	+			−
Immunity						
rheumatoid factor	+				+	−
collagen Type II ab	low	−	+	−		−
antibody response	+	−	+ / −			−
T cell response	+	+	+	+		−
complex formation	+				+	−
Soft Tissues						
cytokines in joint space	+	+		+		?
pannus	+	+	+	+	+	+
periosteal reaction	−	−	−	−		+
Effective Therapies						
NSAIDS	+	+	+ / −	+	−	
corticosteroids	+	+	+	+	+	
methotrexate	+	+	+	+	+	
gold salts	+	+ / −	worsens	−		
penicillamine	+	−	+	−		
cytokine antagonist /antibody			+	+	+ / −	
bisphosphonate		+			+ / −	?

* Table adapted and modified from Reference 50.
** RA = rheumatoid arthritis; other acronyms are the same as in Table 1.

Microscopic examination of the regional lymph nodes and other tissues indicates that the inoculum is transported by the lymphatic system to the circulatory system soon after injection, reaching the lung and liver where a classic granulomatous response occurs. Table 2 shows a clinical and pathologic comparison of adjuvant disease to other models and to rheumatoid arthritis, while Table 3 outlines bony effects.

4. Immunology

The immunological nature of adjuvant arthritis was recognized early by Waksman and Wennersten, who achieved the induction of adjuvant arthritis in rats by the passive transfer of living lymphoid cells from sensitized donors to their normal counterparts. The Lewis strain, which is most commonly used in adjuvant arthritis experiments[13,17] and Wistar strains are particularly susceptible to adjuvant arthritis, while Fischer and BN rats are resistant.

B. COLLAGEN INDUCED ARTHRITIS

Since type II collagen is present only in hyaline cartilage at birth, an immune reaction to type II collagen generates a specific reaction in synovial joints. Antibodies to type II collagen are present

TABLE 3
Reported Effects on Bone in Selected Animal Models of Arthritis — Comparison to Rheumatoid Arthritis*

	RA**	AA	CIA	SCW	AIA	CarA
Radiography						
radiographic lucency	+	+	+	+	+	+
joint space narrowing	+	+	+	+	+	
erosion	+	+	+	+	+	+
bone mineral density	+	+				
Joint						
joint stiffness	+					
tenosynovitis	+				+	
osteophyte formation	+	+	+	+	+	
ankylosis		+	+	fibrous		
instability / dislocation	+					+
Microscopic effects						
mineralization						+
osteopenia	+	+				+
structure						+
bone formation	variable					+
bone resorption	+					+
Mechanical properties	+					+
Bone marrow effects						
bone marrow cells		+	+			+
cytokines of bone	+	+	+			?
Therapies that alter hard tissue response						
NSAIDS					+	
corticosteroids	+	+			+	
methotrexate					+	
cytokine antagonist/antibody				+		
bisphosphonates						+
Altered osseointegration					no effect	

* Table adapted and modified from Reference 50.
** Acronyms are as shown in Tables 1 and 2.

in the synovial fluid of rheumatoid arthritis patients.[18,19] However, antibodies to denatured collagens are also present in RA and in several unrelated diseases, suggesting that antibodies to collagen represent a secondary phenomenon in chronic inflammatory arthritis and are not reflective of the etiology of the disease.

1. Induction of Arthritis

In 1977, Trentham and colleagues recorded that 40% of rats injected intradermally with native type II collagen obtained from human, chick or rat cartilage developed an inflammatory arthritis.[20] Collagen prepared at a concentration of 1 mg/ml in 0.1M acetic acid was emulsified with an equal volume of complete or incomplete Freund's adjuvant; 1 ml of this preparation was injected in the back intradermally in four to six sites in the rat. After 21 days, a booster dose consisting of another 0.5 mg collagen was injected intraperitoneally without adjuvant. Other injection schedules utilizing type II collagen from human, chick or rat cartilage sources and complete or incomplete Freund's adjuvant, but not types I or III, nor the α chains of type II collagen, also are effective.[20]

2. Clinical Observations

In the animals which developed arthritis, the onset occurred acutely at 20 days and persisted for as long as 60 days after immunization. On the first day of arthritis, the arthritic index in one hind limb typically was 4, the highest score. In some cases, both hind limbs were involved.

3. Gross Pathology and Histology

Histologic sections of the joints collected within 24 hours of the onset of arthritis demonstrated a mononuclear synovial infiltrate chiefly composed of T helper cells with few cytotoxic T-cells rather than a neutrophilic infiltrate.[21,22] In the synovial fluid most cells were granulocytes with fewer macrophages, comparable to RA.[22] The synovium and serum contained similar amounts of type II collagen antibody and rheumatoid factor.[22] Samples collected at later stages show proliferation of synovium and fibroblasts, and erosion of cartilage and subchondral bone. Periosteal new bone formation progresses in some cases to joint ankylosis in the carpal, tarsal, metacarpal, metatarsal and interphalangeal regions. Presence of mononuclear cells was observed in the synovium for as long as six months after disease onset. There was no axial skeletal involvement. The incidence of arthritis following injections of Type II collagen did not vary with injections ranging from 0.5 to 2 mg per animal.

Chronic collagen induced arthritis of four to eight months duration resulted in bony ankylosis after total destruction of cartilage and new bone formation.

4. Gender Effects

In humans, the increased risk for females (3×) of developing rheumatoid arthritis has been speculated to be related to sex hormones and to the central nervous system. In collagen induced arthritis, gender differences are opposite in effect for the rat and the mouse.[23]

5. Immunology

Collagen induced arthritis can be transferred passively from one animal to another by activated lymph cells. Nine of 32 naïve rats developed arthritis after receiving pooled spleen and lymph node cells from donors previously injected with type II collagen. Arthritis was not induced by injection of Freund's adjuvant alone, sera from arthritic donors, cells from nonimmunized donors, cells from rat donors which had been injected with type I collagen, cells from nonimmunized donors with solubilized type II collagen, or heat-killed cells from immunized donors. These controls established that disease transfer was not due to transfer of antigen, or chemical components in sera, but rather to transfer of sensitized cells.[24]

The significance of the complement system in collagen induced arthritis is shown by three types of evidence: Affected mice produce high levels of antibodies of the isotypes which are most efficient at activating complement, and increases in C3 correspond with the development of antibody. Administration of cobra venom which depletes complement prevents development of disease. Passive transfer of the disease is successful only in animals which have normal levels of complement.

T-cells have an essential role in collagen induced arthritis. Passive transfer of the complete disease is achieved only when T-cells are involved. Administration of anti-MHC Class II antibodies or anti-CD4 antibodies attenuates the disease, and arthritis cannot be induced in athymic nude rats which lack T-cells.[25,26] Pretreatment with antibodies to the T-cell receptor also significantly reduces the incidence of arthritis.

Comparison of the immunogenetics of RA and collagen induced arthritis is relevant because of the importance of MHC Class II genes in the human disease. Experiments describing the role of the comparable genes in mice in collagen induced arthritis are reviewed by Crofford and Wilder.[3]

A genetic factor was also identified in rats: the dr bb/WOR-UTM strain developed collagen induced arthritis with an incidence of 100%.

Animals injected with other types of collagen raise an immune response which is not arthritogenic. The injection of heterologous type II collagen results in antibody development which is specific to particular epitopes of the molecule. The region which contains these epitopes has been dubbed CB11 as it is contained within the cyanogen bromide fragment.[27] The immune response in collagen induced arthritis does not rely on the production of a single antibody for a particular epitope, but rather a combination of antibodies, which together produce the full arthritis response.[28] Persistent cases of collagen induced arthritis demonstrate interaction of humeral and cellular aspects of immunity.[23] The active induction of arthritis is prevented by the application of specific monoclonal antibodies to T-cells[25] and the passive immunization model of collagen induced arthritis is prevented by depletion of inflammatory fibrocytes.[29]

The concept of oral tolerance in RA and other inflammatory arthritides was developed through studies of collagen induced arthritis and adjuvant arthritis. The role of the T-cell receptor and the interaction of T-cells and B-cells as exemplified by interactions of the CD40 ligand GP39 present on activated CD4 T-cells is reviewed in detail by Durie et al.[23]

6. Significance of the Collagen Induced Arthritis Model

This model was of contemporary importance in demonstrating that homologous tissue could be arthritogenic, independent of complete Freund's adjuvant which contains bacterial cell walls. Features of collagen induced arthritis comparable to those of rheumatoid arthritis include proliferative synovitis in a chronic, polyarticular, erosive disease exhibiting symmetrical involvement of small and medium-sized peripheral joints, sparing of the axial skeleton, erosion of cartilage and subchondral bone at joint margins, and immunity to native Type II collagen[30] (Table 2).

Collagen induced arthritis is distinct from other models of arthritis in that it is induced by an endogenous antigen, and also because arthritis can be induced in naïve animals by the injection of anticollagen antibodies. As has been observed and reported for RA, the cytokines interleukin 1 (IL-1) and tumor necrosis factor (TNF-α) appear to be involved in the pathogenesis of collagen induced arthritis (reviewed by Durie et al.[23]). For example, recent studies employing the IL-1 receptor antagonist (IL-1ra), show that edema, cellular infiltration, and increased proteoglycan levels in synovial fluid during the induction phase of adjuvant arthritis are not inhibited, but IL-1ra has an anti-fibrotic action, decreasing abnormal interstitial collagen deposits and promoting regeneration of normal fat spaces in the synovial lining.[31,32] Collagen induced arthritis differs from RA in that subcutaneous nodules and pulmonary fibrosis are lacking, as are extraarticular manifestations, with the exception of lesions in the hyaline cartilage of the ear.

Adjuvant arthritis and collagen induced arthritis are similar in exhibiting immunologic hypersensitivity to type II collagen. Sera from rats with adjuvant arthritis contains increased amounts of hemagglutinating antibodies specific for type II collagen, and also for the α chain.[33] Also, peripheral blood mononuclear cells collected from rats with adjuvant arthritis respond to homologous type I and also type II collagen by increased incorporation of tritiated thymidine.[20] Trentham and colleagues proposed that adjuvanticity and arthritogenicity are properties derived from different portions of the glycopeptides contained in mycobacterial cell walls. Trentham's group also demonstrated that intact telopeptide regions of the type II molecule are not required, since pepsin-modified molecules were arthritogenic. A significant drawback of collagen induced arthritis is the technically complex preparation of the inducing agent which may become contaminated by other bacterial products, particularly proteoglycans, which have arthritogenic and immunomodulating properties.[34]

Another drawback of collagen induced arthritis is the variable onset from day 20 to 48 after the initial injection.[30]

C. Proteoglycan Induced Arthritis

Proteoglycan can be extracted from cartilage matrix digestion with chondroitinase-ABC and injected with Freund's complete adjuvant to induce arthritis in a susceptible mouse strain, BALB-c. Nine to twelve days after the second booster injection at four weeks, 100% of injected animals develop arthritis which becomes severe 7–8 weeks later.[34] Chronic disease involving cartilage and bone destruction, osteophyte formation and ankylosis, including ankylosis of the spine follows. Both humeral and cell-mediated immunity are involved as demonstrated in experiments with successful transfer of disease by injection of sensitized B- and T-lymphocytes into naïve recipients.

D. Antigen Induced Arthritis

The term "antigen induced arthritis" includes a heterogeneous group of conditions induced by various antigens and methods. The most commonly used antigens are methylated bovine serum albumin and ovalbumin. Early models of antigen induced arthritis were described in the rabbit by Dumonde and Glynn[35] and the mouse[36] and, more recently, the rat. A thorough review which emphasizes lapine antigen induced arthritis and contains many practical details and instructive illustrations was published by Cooke.[5]

Induction of rabbit antigen induced arthritis begins with intradermal or subcutaneous injection of 5 mg BSA or ovalbumin in 1 ml of emulsion at multiple sites in the back. Original protocols followed with one or two booster injections, after 14–21 days, which have been omitted by some recent investigators.[37] Sensitizaton is confirmed by a positive skin test at 21 days, demonstrating presence of a delayed type hypersensitivity response prior to the arthritogenic intraarticular knee injection.[32,37,38] The success of induction varies with species; the rabbit is hyperresponsive so that anaphlylactic shock may occur, while induction in the mouse, and particularly, the dog, is much more difficult.[5]

Within a few hours of injection, monoarticular arthritis begins with acute synovitis, swelling and exudate. Over two weeks, this is replaced by the chronic phase, characterized by invasive pannus and cartilage erosion, which persists as long as 24 weeks.[3]

E. Streptococcal Cell Wall Arthritis

Models of arthritis which employ an arthritogen developed from bacterial components are relevant to arthritis associated with Reiter's syndrome, inflammatory and infectious bowel diseases, rheumatic fever, post-Streptococcal arthritis, and Lyme arthritis all of which appear to be associated with bacterial infection. Although arthritis is inducible in animals by injection of bacterial components from a number of species: *Lactobacillus casei*, *Eubacterium aerofaciens*, *Bifidobacterium* species, and *Peptostreptococcus productus*, the most commonly used agent is *Streptococcus pyogenes*.[3] Streptococcal cell wall arthritis was developed in the 1950s and 1960s from the observation that injection of Streptococcal cell wall material including a peptidoglycan polysaccharide complex which resists biodegradation and persists in tissue, created an inflammatory lesion in the skin of the rabbit.[39]

Arthritis was first induced in rabbits by intraarticular injection and in the rat by intraperitoneal injection. In mice, a similar protocol results in pericarditis rather than arthritis. A whole cell sonicate is prepared from cultures of viable cells by treating them with 90 minutes of ultrasonic vibration, followed by filtration. The rhamnose content of the material is measured and adjusted in order to deliver 60 mcg rh/g body weight intraperitoneally. Alternatively, a heat inactivated whole cell preparation is prepared in phosphate buffered saline by adjusting the turbidity to compare with a previously standardized whole cell sonicate. About 10% of animals develop a mild arthritis which disappears within four days. Ninety percent of animals later develop one of two distinctive forms of disease. In one pattern animals have a period of complete remission followed by severe recurrence after 60 days. In the other pattern, arthritis subsides and then recurs several times over the study period of 130 days, without complete remission. The onset of disease is rapid, within 15–48 hours, marked by the appearance of red, swollen tarsal, carpal and interphalangeal joints.

1. Pathology

The heart, lungs and kidneys show no significant histologic changes. Tissues of the liver, spleen and lymph nodes show infiltration by foamy cells (macrophages) and the intestine shows evidence of peritonitis.

Microscopic changes are typical of an acute exudative inflammatory reaction which changes into a chronic erosive synovitis. Initial changes are observed as early as five hours after injections. Recurrent acute phases last for 10–15 days, with vascular congestion, edema, extensive fibrin deposition and infiltration by neutrophils and mononuclear phagocytes. Not only the synovial membrane, but also joint capsule, periarticular tissue, tendon sheath, muscle bundles, and muscle attachments are affected. Over two weeks the acute phase is replaced by a chronic proliferative synovitis which destroys and replaces subchondral cartilage and bone, and which is characterized by infiltrates of macrophages, lymphocytes and neutrophils, but not lymphoid follicles, in the synovial villi. The severity and duration of arthritis depends on the dose and also the strain of bacteria employed. Group A and B type streptococci are able to induce arthritis without fragmentation, and often with a long latent period. Portions of the bacterial cell which are not able to persist in the host for long periods of time induce only transient arthritis. Animals which have streptococcal cell wall arthritis exhibit other features which are comparable to rheumatoid arthritis, including chronic microcytic anemia, anergy, reduced production of interleukin-2, reduced mononuclear cell proliferation from the spleen in response to mitogens, and responses to several therapeutic agents. Histologic studies of the earliest stages of streptococcal cell wall induced arthritis show that endothelial cells are the first to be damaged.[40]

2. Immunology

The inflamed synovial tissue contains increased amounts of a number of enzymes and other factors which are also upregulated in rheumatoid arthritis, including metalloproteinases, cyclooxygenase D, cytokines, heparin binding growth factor and transforming growth factor-β, neuropeptides such as corticotropin releasing hormone, and the genes c-myc and c-fos. Experiments which show that athymic nude rats can develop acute but not chronic streptococcal cell wall arthritis show that T-cells are required for that phase. Passive transfer from diseased animals to naïve recipients is successful with T-cells. In cell culture, T-cells proliferate not only in response to Streptococcal cell wall, but also in response to extracts prepared from *Mycobacterium* species, suggesting a pathogenetic relationship between streptococcal cell wall arthritis and adjuvant arthritis.[41]

E. CARRAGEENAN INDUCED ARTHRITIS

1. General

Carrageenan is a sulfated mucopolysaccharide which is extracted from seaweeds, the marine algae *Chondrus* spp. and *Gigartina* spp. commonly referred to as Irish moss, or carrageen moss. Multiple spellings of the word Carrageenan appear in the literature. Carrageenan is obtained and used commercially as a thickener and stabilizer in many types of foodstuffs. Carrageenan is a member of the "gel-forming polysaccharides" because κ carrageenan gels upon exposure to potassium ion. Carrageenan has a primary structure of an alternating copolymer comprised of alternating units of sulfated D-galactose and 3, 6-anhydro-D-galactose. The secondary and tertiary structure which accompany this chemical makeup may afford some resistance to digestion by lysosomal enzymes. The potency of carrageenan as an irritant in either acute or chronic inflammatory reactions or in its effect as an anticoagulant is dependent upon both the molecular weight of the carrageenan, and also upon its characteristic fraction, either λ or κ, with the former fraction being more potent.[42]

2. Background

Carrageenan induced footpad inflammation is a widely used model for assaying the antiinflammatory effects of medications. Carrageenan-induced arthritis was described in the rabbit in the 1960s by Gardner, Santer and others,[42–45] after comparison against other irritants for its arthritogenicity in the tibiofemoral joint, where 0.3 ml of 1% carrageenan is injected. Carrageenan particles are actively phagocytosed by macrophages and monocytes where they remain for long periods of time.

Besides the well-documented inflammatory effects associated with the footpad injection, numerous other effects have been reported. Effects on the kinin system include *in vitro* activation of plasmin. *In vivo*, in the rat, carrageenan injection causes intense hypotension which is attributed to not only kinin effects, but also possibly other vasoactive factors which are formed via the complement activation system. Carrageenan also has an effect on the coagulation system: unfractionated, κ, and λ fractions all are anticoagulants which are about 1/15 as effective as heparin. The mechanism for this effect is not completely clear, but may be via the activation of Hagemann factor. Carrageenan also inhibits hemolytic complement, both *in vitro* and *in vivo*.

The mechanisms of action of carrageenan are extensively reviewed.[42–46] Many of the described characteristics of carrageenan are indeterminate because the molecular structure of the carrageenans varies from year to year and also from harvest to harvest, and presumably from species to species. The κ and ι fractions of carrageenan have been demonstrated to induce thrombosis and/or infarction of the tail or digits in mice, rats and guinea pigs. This thrombogenic effect would confound studies involving arthritis; therefore, the λ fraction of carrageenan, which does not have this effect, is recommended.[47]

3. Induction of Arthritis

Carrageenan is convenient for use as an arthritogen because of its chemical structure, allowing the identification in tissue by the periodic acid Schiff reaction, its easy preparation, its resistance to degradation, and the presence of sulfate groups which could be labeled. The rabbit model was extensively characterized by Lowther and Gillard, and coworkers.[43,48] Gardner determined that the optimal concentration of the solution was 1%. In guinea pig joints six hours after a single injection, an influx of polymorphonuclear leukocytes was observed which by 24 hours was being replaced by histiocytes and, at three days, by fibroblasts, subsiding by approximately one month. In the rabbit, arthritis induced in the joint was similar histologically and morphologically to surgical specimens from rheumatoid arthritis patients. A minimum of six injections spaced no more than one week apart resulted in the following changes: increased cellularity of synovial villi with invasion of margins of articular cartilage, loss of articular cartilage, and the formation of pannus. Fibrinoid deposits and osteoclastic bone resorption were occasionally observed. Lymphocytic perivascular cellular infiltrates were dissimilar from RA, where the infiltrate is principally of plasma cells.

4. Species/Strain Differences

As in other models of arthritis there are some strain differences in the reaction to carrageenan. A subset of Wistar rats, which are genetically resistant to the anaphylactic reaction produced by dextran, failed to react to carrageenan.

5. Comparison with Other Models

The carrageenan injection model of arthritis in the rabbit offers the practical advantages of simplicity and reproducibility. Repeated injections maintain the arthritis but do not exaggerate the local response, and the lungs, liver and other viscera are unaffected.[42,46] After five weeks of biweekly carrageenan injection, the inflammatory cellular response results in a villous hypertrophy of the synovium with perivascular collections of lymphocytes and large macrophages with a follicular appearance, and marginal cartilage erosion. Similar evolution of arthritis was reported in the specific pathogen-free pig; however, villus hypertrophy, hyperplasia of the synovium and the presence of fibrin was less than observed in the rabbit.[49] The model has been utilized in the rabbit for numerous orthopaedic research projects.

III. MECHANISTIC APPROACHES TO MODELS OF ARTHRITIS

A. CYTOKINE-INDUCED ARTHRITIS AND INHIBITION WITH ANTI-CYTOKINE ANTIBODIES

Controlled studies of cytokines in animal arthritis are of particular interest because, while upregulation of several pro-inflammatory cytokines in RA is well-known, mechanistic considerations are impossible to resolve from studies of human disease.[50,51] The development of modern techniques which allow the isolation, preparation and use of cytokines *in vivo*, and their positive identification, has stimulated many advances in the study of pathophysiological mechanisms of arthritis. Injection of IL–1 or TNF into joints induces a transient synovitis in normal animals and worsens arthritis in several models of arthritis.[51] Several cytokines of interest in the pathophysiology of arthritis and in bone metabolism (eg IL–1, TNF, TGF-β) which have been studied with mutation-based approaches are reviewed in Ryffel[53] and Brennan.[52]

B. MUTATION-BASED APPROACHES

It has become feasible to manipulate both cellular makeup and cytokine makeup *in vivo*, in the same animal, an approach which has numerous advantages. A number of natural and induced animal mutations afford the opportunity to study how the omission (a.k.a. "gene-knockout") or addition of a particular gene or its transcript affects the evolution of arthritis. These "transgenic" animals (having an altered, but stable genotype arising from a defined genetic manipulation[22]) have been employed in much work associated with arthritis, but little with an orthopaedic approach to arthritis.

Spontaneous "knock-outs" include several strains of osteopetrotic animals (mice, rats, and rabbits) which lack M-CSF (macrophage colony-stimulating factor), and thus demonstrate abnormal bone development. W/W and steel mouse strains have abnormal apoptosis (timed cell death), resulting in lymphoproliferative disease. Other mutations which involve extensive lymphoid abnormality are the aly, nude and SCID (severe combined immunodeficient) mice.[53] Specific gene deletions to develop "knock-out" strains (also symbolized in the literature by "-/-") can be contrived by disrupting genes (homologous recombination) in embryonic stem cells and injecting the cells into blastocytes.[53]

Among natural mutations, one widely used model is the SCID mouse. Recent work implanting human cancellous bone under the skin of SCID mice, demonstrates that osteoblasts remain viable, forming new bone which persists.[54] Also, human synovial membrane from RA patients, implanted into the joint of SCID mice induces arthritis.[55]

IV. EVALUATION METHODS

Evaluation of the specimens from animal models are covered in detail in Chapters 6 through 12. The general principles of evaluation apply as well to animal models of rheumatoid arthritis, but certain methods developed specifically for evaluation of the animal models are discussed briefly below. The reader is directed to the appropriate references for detailed information.

The health of an animal in which inflammatory arthritis is being induced can be monitored by simple methods such as the weight, activity level and quantity of feeding. Chronically ill animals will generally differ in their weight pattern from normal controls, and this may be considered in the interpretation of the results of surgical experiments.

A. Biochemical Assessment of Cartilage and Bone Turnover

There are numerous biochemical tests in serum and urine which reflect the kinetics of formation and resorption of bone and the formation and degradation of articular cartilage components. As surrogates for direct measurement of bone and cartilage, biochemical tests have the advantage of permitting serial studies during the course of disease. Molecules liberated in the process of bone formation or resorption can be followed by ELISA assay in serum or urine. Most useful to describe increased bone formation are osteocalcin, procollagen peptides, and bone-specific alkaline phosphatase. For resorption, elevated levels of calcium, hydroxyproline, tartrate-resistant acid phosphatase, pyridinoline and deoxypyridinoline (products of the degradation of Type I collagen from bone) in plasma or urine are found in RA and animal models of arthritis. It must be recognized that serological measurements in general have inherent inaccuracies related to sampling error, renal function, and tissue distribution. In the case of bone, presence of some markers (for example osteocalcin) in the serum may reflect both formation and resorption, as the marker is released during both processes.[56]

B. Biochemical Assessment of Inflammatory Disease

The activity of inflammatory disease can be followed in animal models with the use of the erythrocyte sedimentation rate and with acute phase reactants. These tests are discussed in Chapter 8. Biochemical assessment of cytokines within synovial fluid effusions can be performed; cytokines within the joint are highly variable and reflect a complex and redundant process of inflammation. Cytokines are released from cells within joint exudates and in synovial membrane, but also from cells in marrow spaces in subchondral bone and from chondrocytes in cartilage. The cytokines implicated in bone and cartilage degradation include PGE2, IL–1α, 6, TNF-α and are known to be elaborated by monocytes, histiocytes, osteoblasts, T-cells and myeloid precursors. Knowledge in this field is expanding rapidly and a current understanding of the subject can be found in recent reviews.[57]

C. Clinical Severity

Scoring systems to date for clinical severity of arthritis are various and subjective. These authors know of none to date that have been formally validated. Jee et al. have recently developed a new, systematic method of assessing joint damage in adjuvant arthritis, using automated histomorphometry.[58]

D. Roentgenographic Methods

Roentgenographic evaluation is useful for following the progression of arthritis non-invasively and sequentially.[59–61] Small animals such as the rat are anesthetized, placed in a supine position, and their hind limbs fixed with tape, one in the lateral, one in the AP position.[60] In each radiograph, the joints of the hind limbs are evaluated together for mineralization, erosions, periostitis, cartilage space, soft tissues and limb alignment.

Radiographic grading systems for evaluating the progression of arthritis require lateral radiographs of the affected limb studied at three-fold magnification. In collagen induced arthritis, radiographs are most likely to demonstrate joint space narrowing, joint space widening, soft tissue swelling, and osteopenia, and unlikely to demonstrate abnormalities in the sacroiliac joints, hips, shoulders, elbows or axial skeleton. Osteopenia is often obscured by periostitis at the bone margins. Changes in the joint space are not reliable because of variability due to early narrowing due to articular cartilage loss, but later apparent widening related to bone changes in the central part of the articular surface.

One of the few reports validating an assessment method for describing the bony effects of inflammatory arthritis employed scintigraphy of the arthritic right compared to left knee, and found that uptake of ^{99}Tc correlated with histological findings, in antigen induced arthritis in the mouse.[62]

Bony changes observed in collagen induced arthritis differ from those of antigen induced arthritis, and periostitis, which is observed in several animal models of inflammatory arthritis, is rarely observed in rheumatoid arthritis, with the exception of juvenile onset disease.

E. Necropsy: Gross Examination

After euthanasia specific gross anatomical findings should be recorded. The presence and grade of joint effusions, fibrinoid debris and "rice bodies" in the synovial fluid, presence and grade of synovitis, area and depth of articular cartilage and bone erosion can be recorded as indices of joint damage. Capsuloligamentous distention and joint instability are also relevant findings in inflammatory arthritis.

F. Histomorphometry: Assessment of Bone Turnover by *In Vivo* Labelling

The nomenclature of bone histomorphometry has been standardized by an international committee of specialists.[63] An introduction to histomorphometry theory and methods, and an extensive review of earlier work have been published.[64] Inflammatory arthritis is characterized in animal models and in human rheumatoid arthritis by abnormalities of bone remodeling kinetics (Chapter 15). These abnormalities can be identified in histological sections of undecalcified bone by *in vivo* labelling of newly formed bone with bone-seeking markers.[65] These labels are incorporated into newly deposited osteoid as it mineralizes so that formation of new mineral can be documented by the primary bone parameters, labeled surface (LS) and mineral apposition rate (MAR), and secondary parameters such as bone formation rate (BFR) can be derived. Pulsed labelling by interval injection of tetracycline, calcein green, xylenol orange and/or other dyes is a standard method of bone histomorphometry, useful in osteonal analysis of cortical bone. Uncommonly used is continuous labelling of bone with markers such as calcein delivered continuously in drinking water. Careful choice of a labelling molecule, with an appropriate administration schedule, will add greatly to the kinetic information that can be obtained through bone histomorphometry. Histomorphometry requires specialized equipment, experience and skill as it is performed on undecalcified, ground sections of bone, and because labels are sensitive to light and processing methods and can be lost from specimens. For these reasons kinetic and morphometric methods, although indispensable in certain experiments are expensive and time-consuming and are often replaced where possible by surrogate measurements.

Assessment of mineralization requires undecalcified histological sections of bone, prepared with silver stains to identify unmineralized osteoid seams on the surfaces of trabecular bone. These stains do not accurately identify gradations in mineralization, which is a gradual process requiring approximately 40 days to complete. Detailed information regarding the state of mineralization of bone can be obtained via density fractionation and by specialized electron microscopic methods.

V. APPLICATIONS OF ANIMAL MODELS

A. Evaluation of the Pathophysiology of Rheumatoid Arthritis

The surgical management of rheumatoid arthritis and the other rheumatic diseases requires special attention to the specific pathology and pathophysiology of these conditions. Rheumatoid involvement of the hip, forefoot and wrist, for example, differ in important ways from degenerative and traumatic disorders at these sites, and informed surgical management reflects an understanding of the unique conditions of joint pathoanatomy and pathomechanics and changes in the soft tissue constraints of the joints. In order to evaluate new therapeutic methods in the surgery of the rheumatic diseases, and to reevaluate old ones, it is preferable to employ appropriate animal models which share features of human disease. Although there are numerous examples of appropriate investigations in animal models of rheumatoid arthritis, some of which are described below, there are many situations where the surgeon seeking guidance has no alternative to generalizing from the results of experiments performed in normal animals or in animal models of degenerative disease. The extensive literature on models of rheumatoid arthritis in Rheumatology and Immunology journals and textbooks contains numerous useful reviews of the immunology and pharmacological management of the condition.

Animal models have been applied extensively to studies of the genetics, biochemistry and pharmacology of rheumatoid arthritis and the rheumatic diseases, but less frequently to studies of bony effects, or to the surgical management of these conditions, even if a broad interpretation of the scope of surgical management is assumed.

Some of the principal areas of orthopaedic investigation where the application of animal models of rheumatoid arthritis is useful are described below.

B. OBSERVATIONS OF BONE MORPHOLOGY AND METABOLISM — EFFECTS IN ANIMAL MODELS OF ARTHRITIS

Abnormal periosteal new bone formation is observed in several of these models adjacent to sites of severe inflammation. In antigen induced arthritis in the mouse, bone apposition at medial and lateral sides of long bones occurs in most animals.[66] Osteophyte formation appears inhibited by NSAIDs (piroxicam) and by corticosteroids (prednisolone), independent of secondary cartilage formation.[67] Osteocyte death and bone marrow degeneration (osteonecrosis) are observed in tibial subchondral bone.[67]

Osteopenia is a hallmark of animal models of inflammatory arthritis and RA. Many properties of cancellous bone of the epiphysis proximal to the arthritic joint are abnormal in carrageenan induced arthritis. The turnover of bone by osteoclasts and osteoblasts is rapid, so that large increases in bone formation are insufficient to compensate for bone loss, documented as a 20% decrease in trabecular bone volume (BV).[68] The trabecular surface is largely occupied by active osteoblasts, osteoid, or osteoclasts instead of a quiescent, resting trabecular surface.[69] The net effect is a rapid remodeling osteopenia. Immobilization of a normal limb in a plaster cast as a model of disuse failed to reproduce this remodeling abnormality, suggesting that arthritis effects, not immobilization are the cause of the remodeling and osteopenic abnormality.[70] Fractal analysis of cross-sectional images shows that the normal anisotropy or organizational arrangement of the subchondral trabecular bone is modified in arthritic specimens.[71] Recent histological work suggested that two bisphosphonates are osteoprotective in carrageenan induced arthritis[72,73] and also may protect the articular cartilage surface[69] from inflammatory arthritis-induced destruction. The stabilizing effects of bisphosphonates on subchondral trabecular bone and on metaphyseal and diaphyseal cortex in inflammatory arthritis (discussed in Chapter 15), are of potential interest to the arthritis surgeon. Bisphosphonates may also affect bone marrow fibroblasts and lymphocytes which are altered in carrageenan induced arthritis. Alterations in bone marrow have also been observed in adjuvant arthritis,[74] collagen induced arthritis,[74] and antigen induced arthritis,[66] as well as in RA.[75,76]

C. FIXATION OF PROSTHETIC IMPLANTS

The fixation of orthopaedic joint implants to bone in patients who have inflammatory arthritis raises specific issues of the bone-prosthesis bond in a condition characterized by abnormalities in bone strength, microanatomy and remodeling kinetics.

Since the late 1970s fixation of metal joint prostheses to the skeleton without the use of acrylic cement has become an option in the knee and the hip, and more recently in the shoulder, elbow, hand and ankle. The fixation of prostheses by this method depends for initial stability upon an interference fit, which requires impaction of an implant with favorable contours into a bone surface precisely prepared within a joint. Beyond the brief phase of initial press-fit stability, which may require protection from weight-bearing, fixation of the implant requires ingrowth of host bone into asperities on the surface of the prosthesis, fabricated by special processes. Ingrowth prostheses have been prepared by bonding fiber metal pads, by sintering metal beads, by arc-bonding of titanium and by other methods.

Extensive research has been devoted to identifying the suitable types of metal surface, especially the size and configuration of voids, that promote rapid and stable bone ingrowth for long-term prosthetic fixation. Another relevant factor is the metallurgy of the implant: titanium and titanium alloys, for example, have been demonstrated in general to promote bone ingrowth compared to cobalt/chromium alloys and stainless steel.

However, there has been less effort to determine the effect of the other side of the prosthesis-metal bond, the host bone which serves as the bed for implantation. This is particularly relevant in rheumatoid arthritis, one of the common indications for joint replacement in the knee, shoulder, and elbow and also the hip. Bone anatomy, strength and remodeling kinetics are key to initial and long term fixation of implants, and the effects of the specific abnormalities of bone in rheumatoid arthritis require elucidation.

The authors have extensively described the abnormalities in juxtaarticular cancellous bone anatomy, remodeling kinetics, and mineralization in the carrageenan injection model of inflammatory arthritis in the rabbit, characterized by osteopenia, rapid remodeling and incomplete mineralization.[68,77] These osteopenic features are discussed in greater detail in Chapter 15. Using a similar model in dogs, characterized by a 20% reduced bone density measured by CT densitometry, Söballe et al. found, after four weeks, a diminished bone ingrowth into titanium alloy porous-coated cylinders implanted into the distal femoral condyle, whereas ingrowth into hydroxyapatite-coated porous cylinders was not significantly different from control, non-arthritic animals. Assessment of shear strength demonstrated a corresponding decrease in the titanium, but not the hydroxyapatite-coated implants. The authors conclude that ingrowth into porous implants was impaired in osteopenic bone in inflammatory arthritis, for titanium alloy implants, but not for hydroxyapatite-coated implants.[78,79] Similar findings were reported by Sennerby and Thomsen[80] who studied bone ingrowth onto threaded pure titanium implants in an antigen induced arthritis model. Ingrowth of bone onto pure titanium implants in arthritic rabbit tibiofemoral joints was diminished compared to control animals.[80] However, Branemark and Thomsen reported in 1997 that bone apposition and biomechanical evaluation was not impaired in collagen-induced arthritis.[81]

Friedman et al. compared bone apposition and ingrowth to implant surfaces and shear strength in a pushout test for cylinders of a pure titanium beaded surface, in the same surface coated with a 50 mm hydroxyapatite (HA) coating and in a grit-blasted, non-beaded titanium alloy surface with HA coating, in rabbits with carrageenan-induced arthritis.[82] Their findings demonstrated a significant diminution of the shear strength of the bone/implant interface in all three groups in inflammatory arthritis compared to normal, possibly due to the diminished trabecular thickness and number induced by the arthritis, observed in all groups, and a thinner layer of apposed bone, observed in the third group. However, bone apposition and ingrowth were not impaired by inflammatory arthritis in these experiments.[83]

There is a need for further studies to determine whether medications commonly used in rheumatoid arthritis (NSAIDs, immunosuppressive and antimetabolic agents) would interfere with bone ingrowth or apposition to orthopaedic implants. Whereas there are reports on the effects of indomethacin and methotrexate into porous metallic implants in normal animals, similar studies in models of inflammatory arthritis would be useful to determine how, in the presence of altered bone remodeling in inflammatory arthritis, drugs known to affect bone formation and resorption affect cementless prosthetic fixation.

D. SYNOVECTOMY

The role of synovectomy in the management of inflammatory arthritis has been controversial for 40 years. Surgical synovectomy of joints was introduced to the management of RA by Vainio and Laine, in the 1950s in Heinola, Finland. Other methods of synovectomy have been applied, including chemical ablation of the synovium with osmic acid, radiosynoviorthesis or radiosynovectomy by injection of radioisotopes of gold, yttrium, rhenium, and strontium, and by less invasive means of surgical synovectomy utilizing the arthroscope. The original focus of joint synovectomy was the knee, but most of the other joints have been treated by synovectomy.

There would be little disagreement among clinicians who treat rheumatoid arthritis that thickened synovial tissue should be removed by the surgeon during surgical procedures performed for other, independent indications. During any surgical procedure for rheumatoid arthritis, such as wrist fusion, total knee arthroplasty, forefoot reconstruction or reconstruction of swan neck deformity, the surgeon will remove synovium with the goal of improving the mechanical function of the joint, diminishing effusion and possibly relieving a source of chronic pain.

The value of synovectomy as an independent operative procedure, performed for indications based on the benefits of synovectomy alone, has been more controversial. A clinical review of synovectomy is beyond the scope of this chapter, but there is evidence that synovectomy has value in relieving pain and improving function in certain sites, but that no effect in retarding the anatomic and radiological progression of the disease has been established.

Animal models have been applied to the controlled study of the effects of synovectomy. Chinol et al.[84] utilized an antigen induced arthritis model to evaluate the suitability of ^{153}Sm- or ^{186}Re-labeled hydroxyapatite as a radiation synovectomy agent. Low leakage rates and satisfactory distribution through the joint were demonstrated. ^{186}Re sulphur colloid also appeared to have satisfactory joint retention in a similar antigen induced arthritis model.[85] In a controlled trial, holmium-laser arthroscopic synovectomy was compared to surgical synovectomy by arthrotomy with a further sham control, in a model of antigen induced arthritis.[86] The laser synovectomy appeared to result in less capsular fibrosis than the open surgical method. Reichel and Weber reported that synovectomy resulted in a reduced exudation of plasma proteins into the knee joint, compared to preoperative controls, in an arthritis model induced by intraarticular injection of human IgG complex in immunized rabbits.[87]

The carrageenan injection model of synovitis in the horse was utilized to evaluate pretreatment with ketoprofen and phenylbutazone in acute joint inflammation. Phenylbutazone was more effective than ketoprofen in reducing lameness, joint temperature, synovial fluid volume and synovial fluid PGE_2.[88] Kim et al. reported that continuous passive motion of the knee in antigen induced arthritis, was associated with significantly greater joint swelling, synovial effusion and histologic synovitis scores compared to immobilized arthritic knees. After six weeks, however, articular cartilage was better preserved in CPM-treated than immobilized knees, as measured by articular cartilage erosion and loss of cellularity.[37]

E. SOFT TISSUE JOINT CONSTRAINTS IN RHEUMATOID ARTHRITIS

A typical characteristic of rheumatoid arthritis is laxity of ligaments supporting joints, and capsuloligamentous distention resulting from synovial effusion, damage to the cells, collagen and proteoglycans of ligaments and suppression of the biosynthesis of type I collagen. This problem contributes greatly to the morbidity of rheumatoid arthritis by creating clinical joint instability, and also by contributing to the destruction of the articular cartilage surface, which depends upon normal articulation of the joint components.

Investigations of the biochemistry, structure and biomechanics of ligament in joint injury models and in models of degenerative or posttraumatic arthrosis have been published. We have not been able to locate published reports of studies of the deficient soft tissue restraints of joints in models of inflammatory arthritis.

Considerable morbidity in RA is related to the involvement of sheaths, with effects on the tendons. A model of tenosynovitis is described by Cooke.[5] In this model, 0.5 ml. of 2% BSA in PBS is injected into the tibialis anterior tendon, an intrasynovial structure which is easily accessible. Within 48 hours, an acute Arthus reaction occurs, with characteristic hemorrhage, hypertrophy of the sheath lining, and influx of polymorphonuclear cells, followed by phagocytic cells and necrosis. The reaction as shown by localization of antigen within the tendon, subsides over two weeks, or six weeks, if challenge is used.

F. BONE STRUCTURE AND STRENGTH

Rheumatoid arthritis patients are at a markedly increased risk of fracture[89] which is devastating to the RA patient when it occurs in the proximal or distal femur.[90] Biomechanical failure at articular surfaces of bones (collapse of joint surfaces, weak prosthetic fixation, difficult juxtaarticular metaphyseal fractures) and in the diaphyses (fracture of long bones during or after prosthetic arthroplasty) contribute to the morbidity of RA. Extensive studies on the morphology, remodeling kinetics, mineralization, microstructure and mechanics of bone have been performed in animal models of rheumatoid arthritis, and are discussed in Chapter 15. These studies are relevant to a number of orthopaedic issues including prosthesis selection and fixation, fixation of fracture implants including screws, understanding fracture risk in RA patients and developing strategies to reduce that risk.

REFERENCES

1. Gardner, D. L., "The experimental production of arthritis: a review," *Ann. Rheum. Dis.*, 19, 297, 1960.
2. Willoughby, D. A., "Human arthritis applied to animal models: towards a better therapy," *Ann. Rheum. Dis.*, 34, 471, 1975.
3. Crofford, L. J. and Wilder, R. L., "Arthritis and autoimmunity in animals," in *Arthritis and Allied Conditions: A Textbook of Rheumatology*, Vol. 2, Koopman, W. J., Ed., Williams & Wilkins, London, 1997, Chapter 29.
4. Zhang, J., Weichman, B. M., and Lewis, A. J., "Role of animal models in the study of rheumatoid arthritis: an overview," in *Mechanisms and Models in Rheumatoid Arthritis*, Henderson, B., Edwards, J. C. W., and Pettipher, E. R., Eds., Academic Press, London, 1995, 363.
5. Cooke, T. D. V., "Antigen-induced arthritis, polyarthritis, and tenosynovitis," in *CRC Handbook of Animal Models for the Rheumatic Diseases*, Vol. 1, Greenwald, R. A. and Diamond, H. S., Eds., CRC Press, Boca Raton, FL, 1988, 53.
6. Taurog, J. D., Argentieri, D. C., and McReynolds, R. A., "Adjuvant arthritis," in *Immunochemical Techniques*. Part 1, Di Sabato, G., Ed., Academic Press, New York, 1988, 339.
7. Pearson, C. M., "Experimental joint disease. Observations on adjuvant-induced arthritis," *J. Chron. Dis.*, 16, 863, 1963.
8. Pearson, C. M. and Wood, F. D., "Studies of arthritis and other lesions induced in rats by the injection of mycobacterial adjuvant. VII. Pathological details of the arthritis and spondylitis," *Am. J. Pathol.*, 42, 73, 1963.
9. Taurog, J. D., Sandberg, G. P., and Mahowald, M. L., "The cellular basis of adjuvant arthritis. I. Enhancement of cell-mediated passive transfer by concanavalin A and by immunosuppressive pretreatment of the recipient," *Cell Immunol.*, 75, 271, 1983.
10. Billingham, M. E. J., "Adjuvant arthritis: the first model," in *Mechanisms and Models in Rheumatoid Arthritis*, Henderson, B., Edwards, J. C. W., and Pettipher, E. R., Eds., Academic Press, London, 1995, 389.

11. Kerwar, S. S., Ridge, S. C., and Oronsky, A. L., "Comparative studies between adjuvant, type II collagen, and streptococcal cell wall-induced arthritis in rats," in *CRC Handbook of Animal Models for the Rheumatic Diseases*, Vol. 1, Greenwald, R. A. and Diamond, H. S., Eds., CRC Press, Boca Raton, FL, 1988, 49.
12. Halloran, M. M., Szekanecz, Z., Barquin, N., Haines, G. K., and Koch, A. E., "Cellular adhesion molecules in rat adjuvant arthritis," *Arthritis Rheum.*, 39, 810, 1996.
13. Swingle, K. F., Jaques, L. W., and Kvam, D. C., "Differences in the severity of adjuvant arthritis in four strains of rats," *Proc. Soc. Exp. Biol. Med.*, 132, 608, 1969.
14. Whitehouse, M. W., Orr, K. J., Beck, F. W. J., and Pearson, C. M., "Freund's adjuvants: relationship of arthritogenicity and adjuvanticity in rats to vehicle composition," *Immunology*, 27, 311, 1974.
15. Herbert, W. J., "Mineral-oil adjuvants and the immunization of laboratory animals," in *Handbook of Experimental Immunology*, Weir, D. M., Ed., Lippincott, Philadelphia, 1978.
16. Whitehouse, M. W., "Adjuvant-Induced Polyarthritis in Rats," in *CRC Handbook of Animal Models for the Rheumatic Diseases*, Vol. 1, Greenwald, R. A. and Diamond, H. S., Eds., CRC Press, Boca Raton, FL, 1988, 3.
17. Muir, V. Y. and Dumonde, D. C., "Different strains of rats develop different clinical forms of adjuvant disease," *Ann. Rheum. Dis.*, 41, 1982, 1982.
18. Trentham, D. E. and Dynesius-Trentham, R., "Collagen-induced arthritis," in *Mechanisms and Models in Rheumatoid Arthritis*, Henderson, B., Edwards, J. C. W., and Pettipher, E. R., Eds., Academic Press, London, 1995, 447.
19. Trentham, D. E., Kammer, G. M., McCune, W. J., and David, J. R., "Autoimmunity to collagen: a shared feature of psoriatic and rheumatoid arthritis," *Arthritis Rheum.*, 24, 1363, 1981.
20. Trentham, D. E., Townes, A. S., and Kang, A. H., "Autoimmunity to type II collagen: an experimental model of arthritis," *J. Exp. Med.*, 146, 857, 1977.
21. Caulfield, J. P., Hein, A., Dynesius-Trentham, R., and Trentham, D. E., "Morphological demonstration of two stages in the development of type II collagen-induced arthritis," *Lab. Invest.*, 46, 321, 1982.
22. Harris, H. E., Liljeström, M., and Klareskog, L., "Characteristics of synovial fluid effusion in collagen-induced arthritis (CIA) in the DA rat; a comparison of histology and antibody reactivities in an experimental chronic arthritis model and rheumatoid arthritis (RA)," *Clin. Exp. Immunol.*, 107, 480, 1997.
23. Durie, F. H., Fava, R. A., and Noelle, R. J., "Collagen-induced arthritis as a model of rheumatoid arthritis," *Clin. Immunol. Immunopathol.*, 73, 11, 1994.
24. Trentham, D. E., Dynesius, R. A., and David, J. R., "Passive transfer by cells of type II collagen-induced arthritis in rats," *J. Clin. Invest.*, 62, 359, 1978.
25. Ranges, G. E., Sriram, S., and Cooper, S. M., "Prevention of collagen-induced arthritis by *in vivo* treatment with anti-L3T4," *J. Exp. Med.*, 162, 1105, 1985.
26. Wooley, P. H., Luthra, H. S., Lafase, P. W., Huse, A., Stuart, J. M., and David, C. S., "Type II collagen-induced arthritis in mice. III. Suppression of arthritis by using monoclonal and polyclonal anti-Ia antisera," *J. Immunol.*, 134, 2366, 1985.
27. Myers, L. K., Rosleniec, E. F., Seyer, J. M., Stuart, J. M., and Kang, A. H., "A synthetic peptide analogue of a determinant of type II collagen prevents the onset of collagen-induced arthritis," *J. Immunol.*, 150, 4652, 1993.
28. Terato, K., Hasty, K. A., Reife, R. A., Cremer, M. A., Kang, A. H., and Stuart, J. M., "Induction of arthritis with monoclonal antibodies to collagen," *J. Immunol.*, 148, 2103, 1992.
29. Fava, R. A., Gates, C., and Townes, A. S., "Critical role of peripheral blood phagocytes and the involvement of complement in tumor necrosis factor enhancement of passive collagen-induced arthritis," *Clin. Exp. Immunol.*, 94, 1993.
30. Jamieson, T. W., De Smet, A. A., Cremer, M. A., Kage, K. L., and Lindsley, H. B., "Collagen-induced arthritis in rats assessment by serial magnification radiography," *Invest. Radiol.*, 20, 324, 1985.
31. Lewthwaite, J., Blake, S. M., Hardingham, T. E., Warden, P. J., and Henderson, B., "The effect of recombinant human interleukin 1 receptor antagonist on the induction phase of antigen induced arthritis in the rabbit," *J. Rheumatol.*, 21, 467, 1994.
32. Lewthwaite, J., Blake, S., Thompson, R. C., Hardingham, T. E., and Henderson, B., "Antifibrotic action of interleukin–1 receptor antagonist in lapine monoarticular arthritis," *Ann. Rheum. Dis.*, 54, 591, 1995.

33. Trentham, D. E., McCune, W. J., Susman, P., and David, J. R., "Autoimmunity to collagen in adjuvant arthritis of rats," *J. Clin. Invest.*, 66, 1109, 1980.
34. Glant, T. T., Mikecz, K., Arzoumanian, A., and Poole, A. R., "Proteoglycan-induced arthritis in BALB/c mice," *Arthritis Rheum.*, 30, 201, 1987.
35. Dumonde, D. C. and Glynn, L. E., "The production of arthritis in rabbits by an immunological reaction to fibrin," *Br. J. Exp. Pathol.*, 43, 373, 1962.
36. Brackertz, D., Mitchell, G. F., and Mackay, I. R., "Antigen-induced arthritis in mice. I. Induction of arthritis in various strains of mice," *Arthritis Rheum.*, 20, 841, 1977.
37. Kim, H. K., Kerr, R. G., Cruz, T. F., and Salter, R. B., "Effects of continuous passive motion and immobilization on synovitis and cartilage degradation in antigen induced arthritis," *J. Rheumatol.*, 22, 1714, 1995.
38. Novaes, G. S., Mello, S. B. V., Laurindo, I. M. M., and Cossermelli, W., "Low dose methotrexate decreases intraarticular prostaglandin and interleukin–1 levels in antigen induced arthritis in rabbits," *J. Rheumatol.*, 23, 2092, 1996.
39. Cromartie, W. J., Craddock, J. G., Schwab, J. H., Anderle, S. K., and Yang, C., "Arthritis in rats after systemic injection of streptococcal cells or cell walls," *J. Exp. Med.*, 146, 1585, 1977.
40. Wilder, R. L., "Streptococcal cell wall-induced arthritis in rats," in *CRC Handbook of Animal Models for the Rheumatic Diseases*, Vol. 1, Greenwald, R. A. and Diamond, H. S., Eds., CRC Press, Boca Raton, FL, 1988, 33.
41. DeJoy, S. Q., Ferguson, K. M., Sapp, T. M., et al., "Streptococcal cell wall arthritis. Passive transfer of disease with a T cell line and crossreactivity of streptococcal cell wall antigens with *Mycobacterium tuberculosis*," *J. Exp. Med.*, 170, 369, 1989.
42. Gardner, D. L., "Production of arthritis in the rabbit by the local injection of the mucopolysaccharide carageenin," *Ann. Rheum. Dis.*, 19, 369, 1960.
43. Santer, V., Sriratana, A., and Lowther, D. A., "Carrageenin-induced arthritis. V. A morphologic study of the development of inflammation in acute arthritis," *Semin. Arthr. Rheum.*, 13, 160, 1983.
44. Lowther, D. A., Gillard, G. C., Baxter, E., Handley, C. J., and Rich, K. A., "Carrageenin-induced arthritis. III. Proteolytic enzymes present in rabbit knee joints after a single intraarticular injection of carrageenin," *Arthritis Rheum*, 19, 1287, 1976.
45. Lowther, D. A. and Gillard, G. C., "Carrageenin-induced arthritis. I. The effect of intraarticular carrageenin on the chemical composition of articular cartilage," *Arthritis Rheum.*, 19, 769, 1976.
46. Amini, D., Daziano, L., Gagnon, J., and Laurin, C. A., "The use of intraarticular osmic acid to produce chemical synovectomy in rabbits," *Clin. Orthop.*, 79, 164, 1971.
47. Bekemeier, H. and Giessler, A. J., "Thrombosis induction by different carrageenins in rats and mice," *Naturwissenschaften*, 74, 345, 1987.
48. Gillard, G. C. and Lowther, D. A., "Carrageenin-induced arthritis. II. Effect of intraarticular injection of carrageenin on the synthesis of proteoglycan in articular cartilage," *Arthritis Rheum.*, 19, 918, 1976.
49. Uruchurtu Marroquin, A. and Ajmal, M., "Carrageenin-induced arthritis in the specific-pathogen-free pig," *J. Comp. Pathol.*, 80, 607, 1970.
50. Henderson, B., Edwards, J. C. W., and Pettipher, E. R., Eds., *Mechanisms and Models in Rheumatoid Arthritis*, Academic Press, Toronto, 1995.
51. Brennan, F. M., "Role of cytokines in experimental arthritis," *Clin. Exp. Immunol.*, 97, 1, 1994.
52. Brennan, F. M., "Transgenic models for arthritis: useful clues to be gained?" *Ann. Med.*, 28, 271, 1996.
53. Ryffel, B., "Gene knockout mice as investigative tools in pathophysiology," *Int. J. Exp. Pathol.*, 77, 125, 1996.
54. Boynton, E., Aubin, J., Gross, A., Hozumi, N., and Sandhu, J., "Human osteoblasts survive and deposit new bone when human bone is implanted in SCID mouse," *Bone*, 18, 321, 1996.
55. Sack, U., Kuhn, H., Kämpfer, I., Genest, M., Arnold, S., et al., "Orthotopic implantation of inflamed synovial tissue from RA patients induces a characteristic arthritis in immunodeficient (SCID) mice," *J. Autoimmunity*, 9, 51, 1996.
56. Kleerekoper, M., "Biochemical markers of bone remodeling," *Am. J. Med. Sci.*, 312, 270, 1996.
57. Miossec, P., "Cytokine abnormalities in inflammatory arthritis," *Ballière's Clin. Rheumatol.*, 6, 373, 1992.
58. Jee, W. S. S., Li, X. J., Ke, H. Z., Li, M., Smith, R., and Dunn, C. J., "Application of computer-based histomorphometry to the quantitative analysis of methylprednisolone-treated adjuvant arthritis in rats," *Bone Miner.*, 22, 221, 1993.

59. Blackham, A., Burns, J. W., Farmer, J. B., Radziwonik, H., and Westwick, J., "An X ray analysis of adjuvant arthritis in the rat: the effect of prednisolone and indomethacin," *Agents Actions*, 7, 145, 1977.
60. Clark, R. L., Cuttino, J. T., Jr., Anderle, S. K., Cromartie, W. J., and Schwab, J. H., "Radiologic analysis of arthritis in rats after systemic injection of streptococcal cell walls," *Arthritis Rheum.*, 22, 25, 1979.
61. Wood, F. D., Pearson, C. M., and Tanaka, A., "Capacity of mycobacterial wax D and its subfractions to induce adjuvant arthritis in rats," *Int. Arch. Allerg. Appl. Immunol.*, 35, 456, 1969.
62. Lens, J. W., van den Berg, W. B., and van de Putte, L. B. A., "Quantitation of arthritis by ^{99m}Tc-uptake measurements in the mouse knee joint: correlation with histological joint inflammation scores," *Agents Actions*, 14, 723, 1984.
63. Parfitt, A. M., "Bone histomorphometry: proposed system for standardization of nomenclature, symbols, and units," *Calcif. Tissue Int.*, 42, 284, 1988.
64. Parfitt, A. M., "The physiological and clinical significance of bone histomorphometric data," in *Bone Histomorphometry, Techniques and Interpretations,* Recker, R., Ed., CRC Press, Boca Raton, FL, 1983, 150.
65. Frost, H. M., "Bone histomorphometry: choice of marking agent and labeling schedule," in *Bone Histomorphometry, Techniques and Interpretations*, Recker, R., Ed., CRC Press, Boca Raton, FL, 1983, 49.
66. Schalkwijk, J., van den Berg, W. B., van de Putte, L. B. A., Joosten, L. A. B., and van der Sluis, M., "Effects of experimental joint inflammation on bone marrow and periarticular bone. A study of two types of arthritis, using variable degrees of inflammation," *Br. J. Exp. Pathol.*, 66, 435, 1985.
67. de Vries, S. J. and van den Berg, W. B., "Impact of NSAIDS on murine antigen induced arthritis. II. A Light Microscopic investigation of antiinflammatory and bone protective effects," *J. Rheumatol.*, 17, 295, 1990.
68. Bogoch, E., Gschwend, N., Bogoch, B., Rahn, B., and Perren, S., "Juxtaarticular bone loss in experimental inflammatory arthritis," *J. Orthop. Res.*, 6, 648, 1988.
69. Bogoch, E. R., Roberts, E., Moran, E., and Fornasier, V. L., "Pamidronate (APD) prevents rapid bone remodelling and trabecular bone loss in experimental inflammatory arthritis," *Trans. Orthop. Res. Soc.*, 20, 226, 1995
70. Bogoch, E. R., Moran, E., Crowe, S., and Fornasier, V., "Arthritis not immobilization causes bone loss in the carrageenin injection model of inflammatory arthritis," *J. Orthop. Res.*, 13, 777, 1995.
71. Caldwell, C. B., Moran, E. L., and Bogoch, E. R., "The fractal dimension of trabecular bone is altered in experimental inflammatory arthritis," *J. Bone Miner. Res.*, in press, 1998.
72. Pysklywec, M. W., Moran, E. L., and Bogoch, E. R., "Zoledronate (CGP 42'446), a bisphosphonate, protects against metaphyseal intracortical defects in experimental inflammatory arthritis," *J. Orthop. Res.*, 15, 858, 1997.
73. Bellingham, C. M., Lee, J. M., Moran, E. L., and Bogoch, E. R., "Torsional properties of the rabbit femoral diaphysis: bisphosphonate (Pamidronate/APD) prevents arthritis-induced loss of fracture toughness," *J. Orthop. Res.*, 13, 876, 1995.
74. Hayashida, K., Ochi, T., Fujimoto, M., Owaki, H., Shimaoka, Y., Ono, K., and Matsumoto, K., "Bone marrow changes in adjuvant-induced and collagen-induced arthritis," *Arthritis Rheum.*, 35(2), 241, 1992.
75. Tomita, T., Kashiwagi, N., Shimaoka, Y., Ikawa, T., Tanabe, M., et al., "Phenotypic characteristics of bone marrow cells in patients with rheumatoid arthritis," *J. Rheumatol.*, 21, 1608, 1994.
76. Owaki, H., Ochi, T., Yamasaki, K., Hakitani, S., Ocamura, M., and Ono, K., "Elevated activity of myeloid growth factor in bone marrow adjacent to joints affected by rheumatoid arthritis," *J. Rheumatol.*, 16, 572, 1989.
77. Lucas, S., Bogoch, E. R., Nespeca, R., and Grynpas, M. D., "Bone changes induced in a rabbit model of experimental arthritis," *Eur. J. Exp. Musculoskel. Res.*, 1, 121, 1993.
78. Søballe, K., Pedersen, C. M., Odgaard, A., Juhl, G. I., Hansen, E. S., et al., "Physical bone changes in carrageenin-induced arthritis evaluated by quantitative computed tomography," *Skeletal Radiol.*, 20, 345, 1991.
79. Søballe, K., "Hydroxyapatite ceramic coating for bone implant fixation: mechanical and histological studies in dogs," *Acta. Orthop. Scand.*, 64(Suppl 255), 1, 1993.
80. Sennerby, L. and Thomsen, P., "Tissue response to titanium implants in experimental antigen-induced arthritis," *Biomaterials*, 14, 413, 1993.

81. Brånemark, R. and Thomsen, P., "Biomechanical and morphological studies on osseointegration in immunological arthritis in rabbits," *Scand. J. Plast. Reconstr. Surg. Hand Surg.*, 31, 185, 1997.
82. An, Y. H., Friedman, R. J., Jiang, M., LaBreck, J. C., Draughn, R. A., et al., "Bone ingrowth to implant surfaces in an inflammatory arthritis model," *J. Orthop. Res.*, 16, 576, 1998.
83. Friedman, R. J., An, Y. H., Jiang, M., Butehorn, H. F., Draughn, R. A., and Bauer, T. W., *Bone Ingrowth into Porous and HA Coated Titanium Implants in Experimental Inflammatory Arthritis*, Fifth World Biomaterials Congress, Toronto, 1996.
84. Chinol, M., Vallabhajosula, S., Goldsmith, S. J., Klein, M. J., Deutsch, K. F., et al., "Chemistry and biological behavior of samarium–153 and rhenium–186-labeled hydroxyapatite particles: potential radiopharmaceuticals for radiation synovectomy," *J. Nucl. Med.*, 34, 1536, 1993.
85. Wang, S., Lin, W., Hsieh, B., Shen, L., Tsai, Z., Ting, G., and Knapp, F. F., Jr., "Rhenium–188 sulphur colloid as a radiation synovectomy agent," *Eur. J. Nucl. Med.*, 22, 505, 1995.
86. Lind, B. M., Moller, K. O., Schramm, U., Baretton, G., Trautmann, C., et al., "Vergleichende experimentelle Untersuchungen zur mechanischen und Holmiumlaser-Synovektomie," *Langenbecks Arch. Chir.*, 378, 273, 1993.
87. Reichel, W. and Weber, K. J., "The stabilizing effect of synovectomy on the synovial membrane in arthritic rabbit knees," *Arch. Orthop. Trauma Surg.*, 105, 11, 1986.
88. Owens, J. G., Kamerling, S. G., Stanton, S. R., Keowen, M. L., and Prescott-Mathews, J. S., "Effects of pretreatment with ketoprofen and phenylbutazone on experimentally induced synovitis in horses," *Am. J. Vet. Res.*, 57, 866, 1996.
89. Hooyman, J. R., Melton III, L. J., Nelson, A. M., O'Fallon, W. M., and Riggs, B. L., "Fractures after rheumatoid arthritis: a population based study," *Arthritis Rheum.*, 27, 1353, 1984.
90. Bogoch, E. R., Ouellette, G., and Hastings, D., "Failure of internal fixation of displaced femoral neck fractures in patients with rheumatoid arthritis," *J. Bone Joint Surg.*, 73B, 7, 1991.

Part V

Animal Models of Joint Replacement and Related Conditions

20 Animal Models for Studying Soft Tissue Biocompatibility of Biomaterials

John A. Jansen

CONTENTS

I. Introduction 393
II. General *In Vivo* Tests 394
    A. Soft Tissue Biocompatibility Assays 394
    B. Commonly Used Animals 394
III. Factors Affecting Wound Healing 395
    A. Wound and Repair Process 395
    B. Animal Considerations 396
    C. Surgical Considerations 397
    D. Material Properties 398
IV. Commonly Used Implant Models 398
    A. Subcutaneous Implant Model 398
    B. Intramuscular Implant Models 399
    C. Percutaneous Implant Models 399
V. Common Evaluation Methods 400
    A. Light Microscopical Preparation 401
    B. Histological and Histomorphometrical Evaluation 402
    C. Other Evaluation Methods 402
References 404

I. INTRODUCTION

Prior to their clinical use biomaterials for surgical implants have to be tested on their biocompatibility. The purpose of this biocompatibility assessment is to exclude a potential toxic and carcinogenic effect of the material. Such a biocompatibility evaluation can occur at three levels,[1] i.e. (1) initial tests; cytotoxicity and mutagenicity; (2) secondary tests; and (3) usage tests.

The initial tests are mainly *in vitro* evaluation procedures. For the secondary or usage tests experimental animals are used.[2] Secondary tests are employed as a screening method for the local *in vivo* compatibility of materials for short and prolonged periods. The objective of the implant usage test is to investigate the functionality of a specific implant design. In general, for the initial secondary tests inexpensive, readily available but still relevant animal models are used. The final functionality evaluation occurs mostly in a different animal.

Considering the above mentioned, we have to notice that for orthopaedic implant materials the biological analysis not only comprises tests for bone compatibility. Also screening of the soft tissue response is desirable, since a lot of orthopaedic materials come in contact with subcutaneous tissue, muscles, fasciae and tendons.

II. GENERAL *IN VIVO* TESTS

The interaction between an implant material and the surrounding tissues can be considered vital for the final clinical performance of implanted artificial medical devices. For example, the promotion of tissue attachment and the concomitant reduction of the highly undesirable chronic imflammatory response and fibrosis around implant materials are of main importance for the biocompatibility of biomaterials.[3]

Roughly, a test material inserted into tissue evokes two types of reactions.[4] First, there is the inflammatory response to the surgical trauma. The following reaction is the tissue response on the biomaterial. Regarding the dynamic character of living tissues, animal models are an essential tool in the evaluation of the biological behavior of implant materials.

In designing an experimental protocol to study the soft tissue biocompatibility of implants, it has to be realized that the healing responses of different animals and tissues can vary considerably. The selection of an appropriate model and implantation site should be such that the obtained results are indicative for use in human subjects.[4-6]

A. Soft Tissue Biocompatibility Assays

Soft tissue compatibility experiments can be performed on two levels:[7] tests for local effects after implantation[8-11] and tests for systemic effects.[12] Considering the scope of this chapter, the significance of systemic reactions will not be discussed. For further information, the reader is referred to Merritt.[13]

Tests on local effects can be classified again according to their duration into short and long term. In short term or acute to subchronic biocompatibility experiments, the period of implantation is less than 30 days. In long term or chronic assays, the implantation time can continue from 30 days until about one year. The exact implantation time is always depending on the purpose of the study.

Since the early introduction of soft tissue implantation tests,[14] new methods and approaches have been introduced continuously to qualify and quantify the tissue response. Characteristic for all methods is that the primary goal is to describe the severity of the inflammatory response. As reflected in the standards of the American Society for Testing and Materials (ASTM),[8-11] the still commonly used tool is histological and histomorphometrical evaluation of retrieved implants and tissue specimens. Suggested scoring indicators are: the general appearance of the tissue reaction and the presence of inflammatory cells.

B. Commonly Used Animals

The selection of a suitable animal model for biocompatibility testing is a complex issue. It is determined by factors, like cost of animals, housing space, technical assistance and experimental objective. For soft tissue purposes, mostly rats and rabbits are used. Animals, such as guinea pigs, goats, dogs, sheep, pigs, calves and monkeys are also used.[15-19]

The advantage of rats is their availability and low cost. Further, breeding programs have resulted in rat species with almost similar intrinsic biologic properties. A disadvantage of the rat model is that the metabolic and wound healing properties are significantly different from bigger animals. We have to emphasize that this can endanger the correct extrapolation of the obtained results. Another problem with rats and all other rodents is that they can only be used for implantation studies shorter than six months. After about 6–8 months of implantation there is the risk of accidental induction of tumors.[20] Consequently, for long term studies other animal models like rabbits have to be used. An additional advantage of the rabbit is that more implant specimens per animal can be tested. This facilitates the statistical design of the experiment and reduces the effect of interanimal variance. Further, bigger animals also allow the use of larger implants.

Although a wide variety of different animal species can be used for soft-tissue biocompatibility tests, it is better to stick to one or two animal models, only a few implantation sites and preferably one operator. The advantage of such a procedure is that an enormous amount of experience is obtained, which assures a good reproducibility and a high intralaboratory validity. This enables a comparative evaluation between different experiments. A disadvantage is that the results of the experiments are hard to extrapolate (to other operators, animals, laboratories, etc.).

III. FACTORS AFFECTING WOUND HEALING

Despite the use of standardized protocols and appropriate animal models, the investigator has to be aware of the fact that there are a lot of other variables which can affect the final tissue response. These influences can be of biological or experimental origin, such as surgical fluctuations (presence of microorganisms, size of incision), health or general condition of the animals (the occurrence of infections after implantation), social behavior of the animals (biting, grating), local properties of the implant site (among species, presence of subcutaneous fat), and implant characteristics (shape, porosity, mechanical properties).

A. Wound and Repair Process

If the integrity of soft tissue is disturbed, e.g., by trauma or surgery, the physiological mechanisms of wound healing start. Two phases in wound healing can be distinguished, i.e. the inflammatory phase and the repair phase.[21] Sometimes even a third intermediate phase, called the proliferative phase is described.[22] In summary, during the inflammatory phase which takes about three days, the following events occur: at disruption of the integrity of the tissue, blood vessels are torn. Subsequently, the wound bed fills with blood from the torn vessels followed by activation of blood coagulation while platelets bind to the exposed collagen. The release of chemotactic substances by platelets, the activation of the complement system by exposed collagen and extracellular ATP attracts inflammatory cells. These cells, mainly polymorphonuclear granulocytes and monocytes differentiate into macrophages, which start to ingest fragments of injured tissue. Furthermore, the macrophages release substances to stimulate replication of fibroblasts and myofibroblasts at the wound edges.

At approximately the third day, formation of collagen fibres by the fibroblasts becomes histologically visible.[22] In addition, a network of capillaries is formed to provide oxygen to support the fibroblast synthesis of collagen. In front of the newly formed collagen matrix, the macrophages still continue to phagocytize the dead material hereby creating an environment for other fibroblasts to settle. This process continues until the wound is completely closed. The tissue formed is called granulation tissue. Already at the sixth day of wound healing, maturation of the collagen fibres starts. By means of collagen synthesis and lysis, remodelling of the collagen network occurs. Meanwhile, myofibroblasts are responsible for wound contraction, hereby reducing the wound surface. Finally, the number of cells will decrease, leaving scar tissue behind. The functional characteristics of this newly formed tissue are less effective compared to the original tissue. The wound strength will never reach its original value and scar tissue is nonelastic.[22-24]

The presence of an implant can provide a continuous inflammatory stimulus. As a result, the acute or inflammatory phase can be prolonged.[25] This will be associated with an additional increase in cellular activity. If this occurs, then also the repair phase will be noticeably delayed and enhanced. The change in timescale and extensivity of wound healing and repair processes are determined by the biocompatibility of the used implant material (Figure 1).

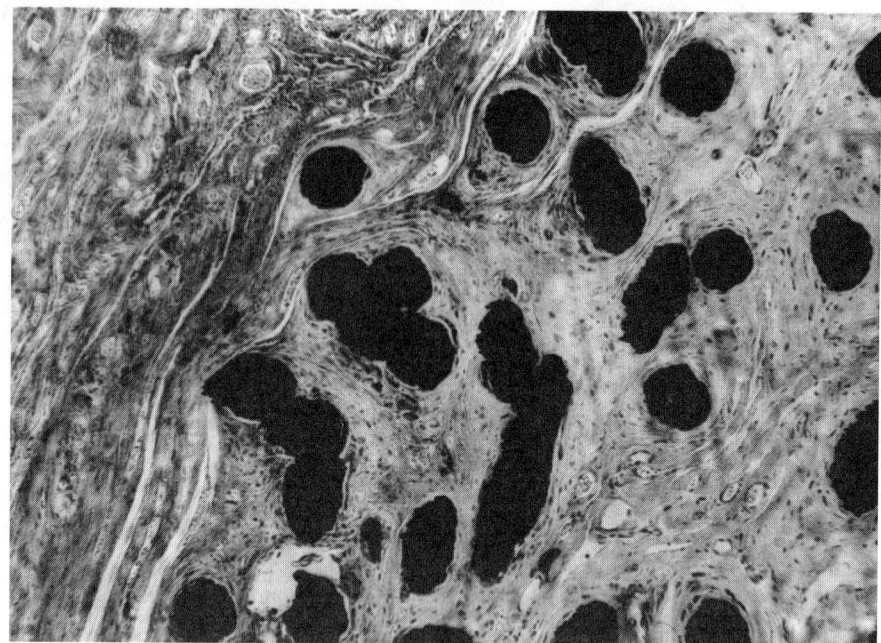

FIGURE 1. Histological section showing the soft tissue response to a porous titanium fibre mesh implant at three months after subcutaneous implantation. The used material evokes a very moderate tissue response, characterized by the presence of a thin fibrous capsule surrounding the mesh, and connective tissue ingrowth into the porosity of the mesh.

B. Animal Considerations

Although the basic process is similar, wound healing can still vary between animal species. Gangjee[18] showed that there are quantitative differences with regard to subcutaneous connective response between rabbits, goats and dogs. In his study at all experimental periods (10, 15, 20 and 30 days) the number of giant cells and polymorphonuclear leucocytes around implants was highest in the rabbit and lowest in the goat. The degree of fibrous capsule maturity showed an inverse relationship. It was highest in the goat and lowest in the rabbit. Therefore, conclusions about biocompatibility tests have also to be related to the experimental animal used. Depending on the final application of the investigated biomaterial, it can even be suggested to repeat the experiment with another type of animal.

Further, the location or tissue, in which the implant is placed, can contribute to the wound healing. Picha[26] observed that when rough surfaced implants are placed close to fatty tissue, the tissue response will be completely different from similar implants placed in a completely muscular or fibrous tissue bed. McGeachie[27] inserted titanium and stainless steel wire into mouse leg muscles. Morphometric analysis showed no difference in muscle reaction between the two metals. Since these results did not corroborate with other studies in which titanium and stainless steel implants were placed in a subcutaneous position, he suggested that probably the skeletal muscle of a mouse has a high tolerance for foreign materials.

Considering the above mentioned, the conclusion appears to be justified that materials only have to be tested in the environment in which they finally will be applied. This is confirmed by Semmelink,[28] who describes the induction of granuloma and plasma cell formation after the subcutaneous implantation of β–whitlockite particles. In contrast, in an osseous environment bone formation is observed without an immuno-response after only one week.

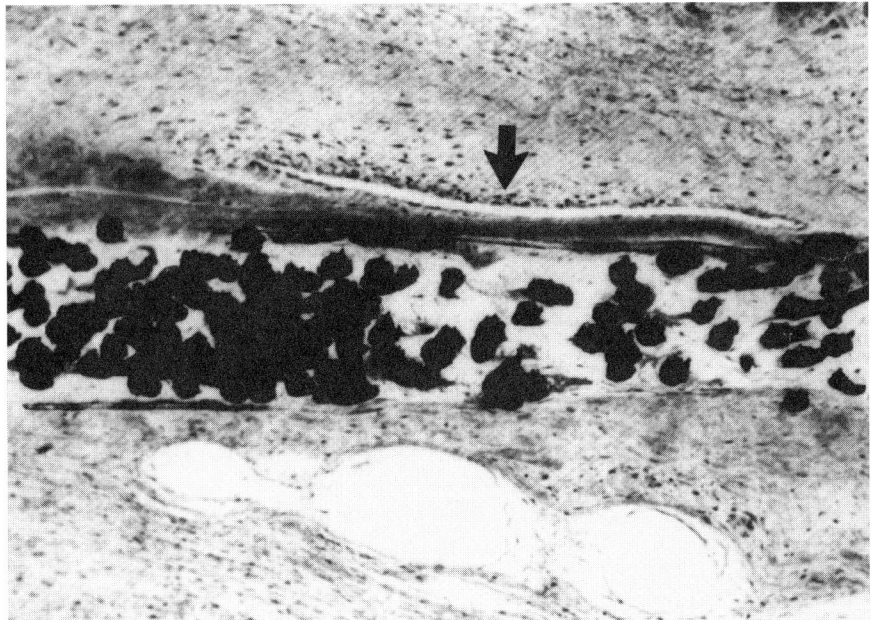

FIGURE 2. A hair (arrow) is visible in the interface between implant and surrounding fibrous tissue capsule. The hair was introduced during the surgical procedure.

C. SURGICAL CONSIDERATIONS

Surgical technique is an important contributing factor in the tissue acceptance of an implant. For example, a careful surgical technique has to be used to limit the damage to the tissue. In addition, implants which do not fit properly into the created space can cause an intolerable pressure on the surrounding tissue structures. This can lead to necrosis and subsequent failure of the implant. Further, it is known that glove powder can create histological complications including serum accumulations. Consequently, it is necessary that the surgical gloves are washed every time before the implants are touched. In addition, provisions have to be met which prevent the possible ingress of hairs or other debris into the implant pocket, like the use of special introducers for the insertion of the experimental specimens (Figure 2).

Another point of concern is migration or mobility of the implants after implantation. It has even been suggested that implants, after insertion, have to be fixed in position with sutures.[29] Whether this is true is difficult to say, since the suture material itself also will evoke an inflammatory reaction. On the other hand, the surgeon has to make a tissue bed in which the sample fits as well as possible. Overextension has to be prevented.

Also care has to be taken in the handling of the animals after implant placement. When implants are placed in the backs of animals, these animals should be treated carefully and not handled at the dorsum skin when they are taken out of their cages.

A final question which arises is the use of antibiotics. To prevent infection, the best approach is to use sterile surgical procedures.[2] Only when this is impossible, the use of antibiotics can be considered. However, it is important to note that antibiotic treatment can also interfere with the final healing response. Therefore, if given, they should be administered for as short a period as possible. A strict standardized pre-and postsurgical administration protocol, like that used for human patients, is advisable.

D. MATERIAL PROPERTIES

When desiging a biocompatibility test, one must consider properties other than the physico-chemical characteristics of the implant which can influence the tissue response. For example, certain geometrical properties of the implant specimen, like shape, size, and surface topography, are known to affect the tissue reaction.[3-5] Therefore, care has to be taken to standardize the geometrical features of the implants for each experiment.

The implants also have to cleaned carefully after their preparation. Any foreign material (chemical matter, debris, etc.) left can alter the tissue response. For polymeric materials a good post-preparation cleansing procedure is first washing in 10% Liquinox solution (Alconox Inc.). Thereafter, the specimens have to be rinsed, cleaned ultrasonically for 30 minutes in a 1% Liquinox solution and given two 15 minute ultrasonic rinses in distilled, deionized water. Subsequently, they have to be given a Soxhlet rinse for 12 hours in distilled, deionized water. Finally, the substrata can be air-dried and sterilized. A sterilization process has to be used that does not change the polymer. For metallic implants, ultrasonic cleaning in 100% ethanol to remove any loose particles, is mostly sufficient. Again, the sterilization procedure has to be selected carefully, since sterilization is not always as clean as supposed.[30] Also the packaging of the specimens after sterilization is important. Especially, in case of rough materials, particles of the wrapping material can stick and be maintained on the specimen surface.

The mechanical properties of the implant material are also important.[2] Similar to bone implants, a mismatch in the mechanical properties between the implant and the surrounding tissue will result in an inadequate stress transfer and distribution at the interface. This can result in a completely different tissue response, i.e. a thicker versus thinner fibrous capsule and more versus fewer inflammatory cells at the implant-tissue interface. Consequently, specimens within one experiment must have similar mechanical properties. Otherwise, the results will be distorted.

IV. COMMONLY USED IMPLANT MODELS

The overall objective of soft-tissue biocompatibility assays is to determine the *in vivo* behavior of materials that are inserted for short or prolonged contact with tissue. Testing of the biological properties is mostly performed by inserting the materials into the subcutaneous or muscle tissues of experimental animals. For some specific applications, like the evaluation of materials used for the fabrication of leads, drains and external fixators, percutaneous implant models also can be used.

A. SUBCUTANEOUS IMPLANT MODELS

For subcutaneous testing, the dorsal subcutis of the experimental animal is the preferred location.[11] Before implantation, the animals are anaesthesized and placed in ventral recumbancy. Then, the back of the animal is shaved, scrubbed with Betadine®, and disinfected with iodine. Paravertebral, between the scapula and the hind limb, a longitudinal incision is made on the left and right sides of the spinal column, through the full thickness of the dorsum skin. Depending on the animal type and size one or more incisions can be made. Subsequently, lateral to the incision a subcutaneous pocket between skin and muscle fascia is created by blunt dissection with a scissors. One implant is inserted in each pocket. When more implants are placed per animal, contact between the specimens after insertion has to be prevented. Finally, the wound(s) are carefully closed with resorbable sutures. Depending on the animal species and housing of the animals, various kinds of suture techniques can be used. For example, in our laboratory we use an intracutaneous suturing technique when the animals are housed together.

At the end of the experiment, the animals are sacrificed and the skin is shaved again. Subsequently, an incision is made through the skin lateral to the implants. Then, the implants are exposed by retracting the skin from the underlying muscle tissue and the implants with their surrounding tissues and the overlying skin are excised. Skin tissue always has to be included into the retrieved sample. This facilitates the final histological comparison between normal and regenerated tissue. For this reason, the animals are shaved after euthanasia. Without complete hair removal, the histological preparation of the samples is hampered, because the embedding material cannot easily penetrate into the sample. Directly after retrieval, the tissue specimens are fixed in 10% buffered formalin. After fixation, the samples can be trimmed to remove excess tissue.

Mostly the specimens are subjected to routine light microscopical examination. When more sophisticated evaluation techniques are used, perfusion fixation is used instead of immersion fixation.[31]

B. INTRAMUSCULAR IMPLANT MODELS

A lot of orthopaedic devices will come in contact with skeletal muscles. Although muscle tissue is highly vascularized, it shows less regenerative capability.[6] In addition, due to intrinsic stress factors related to motion of the muscles, the biocompatibility response of implant materials placed in muscles can differ from subcutaneous tissue. Paravertebral and gluteal muscles are the test sites of first choice for intramuscular implant models.[8]

After anesthesia, the skin over the dorso-lumbar or pelvic regions is shaved and disinfected with Betadine and 75% ethanol. A longitudinal incision is made through the skin and the paravertebral gluteal muscles are exposed. The skin is separated from the underlying fascia with blunt dissection. Subsequently, the fascia is dissected and a small incision is made into the belly of the muscle. An implantation site is created by further separation of the muscle fibers using blunt dissection with a hemostat or rounded scissors. After insertion of the implant, the muscle incision is closed with resorbable sutures. The last step is closure of the skin incision. Depending on the size of the animal, one or more implants can be introduced in the muscle tissue.

At the end of the experiment, the animals are killed and the implants along with a generous zone of surrounding muscle tissue are excised.

C. PERCUTANEOUS IMPLANT MODELS

Percutaneous implants can be placed at different locations. In our laboratory, we implant them into the dorsum, into the tibia and onto the cranium of various types of experimental animals, i.e. guinea pigs, rabbits and goats.[32–34]

For the testing of drains, we prefer the use of goats. For the insertion of the specimens, the animals are anesthesized and the region distal to the costal ridge is shaved, washed and disinfected with iodine. A longitudinal incision is made parallel to the spinal column. Lateral to this incision a subcutaneous pocket is created by blunt dissection with scissors between the subcutaneous fat layer and the musculus obliquus abdominis externus. Centrally in the subcutaneous pocket, the muscle is cleft parallel to the muscle fibers over a distance of about 0.5 cm and a small tunnel is created by blunt dissection. Then, the drain tube can be inserted in this tunnel. Thereafter, the wound is closed using resorbable sutures. To prevent postoperative damage of the wound site, we stable the goats separately with their heads fixed between two vertical bars to prevent the animals from manipulating the percutaneous tubes.

In experiments where percutaneous leads have to be used, percutaneous implants are inserted into the dorsum and on the cranium. The implants are flange-shaped to obtain sufficient subcutaneous stabilization and fixation.

For the cranium implants we prefer the rabbit as experimental animal. After anesthesia, shaving and disinfection, a longitudinal incision is made on the rabbit's skull, approximately 3 cm caudal of the orbita. After exposing the os frontale, the skin is bluntly undermined and a subcutaneous pocket is created. The flange-shaped implant is created and the skin is sutured. Besides separate housing, no special measures have to be taken to prevent damage to the device or percutaneous passage.

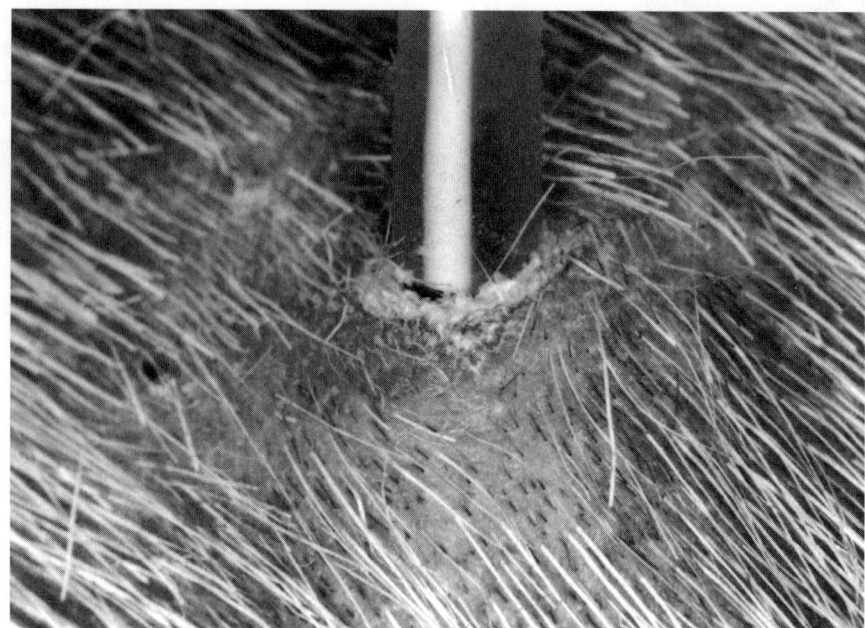

FIGURE 3. A macroscopic picture of a successful percutaneous lead, inserted into the dorsum of a goat, after four months of implantation.

For the dorsum implants we either use rabbits or goats. The surgical technique for placement of the implants always consists of two stages. During the first session of this two-stage procedure, only the subcutaneous part of the percutaneous device is inserted. For installation of the subcutaneous component, a subcutaneous pocket is created lateral to the spinal column. After closure of the wound, the subcutaneous implant is left to heal for a period of at least six weeks before the second stage surgical procedure is performed. At the second session, a small incision is made through the skin over the implant. Subsequently, the percutaneous part of the implant is fixed in the subcutaneous. For this purpose, the subcutaneous component is provided with a special holding element. The last step is closure of the incision with one straight suture. When goats are used, the same protective measures to prevent mutilation of the exit-site have to be taken as described earlier (Figure 3). Rabbits only have to be housed in separate cages.

For the testing of external fixation devices we use the tibia as implantation site. To install the devices a longitudinal incision is made on the medial surface of both legs. After exposing the bone, a hole is drilled through the medial cortex, the medulla and the lateral cortex of the tibia. The implants are inserted in the tibia, so that they clearly protrude above the skin surface, and the incisions are closed. The amount of skin protrusion is determined by the size of the experimental animal.

V. COMMON EVALUATION METHODS

At the end of the implantation period, the experimental animals are sacrificed and the implants with their surrounding tissue retrieved for further evaluation. First inspection consists of a gross examination of the specimens on abnormalities in tissue appearance. Thereafter, further processing is necessary for the histological and histomorphometrical evaluation of the implant-tissue specimens.

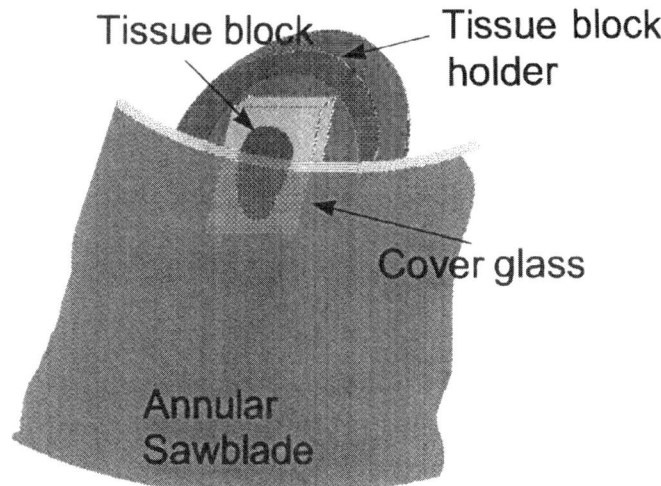

FIGURE 4. Schematic drawing of the sawing technique showing the coverglass fixed to the tissue block. A thin section is cut.

A. LIGHT MICROSCOPICAL PREPARATION

For the correct light microscopical sectioning of implant-containing specimens, only two techniques are suited: (1) the "sawing-grinding" technique as developed by Donath[35] and (2) the modified inner circular "sawing" technique as developed by van der Lubbe and Klein.[36,37] Both methods require embedding of the samples in methylmethacrylate before sectioning.

The equipment necessary for the sawing-grinding method (Exakt-cutting-grinding-system®, Exakt Apparatebau, Germany) consists of a precision guided, diamond-coated band saw and an automatic grinding machine. With the band saw, a first cut is made through the polymerized block. On the exposed tissue-implant surface, a microscope slide is glued. This slide is mounted again in the band saw using a vacuum slide holder. Subsequently, a planoparallel section is cut of a thickness between 50–200 μm. This section is thinned down to a thickness of 5–10 μm with the grinding machine. Finally, the section is stained. For the staining, all the usually employed staining procedures for plastic embedded tissues can be used (see Chapter 7).

For the modified inner circular sawing technique, a horizontal innerlock saw microtome (Fijnmetaal Techniek, Amsterdam, The Netherlands) is used with many adjustments, i.e., freedom of movement of the saw blade, balanced rotation mechanism, thickness of sectioning. To prepare sections, the polymerised tissue/implant block is fixed in the specimen holder and a first cut is made to expose the implant surface. After staining the exposed specimen surface with basic fuchsin, Giemsa or methylene blue, a glass coverslip is fixed on the sample surface with cyanoacrylate-based glue. After drying, the coverslip, with the attached tissue/implant, is sawed off the block using a 1:1 mixture of glycerine and water as cooling liquid and lubricant (Figure 4). The sections obtained have a thickness between 5–10 μm. Finally, a glass slide is glued against the section.

Using the above described methods, it is possible to make thin sections of implant material and surrounding tissue without damaging the interface. It is difficult to express an opinion regarding which method is preferred; both methods have their merits. For example, the sawing-grinding technique results in sections of very high quality. On the other hand, the sawing technique is less elaborate (no grinding or polishing) and allows the preparation of more sections of implant samples with a small diameter.

After the sectioning procedure, the obtained sections can be investigated by light microscopy.

B. Histological and Histomorphometrical Evaluation

For estimation of the soft tissue response to the implants, histological and histomorphometrical evaluation can be performed. Areas of interest for the biological evaluation of soft-tissue implants include: (1) Implant: using the above described techniques, light microscopic sections with the implant *in situ* can be made. The histological appearance of an implant can give information about the stability or degradation behavior of a material. (2) Surrounding tissue: this is the soft tissue capsule surrounding the implant. This capsule is considered to show the inflammatory and healing reaction in response to the surgical trauma and the continued presence of the implant. (3) Interface: this is the type of tissue directly adjacent to the implant surface. The nature of this tissue is determined by the chemical and physical properties of the biomaterial. (4) Interstitial tissue: in case of a porous implant connective tissue will grow into the implant. This is the interstitial tissue. The degree of ingrowth will, in addition to the chemical and physical properties of the material, also depend upon the biomechanical conditions of the biocompatibility test.[38]

Considering these areas of interest the histological evaluation consists of a thorough description of the observed tissue reaction. For the histomorphometry, the following assessment parameters can be used: (1) Epidermal downgrowth (the distance of epidermal migration alongside the implant) and sulcus width (the distance between percutaneous component of the device and the skin). This analysis holds only for percutaneous implant models. (2) A semiquantitative and semiqualitative histological grading scale in which the histological characteristics of the surrounding tissues, interface and interstitium are evaluated by assigning scoring points. Various grading scales are available. The semiquantitative classification of the capsule frequently consists of measurement of the capsule thickness by counting the number of observed fibroblasts. The semiqualitative rating of the capsule, interface and interstitium can be based on numerically rating the tissue morphology (fibrous tissue, maturity, presence of connective tissue or fat tissue) and cellularity (presence of fibroblasts, macrophages, giant cells and other inflammatory cells).[39] An example of a semiquantitative rating system is given in Table 1. (3) The presence and number of blood vessels, plasma cells and inflammatory cells (macrophages, giant cells, polymorphonuclear granulocytes) in the interstititial tissue and surrounding fibrous capsule.

The histomorphometric analyses have to be performed on a sufficient number of representative sections of each implant (at least two sections per implant) and done blindly.

C. Other Evaluation Methods

Light microscopy is especially fitted to obtain in a fast and more or less simple way information about the whole tissue part containing the implant. Occasionally, a more detailed or very specific assessment of the tissue reaction has to occur.

When, for example, accurate information about the tissue changes next to implants is required, immunohistochemical analysis can be used.[40] Besides information about the local immune reactions,[41] immunostaining techniques also offer a possibility to study the presence and distribution of proteins involved in soft tissue remodeling.[42]

Evaluation of biomaterials can also be done by using electron microscopical techniques. Transmission electron microscopy (TEM) provides ultrastructural information about differentiative cellular changes in relation to a specific biomaterial. A disadvantage of TEM is that the preparation of tissue sections is very time consuming. In addition, most biomaterials are too hard to allow the preparation of ultrathin TEM sections. Often, the implant is removed. However, removal of the implant impedes a proper investigation of the tissue-implant interface. Therefore, several methods for preparing ultrathin sections containing intact implant-tissue interfaces have been explored (see Chapter 6). Also, 10 μm sections created by the inner circular "sawing" technique can be used for TEM examination.[37] Scanning electron microscopy (SEM) allows the spatial three-dimensional

TABLE 1
Example of Parameters Used in the Histological Analysis of Soft Tissue Implants

Category	Number or score
General	
Section No.	Independent
Animal No.	Independent
Side	L or R
Site	1, 2, 3…
Implantation period	1, 2, 3, 4, days or weeks
Capsule Localization	
No capsule present	1
Capsule on 1 (dermis) side	2
Capsule on 1 (medial) side	3
Capsule on two sides present	4
Capsule Formation	
No capsule present	1
Loose, fibro-elastic	2
Loose, adipose	3
Loose, fibro-adipose	4
Less dense	5
Dense	6
Capsule Cellular	
— Fibroblast thickness	[1 = 0, 0<2<5, 5 <3<10, 10< 4<30, 5>30]
— Fibroblast contacting surface	1 = YES, 2 = NO
— Acute/chronic inflammatory process	1 = AC, 2 = CHR
— Severity inflammatory process	1 = none, 4 = severe
Inflammatory Cells Location	[1 = non, 2 = end, 3 = middle, 4 = 2 + 3]
— Inflammatory cells contacting surface	1 = YES 2 = NO
macrophages	1 = YES 2 = NO
giant cells	1 = YES, 2 = NO
PMNs	1 = YES, 2 = NO
plasma cells	1 = YES, 2 = NO
— Blood vessels present	1 = YES, 2 = NO
mature/new vessels	1 = MAT, 2 = NW
Capsule Surrounding Tissues	
— Acute/chronic inflammatory process	1 = HC, 2 = CHR
— Severity inflammatory process	1 = none, 4 = severe
macrophages	1 = YES, 2 = NO
giant cells	1 = YES, 2 = NO
PMNs	1 = YES, 2 = NO
plasma cells	1 = YES, 2 = NO
— Blood vessels present	1 = YES, 2 = NO
mature/new vessels	1 = MAT, 2 = NW

examination of tissue-implant specimens. Although, the occurrence of drying artefacts is a well-known phenomenon in SEM samples, the recent development of a so-called "environmental" SEM has almost completely solved this problem.

Finally, electron probe X ray microanalysis (XMRA) can be used to determine changes in elemental composition of the cells and tissues surrounding implants on a microscopic scale.[43] Such changes can occur due to release and accumulation of chemical trace elements from the implanted material.

REFERENCES

1. Langeland, K., "Biocompatibility of dental materials," in *Concise Encyclopedia of Medical and Dental Materials*, Williams, D. F., Ed., Pergamon Press, Oxford, 1990, 59.
2. Lemons, J. E., "Experimental approaches to tooth and bone replacement," in *Animal Models in Dental Research*, Navia, J. M., Ed., The University of Alabama Press, Birmingham, 1977, Chapter 17.
3. von Recum, A. F. and van Kooten, T. G., "The influence of micro-topography on cellular response and the implications for silicone implants," *J. Biomater. Sci. Polymer Edn.*, 7, 181, 1995.
4. Cholvin, N. R., "General compatibility assessment," in *Handbook of Biomaterials Evaluation*, von Recum, A. F., Ed., Macmillan, New York, 1986, Chapter 25.
5. Williams, D. F., "Analysis of the soft tissue response to biomaterials," in *Techniques of Biocompatibility Testing*, Vol. 1, Williams, D. F., Ed., CRC Press, Boca Raton, 1986, Chapter 4.
6. Spector, M., Lalor, P. A., "*In vivo* assessment of tissue compatibility," in *Biomaterials Science*, Ratner, B. D., Hoffman, A. S., Schoen, F. J., and Lemons, J. E., Eds., Academic Press, San Diego, 1996, Chapter 5.3.
7. Van Loon, J., *Biocompatibility Testing of Degradable Polymers*, PhD thesis, University of Leiden, 1995, Chapter 1.
8. The American Society for Testing and Materials, *Standard practice for assessment of compatibility with respect to effect of materials on muscle and bone*, ASTM, F981-93, 1993.
9. The American Society for Testing and Materials, *Standard practice for short-term screening of implant materials*, ASTM, F763-87, 1993.
10. The American Society for Testing and Materials, *Standard practice for selecting generic biological test methods for materials and devices*, ASTM, F748-95, 1995.
11. The American Society for Testing and Materials, *Standard practice for subcutaneous screening test for implant materials*, ASTM, F1408-92, 1996.
12. ISO/DIS 10993-11, Part 11, *Biological evaluation of medical devices. Tests for systemic toxicity*, International Organisation of Standardisation, 1993.
13. Merritt, K., "Systemic toxicity and hypersensitivity," in *Biomaterials Science*, Ratner, B. D., Hoffman, A. S., Schoen, F. J., and Lemons, J. E., Eds., Academic Press, San Diego, 1996, Chapter 4.4.
14. Dixon, C. M. and Rickert, G., "Tissue tolerance to foreign materials," *J. Am. Dent. Assoc.*, 20, 1458, 1933.
15. Jansen, J. A. and de Groot, K., "Guinea pig and rabbit model for the histological evaluation of permanent percutaneous implants," *Biomaterials*, 9, 268, 1988.
16. Jansen J. A., von Recum, A. F., and van der Waerden, J. P. C. M., "Soft tissue response to different types of sintered metal fibre-web materials," *Biomaterials*, 13, 959, 1992.
17. Jansen, J. A., de Ruijter, J. E., Janssen, P. T. M., and Paquay, Y. G. C. J., "Histologic evaluation of a biodegradable polyactive/hydroxyapatite membrane," *Biomaterials*, 16, 819, 1995.
18. Gangjee, T., Colaizzo, R., and von Recum, A. F., "Species-related differences in percutaneous wound healing," *Ann. Biomed. Eng.*, 13, 451, 1985.
19. Grosse-Siestrup, C. and Affeld, K., "Design criteria for percutaneous devices," *J. Biomed. Mater. Res.*, 18, 357, 1984.
20. Woodward, S. C. and Salthouse, T. N., "The tissue response to implants and its evaluation by light microscopy," in *Handbook of Biomaterials Evaluation*, von Recum, A. F., Ed., Macmillan, New York, 1986, Chapter 30.
21. Spector, M., Cease, C., and Tong-Li, X., "The local tissue response to biomaterials," *CRC Critical Rev. Bioeng.*, 5, 269, 1989.
22. Wokalek, H., "Cellular events in wound healing," *CRC Critical Rev. Bioeng.*, 4, 209, 1988.
23. Silver, I. A., "The physiology of wound healing," *Schweiz. Rundschau Med.*, 73, 942, 1984.
24. von Recum, A. F. and Park, J. B., "Permanent percutaneous devices," *CRC Critical Rev. Bioeng.*, 5, 37, 1981.
25. Black, J. B., *Biological Performance of Materials*, Marcel Dekker, New York, 1992.
26. Picha, G. J. and Drake, R. F., "Pillared-surface microstructure and soft-tissue implants: effect of implant site and fixation," *J. Biomed. Mater. Res.*, 30, 305, 1996.

27. McGeachie, J, Smith, E., Roberts, P., and Grounds, M., "Reaction of skeletal muscle to small implants of titanium or stainless steel: a quantitative histological and autoradiographic study," *Biomaterials*, 13, 562, 1992.
28. Semmelink, J. M., Klein, C. P. A. T., Vermeiden, J. P. W., and Althuis, A. L., "Granuloma and plasma cell formation induced by the subcutaneous implantation of β–whitlockite particles," *Biomaterials*, 7, 152, 1986.
29. Morehead, J. M. and Holt, G. R, "Soft-tissue response to synthetic biomaterials," *Otolaryngol. Clin. North Am.,* 27, 195, 1994.
30. Doundoulakis, J. H., "Surface analysis of titanium after sterilization: role in implant-tissue interface and bioadhesion," *J. Prosthet. Dent.,* 58, 471, 1987.
31. Gross, U. M., Powers, D. L., and Clemence-Benson, L., "General aspects of hard tissue processing," in *Handbook of Biomaterials Evaluation*, von Recum, A. F., Ed., Macmillan, New York, 1986, Chapter 36.
32. Jansen, J. A., van der Waerden, J. P. C. M., and de Groot, K., "Epithelial reaction to percutaneous implant materials: *in vitro* and *in vivo* experiments," *J. Invest. Surg.,* 2, 29, 1989.
33. Jansen, J. A., Paquay, Y. C. G. J., and van der Waerden, J. P. C. M., "Tissue reaction to soft-tissue anchored percutaneous implants in rabbits," *J. Biomed. Mater. Res.,* 28, 1047, 1994.
34. Paquay, Y. C. G. J., de Ruijter, J. E., van der Waerden, J. P. C. M., and Jansen, J. A., "A one-stage versus a two-stage surgical technique: tissue reaction to a percutaneous device provided with titanium fiber mesh applicable for peritoneal dialysis," *Am. Soc. Artif. Inter. Organs*, 42, 961, 1996.
35. Donath, K. and Brenner, G., "A method for the study of undecalcified bones and teeth with attached soft tissue," *J. Oral Pathol.*, 11, 318, 1982.
36. Van der Lubbe, H. B. M., Klein, C. P. A. T., and de Groot, K., "A simple method for preparing thin histological sections of undecalcified plastic embedded bone with implants," *Stain Technol.,* 63, 171, 1988.
37. Klein, C. P. A. T., Sauren, Y. H. M. F., Modderman, W. E., and van der Waerden, J. P. C. M., "A new saw technique improves preparation of bone sections for light and electron microscopy," *J. Appl. Biomater,* 5, 369, 1994.
38. Heimke, G., Griss, P., Werner, E., and Jentschura, G., "The effects of mechanical factors on biocompatibility tests," *J. Biomed. Eng.,* 3, 209, 1981.
39. Jansen, J. A. and van't Hof, M. A., "Histological assessment of sintered metal-fibre-web materials," *J. Biomater. Appl.,* 9, 30, 1994.
40. Hunt, J. A., Abrams, K. R., and Williams, D. F., "Modelling the pattern of cell distribution around implanted materials," *Anal. Cell. Pathol.*, 7, 43, 1994.
41. Torgersen, S., Moe, G., and Jonsson, R., "Immunocompetent cells adjacent to stainless steel and titanium miniplates and screws," *Eur. J. Oral Sci.,* 103, 46, 1995.
42. Anselme, K., Bacques, C., Charriere, G., Hartmann, D. J., Herbage, D., and Garrone, R., "Tissue reaction to subcutaneous implantation of a collagen sponge. A histological, ultrastructural, and immunological study," *J. Biomed. Mater. Res.,* 24, 689, 1990.
43. Sigee, D. C., Morgan, A. J., Sumner, A. T., and Warley, A., *X ray Microanalysis in Biology: Experimental Techniques and Applications*, Cambridge University Press, Cambridge, 1993.

21 Animal Models of Bone Ingrowth and Joint Replacement

Dale R. Sumner, Thomas M. Turner, and Robert M. Urban

CONTENTS

I. Introduction 407
II. Bone Ingrowth 408
    A. General Principles of Bone Ingrowth 408
    B. Bone Ingrowth Models 408
        1. Nonweight-Bearing Models 409
        2. Controlled Motion Models 511
        3. Models Communicating with the Joint 411
        4. Weight-Bearing Models 411
        5. Hip Replacement Models 411
        6. Knee Replacement Models 412
    C. Experimental Endpoints 412
    D. Factors Affecting Bone Ingrowth: Recent Highlights 412
III. Joint Replacement 413
    A. Methodological Issues 414
        1. Species Choice 414
        2. Use of the Intact Contralateral Femur as a Control 414
        3. Sample Size 416
        4. Time Points to be Examined 416
        5. Experimental Endpoints 416
    B. Brief Review of Usage of Joint Replacement Models 416
IV. Conclusion 417
Acknowledgment 419
References 419

I. INTRODUCTION

In this chapter, we review animal models related to two areas of orthopaedics: (1) bone ingrowth and (2) joint replacement. Our focus is primarily on considerations pertinent to the use of the models, with less emphasis on specific findings. The chapter, however, should be useful as an entry point to this literature. In most cases we have restricted citations to book chapters and peer-reviewed manuscripts, although in a few instances we have cited peer-reviewed abstracts. The reader should be aware that this is an active area of research and that many new results have not yet been published in peer-reviewed journals. Recent transactions from the Orthopaedic Research Society and the Society for Biomaterials should be consulted for the most recent findings.

II. BONE INGROWTH

The term *bone ingrowth* refers to the development of new bone tissue within an implant. Typically, in orthopaedics the implant receiving the new bone is a porous bone graft substitute or a porous-coated joint replacement component. Our direct experience is primarily with joint replacements and that will be the focus of our discussion, but it is thought that the same general principles that apply to joint replacements also apply to bone graft substitutes.

Research questions involving bone ingrowth include the feasibility of using various porous materials, desirable implant characteristics for bone ingrowth, the effects of interface motion and gaps, the effects of adjuvant therapies used during joint reconstruction, and means of enhancing implant fixation. Our intention is not to review these studies per se as several recent reviews are available,[1-6] but to review briefly the basics of bone ingrowth, discuss the types of models currently available and issues associated with experimental design, and then mention a few highlights from recent studies.

A. GENERAL PRINCIPLES OF BONE INGROWTH

The general principles of bone ingrowth have been identified for some time and comprehensive reviews of the tissue and cell level mechanisms are already available in the literature.[1,4] Bone ingrowth occurs if the implant (1) is made from a biocompatible material, (2) has the appropriate porosity and integrity, (3) is mechanically stable, (4) is in close contact with the host bone and (5) the implantation site is not infected. The process is one of intramembranous bone formation and, thus, resembles gap or defect healing as opposed to endochondral aspects of fracture healing (Figure 1). From a practical point of view, the materials in commercial use are biocompatible and have the appropriate porosity, and with proper surgical technique the risk of infection is low. So, important practical issues involve mechanical stability and proximity of the implant surface to host bone.

Excessive micromotion between the implant and host bone and gaps at the interface are known to inhibit or prevent bone ingrowth. It is generally accepted that relative motion of 150 μm or more leads to failure of fixation by bone ingrowth.[7,8] Motions as small as 40 μm appear to inhibit fixation by bone ingrowth, but interface motion of 20 μm appears to permit fixation by bone ingrowth.[8] Gaps of 0.5–3.0 mm have been shown to inhibit bone ingrowth, with the larger gaps causing more inhibition.[9,10]

B. BONE INGROWTH MODELS

Bone ingrowth has been most thoroughly examined in canine models, although other species have been used. Interestingly, a recent study designed to investigate the effect of ovarian function showed comparable results in canine and primate models.[11] Beyond this study, there is very little direct comparison of species. Based on our own experience with canine and primate models and implants retrieved from patients, it is our opinion that the process, morphology and governing factors operant in humans are well-modeled in canines.

The models can be classified in several ways, including site investigated (cortical v. cancellous bone or metaphysis v. diaphysis), loading status (nonweight bearing v. weight bearing), fit (press-fit v. gap) and host bone status (normal v. altered). The choice of model depends upon the question being asked. If one is interested in testing whether a new material will support bone ingrowth, the simplest approach is to use a nonweight bearing, press-fit model. In contrast, if one is interested in testing a new treatment thought to stimulate bone regeneration, then the most efficient model would be one in which bone ingrowth is inhibited, perhaps progressing from simple to complex models (e.g., a nonweight bearing gap model to a revision total hip replacement [THR] model).

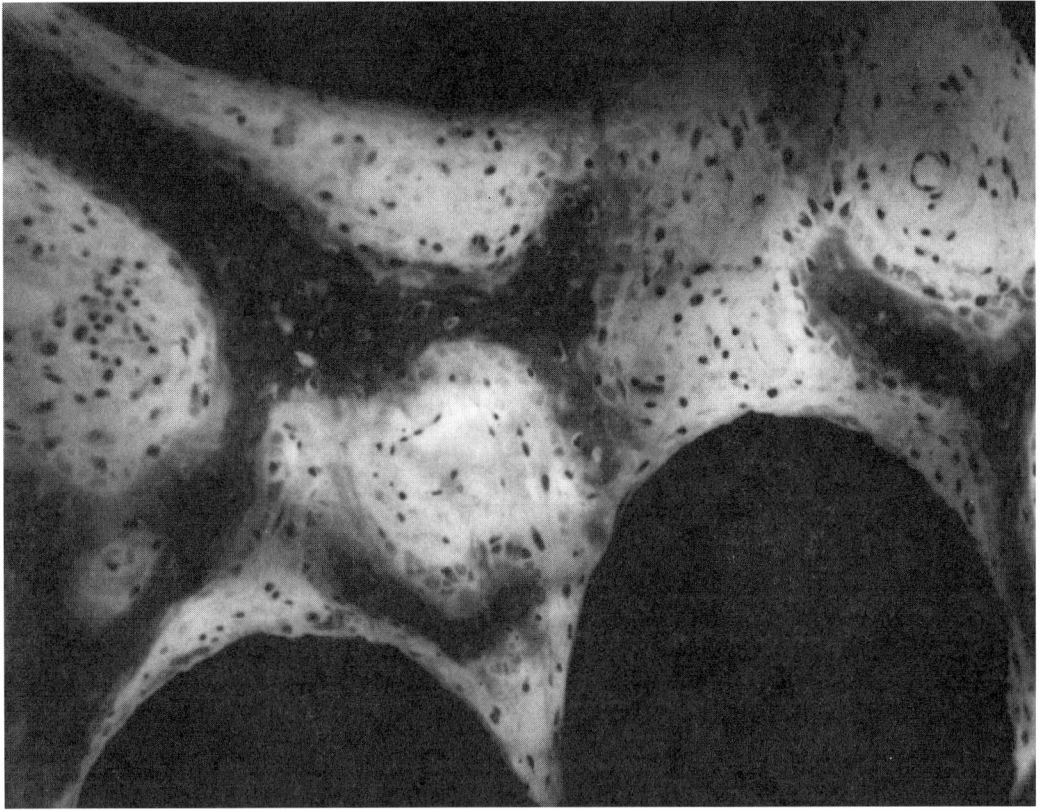

FIGURE 1. Photomicrograph of bone ingrowth at two weeks in a canine model. Note the presence of new intramembranous bone within the void spaces of the porous coating. (from Galante, J. O. and Rivero, D. P., in *Advanced Concepts in Total Hip Replacement,* Harris, W. H., Ed., Slack, Thorofare, NJ, 1985).

1. Nonweight-Bearing Models

Classically in joint replacement research, models may either be categorized as nonweight bearing or weight bearing. The nonweight bearing devices are not directly loaded, are usually implanted for short periods of study (days to weeks) and are used to study implant material or the bone-implant interface isolated from the effects of cyclic weight bearing. In general, these models are used to study implant-related issues (e.g., materials or coatings or surface modifications) or the effects of treatments that may inhibit or enhance bone ingrowth. If a material or surface structure appears promising in a nonweight bearing application, the next consideration is to test the concept under the influence of cyclic weight bearing. This generally necessitates either a segmental replacement or a joint replacement model.

Nonweight bearing models can be classified into those which are applied as a press fit or in which an interfacial gap or defect is created and they can also be classified according to placement location (Figure 2). Placement may be such that the long axis of the implant and bone are aligned ("axial" placement) or so that these two axes are orthogonal ("transcortical" placement). The axial intramedullary devices may be restricted to insertion into only the distal or proximal metaphyseal bone or through all regions of the bone, metaphyseal and diaphyseal. Thus, the shape and size of the device may afford apposition to only metaphyseal trabecular bone or to both metaphyseal trabecular bone and the endosteal cortical surface of the diaphysis.

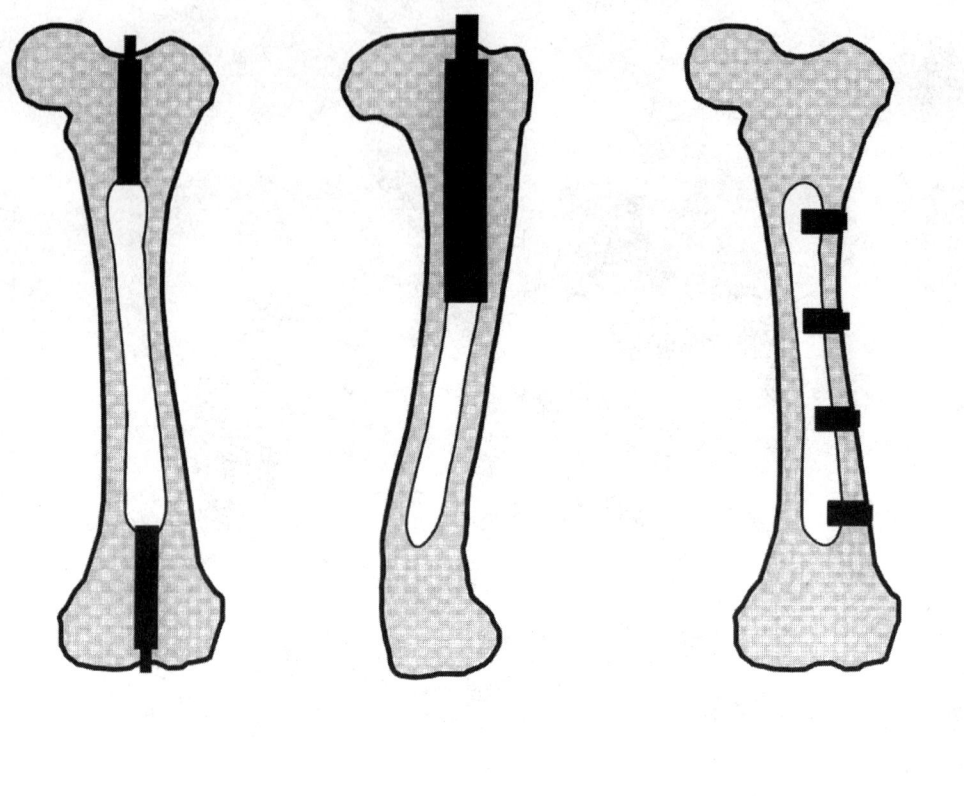

Axial **Axial** **Transcortical**

FIGURE 2. Schematic showing placement of axial and transcortical nonweight-bearing implants for investigation of bone ingrowth.

Transcortical devices are implanted at right angles to the long axis of the bone either in the metaphyseal or diaphyseal region of the bone. Thus, the transcortical implants typically are in contact with cortical bone only (in the case of most diaphyseal sites) or a mixture of cortical and medullary trabecular bone (in the case of metaphyseal sites).

The nonweight bearing devices provide the ability to isolate bone surface as a variable for study, for example, the response of endosteal cortical bone or metaphyseal trabecular bone to an implant. We believe it is most clinically relevant to include a site in trabecular bone or a site adjacent to the endocortical surface because most joint replacement implants are placed within a trabecular bone bed (e.g., the acetabular component in THR and the tibial and femoral components in total knee replacement [TKR]) or within the medullary cavity (e.g., the femoral components in THR). Thus, our bias is that axial and transcortical metaphyseal implants provide more clinically relevant information than transcortical diaphyseal implants.

Nonweight bearing devices also allow for placement of multiple implants within one animal. The test devices are frequently implanted bilaterally, with one side serving as a test material and the other side serving as a paired control. For transcortical implants, sometimes several implants are placed diaphyseally, yielding as many as six to ten implants per animal. From a statistical point of view, each animal, not each implant, constitutes a sample. Thus 10 implants in one animal yields a sample size of one, not 10. Other aspects of sample size determination are described in the section on joint replacement (below).

2. Controlled Motion Models

A disadvantage of nonweight bearing implants is that they cannot replicate the normal load distribution from a weight bearing prosthetic device to the bone nor can they replicate the cyclic loading conditions that a weight bearing prosthetic device undergoes. Recently, this difference between nonweight bearing and weight bearing implants has been bridged to some extent by designing model systems in which the bone-implant interface micromotion can be controlled. In one model system, Søballe and colleagues developed a means to impart controlled motion at the bone-implant interface by having a plunger contact the tibia during the gait cycle.[7,12] In another model Harris' group developed a mechanized means to control the motion of a transcortical metaphyseal implant with respect to the host bone.[8]

3. Models Communicating with the Joint

Another type of implant that bridges some of the differences between nonweight bearing and joint replacement models is the use of devices that communicate with the joint.[13,14] For instance, this type of model has been used to investigate the migration of particles around a device interface without the compounding effects of weight bearing and component movement. Controlled motion can also be imparted to this type of model.[15]

4. Weight Bearing Models

Weight bearing devices have been studied most frequently in the dog; however, other species such as primates, goats or sheep have been utilized. Segmental replacement has been studied, typically centered around a diaphyseal replacement prosthetic device.[16,17] These devices have usually succeeded in having a successful union at the proximal and distal bone-implant junctions but a lesser amount or no bony incorporation in the mid-aspect of the device.

5. Hip Replacement Models

A more common weight bearing model, particularly in recent years, has been hip replacement. This has been applied in two forms. One is a THR with both acetabular and femoral components being inserted. The other is a hip replacement hemiarthroplasty in which only the femoral head is replaced, thereby avoiding the potential complications of an acetabular component. The use of any joint replacement device in an animal model allows that device to experience the cyclic loading of ambulation both for the prosthetic device materials as well as the bone-implant interface. Although different surgical approaches have been utilized for the implantation of hip replacement components, this is a reliable model which can provide very successful clinical function provided proper implantation of the device is achieved.

Variations on the bone-implant interface have also been studied by allowing the presence of only a press fit or, alternatively, the development of defects in the bone adjacent to the implant. Thus, hip arthroplasties may be implanted as a press fit device with the components being impacted into an undersized prepared cavity or as a gap model in which control defects are developed or created adjacent to the prosthetic bone interface. These defect models have been used to test various bone grafts and bone graft substitutes.[18–21]

A further modification of the weight bearing prosthetic joint model is the development of revision models that replicate the bony environment developed in the site surrounding a failed prosthetic device.[22–24] The altered bony environment includes the presence of macrophage-laden granulomas rather than bone marrow at the site of implantation.[23]

6. Knee Replacement Models

Other prosthetic components, notably TKRs have also been used. TKR models are less frequently used, as only a handful of reports have been made[25-31] compared to the large number of THR studies (see below, under the Joint Replacement section). Knee arthroplasties are more complex than hip arthroplasties because of the greater complexity of motion at the knee joint than at the hip joint.

C. EXPERIMENTAL ENDPOINTS

For studies of bone ingrowth, the experimental endpoints typically are morphological (e.g., measurements of the amount of bone ingrowth or bone formation in a gap) or mechanical (e.g., the strength of fixation of the implant to the host bone). There are various morphological instruments. Two we find useful are the "volume fraction" and "extent" of bone ingrowth. The volume fraction refers to the amount of void space within the porous coating occupied by bone and can be measured with the aid of an image analyzer and backscatter scanning electron micrographic images,[32] although point counting of properly imaged and stained ground sections works just as well and in a small-scale study might be preferable. The extent of bone ingrowth is a means to quantify the topographic distribution of bone ingrowth, and can be considered a measure of consistency. Researchers have performed this observation in various ways, but the basic concept is to divide the porous coating into a number of equal size units (e.g., 1 mm fields) and then to determine how many of these fields contain bone ingrowth.[33,34] We have also found it helpful to use measures of trabecular architecture first developed by researchers in metabolic bone disease to characterize the newly formed bone in gaps adjacent to test implants.[10] Other morphological observations have been made, including the presence and thickness of fibrous tissue layers at the interface,[35] bone tissue kinetics,[36] and it can be anticipated that techniques to better understand gene expression (such as *in situ* hybridization and immunohistochemistry) will be used in the future.

Mechanical measures typically have focused on the strength of fixation.[37,38] The interface shear strength is measured by a "pull-out" or a "push-out" test (Figure 3). Typically, pull-out tests have been used for implants placed in an intramedullary site, while push-out tests have been used for both transcortical implants and intramedullary implants. These tests are destructive since the interface is stressed to failure and the peak load is divided by the nominal surface area to calculate the strength of fixation. Recently, we have proposed a nondestructive test,[39] but this has not yet been performed on actual implantations. The basic concept of the test is that it should be possible to measure interface stiffness in a nondestructive way so that both mechanical and morphological observations can be made on the same specimen. Of course, it is possible to measure some aspects of bone ingrowth on an implant that has been forcibly removed from the host bone, but the interface itself is destroyed during destructive testing.

For the weight bearing models, researchers typically measure the amount of bone ingrowth as with the nonweight bearing models. However, for these models, it is more common for the mechanical stability of the implant to be measured than for the strength of fixation to be measured.[29,31,40-45]

D. FACTORS AFFECTING BONE INGROWTH: RECENT HIGHLIGHTS

There are many factors which can affect bone ingrowth such as bone grafts and bone graft substitutes, implant surface treatments, electrical stimulation, growth factors and adjuvant treatments for conditions such as heterotopic ossification, cancer and immune-mediated diseases. Rather than review specific studies on factors affecting bone ingrowth (which have been reviewed in detail elsewhere[1-5]), we felt it would be more useful to highlight a few recent studies.

It is generally thought that the amount of micromotion at the interface influences tissue differentiation, with stable implants permitting bone ingrowth and unacceptable levels of micromotion causing fibrous tissue formation. However, the biology of tissue differentiation at the

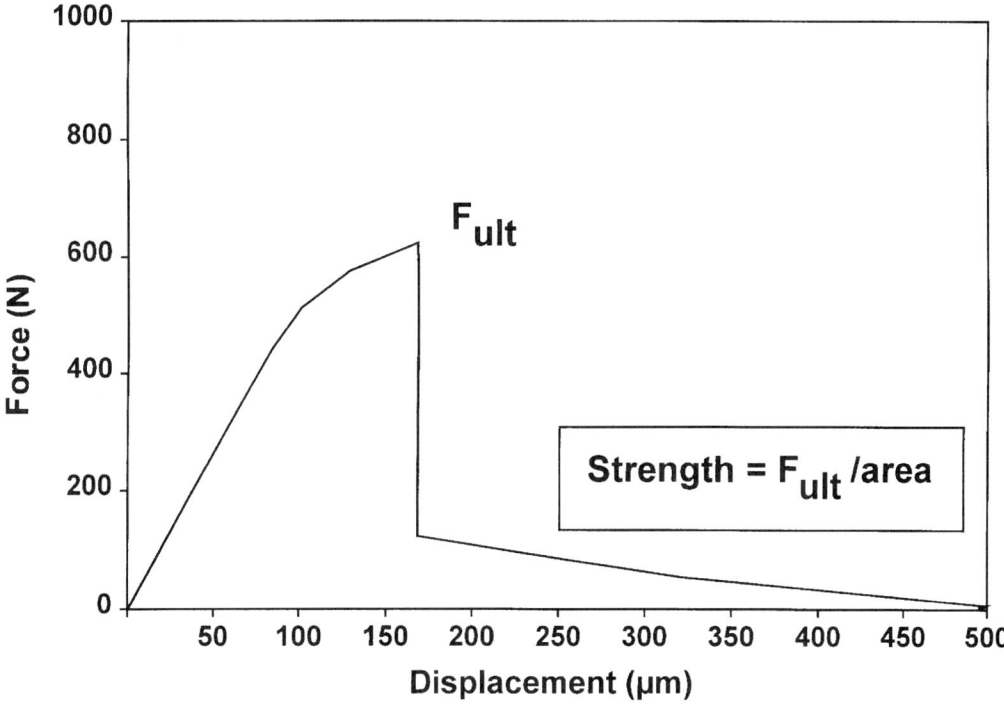

FIGURE 3. An example of a load-displacement curve from a "pull-out" test. Note that the ultimate force can be divided by the nominal surface area of the implant to calculate the strength of fixation. The slope of the curve can be used to calculate the interface stiffness if one accounts for the compliance of the experimental set-up.

interface may be more complex as evidenced by a recent canine study[8] in which 20 μm of initial motion led to increasing interface stiffness during the course of the six week study and continuity of bone ingrowth, 40 μm of initial motion led to decreasing interface stiffness and mixed bone and "fibrocallus," and 150 μm initial motion led to decreasing interface stiffness and a fibrous tissue interface despite the presence of some bone ingrowth. This later finding is quite intriguing because it suggests that bone differentiates, but then possibly undergoes stress fracture and replacement by fibrous tissue in the presence of excessive interface micromotion. Thus, the conventional view, that excessive motion causes direct formation of fibrous tissue, may be incorrect.

We and others have recently shown that transforming growth factor-beta (TGF-β) enhances bone ingrowth or implant fixation strength.[10,46] More recently we have found that there is a positive effect on bone regeneration at the contralateral, non-growth factor-treated, control site as well as at the site directly treated with the growth factor.[47] We have also recently found that bone ingrowth and regeneration are enhanced with local delivery of bone morphogenic protein-2.[48] With both of these growth factors the response is dose-dependent and, in our studies, the lowest of the doses studied to date has been the most effective.

III. JOINT REPLACEMENT

As with the studies of bone ingrowth, the most common animal model of joint replacement is the canine. By far, most of the work has focused on the hip joint with only a few studies of the knee (Figure 4). The research questions are varied, but many are concerned with various aspects of cementless total joint replacement, including implant fixation and adaptive bone remodeling. In this regard, the nature of the interface, materials and implant shape have been investigated. In addition to concern with bone ingrowth and remodeling, other areas investigated include bone cement, the type of bearing surface and wear debris. This latter topic is covered in a separate chapter in this book.

In this section, we focus on methodological issues such as species choice, experimental design, and experimental endpoints. We then briefly review studies on bone remodeling and provide a listing of various studies to provide the reader an entry into the literature. Our listing is not meant to be exhaustive, but in conjunction with some recent reviews,[3,49-52] should be helpful.

A. METHODOLOGICAL ISSUES

The appropriate experimental design depends, of course, upon the research question(s) being asked. Many studies involving total joint replacement can be considered "feasibility" studies. In these studies the question might be, can new material X be used successfully? The first step is to perform a few trial implantations and look for obvious problems. Should none be found, the research may progress to the next stage: how does new material X compare with some standard? At this point, a number of important issues develop, including the type of experimental endpoints, the power of the experiment, and the potential influence of confounding variables.

We can't hope to review all possible research questions so we will review here our own philosophy of experimental design, using some specific examples from our research. Recently, we performed a study to compare bone ingrowth and functional adaptation of the host bone to femoral stems varying in stiffness.[53] The clinical problem being addressed was that of excessive proximal femoral bone loss following cementless THR. Our main endpoint was change in cortical bone. The actual measurement was made by calculating the cortical area (the area enclosed by the subperiosteal and endo-cortical bone surfaces) and the cortical porosity (the relative amount of cortical area occupied by pores larger in size than osteocytic lacunae). The cortical area and porosity of the operated limb were then compared to similar values for the contralateral, unoperated limb to determine the change in cortical bone.

1. Species Choice

The dog is considered the species of choice by many for studies of joint replacement because its bone microstructure is similar to humans, its remodeling kinetics have been extensively studied, the geometry of the proximal femur and mechanical properties of the bone of the distal femur and proximal tibia have been well-characterized, and its size is appropriate for studies of THR and TKR.[54-58] Certainly, other species, most notably sheep and goats, but also pigs and rats have been used, but these models tend to be less well characterized.[59-64] Perhaps, another advantage of the canine model, for THR, at least, is the fact that THRs are performed clinically in the canine.[65-68]

2. Use of the Intact Contralateral Femur as a Control

From a methodological viewpoint, the desire to estimate the change in cortical bone raises certain issues. First, is it appropriate to use the contralateral limb as an intact control? We addressed this by examining the symmetry in geometric properties of the proximal canine femur, finding only very small side-to-side differences.[69] Although it is possible to monitor some aspects of limb function directly,[70,71] we examine the bone mineral content of both tibiae either just at the end or during the entire course of the experiment, based on the assumption that these bone measurements are an adequate reflection of the loads placed on the limb integrated over time. Differences in limb function between the operated and control limbs or among animals in different groups could have a significant confounding effect on the primary endpoints. Previously, we have shown that tibial bone mineral content is symmetrical and that side-to-side differences are likely to be related to clinical function.[72] In addition, another previous study showed that change in tibial bone mineral content of the

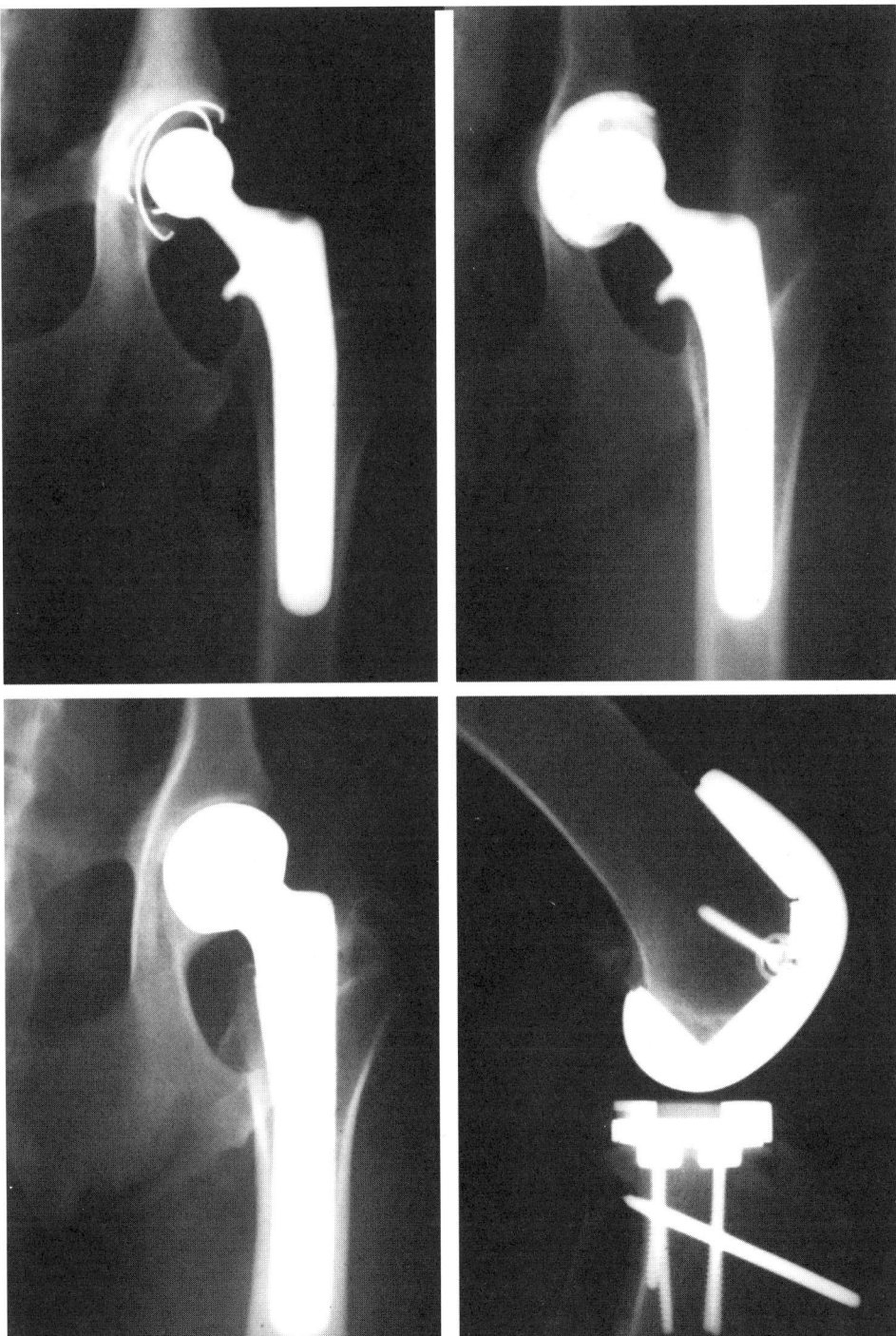

FIGURE 4. *In vivo* radiographs of three canine hip replacement models and one canine knee replacement model used by the authors. (a) a total hip replacement in which a cemented acetabular component is matched with a cementless femoral component, (b) a total hip replacement in which both the acetabular and femoral components are cementless, (c) a hemiarthroplasty in which only a femoral component is used with the head matched to the size of the intact acetabulum, and (d) a cementless knee replacement model with cementless femoral and tibial components.

contralateral femur can occur transiently but is minimal by six months.[73] All of these studies indicate that the contralateral limb is an adequate control. Given sufficient resources, an excellent additional control would be dogs that have not undergone surgery.

3. Sample Size

The next question is what should the sample size be? The answer is not simple, but depends on (1) the magnitude of the effect which the researcher believes is important, (2) the desired probability of showing a significant difference of this magnitude (i.e., the "power" of the significance test), (3) the significance level (usually 0.05) and (4) the variability in the data.[69,74,75] The required sample size is inversely proportional to (1) and directly proportional to (2), (3) and 4). Thus, if the researcher wants a high probability of finding a small difference between two groups for an endpoint that is quite variable, a very large sample size will be needed. More specifically, if the researcher believes a 20% difference between groups is important and plans to use five animals per group, the research design (sample size) would be judged inadequate if one could only reasonably expect to demonstrate a 20% effect with 10 animals per group because of the expected variability in the data.

4. Time Points to be Examined

Another important question is what time points should be investigated? Of course there is no universal answer. However, if one is interested in long-term functional adaptation of bone it is important to choose a time frame which should allow several remodeling cycles to be completed. In the canine, the "activation-resorption-filling" sequence requires approximately two to three months. Interestingly, most studies have shown that most of the adaptation occurs within six months with minor changes occurring thereafter, but it is probably not correct to say that a new steady-state is reached already at six months. In addition, in some circumstances, e.g., if the implant is not fixed to the bone, there can be quite dramatic changes after six months.[76] Our personal opinion is that for functional adaptation studies, six months is a minimum and an additional time point is of value (such as one, two or even more years). To study transient events, such as the process of bone ingrowth, then shorter time points are of interest. We have focused on one month as a "screening" period, but other time points would be needed, depending upon the specific questions to be asked.

5. Experimental Endpoints

A critical issue is what should be observed and measured? Obviously, this depends upon the research question, but it is more involved because there may be unanticipated confounding variables. On the one hand, the researcher cannot possibly measure everything, but on the other hand, it can be devastating to miss something important. This problem is particularly vexing with joint replacement experiments because these experiments are inherently time-consuming and expensive. While one might be able to repeat experiments to make additional observations with rats, this rarely is the case with canines.

B. Brief Review of Usage of Joint Replacement Models

Our review of the THR literature indicates that most of the attention has been on morphological and biomechanical responses with an emphasis on one or more of three primary regions: (1) the interface, (2) the adjacent tissue (usually medullary contents or trabecular bone) or (3) the host cortical bone (Table 1). We have used these two axes (type of question v. type of endpoint) to place a large body of THR studies in context (Table 1). While this classification scheme works for most studies, in some the focus has been elsewhere, such as the articular cartilage after hemiarthroplasty.[77–80] A few studies have focused on biomaterial performance with little or no attention paid to the biological response. Some of these later studies are also listed in Table 1.

TABLE 1
List of *In Vivo* Hip Replacement Models, Sorted According to Type of Question and Type of Endpoint. The Last Part Lists Models in which the Primary Endpoint was Performance of the Biomaterial

Type of endpoint	Type of question		
	Interface	Medullary/trabec. bone	Cortical bone
Morphology	14, 18–24, 40, 42–45, 53, 62, 66, 76, 81, 82, 84, 85, 90–104, 104–127	18–24, 53, 81, 82, 84, 92, 103, 128	22, 23, 44, 51, 53, 66, 76, 81–86, 98, 100, 101, 104, 105, 118, 119, 126, 128–133
Mechanical	24, 39–45, 76, 94, 116, 126, 134–136	137	44, 76, 76, 118, 119, 131, 132, 135, 137–139
Biomaterial performance	34, 77–80, 96, 105, 105, 140, 140–143		

Here, we review the THR models which have focused on morphological endpoints of the cortical bone to serve as a convenient example of the utility of Table 1. We summarize in Table 2 studies of functional adaptation of cortical bone in hip replacement models, basing our review on the appropriate cell of Table 1. Early work showed that the type of porous coating (as long as it allowed bone ingrowth) did not influence the cortical bone response.[81,82] It was also found that the presence of a porous coating caused more bone loss initially, but eventually, animals with uncoated stems had as much bone loss as those with porous coated stems.[76] Work from Bobyn's group suggested that use of proximally porous coated stems caused less cortical bone loss adjacent to the mid-stem than the use of fully coated stems,[83] but the experiment had a small sample size and, hence, low power. Comparison of two of our experiments suggests that even subtle differences in porous coating placement may influence the cortical bone response.[81,84] Specifically, at six months we observed a tendency toward less bone loss with a stem that had the porous coating on the anterior and posterior faces only as opposed to one that had the anterior and posterior porous coating plus a small proximal medial area of porous coating. By far, the most consistent finding has been that cortical bone loss can be reduced by reducing the stiffness of the stem.[53,84–86]

Knee models are much less common and it is not necessary to summarize the studies in tabular format. One of the early studies used a knee model to characterize the mechanical properties of fibrous tissue at the bone-cement-bone interface.[26] Later studies used these models to investigate various interface phenomena, including the potential to establish bone ingrowth fixation.[25,27–29] More recently, knee models have been used to investigate the role of implant design on bone ingrowth and mechanical stability[30,31,87] and wear of articular cartilage.[88]

IV. CONCLUSION

We hope this chapter has shown that there is no perfect model because model choice depends upon the research question. In general, nonweight-bearing models are most useful for studies related to the establishment of implant fixation. Weight-bearing models are useful for issues related to establishment and maintenance of implant fixation. Thus, weight-bearing models are generally utilized for short and long-term studies (weeks to years) while nonweight-bearing models are used in short-term studies (weeks to months).

Since changes in the bony environment develop over time, conditions at the bone-implant interface (e.g., the development of fibrous tissue and migration of particulate debris or interface micromotion) or adjacent to the implant (e.g., loss of cortical bone) must be evaluated over long

TABLE 2
Summary of Femoral Cortical Bone Loss in Canine Models of Cementless Hip Replacement in which Unilateral Implants were in Place for Six Months or Longer

Length of femoral component	Porous coating location[1]	Material[2]	Length of study (months)	Total sample size	Proximal cortical bone loss[3]	Mid-stem cortical bone loss[3]	Reference
73 mm	80%, A+P	Ti alloy	6	9	11%	10%	84
73 mm	80%, A+P	Composite	6	9	4%	6%	84
88 mm	94%, circumferential	Ti alloy	6, 24	22	35%	20%	53
88 mm	94%, circumferential	Composite	6, 24	22	20%	10%	53
73 mm	80%, A+P+prox M	Ti alloy	6, 24	32	15%	15%	81,82
73 mm	80%, circumferential	Ti alloy	6, 24	10	25%	15%	81,82
75 mm	100%, circumferential	CoCr	16, 36	4	20%	25%	83
75 mm	40%, circumferential	CoCr	16, 36	4	25%	2%	83
?	67%, circumferential	CoCr	6, 24	10	—	12%	130

1. Porous coating location; the percent refers to the distal extension of the porous coating. A = anterior stem face, P = posterior stem face, M = medial stem face.
2. Material: Ti = titanium; composite = a relatively low stiffness implant; CoCr = cobalt chrome.
3. Cortical bone loss = difference in cortical area between the operated femur and the contralateral (control) femur. When two time points were available the means or individual data points were averaged because, in general, none of these studies showed a difference in bone loss as a function of time. These are mean values. Some individual cases showed considerably more and some considerably less bone loss.

periods of time. It is apparent that clinically relevant issues, such as adaptive remodeling[52,89] or the changes that occur at the bone implant interface[34] can be well-studied experimentally in weight-bearing models.[14,53,90] The ability to study these issues in animal models allows relatively rapid accumulation of information that otherwise would take years to acquire from human autopsy material.

ACKNOWLEDGEMENT

NIH Grants AR42862, AR16485

REFERENCES

1. Sumner, D. R. and Galante, J. O., "Bone ingrowth," in *Surgery of the Musculoskeletal System*, Evarts, C. M., Ed., Churchill Livingstone, New York, 1990, 151.
2. Sumner, D. R., Kienapfel, H., and Galante, J. O., "Metallic implants," in *Bone Grafts and Bone Substitutes,* Habal, M. B. and Reddi, A. H., Eds., Saunders, Philadelphia, 1992, 252.
3. Callaghan, J. J., "The clinical results and basic science of total hip arthroplasty with porous-coated prostheses," *J. Bone Joint Surg.,* 75A, 299, 1993.
4. Sumner, D. R., "Bone ingrowth: implications for establishment and maintenance of cementless porous-coated interfaces," in *Orthopaedic Knowledge Update: Hip and Knee Reconstruction*, Callaghan, J. J., Dennis, D. A., Paprosky, W. G., and Rosenberg, A. G., Eds., American Academy of Orthopaedic Surgeons, Rosemont, IL, 1995, 57.
5. Kienapfel, H. and Griss, P., "Fixation by ingrowth," in *The Adult Hip*, Callaghan, J. J., Rosenber, A. G., and Rubash, H., Eds., Lippincott-Raven, Philadelphia, 1998, 201.
6. Pilliar, R. M., "Porous-surfaced metallic implants for orthopaedic applications," *J. Biomed. Mater. Res.,* 21(A1), 1, 1987.
7. Soballe, K., Hansen, E. S., B Rasmussen, H., Jorgensen, P. H., and Bunger, C., "Tissue ingrowth into titanium and hydroxyapatite-coated implants during stable and unstable mechanical conditions," *J. Orthop. Res.,* 10, 285, 1992.
8. Bragdon, C. R., Burke, D., Lowenstein, J. D., O'Connor, D. O., Ramamurti, B., et al., "Differences in stiffness of the interface between a cementless porous implant and cancellous bone *in vivo* in dogs due to varying amounts of implant motion," *J. Arthrop.,* 11, 945, 1996.
9. Dalton, J. E., Cook, S. D., Thomas, K. A., and Kay, J. F., "The effect of operative fit and hydroxyapatite coating on the mechanical and biological response to porous implants," *J. Bone Joint Surg.,* 77A, 97, 1995.
10. Sumner, D. R., Turner, T. M., Purchio, A. F., Gombotz, W. R., Urban, R. M., and Galante, J. O., "Enhancement of bone ingrowth by transforming growth factor beta," *J. Bone Joint Surg.,* 77A, 1135, 1995.
11. Shaw, J. A., Wilson, S. C., Bruno, A., and Paul, E. M., "Comparison of primate and canine models for bone ingrowth experimentation, with reference to the effect of ovarian function on bone ingrowth potential," *J. Orthop. Res.,* 12, 268, 1994.
12. Lind, M., Overgaard, S., Ongpipattanakul, B., Nguyen, T., Bünger, C., and Soballe, K., "Transforming growth factor-β1 stimulates bone ingrowth to weight-loaded tricalcium phosphate coated implants," *J. Bone Joint Surg.,* 78B, 377, 1996.
13. Howie, D. W., Vernon-Roberts, B., Oakeshott, R., and Manthey, B., "A rat model of resorption in bone at the the cement-bone interface in the presence of polyethylene wear particles," *J. Bone Joint Surg.,* 70A, 257, 1988.
14. Bobyn, J. D., Jacobs, J. J., Tanzer, M., Urban, R. M., Aribindi, R., et al., "The susceptibility of smooth implant surfaces to periimplant fibrosis and migration of polyethylene wear debris," *Clin. Orthop.,* 311, 21, 1995.
15. Prendergast, P. J., Huiskes, R., and Soballe, K., "Biophysical stimuli on cells during tissue differentiation at implant interfaces," *J. Biomech.,* 30, 539, 1997.

16. Andersson, G. B. J., Gaechter, A., Galante, J. O., and Rostoker, W., "Segmental replacement of long bones in baboons using a fiber titanium implant," *J. Bone Joint Surg.*, 60A, 31, 1978.
17. Heck, D. A., Nakajima, I., Kelly, P. J., and Chao, E. Y., "The effect of load alteration on the biological and biomechanical performance of a titanium fiber-metal segmental prosthesis," *J. Bone Joint Surg.*, 68A, 118, 1986.
18. Russotti, G. M., Okada, Y., Fitzgerald, R. H., Chao, E. Y. S., and Gorski, J. P., "Efficacy of using a bone graft substitute to enhance biological fixation of a porous metal femoral component," in *The Hip*, Brand, R. A., Ed., Mosby, St. Louis, 1987, 120.
19. McDonald, D. J., Fitzgerald, R. H., and Chao, E. Y. S., "The enhancement of fixation of a porous-coated femoral component by autograft and allograft in the dog," *J. Bone Joint Surg.*, 70A, 729, 1988.
20. Kang, J. D., McKernan, D. J., Kruger, M., Mutschler, T., Thompson, W. H., and Rubash, H. E., "Ingrowth and formation of bone in defects in an uncemented fiber-metal total hip-replacement model in dogs," *J. Bone Joint Surg.*, 73A, 93, 1991.
21. Greis, P. E., Kang, J. D., Silvaggio, V., and Rubash, H. E., "A long-term study on defect filling and bone ingrowth using a canine fiber metal total hip model," *Clin. Orthop.*, 274, 47, 1992.
22. Roberson, J. R., Spector, M., Baggett, M. A., and Kita, K., "Porous-coated femoral components in a canine model for revision arthroplasty," *J. Bone Joint Surg.*, 70A, 1201, 1988.
23. Turner, T. M., Urban, R. M., Sumner, D. R., and Galante, J. O., "Revision, without cement, of aseptically loose, cemented total hip prostheses: quantitative comparison of the effects of four types of medullary treatment on bone ingrowth in a canine model," *J. Bone Joint Surg.*, 75A, 845, 1993.
24. Schreurs, B. W., Huiskes, R., Buma, P., and Sloof, T. J., "Biomechanical and histological evaluation of a hydroxyapatite-coated titanium femoral stem fixed with an intramedullary morsellized bone grafting technique: an animal experiment on goats," *Biomatarials*, 17, 1177, 1996.
25. Bobyn, J. D., Cameron, H. U., Abdulla, D., Pilliar, R. M., and Weatherly, G. C., "Biologic fixation and bone modeling with an unconstrained canine total knee prosthesis," *Clin. Orthop.*, 166, 301, 1982.
26. Hori, R. Y. and Lewis, J. L., "Mechanical properties of the fibrous tissue found at the bone–cement interface following total joint replacement," *J. Biomed. Mater. Res.*, 16, 911, 1982.
27. Turner, T. M., Urban, R. M., Sumner, D. R., Skipor, A. K., and Galante, J. O., "Bone ingrowth into the tibial component of a canine total condylar knee replacment prosthesis," *J. Orthop. Res.*, 7, 893, 1989.
28. Walker, P. S., Rodger, R. F., Miegel, R. E., Schiller, A. L., Deland, J. T., and Robertson, D. D., "An investigation of a compliant interface for press-fit joint replacement," *J. Orthop. Res.*, 8, 453, 1990.
29. Stulberg, B. N., Watson, J. T., Stulberg, S. D., Bauer, T. W., and Manley, M. T., "A new model to assess tibial fixation in knee arthroplasty. I. Histologic and roentgenographic results," *Clin. Orthop.*, 263, 288, 1991.
30. Sumner, D. R., Turner, T. M., Dawson, D., Rosenberg, A. G., Urban, R. M., and Galante, J. O., "Effect of pegs and screws on bone ingrowth in cementless total knee arthroplasty," *Clin. Orthop.*, 309, 150, 1994.
31. Sumner, D. R., Berzins, A., Turner, T. M., Igloria, R., and Natarajan, R., "Initial *in vitro* stability of the tibial component in a canine model of cementless total knee replacement," *J. Biomech.*, 27, 929, 1994.
32. Sumner, D. R., Bryan, J. M., Urban, R. M., and Kuszak, J. R., "Measuring the volume fraction of bone ingrowth: a comparison of three techniques," *J. Orthop. Res.*, 8, 448, 1990.
33. Pidhorz, L. E., Urban, R. M., Jacobs, J. J., Sumner, D. R., and Galante, J. O., "A quantitative study of bone and soft tissues in cementless porous-coated acetabular components retrieved at autopsy," *J. Arthrop.*, 8, 213, 1993.
34. Urban, R. M., Jacobs, J. J., Sumner, D. R., Peters, C. L., Voss, F. R., and Galante, J. O., "The bone–implant interface of femoral stems with non-circumferential porous coating: a study of specimens retrieved at autopsy," *J. Bone Joint Surg.*, 78A, 1068, 1996.
35. Soballe, K., Hansen, E. S., Brockstedt-Rasmussen, H., and Bünger, C., "Hydroxyapatite coating converts fibrous tissue to bone around loaded implants," *J. Bone Joint Surg.*, 75B, 270, 1993.
36. Burr, D. B., Mori, S., Boyd, R. D., Sun, T. C., Blaha, J. D., et al., "Histomorphometric assessment of the mechanisms for rapid ingrowth of bone to HA/TCP coated implants," *J. Biomed. Mater. Res.*, 27, 645, 1993.

37. Galante, J., Rostoker, W., Lueck, R., and Ray, R. D., "Sintered fiber metal composites as a basis for attachment of implants to bone," *J. Bone Joint Surg.*, 53A, 101, 1971.
38. Cook, S. D., Walsh, K. A., and Haddad, R. J., "Interface mechanics and bone growth into porous Co-Cr-Mo alloy implants," *Clin. Orthop.*, 193, 271, 1985.
39. Berzins, A., Shah, B., Weinans, H., and Sumner, D. R., "Nondestructive measurements of implant-bone interface shear modulus and effects of implant geometry in pull-out tests," *J. Biomed. Mater. Res.*, 34, 337, 1997.
40. Vanderby, R., Jr., Manley, P. A., Kohles, S. S., and McBeath, A. A., "Fixation stability of femoral components in a canine hip replacement model," *J. Orthop. Res.*, 10, 300, 1992.
41. Jasty, M., Krushell, R. J., Zalenski, E., O'Connor, D., Sedlacek, R., and Harris, W. H., "The contribution of the nonporous distal stem to the stability of proximally porous-coated canine femoral components," *J. Arthrop.*, 8, 33, 1993.
42. Finkelstein, J. A., Anderson, G. I., Waddell, J. P., Richards, R. R., Hearn, T. C., and Schemitsch, E., "A study of micromotion and appositional bone growth to a canine madreporic surfaced femoral component," *J. Arthrop.*, 9, 317, 1994.
43. Heiner, J. P., Manley, P., Kohles, S., Ulm, M., Bogart, L., and Vanderby, R., Jr., "Ingrowth reduces implant-to-bone relative displacements in canine acetabular prostheses," *J. Orthop. Res.*, 12, 657, 1994.
44. Manley, P. A., Vanderby, R., Kohles, S., Markel, M. D., and Heiner, J. P., "Alterations in femoral strain, micromotion, cortical geometry, cortical porosity, and bony ingrowth in uncemented collared and collarless prostheses in the dog," *J. Arthrop.*, 10, 63, 1995.
45. Jasty, M., Bragdon, C. R., Zalenski, E., O'Connor, D., Page, A., and Harris, W. H., "Enhanced stability of uncemented canine femoral components by bone ingrowth into the porous coatings," *J. Arthrop.*, 12, 106, 1997.
46. Lind, M., Overgaard, S., Soballe, K., Nguyen, T., Ongipattanakul, B., and Bünger, C., "Transforming growth factor-β_1 enhances bone healing to unloaded tricalcium phosphate coated implants: an experimental study in dogs," *J. Orthop. Res.*, 14, 343, 1996.
47. Sumner, D. R., Turner, T. M., Urban, R. M., Hawkins, M., Nichols, E. H., et al., "Locally delivered rhTGF-β_2 enhances bone ingrowth and bone regeneration at local and remote sites of skeletal injury," *Trans. Soc. Biomater*, 20, 210, 1997.
48. Sumner, D. R., Turner, T. M., Urban, R. M., Guhl, D., Turek, T., et al., "rhBMP-2 enchances bone ingrowth and gap healing," *Trans. Orthop. Res. Soc.*, 23, 599, 1998.
49. Sumner, D. R. and Turner, T. M., "The effects of femoral component design features on femoral remodeling following cementless total hip arthroplasty," in *Non-Cemented Total Hip Arthroplasty*, Fitzgerald, R. H., Ed., Raven Press, New York, 1988, 143.
50. Galante, J. O., Lemons, J., Spector, M., Wilson, P. D., and Wright, T. M., "The biologic effects of implant materials," *J. Orthop. Res.*, 9, 760, 1991.
51. Jacobs, J. J., Sumner, D. R., and Galante, J. O., "Mechanisms of bone loss associated with total hip replacement," *Orthop. Clin. N. Am.*, 24, 583, 1993.
52. Sumner, D. R., "Bone remodeling of the proximal femur," in *The Adult Hip*, Callaghan, J. J., Rosenberg, A. G., and Rubash, H., Eds., Lippincott-Raven, New York, 1998, 211.
53. Turner, T. M., Sumner, D. R., Urban, R. M., and Galante, J. O., "Maintenance of proximal cortical bone with use of a less stiff femoral component in hemiarthroplasty of the hip without cement," *J. Bone Joint Surg.*, 79A, 1381, 1997.
54. Sumner, D. R., Turner, T. M., Urban, R. M., and Galante, J. O., "Experimental studies of bone remodeling in total hip replacement," *Clin. Orthop.*, 275, 83, 1992.
55. Sumner, D. R., Devlin, T. C., Winkelman, D., and Turner, T. M., "The geometry of the adult canine proximal femur," *J. Orthop. Res.*, 8, 671, 1990.
56. Kuhn, J. L., Goldstein, S. A., Ciarelli, M. J., and Matthews, L. S., "The limitations of canine trabecular bone as a model for human: a biomechanical study," *J. Biomech.*, 22, 95, 1989.
57. Sumner, D. R., Willke, T. L., Berzins, A., and Turner, T. M., "Distribution of Young's modulus in the cancellous bone of the proximal canine tibia," *J. Biomech.*, 27, 1095, 1994.
58. Bloebaum, R. D., Ota, D. T., Skedros, J. G., and Mantas, J. P., "Comparison of human and canine external femoral morphologies in the context of total hip replacement," *J. Biomed. Mater. Res.*, 27, 1149, 1993.

59. Goel, V. K., Drinker, H., Panjabi, M. M., and Strongwater, A., "Selection of an animal model for implant fixation studies: anatomical aspects," *Yale. J. Biol. Med.,* 55, 113, 1982.
60. Thomsen, M., von Strachwitz, B., Loew, M., Cotta, H., Kirsch, S., Schunk, O., and Kubein-Meesenburg, D., "The Gottinger minipig as an animal model in hip endoprosthesis. Anatomy, anesthesia, operation results" (German), *Zeitschrift fur Orthopadie und Ihre Grenzgebiete,* 135, 58, 1997.
61. Powers, D. L., Claassen, B., and Black, J., "The rat as an animal model for total hip replacement arthroplasty," *J. Invest Surg.,* 8, 349, 1995.
62. Phillips, T. W., Gurr, K. R., and Rao, D. R., "Hip implant evaluation in an arthritic animal model," *Arch. Orthop. Trauma Surg.,* 109, 194, 1990.
63. Phillips, T. W., Johnston, G., and Wood, P., "Selection of an animal model for resurfacing hip arthroplasty," *J. Arthrop.,* 2, 111, 1987.
64. Khalily, C., Malkani, A. L., Hellman, E., and Voor, M. J., "Arthroplasty in the goat hip," *J. Invest. Surg.,* 10, 119, 1997.
65. Olmstead, M. L., Hohn, R. B., and Turner, T. M., "A five-year study of 221 total hip replacements in the dog," *J. Am. Vet. Med. Assoc.,* 183, 191, 1983.
66. DeYoung, D. J. and Schiller, R. A., "Radiographic criteria for evaluation of uncemented total hip replacement in dogs," *Vet. Surg.,* 21, 88, 1992.
67. Olmstead, M. L., Hohn, R. B., and Turner, T. M., "Technique for canine total hip replacement," *Vet. Surg.,* 10, 44, 1993.
68. Montgomery, R. D., Milton, J. L., Pernell, R., and Aberman, H. M., "Total hip arthroplasty for treatment of canine hip dysplasia," *Vet. Clin. North Am.,* 22, 703, 1992.
69. Sumner, D. R., Turner, T. M., and Galante, J. O., "Symmetry of the canine femur: implications for experimental sample size requirements," *J. Orthop. Res.,* 6, 758, 1988.
70. Dogan, S., Manley, P. A., Vanderby, R., Jr., Kohles, S., Hartman, L. M., and McBeath, A. A., "Canine intersegmental hip joint forces and moments before and after cemented total hip replacement," *J. Biomech.,* 24, 397, 1991.
71. Page, A. E., Allan, C., Jasty, M., Harrigan, T. P., Bragdon, C. R., and Harris, W. H., "Determination of loading parameters in the canine hip *in vivo*," *J. Biomech.,* 26, 571, 1993.
72. Goethgen, C. B., Sumner, D. R., Platz, C., Turner, T. M., and Galante, J. O., "Changes in tibial bone mass following primary cementless and revision total hip arthroplasty in canine models," *J. Orthop. Res.,* 9, 820, 1991.
73. Smith, A. M., Turner, T. M., and Sumner, D. R., "Unilateral hip replacement causes bilateral changes in tibial bone mineral content in a canine model," *J. Bone Miner Res.,* 11, 693, 1996.
74. Cohen, J., *Statistical Power Analysis for the Behavioral Sciences,* Lawrence Erlbaum Associates, Hillsdale, NJ, 1988.
75. Sokal, R. R. and Rohlf, F. J., *Biometry: the principles and practice of statistics in biological research,* W. H. Freeman & Co., San Francisco, 1981.
76. van Rietbergen, B., Huiskes, R., Weinans, H., Sumner, D. R., Turner, T. M., and Galante, J. O., "The mechanism of bone remodeling and resorption around press-fitted THA stems," *J. Biomech.,* 26, 369, 1993.
77. Cook, S. D., Anderson, R. C., Weinstein, A. M., Skinner, H. B., Haubold, A., and Yapp, R., "An evaluation of LTI carbon and porous titanium hip prostheses," in *Biomaterials and Biomechanics,* Ducheyne, P., Van der Perre, G., and Aubert, A. E., Eds., Elsevier, Amsterdam, 1984, 31.
78. Cook, S. D., Thomas, K. A., and Kester, M. A., "Wear characteristics of the canine acetabulum against different femoral prostheses," *J. Bone Joint Surg.,* 71B, 189, 1989.
79. Cruess, R. L., Kwok, D. C., Duc, P. N., LeCavalier, M. A., and Dang, G.-T., "The response of articular cartilage to weight-bearing against metal," *J. Bone Joint Surg.,* 66B, 592, 1984.
80. Lade, R., Sauer, B., and Doerre, E., "Long term perfomance of high purity aluminum oxide ceramic heads in canine endoprosthetic hip replacement," in *Ceramics in Surgery,* Vincenzini, P., Ed., Elsevier, Amsterdam, 1983, 277.
81. Turner, T. M., Sumner, D. R., Urban, R. M., Rivero, D. P., and Galante, J. O., "A comparative study of porous coatings in a weight-bearing total hip-arthroplasty model," *J. Bone Joint Surg.,* 68A, 1396, 1986.

82. Sumner, D. R., Turner, T. M., Urban, R. M., and Galante, J. O., "Remodeling and ingrowth of bone at two years in a canine cementless total hip-arthroplasty model," *J. Bone Joint Surg.*, 74A, 239, 1992.
83. Bobyn, J. D., Pilliar, R. M., Binnington, A. G., and Szivek, J. A., "The effect of proximally and fully porous-coated canine hip stem design on bone modeling," *J. Orthop. Res.*, 5, 393, 1987.
84. Sumner, D. R., Turner, T. M., Igloria, R., Urban, R. M., and Galante, J. O., "Trabecular bone functional adaptation and bone ingrowth vary as a function of implant stiffness in a canine hip arthroplasty model," *J. Biomech.*, 1997.
85. Maistrelli, G. L., Fornasier, V., Binnington, A., McKenzie, K., Sessa, V., and Harrington, I., "Effect of stem modulus in a total hip arthroplasty model," *J. Bone Joint Surg.*, 73B, 43, 1991.
86. Bobyn, J. D., Mortimer, E. S., Glassman, A. H., Engh, C. A., Miller, J. E., and Brooks, C. E., "Producing and avoiding stress shielding: laboratory and clinical observations of noncemented total hip arthroplasty," *Clin. Orthop.*, 274, 79, 1992.
87. Matthews, L. S. and Goldstein, S. A., "The prosthesis-bone interface in total knee arthroplasty," *Clin. Orthop.*, 276, 50, 1992.
88. LaBerge, M., Bobyn, J. D., Drouin, G., and Rivard, C. H., "Evaluation of metallic personalized hemiarthroplasty: a canine patellofemoral model," *J. Biomed. Mater. Res.*, 26, 239, 1992.
89. Maloney, W. J., Sychterz, C., Bragdon, C., McGovern, T., Jasty, M., Engh, C. A., and Harris, W. H., "Skeletal response to well-fixed femoral components inserted with and without cement," *Clin. Orthop.*, 333, 15, 1996.
90. Kraemer, W. J., Maistrelli, G. L., Fornasier, V., Binnington, A., and Zhao, J. F., "Migration of polyethylene wear debris in hip arthroplasties: a canine model," *J. Appl. Biomater.*, 6, 225, 1995.
91. Amstutz, H. C., Kim, W. C., O'Carroll, P. F., and Kabo, J. M., "Canine porous resurfacing hip arthroplasty," *Clin. Orthop.*, 207, 270, 1986.
92. Chen, P.-Q., Turner, T. M., Ronningen, H., Galante, J., Urban, R., and Rostoker, W., "A canine cementless total hip prosthesis model," *Clin. Orthop.*, 176, 24, 1983.
93. Dai, K. R., Liu, Y. K., Park, J. B., Clark, C. R., Nishiyama, K., and Zheng, Z. K., "Bone-particle-impregnated bone cement: An *in vivo* weight-bearing study," *J. Biomed. Mater. Res.*, 25, 141, 1991.
94. Walenciak, M. T., Zimmerman, M. C., Harten, R. D., Ricci, J. L., and Stamer, D. T., "Biomechanical and histological analysis of an HA coated, arc deposited CPTi canine hip prosthesis," *J. Biomed. Mater. Res.*, 31, 465, 1996.
95. Plenk, H., Jr., Pflüger, G., Böhler, N., Gottsauner-Wolf, F., Grundschober, F., and Schider, S., "Long-term anchorage of cementless tantalum and niobium femoral stems in canine hip-joint replacement," in *Biomaterials and Biomechanics*, Ducheyne, P., Van der Perre, G., and Aubert, A. E., Eds., Elsevier, Amsterdam, 1984, 61.
96. Ducheyne, P., Martens, M., and Burssens, A., "Materials, clinical and morphological evaluation of custom-made bioreactive-glass-coated canine hip prostheses," *J. Biomed. Mater. Res.*, 18, 1017, 1984.
97. Radin, E. L., Rubin, C. T., Thrasher, E. L., Lanyon, L. E., Crugnola, A. M., et al., "Changes in the bone-cement interface after total hip replacement," *J. Bone Joint Surg.*, 64 A, 1188, 1982.
98. Gavens, A. J., Beals, N. B., DeMane, M. F., Davidson, J. A., Roberson, J. R., et al., "Porous polysulfone coated femoral stems," in *Implant Materials in Biofunction*, de Putter, C., de Lange, G. L., de Groot, K., and Lee, A. J. C., Eds., Elsevier, Amsterdam, 1988, 159.
99. Thomas, K. A., Cook, S. D., Haddad, R. J., Kay, J. F., and Jarcho, M., "Biologic response to hydroxyapatite-coated titanium hips: a preliminary study in dogs," *J. Arthrop.*, 4, 43, 1989.
100. Sumner, D. R. and Galante, J. O., "Determinants of stress shielding: design vs. materials vs. interface," *Clin. Orthop.*, 274, 202, 1992.
101. Richards, R. R., Minas, T., Johnston, D. W. C., and Waddell, J. P., "Biologic response to uncemented madreporic canine hip arthroplasty," *Canad. J. Surg.*, 30, 245, 1987.
102. Thornhill, T. S., Ozuna, R. M., Shortkroff, S., Keller, K., Sledge, C. B., and Spector, M., "Biochemical and histological evaluation of the synovial-like tissue around failed (loose) total joint replacement prostheses in human subjects and a canine model," *Biomaterials*, 11, 69, 1990.
103. Ronningen, H., Lereim, P., Galante, J., Rostoker, W., Turner, T., and Urban, R., "Total surface hip arthroplasty in dogs using a fiber metal composite as a fixation method," *J. Biomed. Mater. Res.*, 17, 643, 1983.
104. Paul, H. A. and Bargar, W. L., "Histologic changes in the dog femur following total hip replacement with current cementing techniques," *J. Arthrop.*, 1, 5, 1986.

105. Gitelis, S., Chen, P.-Q., Andersson, G. B. J., Galante, J. O., Rostoker, W., and Andriacchi, T. P., "The influence of early weight-bearing on experimental total hip arthroplasties in dogs," *Clin. Orthop.*, 169, 291, 1982.
106. Griss, P., Greenspan, D. C., Heimke, G., Krempien, B., Buchinger, R., Hench, L. L., and Jentschura, G., "Evaluation of a bioglass-coated Al_2O_3 total hip prosthesis in sheep," *J. Biomed. Mater. Res. Symp.*, 7, 511, 1976.
107. Griss, P., Silber, R., Merkle, B., Haehner, K., Heimke, G., and Krempien, B., "Biomechanically induced tissue reactions after Al_2O_3-ceramic hip joint replacement. Experimental and early clinical results," *J. Biomed. Mater. Res. Symp.*, 7, 519, 1976.
108. Harris, W. H. and Jasty, M., "Bone ingrowth into porous coated canine acetabular replacements: the effect of pore size, apposition, and dislocation," in *The Hip*, Fitzgerald, R. H., Ed., Mosby, St. Louis, 1985, 214.
109. Hedley, A. K., Clarke, I. C., Kozinn, S. C., Coster, I., Gruen, T., and Amstutz, H. C., "Porous-ingrowth fixation of the femoral component in a canine surface replacement of the hip," *Clin. Orthop.*, 163, 300, 1982.
110. Hedley, A. K., Kabo, M., Kim, W., Coster, I., and Amstutz, H. C., "Bony ingrowth fixation of newly designed acetabular components in a canine model," *Clin. Orthop.*, 176, 12, 1983.
111. Itami, Y., Akamatsu, N., Tomita, Y., and Nagai, M., "A cementless system of total hip prosthesis: experimental studies on total hip prosthesis in dogs," *Arch. Orthop. Trauma Surg.*, 100, 183, 1982.
112. Jasty, M., Bragdon, C. R., Haire, T., Mulroy, R. D., and Harris, W. H., "Comparison of bone ingrowth into cobalt chrome sphere and titanium fiber mesh porous coated cementless canine acetabular components," *J. Biomed. Mater. Res.*, 27, 639, 1993.
113. Jasty, M., Bragdon, C. R., Rubash, H., Schutzer, S. F., Haire, T., and Harris, W., "Unrecognized femoral fractures during cementless total hip arthroplasty in the dog and their effect on bone ingrowth," *J. Arthrop.*, 7, 501, 1992.
114. Jasty, M., Rubash, H. E., Paiement, G. D., Bragdon, C. R., Parr, J., and Harris, W. H., "Porous-coated uncemented components in experimental total hip arthroplasty in dogs," *Clin. Orthop.*, 280, 300, 1992.
115. Lembert, E., Galante, J., and Rostoker, W., "Fixation of skeletal replacement by fiber metal composites," *Clin. Orthop.*, 87, 303, 1972.
116. Maistrelli, G. L., Mahomed, N., Fornasier, V., Antonelli, L., Li, Y., and Binnington, A., "Functional osseointegration of hydroxyapatite-coated implants in a weight-bearing canine model," *J. Arthrop.*, 8, 549, 1993.
117. Mendes, D. G., Walker, P. S., Figarola, F., and Bullough, P. G., "Total surface hip replacement in the dog," *Clin. Orthop.*, 100, 256, 1974.
118. Magee, F. P., Weinstein, A. M., Longo, J. A., Koeneman, J. B., and Yapp, R. A., "A canine composite femoral stem," *Clin. Orthop.*, 235, 237, 1988.
119. Cheng, S. L., Davey, J. R., Inman, R. D., Binnington, A. G., and Smith, T. J., "The effect of the medial collar in total hip arthroplasty with porous-coated components inserted without cement," *J. Bone Joint Surg.*, 77A, 118, 1995.
120. Munting, E., "The contributions and limitations of hydroxyapatite coatings to implant fixation: a histomorphometric study of load bearing implants in dogs," *Int. Orthop.*, 20, 1, 1996.
121. Dowd, J. E., Schwendeman, L. J., Macaulay, W., Doyle, J. S., Shanbhag, A. S., et al., "Aseptic loosening in uncemented total hip arthroplasty in a canine model," *Clin. Orthop.*, 319, 106, 1995.
122. Schiller, T. D., DeYoung, D. J., Schiller, R. A., Aberman, H. A., and Hungerford, D. S., "Quantitative ingrowth analysis of a porous-coated acetabular component in a canine model," *Vet. Surg.*, 22, 276, 1993.
123. Spector, M., Shortkroff, S., Hsu, H. P., Lane, N., Sledge, C. B., and Thornhill, T. S., "Tissue changes around loose prostheses: a canine model to investigate the effects of an antiinflammatory agent," *Clin. Orthop.*, 261, 140, 1990.
124. Parvongnukul, K. and Lumb, W. V., "Evaluation of polytetrafluoroethylene-graphite-coated total hip prostheses in goats," *Am. J. Vet. Res.*, 39, 221, 1978.
125. Esslinger, J. O. and Rutkowski, E. J., "Studies on the skeletal attachment of experimental hip prostheses in the pygmy goat and the dog," *J. Biomed. Mater. Res.*, 7, 187, 1973.

126. Oates, K. M., Barrera, D. L., Tucker, W. N., Chau, C. C., Bugbee, W. D., and Convery, F. R., "*In vivo* effect of pressurization of polymethyl methacrylate bone-cement. Biomechancial and histologic analysis," *J. Arthrop.*, 10, 373, 1995.
127. Otsuka, N. Y., Binnington, A. G., Fornasier, V. L., and Davey, J. R., "Fixation with biodegradable devices of acetabular components in a canine model," *Clin. Orthop.*, 306, 250, 1994.
128. Rose, R. M., Martin, R. B., Orr, R. B., and Radin, E. L., "Architectural changes in the proximal femur following prosthetic insertion: preliminary observations of an animal model," *J. Biomech.*, 17, 241, 1984.
129. Bouvy, B. M. and Manley, P. A., "Vascular and morphologic changes in canine femora after uncemented hip arthroplasty," *Veter. Surg.*, 22, 18, 1993.
130. Finkelstein, J. A., Anderson, G. I., Waddell, J. P., Richards, R. R., and Humeniuk, B., "A madreporic-surfaced femoral component in a canine total hip arthroplasty model: bone remodelling response at 6 and 24 months," *Canad. J. Surg.*, 38, 501, 1995.
131. Kohles, S. S., Vanderby, R., Jr., Ashman, R. B., Manley, P. A., Markel, M. D., and Heiner, J. P., "Ultrasonically determined elasticity and cortical density in canine femora after hip arthroplasty," *J. Biomech.*, 27, 137, 1994.
132. Vanderby, R., Jr., Manley, P. A., Belloli, D. M., Kohles, S. S., Thielke, R. J., and McBeath, A. A., "Femoral strain adaptation after total hip replacement: a comparison of cemented and porous ingrowth components in canines," *J. Eng. Med.*, 204, 97, 1990.
133. de Waal Malefijt, J., Sloof, T. J., Huiskes, R., de Laat, E. A., and Barentsz, J. O., "Vascular changes following hip arthroplasty. The femur in goats studied with and without cementation," *Acta. Orthop. Scand.*, 59, 643, 1988.
134. Litsky, A. S., Rose, R. M., Rubin, C. T., and Thrasher, E. L., "A reduced-modulus acrylic bone cement: preliminary results," *J. Orthop. Res.*, 8, 623, 1990.
135. Szivek, J. A., Kersey, R. C., DeYoung, D. W., and Ruth, J. T., "Load transfer through a hydroxyapatite-coated canine hip implant," *J. Appl. Biomater.*, 5, 293, 1994.
136. Heiner, J. P., Kohles, S. S., Manley, P. A., Vanderby, R., Jr., and Markel, M. D., "Stability of proximal femoral grafts in canine hip arthroplasty," *Clin. Orthop.*, 341, 233, 1997.
137. Weinans, H., Huiskes, R., van Rietbergen, B., Sumner, D. R., Turner, T. M., and Galante, J. O., "Adaptive bone remodeling around bonded noncemented total hip arthroplasty: a comparison between animal experiments and computer simulation," *J. Orthop. Res.*, 11, 500, 1993.
138. Cook, S. D., Skinner, H. B., Weinstein, A. M., Lavernia, C. J., and Midgett, R. J., "The mechanical behavior of normal and osteoporotic canine femora before and after hemiarthroplasty," *Clin. Orthop.*, 170, 303, 1982.
139. Lanyon, L. E., Paul, I. L., Rubin, C. T., Thrasher, E. L., DeLaura, R., et al., "*In vivo* strain measurements from bone and prosthesis following total hip replacement," *J. Bone Joint Surg.*, 63A, 989, 1981.
140. Lord, G. A., Hardy, J. R., and Kummer, F. J., "An uncemented total hip replacement: experimental study and review of 300 madreporique arthroplasties," *Clin. Orthop.*, 141, 2, 1979.
141. Claes, L., Burri, C., Neugebauer, R., and Gruber, U., "Experimental investigations on hip prostheses with carbon fibre reinforced carbon shafts and ceramic heads," in *Ceramics in Surgery,* Vincenzini, P., Ed., Elsevier, Amsterdam, 1983, 243.
142. Clarke, I. C., Phillips, W., McKellop, H., Coster, I. R., Hedley, A., and Amstutz, H. C., "Development of a ceramic surface replacement for the hip. An experimental Sialon model, biomaterials," *Med. Dev. Artif. Org.*, 7, 111, 1979.
143. Ferris, B. D., "A quantitative study of the tissue reaction and its relationship to debris production from a joint implant," *J. Exp. Pathol.*, 71, 367, 1990.

22 Animal Models for Investigations of Biomaterial Debris

Martin Lind, Yong Song, and Stuart B. Goodman

CONTENTS

I. Introduction 427
II. Models With Particles without Implants 428
    A. Intramuscular and Intraarticular Injection of Particles................................428
    B. Subcutaneous Pocket Model 429
    C. Subcutaneous Air Pouch Model..429
    D. Rabbit Tibia Model 430
    E. The Bone Harvest Chamber with Particles 430
    F. Model with Particle Application to Mice with Different Degrees of Immunodeficiencies 431
III. Models with Particles and Implants 432
    A. Rat Model with Intraarticular PMMA Plug and Application of PE Particles.............432
    B. Canine Model with Different Implant Textures and PE Particles................433
    C. Total Hip Arthroplasty Models 433
        1. Spector Model 433
        2. Dowd Model..434
    D. Rabbit Tibial Hemiarthroplasty with Fixed and Loose Implant and PMMA Particles 434
    E. Rat Model with Combined Particle Application and Micromotion.............435
IV. Models with Intraarticular Particle Injections and Joint Implants......................435
    A. Rabbit Model...435
    B. Canine Model..436
V. Conclusion..436
VI. Future Research Directions..439
References 439

I. INTRODUCTION

Total joint replacement (TJR) is a very successful procedure for the treatment of end stage arthritis. However, TJR still has problems with regards to wear and late loosening of the components. These problems are becoming more important since the indications for TJR have been extended to a younger patient group. Also the elderly population is expecting an extended and more active lifetime and both of these situations will place increasing demands on TJR both with respect to survival-time and durability.

A key factor for the loosening of prosthetic components is believed to be the generation of and the biologic response to wear particles in periprosthetic tissues. Wear debris typically is generated at the normal articular surfaces of a TJR. A person can perform over 1 million gait cycles in a year and this activity can potentially generate hundreds of millions of wear particles from the metal and polyethylene (PE) articulating surfaces.[23] Particles can also be generated from the interfaces of modular components such as metal backed PE components. The generation of wear debris can be accelerated if third body wear (bone, metallic or other debris) is trapped in the articular surface or if abnormally articulating interfaces exist (such as areas of impingement between the femoral neck and the acetabular cup). The generated particles are typically small (<10 μm) and, therefore, are phagocytosable by macrophages.[23] The macrophages that participate in the phagocytosis become activated which leads to secretion of numerous substances that interact with fibroblasts, osteoblasts, osteoclasts and also other cells from the immunological system.[15,22] This biological reaction modulates the formation and resorption of mesenchymal tissue and eventually leads to some pathological findings in failed TJRs including membrane formation, periimplant osteolysis, and implant loosening.[33,34]

Improved understanding of the biological response to wear debris from TJR is important for revealing the mechanisms of failure of prosthetic components and for possible future interventions for prevention and treatment. Animal models have contributed greatly to the understanding of the biological processes that are involved in the response to wear debris particles in soft and hard tissue.

Animal studies are valuable for a number of reasons. It is possible to develop standardized models in which a particular facet of a complex biological process can be investigated. Particles of known size and material can be implanted at specific locations and at controlled doses. Biological specimens can be harvested at any time point and end point determinations are almost unlimited. Animal models also have the advantage of investigating biological effects in an intact organism where the complex interactions between different tissues are preserved. This is contrary to *in vitro* studies in which single cell cultures lack interactions from other cell-types and extracellular matrix. The disadvantage of animal experiments is the high expense of such studies. Also, it can be difficult to select a model that mimics exactly the surgical situation or biological process in humans that is of interest. Another problem is that different physiology in animals, especially lower species, can cause difficulty in extrapolating data from animal experiments to human situations. This is especially relevant for bone physiology, in which bone remodeling in rodents is very different from humans and healing is faster in animals than in humans.

Despite these disadvantages animal studies have contributed to the understanding of the biological reaction to orthopaedic wear debris. It is the scope of this chapter to review the different animal models that have contributed to the understanding of the biological effects of wear debris and to summarize the data retrieved in these studies. We have chosen to divide the models into two major groups: first models with application of wear debris alone into different tissues, and second models in which wear debris is applied in the interface between bone and an implant.

II. MODELS WITH PARTICLES WITHOUT IMPLANTS

A. Intramuscular and Intraarticular Injection of Particles[29]

In this early study by Stinson, particles were injected into the left gluteal muscle and left knee joint of Hartley guinea pigs. Particles of polymethylmethacrylate (PMMA), PE and nylon with a size range from 0–72 μm were investigated. Animals were sacrificed at two weeks, 1, 6, 12, 24, 30 and 36 months after implantation of particles.

Muscle biopsies were analyzed by qualitative histology and knee joints were investigated by macroscopic evaluation of particle location and joint cartilage status as well as qualitative histology of synovial tissue where particle accumulation was observed.

In muscle tissue, necrotic muscle fibers were seen in relation to particles, which evoked a macrophagic reaction. At later time points up to 12 months, necrosis disappeared and foreign body giant cells predominated. Foci of plasma cells and lymphocytes were very evident at 12 months, typically located adjacent to clumps of particulate material. Giant cells were replaced by fibroblasts at late time points from 12 month for PE particles. Nylon particles were removed and not seen after 18 months. In knee joints, particulate PMMA was accumulated in synovial tissue as early as after two weeks. At two and four weeks the particles were primarily surrounded by macrophages. At later time points particles were surrounded by a fibrous stroma with islands of foreign body giant cells. Small foci of plasma cells and lymphocytes were seen at 18 and 24 months. The joint cartilage was unaffected by particle implantation at all time points. PE and nylon particles gave similar synovial reactions.

B. SUBCUTANEOUS POCKET MODEL

CD rats (four weeks old, Charles River Breeding Labs., MA)[6] and MF1 mice (six weeks old)[25] were used for this model. Bilateral subcutaneous pockets were prepared by blunt dissection over the thoracic area. In the study by Glowacki et al. 50 mg of PMMA, polyethylene and bone particles were placed in the pockets for 12 days.[6] The particle size ranged from 75–250 μm. In the study by Quinn et al.[25] PMMA particles were placed in the pockets for two months. The particle size ranged from 50–300 μm.

Glowacki et al.[6] used qualitative undecalcified histology and semiquantitative histochemistry to evaluate inflammatory and osteoclast response to the different particles. Quinn et al. harvested the granulomas formed at the particle application sites. One part of the granuloma was used for histology and another part was used for cell culture. The granuloma cells were then co-cultured with fibroblastic, osteoblastic and stromal cells on bone slices for evaluation of osteoclastic activity.

Glowacki et al.[6] found that these nonphagocytosable PMMA and PE particles were surrounded by foreign body giant cells in fibrous stroma. The bone particles were surrounded by osteoclasts that showed evidence of surface resorption. Histochemical analyses for an osteoclast marker confirmed a high osteoclastic activity around bone particles. Quinn et al.[25] found mononuclear inflammation with scattered giant cells in response to PMMA particle implantation. These cells stained positive for macrophage markers but not for osteoclast associated tartrate resistant acid phosphatase. The granuloma cells exhibited osteoclastic activity on bone slices after 14 and 21 days of co-culture with osteoblastic or stromal cells but not after co-culture with fibroblastic cells.

C. SUBCUTANEOUS AIR POUCH MODEL[21]

Under anesthesia 20 ml of air was injected into the subcutaneous tissue of a 6–8 week-old SD rat's back to form a single subcutaneous air pouch. At day 5, an additional 10 ml of air was injected together with 5 ml of particle suspension. In this study PMMA particles (11 μm diam.) with and without 10% barium sulfate were injected. Liquid from the air pouch was aspirated at 1, 6, 12, 24, 72 hours and 1 and 2 weeks by injecting 5 ml saline and aspirating 4 ml for analyses.

The air pouch aspirate was analyzed for leukocyte count. Prostaglandin E_2 (PGE_2) levels were measured by enzyme linked immunosorbent assay (ELISA). Tumor necrosis factor (TNFa) was measured by L929 bioassay, and metalloprotease activity was measured by a substrate degradation spectrophotometric assay.

Particulate PMMA with barium sulfate was associated with an earlier influx of leukocytes than PMMA particles without barium sulfate at six hours. At the same time point, increased metalloprotease activity was demonstrated for PMMA particles with barium sulfate. PGE_2 release was highest in the barium sulfate group from 1–24 hours whereas TNFa was increased at 48 and 72 hours. It was concluded that addition of barium sulfate to bone cement for radioopacity could result in an accentuated host inflammatory response to particulate debris from the cement.

D. Rabbit Tibia Model[10-12,14]

Mature NZW rabbits (3-4 kg) were used for this model. After bilateral exposure of the anteromedial aspects of the proximal tibia a 6 mm drill hole was made by a hand drill. A curved curette was used to scoop bone marrow out of the canal underneath the drill hole. After saline irrigation and drying the hole was implanted with standardized amounts of either bulk or particulate material in the right tibia; the left tibia served as a non-implanted sham operated control. Particles of ultrahigh molecular weight polyethylene (UHMWPE), Ti6Al4V alloy, cobalt chromium alloy (CoCr alloy), and PMMA were analyzed. Particle sizes were <1000, 4, 15 and 10–100 µm respectively. Observation time was 16 weeks, but harvesting can occur at any time period(s).

Sterile harvest of bone marrow was used for organ cultures. Weight standardized tissue samples were cultured in Dulbecco modified Eagle's medium for three days and conditioned medium was assayed for PGE_2 levels by radioimmunoassay (RIA). Also, the proximal tibia was harvested and prepared for decalcified histology by conventional techniques. The bone implant interface was analyzed by counting giant cells, histiocytes, lymphocytes, plasma cells, PMN leukocytes, fibroblasts and marrow cells using a standardized system.

Particulate PMMA material exhibited 2.25 fold higher levels of PGE_2 in bone marrow organ cultures compared to controls. Bulk PMMA exhibited identical PGE_2 levels as controls. This response could be inhibited by giving the animals an oral nonsteroidal anti-inflammatory drug, sodium naproxen, during the observation period. Using histological analyses, particulate UHMWPE and PMMA implantation resulted in a bone implant interface with increased cell counts for giant cells and histiocytes when compared to bulk implantations of the same materials. Similar implantation of particulate and bulk titanium (Ti) and CoCr alloy did not demonstrate any increase in giant cells and histiocytes in the particulate material group.

E. The Bone Harvest Chamber with Particles[1,7,9,32]

The bone harvest chamber (Figure (1) is a Ti device that is implanted in the proximal tibial metaphyses of mature male NZW rabbits. The chamber has a 1x1x10 mm pore for tissue ingrowth at the cortical bone level. The top of the chamber can be accessed through a small skin incision and be disassembled for tissue harvest or particle application. In these studies chambers were inserted bilaterally. After osseointegration, repeated harvests of tissue growing into the chamber are possible. The particles investigated were machined high density polyethylene (HDPE), hydroxyapatite (HA), Ti6Al4V alloy and CoCr alloy particles with 4.7, 5.0, 3.0 and 2.7 µm mean diameter respectively. Also diamond and SiC particles with a <20 µm size were investigated. Particles were dissolved in sodium hyaluronate at a concentration of 1×10^8 particles/ml. Initially a six week osseointegration period was applied. Then the carrier material (Healon) was placed bilaterally for three weeks. Tissues were harvested and HDPE particles were placed in the chambers on one side and the contralateral side was left empty as control. After another three weeks tissues were harvested. The previous control side now received titanium alloy (Ti alloy) particles and the contralateral side served as control for a final three weeks.

The tissue specimens harvested were prepared for conventional decalcified histology. Amount of bone and fibrous tissue was quantified as well as cellular parameters of foreign body reaction (histiocytes and giant cells), acute and chronic inflammation (polymorphonuclear leukocytes and lymphocytes/plasma cells).

CoCr alloy and HDPE particles caused 67% and 35% reduction in bone ingrowth into the chamber whereas Ti alloy particles did not exhibit this effect. Also in CoCr alloy and HDPE treated tissue specimens particles where surrounded by mono- and multinucleated histiocytic cells as well as lymphocytes and plasma cells in a fibrous stroma. In tissue specimens receiving Ti alloy particles the cellular composition was similar to control tissue specimens. HA particles increased bone

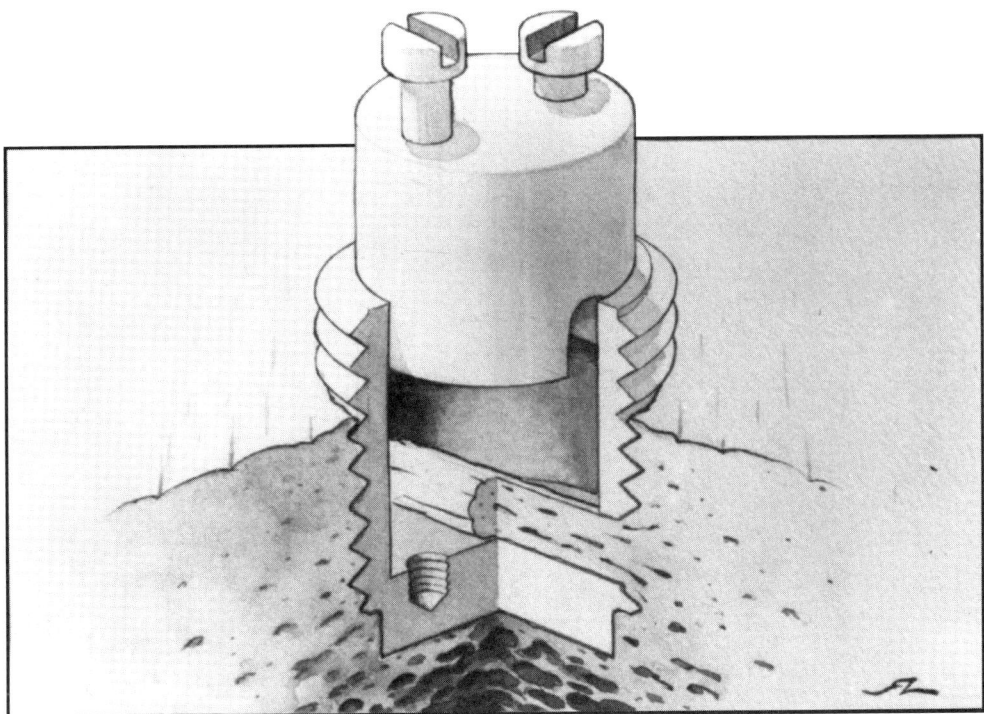

FIGURE 1. The bone harvest chamber. The canal for bone ingrowth at the cortical bone level is illustrated as well as the removable core that enables access to the bone canal for particle application of tissue harvest.

ingrowth at three weeks and no inflammation and granuloma formation was seen. Diamond and SiC particles had no adverse effects in the bone harvest chamber.

F. MODEL WITH PARTICLE APPLICATION TO MICE WITH DIFFERENT DEGREES OF IMMUNODEFICIENCIES[19,20]

Four different strains: normal immunocompetent mice, nu/nu athymic mice deficient in T cells, severe combined immunodeficiency strain (SCID) mice with decreased T and B lymphocyte function, and triple deficient mice (NIH III, nude-beige) lacking T and B lymphocyte function as well as natural killer cells were used.

Using the immunocompetent and immunodeficient mice, Jasty et al.[19] and Jiranek et al.[20] examined the tissue response to injection of PMMA powder. The PMMA particles were 13.6 μm in diameter and approximately 25,000 particles in 0.2 ml aliquots were injected into two ventral and two dorsal subcutaneous sites of each mouse strain. A tissue reaction was then allowed to form in five weeks.

Tissue granulomas formed were analyzed by immunohistochemistry for leukocyte markers: Mac-2 for macrophages, CD3 for T-lymphocytes, CD8 for antigen expressing suppressor T-cells and CD20 for B-lymphocytes. *In situ* hybridization for IL–1b was applied to locate gene expression of this cytokine.

All animal strains demonstrated granulomas when PMMA powder was injected subcutaneously for five weeks. All four mouse strains exhibited the same macrophage accumulation unaffected by the different degrees of deficiencies in immunocompetence. The macrophages in the granuloma expressed messenger RNA for interleukin–1 beta, a marker for macrophage activation, and the

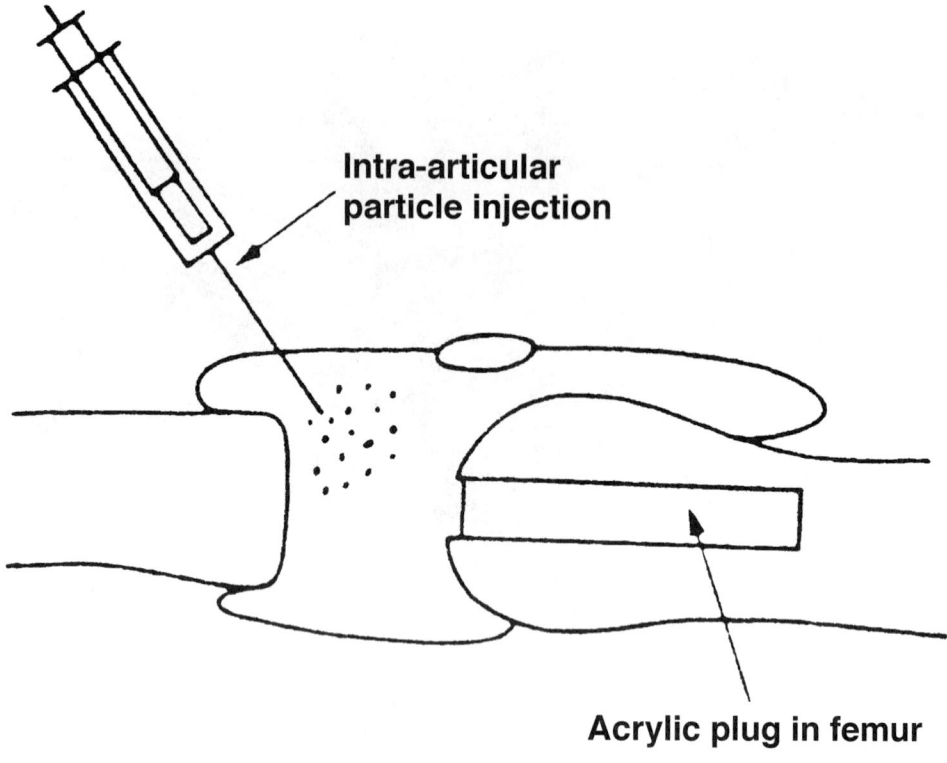

FIGURE 2. The Howie model. The rat knee joint is illustrated with the implanted bone cement plug in the distal femur. Also the site for particle injections is shown.

expression was independent of lymphocyte presence. It was concluded that macrophages are the key cells initiating and maintaining the foreign body response to particulate PMMA, and that lymphocytes are not essential to mounting this response. A similar finding was demonstrated in the rat tibia model using normal and T-cell deficient animals and HDPE particles.[8]

III. MODELS WITH PARTICLES AND IMPLANTS

A. RAT MODEL WITH INTRAARTICULAR PMMA PLUG AND APPLICATION OF PE PARTICLES

In this model by Howie et al.,[18] through a medial knee arthrotomy, a hole (10 mm in length, 1.1 mm diam.) was drilled in the intercondylar notch of mature Lewis rats. A PMMA plug was then inserted into the drill hole so that the distal end of the plug was just below the level of the articular surface. (Figure 2) Post surgery the operated knees were injected with 1 mg UHMWPE particles (20–200 μm diam.) dissolved in 15 ml of a 1:50 serum/saline solution. Injections were performed at two, four, six, and eight weeks postoperatively and the animals were terminated at 10 weeks. A control group received no injections and were terminated at two weeks.

The knee joint and distal femur were harvested for conventional decalcified histological preparation. The interface was investigated by qualitative histology.

In control animals, a complete shell of bone had formed around the implant and a very thin layer of amorphous tissue was located between the cement and bone. No signs of bone resorption were seen. In the animals injected with particles, the bone surrounding the implant had been replaced by a cellular connective tissue composed of macrophages, giant cells and fibroblasts. Bone resorption was most evident close to the joint surface but osteoclasts were also seen on bone surfaces some distance from the joint. Throughout the tissue surrounding the implant, birefringent PE particles were demonstrated using polarized light. This study was the first to demonstrate formation of a peri-implant connective tissue membrane as a result of intraarticular particle exposure.

Using a modification of the Howie model, van der Vis et al.[31] have recently tested numerous different particle types. In their study, the model was modified by placing the particles around the PMMA implant at surgery only and the observation time was four weeks only. The study confirmed that PE particles stimulate peri-implant bone resorption. CoCr alloy, Ti6Al4V alloy and zirconium oxide particles did not have this effect.

B. Canine Model with Different Implant Textures and PE Particles

In this model by Bobyn et al.[3] special cylindrical implants were inserted intraarticularly in the knee joints of mature mongrel dogs. The implants had a smooth texture on half of the surface and a porous coating on the remaining surface. The implants were 9x30 mm and were press-fit in the intercondylar notch of the femur and in the tibial plateau bilaterally. At both locations, implantations were performed parallel to the long bone axis. Initially a 10 week osseointegration period was allowed before intraarticular injections with polyethylene particles began. Thereafter, the animals were injected twice weekly with machined (4.7 μm diam.) polyethylene particles for 10 weeks. Two doses were investigated, 5 and 10 mg per injection in the left and right knees respectively. The particle injection period was followed by another 10 week period before sacrifice. In this period an inflammatory response to the injected particles was allowed to evolve.

After sacrifice the distal femur, proximal tibia and synovial tissue were harvested for conventional undecalcified thin section histology and contact radiography. Four to eight sections at different levels from each implant were analyzed qualitatively for bone ingrowth and periprosthetic membrane formation and for the presence of PE particles within the bone/implant interface.

Synovium from particle-injected animals contained large amounts of PE particles. The smaller particles were seen intracellularly in macrophages and larger particles typically were surrounded by foreign body giant cells. For all implants that were subjected to particles, bone ingrowth was seen at the porous coated surface whereas the smooth surface was surrounded by peri-implant fibrous tissue. Corresponding radiolucent lines were observed at contact radiography in the latter group. The cavities around the smooth implants were observed from the most proximal to the most distal parts of the implant and were of fibrous character. The bone cavities around the smooth implant surfaces were surrounded by a distinct bony shell or neocortex and PE particles were found to have migrated almost exclusively along the smooth surface of implants. This model is the first to show that a porous coated implant surface can prevent particle migration into the bone/implant interface.

C. Total Hip Arthroplasty Models

Two models to investigate the effects of wear particles in a total hip arthroplasty model have been described by Spector et al. and Dowd et al.[4,28]

1. Spector Model[28]

Unilateral implantation of a Charnley like femoral stem and a PE acetabular cup was performed in seven dogs. Prior to implantation a thin layer of bone cement (1–2 mm) was applied to the femoral stem in the laboratory. During surgery the medullary canal was over reamed so that the cement coated stem could be inserted in a loose fit to enable motion of the stem. Before the stem was inserted 250 mg of bone cement particles (500 μm diam.) was placed in the femoral canal. It was hypothesized that abrasion of these bigger particles, between each other and the cement sheath on the stem, would generate smaller phagocytosable particles. Animals were sacrificed at four and seven months.

Evaluation was performed by anteriorposterior and lateral radiographs preoperatively, immediate postoperatively, two, four and seven months postoperatively. Radionucleotide imaging was performed one day prior to sacrifice by intravenous injection of ^{99}mTc-methylene disphosphonate. The femur was excised and the membrane was removed aseptically. Part of the membrane was used for histology and embedded in methylmethacrylate. The other part of the membrane was used for biochemistry. This was accomplished by dicing the tissue and subsequent collagenase treatment to release the cells. Mononuclear cells were isolated by Ficoll gradient centrifugation and subsequently cultured for 72 hours with and without sodium naproxen. Conditioned media were assayed for PGE_2 and interleukin 1 (IL-1) content as well as bone resorbing activity in the fetal rat calvaria organ culture assay.

A radiolucent seam was found at the cement bone interface at four months; at seven months this seam became wider and more irregular. Subsidence of the stem was also seen. Radionucleotide imaging showed increased blood pooling and bone turnover around the operated stem when compared to the non-operated femora. However, no changes in blood flow were observed. Histologically the membrane tissues demonstrated multiple layers of macrophages in a fibrous stroma as well as multinucleated giant cells around PMMA particles. PGE_2 and IL-1 production from the cultured membrane cells could be depressed by naproxen treatment. The medium from the treatment group stimulated bone resorption more than control medium.

2. Dowd Model[4]

Thirty eight dogs received unilateral implantation of cementless femoral stems with a midstem porous coating and porous coated acetabular cup. Three different prosthesis types were used: (1) A control prosthesis with a cementless implant design. (2) A gap prosthesis with a midshaft circumferential 2 mm gap in which different particles could be applied. (3) A motion prosthesis with a midshaft ball joint to enable implant motion. The dogs were randomized to six different groups including, (1) control prosthesis, (2) motion prosthesis, (3) gap prosthesis without particles, (4) with CoCr alloy particles, (5) with Ti6Al4V alloy particles, and (6) with HDPE particles; 100 mg of particles were placed in the gap. The particle sizes ranged from 3–10 µm in diameter. The observation period was 12 weeks.

After sacrifice the proximal femur with implant was harvested as well as synovial tissue from the contralateral hip joint. Part of the membrane was used for histology and graded semiquantitatively with respect to macrophage density. Another part of the membrane was harvested aseptically and used for biochemistry. This was accomplished by dicing the membrane and synovial tissue and subsequently culturing the tissue in organ culture for 72 hours. Conditioned media were assayed for collagenase, gelatinase, PGE_2 and IL-1 activity.

Manual testing demonstrated that the control implants were well-fixed after the 12 week observation period whereas the implants with motion or particles were all manually loose. Histologically, membrane tissue around control implants consisted of very few macrophages. For motion implants a thick membrane was found but with only a moderate number of macrophages. Around the gap implants, a thick membrane was found with high numbers of macrophages in all particle stimulated groups. However, PE particles appeared the stimulate macrophage accumulation less than CoCr alloy and Ti alloy particles. Biochemical analyses of conditioned medium from cultured membrane tissue revealed high levels of collagenase activity and PGE_2 release in metal particle groups and lesser activity in the polyethylene particles group. Gelatinase activity was mainly elevated in the group with motion implants. IL-1 activity was highest in CoCr alloy and PE particle groups.

D. Rabbit Tibial Hemiarthroplasty with Fixed and Loose Implant and PMMA Particles[13,16,17]

Using a medial parapatellar incision, the right knee joints of adult NZW rabbits were accessed. Anterior cruciate ligament and menisci were removed and the tibia plateau cartilage and a small amount of bone were removed by an oscillating saw. Eight animals received fixed implants by conventionally cementing a custom type stemmed and fluted titanium alloy tibial hemiarthroplasty implant in place. Another eight animals received a loose implant. This was accomplished by letting the cement cure *ex vivo* after its application to the implant. When placed in the prepared tibial plateau an additional thin layer of PMMA powder (10–100 μm particles) was applied to the surface of the implant/cement complex to simulate cement debris. To ensure loose fitting, the implant was rotated 90 degrees once. Observation time was three months.

The bone/implant interface was investigated by histology and radiography. Sterile harvest of bone marrow from the proximal 2 cm of the tibia was used for organ cultures. Biopsies from both operated and non-operated tibia were harvested. Weight standardized tissue samples were cultured in Dulbecco modified Eagle's medium for three days and conditioned medium was assayed for PGE_2 levels by RIA and lysosomal enzyme activity.

Wider radiolucent lines and a thick fibrous membrane were observed around the loose implants. Bone marrow from the particle exposed loose implant group produced three times more PGE_2 than non operated marrow. In the fixed group PGE_2 production was less than in non operated marrow. Also lysosomal enzyme activity was increased in the loose implant group.

E. RAT MODEL WITH COMBINED PARTICLE APPLICATION AND MICROMOTION

Male SD rats (350 gm) were used in this model.[2] A 4×13 mm Ti plate was screwed onto the medial surface of the proximal tibia. In the middle of the plate there was a hole into which a circular 2.5 mm test surface could be screwed to contact the cortical bone surface. The test surface protruded 0.5 mm into the cortical bone which was milled at the area to receive the test surface and also to induce bone trauma and hematoma. HDPE particles (machined to 4.7 μm) were applied initially or at later stages by unscrewing the test surface. Implant movement was applied by rotating the test surface 180 degrees 20 times twice daily. Movement was achieved by manually rotating a subcutaneous located wing nut on the test surface by external manipulation. Several investigational groups were included: (1) the cortical bone was allowed to osseointegrate around the test surface for six weeks and then particles were applied for six weeks or particles for six weeks only without any prior osseointegration. (2) osseointegration for two weeks and then motion for two, or six weeks[3] osseointegration for two weeks and then motion and particles for six weeks, or motion for six weeks followed by particles for six weeks.

Histomorphometry was performed on conventional decalcified sections to quantify bone and cartilage fractions at 0.18 and 0.45 mm below the test implant surface. Also the thickness of a soft tissue membrane between implant and bone was measured.

Application of particles alone did not disturb osseointegration and this result was similar for implants with particles from the implantation time or after six weeks osseointegration. Movement created a fibrous membrane that was well established at six weeks and contained small islands of cartilage. When both particles and motion were present, the membrane was similar to the membrane seen with motion alone without signs of inflammation. If particles were applied after six weeks of movement, osseointegration that was otherwise seen in animals rested six weeks after micromotion was prevented.

IV. MODELS WITH INTRAARTICULAR PARTICLE INJECTIONS AND JOINT IMPLANTS

A. RABBIT MODEL

Mature NZW rabbits (3.0–3.5 kg) were used in this model.[5] Through a medial parapatellar incision the knee joint was accessed for intercondylar implantation of a Ti alloy implant (5 × 10 mm) bilaterally. Observation time was 12 weeks. During this period weekly intraarticular injections of 1 mg HDPE particles (7 μm) suspended in hyaluronic acid were performed. Particles were injected on one side and sham injection was performed on the contralateral side.

Synovial tissues were harvested for semiquantitative determinations of particle infiltration and immunohistochemical staining for the macrophage chemoattractant cytokines interleukin 8 (IL–8) and macrophage chemotactic and activating peptide 1 (MCP-1).

In control joints, no signs of inflammation and staining for IL–8 and MCP-1 were found. In joints receiving particles, moderate inflammation with mainly mononuclear cells were seen. Particles were seen beneath the lining layer of the synovial membrane. The synovial tissue demonstrated intensive staining for MCP-1, but moderate IL–8 staining.

B. Canine Model

Using mature Labrador dogs, Ti alloy implants with and without HA coating were inserted bilaterally into the weight bearing part of the femoral condyles surrounded by a 0.75 mm gap.[26] The implants were mounted on a special weight bearing device that ensured implant loading during each gait cycle and also created access to the joint space.[30] Each femur received one uncoated and one HA coated implant. One knee joint was randomly selected for particle injection of 6×10^9 HDPE (2 μm) particles. The contralateral knee had implants inserted but received no particles. Particles were injected weekly from week three and throughout the eight-week observation period.

The implant bone specimens were investigated histologically using decalcified sections for analysis of particle migration along the bone implant interface and undecalcified sections for analysis of bone and fibrous tissue ingrowth.

Particles were found to have migrated along the uncoated implants in large numbers, whereas almost no particles were found around the HA coated implants. Less bone ingrowth and bone formation in the gap along with more fibrous tissue formation were found around uncoated implants when compared with HA coated implants.

V. CONCLUSION

In vivo studies have contributed greatly to the understanding of biological effects of orthopaedic wear debris (Tables 1 and 2). Models without implants used various locations for particle application such as muscular, subcutaneous, bone marrow and periimplant tissue. Particles applied in these locations were able to generate granuloma formation rich in macrophages and giant cells. With longer observation time, from 12 months and longer lymphocyte aggregates typically were seen in these granulomas.[6,25,29] Of the many materials used in orthopaedic implants such as: PE, Ti6Al4V alloy, CoCr alloy and PMMA, PE seems to form the most aggressive granuloma.[31] Particles have to be of phagocytosable size (<20 μm) to elicit an inflammatory response with granuloma formation.[11,12,14] When particles are applied around implants inserted in bone, fibrous tissue membrane formation and bone resorption have been demonstrated in many models.[4,18,28] Some studies suggest that polyethylene causes less accumulation of macrophages around implants than CoCr and Ti6Al4V alloy particles, but which particle material generates the most aggressive granuloma formation seems to depend on the model used.[4] It appears that motion at the bone implant interface produces a more fibrous membrane than particles alone. This membrane is, however, without many macrophages that are seen in particle associated membranes. Implants with a smooth surface texture seem to give access to more particle accumulation at the bone-implant interface than implants with a porous coating or a HA coating.[3,26]

In summary, animal studies have confirmed that application of orthopaedic wear debris particles can lead to radiographically evident osteolysis and formation of macrophage rich fibrous tissue membrane. Furthermore, particle induced granuloma express mediators that participate in accelerated bone resorption such as IL-1b, IL–6, and PGE_2. These characteristics all mimic the pathological

TABLE 1
Animal Models for Effects of Wear Debris Particles (Without Implants)

1st author year[Ref.]	Species	Model, Observation time	Particle type/size	End point
Stinson 1965[29]	Rat	IM and intraarticular particle injections, days–2 years	PMMA, PE, Nylon, <70μm injections every 2nd week	Qualitat. histology
Glowaski 1986[6]	Rat	Subcutaneous pockets, 12 days	PMMA, 75–250 μm	Qualitat. histology. Semiquantative histochemistry
Goodman 1992[10,14]	Rabbit	Rabbit tibia hole with particles, 16 weeks	PMMA, PE, Ti6Al4V, CoCr 10–100 μm	Organ culture marrow from prox. tibia, PGE2 measurements, Quantitat. histology
Quinn 1992[25]	Mouse	Subcutaneous pockets 2 months	PMMA, Bone, PE, 50–300 μm	Qualitat. histology, Co-culture bone lice osteoclast assay, IL-1 synthesis.
Goodman 1992[7,9]	Rabbit	Bone harvest chamber 3 weeks	PE, Ti6Al4V, CoCr, HA, SiC, Diamond, 2–20 μm	Bone ingrowth, Quantitative histology
Jasty 1992,[19] Jiranek 1995[20]	Mouse	SC particles in immunodeficient mice, five weeks	PMMA, 13.6 μm	Immunological cell counts in granuloma
Lazarus 1994[21]	Rat	Air pouch model one hr.–2 weeks	PMMA, 11 μm +/- BaSO4	TNF-a, PGE2 metallo-proteinase in pouch liquid

TABLE 2
Animal Models for Effects of Wear Debris Particles (With Implants)

1st author year[Ref].	Animal	Model, Observation time	Particle type/size, use	End point
Howie 1988[18]	Rat	Intraarticular PMMA rods in knee joint, 10 weeks	PE 20–200 μm, injections every 2nd week	Qualitative histology
Spector 1990[28]	Dog	Total hip arthroplasty 4–7 months	PMMA, 500 μm	Qualitative histology and radiography
Goodman 1992[16,17]	Rabbit	Total knee arthroplasty cemented fixed and loose, 3 months	PMMA, 10-100 μm	Organ culture marrow tissue from proximal tibia, PGE2 measurements
Dowd 1995[4]	Dog	Total hip arthroplasty, 12 weeks	CoCr, Ti6Al4V, PE, 4-10 μm	Quantitative histology, enzymes, PGE2 and IL-1 synthesis
Frøkjær 1995[5]	Rabbit	Intraarticular titanium implants in knee joint, 12 weeks	PE, 7 μm, weekly injections	Immunohistochemistry of synovial tissue for IL-8 and MCP-1
Bobyn 1995[3]	Dog	Titanium implant in knee joint, half smooth, half porous, 30 weeks	PE, 4.7 μm twice weekly in middle 10 weeks	Qualitative histology
Aspenberg 1996[2]	Rat	Tibial implant with micromotion + particles, six weeks	PE, 4.7 μm	Quantitative histology
Rahbek 1997[26]	Dog	Weight-loaded implant with gap in knee joint, 8 weeks	PE, 2 μm, weekly injections	Mechanical tests and quantitative histology

scenario observed clinically in aseptic loosening of prosthetic implants. This large body of experimental evidence points towards a significant role for wear particles in the etiology of aseptic loosening. Further research in this field could lead to future interventional steps that could improve the clinical outcome of prosthetic surgery.

VI. FUTURE RESEARCH DIRECTIONS

Currently the histopathological effects of orthopaedic wear debris particles have been investigated and described in detail. Many of the inflammatory mediators that are expressed and/or regulated during particle induced inflammation have been identified. However, there are still many poorly understood facets of particle induced inflammation and its contribution to prosthetic implant loosening. The clinical picture in patients having a prosthetic implant with known wear debris problems varies. Some patients have well functioning prostheses and are symptom free for decades despite extensive PE wear and particle generation. Other patients develop extensive osteolysis and implant failure within few years. Little is known about this difference in biological response to prosthetic component implantation and the individual response of patients to wear debris particles. Also the possible involvement of the immune system in the reaction to particulate wear has been implicated. Several studies with particle implantation for 12 months and longer have observed lymphocyte accumulation in the particle generated granuloma tissue.

Immunohistochemical studies have demonstrated that some of these lymphocytes express the IL-2 receptor, which is a marker of lymphocyte activation. However, no studies have shown specific lymphocyte reaction or specific antibody production in response to wear particles. New animal models that can detect such lymphocytic activation in long-term studies are needed to answer the question of immune system involvement.

Another important direction for future research is to develop methods and techniques that can prevent the adverse effects of particle induced inflammation in periprosthetic tissues. Such approaches could target the implants, and recent studies have shown that a porous coating or HA coating on a cementless implant can prevent some of the particle migration along the implant-bone interface.[3,26] Another recent study has used a pharmacological approach for treatment of aseptic loosening of prosthetic components. In this study by Shanbhag et al., total hip arthroplasties were performed in dogs. If a bisphosphonate alendronate, was administered, osteolysis could be prevented despite the fact that periimplant membranes still produced the high levels of cytokines.[27]

Finally, recent *in vitro* studies with wear particles and leukocytes have focused on uncovering some of the intracellular signaling pathways involved in particle induced macrophage activation and inflammation.[24] Discovery of specific cascades involved in these processes could pave the way for specific pharmacological target sites to inhibit the adverse expression of bone resorbing and inflammatory cytokines.

REFERENCES

1. Aspenberg, P., Anttila, A., Konttinen, Y. T., et al., "Benign response to particles of diamond and SiC: bone chamber studies of new joint replacement coating materials in rabbits," *Biomaterials,* 17, 807, 1996.
2. Aspenberg, P. and Herbertsson, P., "Periprosthetic bone resorption. Particles versus movement," *J. Bone Joint Surg.,* 78B, 641, 1996.
3. Bobyn, J. D., Jacobs, J. J., Tanzer, M., et al., "The susceptibility of smooth implant surfaces to periimplant fibrosis and migration of polyethylene wear debris," *Clin. Orthop.,* 311, 21, 1995.
4. Dowd, J. E., Schwendeman, L. J., Macaulay, W., et al., "Aseptic loosening in uncemented total hip arthroplasty in a canine model," *Clin. Orthop.,* 319, 106, 1995.

5. Frøkjær, J., Deleuran, B., Lind, M., Overgaard, S., Søballe, K., and Bunger, C., "Polyethylene particles stimulate monocyte chemotactic and activating factor production in synovial mononuclear cells *in vivo*. An immunohistochemical study in rabbits," *Acta Orthop. Scand.,* 66, 303, 1995.
6. Glowacki, J., Jasty, M., and Goldring, S. R., "Comparison of multinucleated cells elicited in rats by particulate bone, polyethylene, or polymethylmethacrylate," *J. Bone Miner. Res.,* 1, 327, 1986.
7. Goodman, S., Aspenberg, P., Song, Y., Knoblich, G., Huie, P., Regula, D., and Lidgren, L., "Tissue ingrowth and differentiation in the bone-harvest chamber in the presence of cobalt-chromium-alloy and high-density-polyethylene particles," *J. Bone Joint Surg.,* 77A, 1025, 1995.
8. Goodman, S., Wang, J. S., Regula, D., and Aspenberg, P., "T-lymphocytes are not necessary for particulate polyethylene-induced macrophage recruitment. Histologic studies of the rat tibia," *Acta Orthop. Scand.,* 65, 157, 1994.
9. Goodman, S. B., Aspenberg, P., Song, Y., Doshi, A., Regula, D., and Lidgren, L., "Effects of particulate high-density polyethylene and titanium alloy on tissue ingrowth into bone harvest chamber in rabbits," *J. Appl. Biomater.,* 6:27, 1995.
10. Goodman, S. B. and Chin, R. C., "Prostaglandin E_2 levels in the membrane surrounding bulk and particulate polymethylmethacrylate in the rabbit tibia. A preliminary study," *Clin. Orthop.,* 257, 305, 1990.
11. Goodman, S. B., Fornasier, V. L., and Kei, J., "The effects of bulk versus particulate polymethylmethacrylate on bone," *Clin. Orthop.,* 232, 255, 1988.
12. Goodman, S. B., Fornasier, V. L., and Kei, J., "The effects of bulk versus particulate ultra-high-molecular-weight polyethylene on bone," *J. Arthroplasty,* 3, S41, 1988.
13. Goodman, S. B., Fornasier, V. L., and Kei, J. "Quantitative comparison of the histological effects of particulate polymethylmethacrylate versus polyethylene in the rabbit tibia," *Arch. Orthop. Trauma Surg.,* 110, 123, 1991.
14. Goodman, S. B., Fornasier, V. L., Lee, J., and Kei, J., "The effects of bulk versus particulate titanium and cobalt chrome alloy implanted into the rabbit tibia," *J. Biomed. Mater. Res.,* 24, 1539, 1990.
15. Goodman, S. B., Huie, P., Song, Y., et al., "Loosening and osteolysis of cemented joint arthroplasties. A biologic spectrum," *Clin. Orthop.,* 337, 149, 1997.
16. Goodman, S. B., Kang, T., and Smith, R. L., "Lysosomal enzyme production at the interface surrounding loose and well-fixed cemented tibial hemiarthroplasties in the rabbit knee," *J. Invest. Surg.,* 6, 413, 1993.
17. Goodman, S. B., Magee, F. P., and Fornasier, V. L., "Radiological and histological study of aseptic loosening using a cemented tibial hemiarthroplasty in the rabbit knee," *Biomaterials,* 14, 522, 1993.
18. Howie, D. W., Vernon-Roberts, B., Oakeshott, R., and Manthey, B., "A rat model of resorption of bone at the cement-bone interface in the presence of polyethylene wear particles," *J. Bone Joint Surg.,* 70A, 257, 1988.
19. Jasty, M., Jiranek, W., and Harris, W. H., "Acrylic fragmentation in total hip replacements and its biological consequences," *Clin. Orthop.,* 285, 116, 1992.
20. Jiranek, W., Jasty, M., Wang, J. T., Bragdon, C., Wolfe, H., Goldberg, M., and Harris, W., "Tissue response to particulate polymethylmethacrylate in mice with various immune deficiencies," *J. Bone Joint Surg.,* 77A, 1650, 1995.
21. Lazarus, M. D., Cuckler, J. M., Schumacher, H. R., Ducheyne, P., and Baker, D. G., "Comparison of the inflammatory response to particulate polymethylmethacrylate debris with and without barium sulfate," *J. Orthop. Res.,* 12, 532, 1994.
22. Maloney, W. J. and Smith, R. L., "Periprosthetic osteolysis in total hip arthroplasty: the role of particulate wear debris," *Instr. Course. Lect.,* 45, 171, 1996.
23. Maloney, W. J., Smith, R. L., Schmalzried, T. P., Chiba, J., Huene, D., and Rubash, H., "Isolation and characterization of wear particles generated in patients who have had failure of a hip arthroplasty without cement," *J. Bone Joint Surg.,* 77A, 1301, 1995.
24. Nakashima, Y., Trindade, M., Sun, D. H., Maloney, W. J., Goodman, S. B., et al., "The effects of signal transduction inhibitors on macrophage activation *in vitro*: requirement for tyrosine kinase activity," *Trans. Orthop. Res. Soc.,* 22, 323, 1997.
25. Quinn, J., Joyner, C., Triffitt, J. T., and Athanasou, N. A., "Polymethylmethacrylate-induced inflammatory macrophages resorb bone," *J. Bone Joint Surg.,* 74B, 652, 1992.

26. Rahbek, O., Overgaard, S., Bendix, K., Bünger, C., Søballe, K., "Sealing effect of hydroxyapatite coating on periimplant particle migration," *Trans. Orthop. Res. Soc.,* 22, 354, 1997.
27. Shanbhag, A. S., Hasselman, C. T., Kovach, C. J., and Rubash, H., "Inhibition of osteolysis by bisphosphonates in a canine total hip arthroplasty model," *Trans. Orthop. Res. Soc.,* 22, 43, 1997.
28. Spector, M., Shortkroff, S., Hsu, H. P., Lane, N., Sledge, C. B., and Thornhill, T. S., "Tissue changes around loose prostheses. A canine model to investigate the effects of an antiinflammatory agent," *Clin. Orthop.,* 261, 140, 1990.
29. Stinson, N. E., "Tissue reaction induced in guinea-pigs by particilate polymethylmethacrylate polythene and nylon of the same size range," *Br. J. Exp. Pathol.,* 46, 135, 1965.
30. Søballe, K., Hansen, E. S., Rasmussen, H. B., Jørgensen, P. H., Bünger, C., Tissue ingrowth into titanium and hydroxyapatite-coated implants during stable and unstable mechanical conditions, *J. Orthop. Res.,* 10, 285, 1992.
31. van der Vis, H. M., Marti, R. K., Tigchelaar, W., Schuller, H. M., and Van Noorden, C. J., "Benign cellular responses in rats to different wear particles in intra-articular and intramedullary environments," *J. Bone Joint Surg.,* 79B, 837, 1997.
32. Wang, J. S., Goodman, S., and Aspenberg, P., "Bone formation in the presence of phagocytosable hydroxyapatite particles," *Clin. Orthop.,* 304, 272, 1994.
33. Willert, H. G. "Reactions of the articular capsule to wear products of artificial joint prostheses," *J. Biomed. Mater. Res.,* 11, 157, 1977.
34. Willert, H. G., Bertram, H., and Buchhorn, G. H., "Osteolysis in alloarthroplasty of the hip. The role of ultra-high molecular weight polyethylene wear particles," *Clin. Orthop.,* 258, 95, 1990.

23 Animal Models of Orthopaedic Prosthetic Infection

Yuehuei H. An and Richard J. Friedman

CONTENTS

   I. Introduction 443
  II. How to Design an Animal Model of Prosthetic Infection 444
      A. Animal Selection 444
      B. Implant Fabrication 446
      C. Bacterial Inoculation 446
      D. Surgical and Necropsy Technique 446
      E. Authors' Preferred Models 446
         1. Subcutaneous Model 446
         2. Diaphyseal Bone Model 447
         3. Cancellous Bone Model 448
         4. Joint Replacement Model 448
 III. Commonly Used Evaluation Methods and Criteria 448
      A. Clinical Features 448
      B. Radiography 451
      C. Laboratory Tests 451
      D. Bacteriological Study 452
      E. Histological Study 452
 IV. Applications of Prosthetic Infection Models 452
      A. Pathogenesis of Bacteria 452
      B. *In vivo* Behavior of Biofilm 453
      C. Effect of Prophylactic and Therapeutic Antibiotics 454
      D. Effect of Biomaterials on Prosthetic Infection Rate 454
      E. Effect of Infection on Biomaterial Surfaces 455
  V. Concluding Remarks 455
Acknowledgements 455
References 456

I. INTRODUCTION

In the United States, more than 200,000 primary hip and 200,000 primary knee arthroplasties are performed each year. Between 0.5% and 2.3% of them will become infected within 10 years (Table 1).[1] Sepsis following total joint replacement can have catastrophic results both physically and psychologically for the patients, leading to failure of the arthroplasty, possible amputation, prolonged hospitalization, and even death.[8] Although the use of prophylactic antibiotics and greatly improved OR techniques have decreased the infection rate from an average of 5.9% in 1975 to 1.2% in 1993,[1] challenges still remain for better preventive and therapeutic measures.

TABLE 1
Infection Rates of Prosthetic Infection

First author, year	Total cases	Infected cases	% of infection*
Josefsson 1993[2]	835	13	1.6
Josefsson 1993[2]	853	9	1.1
Lidwell 1982[3]	5831	34	0.6
Lidwell 1982[3]	2221	52	2.3
Andrews 1981[4]	N/A	68	1.3
Nelson 1980[5]	580	6	1.0
Fitzgerald 1977[6]	3215	42	1.3
Eftekher 1976[7]	800	4	0.5

* numbers of infection cases/total cases ×100 = % of infection.

Animal models of osteomyelitis[9-11] and foreign body infection[12-19] have been well established by using the rabbit, dog, chick, guinea pig, and rat. The experience from these models, especially the models of foreign body infection, has been very helpful in designing an *in vivo* prosthetic infection model. Actually, a prosthetic infection model is an extension of the models of foreign body infection, with more attention to the effect of prosthetic materials, the use of bone tissue (instead of soft tissue), or the imitation of a human total joint replacement.

Several animal models have been reported for the study of orthopaedic prosthetic infection. They include total joint replacement, skeletal implant, and soft tissue model (Table 2). The soft tissue model is actually a foreign body infection model. Animal models have been mainly used for the studies of pathogenesis of bacteria, *in vivo* behavior of bacterial biofilm, effect of biomaterials on prosthetic infection rate, and the effect of infection on biomaterial surfaces.

II. HOW TO DESIGN AN ANIMAL MODEL OF PROSTHETIC INFECTION

A. Animal Selection

A careful selection of an animal model for a study of prosthetic infection is the key for an ultimate result. Theoretically, to imitate a human situation it is better to use a large animal such as a sheep, goat, or dog, especially when attempting to design a joint replacement prosthesis as the implant. A dog femoral model has been used to test the influence of skeletal implants on incidence of infection and the preventive effect of prophylactic antibiotics.[28,29] Animal models using goats or sheep have not been reported. The shortcomings of using large animals include the need for large housing space, difficulty of handling, and high costs.

Rabbit joint replacement models have been reported by using a specially designed femoral head prosthesis,[21] interphalangeal joint prosthesis[22] or the first metatarsophalangeal joint[20] for humans. A rabbit model using bone screws as implants was reported to evaluate the interaction between bacterial biofilm and antibiotics.[24] Implant site infection models by rabbit subcutaneous implantation were also reported.[30–32] More recently, rabbit implant models, such as femoral condyle cylindrical implant,[35] tibial intramedullary nailing,[27] and tibial plating[25] have been reported. Based on the literature and our own experience, rabbits are excellent for implant infection models because (1) the rabbit has a good joint size for even a total joint replacement, (2) it is more easily infected compared to dogs or rats, and (3) it is relatively economical. Small animals (rodents) are good for studying bacterial pathogenesis, implant site infection rate (which normally need large numbers of animals), or antibiotic effects. They have been widely used for foreign body infection models of different purposes.[39] Small animals which have been used include guinea pigs,[19] hamsters,[36,37] and mice.[12-14,16,34]

TABLE 2
Animal Models of Implant or Prosthetic Infection

Models	1st Author (Year)[Ref.]	Animal	Model Description	Inoculation Routes	Bacteria	Number of Inoculated Bactera*	Incubation Time
Joint replacement	Belmatoug 1996[20]	Rabbit	Partial knee arthroplasty (tibial Intraarticular injection plateau) with silastic implant		*S. aureus*	10^{5-8} cfu/ml	—
	Southwood 1985[21]	Rabbit	Prosthetic femoral head replacement	Local inoculation or IV injection	*S. aureus*	10^{6-7} cfu local 10^{8-9} cfu IV	—
	Blomgren 1981[22]	Rabbit	Femur defect filled by cement or total knee replacement	Local inoculation or IV injection	*S. aureus*	10^{8-9} cfu	—
Skeletal model	An 1997[23]	Rabbit	Cylindrical implants inserted into lateral femoral condyle	*In vitro* colonization before implantation	*S. epidermidis*	In suspension of 10^6 cfu/ml	1 hr.
	Isiklar 1996[24]	Rabbit	Femoral intercondylar notch, drill hole, cancellous SS screw	Local inoculation	*S. epidermidis*	At least 10^7 cfu per implant site	—
	Arens 1996[25]	Rabbit	Tibial diaphyseal plating, then, local bacteria injection	Local bacteria injection	*S. aureus*	$4 \times 10^{3-6}$ cfu	—
	Sanzén 1995[26]	Rabbit	Upper tibial implantation of cylindrical Ti and PMMA	Local injection into medul. canal through implants	*S. aureus*	$9 \times 10^{5-7}$ cfu/implant	—
	Melcher 1994[27]	Rabbit	Tibial intramedullary nailing	Local inoculation	*S. aureus*	$2 \times 10^{3} - 4 \times 10^{7}$ cfu	—
	Petty 1985,1988[28,29]	Dog	Cylindrical implants inserted into proximal femoral canal	Local inoculation	Three different bacteria	10^8 cfu	—
Soft tissue model	Nakamoto 1995[30]	Hamster	Coated stainless steel wires placed subcutaneously	Incubation with bacteria before implantation	*S. epidermidis*	In suspension of 10^7 cfu/ml	15 min.
	Chang 1994[31]	Hamster	Subcutaneous cylindrical SS or Ti implants on the back	*In vitro* colonization	*S. epidermidis*	10^7 cfu in TSB†	overnight
				Local inoculation		10^8 cfu/implant	—
	Buret 1991[32]	Rabbit	Silastic placed in subdermal tissue for biofilm study	*In vitro* colonization before implantation	*Pseudomonas aeruginosa*	In suspension of 10^7 cfu/ml	3 hours

* Number of bacteria injected or the concentration of bacteria in the incubation suspension.
† TSB = Trypticase soy broth.

Nevertheless, animal selection is a very important step for an *in vivo* prosthetic infection study. One should be very careful to choose an animal which will fit the purpose of the project.

B. Implant Fabrication

Based on the specific purpose, implants made of different kinds of materials and with different sizes and shapes have been used. Implants can be fabricated into or obtained as a cylinder,[26,28,34,35] a metal wire,[30] a screw,[24] a tissue cage,[39] a bone plate,[25] an intramedullary nail,[27] a small joint prosthesis for human,[20,22] or even a prosthetic joint component.[21] Implants can be made with a smooth surface or porous coated. Most of the implants made for bone ingrowth study are cylindrical and could be used as implants for prosthetic infection.

C. Bacterial Inoculation

Common pathogens isolated from human prosthetic infections should be used for most of the *in vivo* infection models, such as *S. aureus*, *S. epidermidis*, or less frequently *E. coli*, proteus, or some anaerobics. Bacteria can be delivered to the implant surface or its surroundings by (1) colonization of bacteria to the implant *in vitro* before implantation, (2) direct injection into the implant site, or (3) injection of the bacteria into the bloodstream.

The number of bacteria needed to produce an experimental infection varies from one bacterium to another. For example, less slime-producing *S. epidermidis* will be needed to produce an infection than non-slime producers. The route of inoculation makes a large difference too. The relationship between the dose of inoculum and the development of infection after prosthetic replacement has been studied in a rabbit model.[21] Contamination of the implanted wound site with only a few bacteria (less than 50 *S. epidermidis*) likely will result in infection. It is difficult to induce infection when the operation was performed without insertion of a prosthesis (10^4 bacteria), which may suggest that the implant inhibits the defense system for dealing with the insult. It is difficult to produce an infection by inoculating the bacteria intravenously rather than locally and this will be more obvious if this inoculation is given three weeks after operation.[21,22] Based on the numbers in Table 2, at least 10^3 cfu, or an average of 10^{5-8} cfu are needed to produce an implant site infection.

D. Surgical and Necropsy Technique

Surgery should be performed under strict sterile conditions. The skin area of the incision should be shaved carefully and cleaned with 70% alcohol before transfer to the operating room. The skin area should be wiped with 7.5% Povidon-Iodine and 70% alcohol and then properly draped. Standard surgical technique should be employed throughout the operation. The wound should be washed with saline and closed securely in layers. Harvesting specimens also should be done under the exact same sterile conditions in order to obtain a valid microbiological evaluation.

E. Authors' Preferred Models

1. Subcutaneous Model

The subcutaneous model is represented by the work by Chang and Merritt[31] published in 1994. Syrian hamsters and small (2–3 × 10 mm) metal cylinders were used. The implant was inserted into the scapular area through a small incision at the sacral region with the aid of a tube and trochar system. Bacteria were introduced into the implant sites by pre-incubation with the implant or by injection after the implant was placed. Seven days after the implant replacements, the animals were sacrificed and the implants and surrounding tissues excised and placed in test tubes containing normal saline and ground with glass beads to isolate bacteria from the implants and tissues. Then, the resulting suspension was plated on agar plates for bacterial colony counts.

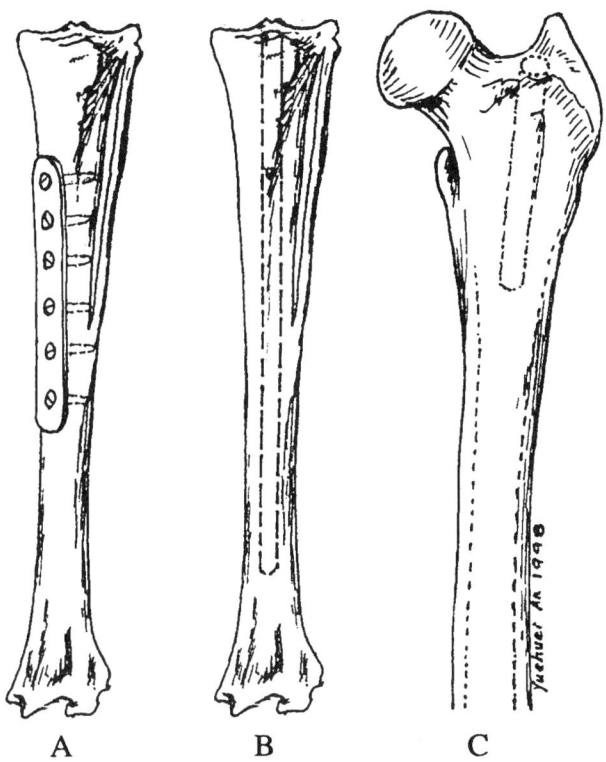

FIGURE 1. Illustrations of diaphyseal implant models for experimental prosthetic infection: (A) a tibia fixed with plate and screws in the rabbit;[25] (B) a tibia inserted with an intramedullary rod in the rabbit;[27] and (C) a canine femur inserted with an intramedullary rod.[28,29]

2. Diaphyseal Bone Model

The rabbit tibial diaphyseal plating model by Arens et al.[25] and the rabbit tibial intramedullary nailing model by Melcher et al.[27] are appropriate for evaluating implant infections because these two fixation devices are the most commonly used implants in orthopaedic surgery. Another diaphyseal infection model is the canine femoral implantation of rods of different materials reported by Petty et al.[28,29]

In the rabbit tibial plating model, the skin over the medial aspect of the midshaft of the tibia was cut under strict sterile conditions. Retraction of the muscles allowed bone exposure and then a bone plate (normally 6-hole plate) was fixed to the medial side of the tibia (Figure 1A). The wound was closed in layers and a plastic catheter was placed into the plate site through an operatively placed needle which was used for injection of bacterial suspension. The number of bacteria needed for a clinical infection ranged from 4×10^3 to 4×10^7 cfu/ml. Twenty-eight days after the surgery, bone samples were taken and ground into small pieces and cultured for bacteria.

In the rabbit medullary nailing model, through an anterior approach, the patellar ligament of the leg was divided and the medullary canal of the tibia opened using a drill (3.0–3.5 mm). The canal was then washed using a metal sucker and an appropriate amount of bacteria suspension was injected into the distal part of the canal. Then a 90 mm long and 3 mm diameter medullary nail was inserted into the tibia (Figure 1B). Twenty-eight days after the surgery, bone samples (distal half of the tibia) were taken and ground into small pieces and cultured for bacteria.

In the canine proximal femoral model,[28,29] the femoral medullary canal was reamed through a drill hole at the greater trochanter. After introduction of bacteria using a Teflon tube, a 4 × 60 mm rod made of stainless steel, PMMA, or polyethylene was inserted into the proximal femur through the drill hole (Figure 1C). Animals were killed 15 days after surgery. Clinical, microbiologic and histologic methods were used to evaluate the occurrence of infection.

If a fracture is introduced in these models as in the clinical condition, the number of bacteria needed for an infection should be fewer than those cited in the reports.

3. Cancellous Bone Model

A cancellous bone model is represented by the animal models described by Isiklar et al.[24] and An et al.[23] They both placed the implants into the femoral condyle cancellous bone with bacteria injected into the implant site adhered to the implant (Figure 2). They are both intraarticular models with the implant inserted through intercondylar notch (screw)[24] or through the lateral condyle (the cylindrical implant).[23] The infection is showed by clear clinical septic arthritis and histologic peri-implant abscess. Another metaphyseal bone model is the rabbit upper tibial implantation of Ti cylinders reported by Sanzén and Linder.[26]

In the screw model, through an anterior approach the distal femoral joint was exposed by dislocating the patellae laterally. A drill hole of 3.5 mm in diameter was created in the intercondylar notch. After bacterial inoculation in the drill hole, a stainless steel screw was placed into the femur. Four weeks later, quantitative bone cultures were performed to evaluate the infection.

In the lateral condyle plug model, a direct lateral approach to the left distal femoral condyle of each animal was made through a lateral incision and a drill hole equal in size to the implant (5 mm diam.) was made 5 mm proximal to the femoral-tibial joint articular surface and 5 mm posterior to the patellofemoral joint surface in a transverse direction. Care should be taken not to penetrate the medial cortex. The implants (with pre-adhered bacteria or without) were then inserted into the holes according to the experimental plan. The animals were sacrificed 28 days after the implantation and infection was diagnosed by clinical examination, histology, and/or a positive culture.

In the upper tibial plug model by Sanzén and Linder,[26] a PMMA or Ti cylindrical implant was implanted. After three weeks, $4 \times 10^{3-8}$ cfu $S.$ $aureus$ was injected into the medullary canal percutaneously through a central hole in the implant to create infection. After another four weeks, the animals were killed and radiographic, bacteriologic, and histologic methods were used to evaluate the infection.

4. Joint Replacement Model

The method described first by Blomgren[22] in 1981 and then by Belmatoug et al.[20] is appropriate as a knee replacement model. They used joint prostheses for human finger joint replacement. The prosthesis used by Blomgren and the silastic nail-shaped implant by Belmatoug et al. have long stems which can be introduced into the diaphyseal part of the bone (Figure 3). Bacteria can be inoculated immediately after the implantation surgery or days or weeks later through local injection into the joint space or intravenously. The latter mimics the hematogenous prosthetic infection. The pathological and radiological characteristics of this model are close to those of human prosthetic joint infection. The model by Southwood et al.[21] is also appropriate for studying joint prosthetic infection, but more effort has to be made for constructing the femoral component.

III. COMMONLY USED EVALUATION METHODS AND CRITERIA

A. Clinical Features

It is difficult to get subjective findings from animals. Several abnormal behaviors may indicate a severe local or systemic infection, such as lethargy, less eating, a lack of weight gain, or weight loss. Physical signs are very important for judging an infection. Postoperatively, temperature should be taken and the wound observed daily for the first week and twice a week thereafter.

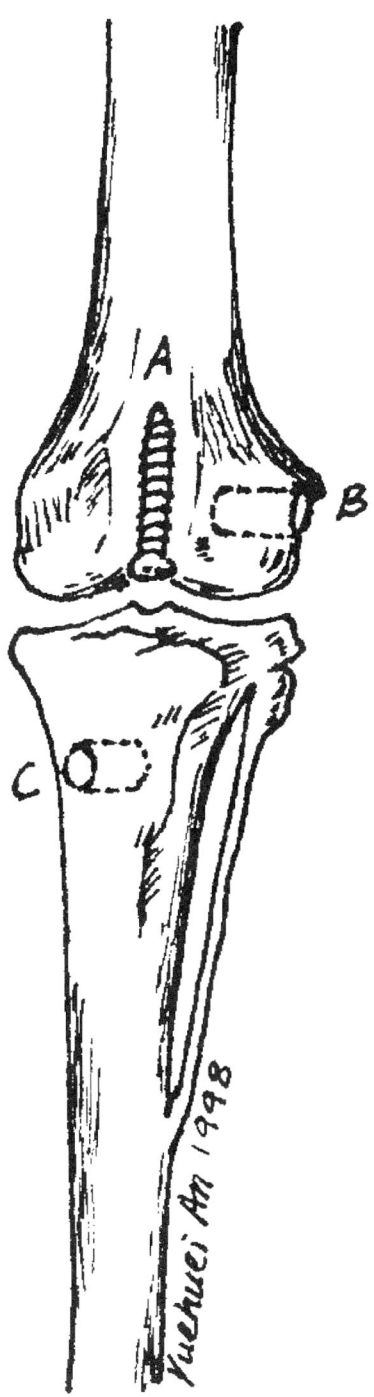

FIGURE 2. Illustration of rabbit cancellous bone implantation models for studying prosthetic infection: (A) an intercondylar implantation of a metal screw;[24] (B) a cylindrical inserted in the lateral femoral condyle of the rabbit;[23] and (C) a cylindrical implant inserted in the upper tibial metaphyseal area.[26]

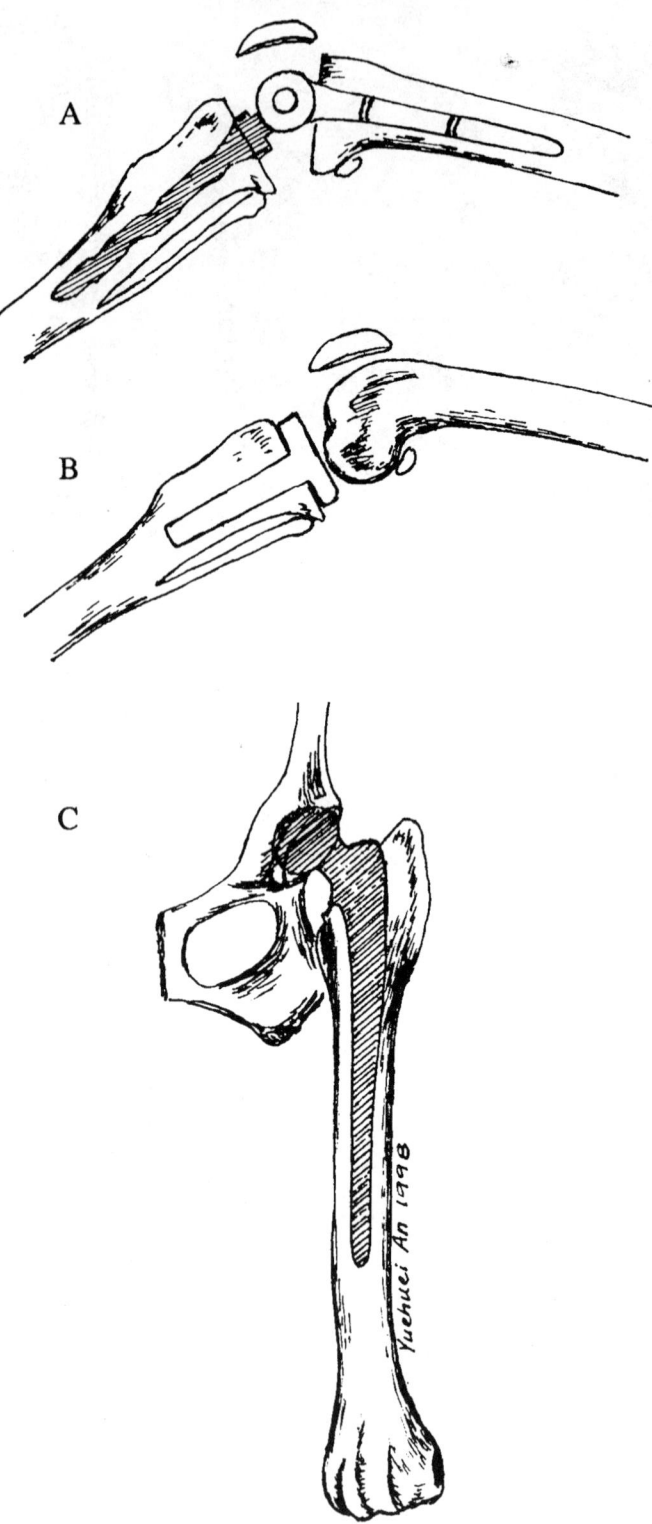

FIGURE 3. Illustrations of joint replacement models: (A) a total knee replacement with a St. Georg Fingermittelgelenk Endoprothese;[22] (B) a partial knee replacement with a Swanson great toe implant HP;[20] and (C) a potential partial hip replacement in the dog for studying prosthetic infection (A rabbit partial hip replacement model was reported by Southwood et al.[21]).

TABLE 3
Appearance of the Wound and Grading

Grade	Appearance	Score	Clinical meaning
A	No abnormal signs	0	No infection
B	Erythema and moderate soft swelling	0.5	Early infection or due to surgery
C	Large soft swelling or pus exudation	1.0	Definite infection
D	Pus exudation and systemic illness	1.5	Definite infection

TABLE 4
Radiographic Criteria

Variable	Definition	Score
Diaphyseal periosteal reaction	+/present	1
	–/absent	0
Osteolysis	+/present	1
	–/absent	0
Sequestrum formation	+/present	1
	–/absent	0
Joint effusion	+/present, widening of joint space	1
	–/absent	0
Soft tissue swelling	+/present	1
	–/absent	0

Temperature change is a sensitive parameter of a local or systemic infection. The appearance of the wound can be recorded and graded using a grading system modified from the one introduced by Petty et al.[28] (Table 3). Normally, infection can be confirmed by soft tissue swelling with pus exudate.

B. Radiography

For an intraosseous model, radiographs should be taken immediately after operation to check the implant position and at two-week intervals thereafter, until the animal is sacrificed. The development and progression of infection can be assessed using five criteria (Table 4), modified on the previous radiological descriptions of osteomyelitis in rabbits.[9] They are (1) diaphyseal periosteal reaction; (2) osteolysis; (3) sequestrum formation; (4) joint effusion; and (5) soft tissue swelling. Using these criteria, a numerical score can be assigned and the six scores added together to give an overall ranking for radiographic severity. Radiography is a very useful method for diagnosing an infection in a diaphyseal area but is less favorable when a prosthetic infection is at an epiphyseal or metaphyseal location.

C. Laboratory Tests

For any abnormal wound exudate or sinus discharge, a swab should be taken for bacterial culture. It is realized that there are no direct relations between results of swab culture and clinical signs of infection. In a recent study by the authors using a rabbit femur implant infection model, there were only two positive swab cultures in 11 histologically diagnosed infections.[23] A culture in broth may yield more positive growth. Blood samples can be collected one day preoperatively and 3, 7, 14, and 21 days post-operatively for blood cell counts and erythrocyte sedimentation rates. Getting

TABLE 5
Histologic Criteria of Prosthetic Infection when Implant is Inserted into Cancellous Bone

Histologic findings	Diagnostic meaning
Inflammation with abscess formation	Infected
Presence of sequestrum (not drilling debris)	Infected, within abscess or near an abscess
Intracellular bacteria (Gram stain)	Infected, when found in an abscess, abscess capsule, or in inflamed tissues (often neutrophil or macrophage infiltrated area)
	Not infected, when found without evident inflammation or abscess
Inflammation with fibrosis	Infected, if bacteria found in inflamed tissues
	Not infected, if no bacteria found

information from blood sampling is sometimes not practical because some of the samples will be clotted (blood aspiration is often not fast enough for animals of this size) and the work is very time consuming.

D. BACTERIOLOGICAL STUDY

The explanted implants can be placed in PBS and agitated on a vortex mixer for five min. or sonicated for 30 min. to harvest bacteria for culture. Serial dilutions will be made and the solutions spread on a tryptic soy agar plate and added to tryptic soy broth for culture. Subcultures will be prepared at 24 hours if needed. The cultures will be considered positive for infection as described by Petty et al.[28] if (1) primary culture or subculture yields any bacteria that had been inoculated or (2) primary or subculture yield any bacteria. This criterion is subject to change according to the individual situation. If the implant is cylindrical, it can also be rolled over an agar surface and cultured overnight for bacterial colonies.

E. HISTOLOGICAL STUDY

The specimens can be embedded in paraffin (if the specimens are bone, they should be decalcified first) and multiple 4 mm thick sections cut. Implants should be removed before embedding. According to the authors' experience, this will not destroy the specimen. We found that abscesses were still intact after the implant was removed.[23] Adjacent soft tissues should be evaluated histologically because soft tissue abscess or drainage tract may exist. Draining lymph nodes also should be evaluated. Selected sections should be stained with Gram stain for detecting the existence of bacteria in the pus, the capsule of the pus, or in any inflamed tissues. The histological parameters in Table 5 can be used to evaluate the samples with implants inserted into cancellous bone. It is useful when an answer of "yes-or-no" is needed. The authors feel only the existence of microabscesses is definite evidence of histological infection (Figure 4).[23] Gram stain is useful for localization of bacteria in abscesses or inflammatory tissues (Figure 5). For the histologic criteria of experimental prosthetic infection involving a diaphyseal area, one should consult the work by Petty et al.[28]

IV. APPLICATIONS OF PROSTHETIC INFECTION MODELS

A. PATHOGENESIS OF BACTERIA

When an animal model is employed, there are always questions, such as how many bacteria can cause prosthetic infection, how are they disseminated, can bacteria stay on the implant surface

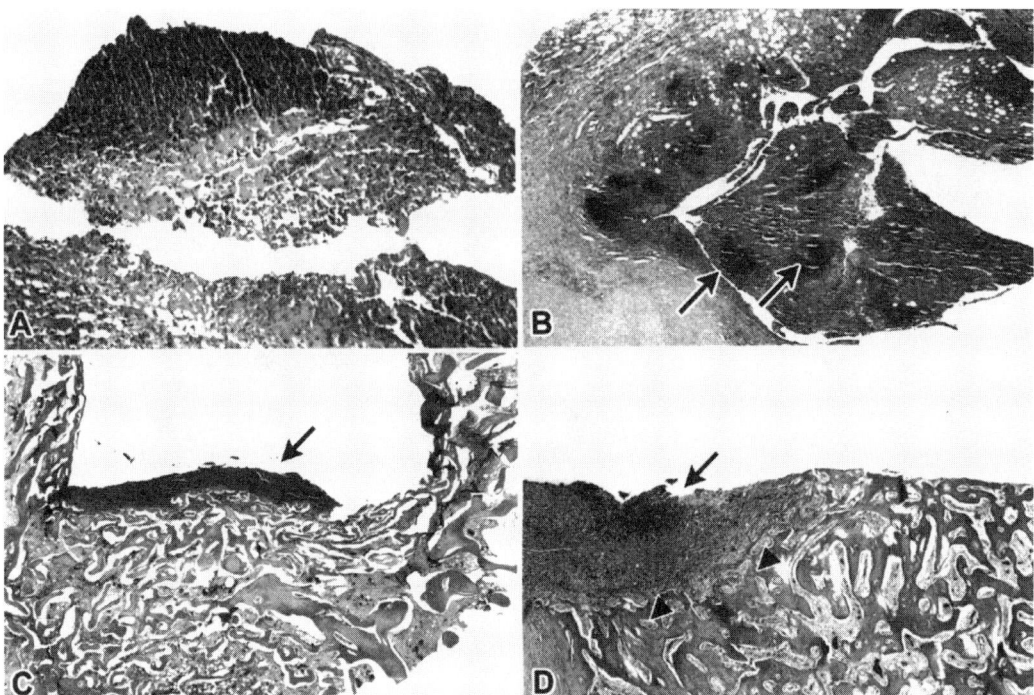

FIGURE 4. The histologic images show the abscesses found near the implant both in bone and soft tissue. (A) An abscess found in the soft tissue next to the lateral end of the implant. The imaginary location of the implant is under the lower part of the image. Part of the pus is separated from the lower part due to the tissue-processing procedure. (B) A sac of inflammatory exudate and pus found in the soft tissue covering the lateral end of the implant. The imaginary location of the implant is on the right side of the image. (C) An abscess found at cancellous bone next to the end of the implant. The implant was removed before sample embedding. (D) An abscess in the cancellous bone next to the side of the implant. The arrowheads indicate the borderline between the abscess and the surrounding bone. This curved borderline was created by erosion by the pus, so the bone surface was excavated away from the interface with the implant. (From An, Y. H., et al., *J. Bone Joint Surg.*, 79B, 816, 1997. With permission.)

for a long time without clinical infection,[33] and are the virulences of different bacteria the same? Zimmerli et al tested the effectiveness of different numbers of colony-forming units of *S. aureus* on the infection rate of a foreign body (tissue cage: a perforated tube).[19] Blomgren verified the possibility of hematogenous bacterial dissemination of a total joint prosthesis and the subsequent infection.[22] He also found that *S. aureus* (Wood 46) and *Propionibacterium acnes* (ATCC-6919) have the same ability to cause hematogenous infection of a total joint replacement. To find the effect of bacterial slime on the infection rate, Christensen et al reported a mouse foreign body infection model. Animals challenged with the slime-producing *S. epidermidis* developed three times as many infections as animals challenged with the strain that did not produce slime.[12] Also, animal models can be used to produce a bacterial biofilm for pathobiological study of implant surfaces.[32]

B. IN VIVO BEHAVIOR OF BIOFILM

Buret and colleagues[32] studied the morphology, ultrastructure, and microbiology of intact biofilm developing on an implant surface which was harvested from implants colonized with *P. aeruginosa* inserted into the peritoneal cavities of rabbits. Also in a rabbit model, Isiklar et al.[24] examined the penetration of antibiotics into biofilm formed by *S. epidermidis* following local and

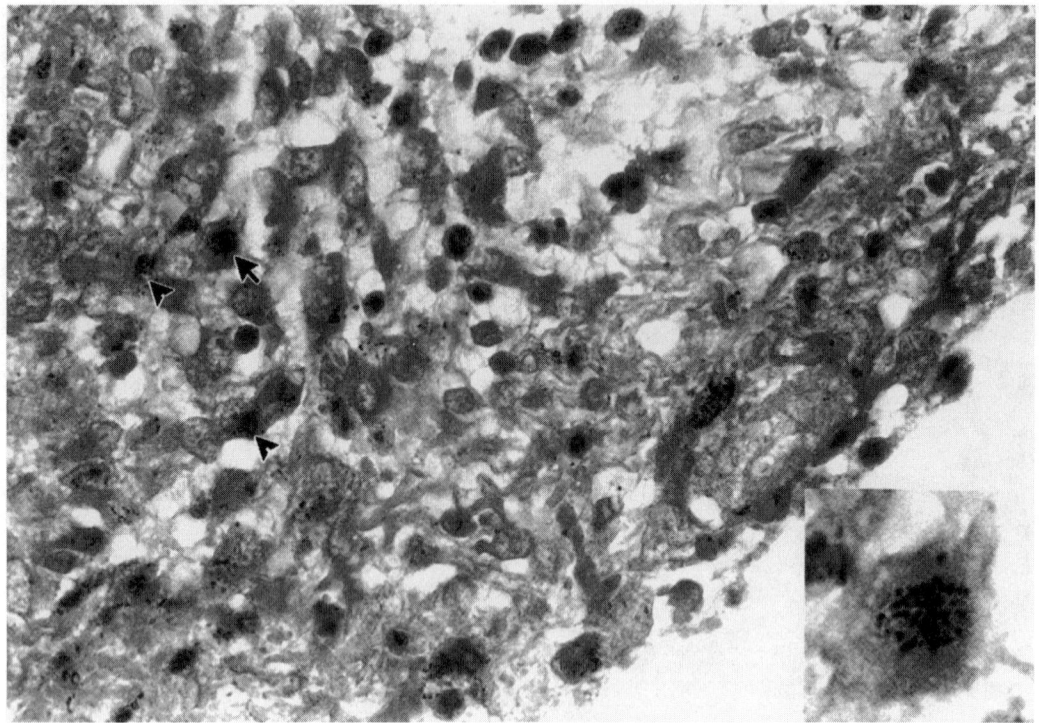

FIGURE 5. Gram stain positive bacteria (the arrow and arrow heads) were found in the cytoplasm of inflammatory cells within an abscess which was composed of polymorph leukocytes (neutrophils), macrophages, and degenerated or dead cells. The cell indicated by the arrow was enlarged to show details in the cell (inset). (From An, Y. H., et al., *J. Bone Joint Surg.*, 79B, 816, 1997. With permission.)

parenteral administration of vancomycin. This method is closer to the human situation because of the use of tibial bone, stainless steel implant, and the introduction of *S. epidermidis* which is a common pathogen for prosthetic infection.

C. Effect of Prophylactic and Therapeutic Antibiotics

Animal models are excellent for studying the effects of prophylactic and therapeutic antibiotics on prosthetic infection because of the homogeneity of the animals for good comparison, reproducibility, and easily controlled time periods for observation.[13,16,24,29] For example, Petty et al.[29] studied the preventive effectiveness of wound irrigation with normal saline, irrigation with saline containing neomycin, antibiotic-impregnated polymethylmethacrylate, and systemic antibiotics administration on prosthetic infection. The result showed that saline irrigation had no effect on infection rate; systemic use of cefazolin and neomycin irrigation slightly reduced the infection rate and the use of bone cement containing gentamicin caused a significant reduction.

D. Effect of Biomaterials on Prosthetic Infection Rate

Several investigations on the effects of orthopaedic implants on the incidence of infection have been reported. Merritt et al.[34] designed a soft tissue model in the mouse and tested the implant site infection rates with porous and dense materials. They found that in the acute model the infection rate with the porous implant was greater, while in the chronic model after tissue invasion the infection rate with the dense materials was greater.

Petty et al.[28] established a dog model to evaluate the influence of skeletal implants on incidence of infection challenged by *S. aureus, S. epidermidis*, and *Escherichia coli*. The results showed that all of the implants (including stainless steel alloy, cobalt-chromium alloy, high density polyethylene, prepolymerized polymethylmethacrylate, and polymethylmethacrylate polymerized *in vivo*) were significantly more likely than the controls (medullary reaming only, without implantation and bacterial challenge) to be associated with infection, and polymethylmethacrylate polymerized *in vivo* was found to be significantly more likely than all other implants to be associated with *S. epidermidis* infection. In another study, the rabbit tibia was plated with steel or titanium plate. Under otherwise identical experimental conditions the rate of infection for steel plates (75%) was significantly higher than that for titanium plates (35%).[25]

Using a rabbit model it was found that the physical configurations of intramedullary nails had significant effect on the implant infection rate. A much higher infection rate was in the group using slotted nails (59%) compared to the group of solid nails (27%).[27]

The use of fibrinolytic agents to coat steel wires to decrease infection was reported recently.[30] In this study both the heparin coated and urokinase coated wires exhibited significantly decreased infection rates compared with uncoated wire. The mechanism of this effect may be the inhibition of bacterial adherence by heparin and fibrinolytic agents.

In the authors' laboratory, it has been demonstrated that albumin coating on titanium surface inhibited bacterial adhesion by 90%.[35] Then, coated (albumin) and uncoated titanium implants were exposed to a suspension of *S. epidermidis* prior to implantation into the lateral femoral condyle of the rabbit. According to the results, animals with albumin coated implants had a much lower infection rate (27%) than those with uncoated implants (62%). This finding may represent a new method for preventing prosthetic infection.[23]

E. Effect of Infection on Biomaterial Surfaces

It has been noted *in vitro* that implant-centered infection has certain effects on the behavior of prosthetic material, such as the corrosion of stainless steel surface,[36] or the destruction of hydroxyapatite coating.[37] Kieswetter and colleagues further studied the destructive effect of *S. aureus, S. epidermidis* or *Proteus* on the integrity of hydroxyapatite coating using an animal model, which consists of subcutaneous implantation of prosthetic materials in hamsters.[38] The results showed that significant destruction of HA coating can occur due to the growth of bacteria. Damage of HA coated implants appeared to be more severe *in vitro* than *in vivo*.

V. CONCLUDING REMARKS

Large amounts of research work have been done with great achievements in understanding the mechanisms of bacterial adhesion and prosthetic infection. Because of the potential tragic results and the large number of prosthetic procedures, prosthetic infection still remains a major challenge to biomedical researchers. For the complex nature of bacteria and their sophisticated interaction with biomaterials and host tissues, the various laboratory methods and animal models are very important and remain the major approach in solving this problem.

ACKNOWLEDGMENTS

The work in this review was supported in part by grants from Arthritis Foundation, Medical University of South Carolina, and Bioengineering Alliance of South Carolina.

REFERENCES

1. An, Y. H. and Friedman, J. R., "Prevention of infection in total joint arthroplasty," *J. Hosp. Infect.*, 33, 93, 1996.
2. Josefsson, G. and Kolmert, L., "Prophylaxis with systematic antibiotics versus gentamycin bone cement in total hip arthroplasty. A ten-year survey of 1688 hips," *Clin. Orthop.*, 292, 210, 1993.
3. Lidwell, O. M., Lowbury, E. J. L., Whyte, W., Blowers, R., Stanley, S. J., and Lowe, D., "Effect of ultraclean air in operating rooms on deep sepsis in the joint after total hip or knee replacement: A randomized study," *Br. Med. J.*, 285, 10, 1982.
4. Andrews. H,J., Arden, G. P., Hart, G. M., and Owen, J. W., "Deep infection after total hip replacement," *J. Bone Joint Surg.*, 63B, 53, 1981.
5. Nelson, J. P., Glassburn, A. R., Talbott, R. D., and McElhinney, J. P., "The effect of previous surgery, operating room environment, and preventive antibiotics on postoperative infection following total hip arthroplasty," *Clin. Orthop.*, 147, 167, 1980.
6. Fitzgerald, R. H., Nolan, D. R., Ilstrup, D. M., and van Scoy, R. E., "Deep wound sepsis following total hip arthroplasty," *J. Bone Joint Surg.*, 59A, 847, 1977.
7. Eftekhar, N. S., Kiernan, H. A., Jr., and Stinchfield, F. E., "Systemic and local complications following low friction arthroplasty of the hip joint," *Arch. Surg.*, 111, 150, 1976.
8. Cheatle, M. D., "The effect of chronic orthopaedic infection on quality of life," *Orthop. Clin. North Am.*, 22, 539, 1991.
9. Norden, C. W., Myerowitz, R. L., and Keleti, E., "Experimental osteomyelitis due to *Staphylococcus aureus* or *Pseudomonas aeruginosa*: a radiographic-pathological correlative analysis," *Br. J. Exp. Pathol.*, 61, 451, 1980.
10. Norden, C. W., "Lessons learned from animal models of osteomyelitis," *Rev. Infect. Dis.*, 10, 103, 1988.
11. Rissing, J. P., "Animal models of osteomyelitis. Knowledge, hypothesis, and speculation," *Infect. Dis. Clin. North Am.*, 4, 377, 1990.
12. Christensen, G. D., Simpson, W. A., Bisno, A. L., and Beachey, E. H., "Experimental foreign body infections in mice challenged with slime-producing *Staphylococcus epidermidis*," *Infect. Immun.*, 40, 407, 1983.
13. Espersen, F., Wilkinson, B. J., Gahrn-Hansen, B., Rosdahl, V. T., and Skinhoj, P., "Experimental foreign body infection in mice," *J. Antimicrob. Chemother.*, 31 (Supp. D), 103, 1993.
14. Gallimore, B., Gagnon, R. F., Subang, R., and Richards, G. K., "Natural history of chronic *Staphylococcus epidermidis* foreign body infection in a mouse model," *J. Infect. Dis.*, 164, 1220, 1991.
15. Mayberry-Carson, K. J., Tober-Meyer, B., Smith, J. K., Lambe, D. W., Jr., and Costerton, J. W., "Bacterial adherence and glycocalyx formation in osteomyelitis experimentally induced with *Staphylococcus aureus*," *Infect. Immun.*, 43, 825, 1984.
16. Mayberry-Carson, K. J., Tober-Meyer, B., Gill, L. R., Lambe, D. W., and Mayberry, W. R., "Effect of ciprofloxacin on subcutaneous abscesses induced with *Staphylococcus epidermidis* and a foreign body implant in the mouse," *Microbios*, 54, 45, 1988.
17. Mayberry-Carson, K. J., Tober-Meyer, B., Gill, L. R., Lambe, Jr. D. W., and Costerton, J. W., "Osteomyelitis experimentally induced with *Bacteroides thetaiotaomicron* and *Staphylococcus epidermidis*. Influence of a foreign-body implant," *Clin. Orthop.*, 280, 289, 1992.
18. Varma, S., Ferguson, H. L., Breen, H., and Lumb, W. V., "Comparison of seven suture materials in infected wounds — An experimental study," *J. Surg. Res.*, 17, 165, 1974.
19. Zimmerli, W., Waldvogel, F. A., Vaudaux, P., and Nydegger, U. E. N., "Pathogenesis of foreign body infection: description and characteristics of an animal mode," *J. Infect. Dis.*, 46, 487, 1982.
20. Belmatoug, N., Cremieux, A. C., Bleton, R., et al., "A new model of experimental prosthetic joint infection due to methicillin-resistant *Staphylococcus aureus*: a microbiologic, histopathologic, and magnetic resonance imaging characterization," *J. Infect. Dis.*, 174, 414, 1996.
21. Southwood, R. T., Rice, J. L., McDonald, P. J., Hakendorf, P. H., and Rozenbilds, M. A., "Infection in experimental hip arthroplasties," *J. Bone Joint Surg.*, 67B, 229, 1985.
22. Blomgren, G. "Hematogenous infection of total joint replacement," *Acta Orthop. Scand. Suppl.*, 187, 7, 1981.

23. An, Y. H., Bradley, J., Powers, D. L., and Friedman, R. J., "An *in vivo* study of preventing prosthetic infection using crosslinked albumin coating," *J. Bone Joint Surg.,* 79B, 816, 1997.
24. Isiklar, Z. U., Darouiche, R. O., Landon, G. C., and Beck, T., "Efficacy of antibiotics alone for orthopaedic device related infections," *Clin. Orthop.,* 332, 184, 1996.
25. Arens, S., Schlegel, U., Printzen, G., Ziegler, W. J., Perren, S. M., and Hansis, M., "Influence of materials for fixation implants on local infection. An experimental study of steel versus titanium DCP in rabbits," *J. Bone Joint Surg.,* 78B, 647, 1996.
26. Sanzén, L. and Linder, L., "Infection adjacent to titanium and bone cement implants: an experimental study in rabbits," *Biomaterials,* 16, 1273, 1995.
27. Melcher, G. A., Claudi, B., Schlegel, U., Perren, S. M., Printzen, G., and Munzinger, J., "Influence of type of medullary nail on the development of local infection. An experimental study of solid and slotted nails in rabbits," *J. Bone Joint Surg.,* 76B, 955, 1994.
28. Petty, W., Spanier, S., Shuster, J. J., and Silverthoene, C., "The influence of skeletal implants on incidence of infection," *J. Bone Joint Surg.,* 67A, 1236, 1985.
29. Petty, W., Spanier, S., and Shuster, J. J., "Prevention of infection after total joint replacement. Experiments with a canine model," *J. Bone Joint Surg.,* 70A, 536, 1988.
30. Nakamoto, D. A., Haaga, J. R., Bove, P., Merritt, K., and Rowland, D. Y., "Use of fibrinolytic agents to coat wire implants to decrease infection. An animal model," *Invest. Radiol.,* 30, 341, 1995.
31. Chang, C. C., Merritt, K., "Infection at the site of implant materials with and without preadhered bacteria," *J. Orthop. Res.,* 12, 526, 1994.
32. Buret, A., Ward, K. H., Olson, M. E., and Costerton, J. W., "An *in vivo* model to study the pathobiology of infectious biofilms on biomaterial surfaces," *J. Biomed. Mater. Res.,* 25, 865, 1991.
33. Smith, M. M., Vasseur, P. B., and Saunders, H. M., "Bacterial growth associated with metallic implants in dogs," *J. Am. Vet. Med. Assoc.,* 195, 765, 1989.
34. Merritt, K., Shafer, J. W., and Brown, S. A., "Implant site infection rates with porous and dense materials," *J. Biomed. Mater. Res.,* 13, 101, 1979.
35. An, Y. H., Stuart, G. W., McDowell, S. J., McDaniel, S. E., Kang, Q., and Friedman, R. J., "Prevention of bacterial adherence to implant surfaces with a cross-linked albumin coating *in vitro*," *J. Orthop. Res.,* 14, 846, 1996.
36. Merritt, K., Brown, S. A., Payer, J. H., and Ryerson, D. H., "Influence of bacteria on corrosion of metallic biomaterials," *Trans. Soc. Biomater.,* 14, 106, 1991.
37. Verheyen, C. C. P. M., Dhert, W. J. A., Petit, P. L. C., Rozing, P. M., and de Groot, K., "*In vitro* study on the integrity of a hydroxyapatite coating when challenged with staphylococci," *J. Biomed. Mater. Res.,* 27, 775, 1993.
38. Kieswetter, K., Merritt, K., and Myers, R., "Effects of infection on hydroxyapatite coating," *Trans. Soc. Biomater.,* 16, 220, 1992.
39. Zimmerli, W., "Experimental models in the investigation of device-related infections," *J. Antimicrob. Chemother.,* 31(Suppl D), 97, 1993.

Part VI

Animal Models for the Study of Ligaments and Tendons

24 Animal Models of Ligament Repair

Jason J. McDougall and Robert C. Bray

CONTENTS

I. Introduction 461
II. Species Commonly Used in Ligament Research 462
    A. Dogs 462
    B. Rabbits..462
    C. Goats and Sheep...462
    D. Rats...463
III. An Overview of Normal Ligament Biology...463
IV. Specific Models of Ligament Healing 464
    A. Ligament Division..464
        1. Mechanical Rupture 464
        2. Surgical Transection 465
        3. Microtrauma by Proxy 466
    B. Ligament Reconstruction and Replacement 467
    C. Other Models..467
V. Evaluation of Healing Responses 467
    A. Histology 468
    B. Biochemistry...468
    C. Biomechanics 469
    D. Vascular Physiology 471
VI. Future Directions...473
References 473

I. INTRODUCTION

Once thought of as inert, vestigial structures, the ligaments of diarthroidal joints are now recognized as complex organs which play a vital role in articular physiology and in the maintenance of joint homeostasis. In addition to a crucial stabilizing function, ligaments have proprioceptive and nociceptive capacities as well as providing an efficient means of transmitting loads between bones with the aim of permitting dynamic and fluid movement. It becomes apparent, therefore, that if these tissues are injured and do not repair effectively, then normal locomotion and future joint structural integrity are compromised. Indeed research has shown that rupture of knee joint ligaments leads to instability with concomitant abnormal loading patterns.[1–3] Over time these abnormal loads could cause degeneration of joint tissues ultimately leading to the development of osteoarthritis.[4–9]

The study of ligament injury and subsequent repair in patients is problematic due to the complex nature of joint trauma. Activity-related injuries rarely involve damage to an isolated ligament but more often result in a number of ligaments being injured with an assortment of insults occurring in other articular tissues. Hence, when investigating ligament repair, it is prudent to use animal models so that the type and degree of ligament injury can be controlled and the confounding influence of extraneous healing responses can be excluded. While vital for the investigation of ligament biology and function, caution must always be observed when interpreting the results of animal research since experimental responses may be species and ligament specific and, therefore, may not always be representative of the human condition. This limitation aside, it is hoped that by repeating experiments in a variety of different species basic trends in joint physiology and pathophysiology will eventually be uncovered. This approach to animal research would render it more persuasive when extrapolating to human conditions.

II. SPECIES COMMONLY USED IN LIGAMENT RESEARCH

The choice of species used in research is usually dictated by non-scientific factors such as cost and the ability of the institution to house certain types of animals. While these are often valid restrictions when choosing an animal model, ideally the investigator should attempt to select the species which is most relevant to the questions posed. Limiting a study to only "tried and tested" models of healing is often unproductive since a result found in one species does not necessarily translate to the same result occurring in all species. A number of different animals have been used in the study of ligament repair of which the most frequently used are outlined here along with their relative advantages and disadvantages.

A. Dogs

Some of the earliest work on ligament healing was performed on dogs[10-12] and this species remains one of the most popular animal models to date.[13] The reason for their prevalence in this area of research is mainly due to the fact that dogs are easy to handle and are generally receptive to various exercise regimens. In addition, since an extensive database has been established from work performed in the veterinary field, researchers have a wealth of published information on which to call. The downside of using dogs in ligament research is the obvious emotive issues involved and the consequent restrictions placed on the investigator by the host institution.

B. Rabbits

Rabbits are widely used in biomedical research due to their docile nature and relatively inexpensive purchasing cost and upkeep. The biochemical and functional properties of rabbit knee ligaments have been well documented (For reviews see references 14-16) and, therefore, this model provides a strong scientific base from which to study ligament healing. The rabbit lends itself to medial collateral ligament (MCL) repair studies since this ligament is well developed and easily accessible in this species. In contrast, rabbit intraarticular ligaments are small and fairly inaccessible making the study of cruciate ligament healing difficult in this particular animal. Several laboratory strains are available; however, it should be noted that some such as the NZW rabbits are prone to obesity during long-term caging and the loads on their joints may be abnormally high.

C. Goats and Sheep

The larger joint sizes in these animals mean that they are ideally suited for cruciate ligament healing studies. Many investigators are now using sheep and goats to assess cruciate ligament reconstructions[17-21] and ligament prostheses.[22-25] Another possible advantage of using these large animals is that the degree of flexion in their stifle joints is less than in rabbits and dogs suggesting a closer analogy to human knee joints. On the downside, the practical and economical considerations are evident with the animals requiring special housing facilities and elaborate post-operative care.

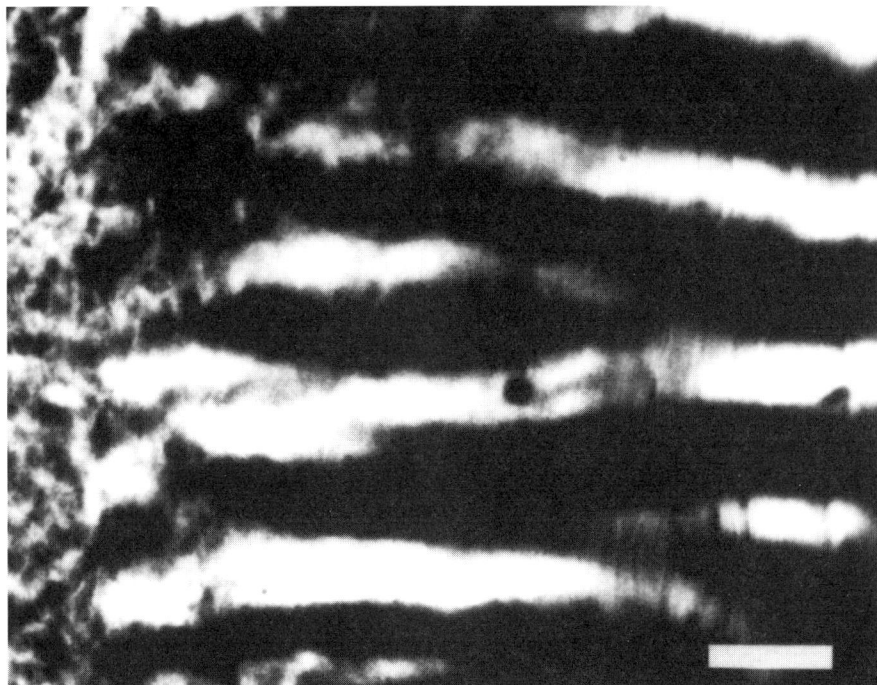

FIGURE 1. Photomicrograph of rat medial collateral ligament taken using polarized light. Longitudinally, the collagen fibres appear to have a sinusoidal repeating pattern often referred to as "crimp." Scale bar = 50μm.

D. Rats

Despite their low cost and ready availability, rats are not generally used in ligament research. In addition to the small size of their joints and the practical encumbrances size imparts, the rat musculoskeletal system continually grows throughout its lifetime such that the bone and articular soft tissues are in a constant state of remodeling. Hence, interpretation of functional and biological outcomes in this species is unreliable.

III. AN OVERVIEW OF NORMAL LIGAMENT BIOLOGY

In order to appreciate ligament healing models more fully, a basic understanding of normal ligament structure and function must first be realized. As defined by *Gray's Anatomy*,[26] ligaments are bundles of white fibrous tissue serving to connect together the articular extremities of bones. Structurally, these tissues are made-up of bundles of collagen fibers which are arranged parallel to the longitudinal axis of the ligament and are mounted in a coarse tissue matrix. Polarized light microscopy reveals a regular sinusoidal pattern along the length of these collagen bundles, a phenomenon which has been described as "crimping" (Figure 1). The significance of this concertina-like pattern is unclear but it is thought to act as a mechanism to accommodate forces, allowing loads to be dissipated before structural damage occurs.[14,27]

Biochemical analyses show that the major component of connective tissues is water accounting for some 70% of the wet weight of the tissue (Table 1). The function of water in ligaments is somewhat moot; however, it has been suggested to be necessary for the intracellular transportation of ions as well contributing to the viscoelastic behavior of the tissue.[28–30] The next most abundant component is collagen, of which type I collagen is the most ubiquitous. Collagen provides the structural mainstay of ligaments by providing a source of stability and mechanical integrity to the tissue. The remaining constituents (elastin, fibronectin and proteoglycans among others) are all

TABLE 1
The Relative Biochemical Constituents of Normal and Scarred Rabbit Medial Collateral Ligament Tissue

Component	Normal ligament proportion (%)	Scar ligament proportion (%)
Water	70	75
Type I collagen	20	12
Other collagen types	3–5	5
Elastin, fibronectin, and glycoproteins	2–4	6
Proteoglycans	<1	1–2

matrix elements whose functions continue to be defined. Proteoglycans are thought to act as "shock absorbers" by altering ligament viscoelasticity via their water binding properties, while elastin may be responsible for the elastic properties of ligaments.[14,15] Further research is required to elucidate the exact role of these substances in normal ligament function so that their levels and chemical composition can be optimized during post-traumatic repair.

IV. SPECIFIC MODELS OF LIGAMENT HEALING

When studying ligament repair *in vivo*, the tissue in question must first be subjected to a certain degree of trauma before it can heal. Therefore, the models used in this area of research also tend to be used in other investigations of soft tissue injury. The limitation in variety and scope of the major models available reflects the rudimentary nature of this type of research and the need for more innovative concepts so that the intricacies of ligament healing can be probed more deeply.

A. Ligament Division

When ligaments fail, they do so by varying degrees of tearing of their collagen bundles. In microfailure the ligament undergoes only partial tearing, often referred to in layman's terms as a sprain, while greater loads result in catastrophic failure in which the entire ligament is severed. Hence, models in which the ligament is divided are often a good approximation to clinical injuries and as such have formed the basis of ligament healing studies. Some of the models of ligament division and repair are summarized in Figure 2.

1. Mechanical Rupture

The seminal work on ligament injuries attempted to replicate the clinical manifestation of trauma by manually subjecting the joint to forces which were in excess of normal physiological limits. Some of the earliest reports of injury induction involved violent abduction of the knee.[31–33] This rather crude injury model resulted in ligament tears which tended to occur near the entheses although the extent and specific location of the rupture varied.[34]

Later studies attempted to refine ligament ruptures by controlling for the site and level of tissue damage with minimal inclusion of other articular structures. Frank et al.[35] produced a "garrotting" model in which a 3-0 braided steel suture was passed under the MCL and then pulled in a firm, upward movement perpendicular to the joint. The resultant level of tissue damage was consistent between animals and was found to be restricted to the site of rupture in the ligament midsubstance. Variations of this technique have included passing a pair of hemostats[36] or a 2.5-mm diameter

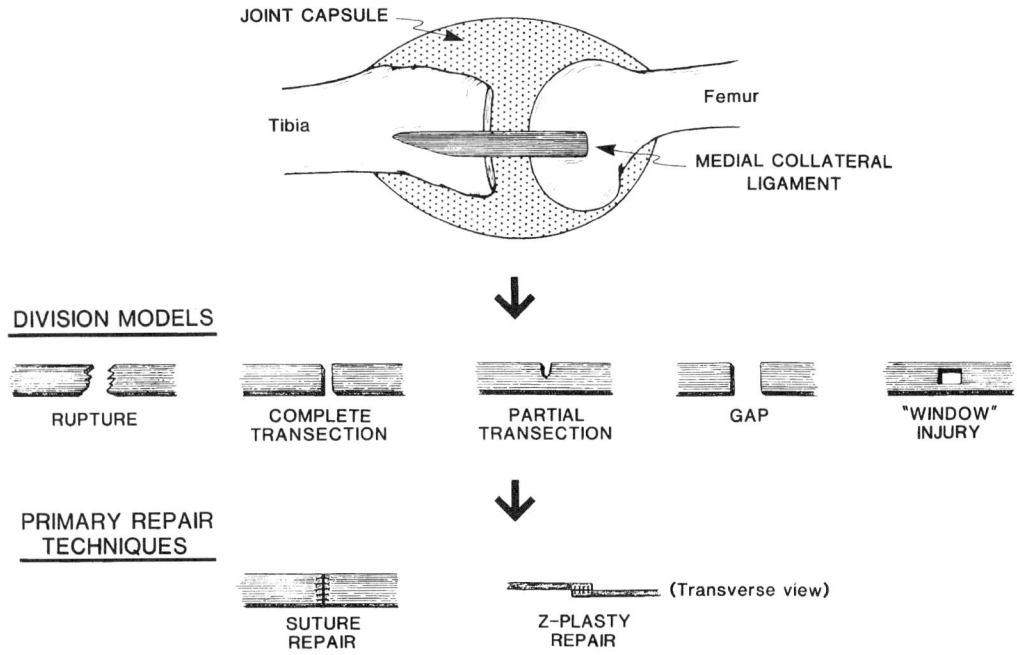

FIGURE 2. Common models of division and repair.

stainless steel rod[37] under the ligament before sharply pulling the implement medially with such force so as to cause a midsubstance rupture.

The morphology of this injury model is similar to the clinical presentation of failure in that the torn ligament ends are frayed and irregular. Although clinically relevant, these injury models still incorporate a significant number of uncontrolled variables such as mechanical disturbance of the ligament in areas remote from the injury site. This inherent complexity of all rupture models means that their use in laboratory investigations should be viewed with a degree of skepticism because of the difficulty in interpreting their effects.

2. Surgical Transection

To obviate some of the concerns associated with rupture models, a less invasive method of injury has emerged, viz., surgical transection of the ligament with a scalpel. Complete transection involves a simple transverse midsubstance division of the ligament resulting in a uniform, regular cut.[10,11] It was originally thought that effective ligament healing could only occur when, as in the rupture model, the separated ends were frayed providing a greater surface area for healing to take place. This belief cast doubt on surgical transection as a viable model of ligament repair. However, Chimich et al.[38] showed that the histological and mechanical properties of a healing MCL were not affected by the configuration of the torn ligament ends, i.e., it didn't matter whether the ends were sheer as in surgical transection, or whether they had the "mopped" appearance of having been ruptured.

It has been postulated that apposition of torn ends may be a prerequisite for the functional recovery of healing ligaments by minimizing the development of potentially inferior scar tissue and maximizing proliferation of dense collagenous material within the healing zone.[12,39,40] To examine this proposition, two distinctive injury-repair models were created in the rabbit MCL.[41] The first was a Z-plasty apposition model in which the ligament was transected in a sagittal plane and the ends overlapped before suturing them together. This procedure maximized the contact area between the two sections of ligament thereby minimizing the amount of scarring. The second model aimed to favor scar production by maintaining separation between the MCL ends. This was achieved by excising a 4-mm segment of tissue from the ligament midsubstance leaving a discrete gap. After 40 weeks of healing, both injury types had remodeled culminating in the formation of a scar, the level of which was, as predicted, greater in the gap model than in the contact state. In addition, gap scars could be distinguished from old ligament whereas Z-plasty scars presented as a mixture of original ligament and neoligamentous scar tissue. Thus gap injury provides a unique model for the study of pure ligament scar without the confounding influence of old tissues infiltrating from surrounding regions.

Numerous other techniques have been developed to promote apposition of ligament ends especially in the anterior cruciate ligament (ACL) where the ligamentous stumps rarely unite through scar tissue and, consequently, atrophy. One of the simplest methods is by primary repair where the ligament ends are brought together by suturing.[42] Although several different approaches have been examined, the ligament ends refuse to bond and healing is eluded. As found in tendon repair studies,[43,44] possible reasons for this lack of ACL healing may be that the type of suture material and stitch pattern used were unsuitable for this tissue. Alternatively, it may be that the forces applied to the ligament during recovery were in excess of what the sutures could handle and ligament stump retraction was inevitable. Therefore, a specialized gap model was created whereby only a mid portion of tissue was removed to produce a window-like injury.[42] This procedure had the advantage of creating a restricted gap injury while retaining mechanical stability. Even though all of these countermeasures were considered, only 5% of these ACLs showed a positive healing response.

Partial ruptures account for nearly a third of all human ACL injuries leading researchers to develop a hemitransection model of injury in which only a limited number of collagen bundles are cut. Evidence exists to suggest that ACL healing with improved functionality can occur in these models although the recovery period is somewhat protracted compared to other ligaments.[12,45,46] In contrast, other studies have been unable to find any sort of healing response in this model even after trying to consolidate the laceration through primary repair.[42,47] The outcome of these studies only reaffirms that a viable and reproducible model of ACL healing still evades the research community.

3. Microtrauma by Proxy

Following ligament rupture, the affected joint becomes lax and unstable leading to an alteration in joint mechanics.[1-3] The forces which are then placed on the joint have to be redistributed throughout the remaining uninjured ligaments leading to the generation of abnormally high loads in these structures. In the absence of any sort of intervention or attempts at repair, the uninjured ligaments may themselves undergo low level failure thereby exacerbating joint instability.[48,49] One of the commonly used models to study secondary microtrauma is to assess the properties of the MCL in an ACL deficient knee. Normally associated with osteoarthritis research, transection of the ACL also provides a unique means of examining MCL integrity in a mechanically unstable environment. One of the first attempts to divide the ACL was performed by Pond and Nuki[5] and involved a blind stab incision to the anterior surface of the joint. A scalpel blade was forced between the femoral and tibial condyles into the joint space in the hope of completely severing the ACL. One of the main drawbacks to this technique is the uncertainty of whether the entire ligament has been transected, not to mention the additional trauma exacted on the patellar ligament and infrapatellar fat pad. A less invasive and more reliable approach to ACL transection is by isolating the ligament following an arthrotomy. This technique entails making a longitudinal incision along the subpatellar region of the joint, reflecting the infrapatellar fat pad, isolating the ACL and then surgically transecting it with a scalpel blade before finally closing the wound with sutures. In this way the investigator may be confident that the whole ligament has been severed with minimal disruption to other articular tissues.

B. LIGAMENT RECONSTRUCTION AND REPLACEMENT

In light of the ineffectiveness of apposition to promote functional recovery of injured ACLs, alternative treatment regimens have been sought. The philosophy that "if it doesn't heal replace it" has gained popularity recently, so much so that reconstruction or replacement of the torn ACL with various biological or prosthetic materials has become common practice. Autografting uses tissues such as patellar ligament, semitendinosus tendon, quadriceps tendon or even meniscus to reconstruct torn cruciates. Since the graft material is harvested from the recipient animal, then there is a certain level of morbidity associated with the donor site. Allografts obviate this problem by removing these tissues from an independent donor animal; however, the disadvantages of this approach include the risk of infection and possible rejection of the graft. Preservation of the substitute ligament following removal from the host has also plagued the clinical use of allografts. Animal studies have shown, however, that freeze-drying the graft for up to a year does not appear to have a detrimental effect on the mechanical viability of the transplanted tissue.[50-54]

Synthetic ligaments have been used to investigate the effect of supplementary support during cruciate ligament repair. One type of artificial ligament is the augmentation device which was designed to allow joint loads to be shared during autologous ligament healing or graft establishment.[55,56] The protective effect the prosthesis imparts is only supposed to be temporary, with it carrying the greatest amount of load immediately after implantation and then gradually transferring load onto the repaired tissue so that it may eventually become a more functionally competent structure.

C. OTHER MODELS

Injured joints are conventionally immobilized to protect the repaired or reconstructed ligament from potentially damaging forces. Immobilization of experimental animals is usually carried out by fitting either a cast, brace or splint to the recovering limb. Although not a model of ligament repair *per se*, joint restraint in conjunction with one of the injury/repair models enables researchers to test the relative benefits of immobilization and remobilization on the postoperative care of healing ligaments.

To our knowledge, *in vitro* models of ligament repair are scarce, reflecting a glaring deficiency in this area of research. One of the few models available was developed by Witkowski et al.[57] to study the migration of ligament cells into a manufactured wound. The technique involves culturing human MCL or ACL fibroblast cells *in vitro* and then streaking them with an inoculating loop thus creating an acellular region. Different mediators can then be assessed for their healing properties under an array of conditions by adding them to the culture medium and monitoring their effect on the migration of cells into the wound.

V. EVALUATION OF HEALING RESPONSES

In order to appraise the effectiveness of the various repair techniques and the influence of the circumstances in which they occur, a multitude of different protocols are employed which span a myriad of disciplines. By performing the evaluations in normal as well as healing ligaments, the level of tissue recovery can be ascertained. This section describes some of the assessment procedures encountered and highlights some of the differences found between a ligament which shows a good healing response (MCL) and one that does not heal effectively (ACL).

A. HISTOLOGY

Ligament structure may be characterized by examining sections of the tissue histologically with haematoxylin and eosin stain. The appearance of ligament cells and collagen organization/size provide a strong indication as to the status of the tissue with respect to metabolic and mechanical parameters. As mentioned earlier, when viewed under polarized light, normal ligaments possess a "crimped" appearance resulting from the coherent organization of collagen fibers. In the scar region of a gap injured MCL, this repeating pattern is lost and the collagen fibers appear disorganized (Figure 3).

In transected ACLs, histological techniques failed to find any sign of healing even following primary repair of the lacerated ligament ends.[42] With the advent of ACL reconstruction, attention has turned to the morphological analysis of healing grafts. In experiments using the patellar ligament autograft, it was found that there was an initial phase of avascular necrosis followed by radical remodeling of the tissue such that the graft eventually had the histological appearance of a normal ACL.[58-60]

Immunohistochemistry has been used to detect neuropeptidergic nerve fibers in normal and healing MCLs.[61,62] Neuropeptides such as substance P (SP) and calcitonin gene-related peptide (CGRP) are pro-inflammatory agents released from the terminals of afferent nerves and which cause vasodilatation and protein extravasation. In addition to their haemodynamic effects, neuropeptides possess trophic properties causing neovascularization and tissue remodeling by promoting fibroblast differentiation. Experiments have shown[62] that six weeks after gap injury, SP and CGRP immunoreactivity is increased in the healing scar tissue although the nerves appear truncated and tangled (Figure 3). Increased presence of peptidergic nerves in the MCL may relate to the sound healing potential of this structure.

B. BIOCHEMISTRY

Organ and cell culture has been used to investigate the biochemical properties of normal and remodeled ligament tissue. The process involves creating an injury to the ligament *in vivo* and then incubating the tissue *in vitro* so that biochemical analyses can be carried out. Examinations usually center on the remodeling processes associated with MCL scar formation, i.e., matrix component synthesis and the formation of various proteases responsible for the removal or structural alteration of these matrix constituents. In ligaments, type I collagen is believed to have a significant role in the biological and biomechanical properties of the tissue[14] and as such a significant amount of biochemical research has been devoted to the mechanisms which control its production. It is thought that the functional outcome of healing ligaments may be improved by promoting the synthesis of the appropriate collagen type and optimizing its organization within the matrix. The specific biochemical parameters which are assessed, therefore, include collagen typing, crosslink analysis, fibril diameter and ground substance content.

Biochemical analysis of MCL scar tissue shows it to have an inferior composition compared to normal ligament, especially with respect to lower type I collagen levels (Table 1). The increased water and elastin content of ligament scars is thought to contribute to the decreased stiffness of the tissue (see below) while collagen fibril diameter does not appear to be applicable since size is unaltered in healing connective tissue.[63]

The effect of injury on ACL biochemistry is still poorly understood; however, some information relating to this subject has started to emerge. Collagenase activity has been found to increase by 82% in transected rabbit ACLs compared to controls[64] providing a possible explanation for the resorption of torn ACL ends. The fact that ACL healing is limited even after apposition of the ligament stumps suggests that there is something intrinsically different about the make-up of this tissue which inhibits remodeling and scar production.

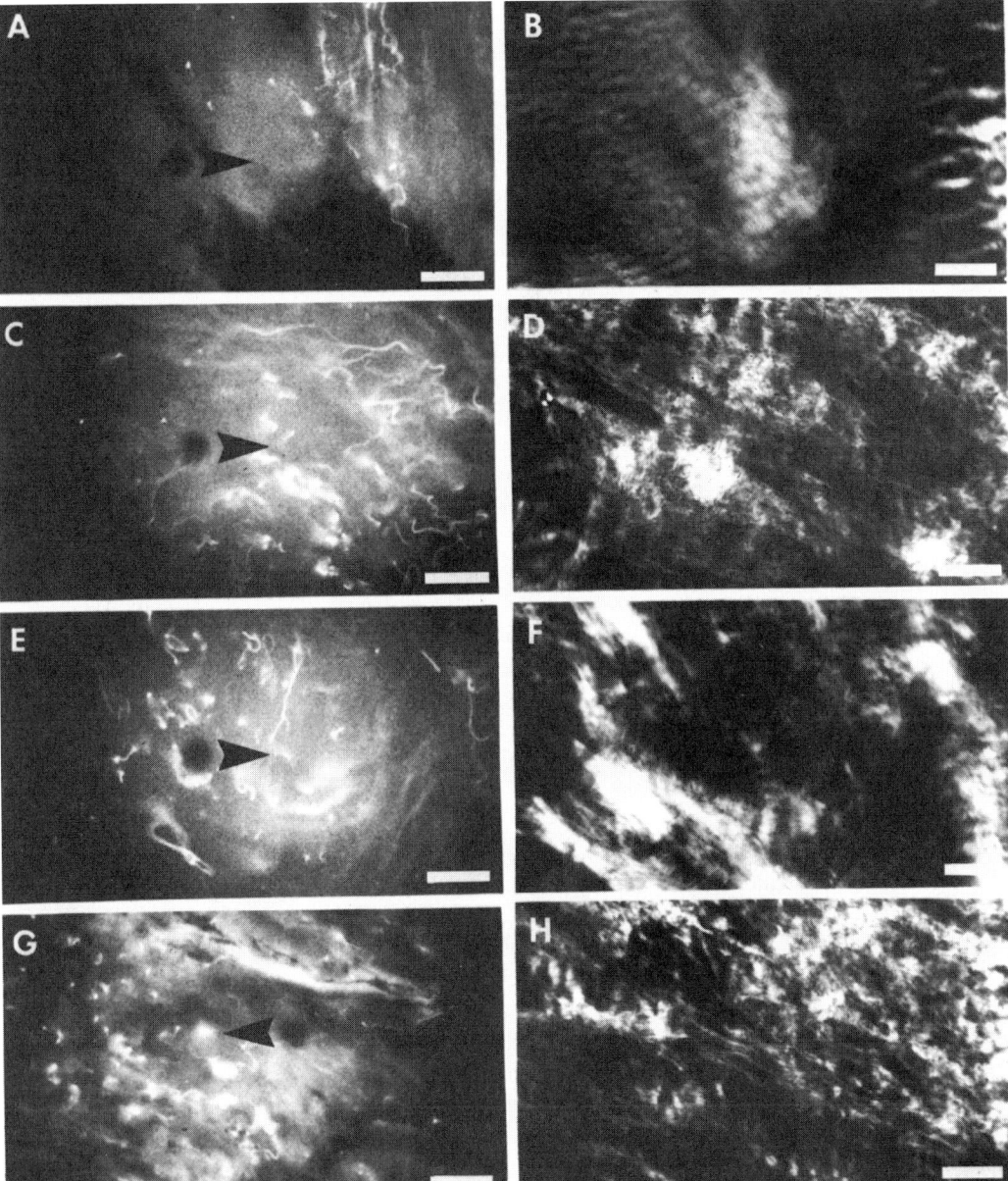

FIGURE 3. Fluorescence micrographs of CGRP- (A,C,E) and SP- (G) like immunoreactivity in the adult rabbit medial collateral ligament six weeks after gap injury. Micrographs B, D, F, and H are corresponding regions taken using polarized light. Note that the "crimping" effect has been replaced by disorganized, amorphous scar tissue. In the immunofluorescent images, the black dots represent suture markers placed to delineate the extremities of the gap injury and the arrows point to the healing zone. The peptidergic nerves are abundant and appear as a tangled mesh of truncated, varicose fibres in the scar tissue. Scale bar = 50μm. (From McDougall et al., *Anat. Rec.*, 248, 29, 1997. With permission.)

C. Biomechanics

Functional assessment is the most widely used measure of post-traumatic recovery in connective tissue research. Innumerable protocols have been developed to test everything from the material properties of ligaments to the mechanical characteristics of bone-ligament-bone complexes. To date, the majority of ligament biomechanics is performed *ex vivo* and, therefore, this evaluation method has little physiological significance. To account for this limitation, researchers have spent many years devising standardized protocols which attempt to control for biological factors such as temperature, specimen orientation and selection of suspension medium. This section attempts to outline some of the more pertinent findings but for an extensive coverage of this subject the reader is directed elsewhere.[16,65,66]

The straightforward approach of merely mounting a section of ligament in a materials testing device is problematic due to the difficulty of securing the ligament ends without altering the properties of the tissue. Furthermore, one is never quite sure whether the clamp-tissue interface is being tested rather than the tissue itself. A more appropriate action is to leave the ligament attached to the bones which can then be clamped into the apparatus. In addition to improved clasping of the tissue, this bone-ligament-bone complex provides a structural unit for testing under tensile loading conditions. Once mounted, load is applied to the ligament and the effect on mechanical behaviors such as ligament stress and strain can be determined. Mathematical analyses of load-deformation plots provides a host of information relating to the functional properties of the ligament. The manner in which the load is applied to the structure (e.g., loading rate, loading level, or whether it is cyclic or not) is also of utmost importance when interpreting mechanical behavior. With regard to developing a sense of tissue biomechanics, cyclic low-load testing may provide the most physiological representation of ligament function under normal conditions.

Load-deformation curves carried out on healing MCLs have shown that the structural properties of bone-healing ligament-bone complexes are initially inferior to control but show progressive improvement at 6 and 12 weeks post-injury.[15,16,66] Interestingly, the material properties of a healing MCL, of which tissue stiffness is an example, were consistently reduced compared to normal even at 14 weeks. This means that under these conditions the bone-healing ligament-bone complex undergoes a certain degree of functional recovery despite a decline in ligament substance integrity. The reason for this apparent recovery paradox is unclear but may be related to compensatory changes occurring at the entheses.

The effect of treatment on injured MCLs has also been studied extensively. Z-plasty repair of torn ligament ends resulted in a significant increase in ligament strength compared to untreated gap injuries[41] reaffirming the benefit of apposition in ligament healing. Immobilization of an injured joint also has significant repercussions on the composition and material properties of neoligamentous scar tissue. Experiments performed on severely injured rabbit knees, i.e., in which there is a composite MCL gap/ACL transection injury, showed that the greater loads associated with joint mobility produce an increase in MCL cross-sectional area; however, load-deformation characteristics of the ligament were the same as in controls.[67] These findings are similar to other reports in which only the MCL was transected.[68,69] The structural properties of healing MCLs which had been subjected to a primary repair with immobilization regimen were found to be poorer than those of both control and untreated ligaments. Taken together these data show that immobilization of injured ligaments is obstructive to the long-term recovery of normal MCL function and that suture repair may only be effective when some modicum of joint stability is maintained via an intact ACL.

Three years after partial transection of the goat ACL, the stiffness and tensile strength of the structure were not significantly different from control values suggesting that functional recovery may be possible in this tissue after all.[46] The reason given for this apparent healing response was that since only the posterolateral bundle was severed, the anteromedial fragment acted as an internal splint providing non-surgical apposition of the ligament stumps. This finding is interesting since numerous attempts at ACL apposition in the complete transection model failed to generate repair responses in this tissue. It is possible that the introduction of ligatures at the site of injury could restrict blood flow to the area thereby preventing the delivery of necessary healing mediators.

When mechanically testing reconstructed ACLs, factors such as graft position, fixation protocol, initial graft tension and postoperative rehabilitative regimens must all be considered. Ultimate load and linear stiffness of newly implanted patellar ligament autografts in dogs have been shown to be about 10% of control bone-ACL-bone complexes.[56,70] Although high-load structural properties of autografts continue to improve with time, complete recovery is never attained even after three years.[71] The use of allograft tissue for ACL reconstruction also exhibits poor functional recovery in other animal knees. As alluded to earlier, cryopreservation of nonautogenous tissue does not seem to affect the material properties of the graft; however, there is some evidence to suggest that some of the techniques used to sterilize the implant may be detrimental.[72]

As an aside, it should be noted that the contralateral limb served as the control in all of the preceding studies and, hence, a possible bilateral effect of joint injury could be prejudicing these results. Reinvestigation using normal animals as controls may be required before any of these results can be fully accepted.

D. Vascular Physiology

Of all of the classic signs of soft tissue healing, none are more conspicuous than those of the microvascular system. Hyperemia, angiogenesis and increased vascular permeability are all synonymous with tissue healing which makes it all the more remarkable that this area of research has been overlooked for so long. The emergence of innovative and readily applicable techniques will only serve to enhance our understanding of joint vascular physiology providing information on one of the most exciting and scientifically pertinent areas of ligament research.

Several methods of measuring ligament blood flow have been developed including hydrogen clearance,[73] microsphere distribution[74,75] and laser Doppler perfusion imaging.[76,77] Bray et al.[75] found that MCL gap injury caused a significant rise in ligament blood flow at three and six weeks post-trauma. This hyperemia may lead to an increase in ligament hydrostatic pressure which, in conjunction with inflammatory extravasation, could explain the high water content of MCL scars and, hence, the decreased stiffness of remodeled ligaments. It follows, therefore, that if the vascular responses to ligament injury could somehow be controlled then the water content of the healing ligament could be moderated, possibly leading to a functionally more normal tissue.

McDougall et al.[77] found that the rabbit knee joint is unable to autoregulate and is wholly dependent upon extrinsic mediators to control its blood supply. Factors known to alter rabbit MCL blood flow include CGRP, adrenaline and the perivascular effects of the sympathetic nervous system.[76,77] When the joint is made unstable through ACL transection, these vasoregulatory mechanisms fail and the MCL is deprived of a balanced blood supply.[78,79] This loss of joint homeostasis may mean that the metabolic needs of the MCL are not being met which could account for the deterioration of the ligament in this model.

To date, measurement of graft revascularization and scar angiogenesis has relied upon qualitatively assessing histological sections of the relevant tissues. This approach is somewhat inadequate since no information is afforded regarding the number of new vessels or whether indeed they are patent with the existing capillary network. In our laboratory we have developed a novel technique which quantifies the volume of vessels supplying various articular structures.[80] By comparing the vascular volumes in a group of normal animals with those found following some sort of intervention, angiogenic activity would be represented as a measurable increase in tissue vascularity. The technique is based on spectrophotometrical analysis of intravascularly injected carmine red dye to give a measure of tissue vascular volume. We have used this methodology to investigate changes in rabbit knee joint vascularity following ACL transection. As shown in Figure 4, compared to sham operated controls, angiogenesis only appears to occur in the ipsilateral MCL of the unstable joint. Thus, microtrauma by proxy to the ipsilateral MCL stimulates vascular remodeling of the tissue which if extensive enough, may cause disorganization of collagen fibrils or even replace a portion of them altogether. This disruption to the collagen crimp pattern could weaken the MCL exacerbating joint instability. Interestingly, the lack of new vessels being formed in the infrapatellar fat pad or synovium may herald a possible explanation for the inadequacies of ACL healing since the bulk of the vasculature which nourishes the ACL arises from these tissues.[81]

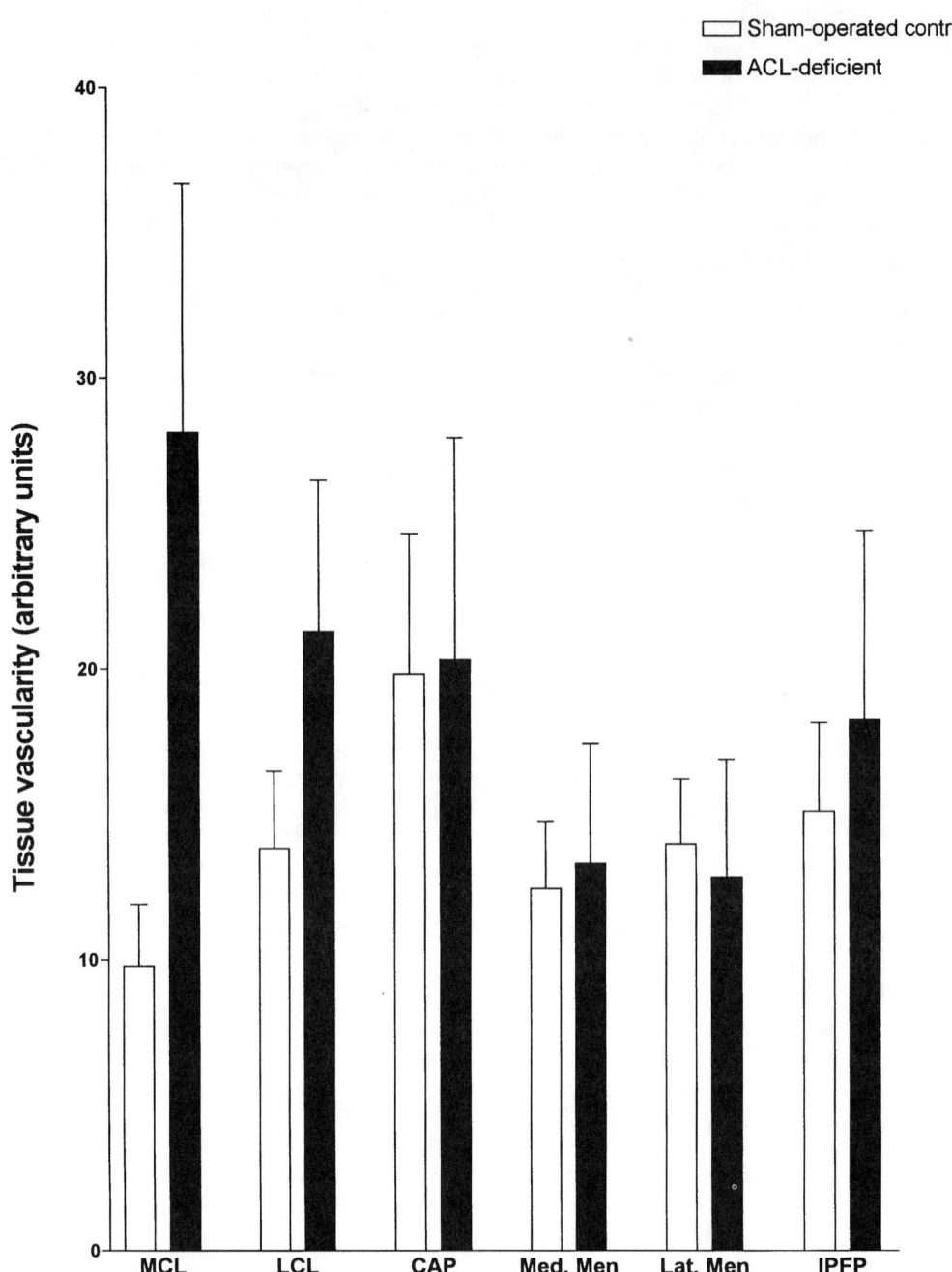

FIGURE 4. The effect of anterior cruciate ligament transection on rabbit knee joint vascularity compared to sham operated controls. The volume of vessels supplying the medial collateral ligament is significantly greater than in controls suggesting that angiogenesis has occurred in this structure as a result of microtrauma by proxy. Vascularity was not significantly altered in any of the other knee joint tissues. (Key: MCL — medial collateral ligament; LCL — lateral collateral ligament; CAP; medial capsule; Med. Men — medial meniscus; Lat. Men — lateral meniscus; IPFP — infrapatellar fat pad).

VI. FUTURE DIRECTIONS

As with all types of animal research, the study of ligament physiology and repair is only as good as the models of the human condition that they have been designed to represent. Can the reason why the MCL heals while the ACL remains latent be due to differences in biochemical composition of the two ligaments or is it related to an ineffective physiological response to injury? Even if a ligament does heal why does it never reach functional normality? The answers to these questions will only be accomplished if we advance existing models of soft tissue healing beyond where the "cut and sew" approach has failed: treatment with unique collagenases to destroy specific types of collagen; chronic administration of vasoactive or vascular remodeling drugs to alter ligament blood supply; transgenic animals which are unable to synthesize certain ligament proteins or which can promote components of abnormal morphology. These are the sorts of models we need to concentrate on if we are ever going to discover the pathways which will lead us to the Utopia of complete ligament healing.

REFERENCES

1. Butler, D. L., Noyes, F. R., and Grood, E. S., "Ligamentous restraints to anterior-posterior drawer in the human knee: A biomechanical study," *J. Bone Joint Surg.*, 62A, 259, 1980.
2. Gollehon, D. L., Torzilli, P. A., and Warren, R. F., "The role of the posterolateral and cruciate ligaments in the stability of the human knee," *J. Bone Joint Surg.*, 69A, 233, 1987.
3. Markolf, K. L., Kochan, A., and Amstutz, H. C., "Measurement of knee stiffness and laxity in patients with documented absence of the anterior cruciate ligament," *J. Bone Joint Surg.*, 66A, 242, 1984.
4. Marshall, J. L. and Olsson, S.-E., "Instability of the knee: A long term experimental study in dogs," *J. Bone Joint Surg.*, 53A, 1561, 1971.
5. Pond, M. J. and Nuki, G., "Experimentally induced osteoarthritis in the dog," *Ann. Rheum. Dis.*, 32, 387, 1973.
6. Jacobsen, K., "Osteoarthrosis following insufficiency of the cruciate ligaments in man: A clinical study," *Acta Orthop. Scand.*, 48, 520, 1977.
7. Hirshman, H. P., Daniel, D. M., and Miyasaka, K., "The fate of unoperated knee ligament injuries," in *Knee Ligaments: Structure, Function, Injury and Repair,* Daniel, D. M., Akeson, W. H., and O'Connor, J. J., Eds., Raven Press, New York, 1990, 481.
8. Brandt, K. D., Braunstein, E. M., Visco, D. M., O'Connor, B., Heck, D., and Albrecht, M., "Anterior (cranial) cruciate ligament transection in the dog: A bona fide model of osteoarthritis, not merely of cartilage injury and repair," *J. Rheumatol.*, 18, 436, 1991.
9. Brandt, K. D., Myers, S. L., Burr, D., and Albrecht, M., "Osteoarthritic changes in canine articular cartilage, subchondral bone, and synovium fifty-four weeks after transection of the anterior cruciate ligament," *Arth. Rheum.*, 34, 1560, 1991.
10. Clayton, M. L. and Weir, G. J., Jr., "Experimental investigations of ligamentous healing," *Am. J. Surg.* 98, 373, 1959.
11. O'Donoghue, D. H., Rockwood, C. A., Jr., Zaricznyj, B., and Kenyon, R., "Repair of knee ligaments in dogs. I. The lateral collateral ligament," *J. Bone Joint Surg.*, 43A, 1167, 1961.
12. O'Donoghue, D. H., Rockwood, C. A., Jr., Frank, G. R., Jack, S. C., and Kenyon, R., "Repair of the anterior cruciate ligament," *J. Bone Joint Surg.*, 48A, 503, 1966.
13. Arnoczky, S. P. and Wilson, J. W., "Experimental surgery of the skeletal system," *Meth. Anim. Exp.* 7, 67, 1986.
14. Amiel, D., Billings, E., and Akeson, W. H., "Ligament structure, chemistry, and physiology," in *Knee Ligaments: Structure, Function, Injury and Repair,* Daniel, D. M., Akeson, W. H., and O'Connor, J. J., Eds., Raven Press, New York, 1990, 77.
15. Frank, C. B., Bray, R. C., Hart, D. A., Shrive, N. G., Loitz, B. J., et al., "Soft tissue healing," in *Knee Surgery,* Fu, F. H., Harner, C. D., Vince, K. G., Eds., Baltimore, Williams & Wilkins, 1994, 189.
16. Woo, S. L.-Y., Smith, B. A., and Johnson, G. A., "Biomechanics of knee ligaments," in *Knee Surgery,* Fu, F. H., Harner, C. D., and Vince, K. G., Eds., Baltimore, Williams & Wilkins, 1994, 155.

17. Jackson, D. W., Grood, E. S., Arnoczky, S. P., Butler, D. L., and Simon, T. M., "Cruciate reconstruction using freeze-dried anterior cruciate ligament allograft and a ligament augmentation device (LAD). An experimental study in a goat model," *Am. J. Sport Med.*, 15, 528, 1987.
18. Jackson, D. W., Grood, E. S., Wilcox, P., Butler, D. L., Simon, T. M., and Holden, J. P., "The effect of processing techniques on the mechanical properties of bone-anterior cruciate ligament-bone allografts. An experimental study in goats," *Am. J. Sport Med.*, 16, 101, 1988.
19. Kasperczyk, W. J., Bosch, U., Oestern, H. J., and Tschcerne, H., "Influence of immobilization on autograft healing in the knee joint. A preliminary study in a sheep knee PCL model," *Arch. Orthop. Trauma Surg.*, 110, 158, 1991.
20. Bosch, U., Decker, B., Kasperczyk, W., Nerlich, A., Oestern, H. J., and Tscherne, H., "The relationship of mechanical properties to morphology in patellar tendon autografts after posterior cruciate ligament replacement in sheep," *J. Biomech.*, 25, 821, 1992.
21. Moeller, H. D., Bosch, U., and Decker, B., "Collagen fibril diameter distribution in patellar tendon autografts after posterior cruciate ligament reconstruction in sheep: Changes over time," *J. Anat.*, 187, 161, 1995
22. Bolton, C. W. and Bruchman, W. C., "The GORE-TEX expanded polytetrafluoroethylene prosthetic ligament," *Clin. Orthop.* 196, 159, 1985.
23. Turner, I. G. and Thomas, N. P., "Comparative analysis of four types of synthetic anterior cruciate ligament replacement in the goat: In vivo histological and mechanical findings," *Biomaterials*, 11, 321, 1990.
24. Huguet, D., Faintreny, A., Mazui, D., Daculsi, G., Passuti, N., and Agado E., "Le mouton: modele animal pour les protheses ligamentaires?" *Chirurgie*, 120, 84, 1994.
25. Durselen, L., Claes, L., Ignatius, A., and Rubenacker, S., "Comparative animal study of three ligament prostheses for the replacement of the anterior cruciate and medial collateral ligament," *Biomaterials*, 17, 977, 1996.
26. Gray, H., "The articulations," in *Anatomy of the Human Body,* 29th ed., Goss, C. M., Ed., Lea & Febiger, Philadelphia, 1973.
27. Diamant, J., Keller, A., Baer, E., Litt, M., and Arridge, R. G. C., "Collagen: Ultrastructure and its relation to mechanical properties as a function of aging," *Proc. Royal Soc. Lond.*, 180, 293, 1972.
28. Frank, C., McDonald, D., Lieber, R., and Sabitson, P., "Biochemical heterogeneity within the maturing rabbit medial collateral ligament," *Clin. Orthop.*, 236, 279, 1988.
29. Chimich, D., Shrive, N., Frank, C., Marchuk, L., and Bray, R., "Water content alters viscoelastic behaviour of the normal adolescent rabbit medial collateral ligament," *J. Biomech.*, 25, 831, 1992.
30. Haut, R. C., "The mechanical and viscoelastic properties of the anterior cruciate ligament and of ACL fascicles," in *The Anterior Cruciate Ligament: Current and Future Concepts*, Jackson, D. W., Arnoczky, S. P., Frank, C. B., Woo, S. L.-Y., and Simon, T. M., Eds., Raven Press, New York, 1993, 63.
31. Miltner, L. J. and Hu, C. H., "Experimental reproduction of joint sprains," *Proc. Soc. Exp. Biol. Med.*, 30, 883, 1933.
32. Miltner, L. J., Hu, C. H., and Fang, H. C., "Experimental joint sprain: Pathologic study," *Arch. Surg.*, 35, 234, 1937.
33. Jack, E. A., "Experimental rupture of the medial collateral ligament of the knee," *J. Bone Joint Surg.*, 32B, 396, 1950.
34. Walsh, S. and Frank, C., "Two methods of ligament injury: A morphological comparison in a rabbit model," *J. Surg. Res.*, 45, 159, 1988.
35. Frank, C., Woo, S. L.-Y., Amiel, D., Harwood, F., Gomez, M., and Akeson, W., "Medial collateral ligament healing: A multidisciplinary assessment in rabbits," *Am. J. Sport Med.*, 11, 379, 1983.
36. Piper, T. L. and Whiteside, L. A., "Early mobilization after knee ligament repair in dogs," *Clin. Orthop.*, 150, 277, 1980.
37. Weiss, J. A., Woo, S. L.-Y., Ohland, K. J., Horibe, S., and Newton, P. O., "Evaluation of a new injury model to study medial collateral healing: Primary repair versus nonoperative treatment," *J. Orthop. Res.*, 9, 516, 1991.
38. Chimich, D., Frank, C., Shrive, N., Bray, R., King, G., and McDonald, D. "No effect of mop-ending on ligament healing: Rabbit studies of severed collateral knee ligaments," *Acta Orthop. Scand.*, 64, 587, 1993.

39. O'Donoghue, D. H., Frank, G. R., Jeter, G. L., Johnson, W., Zeiders, J. W., and Kenyon, R., "Repair and reconstruction of the anterior cruciate ligament in dogs," *J. Bone Joint Surg.*, 53A, 710, 1971.
40. Cabaud, H. E., Rodkey, W. G., and Feagin, J. A., "Experimental studies of acute anterior cruciate ligament injury and repair," *Am. J. Sports Med.*, 7, 18, 1979.
41. Chimich, D., Frank, C., Shrive, N., Dougall, H., and Bray, R., "The effects of initial end contact on medial collateral ligament healing: A morphological and biomechanical study in a rabbit model," *J. Orthop. Res.*, 9, 37, 1991.
42. Amiel, D., Kuiper, S., and Akeson, W. H., "Cruciate ligaments: Response to injury," in *Knee Ligaments: Structure, Function, Injury and Repair*, Daniel, D. M., Akeson, W. H., and O'Connor, J. J., Eds., Raven Press, New York, 1990, 365.
43. Hirsch, D. A., "Tensile properties during tendon healing. A comparative study of intact and sutured rabbit peroneus brevis tendons," *Acta Orthop. Scand. Suppl.*, 153, 1, 1974.
44. Ketchum, L. D., Martin, N. L., and Kappel, D. A., "Experimental evaluation of factors affecting the strength of tendon repairs," *Plast. Reconstr. Surg.*, 59, 708, 1977.
45. Hefti, F. L., Kress, A., Fasel, J., and Morscher, E. W., "Healing of the transected anterior cruciate ligament in the rabbit," *J. Bone Joint Surg.*, 73, 373, 1991.
46. Ng, G. Y. F., Oakes, B. W., McLean, I. D., Deacon, O. W., and Lampard, D., "The long-term biomechanical and viscoelastic performance of repairing anterior cruciate ligament after hemitransection injury in a goat model," *Am. J. Sports Med.*, 24, 109, 1996.
47. Arnoczky, S. P., Rubin, R. M., and Marshall, J. L., "Microvasculature of the cruciate ligaments and its response to injury," *J. Bone Joint Surg.*, 61A, 1221, 1979.
48. Bonamo, J. J., Fay, C., and Firestone, T., "The conservative treatment of the anterior cruciate deficient knee," *Am. J. Sports Med.*, 18, 618, 1990.
49. Finsterbush, A., Frankl, U., Matan, Y., and Mann, G., "Secondary damage to the knee after isolated injury of the anterior cruciate ligament," *Am. J. Sports Med.*, 18, 475, 1990.
50. Cordrey, L. J., McCorkle, H., and Hilton, E., "A comparative study of fresh and preserved homogenous tendon grafts in rabbits," *J. Bone Joint Surg.*, 45B, 182, 1963.
51. Barad, S., Cabaud, H., and Rodrigo, J., "Effects of storage at –80°C as compared to 4°C on the strength of rhesus monkey anterior cruciate ligaments," *Trans. Orthop. Res. Soc.*, 7, 1982.
52. Webster, D. A. and Werner, F. W., "Mechanical and functional properties of implanted freeze-dried flexor tendons," *Clin. Orthop.*, 180, 301, 1983.
53. Curtis, R. J., DeLee, D. C., and Drez, D. J., "Reconstruction of the anterior cruciate ligament with freeze-dried fascia lata allografts in dogs: A preliminary report," *Am. J. Sports Med.*, 13, 408, 1985.
54. Drez, D. J., Jr., DeLee, J., Holden, J. P., Arnoczky, S., Noyes, F. R., and Roberts, T. S., "Anterior cruciate ligament reconstruction using bone-patellar tendon-bone allografts. A biological and biomechanical evaluation in goats," *Am. J. Sports Med.*, 19, 256, 1991.
55. McPherson, G. K., Mendenhall, H. V., Gibbons, D. F., et al., "Experimental mechanical and histologic evaluation of the Kennedy ligament augmentation device," *Clin. Orthop.*, 196, 186, 1985.
56. Yoshiya, S., Andrish, J. T., Manley, M. T., and Kurosaka, M., "Augmentation of anterior cruciate ligament reconstruction in dogs with prostheses of different stiffnesses," *J. Orthop. Res.*, 4, 475, 1986.
57. Witkowski, J., Yang, L., Wood, D. J., and Sung, K.-L. P., "Migration and healing of ligament cells under inflammatory conditions," *J. Orthop. Res.*, 15, 269, 1997.
58. Amiel, D., Kleiner, J. B., Roux, R. D., Harwood, F. L., and Akeson, W. H., "The phenomenon of 'ligamentization': Anterior cruciate ligament reconstruction with autogenous patellar tendon," *J. Orthop. Res.*, 4, 162, 1986.
59. Oakes, B. W., Knight, M., McLean, I. D., and Deacon, O., "Goat ACL autograft collagen remodelling — quantitative collagen fibril analyses over one year," in *Proc. AOSSM*, San Diego, 1992, 45.
60. Panni, A. S., Milano, G., Lucania, L., and Fabbriciani, C., "Graft healing after anterior cruciate ligament reconstruction in rabbits," *Clin. Orthop.*, 343, 203, 1997.
61. Grönblad, M., Korkala, O., Konttinen, Y. T., Kuokkanen, H., and Liesi, P., "Immunoreactive neuropeptides in nerves in ligamentous tissue," *Clin. Orthop. Rel. Res.*, 265, 291, 1991.
62. McDougall, J. J., Bray, R. C., and Sharkey, K. A., "A morphological and immunohistochemical examination of nerves in normal and injured collateral ligaments of rat, rabbit and human knee joints," *Anat. Rec.*, 248, 29, 1997.

63. Frank, C., McDonald, D., Bray, D., Bray, R., Rangayyan, R., Chimich, D., and Shrive, N., "Collagen fibril diameters in the healing adult medial collateral ligament," *Connec. Tissue Res.*, 27, 251, 1992.
64. Amiel, D., Ishizue, K. K., Harwood, F. L., Kitabayashi, L., and Akeson, W. H., "Injury of the ACL: The role of collagenase in ligament degeneration," *J. Orthop. Res.*, 7, 486, 1989.
65. Woo, S. L.-Y., Young, E. P., and Kwan, M. K., "Fundamental studies in knee ligament mechanics," in *Knee Ligaments: Structure, Function, Injury and Repair*, Daniel, D. M., Akeson, W. H., and O'Connor, J. J., Eds., Raven Press, New York, 1990, 115.
66. Gomez, M. A., "The biomechanical properties of normal and healing ligaments," in *The Anterior Cruciate Ligament: Current and Future Concepts*, Jackson, D. W., Arnoczky, S. P., Frank, C. B., Woo, S. L.-Y., and Simon, T. M., Eds., Raven Press, New York, 1993, 227.
67. Bray, R. C., Shrive, N. G., Frank, C. B., and Chimich, D. D., "The early effects of joint immobilization on medial collateral ligament healing in an ACL-deficient knee: A gross anatomic and biomechanical investigation in the adult rabbit model," *J. Orthop. Res.*, 10, 157, 1992.
68. Woo, S. L.-Y., Inoue, M., McGurk-Burleson, E., and Gomez, M. A., "Treatment of the medial collateral ligament injury. II: Structure and function of canine knees in response to differing treatment regimens," *Am. J. Sports Med.*, 15, 22, 1987.
69. Inoue, M., Woo, S. L.-Y., Gomez, M. A., Amiel, D., Ohland, K. J., and Kitabayashi, L. R., "Effects of surgical treatment and immobilization on the healing of the medial collateral ligament: A long-term multidisciplinary study," *Connec. Tissue. Res.*, 25, 13, 1990.
70. Yoshiya, S., Andrish, J. T., Manley, M. T., and Bauer, T. W., "Graft tension in anterior cruciate ligament reconstruction. An *in vivo* study in dogs," *Am. J. Sports Med.* 15, 464, 1987.
71. Ng, G. Y., Oakes, B. W., Deacon, O. W., McLean, I. D., and Lampard, D., "Biomechanics of patellar tendon autograft for reconstruction of the anterior cruciate ligament in the goat: Three year study," *J. Orthop. Res.*, 13, 602, 1995.
72. Newton, P. O., Horibe, S., and Woo, S. L.-Y., "Experimental studies on anterior cruciate ligament autografts and allografts," in *Knee Ligaments: Structure, Function, Injury and Repair*, Daniel, D. M., Akeson, W. H., and O'Connor, J. J., Eds., Raven Press, New York, 1990, 389.
73. Dunlap, J., McCarthy, J. A., Joyce, M. E., Ogata, K., and Shively, R. A., "Quantification of the perfusion of the anterior cruciate ligament and the effects of stress and injury to the supporting structures," *Am. J. Sports Med.*, 17, 808, 1989.
74. Bray, R., Forrester, K., Mc Dougall, J. J., Damji, A., and Ferrell, W. R., "Evaluation of laser Doppler imaging to measure blood flow in knee ligaments of adult rabbits," *Med. Biol. Eng. Comp.*, 34, 227, 1996.
75. Bray, R. C., Butterwick, D. J., Doschak, M. R., and Tyberg, J. V., "Coloured microsphere assessment of blood flow to knee ligaments in adult rabbits: Effects of injury," *J. Orthop. Res.*, 14, 618, 1996.
76. Ferrell, W. R., McDougall, J. J., and Bray, R. C., "Spatial heterogeneity of the effects of calcitonin gene-related peptide (CGRP) on the microvasculature of ligaments in the rabbit knee joint," *Brit. J. Pharmacol.*, 121, 1397, 1997.
77. McDougall, J. J., Ferrell, W. R., and Bray, R. C., "Spatial variation in sympathetic influences on the vasculature of the rabbit knee joint," *J. Physiol.*, 503, 435, 1997.
78. Ferrell, W. R., McDougall, J. J., and Bray, R. C., "Altered neuropeptidergic vasoregulation in the ACL deficient rabbit knee," *Proc. Can. Orthop. Res. Soc., Hamilton*, 1997, 51.
79. McDougall, J. J., Ferrell, W. R., and Bray, R. C., "Sympathetically-mediated constrictor responses in normal and ACL deficient rabbit knees," *Proc. Can. Orthop. Res. Soc., Hamilton*, 1997, 51.
80. Mc Dougall, J. J. and Bray, R. C., "Quantification of vascular volume changes in ACL deficient rabbit knees using a carmine red vascular casting technique," *Trans. Orthop. Res. Soc., New Orleans*, 1998, in press.
81. Arnoczky, S. P., Rubin, R. M., and Marshall, J. L., "Microvasculature of the cruciate ligaments and its response to injury," *J. Bone Joint Surg.*, 61A, 1221, 1979.

25 Animal Models of Tendon Repair

Donald L. Pruitt

CONTENTS

I. Introduction 477
II. Basic Science of Tendon 478
    A. Tendon Healing 478
    B. Tendon Nutrition 478
    C. Tendon Biomechanics 478
III. Animal and Animal Models Selections 479
IV. Commonly Used Models 480
    A. Flexor Tendons 480
        1. Chicken 480
        2. Rabbit 481
        3. Canine 483
        4. Primates 484
        5. Others 485
    B. Achilles Tendon 486
    C. Extensor Tendons 486
    D. General Tendon Models 486
V. Evaluation Methods 486
    A. Morphological Evaluation 486
    B. Biomechanical Evaluation 487
References 488

I. INTRODUCTION

The use of animal models has contributed greatly to our understanding of tendon physiology and biomechanics. Much of the previous investigative work has focused upon flexor tendons, probably because of the challenges clinical flexor tendon repair has presented to surgeons for many years. Flexor tendon models in dogs, chickens, rabbits and primates all have been extensively developed and described. Considerably less use has been made of animal extensor tendon models, but this should not diminish the importance of extensor tendon research. The Achilles tendon model is also well described, particularly in the rats and rabbits.

    This chapter will first provide a historical review of the major animal models, describe each in more detail, and offer some thoughts on how best to choose a model for the specific question the researcher has in mind.

II. BASIC SCIENCE OF TENDON

A. Tendon Healing

Ross[1] characterized soft tissue healing into three phases: the inflammatory, the proliferative, and the organizational or remodeling phases. In 1941, Mason and Allen[2] established a healing curve for divided flexor tendons in a canine model. The early stages after repair were characterized by a profound drop in tensile strength as the tendon ends softened and were held together by a gelatinous clot. They called this period the exudative stage, which lasted for the first 14 days. Thereafter, the formative stage was characterized by increased tissue organization which was directly influenced by the stresses applied to the repair site. For many years, these findings influenced clinicians to immobilize repaired flexor tendons for three weeks before beginning motion.

In 1962, Potenza[3] studied tendon healing within the fibrous sheath in dogs and found no evidence of intrinsic (intratendinous) fibroblast response following tendon injury. He concluded that tendon healing depended upon extrinsic cellular ingrowth. Lindsay et al.[4] noted fibrous adhesions attached to each point of tendon surface disruption in a chicken model. Both Potenza and Lindsay believed adhesions were an integral part of the healing process, essential for nutritional support of the healing tendon. Peacock[5] proposed a one wound concept, which stated that healing was entirely dependent upon cells migrating into the repair site from outside sources.

In 1976, Matthews and Richards[6] performed an experiment in rabbits which showed that the tendon injury itself was not the main stimulus for adhesion formation. Similarly, immobilization alone did not produce adhesions, but the combination of sheath injury and immobilization quickly resulted in massive adhesion formation. Lundborg and Rank[7] showed that lacerated and repaired rabbit tendons, when implanted into the knee joint, healed by an intrinsic cellular response. Manske et al.[8] demonstrated intrinsic healing of lacerated flexor tendons in tissue culture for rabbits, chickens, dogs and monkeys. There was some variation among species, but by 12 weeks rabbit tendons had come closest to a complete repair, followed by chicken, dog and monkey in that order.

B. Tendon Nutrition

For a long time, tendon was considered an avascular tissue. Lundborg et al.[9] described four vascular sources in human cadaver injection studies: longitudinal vessels extending down the tendon from the palm, vessels entering at the synovial reflections, vessels from the long and short vinculae, and vessels entering at the bony insertion. There are two bare areas between the vinculae in which no vessels could be found. It was surmised that these areas might be nourished by diffusion of nutrients. Manske et al. studied the contribution of vascular perfusion and synovial diffusion in chickens[10] and primates.[11] They found that diffusion functions more quickly and completely than perfusion, and is a more important pathway for nutrients.

C. Tendon Biomechanics

Most biomechanical studies of tendon have focused on the ultimate tensile strength and strain, both before and after repair. Tendon ultimate tensile strength has been found to range from 45 to 125 MPa,[12] depending upon the species, type of tendon, preconditioning, and animal age.[13]

Most modern suture techniques have improved purchase in the tendon such that the suture itself breaks before it pulls out of the tendon.[14] Repair strength can be further enhanced with the addition of a peripheral epitenon suture.[15] Gap formation can occur secondary to cyclic forces at the repair site[16] and compromise the final clinical result.[17] The ultimate goal of suture technique research is to develop a repair technique strong enough to allow immediate active motion.

Early attempts to optimize post operative motion protocols were largely empirical; it was hoped that early motion would decrease adhesion formation. Over the past two decades, many experimental studies have helped define the effect of motion on the tendon healing process. Using a canine

TABLE 1
Animal Models for Tendon Repair

Animal Tendon	Useful Tendons per Limb	Tendon Quality*	Cost†	Care	Best Use	Comment
Flexor						
Chicken	1–3	Thin, Flat	$	Easy	Adhesion, Lac. and Repair	Widely Used
Rabbit	2–3	Rubbery	$	Easy	Adhesion	Widely Used
Dog	2–4	Good	$$	Standard	Lac. and Repair	Widely Used
Monkey	4	Excellent	$$$$	Special	Lac. and Repair	Special Projects
Farm Animals	2–4	Variable	$$-$$$	Special	Adhesion, Lac. and Repair	Special Projects
Rat, Mouse	2–4	Small	$	Easy	Tissue Sample	
Extensor						
Chicken	1–3	Fair	$	Easy	All	Small Size
Rabbit	2–3	Fair	$$	Easy	All	Small Size
Dog	3–4	Good	$$	Standard	All	Widely Used
Monkey	4	Good	$$$$	Special	All	Special Projects
Achilles						
Rabbit	1	Good	$$	Easy	All	
Rat	1	Good	$	Easy	All	

* Tendon Quality: Compared to Human tendons
† Costs: $=Least Expensive, $$$$= Most Expensive

model, Gelberman et al.[18] found that carefully applied early protected motion allowed healing not associated with gapping at the repair site and maintenance of a smooth gliding surface. In another study comparing immobilization, delayed mobilization, and early mobilization in repaired canine flexor tendons, the load to failure of immediately mobilized tendons was twice that of immobilized tendons tested at three weeks.[19] These experiments support the concept that early motion improves the quality of the biologic repair process and allows flexor tendon healing within the digital sheath while eliminating the associated adhesion formation. Hitchcock and associates[20] investigated the effects of immediate controlled motion on the strength of flexor tendon repairs in a chicken model and found the initial loss in strength described by Mason and Allen was not inevitable. Similar findings have been reported for extrasynovial flexor tendons in rabbits treated with continuous passive motion.[21]

III. ANIMAL AND ANIMAL MODELS SELECTIONS

Many factors enter into the decision of which animal model to select for a particular tendon research project. First one must consider the general area of tendon research being investigated. If just a small amount of tissue is required, as in *in vitro* tissue culture studies, then rats, mice or chickens may be an appropriate choice. Certain models, such as chickens, are particularly useful for investigating tendon adhesions, while others, such as canine models, are especially useful in laceration and repair studies. Nonhuman primates, such as monkeys, are often used in the final stages of evaluating prosthetic devices for possible human use.[22] Table 1 summarizes selection factors for many of the major tendon models.

There are anatomic differences in tendon morphology between humans and each of the animal models. Mice and rats have very small tendons which run through a short fibroosseous canal (zone 2). Chickens have a more flat and ribbon like flexor profundus tendon compared to man. Chickens have two additional flexor tendons besides the flexor profundus, one going to the middle phalanx and one to the proximal phalanx (Figure 1). Calcification of the tendons above the hock (ankle

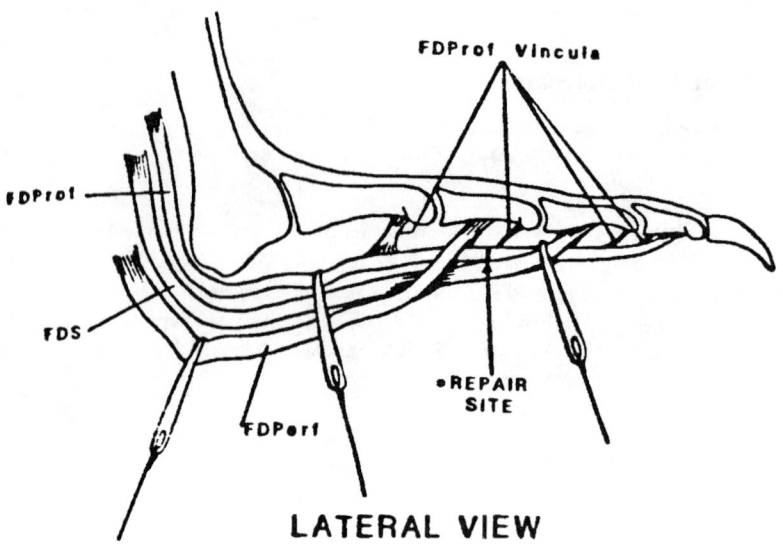

FIGURE 1. Chicken digital flexor system. One tendon inserts onto each phalanx. (From Freehan, L. M., Beauchene, J. G., *J. Hand Surg.*, 15A, 63, 1990, with permission.)

joint) is common. Dogs have a profundus tendon which is more rounded and similar in appearance to humans. Perhaps because dogs bear weight on their front paws, there is an area of fibrocartilage within the flexor tendon.[23] Monkeys, which are the most anatomically similar to humans, have a more developed ulnar side of the hand with smaller thenar muscles and flexor pollicis longus tendons.

Budgetary considerations, such as procurement and housing costs, obviously, enter into the selection process. Mice, rats and chickens are all relatively economical and can be housed in most animal care facilities. Dogs are a little more expensive to procure and house, but can usually be accommodated at most facilities. Sheep, goats, swine and other farm animals often require special housing considerations and are difficult to care for at many animal facilities. Monkeys and other nonhuman primates are the most expensive to both procure and house. Most monkeys need to be quarantined for a period of 6–12 weeks before use to prevent the spread of endemic infectious diseases which they can carry.

IV. COMMONLY USED MODELS

A. FLEXOR TENDONS

1. Chicken

The chicken model has been particularly useful in flexor tendon research. The profundus tendon travels through a long zone 2 compared to the size of the animal. This is an excellent model for studying the formation and prevention of adhesions. In addition, the tendons are usually large enough for use in laceration and repair studies.

The white leghorn chicken is the most commonly used bird. They can be obtained from the usual animal procurement agencies or directly from a farm. Chickens are rather dirty animals, and should be stored in a separate room. It is advisable to spray for mites before placing them in your cages as many animals are infested. It is also recommended that operations be performed in a room other than the usual sterile animal OR to protect the other animals.

Adult white leghorn chickens, usually hens, are used although some investigators have used cockerels.[24] The chickens are initially tranquilized with ketamine (10mg/kg, IM). The birds are next placed into a plastic bag which covers about two thirds of the body with holes for the feet. This helps isolate the cloaca from the operative field. The bird is then wrapped in a towel which is taped circumferentially to prevent the wings from flapping during the operation. The head is left exposed; assesment of the comb's color gives an indication of how the bird is doing. Anesthesia is obtained with pentobarbital (15 mg/kg, IM or SC).[25] Small additional doses of pentobarbital may be given if necessary. Foam blocks placed on each side will help keep the bird in position during surgery. A rubber tourniquet is placed around the leg above the hock. The foot is then sterilely prepped and draped. A cork board is placed beneath the sterile drapes. Sterile needles and rubber bands are used to secure the foot and hold the adjacent toes out of the way. A midline incision along the great toe is the standard surgical approach (Figure 2), although a Brunner (zig-zag) incision may also be used. Magnification is recommended because the structures are quite small. The tendons are usually repaired with a 5-0 or 6-0 core suture and a 7-0 epitenon suture. At the conclusion of the proceedure, sterile dressings and either a soft or hard cast is applied. It is important to make the dressing as waterproof as possible. Antibiotics may be administered either intramuscularly or by mouth via the water supply. Even without antibiotic administration, we have seen relatively few infections despite the amount of dirt the foot is exposed to.

Numerous technical alternatives have been described for the chicken model. Some investigators have performed the proceedure under local anesthesia only (with the animal's head draped to calm it down).[26] Others will first sedate the animal with ketamine or Nembutal and then add a local block.[27] Postoperative dressings have included casts with the toes in flexion,[28] casts with the adjacent toes in hyperextension, soft dressings allowing immediate motion,[29] soft dressings with a tethering splint,[30] (Figure (3) and even attempts at using a continuous passive motion machine.

The chicken is an excellent model for investigating the effects of tendon injury on adhesion formation. This is most effectively performed using a partial tendon laceration. The tendon can be exposed directly, as in a laceration and repair study, or exposed indirectly by making an incision in the sheath just proximal to zone 2, pulling the tendon back into the operative field with traction on the tendon, performing a partial laceration and then allowing the tendon to retract distally back to within the tendon sheath. This technique creates a standard laceration in the tendon without injuring the adjacent sheath.[31] A device has been described which produces a uniform 75% laceration of the tendon.[32] Pharmacological agents may be injected into the tendon sheath by making a small incision over the volar portion of the distal interphalangeal joint, locating the tendon sheath and sliding a blunt tip needle or small IV catheter proximally up the sheath into zone 2.

Euthanasia is performed using an overdose of pentobarbital or intravenous potassium chloride. Good quality veins for injection are readily found in the neck. Additional veins are easily located in the wings, but are usually more fragile than neck veins.

2. Rabbit

The rabbit model has been used most for Achilles tendon and flexor tendon research. The rabbit is an excellent model for partial laceration and adhesion formation studies.[33] Rabbits are inexpensive, clean and docile animals which are easy to work with. There is a small area of zone 2 flexor tendon system within each front paw which is useful in adhesion formation studies. Rabbit tendon, however, has a more elastic or rubbery quality compared to other animals and human tendons. Previous studies have indicated that rabbit collagen is relatively more juvenile when considering collagen cross linking compared to human collagen. Thus it is harder to repair rabbit flexor tendons as cleanly as some other animal models. Despite this drawback, the model is important, especially in adhesion studies[34] and *in vitro* culture experiments.[35]

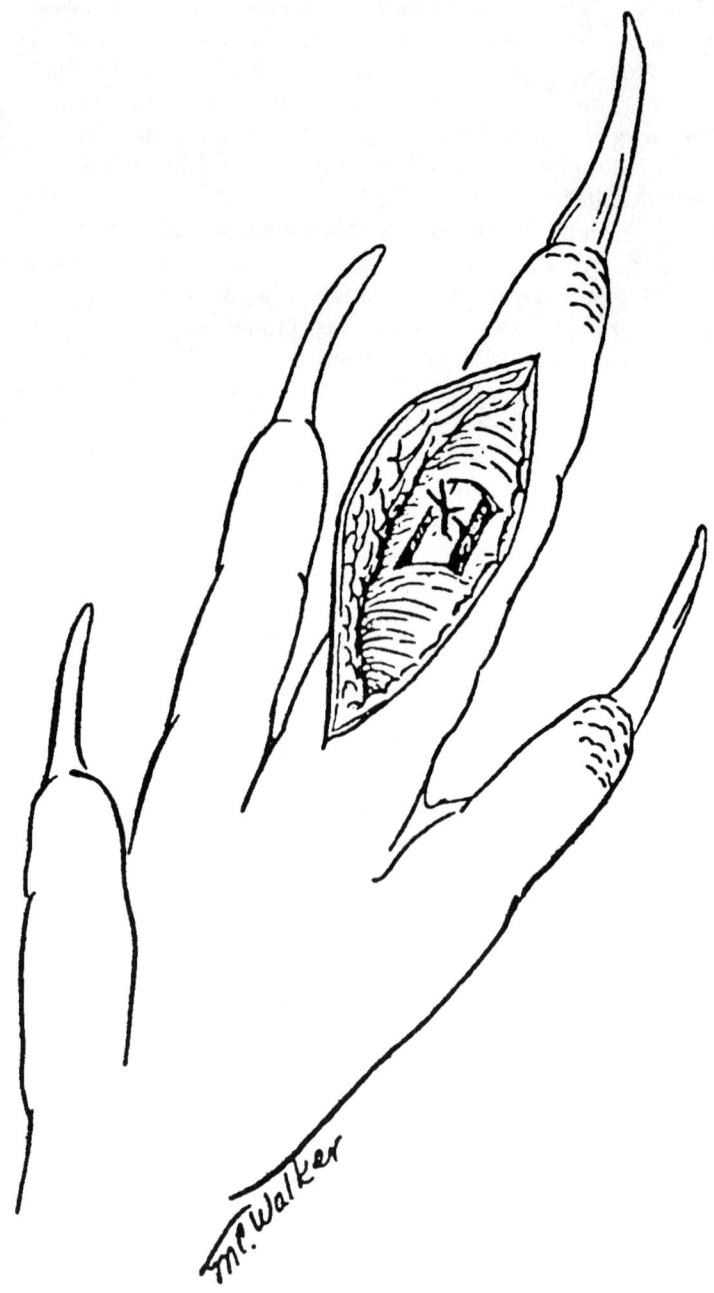

FIGURE 2. Midline approach for exposure of zone 2 of the chicken flexor tendon system. A Brunner zigzag incision also is acceptable. (From Kessler, F. B., et al., "*J. Hand Surg.,* 11A, 241, 1986, with permission.)

Rabbits are rather timid creatures, so it is advisable to first sedate the animals with ketamine (20 mg/kg) given intramuscularly. Full anesthesia is then obtained with Nembutal or pentobarbital given intramuscularly. We have often supplemented this with a local block of 1% lidocaine. A wide Penrose drain wrapped around the limb and secured with a clamp serves as a tourniquet. The limb is shaved and then depiliated with Nair or its equivalent. The forelimb is then prepped and sterilely draped.

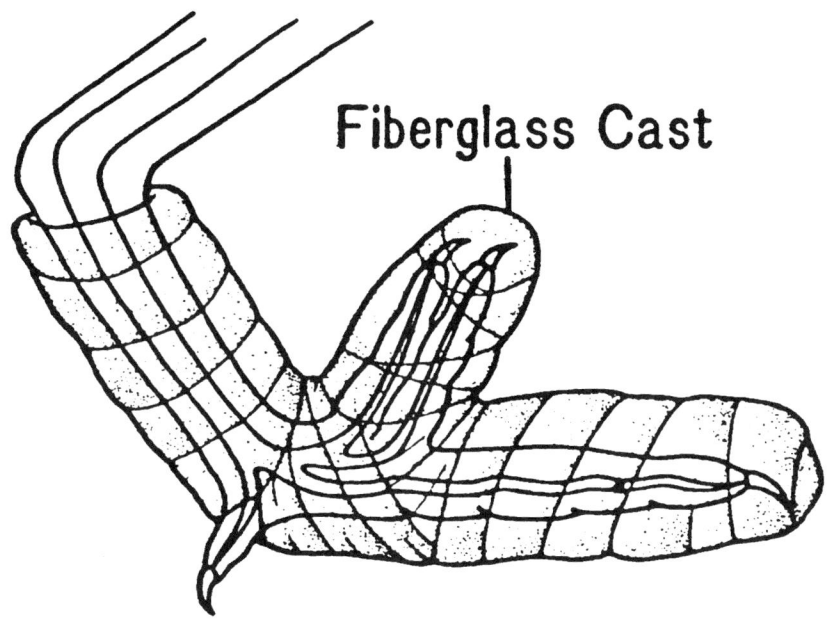

FIGURE 3. Example of the use of a cast which holds the adjacent toes in hyperextension, taking the tension off a flexor tendon repair in the chicken model. (From Hitchcock, T. F., et al., *J. Hand Surg.*, 12A, 590, 1987, with permission.)

A palmar midline or Brunner incision is used to gain acess to the flexor system (Figure 4). The tendons are usually repaired with a 5-0 or 6-0 core suture and a 6-0 or 7-0 epitenon suture. Postoperatively, either a soft dressing is applied (partial laceration studies), or a fiberglass cast (laceration and repair studies) is placed above the elbow. If chewing through the cast is a problem, an Elizabethan collar may be applied. Euthanasia is usually by an overdose of pentobarbital.

3. Canine

The canine has been one of the most important models for flexor tendon research. Mason and Allen used canine wrist flexor and extensor tendons for their classic study to determine the healing rate of tendons. Dogs have also played a prominent role in studies into the formation of adhesions,[36] the effects of controlled motion on flexor tendon healing,[37] tendon suture techniques,[38] and many other topics.

Dogs are commonly obtained in either a conditioned or unconditioned state. Conditioned dogs have undergone veterinary evaluation and are free of worms and other parasites.

The animals are first sedated with ketamine or xylazine. Intravenous access can usually be readily gained from forepaw veins on the upper extremity. Anesthesia is induced with Nembutal or pentobarbital (IM or IV), and maintained with halothane via an endotracheal tube and a Harvard animal ventilator. The dog is placed in a lateral decubitus position with the operative limb below so that the paw pad is facing up. A pediatric tourniquet cuff usually fits well on the upper part of the limb. The extremity is taped into position, shaved and prepped and draped in a sterile manner. Administration of antibiotics, usually penicillin or cephalosporin is advisable.

After draping, the limb is exsanguinated and the tourniquet inflated to 100 mm Hg above systolic blood pressure. Zone II of the flexor tendon system lies beneath the paw pad. Exposing all four tendons requires lifting up the entire paw pad which carries a significant risk of devascularizing

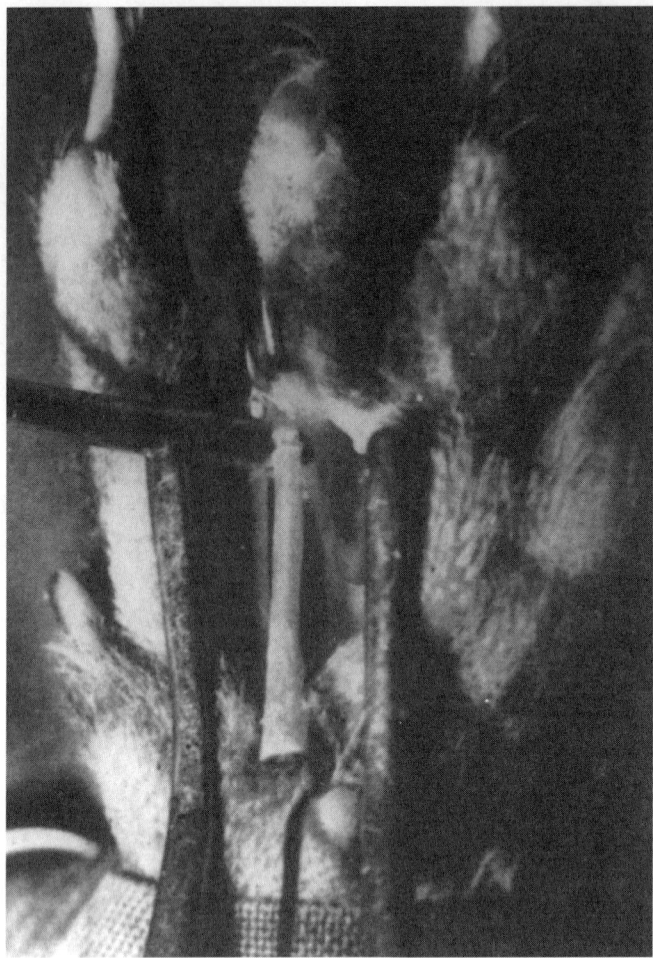

FIGURE 4. Midline exposure of the rabbit flexor tendon system. The profundus tendon has been pulled proximally to expose zone 2 tendon, where a partial laceration has been made. (From Matthews, P., Richards, H., *J. Bone Joint Surg.*, 56B, 618, 1974, with permission.)

the paw pad with subsequent full thickness tissue loss. Most investigators expose only two of the flexor tendons, either the second and fifth through longitudinal incisions on either side of the paw pad, or the third and fourth tendons through a longitudinal incision in the center of the paw pad. The area is quite vascular and having an electrocautery is helpful. The tendons are usually repaired with a 4-0 or 5-0 size core suture and 6-0 epitenon sutures. Wounds are closed with 4-0 nonabsorable sutures.

Postoperative immobilization varies from none to a shoulder spica cast, depending upon the experimental conditions. A "long arm" cast does not work well for dogs, because it moves around enough to rub full thickness sores at the elbow. Single limb shoulder spica casts or orthoses have been extensively used to protect the tendon repair.[19] The area around the paw may be cut out and removed for daily passive motion exercises (Figure 5). We have often made the cast a few days before surgery and bi-valved it for easy reapplication in the operating room.

4. Primates

Monkeys and other nonhuman primates provide the closest approximation of the human flexor tendon system available in an animal model. The anatomic differences between monkey and human

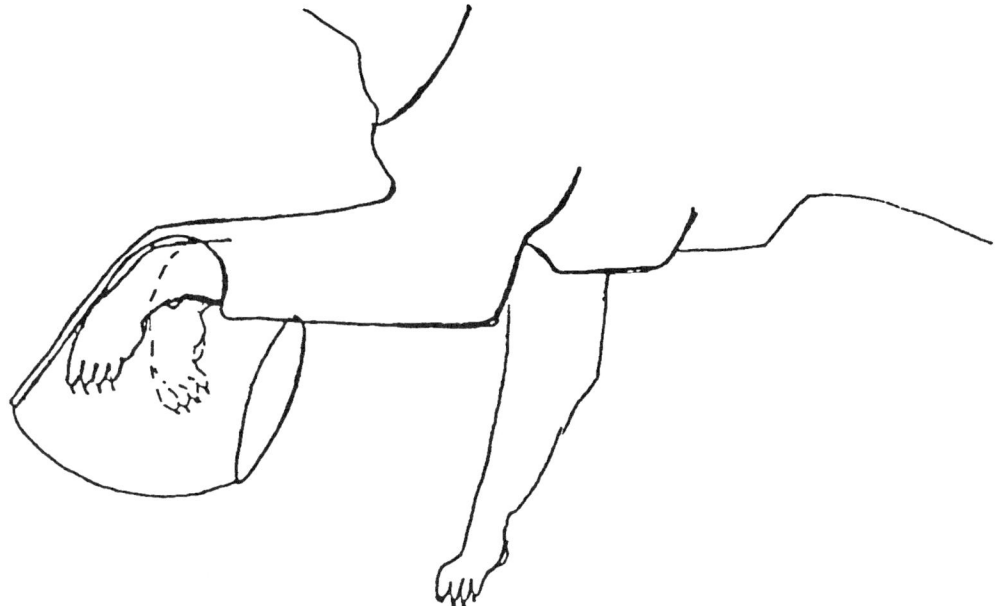

FIGURE 5. Single limb spica cast applied to a dog with the paw area cut out to allow for passive range of motion of the flexor tendons.

hands are small. In general the ulnar side of the monkey hand is more developed and the thumb and associated muscles are less developed. This probably reflects the greater importance of power grasp to the monkey rather than fine prehension.

Primates are only occasionally used in tendon research due to the high cost of both obtaining and maintaining primates. Macaca monkeys are most commonly used; they are plentiful and in no danger of extinction. Most research studies using primates are for evaluating prosthetic materials for possible human use and for projects which would not work in any other model.[22]

Monkeys commonly require at least a six week quarantine period once arriving in the USA to make sure they are not carrying any diseases. They also require special cages with collapsible backs and a separate room away from other animals.

The monkeys are first tranquilized with ketamine (10 mg/kg, IM) while still in their cages. Anesthesia is then induced with sodium pentobarbital (5 mg/kg) and an airway established using an endotracheal tube. Intravenous access is also established. We typically monitor heart rate and blood pressure. Antibiotics are administered. Surgery is performed using a pneumatic tourniquet on the upper arm after shaving the hair from the operative region and a sterile prep and draping. The flexor system is approached using the same incisions as for humans: either a Brunner zig-zag or a midlateral approach. At the conclusion of surgery, sterile dressings and a fiberglass cast are usually applied. Monkeys will often chew on and pick at a cast or dressing until it is completely off, thus a rather strong dressing or splint should be applied.

5. Others

Other animals such as sheep,[39] cows[40] and pigs[41] have been occasionally used in flexor tendon research. These models provide larger tendons to work with and have been useful in evaluating prosthetic materials. However, these animals are more expensive to procure and house at most animal facilities.

B. ACHILLES TENDON

The Achilles tendon is one of the largest extrasynovial tendons in the body. It is commonly involved in traumatic injuries, tendonitis and rupture, making it an important tendon to study in animal models. This model is also commonly used to evaluate the effects of systemic medications,[42] exercise[43] and other stimuli on the musculoskeletal system. The rat[44] and the rabbit[45] are the most commonly used species, although the dog[46] is also occasionally utilized.

Rabbits are prepared for surgery in the same manner as for flexor tendons. A tourniquet is applied to the hind limb above the knee and the extremity is shaved and depiliated. The Achilles tendon is exposed through a posterior midline incision. At the conclusion of surgery, a long leg cast may be applied according to the experimental protocol; a spica cast is usually not necessary.

Rats are anesthetized with sodium pentobarbital (0.5 mg/kg, intraperitoneal), and supplemented during the proceedure with diluted pentobarbital (65 mg/ml) given in 0.1 ml increments. Following induction of anesthesia, the operative leg is shaved and the animal positioned on a cork board. Rubber bands looped around the legs and pinned in place with thumb tacks hold the animal in position. A Penrose drain wrapped around the leg functions as a tourniquet. The Achilles tendon is exposed via a posterior longitudinal incision.

C. EXTENSOR TENDONS

Considerably less experimental work has been carried out on the extensor tendon system compared to the flexor tendons. This may reflect the greater challenges flexor tendon repair has clinically presented to surgeons for so long. Clearly extensor tendons are different from flexors, and have become the subject of more studies in recent years.

At birth, the flexor and extensor tendons of pigs have identical mechanical properties. With growth and aging, both tendons become stronger, stiffer, less extensible and more resilient, but these changes occur to a greater degree in the higher load bearing flexor tendons compared to the flexors.[47] Recent investigations have focused on collagen fibril formation in transected tendons,[48] nutrient pathways to the tendons,[49] and the effects of early motion.[50]

Models to study the extensor tendons have been described in all of the major animal models used to investigate flexor tendons. Models have included rats,[51] rabbits,[50] dogs, chickens,[52] primates and sheep. The same anesthesia, shave, prep and tourniquet placement is used as in the flexor tendon model. The paw is positioned with the dorsal surface up and a midline approach is made to the tendons. Post operative dressings will depend upon the experimental design.

D. GENERAL TENDON MODELS

Rat tail collagen has been noted to have a similar biochemical makeup and crimp pattern when compared to human tendon tissue.[53] Therefore, rat tail tendon is a reasonable source of tissue for ultrastructural, biochemical,[54] and occasionally biomechanical[55] studies. The tail is removed, covered in ice for 10 minutes, the ventral skin is cut for 30% of the tail length and then stripped off. This reveals the dorsal tendon bundles which can be cut free from the vertebrae for use.

Mouse tail tendon tissue may be harvested in a manner similar to the rat model for biochemical and other related experiments.

V. EVALUATION METHODS

A. MORPHOLOGICAL EVALUATION

Standard histologic methods have been used quite extensively to study tendons. For light microscopy, specimens are fixed in 10% neutral buffered formalin, embedded in paraffin, sectioned and stained. Haematoxylin and eosin is the most commonly used stain. Other commonly used stains include Milligan's trichrome, Van Gieson's and Verhoeff's stains. Transmission and scanning electron microscopy are also useful techniques. Specimens for SEM may be fixed in 0.2 N Sorensen's phosphate-buffered glutaraldehyde, dehydrated in an ethanol series and dried in liquid CO_2.[56]

Recent advances in molecular biology techniques permit the study of cellular processes from the beginning of specific gene expression to the synthesis of regulatory and matrix proteins. New techniques such as the polymerase chain reaction (PCR) require only small amounts of tissue allowing experiments to be carried out *in vitro*. *In situ* RNA hybridization localizes gene expression of specific proteins to individual cells within the tendon.[57] The class of regulatory proteins known as growth factors are known to be central to the regulation of wound healing and fracture repair. Their role in the regulation of tendon healing is just beginning to be defined.

B. BIOMECHANICAL EVALUATION

Standard biomechanical testing of tendons includes ultimate tensile strength and elongation to failure. The Instron and MTS mechanical testing machines are most commonly used. Difficulties measuring biomechanical properties of tendons are well recognized, especially in measuring cross sectional area and tendon elongation during testing. Clamp design flaws may allow for tendon slippage (if too loose) or breakage of tendon fibers (if too tight). Placing sandpaper around the tendon helps prevent slippage for lower force experiments. For larger tendons, uneven distribution of forces among tendon bundles may lead to inaccurate strength measurements. A photographic method for evaluating small strain fields within the tendon has been described.[58]

There are two biomechanical tests commonly used to measure tendon gliding function: angular rotation and work of flexion. Angular rotation, most commonly used in dogs and chickens, uses an apparatus to measure angular rotation of the distal joints in response to a small applied load. The specimen is fastened to a metal platform at the level of the proximal phalanx and a precision potentiometer is aligned with the PIP joint. The end of the transected tendon is connected to a weight platform. In dogs, a 150 gm weight was applied for 15 seconds and the angular rotation determined.[59]

The work of flexion is a quantitative representation of all of the forces that resist digital flexion. These forces include the mechanical forces which resist tendon gliding (eg. adhesions), as well as the resistance of the surrounding soft tissues related to joint movement. This parameter has been most commonly used in chickens, but can be used in other animals as well. For chickens, the feet are disarticulated at the knee and the profundus tendon to the long toe is isolated. The feet are secured to a mounting board with the toes down and a counterweight is attached to the toenail to hold the toe in full extension. The tendon end is attached to a mechanical testing machine through a load cell. The testing machine crosshead is advanced and a recording of load cell force vs. excursion is plotted on an X-Y recorder. The area beneath the curve represents the parameter work of flexion (Figure 6).[60]

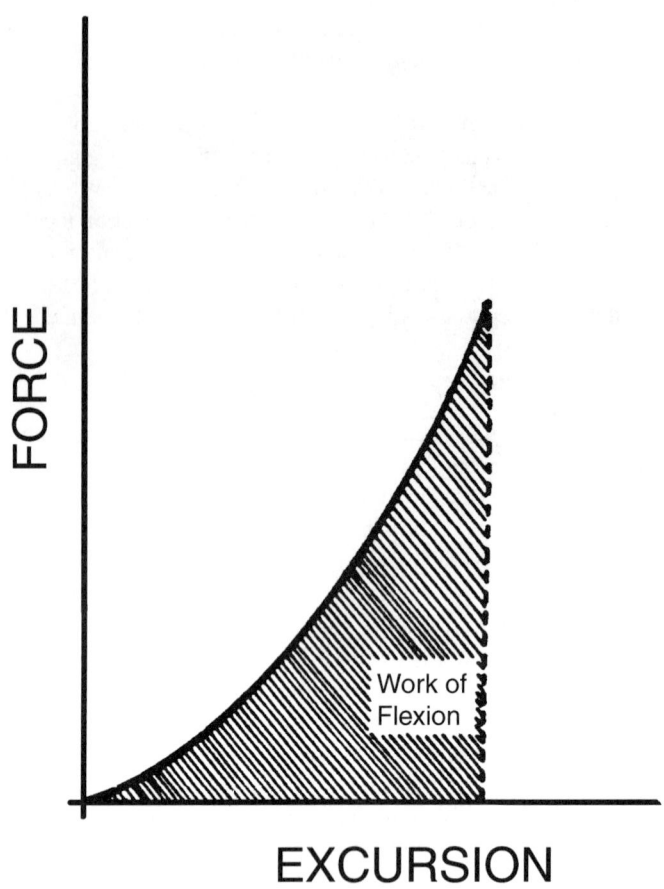

FIGURE 6. Example recording of tendon force plotted against tendon excursion. The area under the curve is the work of flexion. (From Peterson, W. W., et al., *J. Hand Surg.*, 15A, 48, 1990, with permission.)

REFERENCES

1. Ross, R., "The fibroblast and wound repair," *Biol. Rev.*, 43, 51, 1968.
2. Mason, M. L. and Allen, H. S., "The rate of healing of tendons: an experimental study of tensile strength," *Ann. Surg.*, 113, 424, 1941.
3. Potenza, A. D., "Effect of associated trauma on healing of divided tendons," *J. Trauma*, 2, 175, 1962.
4. Lindsay, W. K. and Thompson, H. G., Digital Flexor tendons: "An experimental Study: Part 1. The significance of each component of the flexor mechanism in tendon healing," *Br. J. Plast. Surg.*, 12, 289, 1959–1960.
5. Peacock, E. E., Jr., "Fundamental aspects of wound healing relating to the restoration of gliding function after tendon repair," *Surg. Gynecol. Obstet.*, 119, 241, 1964.
6. Matthews, P. and Richards, H., "The repair potential of digital flexor tendons," *J. Bone Joint Surg.*, 56B, 618, 1974.
7. Lundborg, G. and Rank, F., "Experimental intrinsic healing of flexor tendons based upon synovial fluid nutrition," *J. Hand Surg.*, 3A, 21, 1978.
8. Manske, P. R. and Lesker, P. A., "Histologic evidence of intrinsic flexor tendon repair in various experimental animals: an *in vitro* study," *Clin. Orthop.*, 182, 297, 1984.
9. Lundborg, G., Myrhage, R., and Rydevik, B., "The vascularization of human flexor tendons within the digital sheath region: structural and functional aspects," *J. Hand Surg.*, 2A, 417, 1977.

10. Manske, P. R., Lesker, P. A., and Bridwell, K. H., "Nutrient pathways to flexor tendons of chickens using triated proline," *J. Hand Surg.*, 3, 352, 1978.
11. Manske, P. R. and Lesker, P. A., "Nutrient pathways of flexor tendons in primates," *J. Hand Surg.*, 7A, 436, 1982.
12. Woo, S. L.-Y., "Mechanical properties of tendons and ligaments: I. Quasistatic and nonlinear viscoelastic properties," *Biorheology*, 19, 385, 1982.
13. Viidik, A., "Tensile strength properties of Achilles tendon systems in trained and untrained rabbits," *Acta Orthop. Scand.*, 40, 262, 1969.
14. Urbaniak, J. R., Cahill, J. D., and Mortenson, R. A., "Tendon suturing methods: analysis of tensile strengths," in *AAOS Symposuim on Tendon Surgery in the Hand*, Mosby, St. Louis, 1975, 70.
15. Wade, P. J. F., Muir, I. F. K., and Hutcheon, L. L., "Primary flexor tendon repair: the mechanical limitations of the modified Kessler technique," *J. Hand Surg.*, 11B, 71, 1986.
16. Pruitt, D. L., Manske, P. R., and Fink, B., "Cyclic stress analysis of flexor tendon repair," *J. Hand Surg.*, 16A, 701, 1991.
17. Seradge, H., "Elongation of the repair configuration following flexor tendon repair," *J. Hand Surg.*, 8A, 182, 1983.
18. Gelberman, R. H., Botte, M. J., and Speigelman, J. J., "The excursion and deformation of repaired flexor tendons treated with protected early motion," *J. Hand Surg.*, 11A, 106, 1986.
19. Gelberman, R. H., Woo, S. L.-Y., Lothringer, K., et al., "Effects of early intermittent passsive mobilization on healing canine flexor tendons," *J. Hand Surg.*, 7A, 170, 1982.
20. Hitchcock, T. F., Light, T. R., and Bunch, W. H., "The effect of immediate controlled mobilization on the strength of flexor tendon repairs," *J. Hand Surg.*, 12A, 590, 1987.
21. Salter, R. B., Simmonds, D. F., Malcolm, B. W., et al., "The biological effect of continuous passive motion on the healing of full thickness defects in articular cartilage: An experimental investigation in the rabbit," *J. Bone Joint Surg.*, 62A, 232, 1980.
22. Dunlap, J, McCarthy, J. A., Joyce, M. E., and Manske, P. R., "Biomechanical and histologic evaluations of pulley reconstructions in nonhuman primates," *J. Hand Surg*, 15A, 57, 1990.
23. Nessler, J. P., Amadio, P. C., Breglund, L. J., and An, K. N., "Healing of canine tendons in zones subjected to different mechanical forces," *J. Hand Surg.*, 17B, 561, 1992.
24. Turner, S. M., Powell, E. S., and Ng, C. S., "The effect of ultrasound on the healing of repaired cockerel tendons: is collagen cross linkage a factor?" *J. Hand Surg.*, 14B, 428, 1988.
25. Peterson, W. W., Manske, P. R., Dunlap, J., Horwitz, D. S., and Kahn, B., "Effect of various methods of restoring flexor sheath integrity on the formation of adhesions after tendon injury," *J. Hand Surg.*, 15A, 48, 1990.
26. Freehan, L. M. and Beauchene, J. G., "Early tensile properties of healing chicken flexor tendons: early controlled passive motion versus postoperative immobilization," *J. Hand Surg.*, 15A, 63, 1990.
27. Hitchcock, T. F., Light, T. R., Bunch, W. H., et al., "The effect of immediate constrained digital motion on the strength of flexor tendon repairs in chickens," *J. Hand Surg.*, 12A, 590, 1987.
28. Wray, R. C. and Weeks, P. M., "Experimental comparison of technics of tendon repair," *J. Hand Surg.*, 5, 144, 1980.
29. Kessler, I. and Nissim, F., "Primary repair without immobilization of flexor tendon division within the digital sheath: an experimental and clinical study," *Acta Orthop. Scand.*, 40, 587, 1969.
30. Hitchcock, T. F., Light. T. R., Bunch, W. H., et al., "The effect of immediate constrained digital motion on the strength of flexor tendon repairs in chickens," *J. Hand Surg.*, 12A, 590, 1987.
31. Matthews, P. and Richards, H., "The repair potential of digital flexor tendons," *J. Bone Joint Surg.* 56B, 618, 1974.
32. Hitchock, T. F., Candel, A. G., Light, T. R., and Blevens, A. D., "A new technique for producing uniform partial lacerations of tendons," *J. Orthop. Res.*, 7, 451, 1989.
33. Frykman, E., Jacobsson, S., and Widenfalk, B., "Fibrin sealant in prevention of flexor tendon adhesions: an experimental study in the rabbit," *J. Hand Surg.*, 18A, 68, 1993.
34. Szabo, R. M. and Younger, E., "Effects of indomethacin on adhesion formation after repair of zone II tendon lacerations in the rabbit," *J. Hand Surg.*, 15A, 480, 1990.
35. Mass, D. P., Tuel, R. J., and Labarbera, M., "Effects of constant mechanical tension on the healing of rabbit flexor tendons," *Clin. Orthop.*, 296, 301, 1993.
36. Potenza, A. D., "Effect of associated trauma on healing of divided tendons," *J. Trauma*, 2, 175, 1962.

37. Gelberman, R. H., Vande Berg, J. S., Lundborg, G. M., and Akeson, W. H., "Flexor tendon healing and restoration of the gliding surface: an ultrastructural study in dogs," *J. Bone Joint Surg.*, 65A, 70, 1983.
38. Lin, G. T., An, K. N., Amadio, P. C., and Cooney, W. P., "Biomechanical studies of running suture for flexor tendon repair in dogs," *J. Hand Surg.*, 13A, 553, 1988.
39. Silfverskiöld, K. L. and Anderson, C. H., "Two new methods of tendon repair: an in vitro evaluation of tensile strength and gap formation," *J. Hand Surg.*, 18A, 58, 1993.
40. Koob, T. J. and Vogel, K. G., "Site-related variations in glycosaminoglycan content and swelling properties of bovine flexor tendon," *J. Orthop. Res.*, 5, 414, 1987.
41. Woo, S. L.-Y, Gomez, M. A., Amiel, D., et al., "The effects of exercise on the biomechanical and biochemical properties of swine digital flexor tendons," *J. Biomech. Eng.*, 103, 511, 1981.
42. Inhofe, P. D., Grana, W. A., Egle, D., Min, K. W., and Tomasek, J., "The effects of anabolic steroids on rat tendon. An ultrastructural, biomechanical and biochemical analysis," *Am. J. Sports Med.*, 23, 227, 1995.
43. Simonsen, E. B., Klitgaard, H., and Bojsen-Moller, F., "The influence of strength training, swim training and aging on the Achilles tendon and m. soleus of the rat," *J. Sports Sci.*, 13, 291, 1995.
44. Murrell, G. A., Lilly, E. G., Goldner, R. D., Seaber, A. V., and Best, T. M., "Effects of immobilization on Achilles tendon healing in a rat model," *J. Orthop. Res.*, 12, 582, 1994.
45. Hsu, S. Y., Cheng, J. C., Chong, Y. W., and Leung, P. C., "Glutaraldehyde-treated bioprosthetic substitute for rabbit Achilles tendon," *Biomaterials,* 10, 258, 1989.
46. Badylak, S. F., Tullis, R., Kokini, K., et al., "The use of xenogeneic small intestinal submucosa as a biomaterial for Achilles tendon repair in a dog model," *J. Biomed. Mater. Res.*, 29, 977, 1995.
47. Shadwick, R. E., "Elastic energy storage in tendons: mechanical differences related to function and age," *J. Appl. Physiol.*, 68, 1033, 1990.
48. Matthew, C., Moore, M. J., and Campbell, L., "A quantitative ultrastructural study of collagen fibril formation in the healing extensor digitorum longus tendon of the rat," *J. Hand Surg.*, 12B, 313, 1987.
49. Manske, P. R. and Lesker, P. A., "Nutrient pathways to extensor tendons within the extensor retinacular compartments. An experimental study in dogs," *Clin. Orthop.*, 181, 234, 1983.
50. Wilson, K., Moore, M. J., Rayner, C. R., and Fenton, O. M., "Extensor tendon repair: an animal model which allows immediate post-operative mobilisation," *J. Hand Surg.*, 15B, 74, 1990.
51. Matthew, C., Moore, M. J., and Campbell, L., "A quantitative ultrastructural study of collagen fibril formation in the healing extensor digitorum longus tendon of the rat," *J. Hand Surg.*, 12B, 313, 1987.
52. Ikegami, H., "Experimental study on the effects of tension-reduced early mobilization on extensor tendon healing," *Nippon Seikeigeka Gakkai Zasshi,* 69, 493, 1995.
53. Idler, R. S., "Anatomy and biomechanics of the digital flexor tendons," *Hand Clin.*, 1, 3, 1985.
54. Davison, P. F., "The contribution of labile crosslinks to the tensile behavior of tendons," *Connective Tissue Res.,* 18, 293, 1989.
55. Kastelic, J. and Baer, E., Reformation in tendon collagen: the mechanical properties of biological materials, Presented at the 34th Symposium of the Society for Experimental Biology, 1980.
56. Takasugi, H., Inoue, H., and Akahori, O., "Scanning electron microscopy of repaired tendon and pseudosheath," *Hand,* 8, 228, 1976.
57. Gelberman, R. H., Amiel, D., and Harwood, F., "Genetic expression for type I procollagen in the early stages of flexor tendon healing," *J. Hand Surg.*, 17A, 551, 1992.
58. Amadio, P. C., Berglund, L. J., and An, K. N., "Biochemically discrete zones of canine flexor tendon: Evaluation of properties with a new photographic method," *J. Orthop. Res.*, 10, 198, 1992.
59. Gelberman, R. H., Woo, S. L.-Y, Amiel, D., Horibe, S., and Lee, D., "Influences of flexor sheath continuity and early motion on tendon healing in dogs," *J. Hand Surg.*, 15A, 69, 1990.
60. Peterson, W. W., Manske, P. R., Dunlap, J., Horwitz, D. S., and Kahn, B., "Effect of various methods of restoring flexor sheath integrity on the formation of adhesions after tendon injury," *J. Hand Surg.*, 15A, 48, 1990.

26 Animal Models of Ligament and Tendon Fixation

Franklin A. Young and Yuehuei H. An

CONTENTS

I. Introduction 491
II. Commonly Used Animal Models of Ligament or Tendon Fixation 494
    A. Suture and Suture Anchor 494
    B. Staple 496
    C. Screw/Washer 496
    D. Spiked Bushing, Washer, or Plate 497
    E. Bone or Absorbable Plug 498
    F. Interference Screw 498
    G. Young's Ligament Anchor 498
III. Evaluation Methods 499
IV. Concluding Remarks 499
References 499

I. INTRODUCTION

For successful transplantation or transposition of ligament and tendon, the fixation techniques are very important. The fixation must provide sufficient initial fixation strength during the postoperative period, must not interfere with the soft tissue healing, and must be biocompatible for long term use and ideally bioabsorbable or easily removable.[1]

Common techniques for fixation of ligament or tendon to bone include suture, staple, screw/washer, spiked bushing, washer, or plate, bone plug or block, and interference screw (Figure 1). Several articles on comparative evaluations of different anchoring techniques have been published.[2,3,4,5,6,7] A special tension-adjustable artificial ligament anchor has been reported recently.[8,9] Although there is still a distance between the current version and clinical application, it brings in a new concept for ligament and tendon anchoring.

Large animals such as goats, dogs, sheep, pigs, and monkeys are common species for studies of ligament and tendon fixation. Large bone volume of these animals is the most important factor for the facilitation of bone instrumentation (fixations with screws or implants). Selected animal models from the literature for evaluating ligament or tendon fixation to bone are listed in Table 1.

Table 2 lists mechanical properties (failure load, ultimate strength, stiffness and elastic modulus) of ligament, tendon, or deep fascia of different species. Ideally, a fixation strength should exceed the requirement for normal activity on ligaments or tendons.

The healing of ligament- or tendon-bone interface has been studied histologically.[11] Transplanted autogenous patellar-tendon grafts undergo a process of ischemic necrosis, revascularization, proliferation, and remodeling. At one year, a transplanted graft can have the histological appearance of a normal ligament.[11] Kasperczyk et al.[23] defined a four stage healing process of autogenic patellar

TABLE 1
Selected Animal Models and *In Vitro* Models Used for Evaluating Ligament or Tendon Fixation to Bone

Method	Device or material	Procedure, graft	Animal species	Initial fixation strength (N)	Strength after *in vivo* study	First author, year[Ref.]
Suture	Suture/button	ACLR*, PCLR†, PT‡	Monkey	—	300 (1 yr.)	Clancy 1981[10]
	Stainless steel suture	ACLR, PT	Dog	—	—	Arnoczky 1982[11]
	Suture/button	ACLR, PT	Human knee *in vitro*	248 ± 40	—	Kurosaka 1987[12]
	Suture over bone	ACLR, iliotibial band	Human knee *in vitro*	109 ± 11	—	Kurosaka 1987[12]
	Suture	Infraspinatus tendon	Goat	—	715–824 (12 weeks)	St. Pierre 1995[13]
Suture anchor	Mitek Superanchor	Tibialis anterior tendon	Human cuboid bone	223	—	Fennell 1995[14]
	Absorbable anchor	MCLR‖	Rabbit	—	—	Ono 1997[15]
Staple	Richards CC1A	ACL, MCL, artificial lig.	Sheep	—	160–197 (1 yr.)	Claes 1987[16]
	Staple	ACLR, PT	Human knee *in vitro*	129 ± 16	—	Kurosaka 1987[12]
	Staple	ACLR, iliotibial band	Human knee *in vitro*	137 ± 23	—	Kurosaka 1987[12]
	Metal	ACLR, fascia lata	Goat	—	200–400 (from fig.)	Holden 1988[17]
	Metal	ACLR, PE	Goat	752	1013–1233 (3 mon.)	Powers 1991[18]
Screw/washer	Cortical screw	ACLR, Gore-Tex lig.	Sheep	369	1380 (7 months)	Bolton 1984[19]
	Screw/bushing	ACLR, Kennedy LAD	Goat	364	841 (24 months)	McPherson 1985[20]
	2-mm cortical screw	ACLR, bone-PT-bone	Rabbit	26 ± 5	51 ± 6	Ballock 1989[21]
	Bicortical screw	ACLR, Braided PE	Human knee *in vitro*	160	—	Gillquist 1993[22]
	Cancellous screw	PCLR, bone-PT-bone	Sheep	171 ± 16	708 ± 99 (2 yrs.)	Kasperczyk 1993[23]
	AO screw	ACLR, bone matrix	Goat	73 ± 9 (ligament)	474 ± 146 (1 yr.)	Jackson 1996[24]

Animal Models of Ligament and Tendon Fixation

Fixation	Device	Specimen	Application	Load (in vitro)	Load (in vivo)	Reference
Spiked washer, plate	Plate (Synthes)	Sheep	MCL/ACLR, artificial lig.	—	160–197 (1 yr.)	Claes 1987[16]
	UHMWPE bushing	Goat	ACLR, fascia lata	—	250–400 (from fig.)	Holden 1988[17]
	Cancellous screw	Pig bone	ACLR, bone-PT-bone	309	—	Paschal 1994[25]
	Soft tissue plate	Tendon in vitro	Supraspinatus tendon	170–266	—	Gottsuer-Wolf 1994[26]
	Spiked washer	Tendon in vitro	Supraspinatus tendon	149–514	—	Gottsuer-Wolf 1994[26]
	Tendon anchor	Tendon in vitro	Supraspinatus tendon	399–729	—	Gottsuer-Wolf 1994[26]
	AO resin or metal	Human distal femur	Fascia lata	99 ± 35, 149 ± 43	—	Straight 1994[1]
Bone or PLA plug	Press-fit bone plug	Pig knee in vitro	ACLR, bone-PT-bone	463	—	Rupp 1997[27]
	SR-PLA plug	Bovine knee in vitro	ACLR, bone-PT-bone	1100	—	Tuompo 1996[28]
Interference screw	9-mm interf. screw	Human knee in vitro	ACLR, patellar tendon	476 ± 110	—	Kurosaka 1987[12]
	9-mm (DePuy)	Pig bone	ACLR, bone-PT-bone	535	—	Paschal 1994[25]
	7-mm (Acufex)	Bovine bone in vitro	ACLR, bone-PT-bone	1358 ± 348	—	Kousa 1995[29]
	7-, 9-mm interf. screw	Bovine bone in vitro	ACLR,	1161–1198	—	Shapiro 1995[30]
	7-mm (Linvatec)	Human knee in vitro	ACLR, bone-PT-bone	640 ± 201	—	Pena 1996[31]
	Ti (Arthrex)	Pig knee in vitro	ACLR, bone-PT-bone	769	—	Rupp 1997[27]
	AO cancel. screw	Human knee in vitro	ACLR, PT	208 ± 28	—	Kurosaka 1987[12]
	AO cancellous	Bovine bone in vitro	ACLR, bone-PT-bone	1081 ± 331	—	Kousa 1995[29]
	SR-PLA (Biofix)	Bovine bone in vitro	ACLR, bone-PT-bone	1211 ± 362	—	Kousa 1995[29]
Bioscrew	PLA (Arthrex)	Same	ACLR, bone-PT-bone	418 ± 118	—	Pena 1996[31]
		Pig knee in vitro	ACLR, bone-PT-bone	805	—	Rupp 1997[27]
Cylindrical anchor	Titanium	Goat	ACLR, Dacron lines	—	2000–4000 (2 months)	Young 1995[8,9]

* ACLR = Anterior cruciate ligament reconstruction
† PCLR = posterior cruciate ligament reconstruction
‡ PT = patellar tendon
¶ MCLR = medial collateral ligament reconstruction

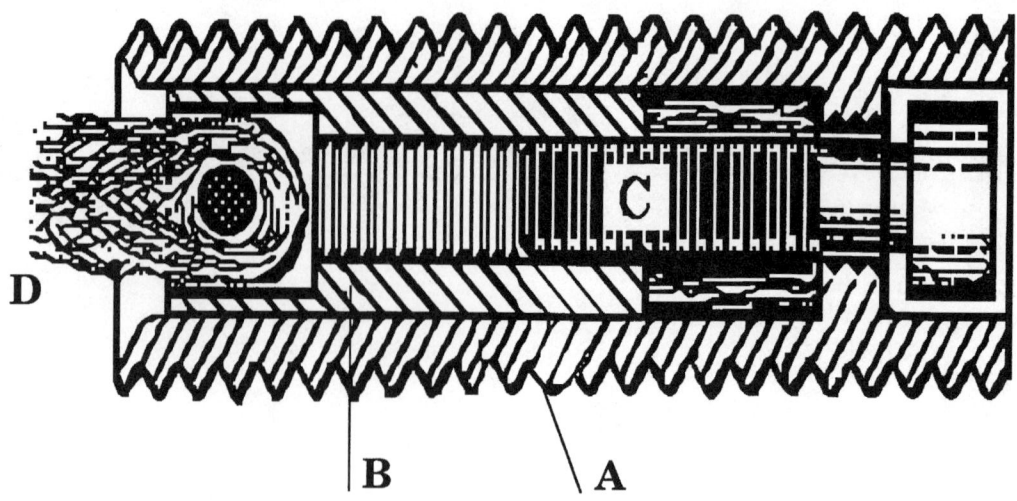

FIGURE 1. Schematic drawing of Young's artificial ligament anchor: (A) exterior cylinder, (B) internal cylinder, (C) tensioning screw, and (D) artificial ligament.

tendon graft, necrosis, revascularization, collagen formation, and remodeling. They also found that the biomechanical data were correlated with the morphological phases.

II. COMMONLY USED ANIMAL MODELS OF LIGAMENT OR TENDON FIXATION

A. SUTURE AND SUTURE ANCHOR

Suture alone and suture-over-button are common techniques for ligament fixation to bone. Common sutures include nonabsorbable suture, bioabsorbable suture, and metal wire. Suture techniques are also commonly used clinically for fixation of tendon to bone especially in hand surgery. They are used mainly for holding the tendon in place for proper healing of the tendon-bone interface. Because of the relatively low initial fixation strength, early vigorous movement is not encouraged.

Clancy et al.[10] examined ACL and PCL replacements in 19 rhesus monkeys. For ACL reconstruction, patellar tendon autografts with bone attached were used. For PCL reconstruction the medial third of the patellar tendon elongated by attached portions of the patella and tibia was employed. Bone tunnels were drilled in the femur and tibia at a location which corrected for changes in ligament joint space exit location because of the size of the tunnels. Fixation was accomplished by sutures through the ligament and tied over a button. Mechanical testing results showed breaking strengths for control (medial one-third) patellar tendon specimens to be 300 N. The same setup was used to test for the grafted ligaments. Graft pull-out strengths, expressed as percentages of the strength of the medial one-third of the patellar tendon at one year, were 81% for ACL and 52% for PCL. Test results from earlier time periods gave lower numbers (e.g. ACL at two months, 34%), and it was concluded that the bone ingrowth into the tunnels was providing the increased fixation.

Arnoczky et al.[11] investigated the patellar tendon healing in ACL reconstruction in dogs. The proximal end of the tendon was fixed to bone by stainless steel suturing. The healing process of the graft was reported. No mechanical testing was used for evaluation.

In an *in vitro* study using human cadaver knee, Kurosaka et al.[12] investigated different fixation techniques in ACL reconstruction using bone-patellar tendon-bone grafts, iliotibial band grafts, and semitendonosus grafts. It was found that fixation strength by the suture-over-button technique was equivalent to that by staples, but was lower than that by cancellous screws and interference screws.

TABLE 2
Selected Mechanical Data of Ligament, Tendon, or Deep Fascia of Different Species

Species	Age	Materials	Failure load (Newton)	Stiffness (N/mm)	Ultimate strength (MPa)	Elastic modulus (MPa)	First author, year[Ref.]
Human	16–26	ACL	1730 ± 660	182 ± 56	—	—	Noyes 1976[32]
	48–86	ACL	1734 ± 283	129 ± 39	—	—	Noyes 1976[32]
	26 ± 6	ACL	1725 ± 269	—	38 ± 4	—	Noyes 1984[33]
	59	ACL	559 ± 47	74 ± 3	—	—	Kurosaka 1987[12]
	42	ACL	2195 ± 427	306 ± 80	—	—	Rowden 1997[34]
	26 ± 6	Bone-PT*-bone	2900 ± 260	—	58 ± 6	306 ± 59	Noyes 1984[33] & Butler 1984[35]
	44	Patellar tendon	—	—	47 ± 16	—	France 1988[36]
	22–35	Bone-PT-bone	2160 ± 157	242 ± 28	—	—	Woo 1991[37]
	40–50	Bone-PT-bone	1503 ± 83	220 ± 24	—	—	Woo 1991[37]
	60–97	Bone-PT-bone	658 ± 129	180 ± 25	—	—	Woo 1991[37]
	26 ± 6	Semitendinosus	1216 ± 50	—	89 ± 5	362 ± 22	Noyes 1984[33] & Butler 1984[35]
	26 ± 6	Gracilis	838 ± 30	—	111 ± 4	613 ± 41	Noyes 1984[33] & Butler 1984[35]
	26 ± 6	Distal iliotibial band	769 ± 99	—	19 ± 3	—	Noyes 1984[33]
	26 ± 6	Fascia lata	628 ± 35	—	79 ± 5	398 ± 17	Noyes 1984[33] & Butler 1984[35]
	44	Fascia lata	—	—	32 ± 14	—	France 1988[36]
	44	Achilles tendon	—	—	61 ± 26	—	France 1988[36]
Rhesus monkey	Young adult	ACL	830 ± 110	194 ± 28	—	—	Noyes 1976[32]
	Young adult	Bone-PT-bone	600 ± 132	—	—	—	Clancy 1981[10]
Goat	Adult	ACL	1691 ± 209	453 ± 120	—	—	Powers 1991[18]
	Adult	ACL	—	259 ± 7	—	487 ± 28	Ng 1996[38]
Sheep	2-year	PCL	950	130	—	—	Kasperczyk 1993[23]
Pig	?	Toe extensor tendon	—	—	47 ± 5	980	Smith 1996[39]
Rabbit	3.0 ± 0.2	ACL	—	—	57 ± 4	600 ± 50	Ishizue 1990[40]
	3.5 ± 0.2 kg	Bone-PT-bone	About 300–400	—	—	—	Ballock 1989[21]
	12 months	Bone-MCL-bone	—	86 ± 1	—	—	King 1995[41]

* PT = patellar tendon

In another study, it was found that the fixation strength by a nonabsorbable suture was equivalent to interference screw and screw/washer.[42]

St. Pierre et al.[13] investigated the healing of tendon attachments in troughs of cancellous bone compared to direct cortical attachment. Twenty goats were subjected to a bilateral tenotomy of the infraspinatus tendon. The tendons were attached using sutures either to the cortical surface of bone or to a trough prepared in the cancellous bone. The techniques were found to provide equivalent fixation both biologically and mechanically.

Recently, different suture anchors have been developed.[43,44] They are made of metal, nonabsorbable,[14] and absorbable materials (such as the expanding suture plug [Arthrex][45]).[15] No foreign body reactions were found by the studies by Barber et al.[45] and Ono et al.[15] Suture anchors are playing an increasingly important role in attaching tendons or ligaments to bone.

B. Staple

Currently many brands of commercial fixation staples are available, such as the Richards type CC1A XSMO staple (spiked).[16] Staples are convenient to use but recorded fixation strength is relatively low.

Claes et al.[16] tested combined replacements of ACL and MCL of four ligament replacement materials in sheep. Some grafts were fixed to bone with a spiked staple (Richards type CC1A XSMO).

Holden et al.[17] measured the strength of fascia lata autograft ACL replacements in 50 goats for periods ranging from zero to eight weeks. The objective was to compare stapled grafts to those fixed with a cancellous bone screw and spiked bushing. The latter control ACL had an average tensile failure load of 2748 N, a value which significantly exceeds that found by other investigators. At time zero, the failure force for the screw/bushing fixed specimens exceeded that for staple fixation. Other time periods yielded no significant differences in strength between the two techniques. The graft failure values were reported only as percentages of the control. At eight weeks, the value for staple fixation was 15% and the screw/bushing fixation was 9% of the control value.

Powers et al.[18] conducted a study of artificial ACL replacements in goats using two tunnels each in the femur and tibia and two ligament strands to simulate the anterolateral and the posterolateral bands. Long chain polyethylene fibers were used for the ligament and staples used to fix them to bone. The increased strength obtained in the three month specimens was deemed to be the result of bone ingrowth into the tunnels providing increased resistance to pullout. Failure modes were not reported.

C. Screw/Washer

Cancellous and cortical screws[22] with or without (commonly for artificial ligament) washers have been used.

Bolton and Bruchman[19] examined the performance of PTFE (Gore-Tex™) artificial ACL replacements in 17 sheep for periods ranging from zero to 369 days. Cortical bone screws placed through eyelets built into the prosthesis were employed to fix the ligaments which were placed in bone tunnels using the "over-the top" technique. Pull-out tests were conducted, and the zero time implants yielded a mean failure strength of 1814 N. Ninety day implants with the bone screws in place yielded a failure value of 2445 N. Screws were removed from one group which averaged 218 days residence and a failure strength of 1379 N. Testing of a control group of ACL specimens yielded a failure strength of 1912 N. Fixation screws pulled out of the bone for the zero time implants, and the increased strengths observed in the experimental groups were attributed to bone growth fixation in the tunnels.

In a goat model, McPherson et al.[20] examined the effects of augmentation by a 6 mm PE braid of a graft ACL replacement consisting of a portion of the rectis femoris tendon, prepatellar tissue, and the central one-third of the patellar tendon. Tensioning was secured by attaching the ligament with a bushing and cortical bone screw to the lateral surface of the femur. Tensile tests were conducted and failure typically was found to occur by pullout of the device from the tibia. The augmented ligaments were found to have a failure strength after initial implantation of 364 N. After two years, the augmented grafts had a strength of 841 N and the unaugmented grafts 528 N. These strengths were compared to a natural goat ACL quoted at 2023 N.

In a sheep model of PCL reconstruction, Kasperczyk et al.[23] investigated the healing of patellar tendon autografts. The graft was fixed to bone by screw/washer. They defined a four stage healing process of autogenic patellar tendon graft, necrosis, revascularization, collagen formation, and remodeling. They also found that the biomechanical data were correlated with the morphological phases. All ligaments failed at the ligament portion, demonstrating the effectiveness of the screw fixation.

In a goat model, Jackson et al.[24] attempted to improve fixation by selecting an ACL replacement material which would foster bone formation in the ligament tunnels. Demineralized bone matrix was used as the ligament and connected to a screw/washer by sutures. Six month and one year experiments were conducted in 10 goats. Seven animals were sacrificed at one year and accelerated bone formation noted in the tunnels. The mean ultimate force to failure for the reconstrructed ligament at one year was 474±146 N compared with the time zero strength of the matrix graft of 73±9 N.

Ballock et al.[21] reported a rabbit model of ACL reconstruction using patellar tendon autografts. They fixed the ligament graft with the spiked bushing, washer, or plate.

D. Spiked Bushing, Washer, or Plate

Currently many brands of commercial spiked washers or plates are available, such as the Synthes type 65.00.11 soft tissue fixation plate,[16] the AO polyacetal resin spiked washer and AO soft tissue fixation plate.[1]

Holden et al. studied the effect of a spiked bushing (with 5-mm diam. shaft) on the fixation of fascia lata grafts for ACL reconstruction in a goat model. The results showed that at eight weeks, the strength of the graft was 9% of the control value, compared to 15% achieved by staple fixation.[17]

Claes et al.[16] tested combined replacements of ACL and MCL of four ligament replacement materials in 30 sheep for one year. Carbon fiber (Lafil), polydioxanone strand, dacron, and a bovine tendon xenograft were employed. The combined replacement technique employed three bone tunnels and a continuous (ACL-MCL) replacement. Both prosthesis ends were anchored on the lateral surface of the femur using either a staple or a spiked fixation plate with screw (Table 1). Tensile tests were conducted for MCL and ACL separately with the staples or fixation plate removed. No ligaments fractured during the tensile tests and failure occurred by pulling the ligament out of the bone tunnel. However, no effects of the fixation techniques on the strength of the fixation and ligament healing were found.

Gottsauner-Wolf et al.[26] tested different methods of fixation of tendons to metal prostheses, including soft tissue fixation plate (Synthes), spiked polyacetal washer (Synthes), and a new enhanced tendon anchor (ETA), a device with spikes designed to interlock both prosthesis and tendon and held in place by two screws. Attachments were made both to the bone blocks of tendon/bone combinations and directly to tendons. Canine supraspinatus tendon was used and 60 "motion units" were tested. The ETA required the largest force to remove, but none of the methods were as strong as the control intact muscle-tendon unit. The use of tendon with attached bone block significantly increased the fixation strength.

E. Bone or Absorbable Plug

Bone plugs or blocks either separately used with artificial ligaments[46] or connected with biological grafts[27] are commonly used clinically for reconstruction of the ACL. The fixation strength of press-fit bone plugs was found to be lower than that found with interference screws.[27] The fixation of bone blocks in the bone tunnel is often reinforced by using an interference screw or sutures tying to a screw/washer on the surface of the bone.

Ligament fixation using a self-reinforced PLA expansion plug was reported by Tuompo et al.[28] in a bovine bone model. The maximum tensile strength of the SR-PLA plug was above 1100 N and the initial strength of the absorbable plug is strong enough for clinical use.

F. Interference Screw

The recent trend toward early motion and aggressive postoperative rehabilitation emphasizes the need for secure initial graft fixation before bony incorporation of the graft.[32,47] Kurosaka et al.[12] in an *in vitro* study demonstrated that the 9-mm interference screw (DePuy) had superior initial fixation strength over a 6.5-mm cancellous screw. One shortcoming of interference screws is their difficulty to remove which makes revision surgery difficult sometimes. Then, bioabsorbable interference screws, such as Bioscrew, appeared.[28,31] The initial fixation strength has been reported to be lower[31] or equivalent to metal ones.[27,29,48]

Paschal et al.[25] compared postfixation (tying to cancellous bone screw) to interference screw fixation in 20 frozen/thawed pig knees using bone-patellar tendon-bone ACL replacement grafts. Displacement of the graft in the bone tunnel by a force of 110 N was measured as well as the load required to pull out the graft from femurs and tibias separately. Displacements were highest in the postfixation and higher in tibias. The strength of the postfixation for femurs was 274 N while for interference it was 543 N. For tibias, the postfixation strength was 343 N and the interference screw strength 527 N. All grafts failed at the point of fixation.

In another study using pig knees, Rupp et al.[27] compared press fitting of the bone block in a bone-patellar tendon-bone graft to the use of a biodegradable (polylactic acid) interference screw. Titanium interference screw fixation served as a control. Pull-out force to failure was measured. The biodegradable screw fixation yielded a load of 805 N and the titanium screw 769 N. The press fit yielded a lower load of only 463 N. All specimens failed at the attachment site.

Kousa et al.[29] also examined interference screw fixation in harvested knees. Bone-patellar tendon-bone ACL replacement grafts were placed in frozen/thawed bovine knees. Interference screws, cancellous bone screws, and fibrillated PLA screws were used. Tensile failure strengths ranged from 1081 N to 1358 N with no statistically significant differences.

Shapiro et al.[30] investigated the screw size on the pullout strength of *in vitro* ACL reconstruction using bovine knee. The results showed there was no significant difference between 7- and 9-mm interference screws. However, in another study, it was found that failed bone plug fixed by 7-mm screw could be refixed successfully with a 9-mm screw.[49] Jomha et al.[50] found that there was a significant weakening of fixation for screw–bone plug angle equal or more than 20 degrees.

G. Young's Ligament Anchor

Young and An reported a new adjustable screw anchor to secure the artificial ACL prosthesis to the femur and tibia (Figure 1).[8,9] Fixation was provided by screw threads on the exterior surface of a hollow cylinder which was placed in the bone tunnels created in the femoral condyle and tibial plateau. The artificial ACL was attached to a sliding portion inside the threaded cylinder, which was adjusted for tension by means of a screw accessed from outside the exterior bone surfaces. Push-out tests of anchors which had functioned for two months in goats gave values of approximately 2000–4000 N.

III. EVALUATION METHODS

Histomorphological analysis is the basic evaluation method for the effect of ligament or tendon fixation to bone.[11,16] The ligament- or tendon-bone interfaces are harvested *en bloc*. The tissue blocks can be embedded in paraffin (for decalcified tissues without fixation devices) or plastic media (for tissues containing fixation devices).

Mechanical testing (tensile test) is the most important method for evaluating the function of the ligament- or tendon-bone interface. Initial[1] and long term mechanical strength of the ligament and tendon fixation should be tested (Table (2) (see Chapter 10). When tensile strength is measured, one has to be aware of where the failure occurs — the fixation device or the ligament itself either at the ligament–device interface or the ligament body. Normally, the ultimate strength of a ligament-device-bone assembly represents the maximum load the fixation device can hold. The ultimate strength of a fixation device or anchor can be tested with an "unbreakable ligament" (stronger than the device holding strength) in a tensile mode or using a pushout test as was the case for Young's ligament anchor.[30,31]

Biochemical evaluation has been used for examining DNA content, content and types of collagens, PGs or GAG synthesis and content in the repair tissues[51] (see Chapter 6). MRI also has been used for evaluating ligament healing in the bone tunnel.[22]

IV. CONCLUDING REMARKS

To summarize, repaired or substituted ligaments and tendons are provided with some sort of mechanical fixation in the form of screws, plates, or sutures to allow healing to take place which will provide the long term survival of the repair. The values obtained by investigators indicate that very long time periods are necessary before bone ingrowth can support either grafts or alloplastic replacements. Initial fixation strength is essential for early movement and rehabilitation programs. The major causes of artificial ligament failure include stretching, loosening and breaking. Future research should focus on initial fixation strength, good long-term bony incorporation, bioabsorbable or easily removable anchors, reduction of artificial ligament stretching, and breakage.

REFERENCES

1. Straight, C. B., France E. P, Paulos L. E., Rosenberg, T. D., and Weiss, J. A., "Soft tissue fixation to bone. A biomechanical analysis of spiked washers," *Am. J. Sports Med.*, 22, 339, 1994.
2. Robertson, D. B., Daniel, D. M., and Biden, E., "Soft tissue fixation to bone," *Am. J. Sports Med.*, 14, 398, 1986.
3. Butler, D. L., Grood, E. S., Noyes, F. R., Zernicke, R. F., and Brackett, K., "Effects of structure and strain measurement technique on the material properties of young human tendons and fascia," *J. Biomech.*, 17, 579, 1984.
4. Amis, A. A., "The strength of artificial ligament anchorages. A comparative experimental study," *J. Bone Joint Surg.*, 70B, 397, 1988.
5. Good, L., Tarlow, S. D., Odensten, M., and Gillquist, J., "Load tolerance, security, and failure modes of fixation devices for synthetic knee ligaments," *Clin. Orthop.*, 253, 190, 1990.
6. Markel, M. D., Rock, M. G., Bergenthal, D. S., Young, D. R., Vanderby, R. Jr., and Chao, E. Y., "A mechanical comparison of gluteus medius attachment methods in a canine model," *J. Orthop. Res.*, 11, 457, 1993.
7. Letsch, R., "Comparative evaluation of different anchoring techniques for synthetic cruciate ligaments. A biomechanical and animal investigation," *Knee Surg. Sports Traumatol. Arthrosc.*, 2, 107, 1994.
8. Young, F. A. and An Y. H., "A new artificial ACL anchor," *Mat. Res. Soc. Symp,* 394, 31, 1995.
9. Young, F. A. and An, Y., Adjustable ligament anchor, U.S. Patent number 5,458,601, 1995.

10. Clancy, W. G., Jr., Narechania, R. G., Rosenberg, T. D., Gmiener, J. G., Wisnefske, D. D., and Lange, T. A., "Anterior and posterior cruciate ligament reconstruction in rhesus monkey," *J. Bone Joint Surg.,* 63A, 1270, 1981.
11. Arnoczky, S. P., Tarvin, G. B., and Marshall, J. L., "Anterior cruciate ligament replacement using patellar tendon. An evaluation of graft revascularization in the dog," *J. Bone Joint Surg.,* 64A, 217, 1982.
12. Kurosaka, M., Yoshiya, S., and Andrish, J. T., "A biomechanical comparison of different surgical techniques of graft fixation in anterior cruciate ligament reconstruction," *Am. J. Sports Med.,* 15, 225, 1987.
13. St. Pierre, P., Olson, E. J., Elliott, J. J., O'Hair, K. C., McKinney, L. A., and Ryan, J., "Tendon-healing to cortical bone compared with healing to a cancellous trough," *J. Bone Joint Surg.,* 77A, 1858, 1995.
14. Fennell, C. W., Ballard, J. M., Pflaster, D. S., and Adkins, R. H., "Comparative evaluation of bone suture anchor to bone tunnel fixation of tibialis anterior tendon in cadaveric cuboid bone: a biomechanical investigation," *Foot Ankle Int.,* 16, 641, 1995.
15. Ono, K., Williams, G. R., Clem, M., Hwa, J. L., Wirth, M. A., and Aufdemorte, T. B., "Repair of soft tissue to bone using a biodegradable suture anchor," *Orthopaedics,* 20, 1051, 1997.
16. Claes, L., Dürselen, L., Kiefer, H., and Mohr, W., "The combined anterior cruciate and medial collateral ligament replacement by various materials: a comparative animal study," *J. Biomed. Mater. Res.,* 21 (A3 Suppl), 319, 1987.
17. Holden, J. P., Grood, E. S., Butler, D. L., Noyes, F. R., Mendenhall, H. V., et al., "Biomechanics of fascia lata ligament replacements: early postoperative changes in the goat," *J. Orthop. Res.,* 6, 639, 1988.
18. Powers, D. L., Jacob, P. A., and Drews, M. J., "Anatomical reconstruction of the anterior cruciate ligament in goats," *J. Invest. Surg.,* 4, 191, 1991.
19. Bolton, C. W. and Bruchman, W. C., "The Gore-Tex™ expanded polytetrafluoroethylene prosthetic ligament," *Clin. Orthop.,* 196, 202, 1985.
20. McPherson, G. K, Mendenhall, H. V., Gibbons, D. F., Plenk, H., Rollmann, W., et al., "Experimental mechanical and histologic evaluation of the Kennedy ligament augmentation device," *Clin. Orthop.,* 196, 186, 1985.
21. Ballock, R. T., Woo, S. L., Lyon, R. M., Hollis, J. M., and Akeson, W. H., "Use of patellar tendon autograft for anterior cruciate ligament reconstruction in the rabbit: a long-term histologic and biomechanical study," *J. Orthop. Res.,* 7, 474, 1989.
22. Gillquist, J. and Good, L., "Load and length changes in an artificial ligament substitute. Ten cases of anterior cruciate ligament reconstruction," *Acta Orthop. Scand.,* 64, 575, 1993.
23. Kasperczyk, W. J., Bosch, U., Oestern, H. J., and Tscherne, H., "Staging of patellar tendon autograft healing after posterior cruciate ligament reconstruction. A biomechanical and histological study in a sheep model," *Clin. Orthop.,* 286, 271, 1993.
24. Jackson, D. W., Simon, T. M., Lowery, W., and Gendler, E., "Biologic remodeling after anterior cruciate ligament reconstruction using a collagen matrix derived from demineralized bone," *Am. J. Sports Med.,* 24, 405, 1996.
25. Paschal, S. O., Seemann, M. D., Ashman, R. B., Allard, R. N., and Montgomery, J. B., "Interference fixation versus postfixation of bone-patellar tendon-bone grafts for anterior cruciate ligament reconstruction," *Clin. Orthop.,* 300, 281, 1994.
26. Gottsauner-Wolf, F., Egger, E. L., Markel, M. D., Schultz, F. M., and Chao, E. Y. S., "Fixation of canine tendons to metals," *Acta Orthop. Scand.,* 65, 179, 1994.
27. Rupp, S, Krauss, P. W, and Fritsch, E. W., "Fixation strength of a biodegradable interference screw and a press-fit technique in anterior cruciate ligament reconstruction with a BPTB graft," *J. Arthrosc.,* 13, 61, 1997.
28. Tuompo, P., "Strength of the fixation of patellar tendon bone grafts using a totally absorbable self-reinforced poly-L-lactide expansion plug and screw. An experimental study in a bovine cadaver," *Arthroscopy,* 12, 422, 1996.
29. Kousa, P, Jarvinen, T. L., Pohjonen, T, Kannus, P., Kotikoski, M, and Jarvinen, M., "Fixation strength of a biodegradable screw in anterior cruciate ligament reconstruction," *J. Bone Joint Surg.,* 77B, 901, 1995.
30. Shapiro, J. D., Jackson, D. W., Aberman, H. M., Lee, T. Q., and Simon, T. M., "Comparison of pullout strength for seven- and nine-millimeter diameter interference screw size as used in anterior cruciate ligament reconstruction," *Arthroscopy,* 11, 596, 1995.

31. Pena, F., Grontvedt, T., Brown, G. A., Aune, A. K., and Engebretsen, L., "Comparison of failure strength between metallic and absorbable interference screws. Influence of insertion torque, tunnel-bone block gap, bone mineral density, and interference," *Am. J. Sports Med.*, 24, 329, 1996.
32. Noyes, F. R. and Grood, E. S., "The strength of the anterior cruciate ligament in humans and rhesus monkeys," *J. Bone Joint Surg.*, 58A, 1074, 1976.
33. Noyes, F. R., Butler, D. L., Grood, E. S., Zernicke, R. F., and Hefzy, M. S., "Biomechanical analysis of human ligament grafts used in knee-ligament repairs and reconstructions," *J. Bone Joint Surg.*, 66A, 344, 1984.
34. Rowden, N. J., Sher, D., Rogers, G. J., and Schindhelm, K., "Anterior cruciate ligament graft fixation. Initial comparison of patellar tendon and semitendinosus autografts in young fresh cadavers," *Am. J. Sports Med.*, 25, 472, 1997.
35. Butler, D. L., "Evaluation of fixation methods in cruciate ligament replacement," *Instr. Course Lect.*, 36, 173, 1987.
36. France, E. P., Paulos, L. E., Rosenberg, T. D., and Harner, C. D., "The biomechanics of anterior cruciate allografts," in *Prosthetic Ligament Reconstruction of the Knee*, Friedman, M. J., Ferkel, R. D., Eds., Saunders, Philadelphia, 1988, Chapter 25.
37. Woo, S. L., Hollis, J. M., Adams, D. J., Lyon, R. M., and Takai, S., "Tensile properties of the human femur-anterior cruciate ligament-tibia complex. The effects of specimen age and orientation," *Am. J. Sports Med.*, 19, 217, 1991.
38. Ng, G. Y., Oakes, B. W., McLean, I. D., Deacon, O. W., and Lampard, D., "The long-term biomechanical and viscoelastic performance of repairing anterior cruciate ligament after hemitransection injury in a goat model," *Am. J. Sports Med.*, 24, 109, 1996.
39. Smith, C. W., Young, I. S., and Kearney, J. N., "Mechanical properties of tendons: changes with sterilization and preservation," *J. Biomech. Eng.*, 118, 56, 1996.
40. Ishizue, K. K., Lyon, R. M., Amiel, D., and Woo, S. L., "Acute hemarthrosis: a histological, biochemical, and biomechanical correlation of early effects on the anterior cruciate ligament in a rabbit model," *J. Orthop. Res.*, 8, 548, 1990.
41. King, G. J., Edwards, P., Brant, R. F., Shrive, N. G., and Frank, C. B., "Intraoperative graft tensioning alters viscoelastic but not failure behaviours of rabbit medial collateral ligament autografts," *J. Orthop. Res.*, 13, 915, 1995.
42. Matthews, L. S., Lawrence, S. J., Yahiro, M. A., and Sinclair, M. R., "Fixation strengths of patellar tendon-bone grafts," *Arthroscopy*, 9, 76, 1993.
43. Barber, F. A. and Deck, M. A., "The *in vivo* histology of an absorbable suture anchor: a preliminary report," *Arthroscopy*, 11, 77, 1995.
44. Barber, F. A., Herbert, M. A., and Click, J. N., "The ultimate strength of suture anchors," *Arthroscopy*, 11, 21, 1995b.
45. Barber, F. A., Herbert, M. A., and Click, J. N., "Suture anchor strength revisited," *Arthroscopy*, 12, 32, 1996.
46. Fujikawa, K., "Clinical study of anterior cruciate ligament reconstruction with the Leeds-Keio artificial ligament," in *Prosthetic Ligament Reconstruction of the Knee*, Friedman, M. J., Ferkel, R. D., Eds., Saunders, Philadelphia, 1988, Chapter 20.
47. Shelbourne, K. D. and Nitz, P. A., "Accelerated rehabilitation after anterior cruciate ligament reconstruction," *Am. J. Sports Med.*, 18, 292, 1990.
48. Caborn, D. N., Urban, W. P., Jr, Johnson, D. L., Nyland, J., and Pienkowski, D., "Biomechanical comparison between BioScrew and titanium alloy interference screws for bone-patellar tendon-bone graft fixation in anterior cruciate ligament reconstruction," *Arthroscopy*, 13, 229, 1997.
49. Kohn, D. and Rose, C., "Primary stability of interference screw fixation. Influence of screw diameter and insertion torque," *Am. J. Sports Med.*, 22, 334, 1994.
50. Jomha, N. M., "Effect of varying angles on the pullout strength of interference screw fixation," *Arthroscopy*, 9, 580, 1993.
51. Sabiston, P., Frank, C., Lam, T., and Shrive, N., "Allograft ligament transplantation. A morphological and biochemical evaluation of a medial collateral ligament complex in a rabbit model," *Am. J. Sports Med.*, 18, 160, 1990.

Part VII

Animal Models of Spinal Conditions

27 Animal Models of Spinal Instability and Spinal Fusion

Harvinder S. Sandhu, Linda E. A. Kanim, Federico Girardi, Frank P. Cammisa, and Edgar G. Dawson

CONTENTS

I. Introduction 505
II. Models of Instability and Internal Fixation 506
   A. Biomechanical Instability 506
   B. *Ex Vivo* Models of Biomechanical Instability and Fixation 506
   C. *In Vivo* Models of Spinal Instability and Fixation 509
   D. Summary of Instability and Internal Fixation 509
III. Animal Models of Spinal Fusion and Biology 511
   A. Posterior Fusion Animal Models 516
   B. Anterior Fusion Animal Models 517
   C. Surgical Approaches and Techniques 518
   D. Vertebral Segments 519
   E. Quadrupedal Animal Models 519
   F. Methods of Evaluation 520
   G. Inhibitors to Spinal Fusion 520
   H. Summary of Animal Models of Spinal Fusion 520
IV. Conclusions 521
References 521

I. INTRODUCTION

Spinal fusion (arthrodesis) is the standard surgical treatment for instability of the spine and has been utilized clinically since 1911. Albee experimentally examined the concept of intersegmental spinal fusion by conducting macroscopic and microscopic studies of posterior spinous process fusions in canines.[1] Since that time, live animal "models" have been routinely used to evaluate individual factors influencing the outcome of spinal surgery. They provide reproducible and quantifiable information which is often difficult or impossible to obtain from human subjects, cadaveric human or animal models, or simulations such as finite element analysis.

    Animal models of the spine are, in many ways, distinct from those used to examine other interventions to the skeletal system. Often, spinal column realignment and stabilization with arthrodesis in a mechanically advantageous arrangement may be the preferred treatment of a spinal disorder. In contrast to arthrodesis of the major joints of the extremities, selective elimination of motion across pathologic vertebral motion segments does not necessarily result in significant functional incapacity for the organism as a whole. Fusions of destabilized vertebral segments are among the most frequently modeled treatments to the musculoskeletal system.

The first section of this chapter will review general concepts of spinal instability and stabilization, and provide a summary of the animal models that have been used previously and the applications for which they are useful. The second section will focus on animal models that have been used to evaluate novel osteoinductive growth factors or implantable, fusion enhancing biomaterials. Models that enable examination of systemic factors influencing the fusion process will also be discussed.

II. MODELS OF INSTABILITY AND INTERNAL FIXATION

Animal models for *ex vivo* and *in vivo* simulation of spinal instability and stability, and of *in vivo* spinal fusion are presented in this section.

A. BIOMECHANICAL INSTABILITY

Although poorly understood and difficult to assess clinically, spinal instability is considered one of the primary causes of chronic spine-related pain and progressive neurologic deficit. It is speculated that dysmorphic and excessive intervertebral motions cause abnormal deformations of ligaments, apophyseal joint capsules, annular fibers and vertebral endplates such that the nociceptors in these respective tissues trigger pain responses.[2] In addition, the increased excursion of the adjacent neural elements is thought to cause compression or stretching of these structures resulting in radicular symptoms and deficits. The precise relationship between dysmorphic motion and clinical symptoms has not been determined. However, accurate assessment of instability is deemed critical to appropriate management of these disorders.

Panjabi's revised definition of instability is based upon the size of the neutral zone, "that part of the range of physiological intervertebral motion, measured from the neutral position, within which the spinal motion is produced with a minimal internal resistance."[3] The neutral zone may increase with injury and incompetence of constraining soft tissue and bony elements, and degeneration and incompetence of the constraining apophyseal joints. It may decrease with paraspinal muscular strengthening, osteophyte formation, internal fixation, and osseous fusion.

A change in the neutral zone may be more sensitive than a change in the corresponding range-of-motion for determining instability. Neutral zone measurements incorporate muscular and peripheral soft tissue contribution to spinal instability whereas range-of-motion measures may not. Historically, biomechanical studies have focused on the mechanics of passive components and, consequently, range-of-motion measures.[4] Thus, implications of active components such as muscular forces in both normal and pathologic conditions are not well understood.

Recently, a simulation of lumbar instability, examining passive and active stabilization components, was developed in an *in vivo* porcine model.[5,6] Sequential surgical injuries in the L3-L4 intersegment were created and sagittal kinematics were measured. Greater axial translation occurred following injury to the disk and annulus; greater sagittal rotation and shear translation occurred following graded injuries to the facet joint. Interestingly, muscular stimulation following each of these selected injuries increased sagittal rotation and shear translation while decreasing axial translation. Paradoxically increasing range-of-motion, muscular stimulation reduced abrupt kinematic behavior in the neutral region, reducing the neutral zone. This methodology provides one of the few experimental demonstrations of the neutral zone theory of instability. However, anatomic dissimilarities between the human and porcine posterior spinal elements and muscular attachment points impose limitations to this model.

B. *EX VIVO* MODELS OF BIOMECHANICAL INSTABILITY AND FIXATION

Several animal models have traditionally been used to evaluate the passive components of spinal stability, examples as shown in Table 1.[7-14] Among *ex vivo* cadaveric models, the bovine spine has

TABLE 1
Ex Vivo Biomechanical Evaluation of Animal Spine Models

Species	Spine Specimen, Preparation	Conditions, Evaluation	First Author, year[Ref.]
Dog	Rib cage-thoracic spine complexes T5–T9	Resection of costovertebral joints; destruction of the rib cage	Oda 1996[7]
Calf	Anterior and middle column defect L2–L5	Anterior fixation with intervertebral body graft after discectomy and endplate excision of L3–L4; anterior fixation only	An 1995[8]
	Destabilization with fixation	TSRH;* bone graft; bone graft + TSRH; BAK†; BAK + TSRH; normal spine	Brodke 1997[9]
Pig and Human	Interspinous lig. and bone specimens	Collagen lig. interrupted with progressive disruption	Dickey 1996[10]
Pig	L3 corpectomy with/fixation	Anterior strut graft with spinal nail fixation	Dawson 1996[11]
	C3–4 discectomy and dissection of posterior longitudinal lig.	Hydroxyapatite with anterior plating; hydroxyapatite; ICBG‡ with anterior plating; ICBG	Takahashi 1997[12]
Sheep	T8–L7 bisegmental specimens	Mechanical testing: extension-flexion, axial rotation, lateral bending T1–T8 three segment specimens	Wilke 1997[13]
Dolphin	Vertebral column	External loads applied to intervertebral segments	Long 1997[14]

* TSRH = Texas Scottish Rite Hospital fixation device
† BAK = intervertebral fixation device (Spine-Tech, Minneapolis, MN)
‡ ICBG = iliac crest bone graft

been the most widely used and validated for the thoracic and lumbar regions. Calf spine specimens, especially, are relatively homoscedastic in bone strength and mineral density and may be preferred over human specimens for mechanically testing particular instrumentation constructs.

Wilke et al. determined range-of-motion, neutral zone, and stiffness properties of thoracolumbar bovine spines (T6-L6) under pure moment loading in flexion and extension, axial left/right rotation, and right/left lateral bending. Similarities in axial rotation and lateral bending were noted between bovine and human cadaveric spines. The bovine spine was suggested to be in a limited way a substitute for the human spine.[13]

The pull-out strength of bone screws placed in bovine vertebral pedicles correlated significantly with the bone mineral density of vertebral bodies. Since the bovine vertebral bodies had significantly higher and less variable mineral density (146 mg/ml, p<0.05) than humans, failures consistently occurred within the implant construct rather than the bone implant interface. The bovine spine is ideal for testing the performance of spinal fixation constructs since there is less interspecimen variability compared with the human cadaveric spine.[15]

Zdeblick et al. have performed L3 corpectomies in calf spines to test stabilization afforded by several anterior fixation devices. In torsion, they found the Kaneda anterior fixation device to be the stiffest, while in axial compression and flexion-extension modes, the Kaneda device and the Texas Scottish Rite Hospital (TSRH) fixation device were the stiffest.[16,17] Earlier, L3 corpectomies in calf spines were used to examine both anterior and posterior fixation devices under mechanical, non-destructive cyclical testing in axial compression, rotation, and flexion.[18] Again the Kaneda device, with fixation one level above and one level below the corpectomy site, was the stiffest anterior fixation system. The Cotrel-Dubousset and Steffee transpedicular fixation systems, incorporating vertebral segments two levels above and below the corpectomy site, were the stiffest posterior fixation systems. Other investigators have used the bovine spine as a model to evaluate instrumentation constructs in the treatment of traumatic or scoliotic conditions.[19,20,21,22]

The bovine model is also an excellent choice for studying the biomechanics of lumbosacral fixation.[23] Although the bovine spine anatomically has smaller vertebral width, six lumbar vertebrae, less lordosis, and larger transverse processes, the range-of-motion at the usually sagittally hypermobile lumbosacral junction was quite similar to the human spine.

The bovine spine was used to biomechanically evaluate the efficacy of posterior instrumentation systems for stabilization of artificial isthmic spondylolisthesis of the lumbosacral joint and found that transpedicular fixation was superior to older distraction constructs.[24]

The bovine spine has also recently been used to evaluate the performance of disc-replacing intervertebral fixation devices using either the anterior or posterior approach for lumbar intervertebral fusion (ALIF, PLIF). *Ex vivo,* an intact bovine lumbar spine was used to demonstrate that PLIF procedures performed with structural iliac crest bone graft (ICBG) were less stiff than intact spines.[9] The use of transpedicular fixation increased initial stiffness significantly (2.5 times the intact spine), and the use of a threaded, fenestrated, cylindrical, and hollow intervertebral fixation device (BAK device; Spine-Tech, Minneapolis, MN) without transpedicular instrumentation, also increased initial stiffness (2 times the intact spine). Other intervertebral fixation devices such as the NOVUS™ fusion device (Sofamor-Danek, Memphis; TN) have been evaluated in the bovine spine with similar results.[25,26,27,28]

Ex vivo animal models of the cervical spine are less frequently used than those of the thoracic and lumbar spines. A bovine model of flexion-distraction cervical injury was used to study stabilization methods in a constrained, nonrepetitive loading environment.[29] Recently, a porcine model was used to evaluate three methods of cervical spine stabilization.[30] In contrast to previous animal and human cadaveric instability models, the specimens were destabilized with a one-level cervical corpectomy and tested under unconstrained and repetitive loading following instrumentation. This investigation demonstrated that anterior grafting combined with posterior lateral mass plating achieved maximum stability. The conclusion was that despite a difference in facet orientation, the porcine spine was ideal for simulation of human cervical spine fixation due to the close size match and otherwise similar geometry.

C. IN VIVO MODELS OF SPINAL INSTABILITY AND FIXATION

In vivo animal models of destabilization and fixation are necessary to examine the effects of skeletal repair and intersegmental fusion on intersegmental instability (Table 2).

An *in vivo* animal model of both anterior and posterior column instability was developed to study the effects of bone arthrodesis and remodeling after instrumentation.[50–52] A destabilizing lesion was created at the L5-L6 intersegment by sectioning both the disc annulus and the posterior elements in adult beagles. Animals that underwent instrumentation had a higher probability of achieving fusion than those that did not. Furthermore, among all successful fusions, those that had instrumentation were stiffer in torsion, axial compression, and flexion following removal. It was also shown, in this model, that osteoporosis was linearly related to the rigidity of the implants in one of the first demonstrations of device related osteoporosis.

In order to simulate an unstable burst fracture, laminectomy, partial facetectomy, and corpectomy of the L5 vertebral segment were performed in canines. Comparison of anterior strut graft with instrumented fixation and without fixation yielded a higher fusion rate in the former. Among the fusions, the spines that had been instrumented were significantly stiffer to torsion.[17]

Rates of fusion may vary considerably within a quadrupedal animal depending upon the segmental level that is being treated.[46] Specifically, posterior intertransverse process fusion (PLSF) without instrumentation was achieved across the interlumbar motion segments of the sheep lumbar spine. In contrast, fusion almost never occurred at the lumbosacral junction with the same treatment. This finding was attributed to increased motion at the lumbosacral junction compared to the intralumbar levels. Similarly, the findings at lumbosacral junction in canines may be attributed to the kinematic behavior at this level.[53] These findings along with studies on kinematic behavior of quadrupedal animals indicate that forces, loads, and treatment results are not equal across segmental levels.[36,54,38,55,56,57,58]

Load-sharing capacity of spinal instrumentation and the posterolateral fusion mass has also been studied in the sheep model following posterolateral fixation and bone grafting with transpedicular fixation.[48] Stability with instrumentation followed by stability after removal of the instrumentation was measured. Fixation provided anterior and middle spinal column support which enabled the spinal intersegments to better resist eccentric loads in the sagittal plane.

Increased motion at spinal levels adjacent to spinal fusion is well documented in clinical literature. *In vivo* motion data from the L2-L3 intervertebral segment were measured before and 12 weeks after posterior spinal instrumentation from L3-L7 in the canine.[39] Vertebral rotations in the coronal plane and excursion of the facet joints increased significantly after the caudal instrumentation. Tissue responses to changes in abnormal joint motion may be investigated using this canine model.

The influence of spinal implants and instrumentation for anterior cervical fusion has been examined in the caprine model.[59,17] A three level anterior cervical discectomy without posterior destabilizing lesions was performed. Several different implant conditions were tested in the discectomy sites including a sham condition. Each discectomy site in a single animal was treated with the same condition.[60] A higher rate of fusion was observed with autograft than with allograft. In a second study, the addition of anterior cervical instrumentation with the autograft did not increase the fusion rate.[61]

D. SUMMARY OF SPINAL INSTABILITY AND INTERNAL FIXATION

Ex vivo animal cadaveric models have generally been used to test instrumentation constructs without the confounding influences of quality of the bone and biologic factors associated with

TABLE 2
In Vivo Animal Models Used for the Biomechanical Evaluation of Spinal Instability

Animal	Type of Procedure, Level	Conditions and Evaluations	Time	1st Author, Year[Ref.]
Rabbits	Injury to facets L1–2, L2–3	Low activity vs. high activity; bilateral facet excision L1–2, L2–3; unilateral facet excision L1–2, L2–3; facets exposed only	2 weeks, 1, 6, 12 months	Stokes 1989[31]
	Posterior fusion of interspace L3–L6	ICBG* + instrumentation with or without decortication; exposure + instrumentation; exposure only	6 weeks	Ishikawa 1994[32]
Dogs	Posterior midline fusion L4–L5	Bone graft + wire fixation+ PMMA; bone graft + wire fixation; bone graft	8 weeks	Feighan 1995[33]
	Distraction hooks and rods L2, L3	Instrumented; removal of instrumentation	2–6 months	Kahanovitz 1984[34]
	Steinman pins, transpedicular L2, L3	Standing, walking and moving from sitting to walking, turning, and moving from a 4-leg stance to hind leg position. documentation of motion	N/A	Wood 1992[35]
	Strain gage, L2–3 facet joints	Loading during static and dynamic activities	3 days	Butteroski 1992[36]
	Posterior facet fusion L3–L4, L4–L5	6.35 mm diam. rods retained and removed 12 weeks; 4.76 mm diam. rods retained & removed after 12 weeks; not operated (control)	12 weeks	Craven 1994[37]
	L3–L7, Steinman pin fixation	In vivo vs. in vitro motion at segment L2–L3	1, 12 weeks	Dekutoski 1994[38]
	Posterior fusion L3–L7, Steinman pins	In vivo motion data collected while animal walked on a treadmill.	1, 12 weeks	Schendel 1995[39]
	Posterior fusion of L1–L5	ICBG* + screws at L1, L3, L5 + rods; ICBG	3, 6 months	Kioschos 1996[40]
	Intervertebral body fusion	Dx cervical spondylomyelopathy with plug for distraction-stabilization	?	Dixon 1996[41]
	Laminar & PLSF‡, fixation L2, L3, L4	Macroporous ceramic, autograft (laminar vs. intertransverse sites)	9 months	Delecrin 1997[42]
	Bilateral laminectomy and facetectomy L5–L6, with fusion	Fusion alone vs. fusion with down sized VSP plates	24 weeks	Edwards 1997[43]
Pigs	Surgical injury to L3–L4 segment	Disc injury; facet joints removed; facet and transverse processes removed	N/A	Kaigle 1995[6]
	Chronic lesion model L3–L4	Anulus injury; facet capsule injury; facet joint slit; facet joint wedge; sham	3 months	Kaigle 1997[5]
	Discectomy and fusion C2–C3, C3–C4, C4–C5, C5–C6	HA: tricortical ICBG	6, 12, 24 weeks	Pintar 1994[44]
Sheep	Posterior fusion L3–L5	3.2, 4.8, 6.4 cm rod, pedicle screws + ICBG; non rodded spines	16 weeks	Johnston 1995[45]
	Posterior laminar fusion/Steinman pins	ICBG across decorticated lamina of L3, L4, L5, L6, sacrum; L6–S1 fusion with decortication, fixation + ICBG; nonoperated	4–43 weeks	Nagel 1991[46]
	PLSF, L3–L4, L4–5	Autograft + decortication + C-D† instrumentation; decortication + C-D instrumentation; C-D instrumentation; non operated	1 year	Guigui 1994[47]
	Posterior fusion of L1–L5	ICBG + transpedicular screw fixation; ICBG removal of fixation	16 weeks	Kotani 1996[48]
	PLSF, L2–L3, L4–5	ICBG; ICBG + fixation; one level with transpedicular screw fixation.	8, 16 weeks	Kanayama 1997[49]
	ALIF§, L4–L5 or L5–L6	ICBG + Ti cage; autograft dowel; sham; nonoperated	24 weeks	Sandhu 1996[25]

* ICBG = iliac crest bone graft
† C-D = Cotrel-Dubousset instrumentation
‡ PLSF = Posterolateral intertransverse process fusion
§ ALIF = anterior lumbar interbody fusion

osseous fusion. In contrast, *in vivo* animal models mainly incorporate the effects of biologic factors such as skeletal repair and remodeling mechanisms and mechanical factors such as the *in vivo* kinematic behavior of specific vertebral segments.

The bovine spine has been the most widely used among animal models for examining the effects of destabilization and internal fixation of the thoracic and lumbar spines *ex vivo*. This has largely been due to the general anatomic similarities and the increased and highly consistent mineral density of the bovine spine compared to the human spine. Both the bovine and porcine models have been used similarly to examine the cervical spine *ex vivo*. Although neither of these models anatomically mimics the human spine with regard to cervical facet joint orientation, the relative similarities of size and geometry were sufficient to warrant their use.

The canine, sheep, and porcine models have been most commonly used to examine both anterior and posterior instability and fixation in the thoracic and lumbar spines *in vivo*. Canines have been particularly popular because of ease of post-operative care in enclosed institutional environments and the ability to treat these spines with conventional forms of internal fixation devices. The alpine goat model is one of the few established for *in vivo* study of the cervical spine primarily due to the ease of approaching the cervical spine surgically, the similarity of vertebral size and geometry with the human, the larger sized intervertebral discs, and the relative ease of post-operative management and evaluation. Finally, the porcine model has been successfully used to examine the active muscular component of spinal column stability, and to examine such stability from the standpoint of the neutral zone.

III. ANIMAL MODELS OF SPINAL FUSION AND BIOLOGY

Established animal models in spinal fusion studies, their specific advantages and limitations, and the significance of the data derived from each are reviewed. Animal models evaluating osteoinductive growth factors are presented in Table 3, and those evaluating biomaterials are presented in Table 4.

Biologic factors influence the success or failure of spinal fusion. The impact of biologic "tools" such as osteoinductive growth factors, which stimulate the biologic processes of skeletal repair, must also be examined rigorously in animal and procedural models wherein successful fusion is difficult to achieve. In order to examine these factors carefully, selective animal models for the study of biologic factors of spinal fusion must be used and interpreted appropriately.

The requirements of an animal model for the study of biologic processes differ from those examining mechanical factors in that the former considers simulation of human geometric relationships and loading characterisics to be relevant, but not critical. The primary endpoint is usually the formation of an osseous fusion mass which connects adjacent vertebral motion segments and exists in space previously occupied by soft tissue. Although local kinematics and mechanical forces may influence success or failure of the fusion attempt, other factors, such as fusion technique, fusion location, evolutionary complexity of the animal, size of the animal,[112] bone architecture,[113] and contents of the bone graft or bone graft substitute may take precedence in determining the outcome of treatment. For example, it is common in lower animals that facet and interlaminar fusions occur simply by surgical exposure and periosteal stripping of the posterior spinal elements.[54] In contrast, the nonhuman primate spine, like the human, often fails to achieve either facet, interlaminar, or intertransverse process fusion despite meticulous decortication and abundant autogenous bone graft along the transverse processes.[149] The speed at which osseous fusion takes place also varies considerably between species and is influenced by the location of the fusion and the type of graft.

Historically, several animal models have been used to examine biologic variables which affect spinal fusion. Rats, rabbits, dogs, guinea pigs, sheep, goats, and nonhuman primates among others, have been treated with facet, interlaminar, intertransverse process, spinous process, anterior interbody and posterior interbody fusions. Study designs have included single level fusions, multiple adjacent level fusions with distinct or similar implants at each level, multiple separated level fusions with distinct or similar implants at each level, and single level fusions with distinct implants on either side.

TABLE 3
Animal Models Employed in the Evaluation of rhBMPs (rhBMP-2, rhBMP-7, etc.)

Animal	Fusion Procedure	Experimental Conditions, Implant	Time (wks)	1st Author, Year[Ref.]
Rats	PLSF,* L4–L5	rhBMP-2 + collagen; ICBG; coillagen only; decortication only	4	Dawson, 1998[95]
Rabbits	PLSF,* L5–L6	rhBMP-2 + collagen; rhBMP-2 + collagen + ICBG†; rhBMP-2 + ICBG; ICBG; or collagen	5	Schimandle 1995[62]
	PLSF, L5–L6	bBP‡-DBM-collagen; DBM alone; autograft alone	5	Boden 1995[63]
	PLSF, L5–L6	rhBMP-2 + collagen; ICBG	4	Hollinger 1996[64]
	Lateral intertransverse process L4–L5	rhBMP-2 + collagen; collagen	10	Boden 1996[65]
	PLSF, L5–L6	bBP + biocoral + collagen; biocoral alone;	5	Boden 1997[66]
	PLSF, L5–L6 with ICBG	Fusion masses separated into outer and central zones, RNA extracted RT/PCR done for BMP	10	Boden 1997[67]
Dogs	Laminar medial transverse process T6–T7, T8–T9, T10–T11, T12–T13	Decortication + autograft (spinous process); BMP/PLA§ or PLA; Decortication only	3, 24	Lovell 1989[68]
	Facet spinous process fusion T13–L1, L2–L3, L4–L5, L6–L7	rhOP-1§ (rhBMP-7) + collagen; collagen; match sticks autograft; implant	6, 26	Cook 1994[69]
	Posterior segmental fusion, L1–L2, L3–L4, L5–L6	Superficial decortication of lamina instrumented: dual plates, rhBMP-2/PLGA;^ PLGA alone; proximal humerus autograft	12	Muschler 1994[56]
	PLSF, L4–5 with decortication	rhBMP-2 + open-pore PLA; open-pore PLA; morselized ICBG	12, 32	Sandhu 1995[70]
	PLSF, L4–5 with decortication	rhBMP-2 + open-pore PLA; open-pore PLA; morselized ICBG	12	Sandhu 1996[71]
	PLSF, L4–5	rhBMP-2 + open-pore PLA with or without decortication	12	Sandhu 1997[72]
	PLSF, L4–5 with no decortication	rhBMP-2 + collagen; rhBMP-2 + open-pore PLA;	12	Sandhu 1996[73]
	PLSF, L4–5 with decortication, bilateral paraspinal approach	rhBMP-2 + collagen; rhBMP-2 + open-pore PLA collagen; corticocancellous rib graft	12	David 1996,[74] 1996[75]
	Spinous processes, lamina, facet fusion, T13–L1, L4–L5, L2–L3, L6–L7	rhBMP-2 + collagen + ICBG; collagen + ICBG + cohesive paste material; ICBG; no implant	6,12	Sheehan 1996[76]
	PLSF, T13–L1, L2–L3, L4–L5, L6–L7	rhBMP-2/collagen + morselized ICBG; ICBG + collagen	16	Helm 1997[77]
	PLSF, T13–L1, L2–L3, L4–L5, L6–L7	rhBMP-2/collagen + ICBG collagen + morselized ICBG autologous morselized ICBG; no implant	12	Li 1996[78]
	PLSF, L4–L5	54, 215, or 860 mg rhBMP-2 + collagen; 215 mg rhBMP-2 + open-pore PLA; collagen; ICBG	12	Sandhu 1996[73]
	Bilateral laminectomy, L5	rhBMP-2/collagen placed on the dura; autogenous bone from laminectomy site placed on dura	12	Kwiatkowski 1997[79]
	PLSF, L1–L2, L3–L4, L5–L6	ICBG alone; ICBG + rhBMP-2, ICBG + collagen "sandwich" + rhBMP-2, ICBG + collagen morsels + rhBMP-2, ICBG + PLGA sponge sandwich + rhBMP-2, and ICBG + open-pore PLA morsels + rhBMP-2	8	Fischgrund 1997[80]

Animal	Procedure/Levels	Materials	N	Reference
Goats	AIF,# C1–C2, C2–C3, C3–C4	Ti BAK + rhBMP-2; HA∞-coated BAK + local reamed graft; BAK + local reamed graft	12	Zdeblick 1995[81]
	AIF, C2–C3, C5–C6	50/50 porous HA-TCP◊ (30, 50, or 70% porosities); 50/50 HA-TCP; autograft	12, 24	Toth 1995[82]
	AIF, C2–C3, C4–C5	rhBMP-2 + HA-TCP + plate; HA-TCP + Ti cage; rhBMP-2 + HA-TCP + Ti cage; HA-TCP + plate; rhBMP-2 + HA-TCP	16	Toth 1997[83]
	AIF, C2–C3, C4–C5	rhBMP-2 + 50% porous 50/50 HA-TCP; HA-TCP; rhBMP-2 calcium carbonate (Bicoral); Bicoral alone	12,16	An 1997[84]
	Anterior cervical fusion C2–C3, C3–C4, C4–C5	rhBMP-2/collagen sponge + plate fixation; saline/collagen sponge + plate fixation; ICBG + plate fixation	11	Bolesta 1997[85]
	Anterior cervical discectomy, fusion, C2–C3, C3–C4, C4–C5	rhBMP-2/collagen sponge + Ti lordotic cages; local bone graft + Ti lordotic cages; rhBMP-2/collagen sponge + porous Ta¶ intervertebral cages; local bone graft + Ta cage	12, 24	Moore 1997[86]
Sheep	ALIF,¢ L4–L5 or L5–L6	Autograft dowel; autograft + Ti cage; rhBMP-2/collagen sponge/Ti cage	24	Sandhu 1996[26]
	ALIF, T5–T6, T7–T8, T9–10	BAK + rhBMP-7/collagen sponge; BAK + ICBG; tricortical ICBG, video assisted thoroscopy	16	Cunningham 1996[87]
	ALIF, L4–L5 or L5–L6	rhBMP-2 with porous Ta cylinder; porous Ta cylinder	24	Sandhu 1996[27]
Rhesus	PLSF, L4-L5	bBP/DBM (groups); DBM alone	12	Boden 1996[88]
Monkeys	Laminectomy & lateral intertransverse fusion, L4–L5	rhBMP-2+collagen; rhBMP-2 in HA-TCP (60:40); collagen, report of technique only	N/A	Boden 1996,[65], 1997[89]
	ALIF, L7–S1	rhBMP-2/collagen with rhBMP-2 soaked freeze dried cortical dowel; ICBG with freeze dried allograft cylinder	24	Boden 1997[89] Hecht 1997[90]
	ALIF, L-S1	rhBMP-2/collagen/TIF£; collagen/TIF, laparoscopic exposure	24	Boden 1997[91]

* PLSF = posterior intertransverse process fusion
† ICBG = iliac crest bone graft
‡ bBP = bovine osteoinductive bone protein
ø PLA = polylactic acid
^ PLGA = polylactide-co-glycolide acid
§ rhOP = recombinant human osteogenic protein, same as rhBMP-7
# AIF = anterior interbody fusion
∞ HA = hydroxyapatite
¶ Ta = tantalum
£ TIF = Tapered Interbody Fusion Device
¢ ALIF = anterior lumbar interbody fusion
◊ TCP = tricalcium phosphate

TABLE 4
Animal Models Used for the Evaluation of Spinal Fusion Implants

Animal	Fusion Procedure(s)	Experimental conditions, implants	Time	1st Author, Year[Ref.]
Mice	PLSF* C1–C2	Autologous tail; ICBG; † transgenetic ICBG; syngenetic donor	4 weeks	Rhee, 1998[92]
Guinea Pigs	L-S (guinea pigs), L4–L5 (dogs)	Dowel graft from iliac crest; H-graft; posterior ICBG (guinea pigs); upper thoracic spinous process graft (dog)	6–72 weeks	Thomas 1975[93]
Rats	PLSF, last three lumbar vertebrae, last 3 thoracic vertebrae	DBM one side + decortication; DBM both sides, spinous processes, interspinous lig., capsule, cartilage of facet joints removed + decortication; bone graft, spinous processes, interspinous lig., capsule, and cartilage of facet joints removed + decortication	3–60 weeks 2, 8, 12wks	Guizzardi 1991[94]
	PLSF, L4–5	Decortication only; ICBG	12 weeks	Dawson 1998[95]
Rabbits	Spinous interspace fusion	DBM; allogenic deep frozen cortical bone; auto cancellous bone	? weeks	Oikarinen 1982[96]
	Spinous interspace fusion, T3–T4, T7–T8	Bone marrow + DBM; DBM; bone marrow	2–20 weeks	Lindholm 1988[97]
	Intervertebral fusion T12–L1, L1–L2	HA microcrystals; rhTGF-β1 + HA microcrystals; vs. powdered T12 rib graft	12 weeks	MacMillan 1997[98]
	PLSF, L5–L6	Healos matrix; Healos with autogenous bone marrow; Healos with heparinized autogenous bone marrow; ICBG	8 weeks	Tay 1997[99]
	AIF‡	Bicoral; Kiel; autogenous rib graft	12 weeks	Tho 1996[100]
	PLSF, L4–L5	Autogenous bone marrow + morselized ICBG; clotted blood + morselized ICBG	12 weeks	Curylo 1997[101]
	PLSF, L5–L6	2.5 cm³ per side of ICBG	6 weeks	Toribatake 1997[102]
	Interfacet interlaminar fusion L1–L2, L3–L4, L5–L6	Collagen + ceramic, 15–30 kDa proteins; ICBG; no graft	12 weeks	Muschler 1993[55]
Dogs	Facet joint fusion T7–T8	Percutaneous approach: HA cancellous bone plugs	6 mo.	Stein 1993[103]
	Unilateral facet fusion, T13–L7	DBM with freeze dried allograft from canine cancellous bone; ICBG	12 weeks	Cook 1995[104]
	Laminar and facet fusion	PLA; PLA + marrow; ICBG; decortication only	24 weeks	Callewart 1995[54]
	Interfacet interlaminar fusion	Bone marrow + collagen ceramic (50:50); autogenous cancellous bone, collagen ceramic (50:50); autogenous bone	? weeks	Muschler 1996[57]
	L1–L2, L3–L4, L5–L6			
	AIF, T7–T8	Calcium carbonate + fixation; calcium carbonate; tricortical ICBG; tricortical ICBG	8 weeks	Fuller 1996[105]
	Interbody fusion, T7–T8	HA: Biphasic (60:40) HA/TCP; Calcium carbonate; ICBG	8 weeks	Emery 1996[106]
	PLSF, L2–L3, L3–L4	40% TCP + 60% HA contact with lamina, facets, transverse processes; 40% TCP + 60% HA in lateral vertebral grooves; ICBG laminar, transverse processes	9 mo.	Delecrin 1997[42]

Animal	Procedure	Details	Duration	Reference
Goats	Anterior discectomy and fusion, C2–C3, C3–C4, C4–C5,	Frozen allograft; autogenous tricortical bone graft; disc excision only; 10/21 fused with tircortical autograft	12 weeks	Zdeblick 1992[60]
	Anterior discectomy and fusion, C2–C3, C3–C4, C4–C5	ProOsteon (Interpore) + cervical plate; ProOsteon (Interpore) + no plate; porous HA; autograft + anterior plate; autograft; no fusion	12 weeks	Zdeblick 1994[61]
	PLIF§, L4-L5	Carbon fiber-reinforced polymer implant-ICBG; allograft bone	6 mo.	Brantigan 1994[107]
Pigs	ALIF¶, L7–S1	Open procedure vs. laproscopic procedure implants: ICBG with Ti cage (MOSS)	12 weeks	Riley 1997[108]
Sheep	PLSF, L3–L4, L4–5	Porous coral or biphasic ceramic + decortication + Cotrel-Dubousset instrumentation	1 yr.	Guigui 1994[47]
	ALIF, L1–L2, L3–L4, L5–L6	DBM; replamineform coral; plaster of Paris; autogenous pelvis bone; autogenous or frozen allograft; Ti cage alone	12 weeks	Nicodemus 1997[109]
	Anterior discectomy and fusion C3–C4, C5–C6	Autograft + Cervi-Lok; autograft + Ti cage (BAK-C); ICBG; discectomy alone; not operated	12 weeks	Goldstein 1997[110]
Horses	Anterior fusion, cervical	Unthreaded bone basket + autograft	8–15 yrs	Bagby 1988[111a], 1997[111b]

* PLSF = Posterolateral intertransverse process fusion
† ICBG = iliac crest bone graft
‡ AIF = anterior intervertebral fusion
§ PLIF = posterior lumbar intervertebral fusion
¶ ALIF = anterior lumbar intervertebral fusion

A. POSTERIOR FUSION ANIMAL MODELS

Canines have historically been the most commonly used animal models to examine factors that influence the spinal fusion process. Albee performed 13 thoracic and lumbar fusions in canines using autograft, allograft ulna, and xenograft ulna to first define the fusion process and the contributions of different graft materials.[1] Hurley, in a classic study in 1959, used the canine to define the role of the paraspinal soft tissues in osteogenesis.[114] Posterior fusions in 37 dogs were performed with either a filter barrier to cells and fluid from the paraspinal tissues, a filter barrier to cells only, or no filter at all. The absence of a filter, or the use of a filter to cells only allowed a solid fusion at L5-L6. However, the use of a filter to cells *and* fluid consistently resulted in nonunions. These findings were among the earliest evidence that the surrounding paraspinal soft tissues provided a source of nutrition and possibly diffusible growth factors necessary for skeletal repair to transpire.

Bone graft enhancers or substitutes have been evaluated in the canine model. Callewart et al. used a lumbar (T13-L7), four level alternating facet and interlaminar anthrodesis in the beagle model.[54] By 24 weeks, facet joints fused spontaneously (77% fusion) simply with decortication of the bone surface of the facets and lamina. Treatment with morselized ICBG resulted in a higher facet fusion rate (100%) compared to decortication alone, or a polylactic acid polymer containing autogenous bone marrow (33%). In dogs, Lovell et al. used an alternating four level, thoracic interlaminar fusion procedure to perform the first investigation of a purified bovine bone morphogenetic protein (bBMP) extract as a bone graft enhancer.[68] Levels that were implanted with bBMP combined with autograft had a higher rate of fusion (71%) radiographically and histologically as compared to the best control sites (29%). In earlier animal studies, BMP had been identified as an osteoinductive morphogen capable of inducing *de novo* bone in extraskeletal sites.[115]

A high fusion rate was achieved with another bBMP extract (15–30 kDa) and with ICBG controls in alternating multilevel lumbar interlaminar fusion performed in canines.[55] Subsequently, a recombinantly produced human BMP (rhBMP-2) was compared to autograft yielding high rates of fusion in both.[56] Another recombinant BMP growth factor (rhBMP-7) was examined via multilevel facet lumbar fusion procedure in which four unilateral fusions in alternate levels of the lumbar spine, T13-L7, were done in canines.[69] Although a high fusion rate was noted with the use of the growth factor, all sites implanted with autograft fused in this facet and laminar fusion model albeit at a slower rate (12 weeks with growth factor and 26 weeks with autograft). Alternating, multilevel (T13-L7) spinous process, lamina, and facet fusions were completed in canines. rhBMP-2 was used as a bone graft enhancer and demonstrated a higher rate of spinous process, lamina, facet, fusion with the growth factor as well as a greater volume and greater stiffness in the fusion mass compared to control conditions.[76] The criticism of each of these studies, is that facets and lamina fuse too readily in canines to enable valid comparison with experimental conditions.

Sandhu et al also used the canine model to examine the performance of rhBMP-2 but refined the fusion technique to increase the difficulty of achieving fusion.[70] Rather than facet and laminar fusions, in a series of experiments single level L4-L5 bilateral intertransverse process fusions (PLSF) were performed. Successful fusion was defined as solid osseous bridging between the transverse processes. Dogs implanted with morselized ICBG along the transverse processes in this single level L4-L5 fusion model failed to achieve successful fusion within three months, but did so after six months. However, all animals (100%) that had been treated with rhBMP-2, ranging in dose from 58 mg to 2300 mg, went on to successful fusion within the three month time period.[71] A further study, using a similar single level fusion procedure only *without decorticating* the posterior elements, demonstrated that an 89% fusion rate could be achieved with rhBMP-2.[72]

A bilateral paraspinal approach followed by transverse process decortication was used to also examine rhBMP-2 in a fibrillar collagen delivery vehicle for PLSF.[74,75] They reported a 100% fusion rate with the growth factor and a 33% fusion rate with autograft.

Schimandle and Boden have argued that the rabbit model is more cost-effective, easier to manage, allows adequate ICBG harvest, and simulates the human condition as well as the larger quadrupedal animals do.[116,117] They used a single level (L5-L6) PLSF technique in rabbits to investigate rhBMP-2 with a collagen carrier and found 100% fusion with the growth factor and 42% fusion with autograft within four weeks of implantation.[62] Since the fusion rate observed in the control condition was similar to that seen in human studies of uninstrumented PLSF, this animal model combined with surgical technique was suggested as an appropriate simulation of the human condition.

Subsequently, Feiertag et al. modified the above model to study nonunion of spinal arthrodesis.[118] Of 35 rabbits that had undergone L5-L6 PLSF with ICBG, 23 were subjected to a "lifting" protocol to induce motion at the fusion site. The lifted animals exhibited a 13% rate of fusion compared to a 50% rate of fusion in control animals that had not been lifted. A relatively high 17% mortality rate secondary to anesthetic and surgical complication was noted in this study and was consistent with other reports using this model.

Lower mammalian models have been criticized with regard to biologic investigations since the biology may differ sufficiently and, therefore, findings may not be extrapolated to higher animals such as nonhuman primates and humans. Aspenberg et al. have asserted that known osteoinductive implants such as demineralized bone matrix (DBM) or BMP reliably induced extraskeletal bone formation in rodents but did not reliably do so in higher animals and in primates.[119] DBM implants augmented with rhBMP-2 *did* induce intramuscular bone in squirrel monkeys but the quantity of bone was less than that seen in lower animals. They presume that the higher the evolutionary complexity of the species, the fewer BMP-2 receptor expressing cells there are available in the surrounding extraskeletal tissues. For this reason, studies of bone inducing growth factors that have used higher animal models such as nonhuman primates have been considered even more relevant to the human condition.

Boden et al has used PLSF in rhesus monkeys to further explore implants previously examined in the rabbit model.[88] Recently, they demonstrated successful dose-dependent L4-L5 intertransverse process spinal fusions in this model using a purified bovine osteoinductive bone protein extract. Although the number of animals tested was limited, those implanted with higher doses of the osteoinductive implant achieved fusion whereas those implanted with lower doses or the carrier only did not fuse. Lately, Boden has explored a less invasive approach to the fusion site.[91]

B. Anterior Fusion Animal Models

Anterior interventions on the spinal column have different requirements than posterior interventions. Animal models employed for the study of anterior spinal fusion generally are larger and capable of safely undergoing transperitoneal, retroperitoneal, transthoracic, or anterior cervical approaches to the spine. Since fusion entails obliteration of the disc space and osseous bridging across the intervertebral gap, anatomic relationships that mimic the human condition including relatively large disc spaces and parallel vertebral endplates are preferable. Most commonly they are goat, sheep, pig, and rhesus monkey models and less commonly the canine.

An anterior intervertebral fusion, T7-T8, was performed in beagles to examine calcium carbonate ceramics in a load bearing area of the spine.[105] Discectomy was followed by decortication of the adjacent endplates. The implant was placed into the discectomy site. A lower fusion rate was found with the bone substitute (5%) than with tricortical ICBG (75%). ICBG implants were superior to hydroxyapapite, tricalcium phosphate, and calcium carbonate ceramic implants in another study employing the same surgical procedure and animal model.[106]

Lumbar intervertebral fusions were performed on Spanish goats. Spines implanted with the carbon-fiber reinforced polymer implant containing autogenous bone graft achieved a quicker and more reliable fusion than those with ethylene oxide-sterilized allograft bone.[107]

A retroperitoneal approach was used to perform anterior interbody fusions (ALIF), L5-L6, in sheep. A cylindrical, fenestrated titanium cage was either packed with ICBG or filled with the growth factor, rhBMP-2, on a collagen sponge and implanted in the fusion site. Greater ingrowth of *de novo* bone through the cage and a higher fusion rate was found with rhBMP-2 compared to autograft (100% vs 33%).[25,26] A subsequent study using this same surgical technique and animal model found a higher histologic fusion rate for a threaded porous tantalum cylinder implanted with rhBMP-2/collagen than without it (100% vs 17%).[27]

Rhesus monkeys were implanted with cylindrical allograft bone dowels combined with either rhBMP-2 on a collagen sponge or ICBG using the same ALIF, L7-S1, surgical open approach.[90] The primates implanted with rhBMP-2 achieved fusion as early as six weeks, a finding not observed in animals implanted with ICBG. More remarkably, the allograft dowels with rhBMP-2 underwent complete resorption and substitution by 12 weeks suggesting an acceleration of the remodeling process by the addition of rhBMP-2.

Cervical fusion techniques and implants have primarily been evaluated in the caprine and sheep models. An adjacent three level anterior cervical discectomy, decortication of the endplates, and fusion of the subaxial cervical spine performed in alpine goats to compare tricortical ICBG, tricortical allograft, and no graft,[60,61] Segments implanted with ICBG had a higher rate of fusion than those implanted with either allograft or no graft. Addition of the internal fixation did not increase the fusion rates.[61] Subsequently, coral hydroxyapatite bone substitute, autogenous tricortical bone graft, and fresh-frozen tricortical allograft with and without internal fixation were examined using the same three level surgical fusion procedure in the caprine model.[61] A significant rate of collapse was noted with the ceramic but some evidence of early creeping substitution was present. In this study, internal fixation prevented implant extrusion and improved the fusion rate in the group receiving the ceramic. Recently, the alpine goat model was employed to evaluate titanium intervertebral cages designed for the cervical spine. There was a 100% fusion rate for sites implanted with rhBMP-2 in collagen compared to 86% fusion rate with autograft.[81]

Two level (C2-C3, C4-C5) anterior discectomy and fusion procedure with ceramics as bone substitutes placed in the discectomy site was also examined in the caprine model. Sixty seven percent of hydroxyapatite-tricalcium phosphate (HA-TCP) implanted sites fused compared to 50% of the autograft implanted sites. By the time of explantation, 50% of the implanted ceramics fractured. The rate of fracture was unrelated to the porosity of the ceramic.[82] The fusion rate and fracture rate did not change with the addition of rhBMP-2 to HA-TCP or with an alternate ceramic (bicoral calcium carbonate ceramic).[83]

The three level cervical fusion procedure in the caprine model was recently used again. The effect of rhBMP-2 soaked collagen sponge inserted into the discectomy site with adjacent vertebrae fixed with an anterior plate was compared to tricortical ICBG without internal fixation.[85] Fusion rates were not statistically different possibly due to the small sample size. Both titanium and tantalum intervertebral fusion cages with and without rhBMP-2 were evaluated in the above model, with the greatest mechanical stiffness following fusion noted with the rhBMP-2-implanted tantalum cages.[86]

C. Surgical Approaches and Techniques

Although posterior lateral intertransverse process fusion (PLSF) is the most common surgical technique performed in clinical practice, historically, facet and interlaminar fusions have been the most often performed techniques in animal models. There is sufficient evidence that success rates of these distinct methods vary considerably and that results of one method cannot be extrapolated to predict results of another. For example, facet and interlaminar fusion following decortication and autogenous bone grafting is easily achieved in the beagle whereas intertransverse process fusion is far more difficult in the same animal model.[54,70,73]

There are several variations of the posterior intertransverse process fusion surgical technique. A bilateral paraspinal approach is used by some investigators ('lateral transverse process fusion')[75] whereas a midline incision is used by others.[70,65] Variations also exist in the amount and anatomical sites of decortication executed in similar fusion attempts.[32, 72]

A minimally invasive modification to the common open surgical approach of intertransverse process fusion procedure was developed in the rabbit and rhesus monkey model.[65] rhBMP-2 carried by a fibrillar collagen vehicle was implanted in four rhesus monkeys.

Laparoscopic, endoscopic, dorsal, and open approaches have been used to accomplish an anterior interbody fusion implanting cages filled with growth factors in sheep and rhesus monkeys.

The sheep model was first used to examine multilevel anterior interbody fusions, T5-T10, using video-assisted endoscopic technique (VATS) vs. the open technique to implant the BAK device either with ICBG, tricortical ICBG, or in combination with an anterior fixation plate.[87] Laparoscopic ALIF was performed in the rhesus monkey using titanium intervertebral fusion cages which contained either rhBMP-2 with a collagen carrier or the collagen carrier alone.[91] All primates implanted with rhBMP-2 achieved successful fusion whereas none of the animals implanted with carrier alone did.

D. Vertebral Segments

The location of a fusion procedure has significant bearing on the outcome. In the canine and sheep models, the fusion rate at the lumbosacral junction, which is likely influenced by the local kinematics, is far lower than that of adjacent levels.[46,53,54] Although multilevel fusions in the same animal represent a cost-effective design methodology, valid comparisons between the treated vertebral segments require counterbalancing the scheme of treatment conditions so that treatments are equally influenced by the inherent motion at each of the treatment levels.

To internally control for both the location of the fusion level and the idiosyncratic variability of the animal, a within-animal experimental design is appealing, However, a within-animal design of comparative treatments at the same level (i.e. experimental treatment on the left side and control treatment on the right side of L4-L5) uses a methodology that is flawed. The effect of one of the treatments on the motion segment will invariably influence the outcome of the other treatment. If one side is fused, the other is more likely to also be fused independent of any treatment effect.

E. Quadrupedal Animal Models

There are two significant considerations in the lower mammalian spine that further distinguish these models from the human condition. First, in quadrupedal models, the loads applied to the ambulating spine are quite different from those applied on the ambulating human spine (bipedal). Specifically, posterior elements of the cervical and lumbar portions of the spine are under tension in the quadruped as opposed to compression in the bipedal human. Since skeletal repair is influenced by these distinct mechanical forces, the biologic simulation of the human condition is, to some extent, undermined.

Second, the control condition in these models, to which all experimental conditions are compared, requires the harvest and transplantation of the current clinical standard of graft material, autogenous cortico-cancellous bone from the iliac crest (ICBG). This control is selected simply because it is the technical standard in clinical practice. However, the quality and quantity of retrievable ileum may be quite different in the animal models compared to the human. For example, in the lower mammalians, the relative proportion of cortical to cancellous bone is higher than that of humans, and often the transplanted graft consists largely of cortical strips with sparse cancellous bone. Theoretically, the control condition in these models should not be considered a precise reflection of the standard of clinical practice.

F. Methods of Evaluation

From a mechanical standpoint, the intent of an intervertebral spinal fusion is to create an osseous mass of bone consolidating portions of adjacent vertebral segments such that the intervertebral neutral zone is substantially reduced to acceptable levels. Noninvasive methods employed to determine success of fusion in the animal model are similar to those used clinically. Interval plain radiographs and even computed tomography scans can be useful in determining the quantity of *de novo* bone formation, the continuity between the fusion mass and the elements of the spinal column, and dynamic instability of the fusion segment by using positional radiographs.

The advantage of the animal model is that additional methods can be used for more detailed evaluation. Typically, the spines are explanted at predetermined time points and evaluated with gross manipulation, nondestructive and destructive mechanical testing, and microscopic analysis. Manual palpitation and gross manipulation of the fusion site addresses the question of fusion (no motion) or no fusion (motion). Mechanical testing can objectively measure stiffness by applying loads and measuring the corresponding displacements. The most useful data, however, is often gleaned from histologic analysis which determines not only *de novo* bone formation, but also describes the extent of remodeling and host reaction to the intervention. Manual and mechanical testing, as well as histologic analysis provide data on performance of alternative implants in animal models. Such analysis, of course, is not available clinically.

Given the same animal model and spinal fusion procedure, various time intervals for observation have been employed. Comparison among results from different studies may be difficult because of distinct end points of evaluation.

G. Inhibitors to Spinal Fusion

The intertransverse process spinal fusion procedure in the rabbit model has been used to study the effects of nicotine exposure,[120] critical period for removal of nicotine,[121] and anti-inflammatories. Other factors and substances for the promotion and inhibition of spinal fusion may be screened using lower vertebrate animal models.

H. Summary of Animal Models of Spinal Fusion

In summary of spinal fusion techniques and approaches, the posterior intertransverse process fusion (PLSF) is the most relevant and useful of the posterior fusion techniques for extrapolating data to the human condition. Facet and interlaminar fusions are achieved too easily in most lower animal models to be a useful test of spinal fusion implants. Challenging interventions such as the PLSF technique applied to lower animal models, may provide a useful and very inexpensive method to examine potential fusion enhancing biomaterials. The canine model has been the most widely used, is small enough to be cost effective and manageable, and is large enough to allow internal fixation and adequate iliac crest bone for harvest. For primarily biologic questions, however, the rabbit is cost effective and manageable, and still permits adequate iliac bone for harvest. The intermediate fusion rate associated with ICBG condition in this model is useful to the extent that it allows valid comparison of both positive and negative experimental conditions.

For anterior fusions, the sheep, goat, and dog have been used successfully. The larger disc spaces and vertebral bodies characterizing the sheep and caprine models are useful in the sense that sufficiently large implants or anterior fixation devices may be placed in the discectomy space.

The cervical spine models are essentially anterior cervical discectomy and fusion models and, in this regard, the alpine caprine model has been most widely used and validated. Most investigators using this model have preferred to use either an alternating two level fusion model or a contiguous three level fusion model, and a balance design to eliminate the influence of location.

For both anterior and posterior fusions, the most valid biologic model is the nonhuman primate, with its biology and the mechanics most closely simulating that of the human. This model, however, is highly cost ineffective, challenging to manage, and, because of its evolved mental capacity, the most ethically difficult to use. Several nonhuman primate studies have demonstrated efficacy of rhBMP-2.[90,91] Studies using these highly evolved mammals are conducted on a very selective basis and usually just prior to embarking on human clinical studies.

IV. CONCLUSIONS

The spine is one of the most frequently modeled parts of the skeletal system, in part because of its complexity, and in part because of its simplicity. The complex coupled motions of different segments of the spinal column, the assessment of spinal instability, and the effects of spinal fixation are not completely understood and continue to be rigorously investigated in a variety of selected animal models. On the other hand, because of the anatomic and kinematic consistency of selective spinal segments, interventions to the spinal column are often reliably predictable. In fact, creation of a spinal fusion is often a useful *preliminary* test of the biologic characteristics of an experimental implant meant for general skeletal repair.

As discussed in this chapter, numerous animal models have been employed to explore various aspects of spinal instability, spinal fixation, and osseous spinal fusion. Each animal model has its particular benefits and limitations. Still others are left to be explored. With the continued evolution of spinal fixation technology and our entrance into the era of biologic control of osteogenesis, the selection of appropriate and relevant animal models and surgical techniques for the study of adjuvant therapies in spinal fusion surgery becomes increasingly important.

REFERENCES

1. Albee, F., "An experimental study of bone growth and the spinal bone transplant," *JAMA* 60, 1044, 1913.
2. Roberts, S., Eisenstein, S. M., Menage, J., Evans, E. H., and Ashton, J. K., "Mechanoreceptors in intervertebral discs. morphology, distribution, and neuropeptides," *Spine*, 20, 2645, 1995.
3. Panjabi, M. M., "The stabilizing system of the spine, Part II: neutral zone and instability hypothesis," *J. Spinal Disord.*, 5, 390, 1992.
4. Adams, M. A., "Spine update: mechanical testing of the spine: an appraisal of methodology, results, and conclusions," *Spine*, 20, 2151, 1995.
5. Kaigle, A. M., Holm, S. H., and Hansson, T. H., "1997 Volvo Award in biomechanical studies. Kinematic behavior of the porcine lumbar spine: a chronic lesion model," *Spine,* 22, 2796, 1997.
6. Kaigle, A. M., Holm, S. H., and Hansson, T. H., "Experimental instability in the lumbar spine," *Spine*, 20, 421, 1995.
7. Oda, I., Abumi, K., Duosai, L., Shono, Y., and Kaneda, K., "Biomechanical role of the posterior elements, costovertebral joints, and rib cage in the stability of the thoracic spine," *Spine*, 21, 1423, 1996.
8. An, H. S., Lim, T. H., You, J. W., Hong, J. H., Eck, J., and McGrady, L., "Biomechanical evaluation of anterior thoracolumbar spinal instrumentation," *Spine,* 20, 1979, 1995.
9. Brodke, D. S., Dick, J. C., Kunz, D. N., McCabe, R., and Zdeblick, T. A., "Posterior lumbar interbody fusion: a biomechanical comparison, including a new threaded cage," *Spine,* 22, 26, 1997.
10. Dickey, J. P., Bednar, D. A., and Genevieve, A. D., "New insight into the mechanics of the lumbar interspinous ligament," *Spine*, 21, 2720, 1996.
11. Dawson, J. M., DeBoer, D. K., Spengler, D. M., and Schwartz, H. S., "The spinal nail: a new implant for short-segment anterior instrumentation of the thoracolumbar spine," *J. Spinal Disord.,* 9, 299, 1996.
12. Takahashi, T., Tominaga, T., Yoshimoto, T., Koshu, K., Yokobori, A. T., Jr., and Aizawa, Y., "Biomechanical evaluation of hydroxyapatite intervertebral graft and anterior cervical plating in a porcine cadaveric model," *Biomed. Mater. Eng.,* 7, 121, 1997.
13. Wilke, H. J., Kettler, A., and Claes, L. E., "Are sheep spines a valid biomechanical model for human spines?" *Spine,* 22, 2365, 1997.

14. Long, J. H., Jr., Pabst, D. A., Shepherd, W. R., and McLellan, W. A., "Locomotor design of dolphin vertebral columns: bending mechanics and morphology of *Delphinus delphis*," *J. Exp. Biol,* 200, 65, 1997.
15. Wittenberg, R. H., Shea, M., Edwards, W. T., Swartz, D. E., White, A. A., III, and Hayes, W. C., "A biomechanical study of the fatigue characteristics of thoracolumbar fixation implants in a calf spine model," *Spine,* 17, S121, 1992.
16. Zdeblick, T. A, Warden, K. E., Zou, D., McAfee, P. C., and Abitol, J. J., "Anterior spinal fixators. A biomechanical *in vitro* study," *Spine,* 18, 513, 1993.
17. Zdeblick, T. A., Shirado, O., McAfee, P. C., deGroot, H., and Warden K. E., "Anterior spinal fixation after lumbar corpectomy. A study in dogs," *J Bone Joint Surg,* 73A, 527, 1991.
18. Gurr K. R., McAfee, P. C., and Shih, C. M., "Biomechanical analysis of anterior and posterior instrumentation systems after corpectomy: a calf-spine model," *J. Bone Joint Surg.,* 70A, 1182, 1988.
19. Bone, L. B., Ashman, R. B., Roach, J. W., and Johnson II, C. E., "Mechanical comparison of anterior spine instrumentation in a burst fracture model," *Orthop. Trans.,* 11, 87, 1987.
20. Johnson II, C. E., Ashman, R. B., Sherman, M. C., Eberle, C. F., Herndon, W. A., et al., "Mechanical consequences of rod contouring and residual scoliosis in sublaminar segmental instrumentation," *J. Orthop. Res.,* 5, 206, 1987.
21. Wenger, D. R., Carollo, J. J., Wilkerson, J. A., Jr., Wauters, K., and Herring J. A., "Laboratory testing of segmental spinal instrumentation versus traditional Harrington instrumentation for scoliosis treatment," *Spine,* 7, 265, 1982.
22. Wenger, D. R., Wauters, K., Herring J. A., and Carollo, J. J., "Comparative mechanics of segmental spinal instrumentation versus traditional Harrington instrumentation in scoliosis treatment — laboratory analysis," *Orthop. Trans.,* 5, 16, 1981.
23. McCord, D. H., Cunningham, B. W., Shono, Y., Myers, J. J., and McAfee, P. C., "Biomechanical analysis of lumbosacral fixation," *Spine,* 17, S235, 1992.
24. Shirado, O., Zdeblick, T. A., McAfee, P. C., Cunningham, B. W., DeGroot, H., and Warden, K. E., "Quantitative histologic study of the influence of anterior spinal instrumentation and biodegradable polymer on lumbar interbody fusion after corpectomy: a canine model," *Spine* 17, 795, 1992.
25. Sandhu, H. S., Turner, S., Kabo, J. M., Kanim, L. E. A., Liu, D., et al., "Distractive properties of a threaded interbody fusion device: an *in vivo* model," *Spine,* 21, 1201, 1996.
26. Sandhu, H. S., Turner, S., Kanim, L. E. A., Kabo, J. M., Toth, J. M., and Dawson E. G., Augmentation of threaded titanium fusion cages with rhBMP-2 for experimental anterior lumbar fusion, 31st Annual Meeting Scoliosis Research Society, Ottawa, ON, Canada, Sept. 25–28, 1996.
27. Sandhu, H. S., Toth, J. M., Turner, S., Kabo, J. M., Kanim, L. E. A., and Dawson, E. G., Porous tantalum metal as a biologically advantageous spinal fixation material, 31st Annual Meeting Scoliosis Research Society, Ottawa, ON, Canada, Sept. 25–28, 1996.
28. Magin, M. N., "The dorsal interbody fusion technique," *Proc. Int. Conf. on BMPs,* 2, 37,1997.
29. Sutterlin, C. E., McAfee, P. C., Warden, K. E., Rey, R. M., Farey, I. D., et al., "A biomechanical evaluation of cervical spinal stabilization methods in a bovine model. Static and cyclical loading," *Spine,* 13, 795, 1988.
30. Richman, J. D., Daniel, T. E., Anderson, D. D., Miller, P. L., and Douglas, R. A., "Biomechanical evaluation of cervical spine stabilization methods using a porcine model," *Spine,* 20, 2192, 1995.
31. Stokes, I. A. F., Counts, D. F., and Frymoyer, J. W., "Experimental instability in the rabbit lumbar spine," *Spine,* 14, 68, 1989.
32. Ishikawa, S., Shin, H. D., Bowen, J. R., and Cummings, R. J., "Is it necessary to decorticate segmentally instrumented spines to achieve fusion?" *Spine,* 15, 1686, 1994.
33. Feighan, J. E., Stevenson, S., and Emery, S. E., "Biologic and biomechanic evaluation of posterior lumbar fusion in the rabbit: the effect of fixation rigidity," *Spine,* 20, 1561, 1995.
34. Kahanovitz, N., Arnoczky, S. P, Levine D. B., and Otis, J. P., "The effects of internal fixation on the articular cartilage of unfused canine facet joint cartilage," *Spine,* 9, 268, 1984.
35. Wood, K. B., Schendel, M. J., Pashman, R. S., Butterman, G. R., Lewis, J. L., et al., "*In vivo* analysis of canine intervertebral and facet motion," *Spine,* 17, 1180, 1992.
36. Buttermann, G. B., Schendel, M. J., Kahmann, R. D., Lewis, J. L., and Bradford D. S., "*In vivo* facet joint loading of the canine lumbar spine," *Spine,* 17, 81, 1992.

37. Craven, T. G., Carson, W. L., Asher, M. A., and Robinson, R. G., "The effects of implant stiffness on the bypassed bone mineral density and facet fusion stiffness of the canine spine," *Spine,* 15, 1664, 1994.
38. Dekutoski, M. B., Schendel, M. J., Ogilvie, J. W., Olsewski, J. M., Wallace, L. J., and Lewis J. L., "Comparison of *in vivo* and *in vitro* adjacent segment motion after lumbar fusion," *Spine,* 19, 1745, 1994.
39. Schendel, M. J., Dekutoski, M. B., Ogilvie, J. W., Olsewski, J. M., and Wallace, L. J., "Kinematics of the canine lumbar intervertebral joint: an *in vivo* study before and after adjacent instrumentation," *Spine,* 20, 2555, 1995.
40. Kioschos, H. C., Asher, M. A., Lark, R. G., and Harner, E. J., "Overpowering the crankshaft mechanism: the effect of posterior spinal fusion with and without stiff transpedicular fixation on anterior spinal column growth in immature canines," *Spine,* 21, 1168, 1996.
41. Dixon, B. C., Tomlinson, J. L., and Kraus, K. H., "Modified distraction-stabilization technique using an interbody polymethyl methacrylate plug in dogs with caudal cervical spondylomyelopathy," *J. Am. Vet. Assoc.,* 208, 61, 1996.
42. Delecrin, J., Aguado, E. N., Guyen, J. M., Pyre, D., Royer, J., and Passuti, N., "Influence of local environment on incorporation of ceramic for lumbar fusion: comparison of laminar and intertransverse sites in a canine model," *Spine,* 22, 1683, 1997.
43. Edwards, W. T., Itoh, M., Fay, L. A., and Yuan, H. A., "Neutral axis measurement demonstrates delayed fusion in the spine," *Trans. Orthop. Res. Soc.,* 22, 366, 1997.
44. Pintar, F. A., Maiman, D. J., Hollowell, J. P., Yoganandan, N., Droese, K. W., et al., "Fusion rate and biomechanical stiffness of hydroxyapatite versus autogenous bone grafts for anterior discectomy: an *in vivo* animal study," *Spine,* 19, 2524, 1994.
45. Johnston II, C. E., Welch, R. D., Baker, K. J., and Ashman, R. B., "Effect of spinal construct stiffness on short segment fusion mass incorporation," *Spine,* 20, 2400, 1995.
46. Nagel, D. A., Kramers, P. C., Rahn, B. A., Cordey, J., and Perren, S. M., "A paradigm of delayed union and nonunion in the lumbosacral joint: a study of motion and bone grafting of the lumbosacral spine in sheep," *Spine,* 5, 553, 1991.
47. Guigui P., Plais P. Y., Flautre, B., Viguier E., Blary M. C., et al., "Experimental model of posterolateral spinal arthrodesis in sheep. Part 1, experimental procedures and results with autologous bone graft," *Spine,* 19, 2791, 1994.
48. Kotani, Y., Cunningham, B. W., Cappuccino, A., Kaneda, K., and McAfee, P. C., "The role of spinal instrumentation in augmenting lumbar posterolateral fusion," *Spine,* 21, 278, 1996.
49. Kanayama, M., Cunningham, B. W., Weis, J. C., Parker, L. M., Kaneda, K., and McAfee, P. C., "Maturation of the posterolateral spinal fusion and its effect on load-sharing of spinal instrumentation: an *in vivo* sheep model," *J. Bone Joint Surg.,* 79A, 1710, 1997.

50–51. McAfee, P. C., Farey, I. D., Sutterlin, C. E., Gurr, K. R., Warden, K. E., and Cunningham, B. W., "Device-related osteoporosis with spinal instrumentation," *Spine,* 16, S190, 1991.

52. McAfee, P. C., Farey, I. D., Sutterlin, C. E., Gurr, K. R., Warden, K. E., and Cunningham, B. W., "1989 Volvo Award in basic science. Device-related osteoporosis with spinal instrumentation," *Spine,* 14, 919, 1989.
53. Nagata, H., Schendel, M. J., Transfeldt, E. E., and Lewis, J. L., "The effects of immobilization of long segments of the spine of the adjacent and distal facet force and lumbosacral motion," *Spine,* 18, 2471, 1993.
54. Callewart, C. C., Kanim, L. E. A., Seeger, L. L., and Dawson, E. G., "Variable fusion rates in the canine model," *Orthop. Trans.,* 19, 601, 1995.
55. Muschler G. F., Huber B., Ullman, T., Barth, R., Easley, K., et al., "Evaluation of bone-grafting materials in a new canine segmental spinal fusion model," *J. Orthop. Res.,* 11, 514, 1993.
56. Muschler, G. F., Hyodo, A., Manning, T., Kambic, H., and Easley, K., "Evaluation of human bone morphogenetic protein-2 in a canine spinal fusion model," *Clin. Orthorp.,* 308, 229, 1994.
57. Muschler, G. F., Negami, S., Hyodo, A., Gaisser, D., Easley, K., and Kambic, H., "Evaluation of collagen ceramic composite graft materials in a spinal fusion model," *Clin. Orthop.,* 328, 250, 1996.
58. Schendel, M. J., Dekutoski, M. B., Ogilvie, J. W., Olsewski, J. M., and Wallace, L. J., "Kinematics of the canine lumbar intervertebral joint: an *in vivo* study before and after adjacent instrumentation," *Spine,* 20, 2555, 1995.

59. Zdeblick, T. A, Cooke, M. E., Wilson, D., Kunz, D. N., and McCabe, R., "Anterior cervical discectomy, fusion, and plating. A comparative animal study," *Spine*, 18, 1974, 1993.
60. Zdeblick, T. A, Wilson, D., Cooke, M. E., Kunz, D. N., McCabe, R., et al., "Anterior cervical discectomy and fusion. A comparison of techniques in an animal study," *Spine*, 17, S418, 1992.
61. Zdeblick, T. A., Cooke, M. E., Kunz, D. N., Wilson, D., and McCabe, R. P., "Anterior cervical discectomy and fusion using a porous hydroxyapatite bone graft substitute," *Spine*, 19, 2348, 1994.
62. Schimandle, J. H., Boden, S. D., and Hutton, W. C., "Experimental spinal fusion with recombinant human bone morphogenetic protein-2," *Spine*, 20, 1326, 1995.
63. Boden, S. D., Schimandle, J. H., and Hutton, W. C., "Lumbar intertransverse-process spinal arthrodesis with use of a bovine bone-derived osteoinductive protein: a preliminary report," *J. Bone Joint Surg.*, 77A, 1404, 1995.
64. Hollinger, E. H., Trawick, R. H., Boden, S. D., and Hutton, W. C., "Morphology of the lumbar intertransverse process fusion mass in the rabbit model: a comparison between two bone graft materials — rhBMP-2 and autograft," *J. Spinal Disord.*, 9, 125, 1996.
65. Boden, S. D., Moskovitz, P. A., Morone, M. A., and Toribitake, Y., "Video-assisted lateral intertransverse process arthrodesis, validation of a new minimally invasive lumbar spinal fusion technique in the rabbit and nonhuman primate (rhesus) models," *Spine*, 21, 2689, 1996.
66. Boden, S. D., Schimandle, J. H., Hutton, W. C., Damien, C. J., Benedict, J. J., et al., "*In vivo* evaluation of a resorbable osteoinductive composite as a graft substitute for lumbar spinal fusion," *J. Spinal Disord.*, 10, 1, 1997.
67. Boden, S. D., Morone, M. A., Hair, G., and McCuaig, K., Gene expression during experimental posterolateral lumbar spinal fusion, 12th Annual Meeting of North American Spine Society, New York, Oct. 22–25, 1997.
68. Lovell, T. P., Dawson, E. G., Nilsson, O. S., and Urist, M. R., "Augmentation of spinal fusion with bone morphogenetic protein in dogs," *Clin. Orthop.*, 243, 266, 1989.
69. Cook, S. D., Dalton, J. E., Tan, E. H., Whitecloud, T. S., and Rueger, D. C., "*In vivo* evaluation of recombinant human osteogenic protein (rhOP-1) implants as a bone graft substitute for spinal fusions," *Spine*, 19, 1655, 1994.
70. Sandhu, H. S., Kanim, L. E. A., Kabo, J. M., Toth, J. M., Zeegen, E. H., et al., "Evaluation of rhBMP-2 with an OPLA carrier in a canine posterolateral (transverse process) spinal fusion model," *Spine*, 20, 2669, 1995.
71. Sandhu, H. S., Kanim, L. E. A., Kabo, J. M., Toth, J. M., Zeegen, E. N., et al., "Effective doses of recombinant human bone morphogenetic protein-2 in experimental spinal fusion," *Spine*, 21, 2115, 1996.
72. Sandhu, H. S., Kanim, L. E. A., Toth, J. M., Kabo, J. M., Liu, D., et al., "Experimental spinal fusions with recombinant human bone morphogenetic protein-2 without decortication of osseous elements," *Spine*, 22, 1171, 1997.
73. Sandhu, H. S., Luppen, C. A., and Kanim, L. E. A., "Spinal applications for recombinant bone morphogenetic protein," *Spine*, 10, 255, 1996.
74. David, S. M., Murakami, T., Tabor, O. B., Meyer, R. A., Jr., Howard, B., et al., "Lumbar spinal fusion using recombinant human bone morphogenetic protein (rhBMP-2): a randomized, blinded and controlled study," *Orthop. Trans.*, 20, 872, 1996–1997.
75. David, S. M, Gruber, H. E., Murakami, T., Tabor, O. B., Meyer Jr., et al., "Lumbar spinal fusion using recombinant human bone morphogenetic protein (rhBMP-2): a randomized, blinded and controlled study," *Trans. Orthop. Res. Soc.*, 21, 119 1996.
76. Sheehan, J. P., Kallmes, D. F., Sheehan, J. M., Jane, J. A., Jr., Fergus, A. H., et al., "Molecular methods of enhancing lumbar spine fusion," *Neurosurgery*, 39, 548, 1996.
77. Helm, G. A., Sheehan, J. M., Sheehan, J. P., Jane, J. A., Jr., diPierro, C. G., et al., "Utilization of type I collagen gel, demineralized bone matrix, and bone morphogenetic protein-2 to enhance autologous bone lumbar spinal fusion," *J. Neurosurg.*, 86, 93, 1997.
78. Li, S. T., Bolton W., Helm, G., Gillies, G., and Frenkel, S., "Collagen as a delivery vehicle for bone morphogenetic protein (BMP)," *Trans. Orthop. Res. Soc.*, 21, 647, 1996.
79. Kwiatkowski, T. C., Meyer, R. A, Jr., Gruber, H. E., Tabor, O. B., Jr., Murakami, T., et al., "Spinal laminectomy with recombinant human bone morphogenetic protein (rhBMP-2) in the canine: a safety study," *Trans. Orthop. Res. Soc.*, 22, 189, 1997.

80. Fischgrund, J. S., James, S. B., Chabot, M. C., Hankin, R., Herkowitz, H. N., et al., "Augmentation of autograft using rhBMP-2 and different carrier media in the canine spinal fusion model," *J. Spinal Dis.*, 10, 467, 1997.
81. Zdeblick, T. A, Cooke, M. E., Rapoff, A. J., Swain, C., Markel, M., and Ghanayem, A. J., Cervical interbody fusion cages, an animal model with and without BMP, 10th Annual Meeting North American Spine Society, Washington, DC, Oct. 18–21, 1995.
82. Toth, J. M., An, H. S., Lim, T. H., Ran, Y., Weiss, N. G., et al., "Evaluation of porous biphasic calcium phosphate ceramics for anterior cervical interbody fusion in a caprine model," *Spine*, 20, 2203, 1995.
83. Toth, J. M., Lim, T. H., An, H. S., Yoshida, H., Roh, J., et al., "Enhancement of cervical spine fusion with reinforced porous 50/50 HA/TCP loaded with rhBMP-2 in a caprine model," *Trans. Orthop. Res. Soc.*, 21, 188, 1997.
84. An, H. S., Toth, J. M., Beller, R. A., You, J.-W., Kho, A., et al., Calcium phosphate ceramics for cervical interbody fusion, 12th Annual Meeting of North American Spine Society, New York, Oct. 22–25, 1997.
85. Bolesta, M. J., Welch, R. D., and Viere, R. G., Healing of anterior interbody fusions with recombinant human bone morphogenetic protein, 12th Annual Meeting of North American Spine Society, New York, Oct. 22–25, 1997.
86. Moore, D. K., Deguchi, M., Rapoff, A. J., O'Brien, T. J., Swain, C., et al., Lordotic cervical porous tantalum interbody fusion cages in a goat model, 12th Annual Meeting of North American Spine Society, New York, Oct. 22–25, 1997.
87. Cunningham, B. W., Kotani, Y., McNulty, P. S., Cappuccino, A., Fedder, I. L., and McAfee, P. C., "Minimally invasive thoracoscopy versus open thoracotomy in the sheep thoracic spine: a comparative *in vivo* spinal fusion study," *Trans. Orthop. Res. Soc.*, 21, 656, 1996.
88. Boden, S. D., Schimandle, J. H., and Hutton, W. C., "Evaluation of a bovine-derived osteoinductive bone protein in a nonhuman primate model of lumbar spinal fusion," *Trans. Orthop. Res. Soc.*, 21, 118, 1996.
89. Boden, S. D., Moskovitz, P. M., Morone, M. A., Martin, G., and Toribitake, Y., "Video-assisted lateral intertransverse process arthrodesis validation of a new minimally invasive spinal fusion technique," *Trans. Orthop. Res. Soc.*, 22, 378, 1997.
90. Hecht, B. P., Fischgrund, J. S., Herkowitz, H, Penman, L., and Toth, J., The use of rhBMP-2 to promote spinal fusion in a nonhuman primate anterior interbody fusion model utilizing freeze-dried allograft cylinder, 12th Annual Meeting of North American Spine Society, New York, Oct. 22-25, 1997.
91. Boden, S. D., Horton,, W. C., Martin, G., Truss, T., and Sandhu, H. S., "Laparoscopic anterior spinal arthrodesis with rhBMP-2 in a titanium interbody threaded cage," *N. Am. Spine Soc.*, 272, 1997.
92. Rhee, J. M., Tay, B. K-B., Gould, S. E., and Bradford, D. S., Mouse spinal fusion model, Personal Communication, 1998.
93. Thomas, I., Kirkaldy-Willis, W. H., Singh, S., and Paine, K. E. W., "Experimental spinal fusion in guinea pigs and dogs: the effect of immobilization," *Clin. Orthop.*, 112, 363, 1975.
94. Guizzardi, S., Di Silvestre, M., Scandroglio, R., Ruggeri, A., and Savini, R., "Implants of heterologous demineralized bone matrix for induction of posterior spinal fusion in rats," *Spine*, 17, 701, 1991.
95. Dawson, E. G., Kanim, L. E. A., and DeGrange, D. A., Personal Communication, 1998.
96. Oikarinen, J., "Experimental Spinal fusion with decalcified bone matrix and deep-frozen allogeneic bone in rabbits," *Clin. Orthop.*, 162, 210, 1982.
97. Lindholm, T. S., Ragni, P., and Lindholm, T. C., "Response of bone marrow stroma cells to demineralized cortical bone matrix in experimental spinal fusion in rabbits," *Clin. Orthop.*, 230, 296, 1988.
98. MacMillan, M., May, D., Suri, M., Shrestha, T. B., Miller, G. J., and Strates, B. S., "Stimulation of intervertebral fusion by TGF-beta with microcrystals of a biodegradable hydroxyapatite," *Trans. Orthop. Res. Soc,* 22, 187, 1997.
99. Tay, B. K-B., Le, A. X., Helman, M., Lotz, J., and Bradford, D. S., "Use of a collagen hydroxyapatite matrix in spinal fusion: a rabbit model," *Trans. Scoliosis Res. Soc.*, 97, 1997.
100. Tho, K. S. and Krishnamoorthy, S., "Use of coral grafts in anterior interbody fusion of the rabbit spine," *Ann. Acad. Med. Singapore,* 25, 824, 1996.
101. Curylo, L. J., Johnstone, B., Petersilge, S., Janicki, J., and Yoo, J., "Autogenous bone marrow augmentation of spinal fusion in a rabbit posterolateral fusion model," *Trans. Orthop. Res. Soc.*, 22, 356, 1997.

102. Toribatake, Y., Hutton, W. C., Boden, S. D., and Morone, M. A., "Revascularization of the fusion mass in a posterolateral intertransverse process fusion," *Trans. Orthop. Res. Soc.*, 22, 192, 1997.
103. Stein, M., Elliott, D., Glen, J., and Morava-Protzner, I., "Young Investigator Award. Percutaneous facet joint fusion: preliminary experience," *J. Vasc. Intervent. Radiol.*, 4, 69, 1993.
104. Cook, S. D., Dalton, J. E., Prewett, A. B., and Whitecloud III, T. S., "*In vivo* evaluation of demineralized bone matrix as a bone graft substitute for posterior spinal fusion," *Spine*, 20, 877, 1995.
105. Fuller, D. A., Stevenson, S., and Emery, S. E., "The effects of internal fixation on calcium carbonate: ceramic anterior spinal fusion in dogs," *Spine*, 21, 2131, 1996.
106. Emery, S. E., Fuller, D. A., and Stevenson, S., "Ceramic anterior spinal fusion: biologic and biomechanical comparison in a canine model," *Spine*, 21, 2713, 1996.
107. Brantigan, J. W., McAfee, P. C., Cunningham, B. W., Wang, H., and Orbegoso, C. M., "Interbody lumbar fusion using a carbon fiber cage implant versus allograft bone: an investigational study in the Spanish goat," *Spine,* 19, 1436, 1994.
108. Riley L. H., III, Eck, J. C., Yoshida, H., Toth, J. M., Nguyen, C., et al., "Laparoscopic assisted fusion of the lumbosacral spine: a biomechanical and histologic analysis of the open versus laparoscopic technique in an animal model," *Spine,* 22, 1407, 1997.
109. Nicodemus C., Simmons, J. W., Hadjipavlou, A. G., Yang, J., and Simmons, D., Lumbar spine stabilization using various graft materials in mature sheep, 12th Annual Meeting of North American Spine Society, New York, Oct. 22-25, 1997.
110. Goldstein, J. A., Cunningham, B. W., Kanayama, M., Stewart, G., Ng, J. T. W., et al., Anterior cervical fusion: comparison of autograft, locking plates and a fusion cage using an *in vivo* sheep model, 12th Annual Meeting of North American Spine Society, New York, Oct. 22-25, 1997.
111a. Bagby, G. W., "Arthrodesis by the distraction-compression method using a stainless steel implant," *Orthopaedics,* 11, 931, 1988.
111b. Bagby, G. W., Cunningham, B. W., McAfee, P. C., Grant, B. D., Reed, S. M., and Piercy, R.. J. Bone maturation study within the self-contained implant for spinal arthrodesis, 12th Annual Meerting of the North American Spine Society, New York, Oct. 22–22, 1997.
112. Schmidt-Nielsen, K., *Scaling: Why Is Animal Size So Important?* Cambridge Univ. Press, Cambridge, 1984.
113. Mullender, M. G., Huiskes, R., Versleyen, H., and Buma, P., "Osteocyte density and histomorphometric parameters in cancellous bone of four mammalian species," *Trans. Orthop. Res. Soc.*, 22, 588, 1997.
114. Hurley, L. A., Stinchfield, F. E., Bassett, C. A. L., and Lyon, W. H., "The role of soft tissues in osteogenesis," *J. Bone Joint Surg.*, 41A, 1243, 1959.
115. Urist, M. R. "Bone formation by autoinduction," *Science*, 150, 893, 1965.
116. Schimandle, J. H. and Boden, S. D., "Spine update: the use of animal models to study spinal fusion," *Spine,* 19, 1998, 1994.
117. Schimandle, J. H. and Boden, S. D., "Spine update: animal use in spinal research," *Spine,* 19, 2474, 1994.
118. Feiertag, M. A., Boden, S. D., Schimandle, J. H., and Norman, J. T., "A rabbit model for nonunion of lumbar intertransverse process spine arthrodesis," *Spine,* 21, 27, 1996.
119. Aspenberg, P., Wang, E., and Thorngren, K. G., "Bone morphogenetic protein induces bone in squirrel monkey, but matrix does not," *Trans. Orthop. Res. Soc.*, 39, 101, 1993.
119b. Aspenberg, P. and Turek, T., "BMP-2 for intramuscular bone induction: effect in squirrel monkeys is dependent on the implantation site," *Acta Orthop. Scand.*, 67, 3, 1996.
120. Silcox III, D. H., Daftari, T., Boden, S. D., Schimandle, J. H., Hutton, W. C., and Whitesides, T. E., Jr., "The effect of nicotine on spinal fusion," *Spine,* 20, 1549, 1995.
121. Wing, K. J., Fisher, C. G., O'Connel, J. X., and Wing, P., Stopping nicotine exposure prior to surgery, 12th Annual Meeting of North American Spine Society, New York, Oct. 22–25, 1997.

28 Animal Models of Spinal Cord Compression

Shimpei Miyamoto, Kazuo Yonenobu, and Keiro Ono

CONTENT

I. Introduction 527
II. Selection Of Animal Models 528
III. Commonly Used Models 528
    A. Classical Models 528
    B. Herniated Intervertebral Disc 529
    C. Tumor 529
    C. Screw 530
    D. Spondylosis 530
    E. Vertebral Hyperostosis 532
IV. Concluding Remarks 536
References 536

I. INTRODUCTION

Myelopathy due to chronic compression of the spinal cord is often encountered in cases of cervical spondylosis, ossification of the posterior longitudinal ligament (OPLL), cervical disc herniation, and spinal tumor, but the pathogenesis and pathophysiology of chronic compression myelopathy remain unclear or controversial.[1] The availability of experimental animal models that mimic the human diseases and consistently reproduce the diseases would materially facilitate basic investigations of the pathogenesis and pathophysiology. They could provide the basis for a more comprehensive analysis than is possible with humans.

    Models of spinal cord compression were classified by Fehlings and Tator[2] as kinetic (acute) compression and static (chronic or subacute) compression according to biomechanics of the applied forces. Acute (kinetic) compression models involve rapid compression of the spinal cord in less than one second.[2] The applied force compresses the spinal cord with increasing velocity (with acceleration) to the point of maximal compression. Allen started modern research of acute spinal cord injury.[3] He introduced a model of acute spinal cord injury which could be quantified and standardized. He dropped weights from various heights onto the surgically exposed spinal cords of dogs, and expressed the force of the contusion injuries in gram centimeters.[3] This weight-dropping method became one of the most frequently used acute compression models, and numerous modifications have been developed. Fehlings and Tator[2] classified acute compression models as follows: weight-drop method, extradural balloon compression method, clip compression method, and vertebral dislocation method. A review of these models is described elsewhere.[2] Acute compression models attempt to simulate the biomechanical features of most types of acute spinal cord injuries, and do not seem to simulate myelopathy due to chronic compression of the spinal cord.

In contrast, chronic (static) compression models use forces which slowly compress and injure the spinal cord at an approximately constant velocity (without acceleration). Chronic compression models attempt to simulate mass lesions which gradually compress the spinal cord such as osteophytes, OPLL, herniated intervertebral discs, and spinal tumors. According to Fehlings and Tator,[2] two types of lesion-making methods can be considered in this category: the technique of graded addition of compression, and the use of slowly expanding mass lesions. In this chapter, these chronic compression models will be discussed.

II. SELECTION OF ANIMAL MODELS

Fehlings and Tator[2] called attention to the fact that many experiments of spinal cord injury from the literature contained serious errors in the experimental design. The main flaws were selection of an inappropriate model and/or species, inadequate sample size, and lack of objectivity.

The model of chronic spinal cord compression selected should be highly reproducible and appropriate for the particular animal species to be studied. Selection of an appropriate method for a particular species is sometimes difficult, and depends on intricacies of the methods and body size of the animal. For example, although advancement of a screw to produce spinal cord compression reported by Hukuda and Wilson[4] may be reproducible and acceptable for experiments on larger animals such as dogs or primates, this method may be too intricate to be applied to smaller animals such as rats or mice.

Methods used to evaluate outcome of experimental spinal cord compression also should be considered when the model and species are selected. Clinical neurological evaluation such as Tarlov's scale[5] or its modifications may be easier to perform on larger animals. Fehlings and Tator,[2] however, criticize the use of Tarlov's scale and its modifications in species other than primates because dogs and cats with complete transection of the spinal cord are capable of "spinal walking" and other forms of reflex limb movement, and these interfere with clinical neurologic evaluation of limb function. Functional clinical tests such as inclined plate,[6] which seem to be consistent and quantifiable, may be more easily applied to smaller animals such as rats or mice. Histopathology of the injured spinal cord may be essential and can be applied to any of the models and species. Spinal cord blood flow (SCBF) can be measured in either large animals such as dogs and monkeys[1,7] or smaller animals such as rats.[8] Magnetic resonance imaging (MRI) may be capable of resolving the finer pathoanatomical features of the injured spinal cords of the larger animals.[1] Neurophysiologic tests such as measurements of somatosensory evoked potentials (SEP) or motor evoked potentials (MEP) may be very difficult and intricate on smaller animals.

III. COMMONLY USED MODELS

Selected reports on chronic compression models are listed in Table 1.

A. Classical Models

In the 1950s, Tarlov developed a method of gradual spinal cord compression in dogs by slowly inflating an extradural balloon.[5,9–11] The balloon was introduced through a laminectomy defect at the level of T12, and threaded extradurally to the mid-thoracic region. The animals became paraparetic within 45 minutes and paraplegic within two hours. Doppman et al. described a percutaneous technique for producing intraspinal mass lesions in dogs and monkeys.[7] Small balloon catheters introduced through needles into the spinal canal are positioned under fluoroscopic control to simulate epidural masses. Selective spinal cord angiography and silicone perfusion studies demonstrate the effect of such masses on SCBF. Croft et al. applied graded pressure by placement of weights directly

TABLE 1
Selected Reports on Chronic Spinal Cord Compression Models

1st Authors, Year[Ref.]	Animal Species	Method of Cord Compression
Tarlov 1953–1957[5,9–11]	Dog	Balloon Inflation
Doppman 1973[7]	Dog, monkey	Balloon Inflation
Croft 1972[12]	Cat	Weights
Eidelberg 1976[13]	Ferret	Weights
Bennett 1977[14]	Cat	Casein Plastic Mass
Olsson 1958[15]	Dog	Disc Protrusion
Coman 1951[17]	Rat	Tumor Growth
Ushio 1977[18]	Rat	Tumor Growth
Aoki 1997[19]	Rat	Tumor Growth
Hukuda 1972[4]	Dog	Metal Screw
Hukuda 1988[20]	Dog	Metal Screw
Schramm 1983[21]	Cat	Implantable Screw
Al-Mefty 1993[1]	Dog	Teflon Screw and Washer
Miyamoto 1991[22]	Mouse	Osteophyte and Disc
Miyamoto 1992[34]	Mouse	Ossified Ligamentum Flavum
Saito 1992[40]	Rabbit	Ossified Ligamentum Flavum

onto the spinal cord of cats exposed through laminectomy to produce reversible blocking of MEP and SEP as a means of estimating spinal cord damage.[12] Eidelberg et al. applied weights sequentially to the thoracic spinal cords of ferrets to produce a model of incomplete cord injury.[13] Bennett and McCallum inserted a casein plastic mass into the epidural space at cervical spine levels of cats.[14] The casein plastic mass slowly absorbed water and increased in weight and volume from 50 to 100% during the five to 18 days they were implanted, and produced spinal cord compression severe enough to result in loss of SEP.

B. HERNIATED INTERVERTEBRAL DISC

It is a well-known fact that certain chondrodystrophic breeds of dogs have a greater tendency towards degeneration and herniation of the intervertebral discs than other breeds.[15] Olsson, a Swedish veterinarian, studied approximately 1300 clinical cases of disc herniations in dogs during an eight-year period, and found symptomatic cervical disc herniation in only 40 of these cases.[15] This might mean that even in the chondrodystrophic breeds of dogs symptomatic cervical disc herniation is exceptional, and is not easily available as an experimental model.

C. TUMOR

According to Batson, the human vertebral venous system lacks valves and acts as a pathway of cancer metastasis.[16] Based on this theory, Coman and DeLong injected a suspension of Walker 256 carcinoma cells into the tail veins of rats while applying abdominal pressure.[17] Tumors grew in the vertebral venous system in twelve of fourteen rats, and six rats developed paraplegia.

To simulate extradural tumors in humans, Ushio et al. produced epidural spinal cord compression in rats by injection of Walker 256 carcinoma cell suspension anterior to the T12 or T13 vertebral body.[18] Tumors grew through the intervertebral foramina to compress the spinal cord and produce paraplegia in three to four weeks. They could assess the effects of several treatments such as dexamethasone, radiation, laminectomy, and cyclophosphamide on clinical symptoms. This animal model appears to be useful for studying the treatment of human spinal cord compression produced by epidural neoplasms.

Aoki et al. placed 2-mm cubes of c-SST-2 mammary carcinoma in the epidural spaces of rats through a laminectomy defect.[19] Forty-one of 45 rats developed paralysis of the hind legs. Paraparesis occurred 6–16 days after implantation of the tumor cells, and paraplegia occurred 6–21 days after implantation.

C. Screw

Hukuda and Wilson performed a series of experiments to determine the pathogenesis of cervical spondylotic myelopathy.[4] Their hypothesis was that vertebral osteophytes impinged on the spinal cord, and produced not only compression but also local ischemia of the spinal cord. This local vascular insufficiency was enhanced by disturbance in blood supply from systemic circulation to the spinal cord, and might then produce irreversible changes in the spinal cord, namely, myelopathy. They produced maximal tolerable compression of the spinal cord in dogs by advancing a screw through the anterior portion of C5 vertebral body into the spinal canal until limb weakness occurred. Chronic vascular insufficiency was established in the cervical spinal cord by blocking or ligating the anterior spinal artery, the vertebral arteries, and their branches in various combinations. From neurologic, microangiographic and, histopathologic findings, they concluded that the effects of vascular insufficiency and compression were synergistic.

Using a modified model of cervical spondylotic myelopathy, Hukuda et al. investigated the effects of several types of abnormalities such as systemic arterial hypotension, systemic arterial hypertension, cervical hyperflexion, cervical hyperextension, and cervical instability on histopathology of the spinal cord.[20] They produced the modified model of cervical spondylotic myelopathy using the following method. A dual screw was inserted into a drill hole piercing the C5 vertebral body of dogs, and under X ray control, an inner screw was gradually advanced into the spinal canal until it reduced the anteroposterior diameter of the canal by 45%. In addition, the vertebral arteries were obliterated bilaterally by inserting a catheter from C6 to the C2 level. They found pathologic changes of the spinal cord which were characteristic of each type of abnormality: peripheral necrosis of the central gray matter in systemic arterial hypotension, capillary congestion and subarachnoid hemorrhage in systemic arterial hypertension, and linear necrosis of the central gray matter and occluded anterior spinal artery in cervical hyperflexion. The pathologic severity was proportional to the number of applied abnormalities.

Schramm et al. applied implantable compression screws dorsally against L1 vertebral bodies of cats, advanced the screws by stepwise tightening at intervals of four to seven days, and used cortical and spinal evoked potentials to monitor the effect of the chronic cord compression.[21] They found that neurological alternations appeared later than alterations in spinal evoked response but earlier than alterations in cortical evoked response.

Al-Mefty et al. achieved subclinical cervical cord compression in dogs by placing a Teflon screw anteriorly and a Teflon washer posteriorly, producing and average of 29% stenosis of the spinal canal.[1] They reported that twelve of fourteen dogs developed delayed and progressive signs of myelopathy, with a mean latent period to onset of myelopathy of seven months. It is noteworthy that this model of cervical spondylotic myelopathy allowed control of the spinal cord compression, an assessment of neurologic deficits, imaging evaluations like MRI, SEP recordings, SCBF measurements, and postmortem histopathologic examinations.

D. Spondylosis

The term "spondylosis" describes chronic degenerative lesions of multiple or single intervertebral discs and the consequent osteophytosis of related vertebral bodies. Spondylosis is an important cause of musculoskeletal disability in humans: the disc degeneration leads to the clinical syndromes of radiculopathy and myelopathy. Miyamoto et al. established an experimental model of cervical spondylosis in rodents with the working hypothesis that mechanical instability in the spine would induce disc degeneration and spondylosis.[22]

Fifty-seven adult (six-month-old) male ICR strain mice weighing approximately sixty grams were used. They were divided into two groups: thirty mice for the experimental group and twenty-seven mice for the control group. Each group was further divided into three subgroups as described below. Thirty mice in the experimental group were surgically treated with the following procedures. The back paravertebral muscles of the mice were detached from the spinous processes, laminae and facets of the cervical, thoracic and lumbar vertebrae. Next, the spinous processes together with supraspinous and interspinous ligaments of the cervical, thoracic and lumbar vertebrae were all resected. The skin was closed without reattachment of the paravertebral muscles so that these muscles could not work as they used to. After recovery, the animals were allowed to move freely about their cages. The 27 mice in the control group were not treated at all in order to observe the aging process of the spine.

Ten of the thirty mice in the experimental group were sacrificed two months after operation (E2M group) to observe short-term effects of the surgical intervention, and nine of the twenty-seven mice in the control group were sacrificed and examined at the same time (C2M group). Another ten mice in the experimental group were sacrificed six months after operation (E6M group) to observe mid-term effects of the surgical intervention while another nine mice in the control group were sacrificed simultaneously (C6M group). The remaining ten mice in the experimental group were sacrificed twelve months after operation (E12M group) to observe long-term effects while the remaining nine mice in the control group were also sacrificed (C12M group). All the animals were killed by carbon dioxide or ether inhalation.

Radiographic studies were performed to examine disc-space narrowing, osteophyte formation and changes in the spinal alignment. The entire vertebral column was quickly dissected free and the cervical and upper thoracic spine was examined radiographically along the anterior-posterior and lateral planes. The radiographic examinations showed that in the C2M, C6M and C12M groups no disc-space narrowing, osteophyte formation or abnormal spinal alignment was observed. No evidence of osteophyte formation was seen in the E2M and E6M groups. In all the cervical vertebral columns in the E12M group, however, lesions were characterized by disc-space narrowing and/or anterior osteophyte formation most prominently seen at C4-5, C5-6 and C6-7 levels. Pathologic cervical kyphosis was also observed in half of the specimens in the E6M and E12M group.

Histologic studies were performed to examine the degeneration process of the intervertebral discs, osteophyte formation, and spinal cord compression. The entire dissected vertebral column of each specimen was fixed in neutral buffered formalin, decalcified in formic acid, split mid-sagittally, embedded in paraffin, sectioned, and stained for the light microscopic examination. Histologic examinations revealed that most of the intervertebral discs appeared normal in the C2M and C6M groups. In contrast, most of the intervertebral discs in the C12M and E2M groups exhibited proliferation of cartilaginous tissue and loss of lamellar structure in the anterior annulus fibrosus. Shrinkage or disappearance of the nucleus pulposus and clefts or fissures in the annulus fibrosus were observed in several discs. In the E6M group, disc degeneration was more advanced and the lesions were characterized by shrinkage or disappearance of the nucleus pulposus and fissures in the annulus fibrosus. Herniation of disc materials and anterior osteophyte formation were observed in several discs. In the E12M group, disc degeneration was most advanced and the lesions were characterized by herniation of disc materials and anterior osteophyte formation. Posterior herniation of disc materials led to impingement on and deformity of the contiguous spinal cord (Figure 1).[22]

Although the degree of spinal cord compression was not severe and neurological deficits were evident in none of the animals, the authors believe that this experimental model could be valuable in helping to understand the pathoanatomy and pathophysiology of myelopathy due to static compression of the spinal cord caused by cervical spondylosis or disc herniation.

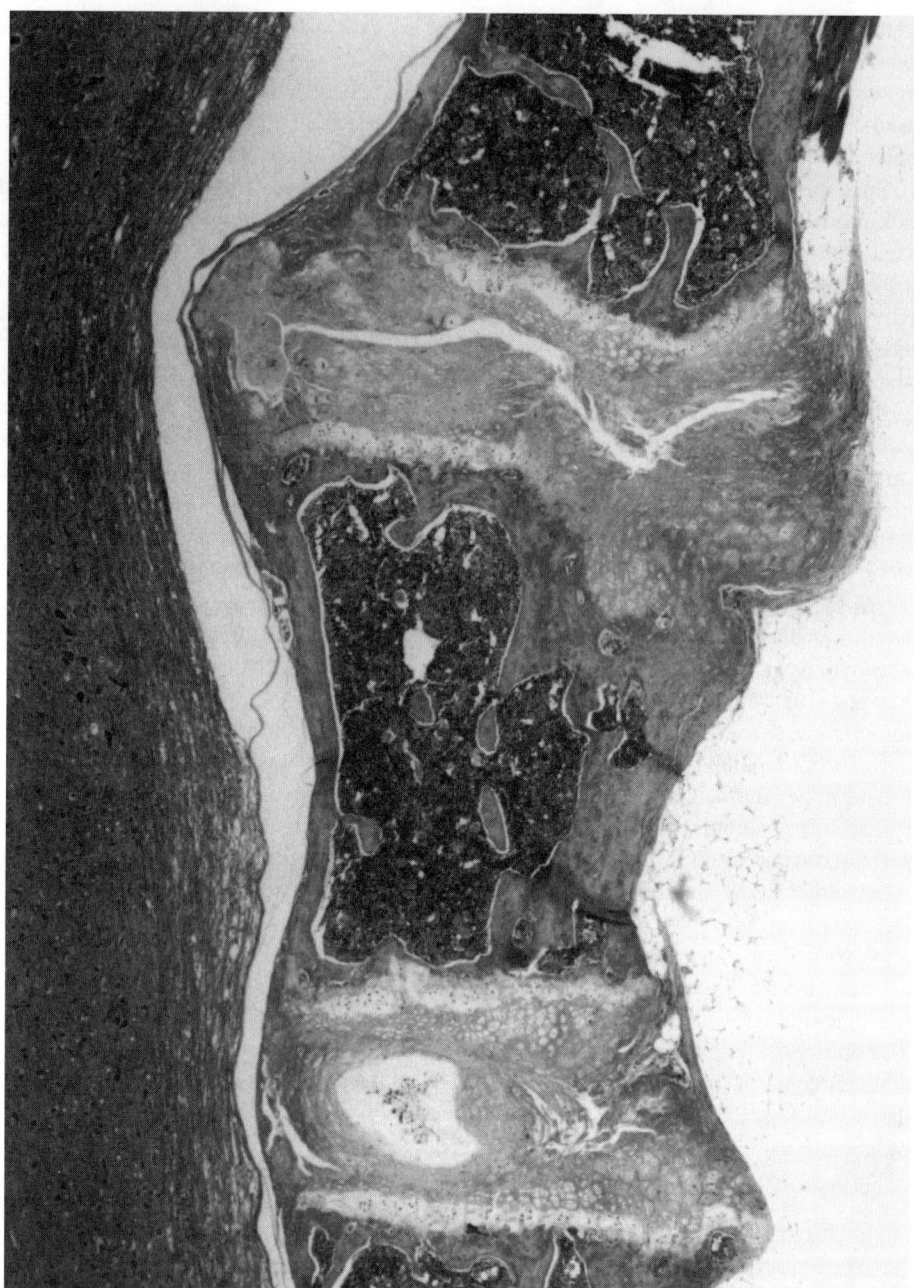

FIGURE 1. A mid-sagittal section shows that herniation of disc tissue was compressed and deformed in this model (Masson stain). Neurons in the gray matter at the compressed level seem to be reduced in number.

E. Vertebral Hyperostosis

Both ankylosing spinal hyperostosis (ASH)[23] and diffuse idiopathic skeletal hyperostosis (DISH)[24] are well-known conditions in which hyperostosis is associated with ossification of the ligaments of the spine. Patients with ASH or DISH often develop ossification of the posterior longitudinal ligament (OPLL)[25,26] and/or ossification of ligamentum flavum (OLF)[27,28] causing serious neurological complications. Protrusion of the thickened, hypertrophied and ossified ligaments into the spinal canal leads to compression and deformation of the contiguous spinal cord and nerve roots, which cause myelopathy and radiculopathy. Histopathologic studies of OPLL[25] and OLF[28] have revealed that the development and growth of OPLL and OLF are based on hypertrophy of the ligamentous tissues with proliferation of fibrocartilaginous cells. The regular fibrous matrices and frame work are disrupted in OPLL and OLF, and collagen fibers increase in number and size.[25,28] Numerous fibrocartilaginous cells are found in the increased collagenous matrices. Some parts of the ossified mass are almost always in continuity with the posterior cortex of the vertebral body in OPLL, and with the lamina in OLF. The ossification extends mainly along the superficial layer of the hypertrophied posterior longitudinal ligament and ligamentum flavum. There are many cartilaginous cells with abundant matrices and nutritional vessels at the area of the ossification front. These histologic findings may indicate that OPLL and OLF have developed through a process of endochondral ossification. In some parts, bone apposition toward the spinal canal is found on the surface of the ossified ligaments.[25] There is no evidence of inflammation in the hypertrophied or ossified ligaments. In the pathogenesis of ossification at the juxta-skeletal sites such as OPLL and OLF, growth factors which can initiate and stimulate new cartilage and bone formation may be important.[29]

During the past decade, studies have shown that a number of growth factors regulate the development, growth and maintenance of cartilage and bone tissues.[29,30] Among them, bone morphogenetic proteins (BMPs) and transforming growth factor-βs (TGF-βs) may have more important roles on the pathogenesis of OPLL and OLF, because BMPs initiate cartilage and bone differentiation and induce new cartilage and bone formation *in vivo*, whereas TGF-βs stimulate cartilage and bone formation by determined chondroprogenitor and osteoprogenitor cells *in vivo*.[31] BMPs were originally defined as bone-inducing substances responsible for new cartilage and bone formation at the extra-skeletal sites by intramuscular or subcutaneous implantation of demineralized bone matrix.[29,32] The process of new cartilage and bone formation induced by the implantation of BMPs is a sequential cascade of several events: activation and migration of undifferentiated mesenchymal stem cells; attachment of the cells to collagenous matrices; proliferation of the mesenchymal stem cells; differentiation of the stem cells into chondrocytes; maturation and hypertrophy of the chondrocytes and mineralization of the cartilage matrix; angiogenesis and vascular invasion; new bone formation via endochondral ossification; remodeling of the bone and hematopoietic marrow formation in the ossicle.[33]

Because some BMPs have bone-inducing potentials as described above, they may be causative factors in the development and growth of OPLL and OLF. This idea led Miyamoto and coworkers to determine whether BMPs could experimentally induce OLF and secondary spinal cord compression in animals.[34]

Forty-eight adult (24-wks-old) male ICR strain mice were used for the experiment. These mice were divided into two groups; 30 mice in the experimental group and 18 in the control group. Partially-purified BMP fraction was prepared from a murine osteosarcoma (Dunn) as previously described.[35] This BMP-active fraction was confirmed to contain murine BMP-4.[35,36] The partially-purified murine BMP-4 was mixed with telopeptide-depleted bovine skin type I collagen in a 0.01 M hydrochloride solution and then lyophilized. The collagen was used as a delivery system for murine BMP-4 which prevented rapid outward diffusion and permitted sustained release.[37]

Thirty mice in the experimental group were surgically treated with the following method. A longitudinal skin incision was made over the lumbar vertebral column, and the dorsal paravertebral muscles were detached from the spinous processes and laminae. The supraspinous and interspinous ligaments of L2-3 or L3-4 were resected. A mid-sagittal slit was made between the right and left ligamenta flava. Using a surgical microscope, two tiny and thin sheets of the BMP-4/collagen composite (100 mg) were inserted into the posterolateral epidural space through the slit, and thereby, in the epidural space, one of the BMP-4/collagen sheets was on the surface of the right ligamentum flavum and the other the left. Eighteen mice in the control group were surgically treated with the same procedure except that collagen (100 mg) without the murine BMP-4 fraction was inserted into the same location in the epidural space.

The 30 mice in the experimental group were sacrificed four, six and eight weeks (10 mice for each time period) after implantation of the BMP-4/collagen composites. They were named E4, E6 and E8 groups, respectively. The 18 mice in the control group were also sacrificed at the same intervals (six mice for each time period). Blocks of whole lumbar spinal columns containing the implanted site were resected together with the paravertebral muscles. They were radiographically examined, and then the specimens were fixed, decalcified, cut sagittally or transversely, embedded, and sectioned. Serial sections were stained with hematoxylin and eosin, toluidine blue, van Gieson's solution, and Kluver-Barrera's solution for light microscopic examination.

Radiographic examination of the spinal columns showed that, in all the mice of the experimental group, a pair of beak-like calcified prominences arose from the laminae at the BMP-4/collagen-implanted segment and protruded into the spinal canal on a lateral radiograph.[34] The cephalic and caudal parts of these bony prominences did not unite completely, even the largest. The degree of narrowing of the spinal canal caused by the bony prominences was obtained by dividing the width of the bony prominences by the diameter of the spinal canal measured on a lateral radiograph. By this calculation, the degree of narrowing was 34.7±10.9% (mean±SD) in the E4 group, 42.8±12.3% in E6, and 46.4±13.9% in E8. Between the E4 and E8 groups, there was a significant difference in the degree of narrowing of the spinal canal caused by the bony prominences ($p<0.05$, Student's t test). This indicated that the size of the bony prominences had increased gradually with the lapse of time. In the control group, there was no abnormal shadow on radiographs.

Histologic examinations of the ligamenta flava showed that, in all the mice of the E4 group, the ligamentum flavum became hypertrophied and a pair of newly formed bony prominences protruded into the spinal canal from the ventral (canal-side) surfaces of the contiguous laminae. The bony prominences adhered to the dura mater and led to compression of the contiguous spinal cord. Fibrous or cartilaginous tissue was found to intervene between the cephalic and caudal parts of the ossified loci. The ossification appeared to extend along the ventral (superficial) layer of the hypertrophied ligament because the ossification was more advanced in that layer. The ventral (superficial) layer was gradually replaced by newly formed bone and cartilage tissues, whereas in the dorsal (deep) layer, collagenous fibers were irregularly hyalinized and elastic fibers were partially disappeared and disrupted. In all the mice in the E6 and E8 groups, endochondral ossification was more advanced. The ossified areas extended further along the hypertrophied ligament but their cephalic and caudal parts did not unite completely with the intervening fibrous and cartilaginous tissue. A small amount of original ligamentous tissue, in which elastic fibers were decreased in number and scattered within newly formed fibrocartilaginous tissue, remained in the dorsal (deep) layer. In the control group, the fibrous constructs of the ligament appeared intact and no new bone or cartilage formation was observed.

Histologic examinations of the spinal cords showed that, in the E4 group, the spinal cord exhibited deformation secondary to compression by the protruded ossified ligamentum flavum (See Figure 5 in Ref. 34). In spite of this mild deformation, little or no degenerative change was evident in the spinal cord. In the E6 and E8 groups, the spinal cord exhibited marked deformation. In those cases with moderate deformation of the spinal cord, demyelination was evident in the posterior and lateral white columns while in the anterior white columns it was not.[34] The gray matter appeared intact; neuronal loss or chromatolysis in the anterior horn was not observed.[34] In those cases with severe deformation of the spinal cord, both the white and the gray matter degenerated (Figure 2).[34] Demyelination and loss of axonal fibers in the posterior and lateral white columns were marked, while little demyelination was observed in the anterior white column (Figure 2).[34] Neuronal loss and chromatolysis in the anterior horn were also evident (Figure 2).[34] These degenerative changes in the spinal cord were, on the whole, more severe in the E8 than in the E4 group. In the control group, no spinal cord was deformed or degenerated.

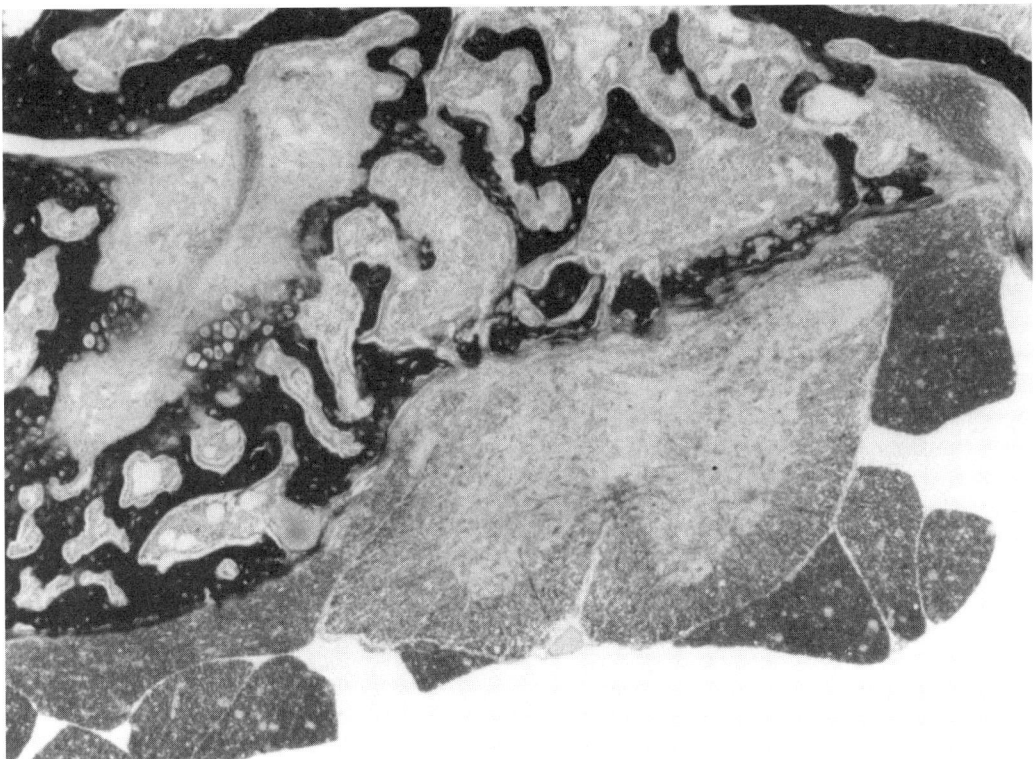

FIGURE 2. A transverse section shows severe deformation of the spinal cord by ossified ligamentum flavum (toluidine blue stain). Demyelination and loss of axonal fibers are predominant in the posterior and lateral columns.

The authors concluded that OLF could be experimentally induced in mice by the bone-inducing effect of murine BMP-4. This experimental study has provided direct evidence that the stem cells which can respond to the cartilage-inducing and bone-inducing signals of murine BMP-4 and, consequently, differentiate into chondrocytes and osteoblasts do exist in or around the ligamenta flava of mice. Another important outcome is that the pathologic findings in this model bear close resemblances to those reported in clinicopathologic studies of OLF.[28,38,39] In both the present experimental study and the clinicopathologic studies, the following have been found: ossification is accompanied by degeneration and hypertrophy of the ligament; the bony prominences arise from the ventral (superficial) surface of the laminae and extend along the ventral (superficial) layer of the hypertrophied ligament; the cephalic and caudal parts of the ossified areas do not unite completely with the intervening fibrous or cartilaginous tissue; the ossified ligament develops through a process of endochondral ossification; the ossified ligamentum flavum increases in size over time and causes gradual compression and deformation of the spinal cord leading to the pathologic changes. These results may indicate that BMPs play an important role in the development and growth of not only OLF but also OPLL. However, because the murine BMP-4 fraction used in this experimental study was not completely pure, it might have contained minor contaminants such as TGF-βs. It remains a possibility that TGF-βs may also play a role in the development and growth of the ossification.

In addition, this experimental model could be useful for the pathoanatomical and pathophysiological study of myelopathy due to gradual chronic compression of the spinal cord caused by OPLL or OLF. Saito et al. used rabbits and a crude fraction of unidentified BMP-like substances to simulate OLF, and showed similar outcomes.[40]

IV. CONCLUDING REMARKS

The availability of an experimental animal model that mimics the human disease and consistently reproduces the disease would facilitate basic investigations of pathogenesis and pathophysiology, and moreover, would be useful to test or compare effectiveness of various treating modalities. It could provide the basis for a more comprehensive analysis than is possible with humans. Although many models of spinal cord compression have been reported so far, there have been few reports on the test or comparison of various treatment modalities using those models.[17] Development of more reproducible and less complicated models might be required in order to test or compare the surgical or non-surgical treatment modalities.

Although cervical spondylosis, OPLL, cervical disc herniation, spinal tumor, and spinal cord tumor may be all similar in terms of spinal cord compression, each has its own characteristic features. For example, hard osteophytes produce multi-segmental cord compression in most cases of cervical spondylosis whereas soft discs cause single-level compression in many cases of cervical disc herniation. Does a model of spinal cord compression by a single metal or plastic screw simulate cervical spondylosis or disc herniation? It may be necessary to develop a specific model which has characteristic features of a specific human disease causing spinal cord compression.

Evaluation of outcome is an extremely important aspect of models of spinal cord compression. Methods for the evaluation include clinical, anatomical, physiological, pathological, and radiological techniques. These outcome parameters should be objective, consistent, and quantifiable. Great care should be taken in the experimental design and the analysis of data.

REFERENCES

1. Al-Mefty, O., Harkey, H. L., Marawi, I., Haines, D. E., Peeler, D. F., et al., "Experimental chronic compressive cervical myelopathy," *J. Neurosurg.*, 79, 550, 1993.
2. Fehlings, M. G. and Tator, C. H., "A review of models of acute experimental spinal cord injury," in *Spinal Cord Dysfunction: Assessment*, Illis, L. S., Ed., Oxford University Press, Oxford, 1988, Chapter 1.
3. Allen, A. R., "Surgery of experimental lesions of spinal cord equivalent to crush injury of fracture dislocation: preliminary report," *J. Am. Med. Assoc.*, 57, 878, 1911.
4. Hukuda, S. and Wilson, C. B., "Experimental cervical myelopathy: effects of compression and ischemia on the canine cervical cord," *J. Neurosurg.*, 37, 631, 1972.
5. Tarlov, I. M., *Spinal Cord Compression: Mechanism of Paralysis and Treatment*, Thomas, Springfield, 1957.
6. Rivlin. A. S. and Tator, C. H., "Objective clinical assessment of motor function after experimental spinal cord injury in the rat," *J. Neurosurg.*, 47, 577, 1977.
7. Doppman, J. L., Ramsey, R., and Thies, R. J., II, "A percutaneous technique for producing intraspinal mass lesions in experimental animals," *J. Neurosurg.*, 38, 438, 1973.
8. Haining, J. L., Turner, M. D., and Pantall, R. M., "Measurement of local cerebral blood flow in the unanesthetized rat using a hydrogen clearance method," *Circ. Res.*, 13, 313, 1968.
9. Tarlov, I. M., Klinger, H., and Vitale, S., "Spinal cord compression studies I: experimental technique to produce acute and gradual compression," *Arch. Neurol. Psych.*, 70, 813, 1953.
10. Tarlov, I. M. and Klinger, H., "Spinal cord compression studies II: time limits for recovery after acute compression in dogs," *Arch. Neurol. Psych.*, 71, 271, 1954.
11. Tarlov, I. M., "Spinal cord compression studies II: time limits for recovery after gradual compression in dogs," *Arch. Neurol. Psych.*, 71, 588, 1954.
12. Croft, T. J., Brodkey, J. S., and Nulsen, F. E., "Reversible spinal cord trauma: a model for electrical monitoring of spinal cord function," *J. Neurosurg.*, 36, 402, 1972.
13. Eidelberg, E., Staten, E., Watkins, J. C., McGraw, D., and McFadden, C., "A model of spinal cord injury," *Surg. Neurol.*, 6, 35, 1976.
14. Bennett, M. H. and McCallum, J. E., "Experimental decompression of spinal cord," *Surg. Neurol.*, 8, 63, 1977.

15. Olsson, S. E., "The dynamic factor in spinal cord compression: a study on dogs with special reference to cervical disc protrusions," *J. Neurosurg.,* 15, 308, 1958.
16. Batson, O. V., "The function of the vertebral veins and their role in the spread of metastasis," *Ann. Surg.,* 112, 138, 1940.
17. Coman, D. R. and DeLong, R. P., "The role of vertebral venous system in the metastasis of cancer to the spinal column," *Cancer,* 610, 1951.
18. Ushio, Y., Posner, R., Kim, J. H., Shapiro, W. R., and Posner, J. B., "Treatment of experimental spinal cord compression caused by extradural neoplasms," *J. Neurosurg.,* 47, 380, 1977.
19. Aoki, Y., Maruo, S., Arakawa, A., Sasaki, S., and Hori, S., "Experimental study of spinal cord compression by epidural tumors using electron probe X ray microanalysis and confocal laser scanning microscopy," *J. Orthop. Sci.,* 2, 434, 1997.
20. Hukuda, S., Ogata, M., and Katuura, A., "Experimental study on acute aggravating factors of cervical spondylotic myelopathy," *Spine,* 13, 15, 1988.
21. Schramm, J., Shigeno, T., and Brock, M., "Clinical signs and evoked response alterations associated with chronic experimental cord compression," *J. Neurosurg.,* 58, 734, 1983.
22. Miyamoto, S., Yonenobu, K., and Ono, K., "Experimental cervical spondylosis in the mouse," *Spine,* 16, S495, 1991.
23. Forestier, J. and Rotes-Querol, J., "Senile ankylosing hyperostosis of the spine," *Ann. Rheum. Dis.,* 9, 321, 1950.
24. Resnick, D. and Niwayama G., "Radiographic and pathologic features of spinal involvement in diffuse idiopathic skeletal hyperostosis (DISH)," *Radiology.,* 119, 559, 1976.
25. Ono, K., Ota, H., Tada, K., Hamada, H., and Takaoka, K., "Ossified posterior longitudinal ligament: a clinicopathologic study," *Spine,* 2, 126, 1977.
26. McAfee, P., Regan, J. J., and Bohlman, H. H., "Cervical cord compression from ossification of the posterior longitudinal ligament in non-Orientals," *J. Bone Joint Surg.,* 69B, 569, 1987.
27. Yonenobu, K., Ebara, S., Fujiwara, K., Yamashita, K., Ono, K., and Yamamoto, T., "Thoracic myelopathy secondary to ossification of the spinal ligament," *J. Neurosurg.,* 66, 511, 1987.
28. Okada, K., Oka, S., Tohge, K., Ono, K., Yonenobu, K., and Hosoya, T., "Thoracic myelopathy caused by ossification of the ligamentum flavum: clinicopathologic study and surgical treatment," *Spine,* 16, 280, 1991.
29. Urist, M. R., DeLange, R. J., and Finerman, G. A. M., "Bone cell differentiation and growth factors," *Science,* 220, 680, 1983.
30. Baylink, D. J., Finkelman, R. D., and Mohan, S., "Growth factors to stimulate bone formation," *J. Bone Min. Res.,* 8 (Supp. 2), S565, 1993.
31. Reddi, A. H., "Regulation of cartilage and bone differentiation by bone morphogenetic proteins," *Curr. Opin. Cell Biol.,* 4, 850, 1992.
32. Urist ,M. R., "Bone: formation by autoinduction," *Science,* 150, 893, 1965
33. Reddi, A. H., "Cell biology and biochemistry of endochondral bone development," *Collagen Relat. Res.,* 1, 209, 1981.
34. Miyamoto, S., Takaoka, K., Yonenobu, K., and Ono, K., "Ossification of the ligamentum flavum induced by bone morphogenetic protein: an experimental study in mice," *J. Bone Joint Surg.,* 74B, 279, 1992.
35. Takaoka, K., Yoshikawa, H., Hashimoto, J., Miyamoto, S., Masuhara, K., et al., "Purification and characterization of a bone-inducing protein from a murine osteosarcoma (Dunn type)," *Clin. Orthop.,* 292, 329, 1993.
36. Takaoka K, Yoshikawa H, Hashimoto J, Masuhara K, Miyamoto S, et al., "Gene cloning and expression of a bone morphogenetic protein derived from a murine osteosarcoma," *Clin. Orthop.,* 94, 344, 1993.
37. Takaoka, K., Koezuka, M., and Nakahara, H., "Telopeptide-depleted bovine skin collagen as a carrier for bone morphogenetic protein," *J. Orthop. Res.,* 9, 902, 1991.
38. Hattori, A., Endoh, H., Suzuki, K., and Kaneda, M., "Ossification of the thoracic ligamentum flavum with compression of the spinal cord: a report of six cases," *J. Jap. Orthop. Assoc.,* 50, 1141, 1976.
39. Yoshizawa, H., Ohiwa, T., Iwata, H., Nishizawa, K., and Nakamura, H., "High thoracic myelopathy due to ossification of the ligamentum flavum," *Neuro-Orthop.,* 5, 36, 1988.
40. Saito, H., Mimatsu, K., Sato, K., and Hashizume, Y., "Histopathologic and morphometric study of spinal cord lesion in a chronic cord compression model using bone morphogenetic protein in rabbits," *Spine,* 17, 1368, 1992.

29 Animal Models for Reconstruction of Vertebral Column and Intervertebral Disc

Hiromi Matsuzaki and Ken Wakabayashi

CONTENTS

I. Introduction 539
II. Animal Models and Animal Selections 540
III. Spinal Column Reconstruction without Intervertebral Disc 540
IV. Dynamic Spinal Reconstruction 541
    A. Allograft 541
    B. Autograft..545
    C. Artificial Intervertebral Discs..545
IV. Discussion and Conclusion 546
References 547

I. INTRODUCTION

The development of spinal surgery has offered a variety of solutions for the treatment of various spinal diseases or conditions. For segmental spinal defects (due to tumors or other pathological conditions), replacement of vertebral body with autogenic bone, allogenic bone,[1] bone cement, metal, or other biomaterials such as hydroxyapatite or ceramics[2,3] has been reported (see Chapter 27 for spinal fusion methods). However, complications are found associated with spine fusion such as the degeneration of adjacent segments due to overloading. Therefore, the necessity of spinal reconstruction with physiological mobility having intervertebral disc function (dynamic spinal reconstruction) has been realized and several animal experiments have been conducted to evaluate the efficacy of different biological grafts or prostheses.

In order to restore the mobility of a natural spinal column with functional intervertebral disc, allografts, autografts, and artificial intervertebral disc prostheses have been investigated in animal models. The ultimate purpose is to search for reliable spinal prostheses leading to future clinical application. Clinically, several disc prostheses have been used for human patients with predominantly degenerative disc disease. Several prostheses have been reported, such as the Charite SB disc.[4]

It has been noticed that animals useful for experiments are those of bipedalism because their weight is loaded on the spine, while experimentation using quadrupedal animals cannot provide sufficient data. However, at present, it is extremely difficult to conduct an animal experiment using bipedal animals (nonhuman primates), so most researchers are using quadrupedal animals, such as dogs or sheep. In this chapter, spinal reconstruction without mobility and dynamic spinal reconstruction having intervertebral disc function will be described.

TABLE 1
Animal Models of Vertebral Body and Intervertebral Disc Replacement

Procedure	1st Author, Year[Ref.]	Animal	Prosthesis	Level	Time Period
Vertebral body	Waku 1990[5]	Dogs	Ti plate + cement	L4	16 wks.
Allogenic disc unit	Olson 1991[6]	Dogs	Stored allograft	T7–9	18 mo.
	Katsuura 1994[7]	Dogs	Stored allograft	Middle L	48 wks.
	Matsuzaki 1996[8]	Dogs	Stored allograft	L4–6	3 yrs.
Autogenic disc unit	Frick 1994[9]	Dogs	Fresh autograft	L2/3,L4/5	4 mo.
	Luk 1997[10]	Monkeys	Fresh autograft	L3/4,L4/5	12 mo.
Disc prosthesis	Hou 1991[11]	Monkeys	Artificial disc	L4/5	15 mo.
	Vuono-Hawkins 1994[12]	Dogs	Artificial disc	L2/3,L5/6	12 mo.
	Nakamura 1997[13]	Dogs	Artificial disc	L3/4,L5/6	3 mo.
	Kostuik 1997[14]	Sheep	Artificial disc	Lumbar	6 mo.

II. ANIMAL MODELS AND ANIMAL SELECTIONS

Large animals such as dogs, sheep, and monkeys, have been used for experimentation of vertebral column and intervertebral disc graft or prosthesis (Table 1). In an immobilized spinal reconstruction, only the replacement of a vertebral body is done, so quadrupedal animals such as dogs or sheep may be appropriate. The evaluation of the function of intervertebral disc is not applied in this case, and only the healing of the graft or prosthesis to the adjacent vertebral body is concerned. Goats and pigs may serve as experimental animals for reconstruction or replacement of spinal columns. Because instrumentation (internal fixations such as plate, screw, etc.) is needed for spinal reconstructions, small animals, such as mice, rats and rabbits, are inappropriate simply due to their small bone volume.

In dynamic spinal reconstruction, the long term results of intervertebral disc function under similar weight-bearing conditions to humans is essential. Therefore, bipedal animals, such as monkeys, should be selected. However, due to their cost and limited availability, quadrupedal animals have been used. Although chickens are bipedal, they are too small to qualify.

III. SPINAL COLUMN RECONSTRUCTION WITHOUT INTERVERTEBRAL DISC

Using seven adult mongrels, Waku replaced the vertebral body alone using a special plate prosthesis made of a titanium alloy.[5] The prosthesis was used to sustain the vertebral body with processes near the upper and lower ends. By intruding forward, L4 and its upper and lower intervertebral discs are excised, preparing furrows at the front of the L3 and L5, the prosthetic plate was put into it and filled with bone cement. Animals were killed four and 16 weeks postoperatively, L3 and L5 were taken out as en bloc and radiographically and histologically examined.

Although dislodgment of artificial vertebral body was not observed, submerging of artificial vertebral body into fixed vertebral body was found in all cases. However, when periodically examined, at the axial loading point of the end of vertebral body, the foregoing endplate was absorbed and by a large amount of new bone formation, an endplate thicker than normal was formed. The newly formed bone remained premature at four weeks and became a layer of trabecular structure at 16 weeks. The results indicated that by an appropriate loading, the remodeling of bone became marked.[5]

IV. DYNAMIC SPINAL RECONSTRUCTION

For restoring normal spinal function, allogenic and autogenic grafts of intervertebral disc with the adjacent vertebral bodies (a part of the vertebral body, about 1.0–1.5 cm thick on each side of the disc to be grafted), namely vertebral body/intervertebral disc complex, have been reported.[7,8] An alternative to allografts is the fast development of artificial intervertebral disc prostheses. Major animal models for the above-mentioned applications are described as the following.

A. ALLOGRAFT

1. Multi-Segment Graft

Using 15 large adult mongrels (20–25kg), Olson et al.[6] freeze stored large spinal graft units from T7 to T9 at 80°C for two weeks, excised the 8th rib and reached the vertebral body by transthoracic approach, cutting the vertebral bodies from T7 to T9 into a triangular form, and grafting the unit so it would not dislodge forward. In this procedure, no instrument was used to stabilize the unit. After 18 months, there was no biomechanical difference from normal cases. However, bone strength of the grafted area decreased significantly as compared with that of the normal area ($p<0.05$). The bone union was obtained histologically, but at the middle vertebral body (T8), an incomplete revascularization was noted and the stratified structure of the intervertebral disc disappeared, showing progress of its denaturation.[6]

2. Katsuura's Method

In 13 adult mongrels, Katsuura et al.[7] used allogenic grafting units attaching parts of the vertebral body to lumbar intervertebral discs which were stored at –80°C for four weeks in a programmed freezing system. Intruding transperitoneal abdominally, and after excising parts of the intervertebral disc and vertebral body, the unit was grafted. The unit was fixed by fixing the partial vertebral body plates to the upper and lower vertebral bodies. During three months until the evulsion of the plate was made, there was no mobility of the grafted intervertebral disc, so that there was a risk of marked denaturation of the disc. The metabolic activity of the intervertebral disc after four weeks of frozen storage decreased up to 44% of the fresh cases in terms of ^{35}S-sulfate incorporation. This means that because of freezing, the function of the intervertebral disc cell markedly decreased. Further, radiographically, narrowing of the intervertebral disc cavity progressed. These were closely similar to our results.[1]

3. Matsuzaki's Method

The authors reported a basic procedure of allogenic intervertebral disc grafting in dogs.[8] The lumbar spine (L1–7) was removed using a sterile technique from 11 mongrel dogs (15 kg average BW) to prepare disc units for transplantation. In order to fix a disc in the recipient lumbar spine without fail, it was removed with adjacent vertebrae transected about 1 cm apart from the surfaces facing the disc so that the cut surfaces could be fixed to the cut surfaces of the recipient vertebrae by bone fusion (Figure 1). Eight disc units were immersed for impregnation in 10% dimethyl sulfoxide (DMSO) at 4°C for one hr, and stored at –80°C for 1–16 weeks (mean, eight weeks). Another group of eight disc units were immersed for impregnation in 10% DMSO at 4°C for three hours, and frozen at –30 for one hr and further at –80°C for one hr, and then stored in liquid nitrogen at –196°C for 1–6 weeks (mean, four weeks) until transplantation.

Recipient animals were anesthetized intravenously with Nembutal, and the lumbar vertebrae were exposed through a peritoneal approach. The frozen disc unit was promptly thawed at 37°C in a thermostatic chamber and immediately transplanted. The disc unit was inserted mainly in the L4/5 or L5/6 intervertebral space. The disc to be replaced was removed with the adjacent vertebral

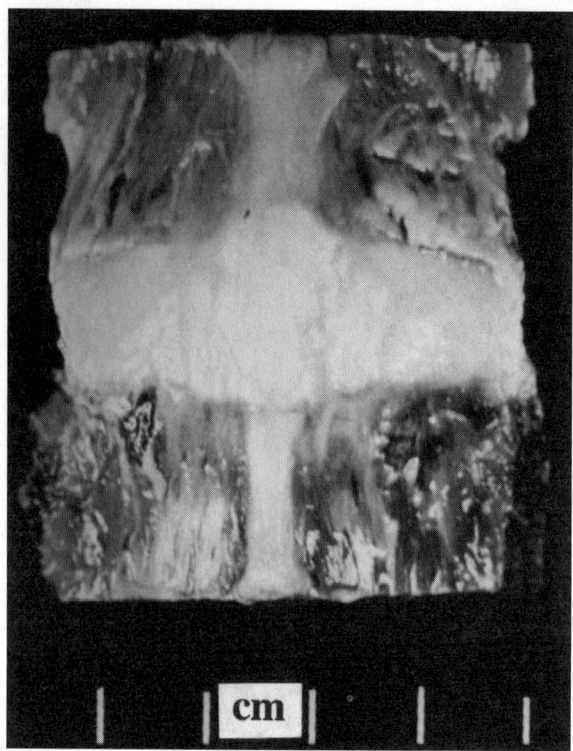

FIGURE 1. A donor disc unit removed with the adjacent vertebrae transected about 1 cm apart from the surfaces facing the disc.

bodies transected at levels about 1 cm apart from their surface facing the disc to be removed. The recipient bed was prepared carefully particularly so as not to injure the dura. The remaining vertebral bodies were cut so that the space between them could match the disc unit to be inserted. An AO mini plate was fixed to the inserted disc with two screws, leaving an about 1-cm long segment of the plate above the transected surface of the graft vertebral body so that it could be fixed to the recipient vertebral body. The recipient bed was trimmed, and the plate was fixed to the recipient vertebral body with screws (Figure 2).

At six months, bone fusion was accomplished in all animals according to radiographic evaluation. The fusion did not vary depending on the length of storage of the units. The disc units stored at –196°C showed less bone resorption than those stored at –80°C, and we had an impression that they achieved bone fusion somewhat faster. The disc space was not changed at six months, and its apparent narrowing appeared at around 12 months in both groups. The narrowing progressed gradually. At three and five years, the space became remarkably narrow with growth of osteophytes, indicating that the disc unit lost the normal disc function, although the disc space remained. Radiographs obtained with soft X rays also showed an adequate intervertebral space.

MRIs were obtained with lumbar spines removed from the recipients at six and 36 months. With specimens removed at six months, high signal areas representing the transplanted discs were demonstrated by T2 weighted, but the recipient discs and fresh discs also had the similar signal intensity to one seen in the transplanted disc. In the vertebral components of the unit, areas of high and low signal intensities were mixed (Figure 3). In other words, bone was not adequately mature in the transplanted unit. At 36 months, as at six months, the transplanted disc had the same high signal intensity as the adjacent discs. The bony portions of the transplanted vertebral component showed no change in signal intensity as observed at six months, indicating that the transplanted bone had been completely replaced with newly formed bone.

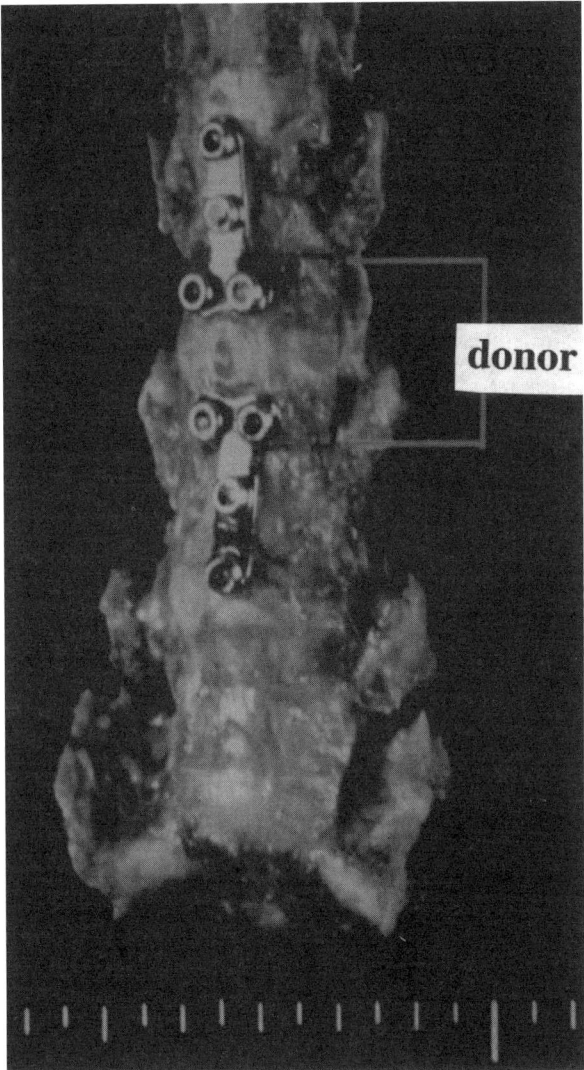

FIGURE 2. The fixation of the disc unit was made with two plates and screws. The disc in the donor unit (between the two lines) remains mobile.

Before transplantation, for the disc units stored at both temperatures, vacuolation and other features of degeneration were observed in the nucleus pulposus using histological examination. However, the anulus fibrosus was morphologically well preserved, particularly for the units stored at −196°C. The changes were not dependent on storage times in both groups.

Discs removed from recipients at six months were decalcified, stained with HE, and examined. With disc units stored at −80 and −196°C, the graft-recipient bone interface showed excellent bone fusion. Cells decreased markedly in number, however, with fibrosis in bone marrow of the disc unit. No lymphocyte was observed, nor was there any evidence suggestive of rejection.[15] Regardless of store temperatures, the normal structure of the nucleus pulposus was maintained in the disc unit, but cells without the nucleus were distributed evenly, practically without infiltration of lymphocytes. The disappearance of the nucleus was remarkable in the annulus fibrosus, which was slightly edematous, but had the normal lamelliform structure. Storage at −196°C is more suitable for the preservation of the annulus fibrosus than that at −80°C.

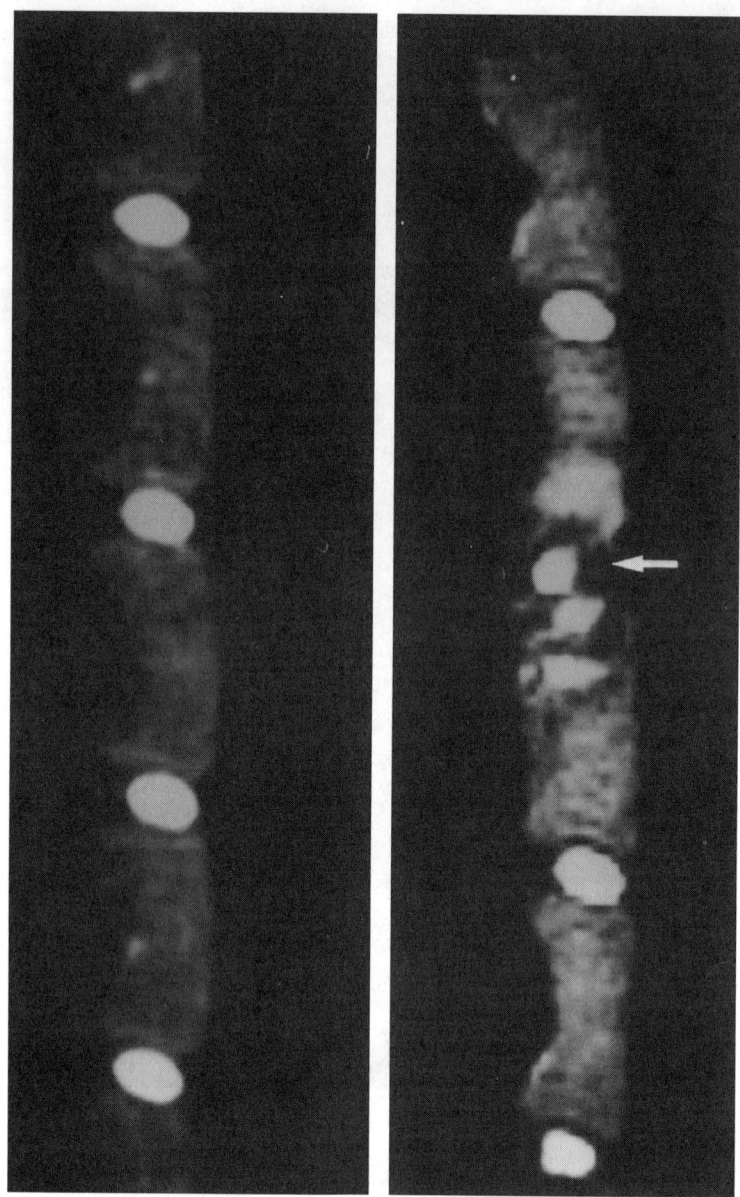

FIGURE 3. MRI images of a fresh disc (A) and at six months after the transplantation (B). The high signal area represents the transplanted disc (arrow).

The cell activity in the transplanted intervertebral disc was also assessed at three and six months. In disc units stored at −80 and −196°C, both incorporation of ^{35}S-sulfate and ^{3}H into cells and their liberation from the cells into the supernate were markedly decreased. When assessed by incubation of the cells for 12 hours, the cell activity of the transplanted discs was 1/5 to 1/20 of that of fresh intervertebral discs. With an expectation that their incorporation and liberation would increase if cells were active, the cells were cultured further for 24 hours, but neither ^{35}S-sulfate or ^{3}H-proline was incorporated in, and liberated from, the cells any more, suggesting that the cells were not active or dead.

B. Autograft

Similar procedure was employed in the several investigations of autogenic transplantation of intervertebral disc units. Because of their autogenic nature, there was no immunological reaction, therefore, pure changes in intervertebral disc grafting can be evaluated.

Using eight adult mongrel dogs (23–30kg), Frick et al.[9] performed an autogenic intervertebral disc transfer. Grafting was done by exchanging L2/3 and L4/5 discs. The discs were excised and thin sections (2 to 4 mm in depth) were prepared for grafting. The grafted intervertebral discs were fixed with wire so that mobility of the disc was preserved. Animals were killed after four months. Their spines were longitudinally incised, and mobility, histological change, and biochemical parameters were examined. The synthesis (^{35}S) of proteoglycan was preserved in the annulus fibrosus, but decreased in the nucleus pulposus ($p<0.05$). However, no change in DNA content was observed. Histologically, revascularization and remodeling were observed in the donor bone. The structure of the annulus fibrosus was preserved but the endplate changed irregularly.[9] From these facts, it was suggested that the intervertebral disc once cut off does not completely recover to normal, and the trend is high in the nucleus pulposus. The reason that the result was somewhat unfavorable in comparison with that of Luk et al.[10] may be possibly due to differences in experimental animals and grafted intervertebral disc units. However, these tests are said to be of value for discussing the fate of grafted sections.

Luk et al.[10] performed an autografting of intervertebral disc in rhesus monkey. The rhesus monkey is a bipedal animal, so the results are interesting. To the upper and lower parts of a lumbar intervertebral disc, a thin section vertebral body of 1.5 mm in depth was attached and excision of intervertebral disc was conducted, thereafter, replaced at the same site followed by 12-month observation. The bone union was obtained within two to four months, and though narrowing of the intervertebral disc space was temporarily observed, it recovered normally after 12 months. Histologically, only a mild denaturation was observed and the decrease in proteoglycan of the nucleus pulposus in the early period was gradually re-accumulated. The large mobility of the grafted intervertebral disc in the early stage of grafting gradually stabilized. From these facts, it was suggested that cut-off intervertebral disc decreases its function in the early stage because of ischemia but thereafter recovers. However, autograft of the intervertebral disc is inappropriate for clinical application.

C. Artificial Intervertebral Discs

Compared to allogenic and autogenic grafts, intervertebral disc prostheses have shown their bright future because of their availability. Over the last several decades, a tremendous effort has been made to develop an artificial disc to replace a degenerated disc. The goal is the restoration of the natural mechanics of the segment after disc excision, thus relieving pain and preventing further degeneration of adjacent segments.

Hou et al.[11] placed artificial intervertebral discs made of silicon rubber in lumbar spines of four adult monkeys. After operation, intervertebral disc space was preserved. Histologically, the periphery of the artificial intervertebral disc was covered with fibrous tissue. It is highly possible that, after a long period, problems such as breakdown of silicone rubber may occur.

Vuono-Hawkins et al.[12] prepared 3-layer elastomeric intervertebral disc spacers for grafting into 12 dogs (20–30kg). The surface layers of the spacers were composed of hydroxyapatite. The plan was to replace the inner layers of the anulus fibrosus and nucleus pulposus, not completely replace the intervertebral disc. Under mechanical testing, compressive and torsional stiffness decreased to 25–42% of levels seen in control animals. The bone ingrowth was unfavorable on the surface of the hydroxyapatite. Much connective tissue was noted. Early migration was observed in 5 of 12 dogs.

In an experiment by Nakamura et al.[13] an artificial intervertebral disc was developed and used for lumbar fixation in beagles. The intervertebral disc spaces were kept at favorable conditions at least one year. The artificial disc was composed of annulus fibrosus, nucleus pulposus and hydrous polyvinyl alcohol hydrogel (PVA-H). The area that affixed to the bed was a titanium fiber mesh with 70% porosity. The mesh was impregnated with PVA-H and gelatinized by low temperature crystallization. Infiltration of neogenetic bone was observed three months after the operation. Results of compression and torsion testing indicated a 2/3 decrease compared to a normal intervertebral disc. PVA-H is the most likely material for artificial intervertebral discs. By adjusting water content, it is possible to develop an artificial disc similar to a normal one. It is of great significance that the device can attach tightly to the vertebral body bed with titanium mesh.

Kostuik et al.[14] inserted artificial intervertebral discs composed of metal and springs into lumbar spines of six ewes. Six months after the replacement, the device combined with the bed of the vertebral body and mobility of the disc was observed. The device was durable mechanically, but use of the spring carries the risk of debris accumulation.

IV. DISCUSSION AND CONCLUSION

Spinal reconstructions are important procedures for many pathological conditions. For the vertebral body, autografts, allografts and biomaterials can be used for clinical reconstruction of spinal column defects due to the removal of tumors. For intervertebral disc replacement, autografts, allografts and artificial intervertebral discs have been studied. It has been realized that a better method for spinal reconstruction is replacement of the intervertebral disc, aiming for dynamic spinal function. This is a relatively new area with only a limited number of experimental reports, so many questions remain unanswered.

Autografting an intervertebral disc has no clinical application because there are no donor sites in the human body. Therefore, the alternatives are restricted to allograft or artificial intervertebral disc. It is difficult to preserve the intervertebral disc function of allograft for a long period because of the decrease in cell function (including cell death) due to freezing and progress of degeneration after grafting. In addition, immunorejection may play an important role in the degeneration process. Therefore, intervertebral disc allografts are only useful as spacers for a limited period of time and are unlikely to be functional motion segments.[8,9]

For artificial disc replacement, different materials have been used including polyethylene and metal,[16,17] titanium (Kostuik prosthesis),[14] polyurethane elastomer reinforced with Dacron fibers,[18] silicone rubber,[19] a hyaluronan-containing device,[20] and a hydrogel.[21] Although a certain number of clinical reports have appeared, there are many recognized and potential problems and questions, for example, durability of materials,[14] wear debris,[14,22] toxicity (Link prosthesis),[4] stability of the fixation (migration or dislocation),[23] or collapse of the prosthesis into the vertebral body.[16] More basic research and animal experimentation are needed for the development of better materials and better prosthetic designs.[19,21,24]

REFERENCES

1. Griss, P. and Pferffer, M., "Vertebral body replacement with homologous femoral head transplants," *Int. Orthop.*, 15, 65, 1991.
2. Matsui, H., Tatezaki, S., and Tsuji, H., "Ceramic vertebral body replacement for metastatic spine tumors," *J. Spinal Disord.*, 7, 248, 1994.
3. Hosono, N., Yonenobu, K., Fuji, T., Ebara, S., Yamashita, K., and Ono, K., "Vertebral body replacement with a ceramic prosthesis for metastatic spinal tumors," *Spine*, 20, 2454, 1995.
4. Cinotti, G., David, T., and Postacchini. F., "Results of disc prosthesis after a minimum follow-up period of 2 years," *Spine*, 1996, 21, 995, 1996.
5. Waku, S., "An experimental study on a new artificial vertebral body," *Nippon Seikeigeka Gakkai Zasshi*, 64, 409, 1990.
6. Olson, E. J., Hanley, E. N. Jr., Rudert, M. J., and Baratz, M. E., "Vertebral column allografts for the treatment of segmental spine defects. An experimental investigation in dogs," *Spine*, 16, 1081, 1991.
7. Katsuura, A. and Hukuda, S., "Experimental study of intervertebral disc allografting in the dog," *Spine*, 19, 2426, 1994.
8. Matsuzaki, H., Wakabayashi, K., Ishihara, H., Ishikawa, H., and Ohkawa, A., "Allografting intervertebral discs in dogs: a possible clinical application," *Spine*, 21, 178, 1996.
9. Frick, S. L., Hanley, E. N., Jr., Meyer, R. A., Ramp, W. K., and Chapman, T. M., "Lumbar intervertebral disc transfer. A canine study," *Spine*, 19, 1826, 1994.
10. Luk, K. D., Ruan, D. K., Chow, D. H., and Leong, J. C., "Intervertebral disc autografting in a bipedal animal model," *Clin. Orthop*, 337, 13, 1997.
11. Hou, T. S., Tu, K. Y., Xu, Y. K., Li, Z. B., Cai, A. H., and Wang, H. C., "Lumbar intervertebral disc prosthesis. An experimental study," *Chin. Med. J. (Engl)*, 104, 381, 1991.
12. Vuono-Hawkins, M., Zimmerman, M. C., Lee, C. K., Carter, F. M., Parsons, J. R., et al., "Mechanical evaluation of a canine intervertebral disc spacer: *in situ* and *in vivo* studies," *J. Orthop. Res.*, 12, 119, 1994.
13. Nakamura, M., Sakaguchi, K., Oka, M., Yura, S., Gen, S., et al., "Mechanical prosthesis of canine artificial intervertebral disc," *J. Jpn. Soc. Clin. Biomech.*, 18, 437, 1997.
14. Kostuik, J. P., "Intervertebral disc replacement. Experimental study," *Clin. Orthop.*, 337, 27, 1997.
15. Langer, F., Czitrom, A., Pritzker, K. P., and Gross, A., "The immunogenicity of fresh and frozen allogenic bone," *J. Bone Joint Surg.*, 57A, 216, 1975.
16. Buttner-Janz, K., Schellnack, K., and Zippel, H., "Biomechanics of the SB Charite lumbar intervertebral disc endoprosthesis," *Int. Othop.*, 13, 173, 1989.
17. Steffee, A., "The Steffee artificial disc," in *Clinical Efficacy and Outcome in the Diagnosis and Treatment of Low Back Pain*, Weinstein, J. N., Ed., Raven Press, New York, 1992, 245.
18. Lee, C. K., Langrana, N. A., Parsons, J. R., and Zimmerman, M. C., "Development of a prosthetic intervertebral disc," *Spine*, 16 (Suppl), S253, 1991.
19. Schneider, P. G. and Oyen, R., "Surgical replacement of the intervertebral disc. First communication: replacement of lumbar discs with silicon-rubber. Theoretical and experimental investigations," *Z. Orthop. Ihre. Grenzgeb.*, 112, 1078 , 1974.
20. Ray, C. D., "The artificial disc," in *Clinical Efficacy and Outcome in the Diagnosis and Treatment of Low Back Pain*, Weinstein, J. N., Ed., Raven Press, New York, 1992, 205.
22. Hellier, W. G., Hedman, T. P., and Kostuik, J. P., "Wear studies for development of an intervertebral disc prosthesis," *Spine,* 17(Supp.), S86, 1992.
23. Griffith, S. L., Shelokov, A. P., Buttner-Janz, K., LeMaire, J. P., and Zeegers, W. S., "A multicenter retrospective study of the clinical results of the LINK SB Charite intervertebral prosthesis. The initial European experience," *Spine*, 19, 1842, 1994.
24. Lemaire, J. P., Skalli, W., Lavaste, F., Templier, A., Mendes, F., et al., "Intervertebral disc prosthesis. Results and prospects for the year 2000," *Clin. Orthop.*, 337, 64, 1997.

30 Animal Models of Scoliosis

Noriaki Kawakami, Masao Deguchi, and Tokumi Kanemura

CONTENTS

I. Introduction	549
II. Animal Selection	550
III. Animal Models of Scoliosis	550
A. Dietary Methods	551
B. Immobilization	551
C. Systemic Methods	552
1. Pinealectomy	552
D. Local Methods	554
1. Intercostal Nerve Resection	554
2. Rib Resection	555
3. Tethering between Scapula and Pelvis	556
4. Removal of Transverse Processes or Ligaments	557
5. Muscle Imbalance	558
6. Other Methods	558
IV. The Use of Scoliosis Models in Spinal Research	229
A. Pathogenesis of Scoliosis	559
B. Exploration for New Treatment	559
V. Future Directions of Research	559
References	560

I. INTRODUCTION

Spinal deformity was first described by Hippocrates.[1] Galen introduced the terms *scoliosis*, *kyphosis*, and *lordosis* into medical terminology in the second century.[2] It is not too much to say that the battle of research and treatment of scoliosis has been lasting for more than two thousand years. Still there remain many unknown points as to its causes and pathology in spite of many extensive experimental and clinical studies done by numerous researchers.

Experimental scoliosis has been produced using different animals since the beginning of the 20th century. The fundamental goals of experimental scoliosis are to produce ideal models of scoliosis, which are comparable to human idiopathic scoliosis and to clarify the etiology and develop new therapeutic methods. According to the literature, surgical procedures performed directly on the spine and its vicinity been used to produce scoliosis as have dietary feeding and injection of pharmacological agents. It has been a process of "trial and error" and each of the animal models has its own characteristics. In this chapter, experimental models of scoliosis from the literature are summarized, with emphasis on the experimental procedures for producing scoliosis and the effects of species, age and size of the experimental animal.

TABLE 1
Animal Selection

Procedures	Animals
Prenatal procedures	Lamb, mouse, rat, chicken egg
Immobilization using external fixation	Rat, rabbit, dog
Pinealectomy	Chicken, rat
Direct injuries of epiphysial plate	Pig, goat, dog
Resection and/or incision of ribs, transverse processes, and/or ligaments	Rabbit, pig, monkey, baboon
Rhizotomy, intercostal nerve resection, or spinal cord injuries	Rabbit, dog
Tethering	Rabbit, dog, rat
Excision of paravertebral muscle	Rabbit, rat, mouse, monkey
Electrostimulation	Rabbit, dog
Irradiation	Rabbit
Magnet implantation	Rat

II. ANIMAL SELECTION

Many animal species including mammals and birds have been used (Table 1). Rabbits and rats are selected most frequently among mammals. Dogs, pigs, and monkeys also have been used. However, most of them are quadrupedal except monkeys. In these animals, the anatomic structure of the spine is different from that of human. Therefore, the models in quadrupedal mammals have certain limits for researching the etiology of scoliosis that seem to be influenced by gravity. For this reason, Yamamoto[3] and Machida[4] produced scoliosis in bipedal rats using Goff's method[5] and studied the effect of gravity on the occurrence of scoliosis. Birds are bipedal animals and better suited for these circumstances. Chickens have been reported to develop spontaneous scoliosis without any congenital vertebral anomalies.[6] Chickens have also been used to produce scoliosis by pinealectomy because the location of the pineal gland can be easily approached.[7-15] However, the chicken spine has no intervertebral discs and tends to fuse spontaneously with only two levels remaining unfused in the thoracic portion. This anatomic difference from humans limits the use of the model for investigating the etiology of scoliosis. In addition to the anatomic features, age and size of the animal are also very important because the onset and progression of scoliosis are strongly related to growth. Each animal species has its own rate of growth. For most studies young animals are selected, such as 1–10-week-old rabbits, 2–20-week-old cats, or 6–20-week-old dogs.

III. ANIMAL MODELS OF SCOLIOSIS

Animal models of scoliosis are divided into two groups, spontaneous and induced scoliosis (Table 2). Several species of chicken,[6] quail,[16] duck,[17] rat,[18] rabbit[19,20] and horse[21] has been reported to show spontaneous occurrence of scoliosis, although congenital vertebral anomalies were involved in some of them. Experimental methods for induction of scoliosis in the prenatal periods include maternal exposure to reduced oxygen (hypoxia) in mice,[22] and injection of insulin,[23] 6-aminonicotinamide,[24] azaserine,[24] or thiadiazole[24] into chicken eggs. Kent and Zingg[25] also reported surgical procedures through a hysterotomy in a lamb. Most methods of creating experimental scoliosis involve induction during the postnatal period. Although many procedures have been reported, they are mainly classified into dietary deficiency, immobilization, systemic procedures (such as pinealectomy), and local procedures (Table 2). Local procedures are those applied directly to the (1) spinal column such as unilateral damage of the epiphysis by irradiation[26,27] or surgical resection,[28,29] unilateral resection of the transverse processes and laminae,[30] and unilateral fusion of laminae and transverse processes;[31,32] (2) the surrounding tissues, such as unilateral resections

TABLE 2
Types of Experimental Scoliosis

I. Spontaneous occurrence
II. Artificial production
    A. Prenatal induction (hypoxia, hypovitaminosis, insulin injection, etc.)
    B. Postnatal induction
        1. Dietary deficiency (lathyrism, etc.)
        2. Immobilization (cast, band, or splint)
        3. Systemic procedures (such as pinealectomy)
        4. Local procedures
            — Damage to spine (epiphysis, lamina, spinous or transverse process)
            — Damage to surrounding tissues (ligament, rib, or muscle)
            — Damage to neural tissues (spinal cord, root, or intercostal nerve)

of transverse ligaments,[30,33] ribs[30,34–40] and paravertebral muscles;[30,34,41–44] and (3) neural tissues, such as surgical damage of spinal cord, roots and intercostal nerves.[45–51]

A. DIETARY METHODS

Experimental diets containing *Lathyrus odoratus* induced scoliosis.[3,52,53] Geiger et al.[52] investigated the effect of feeding lathyrus peas and succeeded in inducing lathyrism in young and adult rats. He noticed growth retardation of young animals with other symptoms, such as lameness, spinal curvature, sternal curvature, etc. Autopsy showed extreme curvature that was mostly ventral in the thoracic region. Ponseti[53] studied characteristics of scoliosis induced by lathyrism in rats to explore the pathogenesis of scoliosis and reported hitherto unrecognized lesions in the epiphyseal plates. Yamamoto[2] fed bipedal rats on modified Steenbock's diet that included *Lathyrus odoratus* to analyze the influence of the upright position on the development of scoliotic deformity due to dietary feedings. He compared them with rachitic quadrupedal, control bipedal, and control quadrupedal rats, and reported that marked scoliosis developed only in rachitic bipedal rats. This result indicated that the upright position played an important role to produce marked scoliosis even in rachitic rats. Compared with the relatively large amount of *Lathyrus odoratus* that was needed to produce scoliosis, feeding a small amount of B-aminopropionitrile (BAPN) has also been proved to produce scoliosis and other skeletal deformities by Lalich and Angevine.[54]

Vitamin deficiency is another way of dietary feeding to produce scoliosis in animals.[55,56] Kitamura et al.[55] reported that rainbow trout developed scoliosis or lordosis during feeding a synthetic diet with a lack of vitamin C. Lim and Lovell[56] studied the effects of dietary feeding with vitamin C deficiency on the spinal columns of channel catfish that presented scoliosis or lordosis with the frequency of 60.9% 22 weeks after feeding.

B. IMMOBILIZATION

Immobilization has proven to be one of the ways that induces scoliosis in animals. Cast, strip, and splint were reported as tools to immobilize animals into a scoliotic position.[57–59] Wullstein[59] first reported experimental scoliosis in two dogs provoked by immobilizing them into a scoliotic position with strips of leather for seven to 13 months. The scoliosis produced in his study was slight — a long C curve involving the whole spine. Hakkarainen[57] used a three-point plaster-of-Paris corset with a thin layer of foam rubber for immobilizing rabbits in a scoliotic position for two to 5 weeks. He observed regression of scoliosis less than 30 degrees at the time of removal of immobilization, whereas progression of scoliosis occurred when the initial curve exceeded

30 degrees. Poussa et al.[58] investigated the effect of forced lordotic position by fitting extension splints in rabbits. Their study was based on the concept of rotated lordosis in the pathological mechanism of scoliosis introduced by Somerville,[60] Roaf,[60-62] and Dickson,[63] and 53.5% of rabbits used in this experiment demonstrated scoliosis greater than 10 degrees with a mean of 47 degrees. An external fixation apparatus developed by Stokes, et al.[64] was another device to produce scoliosis due to vertebral wedging. Mente et al.[65] applied it to rat tails for establishing if scoliosis progression could be explained in terms of mechanical forces causing vertebral wedging. Not only did they produce vertebral wedging and scoliosis, but also they showed the reversal of an induced vertebral wedging by distraction instead of compression.[66]

C. Systemic Methods

1. Pinealectomy (Figure 1)

Thillard[14] was the first to report experimental scoliosis using pinealectomy. She produced scoliosis by removing the pineal glands of young chickens (2–3 days old). Dubousset et al.[9] subsequently reported that the experimental removal of the pineal gland with its stem and part of the choroid plexus of the third ventricle in young chickens (1–5 days old) produced a high rate of severe or moderate deformity of the spine, but they failed to produce scoliosis in 20 quadrupedal rats with the same procedure. Their study also indicated that the scoliotic deformity in the chicken has the same anatomical characteristics (vertebral rotation and rib hump) as those found in human idiopathic scoliosis.

Machida et al.[10,67] advanced Dubousset's experimental work and performed an electrophysiological study on experimental scoliosis of pinealectomized chickens. They found no pathologic change in the brain in either the pinealectomized scoliosis group or non-scoliosis group. However, the somato-evoked potential (SEP) data showed significant abnormalities in the scoliosis group. By electrophysiological examination, cortical potentials in the scoliosis group were found delayed, suggesting conduction disturbance rostral to the brain stem. Their results strongly suggested that a certain yet unidentified defect of neurotransmitters or neurohormonal systems in the pineal body must play a major role in this model.

Machida et al.[11,12] also reported the role of the deficiency of melatonin and serotonin. They compared the occurrence and severity of scoliosis in each group treated with melatonin and serotonin. The result showed that scoliosis developed in 73% in the serotonin group, and 20% in the melatonin group. The melatonin-treated chickens with scoliosis had less severe spinal deformity than those in the serotonin-treated group. Moreover, in their serial study, they were able to prevent development of scoliosis by using 5-hydroxytryptophan (5-HTP), a precursor of serotonin, which can pass through the blood-brain barrier. They proposed that a serotonin deficit secondary to a defect of melatonin may have disturbed postural muscle tone or postural equilibrium resulting in scoliosis in the pinealectomized chicken. Consequently they implied that prevention of the development of scoliosis or its progression in chickens treated with 5-HTP suggested that serotonin may have potential therapeutic value. Finally, Machida et al.[4] performed pinealectomy in quadrupedal and bipedal rats. They reported that the scoliosis developed only in pinealectomized bipedal rats, not in quadrupedal rats, and suggested that the bipedal condition was important for the etiology of human scoliosis.

Using the pinealectomized chicken model, Coillard et al.[7] defined three types of morphological deformations of the vertebrae. Vertebral deformity type 1 was characterized by three-dimensional corporal torsion, which defined the horizontal disorientation of the curve. Vertebral deformities types 2 and 3 were the transitional vertebrae in the curve and defined as lateral imbalance in the election plane of the curve. They compared the result with human idiopathic scoliosis and suggested that the determination of morphoanatomical criteria was possible to force a new therapeutic strategy based on real vertebral and spinal deformation.

Animal Models of Scoliosis

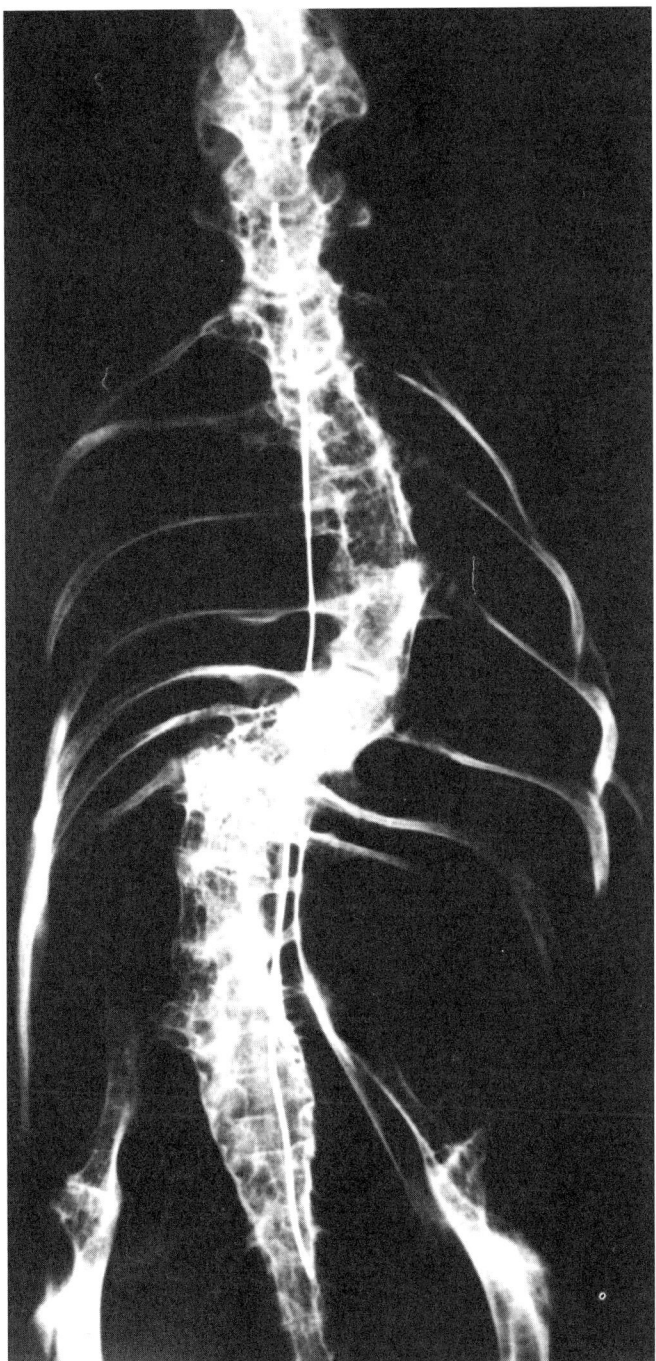

FIGURE 1. Scoliosis in a pinealectomized chicken. The pineal gland was removed at three days of age. Scoliosis developed and progressed to 83 degrees at 16 weeks of age. The wire that marks the tips of the spinous processes clearly demonstrates rotation of the vertebral column.

Wang et al.[15] characterized the scoliosis produced in young chickens after pinealectomy and compared these characteristics with those seen in human adolescent idiopathic scoliosis. The results of their study showed that 60% of chickens that received pinealectomies three days after hatching acquired scoliosis. They reported that similarities included development of single and double curves, degree of curvature, stability of the curve, numbers of vertebrae involved, direction of rotation, and progression characteristics and that differences included wedged vertebrae in the chickens, in conjunction with curve development and increased variability in vertebrae involved.

Kanemura et al.[13] clarified the natural course of scoliosis after pinealectomies in chickens. They found that 85% of animals in the pinealectomy group developed scoliosis (7–85 degrees) featuring a three-dimensional spinal deformity consisting of both lateral curvature and vertebral rotation with rib humps. They found that the scoliotic curvature progressed as the pinealectomized chickens grew older at least until 16 weeks of age (equivalent to the age of puberty in humans). The natural course of scoliosis in the pinealectomized chickens followed one of three patterns; (1) scoliosis progressed rapidly and finally reached more than 84 degrees at the age of 16 weeks, (2) scoliosis progressed more slowly and was usually mild at the age of 16 weeks, and (3) no spinal deformity was seen during the observation period. They could predict that scoliosis exceeding 20 degrees within four weeks after pinealectomy would show progression. Their study also showed pinealectomized chickens had several other differences from normal chickens, such as poor weight gain, underdeveloped cockscombs, and the late onset of egg laying. Consequently the scoliosis developing in chickens after pinealectomy was similar to human idiopathic scoliosis and thus seems to be a useful model of idiopathic scoliosis.

D. LOCAL METHODS

1. Intercostal Nerve Resection

Intercostal nerve resection, including rhizotomy, is one of the procedures that produce scoliosis constantly. Historically, these procedures have been studied by many investigators.[28,47–51,68] However, the developmental mechanisms of scoliosis are still controversial although these procedures are known to cause a condition of muscle imbalance.

Liszka[47] compared the effects of unilateral transaction of four anterior and posterior roots with those of five posterior roots on the spine. Both groups demonstrated scoliosis with convexity toward the affected side although rabbits with resections of five posterior spinal roots showed more marked scoliosis. He assumed that elimination of sensory impulses was more detrimental to the stability of the spinal column than motor paralysis of the muscles. MacEwen[69] repeated these experiments and reached the assumption that severance of the reflex arc was responsible for the development of scoliosis.

Alexander et al.[48] reported the effects of a number of roots transacted on the severity of scoliosis, and found that the magnitude of scoliosis depended upon the number of roots severed. They also studied the histological change of the spinal cord after the division of posterior roots. Interestingly, this study demonstrated not only dorsal column degeneration but also anterior horn cell chromatolysis in spite of preservation of anterior roots. Consequently, they concluded that scoliosis induced by rhizotomy was not due to interruption of the sensory feedback, but was caused by damage of the anterior horn cells and subsequent motor paralysis, which was contrary the assumption described by Liszka[47] and MacEwen.[70]

Suk et al.[50] performed the same procedures in rabbits to confirm if scoliosis is produced by only anterior rhizotomy, posterior rhizotomy, or a combination. They found that three groups demonstrated almost the same severity of scoliosis. They concluded that scoliosis might be induced by selective posterior root paralysis as well as anterior root paralysis.

According to the these investigations, the experimental scoliosis is different from human idiopathic scoliosis, and similar to paralytic scoliosis. Although the pathogenesis of scoliosis induced by intercostal nerve resection remains unclear, these animal models of scoliosis are reproducible and constant.

Animal Models of Scoliosis 555

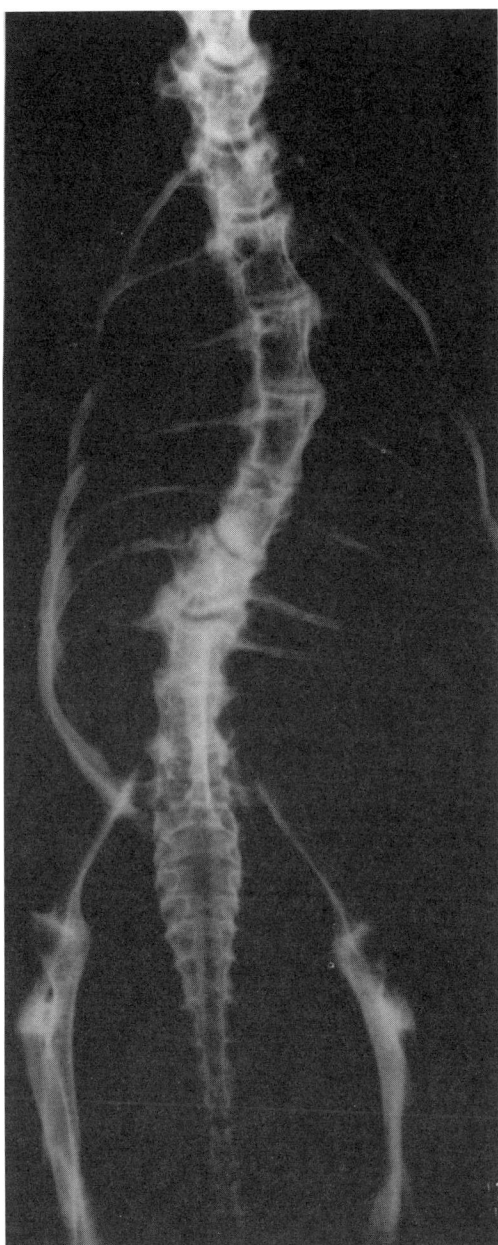

FIGURE 2. Scoliosis due to rib resection in a chicken. Three consecutive ribs on the right (T3, T4. and T5) were resected for a length of 3 mm just lateral to the costotransverse joint at the age of four weeks. Scoliosis of 42 degrees with the convexity to the operated side developed at 20 weeks of age.

2. Rib Resection (Figure 2)

Among unilateral operations on structures in the vicinity of the spine, posterior rib resection lateral to the transverse process is recognized as a procedure which constantly produces scoliosis.[33,34,71,72] The rib cage has several important biomechanical functions related to the spine. It stiffens and strengthens the spine, thus providing greater resistance to displacement. The vertical stability of the thoracic spine is maintained by equal support of the ribs on both sides. Since

Bisgard[46] first reported experimental thoracogenic scoliosis, this type of experimental model has been used for investigation of the etiology and pathology of scoliosis.[29,35,38,40,43,51] The results of these studies indicated that scoliosis might be caused by the weakness or absence of a structure on the convex side of the curve or the over activity of its antagonist on the concave side.

Pal and Bhatt[30] and Sevastik et al.[40] studied the mechanism of production of experimental scoliosis by rib resection in rabbits and developed the following hypothesis. After the resection of a few ribs on one side, the intact ribs on the opposite side exert more load against the vertebral column and this imbalance causes the vertebral column to move laterally and rotate. Scoliosis following rib resection is usually progressive until skeletal maturity is obtained. The more ribs that are resected, the longer the scoliosis becomes. The apex of the scoliosis is usually at the middle of the vertebrae from which the ribs are resected, but it varies to the upper or lower segment in some cases. The convexity of the curvature is always toward the operated side and the apical vertebral body is rotated toward the convex side of the curvature as it occurs in human idiopathic scoliosis.

While factors which affect the severity of scoliosis induced by rib resection have been suggested by DeRosa[73] and Langenskiold and Michelsson,[33,34,71,73] Deguchi et al.[74,75] reported the most detailed study on the relationship between the severity or progression of scoliosis and the number of ribs resected or the effect of age. The study showed that a younger age at operation and the resection of more ribs were important factors in producing severely progressive scoliosis. Interestingly, chickens that underwent rib transaction showed bone healing at the transacted sites several weeks after surgery and no scoliosis developed. Furthermore, the authors clearly showed a negative correlation between the number of healed ribs and the severity of the scoliosis. As the operation creates a gap in the bony skeleton along one side of the spine, mechanical pressure from the contralateral intact ribs may push the mobile column into the gap, and subsequent asymmetrical rib growth may fix and increase the deformity. However, when some of the operated ribs fuse, asymmetrical load transmission to the spine is reduced and progression subsequently may be slower. Therefore, to induce severe scoliosis by rib resection, a younger age at operation, the resection of more ribs, and little bone regeneration at the site of resection are important factors.

Deguchi et al.[75] also examined the alterations of the rib cage and vertebrae in the transverse plane using CT scanning. While there was a significant positive correlation between the Cobb angle and the apical vertebral rotational angle as in human idiopathic scoliosis, curvature following rib resection did not completely resemble the three-dimensional feature of human idiopathic scoliosis. The apical vertebra actually rotated to the convex side of the curvature and a rib hump developed on the concave side of the curve. The direction to which the anterior midline of the body deviated was different between human idiopathic scoliosis and experimental scoliosis by rib resection; it was to the opposite side of the convexity of the curvature in humans, which tended to be the convex side in scoliosis following rib resection. This is because rib resection is a localized surgical interference that influences only the operated area or its vicinity, which causes a distortion of the thoracic cage. The treated ribs result in asymmetric rib growth, which makes the midline of the body deviate to the side and prevents rib humping, which should have developed on the convex side from developing on the physiological side. Therefore, a rib hump developing on the concave side of the curve might be termed "pseudohump." Despite such problems, rib resection is one of the standardized procedures which constantly produce identical scoliosis.

3. Tethering between Scapula and Pelvis

From the view point of the direction of a tether to the spinal column, tethering procedures for producing scoliosis in animals were divided into two groups: a posterior tether and a unilateral tether. There was another classification of tethering such as external and internal. Some of immobilization tools,[57-59] which were already described in this chapter, can be attributed to an external tether. A posterior tether which succeeded in producing scoliosis was induced by not only an extension splint[58] as an external tether but also an approximation of the spinous process by nylon suture[76] and sutures of the three lumbar spinous processes.[77] These experiments were conducted to prove the concept of rotated lordosis in the pathological mechanisms of scoliosis.

On the other hand, a unilateral tether was reported as one of the successful methods to produce marked scoliosis such as unilateral fixation of ribs by Dacron thread,[35] placement of coil splint between articular processes,[78] and suture of the inferior angle of the scapula to the pelvis.[79,80] According to the purposes of experimental scoliosis, it is important that experimental models of scoliosis need to be easily reproducible and remain intact on the spine and the tissues around the spine. From this view point, Sarwark's model may be appropriate to use in spinal research.

Sarwark et al.[79] reported an experimental model of scoliosis by suturing the inferior angle of the scapula to the pelvis at the base of the tail in 15-21 day old male SD rats. They released the tethers in 1–12 weeks and found that a minimum period of six weeks was needed for production of permanent structural curves. They also noted that the longer the period of tethering is, the more severe the scoliotic curves become. The characteristics of the scoliotic curves were long C shaped thoracolumbar curves with vertebral rotation and apical wedging. In addition to these morphological investigations, they studied the effects of the procedures on neurologic function. They found an increased incidence of a mild spasticity in the ipsilateral hindlimb with curvature greater than 40 degrees. The conclusions of this model included no direct trauma to the spine, existence of similar changes to human idiopathic scoliosis, technical easiness in producing scoliotic curvatures in a large number of animals in a relatively short period of time, and appropriate models for study of scoliosis.

4. Removal of Transverse Processes or Ligaments

As already described, rib cage deformity or rib resection was reported to disturb the balance of the spine and caused spinal deformity. A morphometric and morphological study on the thoracic cage indicated ribs played some role of transmission of considerable load from the sternum to the spinal column. The transverse processes and ligaments around them were important structures as load transmitters between ribs and the spinal column. Langenskiold and Michelsson[34] paid attention to the importance of costo-transverse ligaments and succeeded in producing scoliosis by section of anterior and posterior costo-transverse ligaments at four or five levels in rabbits and pigs. However, severe progressive scoliosis was only seen in a few of 47 rabbits. They speculated this was because scar formation around the incised ligaments might have prevented the progression of scoliosis. The same procedures in pigs revealed a scoliosis of 30–45 degrees with the rate of occurrence at 90%. They concluded that posterior costo-transverse ligament was of decisive importance in its equilibrium and symmetrical growth by transmitting the effect of the normal muscle tone to the spine. Karaharju[35] also severed dorsal ligaments of the 8th to 12th ribs in three pigs and produced scoliosis of 12–35 degrees.

Pal et al.[30] studied the route of load transmission from ribs to the vertebral body. He pointed out the various sites, such as ribs, costo-transverse joints, transverse process, lamina, and facet joints that have possibilities in producing asymmetry in load transmission. Because posterior rib resection and hemilaminectomy have already been proved to produce scoliosis in animals, they removed unilateral transverse process and facet joints without interference of ribs in rabbits. The resection of the transverse processes from T3 to T6 showed no scoliosis or mild scoliosis. The extended resection of them from T3 to T9 also demonstrated mild scoliosis. However, removal of seven unilateral segments of the transverse processes in addition to the successive four segments of the facet joints revealed rapidly progressive lordoscoliosis with the convexity toward the operated side. The model described by Pal et al.[30] has proved that the mechanical asymmetry in load transmission from the ribs to the spine resulted in scoliosis, although Smith and Dickson[80] reported that the previous experiments of rib resection to produce scoliosis caused spinal cord damage due to damages of segmental vessels at the time of rib resection and that scoliosis was induced by the spinal cord damage.

5. Muscle Imbalance

Since Arnd[41] first reported the production of scoliosis in rabbits by unilateral excision of the deep back muscles, several studies dealing with muscle imbalance due to incision or excision of the back muscles was conducted to produce scoliosis in animals.[30,33,41–44] Because Arnd's results showed scoliosis with convexity of the curve toward the unoperated side, he suggested the importance of contracture of the scar tissues for the induction of scoliosis. On the contrary, Schwartzmann and Miles[43] studied the effects of muscle imbalance on the alignment of the vertebral column by excision of muscles in rats and mice. They found that muscle imbalance produced not only by excision of unilateral back muscles but also unilateral muscle release with inert materials to prevent muscle reattachment caused scoliosis experimentally. They stressed the importance of muscle imbalance for producing scoliosis through the results of failure by muscle excision and release that did not produce imbalance. Furthermore, Miles[42] advanced this experiment and found the difference of curve patterns between muscle resection and nerve resection.

Stilwell[44] used monkeys to study radiographic and histological changes of vertebrae during curve progression produced by resection of one or both sacrospinalis muscles from the sacrum to the upper thorax. All of 11 monkeys that underwent this procedure showed marked scoliosis with intervertebral disc distortion and vertebral wedging. However, his procedures included incision of the interspinous and flaval ligaments because his main concerns were not the effects of muscle excision on the spine but the mechanisms of bone modeling in a weight bearing spine under conditions of scoliosis progression. In an additional study, he failed to produce significant structural deformity or persistent curve by denervation or resection of a variety of asymmetrical muscles.

As already noted, muscle imbalance induced by unilateral excision of the back muscle or by denervation due to rhizotomy or intercostal nerve resection has been recognized to cause scoliosis in animals. Clinically, it is well known that neuromuscular disorders have many features including scoliosis. On the basis of these findings, some attempts at electrical stimulation of unilateral back muscles in animals have been done since 1970s.[81–84] Olsen et al.[83] reported production of scoliosis of 6.8 degrees with its concavity toward the stimulated side by stimulating unilateral paravertebral muscles through implanted electrodes in dogs.

Monticelli et al.[82] produced thoracolumbar scoliosis of 15–36 degrees by almost the same procedures in rabbits. Joe[84] studied the changes of muscles caused by electrical stimulation histologically. He noted that the diameter of type I fibers increased on the stimulated side, which might be the cause of muscle imbalance. These changes returned to the same level as that of the control groups three weeks after the termination of the stimulation although scoliosis remained almost the same as before. These experimental attempts generally produced mild scoliosis in animals and seemed to have some limits for producing severe progressive scoliosis.

6. Other Methods

Besides the common procedures which produce scoliosis in animals described above, several other experimental models have been reported in the literature.[85,86] Ehara et al.[85] implanted a few magnets on the unilateral sides of the lumbar spines and succeeded in producing scoliosis in rats. Scoliosis induced by three magnets implanted tended to progress more severely than scoliosis with two magnets, with their average Cobb angles 24 degrees, and 14.2 degrees, respectively. Their final goals are clinical application for the treatment of scoliosis. However, the use of magnets basically needs further work for toxicity determination.

An interesting study has been presented by Chuma et al.[86] The development of MRI makes clear that some scoliotic patients have a spinal cord lesion, syringomyelia. It seems that scoliosis is strongly related to syringomyelia, while the pathogenesis of scoliosis due to syringomyelia is still unclear. However, Chuma reported experimental models of scoliosis due to syringomyelia which was produced by a kaoline injection into the cisterna magna in dogs. In spite of the relatively low rate of occurrence of scoliosis (3/11) this study may have just opened the door to a new world of experimental research on scoliosis, and pinealectomy.

IV. THE USE OF SCOLIOSIS MODELS IN SPINAL RESEARCH

A. Pathogenesis of Scoliosis

Scoliosis resulted from a wide variety of pathological conditions that have been classified by the Scoliosis Research Society.[87] However, sixty or seventy percent of scoliosis cases are diagnosed as idiopathic scoliosis. Because of this, one of the main purposes of the experiments for production of scoliosis in animals, in which many investigators have been deeply involved as mentioned above, was to elucidate real pathogenesis and mechanisms of development of human idiopathic scoliosis. Those animal models were evaluated and compared with human idiopathic scoliosis not only morphologically by X ray or dissection, but also biochemically. The existence of rotation associated with lateral curvature and apical vertebral wedging were main concerns in morphological study. Histology observations about bone remodeling, epiphyseal changes, the spinal cord damage, or difference between convex and concave side of the paravertebral muscle were also reported in each experimental model. As for biochemical analysis in experimental animal models, some research has been done by using human blood, urine, or intervertebral discs resected at surgery.[88] Machida et al.[11] estimated serum melatonin in the control group and the pinealectomized group in chickens for investigating the relation between melatonin and scoliosis.

B. Exploration for New Treatment

The other purpose of the experiments of producing scoliosis models in animals was for providing new methods of treatment of already established scoliosis. Some of the experiments were done to evaluate the influences of the same procedures as those which induced scoliosis on the opposite side of the spine.[31,34] Others were conducted to determine whether some new methods of treatment of scoliosis were effective or not (Table 3).[8,39,57,85,89–92] Because surgical correction of scoliosis needs direct approach to the spine, it is necessary to keep the spine and its vicinity intact at the time of producing scoliosis. From this view point, the models produced by the tethering of the scapula to the ipsilateral pelvis or pinealectomy may be ideal for searching a new procedure for treatment of scoliosis, as reported by Dabney et al.,[89] Salzman et al.,[93] Glassman et al.[90] and Deguchi et al.[8]

V. FUTURE DIRECTIONS OF RESEARCH

Most of the experiments related to the animal models of scoliosis were conducted early in this century by analyzing morphological changes of the spine due to each procedure. Because of limitations of the methodological approach, such as with X ray or histology at that time, it was always controversial to make a conclusion as to whether the results obtained from the experiments were primary or secondary. On the other hand, there were some experiments whose data contradicted each other in spite of use of almost the same procedures. This kind of contradiction may be possibly happen because development of idiopathic scoliosis is evoked by more than one etiologic factor as Ponseti et al.[53] described. Thus, it is now widely accepted that more than one pathological mechanism is responsible for idiopathic scoliosis and that idiopathic scoliosis is a group of disorders that show some common parameters. From this viewpoint, many original ideas and experiments using animal models will be reported and studied in the future. It cannot be definitely said in which

TABLE 3
The Use of Scoliosis Models in Spinal Research

Procedures used for induction of scoliosis	Procedures used for correction of scoliosis	First author, year[Ref.]
Placement of a staple on the lumbar spine and intercostal muscle	Staple on the opposite side	Nachlas 1951[31]
Hemilaminectomy, incision of ligaments	Cut of the posterior and anterior costo-transverse ligament on the concave side	Langenskiold 1962[34]
Immobilization by cast	Soft tissue release on the concave side	Hakkarainen 1981[57]
Tethering of the scapula to the ipsilateral pelvis with suture	Distraction with spinal instrumentation	Dabney 1988[89], Salzman 1991[92] Glassman 1995[90]
Resection of three intercostal nerves	Rib shortening on the concave side	Sevastik 1990[39]
Unilateral removal of back muscles	Direct attachment of two or three magnets on concave side of the lumbar spine	Ehara 1992[85]
Unilateral resection of five ribs	Spinal instrumentation (shape memory alloys)	Sanders 1993[91]
Pinealectomy	Rib resection on the concave side of the scoliotic spine	Deguchi 1996, 1997[8,75]

direction experimental research of scoliosis will advance toward in the future. However, there are three or four possible clues for the future experimental research of scoliosis right now.

First, as three dimensional analysis by computers is becoming popular in clinical research of idiopathic scoliosis, experimental works will be conducted for the proof of hypothesis or results that are derived from biomechanical analysis by computers, and vice versa. Scoliosis can be said to be a disorder of misalignment of vertebrae happening during the growth period. Thus, a biomechanical approach for experimental scoliosis is absolutely important with the help of computer analysis.

Second, biochemical analysis of experimental scoliosis will be advanced further. Biotechnological skills and machines have developed recently and will continue to develop more in the future. These developments will make microanalysis of hormones and other substances possible in the blood, urine, and tissues not only *in vitro* but also *in vivo*. Pathogenesis of scoliosis in pinealectomized chickens will be studied continuously biochemically and may open the door to a new treatment of human idiopathic scoliosis without braces or surgical operations.

Third, a gene analysis will be applied for the study of the experimental scoliosis in animals. Particularly, spontaneous occurrence of scoliosis in a highly inbred line of chickens reported by Taylor[6] and in Ishibashi rats,[18] may be a good pathology for a gene analysis. On the other hand, it may be possible to produce experimental models of Recklinghausen disease, Marfan syndrome, etc., which are recognized as hereditary diseases by using the method of genetic recombination.

Finally, the studies and analysis of new surgical or physiological treatments using some kinds of the animal models of scoliosis will be continued in the same trend as those reported in the literature.

REFERENCES

1. Ogilvie, J. W., "Histological aspects of scoliosis," in *Moe's Text Book of Scoliosis and Other Spinal Deformities,* 3rd edition, Lonstein, J. E., Bradford, D. S., Winter, R. B., and Ogilvie, J. W., Eds., W. B. Saunders, Philadelphia, 1995.
2. Huebert, H. T., "Scoliosis: A brief history," *Manitoba Med. Rev.,* October, 452, 1967.

3. Yamamoto H., "Experimental scoliosis in rachitic bipedal rats," *Tokushima J. Exp. Med.,* 13, 1, 1966.
4. Machida, M., Murai, I., Miyashita, J., Dubousset, J., Yamada, T., and Kimura, J., Pathogenesis of idiopathic schoolhouses: experimental study in rats, 32nd Annual Meeting, Scoliosis Research Society, St. Louis, MO, Sept. 25–27, 1997.
5. Goff, C. W. and Landmesser,W., "Bipedal rats and mice," *J. Bone Joint Surg.,* 39A, 616, 1957.
6. Taylor, L. W., "Kyphoscoliosis in a long-term selection experiment with chickens," *Avian Dis.,* 1971; 15, 376.
7. Coillard, C. and Rivard, C. H., "Vertebral deformities and scoliosis," *Eur. Spine J.,* 5, 91, 1996.
8. Deguchi, M., Kawakami, N., and Kanemura, T., "Correction of scoliosis by rib resection in pinealectomized chickens," *J. Spinal Disorders,* 9, 207,1996.
9. Dubousset, J., Queneau, P., and Thillard, M. J., "Experimental scoliosis induced by pineal and diencephalic lesions in young chickens: its relation with clinical findings in idiopathic scoliosis," *Orthop. Trans.,* 7, 7, 1983.
10. Machida, M., Dubousset, J., Imamura, Y., Iwaya, T., Yamada, and T., Kimura, J., "An experimental study in chickens for the pathogenesis of idiopathic scoliosis," *Spine,* 18, 1609, 1993.
11. Machida, M., Dubousset, J., Imamura,. Y., Iwaya, T., Yamada, T., and Kimura, J., "Role of melatonin deficiency in the development of scoliosis in pinealectomised chickens," *J. Bone Joint Surg.,* 77B, 134, 1995.
12. Machida, M., Miyashita, Y., Murai, I., Dubousset, J., Yamada, T., and Kimura, J., "Role of serotonin for scoliosis deformity in pinealectomized chickens," *Spine,* 22, 1297,1997.
13. Kanemura, T., Kawakami, N., Deguchi, M., Mimatsu, K., and Iwata, H., "Natural course of experimental scoliosis in pinealectomized chickens," *Spine,* 22, 1563, 1997.
14. Thillard, M. J., *Deformations de colonne vertebrale consecutives a l'epiphysectomie ches le poussin,* Extrait des Comptes Rendus de l'Association des Anatomistes, 248, 1238, 1959.
15. Wang, X., Jiang, H., Raso, J., Moreau, M., Mahood, J., Zhao, J., and Bagnall, K., "Characterization of the scoliosis that develops after pinealectomy in the chicken and comparison with adolescent idiopathic scoliosis in humans," *Spine,* 22, 2626, 1997.
16. Hijikata, S., Ichihara, M., Nakai, S., and Nakayama, K., "SQHM as an experimental model of idiopathic scoliosis," *J. Tokyo Electr. Hosp.,* 8, 35, 1978.
17. Rigdon, R. H. and Mack, J., "Spontaneously occurring scoliosis in the white Pekin duck," *Am. J. Vet. Res.,* 29, 1081, 1968.
18. Ishibashi, M., "'Ishibashi rat' as an experimental model of congenital spinal deformity," *Exp. Anim.,* 25, 320, 1976.
19. Kin, A., "Radiographical and histological studies on the spinal deformities in the hereditary lordoscoliotic rabbit," *J, Jpn. Orthop. Assoc.,* 68, 458, 1994.
20. Sawin, P. B. and Crary, D. D., "Genetics of skeletal deformities in the domestic rabbit (*Oryctolagus Cuniculus*)," *Clin. Orthop.,* 33, 71, 1964.
21. Roony, J. R., "Congenital equine scoliosis and lordosis," *Clin. Orthop.,* 62, 25, 1969.
22. Ingalls, T. H. and Curley, F. J., "Principles governing the genesis of congenital malformations induced in mice by hypoxia," *New Eng. J. Med.,* 257, 1121, 1957.
23. Duraiswami, P. K., "Experimental causation of congenital skeletal defects and its significance in orthopedic surgery," *J. Bone Joint Surg.,* 34B, 464, 1952.
24. Murphy, M. L., Dagg, C. P., and Karnofsky, D. A., "Comparison of teratogenic chemicals in the rat and chick embryos," *Proc. Am. Acad. Pediatr.,* 19, 701, 1957.
25. Kent, G. M. and Zingg, W., "Experimental scoliosis in fetal lambs," *Surg. Forum,* 25, 75, 1974.
26. Arkin, A. M. and Simos, N., "Radiation scoliosis," *J. Bone Joint Surg.,* 32A, 396, 1950.
27. Engel, D., "Experiments on the production of spinal deformities by radium," *Am. J. Roentgenol.,* 42, 217, 1939.
28. Bisgard. J. D., and Musselman, M. M., "Scoliosis — its experimental production and growth correction; growth and fusion of vertebal bodies," *Surg. Gynecol. Obst.* 70, 1029, 1940.
29. Haas, S. L., "Experimental production of scoliosis," *J. Bone Joint Surg.,* 21, 963, 1939.
30. Pal, G. P., Bhatt, R. H., and Patel, V. S., "Mechanism of production of experimental scoliosis in rabbits," *Spine,* 16, 137, 1991.
31. Nachlas, I. W. and Borden, J., "The cure of experimental scoliosis by directed growth control," *J. Bone Joint Surg.,* 66A, 24, 1951.

32. Nishiike, A., "Experimental study of scoliosis — unilateral posterior fusion of the spine," *J. Jpn. Orthop. Assoc.,* 40, 1041, 1966.
33. Langenskiold, A. and Michelsson, J. E., "Experimental progressive scoliosis in the rabbit," *J. Bone Joint Surg.,* 43B, 116, 1961.
34. Langenskiold, A. and Michelsson, J. E., "The pathogenesis of experimental progressive scoliosis," *Acta Orthop. Scand. Suppl.,* 59, 1, 1962.
35. Karaharju, E., "Deformation of vertebrae in experimental scoliosis," *Acta Orthop. Scand.,* 1967; Supp. 105.
36. Piggot, H., "Posterior rib resection in scoliosis — a preliminary report," *J. Bone Joint Surg.,* 53B, 663, 1971.
37. Thomas, S. and Dave, P. K., "Experimental scoliosis in monkeys," *Acta Orthop. Scand.,* 56, 43,1985.
38. Sevastikoglou, J. A., Aaro, S., Lindholm, T. S., and Dahlborn, M., "Experimental scoliosis in growing rabbits by operations on the rib cage," *Clin. Orthop.,* 136, 282, 1978.
39. Sevastik, J., Agadir, M., and Sevastik, B., "Effects of rib elongation on the spine II. Correction of scoliosis in the rabbit," *Spine,* 15, 826, 1990.
40. Sevastik, J., Agadir, M., and Sevastik, B., "Effects of rib elongation on the spine I. Distortion of the vertebral alignment in the rabbit," *Spine,* 15, 822, 1990.
41. Arnd, C., "Experimental Beitrage Zur Lehre der Skoliose. Der Einfluss des Musculus erector trunci auf die Wirbelsaule," *Arch. F. Orthop.,* I, 145, 1903.
42. Miles, M., "Vertebal changes following experimentally produced muscle imbalance," *Arch. Phys. Med. Rehabil.,* 28, 284, 1947.
43. Schwartzmann, J. R. and Miles, M., "Experimental production of scoliosis in rats and mice," *J. Bone Joint Surg.,* 27, 59, 1945.
44. Stilwell, D., "Structural deformities of vertebrae," *J. Bone Joint Surg.,* 44A, 611, 1962.
45. von Lesser, L., "Experimentelies und Klinishes uber Skoliose," *Arch. F. Path. Anat.,* 113, 10, 1888.
46. Bisgard, J. D., "Experimental thoracogenic scoliosis," *J. Thorac. Surg.,* 4, 435, 1935.
47. Liszka, O., "Spinal cord mechanisms leading to scoliosis in animal experiments," *Acta Medica Polona,* 2, 45, 1961.
48. Alexander, M. A., Bunch, W. H., and Ebbesson, S. O. E., "Can experimental dorsal rhizotomy produce scoliosis?" *J. Bone Joint Surg.,* 54A, 1509, 1972.
49. De Salis, J., Beguiristain, J. L., and Canadell, J., "The production of experimental scoliosis by selective arterial ablation," *Int. Orthop.,* 3, 311, 1980.
50. Suk, S. I., Song, H. S., and Lee, C. K., "Scoliosis induced by anterior and posterior rhizotomy," *Spine,* 14, 692, 1989.
51. Agadir, M., "Induction of scoliosis in the growing rabbit by unilateral rib-growth stimulation," *Spine,* 13, 1065, 1988.
52. Geiger, B. J., Steenbock, H., and Parsons, H. T., "Lathyrism in the rat," *J. Nutr.,* 6, 427, 1933.
53. Ponseti, I. V., "Lesions of the skeleton and of other mesodermal tissues in rats fed sweet-pea (*Lathyrus odoratus*) seeds," *J. Bone Joint Surg.,* 36A, 1031, 1954.
54. Lalich, J. J. and Angevine, D. M., "Dysostosis in adult rats after prolonged B-aminopropyonitrile feeding," *Arch. Path.,* 90, 22, 1970.
55. Kitamura, S., Ohara, S., Suwa, T., and Nakagawa, K., "Studies of vitamin requirements of rainbow trout, *Salmo gairdneri*. I. Ascorbic acid," *J. Jpn. Soc. Fish,* 31, 818, 1965.
56. Lim, C. and Lovell. R. T., "Pathology of the vitamin C deficiency syndrome in channel catfish (*Ictalurus punetatus*)," *J. Nutr.,* 108, 1137, 1978.
57. Hakkarainen, S., "Experimental scoliosis: production of structural scoliosis by immobilization of young rabbits in a scoliotic position," *Acta Orthop. Scand. Suppl.,* 192, 1, 1981.
58. Poussa, M., "Scoliosis in growing rabbits induced with an extension splint," *Acta Orthop. Scand.,* 62, 136, 1991.
59. Wullstein, L., "Die Skoliose in ihrer Behandlung und Entstehung nach Klinischen und experimentallen Studien," *Z. Orthop. Chir.,* 10, 177, 1992.
60. Somerville, E. W., "Rotational lordosis: the development of the single curve," *J. Bone Joint Surg.,* 34B, 423, 1952.
61. Roaf, R., "Rotation movements of the spine with special reference to scoliosis," *J. Bone Joint Surg.,* 57B, 500, 1958.

62. Roaf, R., "Vertebral growth and its mechanical control," *J. Bone Joint Surg.,* 42, 40, 1960.
63. Dickson, R. A., Lawton, J. O., Butt, W. P., "The pathogenesis of idiopathic scoliosis," in *Management of Spinal Deformities*, Dickson, R. and Bradford, D. S., Eds., Butterworth, London, 1984, 1.
64. Stokes, I. A. F., Spence, H., Aronsson, D. D., and Kilmer, N., "Mechanical modulation of vertebral body growth," *Spine,* 1, 1162, 1996.
65. Mente, P. L., Stokes, I. A. F., Spence, H., and Aronsson, D., "Progression of vertebral wedging in an asymmetrically loaded rat tail model," *Spine,* 22, 1292, 1997.
66. Mente, P. L., Aronsson, D. D., Stokes, I. A. F., Spence, H., and Iatridis, J. C., Reversal of an induced vertebral wedge deformity, 32nd Annual Meeting, Scoliosis Research Society, St. Louis, MO, Sept. 25–27, 1997, 42.
67. Machida, M., Dubousset, D., Imamura, Y., Iwaya, T., Yamada, T., Kimura, J., and Toriyama, S., "Pathogenesis of idiopathic scoliosis: SEPs in chicken with experimentally induced scoliosis and in patients with idiopathic scoliosis," *J. Pediat. Orthop.,* 14, 329, 1994.
68. Barrios, C., Tunon, M. T., De Salis, J. A., Beguiristain, J. L., and Canadell, J., "Scoliosis induced by medullary damage: an experimental study in rabbits," *Spine,* 12, 433, 1987.
69. MacEwen, G. D., "Experimental scoliosis," in *Proceedings Second Symposium Scoliosis*, Zorab, P. A., Ed., Churchill-Livingstone, London, 1968, 18.
70. MacEwen, G. D., "Experimental scoliosis," *Israel J. Med. Sci.,* 9, 714, 1973.
71. Michelsson, J. E., "The development of spinal deformity in experimental scoliosis," *Acta Orthop. Scand.,* 81, 1, 1965.
72. MacEwen, G. D., "Experimental scoliosis," *Clin. Orthop.,* 93, 69, 1973.
73. DeRosa, G. P., "Progressive scoliosis following chest wall resection in child," *Spine,* 10, 618, 1985.
74. Deguchi, M., Kawakami, N. Kanemura ,T., Mimatsu, K., and Iwata, H., "Experimental scoliosis induced by rib resection in chickens," *J. Spinal Disorders,* 8, 179, 1995.
75. Deguchi, M., Kawakami, N., and Kanemura, T., "Correction of experimental scoliosis by rib resection in the transverse plane," *J. Spinal Disorders,* 10, 197, 1997
76. Lawton, J. O. and Dickson, R., "The experimental basis of idiopathic scoliosis," *Clin. Orthop.,* 210, 9, 1985.
77. Kasuga, K., "Experimental scoliosis in the rat spine induced by binding the spinous processes," *J. Jpn. Orthop. Assoc.,* 68, 798, 1994.
78. Takaoka, R., "Studies on experimental scoliosis by means of elastic stainless steel (Co-elinvar)," *J. Jpn. Orthop. Assoc.,* 42, 29, 1958.
79. Sarwark, J. F., Dabney, K. W., Salzman, S. K., et al., "Experimental scoliosis in the rat I. Methodology, anatomic features and neurologic characterization," *Spine,* 13, 466, 1988.
80. Smith, R. M. and Dickson, R. A., "Experimental structural scoliosis," *J. Bone Joint Surg.,* 69B, 576, 1987.
81. Bobechko, W. P., "Spinal pacemakers in scoliosis," *J. Bone Joint Surg.,* 55B, 232, 1973.
82. Monticelli, G., Ascani, E., Salsano, V., and Salsano, A., "Experimental scoliosis induced by prolonged minimal electrical stimulation of the paravertebral muscles," *Italian J. Orthop. Trauma,* 1, 39, 1975.
83. Olsen, G. A., Rosen, H., Stoll, S., and Brown, G., "The use of muscle stimulation for inducing scoliotic curves. A preliminary report," *Clin. Orthop.,* 113, 198, 1975.
84. Joe, T., "Studies of experimental scoliosis produced by electrical stimulation, With special reference to the histochemical properties of the muscle," *J. Jpn. Med. School.,* 57, 416, 1990.
85. Ehara, S., Wada, E., Ono, K., Hamamura, A., Endoh, M., Matsuura, Y., and Ishibashi, M., "Experimental scoliosis in the rats using magnetic force," *J. Jpn. Scoliosis Soc.,* 7, 17, 1992.
86. Chuma A, Kitahara, H., Minami, S., Goto, S., Takaso, M., and Moriya, H., "Structural scoliosis model in dogs with experimentally induced syringomyelia," *Spine,* 22, 589, 1997.
87. Society, T.C.S.R., "A glossary of scoliosis terms," *Spine,* 1, 57, 1976.
88. Sterns, G., Chen, J. T., McKinley, J. B., and Ponseti, I. V., "Metabolic studies of children with idiopathic scoliosis," *J. Bone Joint Surg.,* 37A, 1028, 1955.
89. Dabney, K. W., Salzman, S. K., Wakabayashi, T., et al., "Experimental scoliosis in the rat II. Biomechanical analysis of the forces during Harrington distraction," *Spine,* 13, 472, 1988.
90. Glassman, S. D., Zhang, Y. P., Shields, C. B., Linden, R. D., and Johnson, J. R., "An evaluation of motor-evoked potentials for detection of neurologic injury with correction of an experimental scoliosis," *Spine,* 20, 1765, 1995.

91. Sanders, J. O., Sanders, A. E., More, R., and Ashman, R. B., "A preliminary investigation of shape memory alloys in the surgical correction of scoliosis," *Spine,* 18, 1640, 1993.
92. Salzman, S. K., Mendez, A. A., Dabney, K. W., et al., "Serotonergic response to spinal distraction trauma in experimental scoliosis," *J. Neurotrauma,* 8, 45, 1991.
93. Salzman, S. K., Dabney, K. W., Mendez, A. A., et al., "The somatosensory evoked potential predicts neurologic deficits and serotonergic pathochemistry after spinal distraction injury in experimental scoliosis," *J. Neurotrauma,* 5, 173, 1988.

Part VIII

Microsurgical Technique

31 Microsurgery and Orthopedic Animal Models

Yuehuei H. An

CONTENTS

I. Introduction 567
II. Training of Microsurgical Techniques 568
III. Basic Requirements for Experimental Microsurgery 569
IV. Basic Techniques 570
    A. Vessel Anastomosis 570
        1. End-to-End Anastomosis 570
        2. End-to-Side Anastomosis 572
        3. Telescoping or Sleeve Technique 573
        4. Precautions 573
    B. Nerve Repair 573
        1. Basic Anatomy 573
        2. Suture Materials 574
        3. Animal Selection 574
        4. Types of Repair: Epineural or Perineural 574
        5. Precautions 575
V. Microsurgery and Orthopedic Animal Models 575
    A. Reimplantation and Transplantation of Severed Limbs 575
    B. Vascularized Bone Graft or Whole Joint Transplantation 576
    C. Vascularized Periosteum Graft 576
    D. Vascularized Muscle Transfer 577
    E. Revascularization of Avascularized Bone 577
    F. Vascularized Tendon Graft 577
    G. Vascularized Nerve Graft 578
References 578

I. INTRODUCTION

The microsurgical technique has been well developed for more than 20 years. The "dawning of microsurgery" in the early 1960s arose from the development of the operating microscope,[1] the availability of heparin,[2] the perfection of microinstruments, the previous work on vessel anastomoses and replantation and transplantation of animal and human organs.

From the mid 1950s to the early 1960s several groups in Russia,[3] the United States,[4] and Japan,[5] engaged in the investigation of replanting animal legs. Due to the lack of established microsurgical techniques, there were more failures than successes. By the early 1960s, successful replantation of canine legs had been reported. With the development of microinstruments and the use of the microscope, especially after the availability of a new microvascular method by Jacobson and Suarez in 1960,[6] the era of microsurgery began. Using the method by Jacobson and Suarez, vessels 1 mm in diameter could be anastomosed. In 1961, Lee and Fisher published their first paper on the portacaval shunt in the rat.[7]

In May 1962, Malt and McKhann in Boston achieved the world's first arm replantation which was reported in *JAMA* in 1964.[8] In January 1963, Zhong-Wei Chen et al. in Shanghai successfully reimplanted a complete amputation of the forearm which was reported in *Chinese Med. J.*[9] In 1963, Inoue et al. succeeded in replanting a severed hand.[10] In July 1965, Komatsu and Tamai achieved a replantation of an amputated thumb.[11] In Shanghai, Chen and Yang in 1967 first performed successful toe-to-hand transfer.[12,13] Similar work was first reported in English literature by Cobbett in 1969.[14]

From the mid 1960s to the early 1970s, more achievements were reported in experimental microsurgery. The first model of free skin flap in the dog was reported by Krizek et al. in 1965.[15] The first muscle transfer in the dog was reported in 1968[16] and vascularized bone graft by Buncke et al. in 1967.[17] By the early 1970s, Lee's group had established numerous rat models of organ transplantation.[18]

In 1973, the first human application of free skin flap was reported by Daniel and Taylor[19,] and the first case of a pectoralis major muscle transplantation was done by a Shanghai group.[20] Also in 1973 (published in 1983), Ueba and Fujikawa successfully transferred a vascularized fibular bone graft for the treatment of neurofibromatosis of the ulna.[21] Following this were reports on free vascularized fibular graft transfer to tibial defects by Taylor et al.,[22] and free groin-iliac osteocutaneous flap by Taylor and Watson,[23] and Tamai.[24] In 1976, Baudet et al. proposed the term *musculocutaneous flap* and emphasized the usefulness of the latissimus dorsi musculocutaneous flap.[25]

By the late 1970s, the development of microsurgery had already come into a mature stage. Clearly, the current microsurgical techniques depend mostly on the extensive research work using animal models.

II. TRAINING OF MICROSURGICAL TECHNIQUES

Self confidence is the key in learning microsurgical technique. Most beginners are able to complete a successful vascular anastomosis with 1-3 days of training in the laboratory. The learner has to realize that it is a hard and continuous training process with an initial intensive exercise period of six hours per day five days per week for at least four weeks and refreshment exercises (several anastomoses per time) several times per year.

Proper training in microsurgical techniques could not be done without using animals, for no other alternatives can mimic a living vessel which possesses a contractible wall and the ability to clot. However, well prepared beginners will need fewer animals to complete their training, for they have practiced by stitching rubber gloves or silicone tubes and anastomosing vessels harvested from commercially available meat.[26] Another training model was reported by Kim et al., which uses cold stored vessels harvested from sacrificed animals used in other projects.[27]

Although rabbits and cats have been used, rats have been the dominant animal subject at most training centers. Rats are anesthetized by intraperitoneal injection of pentobarbital. Rat abdominal aorta and carotid arteries are the first choice for a beginner because they are larger and stiffer than the femoral artery. They are easier for beginners to get the first functioning anastomoses, which is very important for building up confidence for mastering the technique. Also these vessels are located deep in the body, and require more work to expose them. Therefore, techniques of microdissecting of tissues, ligating vessel branches, and tying knots can be exercised before the finer vascular work. For a careful and patient operator, a carotid artery of a 400 to 500 gm SD rat could be cut and anastomosed up to eight times.

Rat femoral vessels are the second choice for practice. They are smaller than the aorta and carotid vessels, and they contain thinner vessel walls. One needs at least several days practice on

FIGURE 1. A Wild (Type 308795, Heerbrugg, Switzerland) table top operating microscope.

aorta or carotid artery before working on a femoral vessel. Toward the end of a training program, one may want to repair a rat tail vessel.[28] However, the femoral and tail arteries are not the first choice for beginners.

Also, exercise on vein anastomoses is better started at the later stage of the training program because it is more challenging than arterial repair.

III. BASIC REQUIREMENTS FOR EXPERIMENTAL MICROSURGERY

A genuine scientific interest, a positive attitude, responsibility and patience are essential for the surgeon performing the surgery. All personnel should be well prepared and highly motivated. Team work is important for surgical performance as is collaborating with scientists of other specialties.

A normal operating room for animal surgery should be equipped with an operating microscope. A steady and adjustable operating table is ideal. The housing facility needs to be equipped with intensive monitoring devices such as a skin thermometer and a Doppler vessel blood flow monitor. A qualified full-time technical assistant is necessary to achieve optimal postoperative care and monitoring of the animals.

A table top microscope is economical and sufficient for most procedures on rats and rabbits (Figure 1). Only a simple and personalized set of microinstruments is necessary. This set should include two microsurgical forceps, a vessel dilator, a needle holder, dissecting scissors, straight scissors, several single vessel clamps and a vessel approximator (twin clamps). Proper maintenance of instruments is necessary (Figure 2).

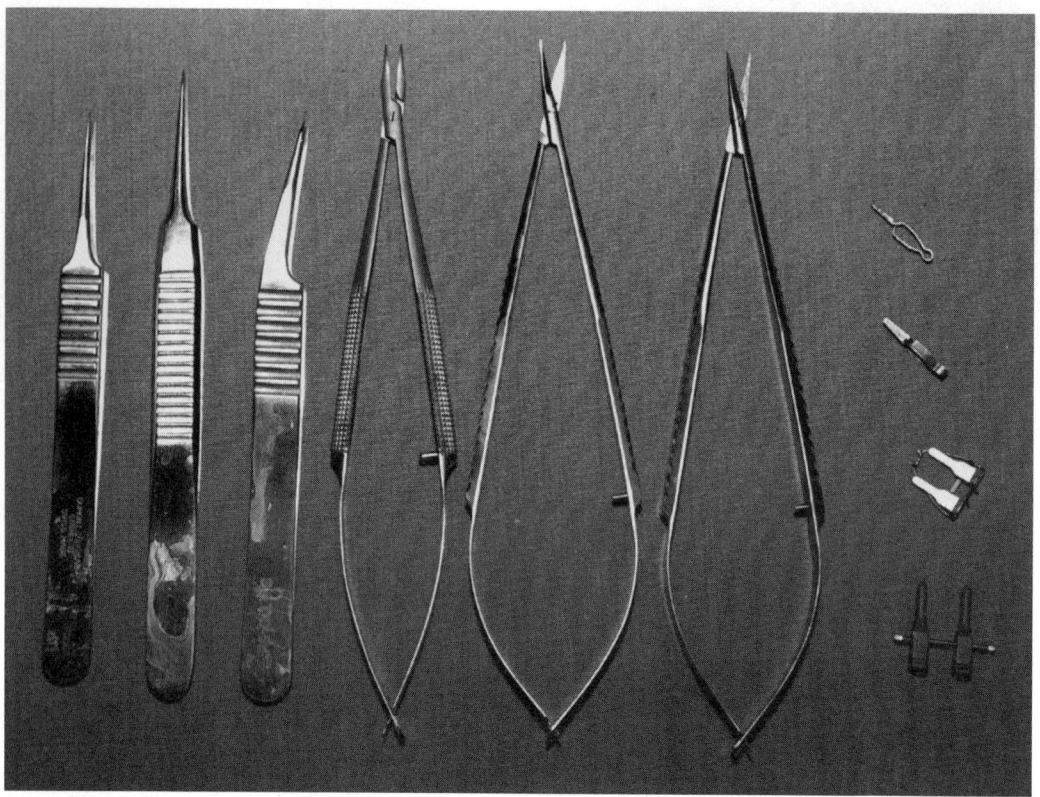

FIGURE 2. Essential microsurgical instruments (left to right): two microsurgical forceps, a vessel dilator, a needle holder, dissecting scissors, straight scissors, two single vessel clamps, and a vessel approximator (twin clamps).

IV. BASIC TECHNIQUES

Microsurgical techniques include micromanipulation of tissues, microdissecting, and microsuturing techniques for different types of tissues such as vessels and nerves. For the purpose of this book, only the suturing techniques for vessel and nerve repair are described. One should read the books by Lee,[18] Pho,[12] and Mehdorn and Müller[29] for more details.

A. Vessel Anastomosis

1. End-to-End Anastomosis

End-to-end anastomosis is the conventional technique for repairing both small arteries and veins.[12,18,29,30] After the artery is freed of surrounding soft tissues and branches ligated, a vessel approximator is applied before the cut is made with a pair of straight scissors. A light colored background plastic sheet (commonly yellow) could be used underneath the vessel to give a sharper and clearer view of the field. The loose soft tissue around the end of vessel can be trimmed off using the "pull-and-cut" method (Figure 3A). Spasm of the vessel opening is common which could be solved by warm saline irrigation or by using a vessel dilator (Figure 3B,C). For bigger vessels such as abdominal aorta or carotid artery, a curved needle holder is more efficient for dilating the vessel opening (Figure 3C).

If a carotid artery is used for anastomoses, 8–10 stitches should be placed in the sequence shown in Figure 4A. For the first and second sutures a tail is needed for retraction. There are

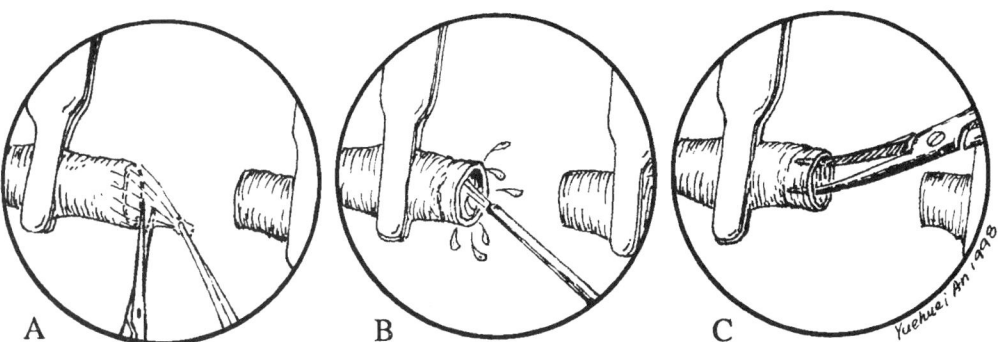

FIGURE 3. The procedure of trimming loose connective tissues around the vessel opening (A), saline irrigation (B), and vessel dilation with a curved needle holder (C).

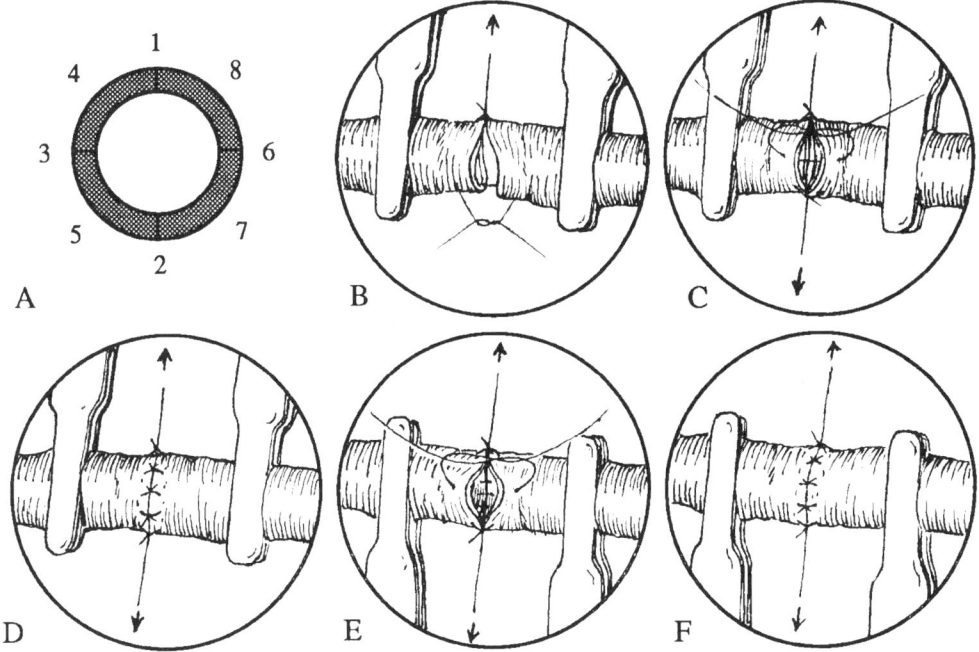

FIGURE 4. Procedures of end-to-end anastomoses of small vessels: (A) the sequence of suture placement; (B) the 1st and 2nd sutures; (C) the 3rd suture; (D) the 4th and 5th sutures (the front side is complete now); (E) the 6th suture after the vessel-clamps assembly is flipped 180°; and (F) the 7th and 8th sutures (the anastomosis is completed).

commercially available weights for use in retraction, but a single heavier vessel clamp will serve the same purpose (Figure 4B,C). After one side of the vessel is sutured the whole assembly, including the vessel and the clamps, will be turned over to expose the other side of the vessel for completion of the anastomoses (Figure 4D,E). For testing the patency of the anastomoses, a full and pulsating artery is the best test. For vein repair, the milking test could be used (Figure 5). However, it is not recommended in clinical settings to use this test for its potential traumatic effect. Minor leaks are often seen after releasing the vessel clamps. A cotton ball can be applied to the anastomotic site for a short period (3–10 min.), thus stopping most leaks.

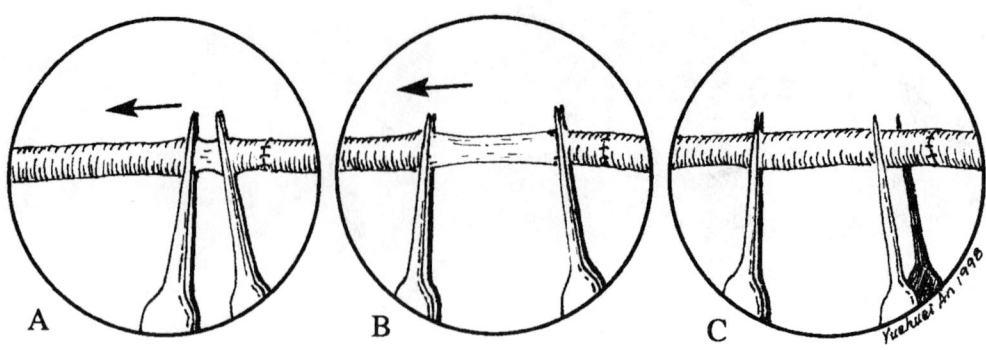

FIGURE 5. A milking test for testing vessel patency.

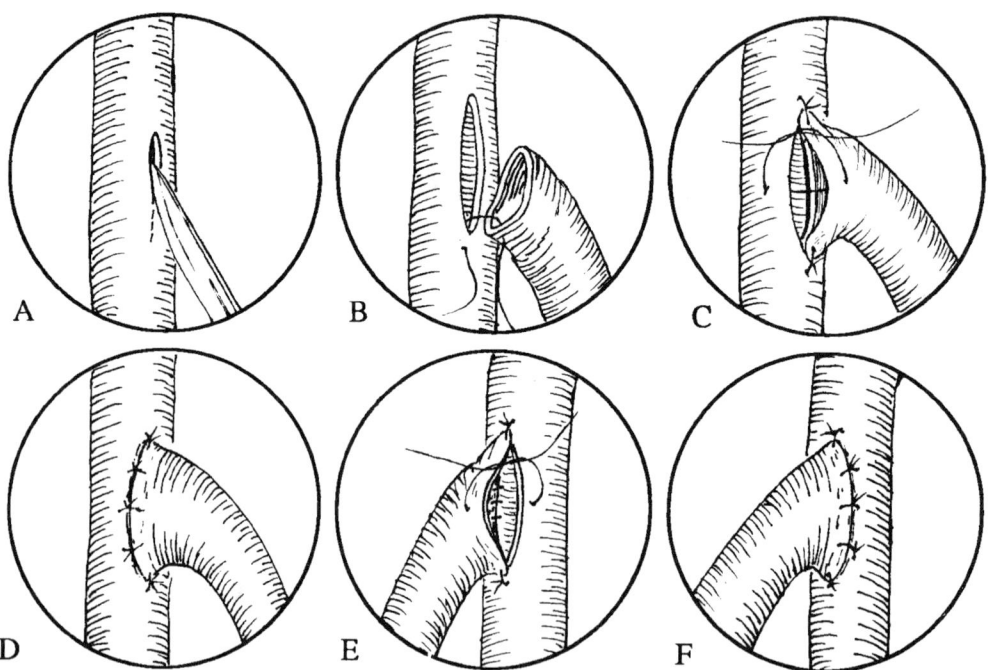

FIGURE 6. Procedures of end-to-side anastomoses of small vessels: (A) A longitudinal incision is made using a small knife. (B,C) Two angle stay sutures are placed to connect the two vessels and the 3rd suture is placed on the front side. (D) The 4th and 5th sutures are used to complete the front side. (E,F) The whole assembly is flipped 180° to the right side for suturing the back side.

2. End-to-Side Anastomosis

Compared to end-to-end technique, end-to-side anastomosis is much less used and mainly for one-vessel recipient area or for overcoming unequal vessel size. This can be practiced on various vessels such as arterio-arterial or arterio-venous anastomoses. Basically, a longitudinal incision is made using a small knife and extended with scissors. Two angle stay sutures are placed to connect the two vessels. Finally, the anterior and posterior walls are sutured (Figure 6).

3. Telescoping or Sleeve Technique

Lauritzen introduced the telescoping or sleeve technique for microvascular repair in 1978.[31] It is used when the upstream vessel is smaller than the downstream vessel. For example, in an artery repair, a smaller proximal end (donor artery) is inserted in the distal artery (recipient artery). An artery cuff technique described by Hung et al. was based on the same principle.[32] Saitoh et al. found that this technique gave very high patency rates when used at both ends of vein grafts interposed for venous defects. No difference was noted when comparing it to conventional end-to-end technique in the degree of stenosis.[33] Further modifications of this technique include a "sutureless sleeve,"[34] a three-stitch sleeve,[35] and various other sleeve methods.[36]

4. Precautions

Factors affecting success of vessel repair include stitching technique, tension at the anastomotic site, excessive injury of the vessels openings by careless practice, or thermal conditions (should be kept close to 37° C). A careful and protective manner is essential for successful anastomoses. Identical stitch intervals are more important than the number of stitches. No more than 10 stitches for rat aorta or carotid artery and eight for femoral artery are necessary.[37] Steady and firm penetration of the vessel wall, using a proper size needle, is essential for avoiding unnecessary damage and leaking. Operators should do their best to avoid introducing any adventitia, loose fibrin, or foreign materials into the vessel lumen. Excessive tension can be solved by freeing more soft tissues around the vessel or by using a vessel graft when necessary. A reasonable thermal condition can be maintained by using warm saline irrigation or sponges soaked with warm saline, keeping the whole anastomotic procedure "under water."

B. Nerve Repair

1. Basic Anatomy

A peripheral nerve is composed of one or more fascicles, covered by an epineurium (Figure 7). Each fascicle is formed by a conglomeration of thousands nerve fibers (axon with surrounding Schwann cell sheath). The connective tissue surrounding the fascicle is called perineurium. A mesoneurium is attached to one side of the nerve trunk for blood supply of the nerve.

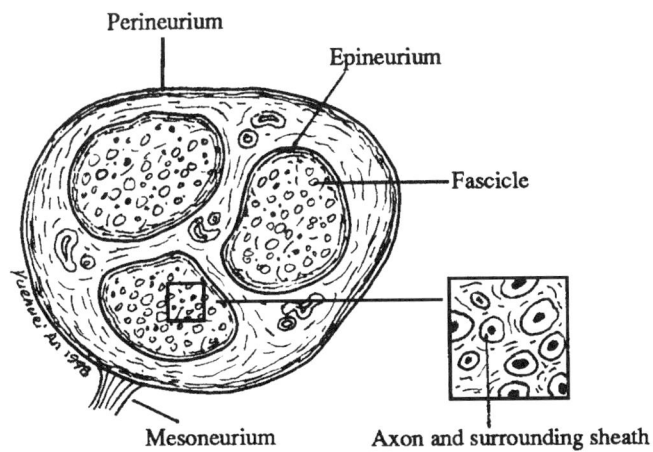

FIGURE 7. Transverse section of a peripheral nerve.

FIGURE 8. Rat sciatic nerve is made up of three fascicles

2. Suture Materials

The rule is that the finest possible suture material should be used in the smallest possible quantity. In most cases, 9/0, 10/0 or 11/0 monofilament nylon threads on tapered needles should be used. A newer product, polylactic acid suture, can make remarkable union of the severed nerve with no inflammatory reaction upon absorption of the sutures.

3. Animal Selection

Rat sciatic nerve, 1.0 mm in diameter (for rats with 400 gm body weight), runs behind the biceps femoris and divides into tibial, peroneal, and sural nerves when it reaches the knee level (Figure 8). The sciatic nerve is made up of three fascicles which enables both epineural and perineural anastomoses (Figure 9). Rabbit sciatic nerve is much thicker, 2.5–3.0 mm in diameter (for rabbits with 4 kg body weight) and is longer than rat sciatic nerve. It also runs behind the biceps femoris and divides into tibial, peroneal, and sural nerves at the knee level. Rabbit sciatic nerve is easier for vascularized nerve graft.[38]

4. Types of Repair: Epineural or Perineural

Epineural repair means the sutures are placed into the circumference of the nerve (epineurium) following the basic principles of microsurgery as for vessel anastomoses. For a peripheral nerve of 1-2 mm in diameter, 4–6 stitches are sufficient (Figure 9B). If the fascicles or funiculi in the nerve can be dissected clear and exact adaptations can be achieved, a perineural suture technique

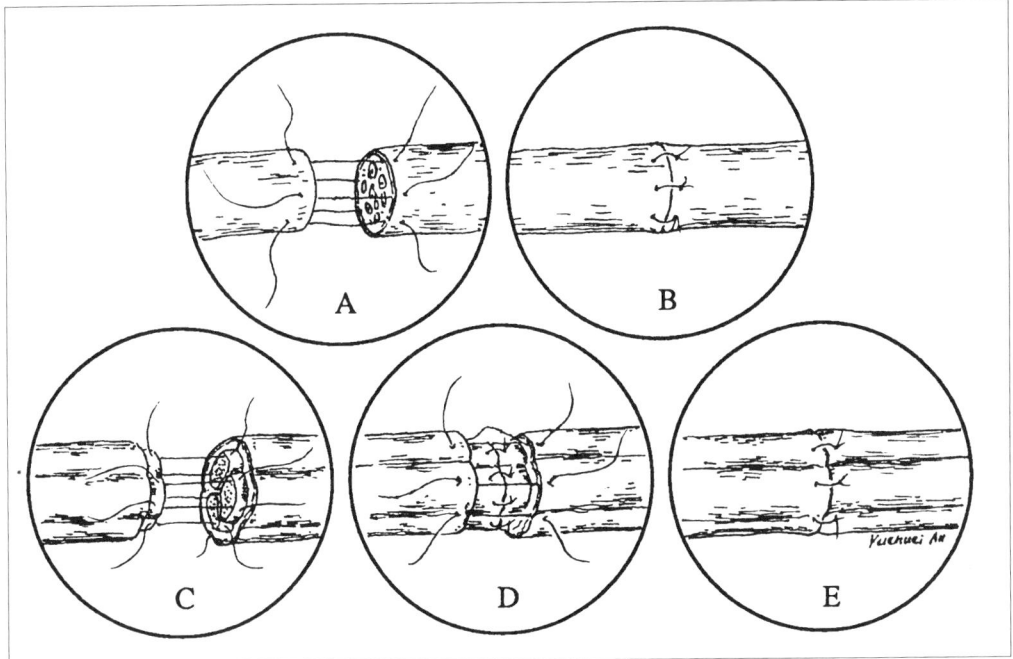

FIGURE 9. Epineural (B) and perineural (C) anastomoses.

(or interfascicular or funicular suture technique) should be used (Figure 9C). Two or three stitches are sufficient for each fascicle. After the completion of interfascicular repair, the epineural sheath should be sutured interruptedly to ensure a tension free repair.

The interfascicular suture technique was first introduced for clinical application by Smith in New York[39] in 1964 and Ito et al.[40] in Japan also in 1964 (according to Tamai[13]). Few years later, the work by Bora[41] and Hakstian[42] also contributed in the fascicular orientation and repair. Reported by Millesi et al., interfascicular nerve grafting can produce recovery of useful function in 80% of their patients.[43]

5. Precautions

The first step of nerve repair is exposure of a nerve lying in scar tissue (neurolysis). Repair of the nerve should not be started unless the corresponding fascicles and/or group of fascicles have been identified in the best possible way. Tension should be avoided because the greater tension the more scar tissue will form. Tension also reduces the precision of fascicle adaptation. The mesoneurium must be freed enough to obtain an adequate length of the nerve for repair. If a segment of the nerve is resected, a nerve graft should be considered.

V. MICROSURGERY AND ORTHOPEDIC ANIMAL MODELS

A. REIMPLANTATION AND TRANSPLANTATION OF SEVERED LIMBS

Most clinical achievements in microsurgery were based on animal research.[44] In the late 1950s in Japan, Tamai's group tried to replant incompletely amputated thighs in two human cases.[13] Due to the lack of microsurgical technique, the results were unfavorable. In 1961 they started extensive animal research, working on replantation of canine thighs and succeeded after two years of hard work.[5] At the same time or earlier, two other groups, Lapchinsky in Russia[3] and Snyder et al.[4] in the United States reported their results on canine thigh replantation. Due to the technique of vessel anastomoses and the use of an operating microscope, the dream of replanting severed limbs finally became a reality in 1962.[8]

An early report of animal study of digital replantation was done by Buncke et al. in 1965.[45] Most of these cases failed. A successful replantation of a totally amputated thumb in human had already been achieved in 1965 by Komatsu and Tamai.[11] Today, digital replantation has become a common procedure in many major centers. Cheng replanted 304 severed digits from 1978 to 1983 with 280 survivals (92.1%).[46] Cheng once treated a case of traumatic amputation of all ten fingers, of which nine fingers, suitable for replantation, survived.[47]

Experimental toe-to-hand transplantation was fist reported by Buncke et al. in 1966.[13] Although the human application of this technique was first reported in English literature by Cobbett in 1969,[14] Chen and Yang in 1967 had already performed successful toe-to-hand transfer.[12,13]

Unlike other organs (such as kidneys or lungs), limb homotransplantation has not been very successful. Canine hind limb homotransplantation is a traditional model because the pre-microsurgical vascular technique (early 1960s or earlier) did not allow the possibility of using small animals.[48,49] After the perfection of microsurgical technique in the 1960s, rats became an ideal model for studying immune rejection.[50] Rats can be easily paired according to their histocompatibility and are economical compared to dogs.[51] Furnas et al. were among the first to perform a successful rat leg homotransplantation using an immunosuppressant.[52] Animal models of limb homotransplantation will continue to play an important role for conquering immunorejection.

B. Vascularized Bone Graft or Whole Joint Transplantation

The first vascularized autogenous joint graft was reported by Buncke et al. in 1967.[17] Both the short-term rat knee model and long-term monkey model showed the transplanted joints survived completely with preservation of normal cellular architecture and function. Tamai et al. in 1971 found that a reimplanted canine knee joint showed no degenerative changes even after a year.[53] The experimental vascularized rib graft was reported by Strauch et al. in 1971.[54] Their work was followed by many researchers, leading to extensive use of free vascularized bone grafts in humans for the treatment of bone defects.[21,22,24]

Vascularized bone or whole joint allograft has been a great challenge since Reeves described the first allograft of a vascularized knee joint in the orthotopic position in dogs using immunosuppression.[55] Like the case in the whole limb homotransplantation, nothing dramatic will happen without a break-through in immunology.

A creative method, "molded vascularized osteogenesis" was reported by Nettelblad et al. in 1984.[56] They showed that corticocancellous bone chips placed in a titanium chamber with a vascular pedicle running through it resulted in a suitable size and shape of vascularized bone graft (Figure 7). This concept was supported by Mizumoto et al. who implanted a bone-chip containing polyethylene chamber with saphenous vessels running through it in rabbit thigh muscles.[57] Recently, two similar models have been reported in which the positive effect of DBM and growth factors (TGF-β and bFGF) on osteogenesis and angiogenesis were found.[58,59]

C. Vascularized Periosteum Graft

Finley et al. in 1978 reported autogenous periosteum grafts with vascular anastomoses for the repair of bony defect in an animal model.[60] In 1979, the first clinical use of free periosteal graft for the treatment of congenital pseudoarthrosis of the tibia was successfully conducted in Beijing.[61] Although there has been some concern about the osteogenic properties of vascularized periosteal grafts,[62,63] several clinical applications of this technique for repairing bone defects have shown promising results[33,64] Recently, researchers have paid attention to periosteum allograft transplantation. Unlike whole limb or whole joint homotransplantation, this new approach seems to work with the help of immunosuppression. The periosteum implanted in muscles formed bone. The one implanted in the radius healed the defect.[65,66]

D. Vascularized Muscle Transfer

The first muscle transfer (rectus femoris) in the dog was reported by Tamai in 1968.[16] The transplantation was proven to be successful by electromyography and light and electron microscopy. In 1973, a Shanghai group successfully treated a severe Volkmann's ischemic contracture of the forearm using a free pectoralis major muscle transfer.[20] Today, free muscle transfer to restore functional defects in the extremities or to reconstruct facial animation has been well-established.[67] Rats are economical animals for studying free muscle transfer. Microvascular transfer of gracilis muscles in rats has been reported.[68]

E. Revascularization of Avascularized Bone

Dickson and Duthie[69] and Boyd and Ault[70] implanted arteries into the femoral shaft and head in canine models. They found prolonged blood flow by cineradiographic arteriograms. Hori et al. further studied the effect of vessel transplantation into bone and documented the growth of new blood vessels and new bone formation.[71] Using vascular bundle transplantation, Hori et al. treated nine patients with Kienboeck's disease and one with scaphoid necrosis. The results were promising after a three year follow-up.[71] A rat model of revascularization of ischemic femoral head was also reported.[72]

Another approach for treatment of osteonecrosis is a vascularized bone graft. Femoral head osteonecrosis has been treated using vascularized fibula graft[73,74] and vascularized iliac crest graft.[75] The clinical application was started seven years prior to the first experimental study in a canine femoral head model using vascularized ribs transfer.[76] Recently, a vascularized periosteal bone graft has been reported to treat osteonecrosis of talus.[64] Although most cases of femoral head necrosis have been treated with femoral head replacement or total hip arthroplasty, vascularized bone graft still remains as an alternative treatment, especially when the collapse is not severe.[77,78]

F. Vascularized Tendon Graft

There have been two different kinds of vascularized tendon grafts, the composite skin-tendon grafts originally described by Taylor and Townsend,[79] and the free or pedicle tendon transfer by Morrison.[80] The former has been often used to treat complicated soft tissue defect involving both tendon(s) and skin. Successful clinical applications have been reported, including the composite grafts of extensor hallucis brevis in dorsal foot flap,[79] external oblique aponeurosis strips in groin-epigastric flap,[79] palmaris longus or flexor carpi radialis in Chinese forearm flap,[81,82] and partial triceps tendon in upper lateral arm flap.[83]

The free vascularized tendon grafts feature small sizes suitable for reconstruction of flexor tendons in "Zone 2." There is a less likely chance of adhesion, and it promotes gliding due to an intact tendon sheath and vascularity. It also facilitates early healing at repair junctures due to their vascularized nature, and can even improve the vascularity of the tendon bed and the finger.[80,84] The free vascularized tendon grafts described in the literature include the extensor hallucis longus or brevis, the extensor indicis proprius, the palmaris longus, and the sublimus tendons of the ring and little fingers.[84,85]

Animal models on this subject were developed after clinical application of the monkey "Zone 2" model by Morrison's group[86] and the tibialis tendon model in the rabbit by Moriyama.[87]

G. Vascularized Nerve Graft

The first free vascularized nerve graft was reported by Taylor and Ham in 1976.[88] They tested the new technique using a pig femoral neurovascular graft model before applying it to a patient with a median nerve defect. A 24 cm long vascularized superficial radial nerve was grafted into the defect using interfascicular repair technique. Postoperative follow-up showed that the nerve had grown 26 cm distally in six months.

The vascularized nerve grafting has the advantages of (1) early revascularization which may increase the speed of axon and myelin sheath regeneration (average speed is 1 mm per day compared to 1.5 mm per day in Taylor's case) and (2) early revascularization which may decrease the extent of fibroblast infiltration and endoneural scarring. The indication for a vascularized nerve grafting may include a severely compromised recipient bed with a long defect. Usable donor nerves are limited. There have been many reports of donor nerves being used clinically, such as superficial radial nerve,[88] sural nerve,[89] ulnar nerve,[90,91] deep peroneal nerve[92] and external popliteal sciatic nerve.[93]

Rabbits are the most commonly used animals for nerve repair, of which vascularized sciatic and median nerve have been reported.[94-96] Rat femoral nerves have also been used.[97]

REFERENCES

1. Holmgren, "G., Some experiences in surgery of otosclerosis," *Acta Otolaryngol.*, 5, 460, 1923.
2. Charles, A. F. and Scott A. A., "Studies on heparin I. The preparation of heparin," *J. Biol. Chem.*, 102, 425, 1933.
3. Lapchinski, A. G., "Recent results of experimental transplantation of preserved limbs and kidneys and possible use of this technique in clinical practice," *Ann. N. Y. Acad. Sci.*, 87, 539, 1960.
4. Snyder, C. C., Knowles, R. P., Mayer, P. W., and Hobbs, J. C., II., "Extremity replantation," *Plast. Reconst. Surg.*, 26, 252, 1960.
5. Onji, Y., Murai, Y., Tamai, S., Hashimoto, T., Yamaguchi, T., Akiyama, H., and Tsujimoto, A., "Experimental surgery on resuscitation and reunion of amputated or nearly amputated extremity," *Plast. Reconstruct. Surg.*, 31, 151, 1963.
6. Jacobson, J. H. and Suarez, E. I., "Microsurgery in anastomosis of small vessels," *Surg. Forum.*, 11, 243, 1960.
7. Lee, S. and Fisher, B., "Portacaval shunt in the rat," *Surgery*, 50, 668, 1961.
8. Malt, R. A. and McKhann, C. F., "Replantation of severed arms," *J. Am. Med. Assoc.*, 189, 114, 1964.
9. Chen, Z. W., Chen, Y. C., and Bao, Y. S., "Salvage of the forearm following complete traumatic amputation: report of a case," *Chin. Med. J.*, 82, 632, 1963.
10. Inoue, T., Toyoshima, Y., Fukusumi, H., et al., "Factors necessary for successful replantation of upper extremities," *Ann. Surg.*, 165, 225, 1967.
11. Komatsu, S. and Tamai, S., "Successful replantation of a completely cut-off thumb," *Plast. Reconstr. Surg.*, 42, 374, 1968.
12. Pho, R. W. H., *Microsurgical Technique in orthopedics*, Butterworths, London, 1988.
13. Tamai, S., "History of microsurgery — from the beginning until the end of the 1970s," *Microsurgery*, 14, 6, 1993.
14. Cobbett, J. R., "Free digital transfer. Report of a case of transfer of a great toe to replace an amputated thumb," *J. Bone Joint Surg.*, 51B: 677, 1969.
15. Krizek, T. J., Tani, T., Desprez, J. D., and Kiehn, C. L., "Experimental transplantation of composite grafts by microsurgical vascular anastomoses," *Plast. Reconstr. Surg.*, 36, 538, 1965.
16. Tamai, S., Komatsu, S., Sakamoto, H., Sano, S., and Sasauchi. N., "Free muscle transplants in dogs, with microsurgical neurovascular anastomoses," *Plast. Reconstruct. Surg.*, 46, 219, 1970.
17. Buncke, H. J. Jr., Daniller, A. I., Schulz, W. P., and Chase, R. A., "The fate of autogenous whole joints transplanted by microvascular anastomoses," *Plast. Reconstr. Surg.*, 39, 333, 1967.
18. Lee, S., *Experimental Microsurgery*, Igaku-Shoin, New York, 1987.

19. Daniel, R. K., and Taylor, G. I., "Distant transfer of an island flap by microvascular anastomoses," *Plast. Reconstr. Surg.*, 52, 111, 1973.
20. Research Laboratory for Replantation of Severed Limbs, Shanghai Sixth People's Hospital, "Free muscle transplantation by microsurgical neurovascular anastomoses. Report of a case," *Chin. Med. J.*, 2, 47, 1976.
21. Ueba, Y. and Fujikawa, S., "Vascularized fibular graft to neurofibulomatosis of the ulna — a 9-year follow up," *Orthop. Surg. Tramatol.*, 26, 595, 1983 (in Japanese).
22. Taylor, G. I., Miller, G. D., and Ham, F. J., "The free vascularized bone graft. A clinical extension of microvascular techniques," *Plast. Reconstr. Surg.*, 55, 533, 1975.
23. Taylor, G. I. and Watson, N., "One-stage repair of compound leg defects with free, revascularized flaps of groin skin and iliac bone," *Plast. Reconstr. Surg.*, 61, 494, 1978.
24. Tamai, S., "Osteocutaneous transplantation. Iliac osteocutaneous neurovascular flap," in *Microsurgical Composite Tissue Transplantation*, Serafin, D. and Bunke H. J., Eds., Mosby, St Louis, 1979, 391.
25. Baudet, J., Guimberteau, J. C., and Nascimento, E., "Successful clinical transfer of two free thoraco-dorsal axillary flaps," *Plast. Reconstr. Surg.*, 58, 680, 1976.
26. Steffens, K., Koob, E., and Hong, G., "Training in basic microsurgical techniques without experiments involving animals," *Arch. Orthop. Trauma Surg.*, 111, 198, 1992.
27. Kim, D. C., Hayward, P. G., and Morrison, W. A., "Training model for microvessel anastomosis," *Microsurgery*, 15, 820, 1994.
28. Bao, J. Y., "Rat tail: a useful model for microvascular training," *Microsurgery*, 16, 122, 1995
29. Mehdorn, H. M. and Müller, G. H., *Microsurgical Exercises*, Georg Thieme Verlag Stuttgart, New York, and Thieme Medical Publishers, New York, 1989.
30. Gahankari, D. R., Lalwani, N. R., and Phatak, A. M., "Classification and comparison of five techniques of end-to-end microarterial anastomoses in rats: a new proposed technique," *Microsurgery*, 16, 793, 1995
31. Lauritzen, C., "A new and easier way to anastomose microvessels," *Scand. J. Plast. Reconstr. Surg.*, 12, 291, 1978.
32. Hung, L. K., Au, K. K., and Ho, Y. F., "Comparative study of artery cuff and fat wrap in microvascular anastomosis in the rat," *Br. J. Plast. Surg.*, 41, 278, 1988.
33. Saitoh, S., Kitagawa, E., and Nakatsuchi, Y., "Comparison of the telescoping anastomotic technique with the end-to-end technique utilizing vein grafts for venous defects: short- and long-term results," *Microsurgery*, 16, 631, 1995.
34. Duarte, A., Valauri, F. A., and Buncke, H. J., "Microvascular thermic sleeve anastomosis: a sutureless technique," *J. Reconstr. Microsurg.*, 4, 53, 1987.
35. Chen, Z. W., Zhang, L., and Wang, M. J., "Experimental investigation of telescoping anastomosis of small arteries," *Shanghai Med. J.*, 6, 282, 1983.
36. Zhang. L., Moskovitz, M., Baron, D. A., and Siebert, J. W., "Different types of sleeve anastomosis," *J. Reconstr. Microsurg.*, 11, 461, 1995.
37. Colen, L. B., Gonzales, F. P., and Buncke, H. J., "The relationship between the number of sutures and the strength of microvascular anastomoses," *Plast. Reconstr. Surg.*, 64, 325, 1979.
38. Amillo, S., Yanez, R., and Barrios, R. H., "Nerve regeneration in different types of grafts: experimental study in rabbits," *Microsurgery*, 16, 621, 1995.
39. Smith, J. W., "Microsurgery of peripheral nerves," *Plast. Reconstruct. Surg.*, 33, 317, 1964.
40. Ito, T., Hirotani, H., and Yamamoto, K., "Peripheral nerve repairs by the funicular suture technique," *Acta Orthop. Scand.*, 47, 283, 1976.
41. Bora, F. W., "Peripheral nerve repair in cats. The fascicular stitch," *J. Bone Joint Surg.*, 49A, 659, 1967.
42. Hakstian, R. W., "Funicular orientation by direct stimulation. An aid to peripheral nerve repair," *J. Bone Joint Surg.*, 50A, 1178, 1968.
43. Millesi, H., Meissl, G., and Berger, A., "Further experience with interfascicular grafting of the median, ulnar, and radial nerves," *J. Bone Joint Surg.*, 58A, 209, 1976.
44. Buncke, H. J., Jr., "Microsurgical research — a personal experience," *Microsurgery*, 16, 186, 1995.
45. Buncke, H. J., Jr., and Schulz, W. P., "Experimental digital amputation and replantation," *Plast. Reconstr. Surg.*, 36, 62, 1965
46. Chen, Z. W. and Yu, H. L., "Current procedures in China on replantation of severed limbs and digits," *Clin. Orthop.*, 215, 15, 1987.

47. Cheng, G. L. and Pan, D. T., "Successful replantation of nine digits in a single case," *Chinese. J. Surg.*, 22, 681, 1984.
48. Goldwyn, R. M., Beach, P. M., Feldman, D., and Wilson, R. E., "Canine limb homotransplantation," *Plast. Reconstr. Surg.*, 37, 184, 1966.
49. Lance, E. M., Inglis, A. E., Figarola, F., and Veith, F. J., "Transplantation of the canine hind limb. Surgical technique and methods of immunosuppression for allotransplantation. A preliminary report," *J. Bone Joint Surg.*, 53A, 1137, 1971.
50. Buncke, H. J. and Murray, D. E., "Small vessel reconstruction," in *Microneurosurgery*, Rand, R., Ed., Mosby, St. Louis, 1974, 183.
51. Doi, K., "Homotransplantation of limbs in rats. A preliminary report on an experimental study with nonspecific immunosuppressive drugs," *Plast. Reconstr. Surg.*, 64, 613, 1979.
52. Furnas, D. W., Black, K. S., Hewitt, C. W., and Achauer, B. M., "Cyclosporine and long-term survival of composite tissue allografts (limb transplants) in rats," *Transplant. Proc.*, 15 (Suppl), 3063, 1983.
53. Tamai, S., Sasauchi, N., Hori, Y., Tatsumi, Y., and Okuda, H., "Microvascular surgery in orthopedics and traumatology," *J. Bone Joint Surg.*, 54B, 637, 1972.
54. Strauch, B., Bloomberg, A. E., and Lewin, M. L., "An experimental approach to mandibular replacement: island vascular composite rib grafts," *Brit. J. Plast. Surg.*, 24, 334, 1971.
55. Reeves, B., "Orthotopic transplantation of vascularized whole knee joints in dogs, *Lancet*, I, 500, 1969.
56. Nettelblad, H., Randolph, M. A., Ostrup, L. T., and Weiland, A. J., "Molded vascularized osteoneogenesis: a preliminary study in rabbits," *Plast. Reconstr. Surg.*, 76, 851, 1985.
57. Mizumoto, S., Inada, Y., and Weiland, A. J., "Pre-formed vascularized bone grafts using polyethylene chamber," *J. Reconstr. Microsurg.*, 8, 325, 1992.
58. Mizumoto, S., Inada, Y., and Weiland, A. J., "Fabrication of vascularized bone grafts using ceramic chambers," *J. Reconstr. Microsurg.*, 9, 441, 1993.
59. Weiss, A. C., Olmedo, M. L., Lin, J. C., and Ballock, R. T., "Growth factor modulation of the formation of a molded vascularized bone graft *in vivo*," *J. Hand Surg.*, 20A, 94, 1995.
60. Finley, J. M., Acland, R. D., and Wood, M. B., "Revascularized periosteal grafts — A new method to produce functional new bone without bone grafting," *Plast. Reconstr. Surg.*, 61, 1, 1978.
61. Cheng, H. H., Yin, D. Q., Chia, S. L., and Shen, C., "The treatment of congenital tibial pseudoarthrosis using free periosteal graft — Report of a case," *Chin. J. Exp. Surg.*, 4, 35, 1987.
62. Puckett, C. L., Hurvitz, J. S., Metzler, M. H., and Silver, D., "Bone formation by revascularized periosteal and bone grafts, compared with traditional bone grafts," *Plast. Reconstruct. Surg.*, 64, 361, 1979.
63. van den Wildenberg, F. A., Goris, R. J., and Tutein Nolthenius-Puylaert, M. B., "Free revascularised periosteum transplantation: an experimental study," *Br. J. Plast. Surg.*, 37, 226, 1984.
64. Doi, K. and Sakai K., "Vascularized periosteal bone graft from the supracondylar region of the femur," *Microsurgery*, 15, 305, 1994.
65. Liu, J. Y., Wang, D., and Cheng, H. H., "Experimental study of the osteogenic capacity of periosteal allografts: A preliminary report," *Microsurgery*, 15, 87, 1994.
66. Liu, J. Y., Wang, D., and Cheng, H. H., "Use of revascularized periosteal allografts for repairing bony defects: an experimental study," *Microsurgery*, 15, 93, 1994.
67. McKee, N. H. and Kuzon, W. M., "Functioning free muscle transplantation: factors affecting success," *Ann. Plast. Surg.*, 23, 249, 1989.
68. Yim, K. K., Lineaweaver, W. C., Siko, P. P., and Buncke, H. J., "Microvascular transfer of anterior and posterior gracilis muscles in rats," *Microsurgery*, 12, 262, 1991.
69. Dickson, R. C. and Duthie, R. B., "The diversion of arterial blood flow to bone," *J. Bone Joint Surg.*, 45A, 356, 1963.
70. Boyd, R. J. and Ault, L. L., "An experimental study of vascular implantation into the femoral head," *Surg. Gynecol. Obstet.*, 121, 1009, 1965.
71. Hori, Y., Tamai, S., Okuda, H., Sakamoto, H., Takita, T., and Masuhara, K., "Blood vessel transplantation to bone," *J. Hand Surg.*, 4A, 23, 1979.
72. Saldana, M. J., Niebauer, J. J., Brown, R., McCarroll, R., and Lichtman, D. M., "Microsurgical revascularization of ischemic rat femoral heads," *J. Hand Surg.*, 15A, 309, 1993.
73. Fujimaki, A. and Yamaguchi, Y., "Vascularized fibular grafting for treatment of aseptic necrosis of the femoral head — preliminary results in four cases," *Microsurgery*, 4, 17, 1983

74. Urbaniak, J. R., "Aseptic necrosis of the femoral head treated by vascularized fibular graft," in *Microsurgery for Major Limb Reconstruction*, Urbaniak, R. J., Ed., Mosby, St Louis, 1987, 178.
75. Chen, Z. W., Zhang, G. L., and Qiu, H. B., "Vascularized pedicle iliac crest graft as treatment for aseptic necrosis of femoral head," in *Microsurgery for Major Limb Reconstruction*, Urbaniak, R. J., Ed., Mosby, St Louis, 1987, 200.
76. González del Pino, J., Gomez Castresana, F., Benito, M., and Weiland, A. J., "Role of free vascularized bone grafts in the experimentally-induced ischemic necrosis of the femoral head," *J. Reconstr. Microsurg.*, 6, 151, 1990.
77. Malizos, K. N., Soucacos, P. N., and Beris, A. E., "Osteonecrosis of the femoral head. Hip salvaging with implantation of a vascularized fibular graft," *Clin. Orthp.*, 314, 67, 1995.
78. Leung, P. C., "Femoral head reconstruction and revascularization. Treatment for ischemic necrosis," *Clin. Orthop.*, 323, 139, 1996.
79. Taylor, G. I. and Townsend, P., "Composite free flap and tendon transfer: an anatomical study and a clinical technique," *Brit. J. Plast. Surg.*, 32, 170, 1979.
80. O'Brien, B. McC. and Morrison, W. A., *Reconstructive Microsurgery*, Churchill Livingstone, Edinburgh, 1987, 340.
81. Reid, C. D. and Moss, L. H., "One-stage flap repair with vascularised tendon grafts in a dorsal hand injury using the 'Chinese' forearm flap," *Brit. J. Plast. Surg.*, 36, 473, 1983.
82. Yajima, H., Inada, Y., Shono, M., and Tamai, S., "Radical forearm flap with vascularized tendons for hand reconstruction," *Plast. Reconstr. Surg.*, 98, 328, 1996.
83. Hou, S. M. and Liu, T. K., "Vascularized tendon graft using lateral arm flap. Five microsurgery cases," *Acta Orthop. Scand.*, 64, 373, 1993.
84. Morrison, W. A. and Cleland, H., "Vascularized flexor tendon grafts," *Ann. Acad. Med. Singapore*, 24, 26, 1995.
85. Vermeylen, J. and Monballiu, G., "The use of the extensor indicis proprius as a vascularised tendon graft. A preliminary report," *J. Hand Surg.*, 16, 185, 1991.
86. Singer, D. I., Morrison, W. A., Gumley, G. J., et al., "Comparative study of vascularized and nonvascularized tendon grafts for reconstruction of flexor tendons in Zone 2: an experimental study in primates," *J. Hand Surg.*, 14A, 55, 1989.
87. Moriyama, M., "Vascularized tendon grafting in the rabbit," *J. Reconstr. Microsurg.*, 8, 83, 1992.
88. Taylor, G. I. and Ham, F., "The free vascularized nerve grafts. A further experimental and clinic application of microvascular technique," *Plast. Reconstruct. Surg.*, 57, 413, 1976.
89. Franchinelli, A., Masquelet, A., Restrepo, J., and Gilbert A., "The vascularized sural nerve," *Int. J. Microsurg.*, 3, 57, 1981.
90. Birch, R., Dunkerton, M., Bonney, G., and Jamieson, A. M., "Experience with the free vascularized ulnar nerve graft in repair of supraclavicular lesions of the brachial plexus," *Clin. Orthop.*, 237, 96, 1988.
91. Gu, Y., Zhang, G. M., Chen, D. S., Yan, J. G., Cheng, X. M., and Chen, L., "Seventh cervical nerve root transfer from the contralateral healthy side for treatment of brachial plexus root avulsion," *J. Hand Surg.*, 17B, 518, 1992.
92. Koshima, I., Okumoto, K., Umeda, N., Moriguchi, T., Ishii, R., and Nakayama, Y., "Free vascularized deep peroneal nerve grafts," *J. Reconstr. Microsurg.*, 12, 131, 1996.
93. Oberlin, C. and Alnot, J. Y., "Use of the external popliteal sciatic nerve as a vascularized graft. Anatomical study and clinical applications," *Rev. Chir. Orthop.*, 71, 94, 1985.
94. Restrepo, Y., Merle, M., Michon, J., Folliguet, B., and Barrat, E., "Free vascularized nerve grafts: an experimental study in the rabbit," *Microsurgery*, 6, 78, 1985.
95. Shibata, M., Tsai, T. M., Firrell, J., and Breidenbach, W. C., "Experimental comparison of vascularized and nonvascularized nerve grafting," *J. Hand Surg.*, 13A, 358, 1988.
96. Ozcan, G., Shenaq, S., and Spira, M., "A new vascularized nerve graft model in the rabbit," *J. Reconstr. Microsurg.*, 8, 35, 1992.
97. Ozcan, G., Shenaq, S., and Spira, M., "Study of microcirculation of rat femoral nerve and development of a new vascularized nerve graft model," *J. Reconstr. Microsurg.*, 7, 133, 1991.

Appendix 1
Abbreviations

ASTM	American Society for Testing and Materials
BMC	Bone mineral content
BMD	Bone mineral density
BMP	Bone morphogenetic protein
BSE	Backscattered electron microscopy
CSLM	Confocal laser scanning microscopy
CoCr	Cobalt chromium alloy
CSD	Critical size defect
DBM	Demineralized bone metrix
ECM	Extensive extracellular matrix
DEXA	Dual energy X ray absorptiometry
ELISA	Enzyme linked immunosorbent assay
ESAF	Endothelial cell-stimulating angiogenesis factor
FGF	Fibroblast growth factor
GAG	Glycosaminoglycans
GLP	Good laboratory practice
GFs	Growth factors
HA	Hydroxyapatite, hyaluronic acid
IGF	Insulin-like growth factor
IHCS	Immunohistochemical staining
IM	Intramuscularly
IP	Intraperitoneally
ISH	*In situ* hybridization
KS	Keratan sulfate
MBT	Molecular biological technique
NGF	Nerve growth factor
NSAIDs	Nonsteroidal antiinflammatory drugs
NZW	New Zealand White
OA	Osteoarthritis
OM	Osteomyelitis
ON	Osteonecrosis
OVX	Ovariectomy
PDGF	Platelet-derived growth factor
PGs	Proteoglycans
PGA	Polyglycolic acid
PLA	Polylactic acid
PMMA	Polymethylmethacrylate
QRD	Quantitative roentgenographic densitometry
QUS	Quantitative ultrasound
rhbFGF	Recombinant human basic fibroblast growth factor
RIA	Radioimmunoassay
SC	Subcutaneously

SC tissue	Subcutaneous tissue
SPA	Single-photon absorptiometry
SD rat	Sprague-Dawley rat
TIMP	Tissue inhibitor of metalloproteinase
Tb.N	Number of trabeculae
Tb.Sp	Trabecular spacing or trabecular separation
Tb.Th	Thickness of trabeculae
TGF	Transforming growth factor-β
TJR	Total joint replacement
TCP	Tricalcium phosphate
TBV	Trabecular bone volume
UHMWPE	Ultra high molecular weight polyethylene

Appendix 2
Useful Journals

Acta Orthop. Scand.
Am. J. Vet. Res.
Arch. Orthop. Trauma Surg.
Arthritis Rheum.
Biomaterials
Calcif. Tissue Int.
Calcif. Tissue Res.
Clin. Orthop.
J. Appl. Biomater.
J. Biomech.
J. Biomed. Mater. Res.
J. Bone Joint Surg. (Am)
J. Bone Joint Surg. (Br)
J. Bone Miner. Res.
J. Hand Surg. (Am)
J. Hand Surg. (Br)
J. Invest. Surg.
J. Orthop. Res.
Lab. Anim.
Lab. Anim. Sci.
Orthop. Clin. North Am.
Osteoarthritis Cartilage
Plast. Reconstr. Surg.

Useful Books

Handbook of Biomaterials Evaluation, von Recum, A., Ed., Macmillan, New York, 1986.
Handbook of Laboratory Animal Science, Svendsen, P., Hau, J., Eds., CRC Press, Boca Raton, 1994.

Appendix 3
Major Sources of Laboratory Animals

Major vendors of laboratory animals include Charles River (www.criver.com), B & K Universal, Taconic (www.taconic.com), and Harlan (www.harlan.com/home.htm). They have branches in Europe, North America, Asia, and other continents. For addresses and phone numbers, go to the internet (search for laboratory animal vendors) or see *Laboratory Animal Buyers' Guide 1996* (along with the 1996 journal issues of *Laboratory Animals*), published for Laboratory Animal Ltd. by PRC Associates, Great Britain.

Index

A

Abbreviations, 583–584
Abdominal aorta, 49
Acetaminophen, 65
Achilles tendon, 477, 486
Achromycin, 118
ACL, see Anterior cruciate ligament
Activated partial thromboplastin time (APTT), 270
Adhesion formation, 481
Adrenaline, 471
AIDS, 337
Alcian blue, 90
Alcohol abuse, 261
Alkaline phosphatase (ALP), 91
Allograft, 252, 318
ALP, see Alkaline phosphatase
American Academy of Orthopædic Surgeons, 3
American Society for Testing and Materials (ASTM), 394
Amputation, 22, 443
Analgesics, commonly recommended, 66
Analysis of variance (ANOVA), 29–30, 171, 317
Anchorin, 311
Anesthesia, see Surgical techniques, surgical facilities, peri-operative care, anesthesia, and
Angiogenesis, 229, 471
Animal(s)
  basic rights of, 4
  models, survey of, 42
  purchasing, 42
  replacement of, 5
  research, ethical issues of, 4
  species, used in orthopædic research, 43
  variance, 20
  Welfare Act (AWA), 5
  welfare, worldwide concern for, 10
Animals selections, in orthopædic research, 39–57
  choosing animal models, 40–50
    available background data of animal, 42
    commonly used animal models, 46–49
    commonly used animals, 42–46
    ethical and general considerations, 30–42
    FDA guidelines, 50
  reason for using animal models, 39–40
Ankylosing spinal hyperostosis (ASH), 532
ANOVA, see Analysis of variance
Anterior cruciate ligament (ACL), 178, 466
  reconstruction, 41, 45, 48, 494
  resection, knee instability of, 331
Anterior interbody fusions, 517
Antibiotic
  therapeutic, 454
  toxicity, 6
Antibody localization method, 127
Anti-Factor VIII immunohistochemistry, 129
Anti-osteoporosis agents, 281
Anuerysms, 101
APTT, see Activated partial thromboplastin time
Arachidonic acid metabolism, alteration of, 73
Arthritis
  adjuvant, 372, 376
  antigen induced, 377
  carageenan-induced, 292, 378
  collagen induced, 373
  induction of, 372, 379
  model, 293
  proteoglycan induced, 377
  streptococcal cell wall, 377
Arthroplasty, 443
Arthroscopy, 335, 355
Articular cartilage, 165, 173
  deformation of, 168
  degeneration of, 172
  lubrication of, 168
  properties of, 172
  tests performed on, 173
Articular cartilage defect, animal models of, 309–325
  animal models for cartilage repair, 311–312
    cartilage defect models, 312
    heterotopic models of chondrogenesis, 311–312
  author preferred animal models, 312–314
    heterotopic models, 312
    rabbit distal femoral joint defect, 312–314
    second choice, 314
  evaluation of cartilage defect repair, 314–317
    histology and histomorphometry, 315–317
    macro findings at necropsy, 314–315
    mechanical testing, 317
    other methods, 317
  grafts investigated for cartilage repair, 318–319
    autograft, allograft, and xenograft, 318
    biomaterials, 318
    chondrocytes or chondrogenic cells grafting, 319
    stimulus for chondrogenesis, 318–319
  normal articular cartilage and repair process, 310–311
    normal cartilage, 310–311
    repair process of cartilage defect, 311
ASH, see Ankylosing spinal hyperostosis
Aspirin, 65
ASTM, see American Society for Testing and Materials
Atropine, 63

Authorship, 34
Autograft, 252, 545–546
AWA, *see* Animal Welfare Act

B

Backscattered electron microscopic (BSEM) images, 92
Bacteria, pathogenesis of, 453
Bacterial adhesion, albumin coating on, 18
B-aminopropionitrile (BAPN), 551
BAPN, *see* B-aminopropionitrile
Barbiturate overdose, 72, 74
Barium sulfate, addition of to bone cement, 429
bBMP, *see* Bovine bone morphogenetic protein
BCF, *see* Blood cell flux
Bending test, 151, 251
BFR, *see* Bone formation rate
Bicortical plug models, 23
Bilateral models, 22
Bioabsorbable materials, 219
 development of, 40
 degradation of, 221
 fixation of osteotomies using, 226
 fixation of osteotomy using, 231, 228
Bioabsorbable materials, animal models for testing, 219–240
 basics of bioabsorbable materials, 220
  biodegradtaion, 220
  common bioabsorbable materials, 220
  mechanical properties of bioabsorbable materials, 220
 commonly used animal models, 220–234
  biocompatibility and biodegradation test in soft tissues, 220–222
  biodegradation in bone tissue, 222–224
  bone replacement, 229
  drug delivery, 239
  fixation or fracture and osteotomy, 225–229
  repair of cartilage defect, 230
  repair of meniscus, 230–234
  repair of tendons and ligaments, 234
  small blood vessel and nerve regeneration, 234
 evaluation methods, 234
Biocompatibility test, designing of, 398
Biodegradation test, 220
Bioimplants, 60
Biological markers, 209
Biomaterial debris, animal models for investigation of, 427–441
 models with intraarticular particle injections and joint implants, 435–436
  canine model, 436
  rabbit model, 435–436
 models with particles and implants, 432–435
  canine model with different implant textures and PE particles, 433
  rabbit tibial hemiarthroplasty with fixed and loose implant, 434–435
  rat model with combined particle application and micromotion, 435
  rat model with intraarticular PMMA plug and application of PE particles, 432–433
  total hip arthroplasty models, 433–434
 models with particles without implants, 428–432
  bone harvest chamber with particles, 430–431
  intramuscular and intraarticular injection of particles, 428–429
  model with particle application to mice with different degrees of immunodeficiencies, 431–432
  rabbit tibia model, 430
  subcutaneous air pouch model, 429
  subcutaneous pocket model, 429
Biomechanical instability, 506
Biomedical literature, 16
Biomedical research, organizations engaged in, 3
Biotechnology, development of techniques in, 34
Bisphosphonate, 285, 373–374
Blood
 cell counts, 451
 cell flux (BCF), 275
 viscosity, 270
BMC, *see* Bone mineral content
BMD, *see* Bone mineral density
BMP, *see* Bone morphogenic protein
Bone(s)
 avascularized, 577
 bone ingrowth in osteopenic, 17
 breaker, 204
 bridges, 272
 cancellous, 28, 143
 cement, 539
 -cement interface, 159
 circulation, 44
 composition of, 140
 cortical, 140, 292
 cutter, 203
 density, 19, 148
 evaluation of histologic changes in, 125
 formation rate (BFR), 382
 fracture, *see* Osteotomy, animal models of bone fracture or
 fragments, 119
 grafting, 209, 252
 -HA interface, 159
 harvest chamber, 430
 histologic study of, 117
 -implant interface, 17, 128, 159
 Kiel, 252
 labeling, 117
 lengthening, 24
 -ligament-bone complex, 176, 180
 loading points of, 152
 loss, 417
  estrogen-depletion, 281
  femoral cortical, 418
 marrow, 242
  extravasation of erythrocytes in, 265
  fresh necrosis of, 262, 263
  immune complexes in, 269
 mechanical testing of, 140

Index

mineral content (BMC), 98
mineral content, measuring, 148
mineral density (BMD), 22, 264
mineralization, 117
morphogenic protein (BMP), 253, 533
nasal, 227
nonunion, animal models of, 207
plugs, 498
regeneration, guided, 229
remodeling, 249
repair, 46
replacement, 219
scan, 96, 209
specimens, method for storing, 145
substitute, subcutaneous, 45
tendon fixation to, 491, 492–493
tissue
   biodegradation in, 222
   implant degradation in, 223
trauma, 435
tumors, 116
union, 208, 209
volume (BV), 383
zygomatic, 230
Bone, mechanical properties and testing methods of, 139–163
  composition and structure of, 140
  general considerations of mechanical testing of bone, 144–151
    measurement of bone densities and mineral content, 148–149
    mechanical testing and data collection, 149–151
    sample preparation, 145–147
    specimen harvesting and storage, 145
  mechanical properties of bone, 140–144
    mechanical parameters, 140–141
    mechanical properties of cancellous bone, 143–144
    mechanical properties of cortical bone, 142–143
  mechanical testing techniques, 151–157
    bending test, 151–152
    compression and tensile test, 153–154
    indentation test, 155
    screw pullout test, 155
    strain gauge, 156–157
  torsional test, 155
    ultrasonic methods, 157
  testing of bone-implant interface, 157–159
    other tests, 159
    pushout and pullout test, 157–158
    removal torque, 159
    screw pullout test, 158–159
Bone defect repair, animal models of, 241–260
  author preferred models, 243–248
    calvarial defect models, 243–247
    long bone defect model, 247–248
    rat subcutaneous model, 243
    second choice, 248
  bone defect models and animal selection, 243
  bone substitutes and future directions of research, 252–253
    autograft, allograft, and xenograft, 252
    biomaterials, 252–253
    bone marrow, 253
    DBM, 252
    growth factors, 253
    mechanisms of bone repair by bone grafting, 252
    tissue engineered composite graft, 253
  evaluation of bone defect repair, 248–251
    histology and histomorphometry, 250–251
    mechanical testing, 251
    radiography, 248–249
  heterotropic models of osteogenesis, 242–243
Bone ingrowth and joint replacement, animal models of, 407–425
  bone ingrowth, 408–413
    experimental endpoints, 412
    factors affecting, 412–413
    general principles of, 408
    models, 408–412
  joint replacement, 413–417
    methodological issues, 414–416
    usage of joint replacement models, 416–417
Books, useful, 585
Bovine bone morphogenetic protein (bBMP), 516
Bradycardia, 63
BSEM images, *see* Backscattered electron microscopic images
Buprenorphine, 66
Butorphanol, 66
BV, *see* Bone volume

C

Cadavers, 168
Calcitonin gene-related peptide (CGRP), 468
Callus formation, 209–210, 249
Calvarial defect models, 243, 245
Cancellous bone, 28
  implant inserted into, 452
  material properties of, 144
  mechanical properties of, 143, 146–147
  model, 448
  structural properties of, 144
Cancer, 412, 529
Canine supraspinatus tendon, 4967
Capillary congestion, 530
Carbon dioxide, 73
Cardiorespiratory system, 67
Caretakers, 10
Cartilage, *see also* Articular cartilage
  articular, 165
  behavior of, 166
  biology, 45
  defect(s), 87
    models, 312–313
    repair of, 230, 233
  deformation of articular, 168
  degeneration, 343
  dehydration, prevention of, 170
  evaluation of histologic changes in, 125
  hardness of, 172

histologic study of, 117
hyaline, 316
photography of repaired, 315
regeneration potential of damaged, 309
surface, 165, 167
testing, set-ups proposed for, 165
thickness, 168, 170
tissue, mechanical testing of, 317
viscoelastic behavior of, 166
Cartilage, mechanical testing of, 165–174
   in vivo indentation testing, 172–173
   indentation theory for compliant layers, 166
   lubrication of articular cartilage and friction measurement, 168–169
   specifications of articular cartilage indentation, 167–168
   use of indentation and friction measurements in orthopædic animal model, 169–172
CCD camera, 179
CDC, *see* Centers for Disease Control
Cefazolin, 200
Cell
   adhesion, 39
   -seeding, 343
Centers for Disease Control (CDC), 75
Cephalosporin, 483
Cervical disc herniation, 536
Cervical spondylosis, 536
CGRP, *see* Calcitonin gene-related peptide
Chi-square test, 31, 316
Cholesterol, 266
Chondrocalcin, 311
Chondrocytes, 310
   apoptotic, 128
   difficulty of engineering, 314
Chondroitin sulfate, 360
Chondronectin, 344
Ciprofloxcin, 229
Circulatory collapse, 74
Clip compression, 527
Clodronate, 199
CNS injury, 288
CoE, *see* Council of Europe
Coefficient of variation (CV), 28
Collagen 140
   activation of complement system by exposed, 395
   fibrils, longitudinal arrangement of, 175
   induced arthritis, 373, 375
   mutations, types IX and XI, 350
   network, 165
      layered, 165
      remodeling of, 395
   type I, 95, 340–341, 464
   type II, 340
Commercial fixation staples, 496
Competence, 10
Compression
   plating, 211
   test ranges, 142
   loading, 166

Compressive tests, longitudinal modulus tested by, 143
Computed tomography (CT), 88, 355
Computer software, custom, 169
Confocal laser scanning microscopy (CSLM), 94
Continuous passive motion (CPM), 318
Control
   data, differences between experimental and, 28
   groups
      commonly used, 22
      types of, 21
   plug, 20
Correlation analysis, 26
Cortex remodeling, 251
Cortical bone, 140
   bending properties of, 142
   fracture risk in, 292
   mechanical properties of, 140
Corticosteroids, 261, 285
Council of Europe (CoE), 9
CPM, *see* Continuous passive motion
Cranial defects, 44
Crimping, 463
Critical size of defect (CSD), 243
CSD, *see* Critical size of defect
CSLM, *see* Confocal laser scanning microscopy
CT, *see* Computed tomography
CV, *see* Coefficient of variation
Cyclic stress relaxation, 183–184
Cyclophosphamide, 529
Cytokines, controlled studies of, 380
Cytolysis, 266, 267

D

Data
   analysis, *see* Research ethics, experimental design, evaluation methods, data analysis, publication, and
   collection error, 26, 27
   common types of, 125
   entry, 27
   validity of, 33
DBM, *see* Demineralized bone matrix
DEA, *see* Dual energy absorptiometry
Decalcification
   improper, 120
   methods, 132
Decapitation, 72, 73
Declomycin, 118
Dehydration
   effects of, 183
   schedule, for small specimens, 133
Demineralized bone matrix (DBM), 517
Denervation, 353
Dermatan sulfate, 310
Descriptive statistics, 28
Desflurane, 64
Designer bone disease, 287
DEXA, *see* Dual-energy X ray absorptiometry
Diaphyseal fracture models, 229

Diaphyseal models, 225
Diffuse skeletal hyperostosis (DISH), 532
Diffusion chamber, 242
Disc-space narrowing, 531
DISH, see Diffuse skeletal hyperostosis
Disk implant, 247
Dissecting microscopy, 88
Distal osteomy union, 251
Distribution-free tests, 31
Disuse osteopenia studies, 288
DNA
  cloned, 100
  nucleic acid sequences, localization of, 127
Donor site morbidity, 252
Dowd model, 434
Drug(s)
  delivery, 219, 229
  paralytic, 8
  systemic use of, 210
Dual energy absorptiometry (DEA), 98
Dual-energy X ray absorptiometry (DEXA), 98, 209, 282
DXA, see Dual energy X ray absorptiometry
Dynamic spinal reconstruction, 541

E

ECG monitoring, 67
ECM, see Extracellular matrix
EDTA method, of decalcification, 132
EGF, see Epidermal growth factor
Elastin, 463
Electrical stimulation, 210
ELISA, see Enzyme linked immunosorbent assay
Embedding, selected protocols for staining and, 132–136
  glycol methacrylate protocols for smaller bones and cartilage, 135–136
  methyl methacrylate embedding of large specimen, 135
  methyl methacrylate embedding of small to medium size specimens, 132–134
  paraffin embedding, 132
  preparation of methacrylate solutions, 134–135
  staining and enzyme localizations, 136
End-to-side anastomosis, 572
End-stage disease, 349
Enflurane, 64
Enhanced Tendon Anchor (ETA), 497
Enzyme histochemistry, 132, 335
Enzyme linked immunosorbent assay (ELISA), 95, 429
Enzyme localization, 136
Epidermal growth factor (EGF), 253
Epiphysis, damage of by irradiation, 550
Epiphysometaphyseal implantation models, 225
Error, common sources of, 26
ERT, see Estrogen replacement therapy
Erythema, 86
Erythrocyte sedimentation rate, 451
Esophageal grafting, 319
Estrogen replacement therapy (ERT), 280–281
ETA, see Enhanced Tendon Anchor
Ethanol, as primary fixative 119
Ethics and regulations, for care and use of laboratory animals, 3–14
  animal research and ethics, 4–5
    importance of animal research, 4
    three Rs and alternatives in orthopædic research, 5
  good laboratory practice, 11–12
  legislation and guidelines, 5–11
    European legislation, 9–10
    legislation and guidelines of other countries, 10–11
    U.S. legislation and guidelines, 5–9
  obtaining approval from IACUC, 11
EU, see European Union
European legislation, 9
European Union (EU), 9
Euthanasia and necropsy, 71–81
  common sampling procedures, 76–79
    joint fluid collection, 77
    sampling for cultivation, 78–79
    sampling for histopathological evaluation, 76
    sampling for mechanical testing, 77
    sampling for serology, 77–78
  euthanasia methods, 72–75
    barbiturate overdose, 74–75
    carbon dioxide, 73–74
    decapitation or cervical dislocation, 73
    gas anesthetic overdose,
    KCL or exsanguination, 74
  necropsy, 75–76
Evaluation, methods of in orthopædic animal research, 85–114
  biochemical methods, 95–96
  clinical observation, 86
  detecting biochemicals in tissues, 95–96
  macro observation at necropsy, 87
  mechanical testing, 96
  molecular biological techniques, 99–102
    basic terminology and methods in molecular biology, 99–100
    diagnosis and prognosis, 101
    etiology and histopathogenesis, 100–101
    gene therapy, 101–102
  radiography, 86–87
    high resolution radiography and microradiography, 87
    plain radiography, 86–87
  special techniques, 96–99
    arthroscopy, 99
    autoradiography, bone scan, scintigraphy, 96–97
    computed tomography, 97
    magnetic resonance imaging, 97–98
    measuring tissue blood flow, 99
    single-photon absorptiometry, single X ray absorptiometry, and dual-energy absorptiometry, 98
    special techniques, ultrasound, 98–99

structural and morphological evaluation, 87–94
  confocal microscopy, 94
  dissecting microscopy, 88–89
  electron microscopy, 93–94
  histology and histomorphometry, 89–93
  measurement of length and area, 87–88
Exercise physiology, 6
Experimental design, see Research ethics, experimental design, evaluation methods, data analysis, publication, and
Experimental endpoints, 416
Experimental plug, 20
Experimental protocol, 32
Exsanguination, 74
Extensor tendon models, 477
Extracellular matrix (ECM), 310
Extradural balloon compression, 527

F

FATC, see Femur-anterior cruciate ligament–tibia complex
FDA, see Food and Drug Administration
Feeding patterns, 86
Femoral head necrosis, 264
Femoral tunnel, anterior placement of, 188
Femur-anterior cruciate ligament-tibia complex (FATC), 180–181
Femur-MCL-tibia complex (FMTC), 181
FGF, see Fibroblast growth factor
Fibrin clot, 328
Fibroblast(s)
  growth factor (FGF), 242, 253
  proliferation of, 375
Fibrocallus, 413
Fibrocartilage, 272, 316, 339
  formation, 340
  meniscal, 328
  repair of meniscal lesions with, 337
Fibronectin, 311, 344
FISH, see Fluorescent in situ hybridization
Fisher exact test, 316
Fixation
  period, safe, 124
  protocol, 470
Flexor tendons, 480
Fluid analyses, 357
Fluorescent in situ hybridization (FISH), 100
FMTC, see Femur-MCL-tibia complex
Food and Drug Administration (FDA), 11
  commissioner of, 12
  guidelines, 286
Foreign body
  infection, 444
  reaction, 234, 340
Formic-citrate method, of decalcification, 132
Fracture(s)
  callus, 155
  fixation of, 219
  fixation, 197
  healing, 208
  abnormal, 197
  effect of systemic drugs on, 211
  endochrondral aspects of, 408
  modes of, 198
  normal, 198
lines, 87
model, 203
  diaphyseal, 199–201
  rat, 202, 211
  radial, 47
  reproducible, 197
  risk, 292
  stability, 201
Fracturing forceps, 198
Fraud, 35
Friction test, 170

G

GAG, see Glycosaminoglycan
Gap prosthesis, 434
Gas anesthetic overdose, 72, 74
GC, see Glucocorticoid
GFs, see Growth factors
Giant cells, 403, 432
GLP, see Good laboratory practice
Glucocorticoid (GC), 290
Glutamic-pyruvic transaminase (GPT), 266
Glutamine-oxaloacetic transaminase (GOT), 266
Gluteraldehyde, buffered, 76
Glycol methacrylate, 134
Glycosaminoglycan (GAG), 140, 310
Gold salts, 373
Goldner's stain, 123
Good laboratory practice (GLP), 11
GOT, see Glutamine-oxaloacetic transaminase
GPT, see Glutamic-pyruvic transaminase
Graft position, 470
Grafting, 48
  esophageal, 319
  skin, 319
Granuloma, particle induced, 436
Gravity bottle, 148
Growth
  factors (GFs), 210, 253, 318
  hormone, 200
  plate abnormalities, 263
Guillotine-like fracture apparatus, 198

H

HA, see Hyaluronic acid; Hydroxyapatite
HA-TCP, see Hydroxyapatite-tricalcium phosphate
Hagemann factor, activation of, 379
Halothane, 65
Haversian remodeling
  in adult dog, 286
  disuse effects on, 288
  effect of GCs on, 291
  low levels of, 285
HDPE, see High density polyethylene

Healing
    direct early, 198
    rapid, 202
Heart block, 63
Hematoma, 435
Hematoxylin/eosin, 122–123
Heparin, 455
Hepatitis, 337
Hertz theory, 166
High density polyethylene (HDPE), 430
High performance liquid chromatography (HPLC), 317
High-pressure liquid chromatography (HPLC), 95
High resolution radiography, 87
Hindlimb suspension, 290
Hip
    joint, intracapsular tamponade in, 261
    replacement models, 411
Histiocytes, 381
Histological scoring system, 251
Histological study, in orthopædic animal research, 115–138
    experimental design, 116–118
        bone labelling, 117–118
        decalcify or not decalcify, 117
        desired endpoints for examination, 116–117
    fixation and transportation, 119
    histological evaluation and histomorphometry, 125–125
    other important procedures, 126–129
        immunohistochemisty, 126–127
        in situ hybridization, 127
        tissues containing implants, 128–129
    specimen harvesting, 118–119
    specimen processing and sectioning, 119–122
        decalcified specimens, 119–120
        soft tissues, 121–122
        undecalcified specimens, 120–121
    staining techniques, 122–125
        histochemical staining, 123–125
        staining of bone and cartilage, 122–123
        staining of soft tissues, 123
Histology laboratory, equipment and tools for orthopædic, 137
Histomorphometry system, 126
HIV transmission, risk of, 343
Hormones, systemics use of, 210
HPLC, *see* High performance liquid chromatography; High-pressure liquid chromatography
5-HTP, *see* 5-Hydroxytryptophan
Human autopsy material, 419
Hyaline cartilage, 316
Hyaluronic acid (HA), 95
Hydroxyapatite (HA), 140
Hydroxyapatite-tricalcium phosphate (HA-TCP), 518
5-Hydroxytryptophan (5-HTP), 552
Hyperemia, 471
Hysteresis, 165

I

IACUC, *see* Institutional Animal Care and Use Committee
Ibuprofen, 199
ICBG, *see* Iliac crest bone graft
IGF, *see* Insulin-like growth factor
Iliac crest bone graft (ICBG), 508
Image
    analysis systems, PC-based interactive, 125
    processing system, 179
    software, 88
Immobilization, 353
    methods, in rats, 288
    permanent, surgical, 289
    postoperative, 484
    regimen, 470
    reversible forelimb, in dogs, 288
    scoliosis induced by, 551
    tools, 556
Immune-mediated diseases, 412
Immunocytochemistry, 126
Immunohistochemistry, 126
    anti-Factor VIII, 129
    problems encountered with, 127
Implant(s)
    -tissue interface, 94
    albumin coated, 455
    biomaterials for surgical, 393
    bone specimens, 436
    -containing specimens, 401
    degradable polyurethane, 342
    dorsal, 400
    fabrication, 446
    influence of skeletal, 444
    materials, 173
    models, 445
    nondegradable, 338
    site infection rate, 18
    soft tissue
        biocompatibility of, 394
        response to, 402
    spinal fusion, 514–515
    tissue acceptance of, 397
Indentation
    depths of, 172
    test, 145, 155
        measurements, 167
        Poisson's ratio of cartilage from, 167
    theory, 166
India ink, vascular injections of, 90
Indomethacin, 211, 229
Inflammatory arthritis, 291
Information, sources of, 138
Institutional Animal Care and Use Committee (IACUC), 7, 23, 72
Insulin-like growth factor (IGF), 253
Intellectual property, 34
Interleukin 1, 434
International property rights, protection of, 34

Intervertebral disc
  artificial, 545–546
  grafting, allogenic, 541
  replacement, 540
  transplanted, 544
Intra-specimen variability, 183
Intramedullary nailing, 211

J

Jogging, cyclic loading applied when, 184
Joint
  disease, models of inflammatory, 370
  fluid collection, 77
  geometry, 350
  infections, 44
  injection, chemical models with, 354
  mechanics, 5
  osteoarthritis models for, 354
  replacement, 407, 413, *see also* Bone ingrowth and joint replacement, animal models of
    models, 416
    research, 409
  stability, 336
  surface drilling, 314
  tissue, 350
  transplantation, whole, 576
Journals, useful, 585

K

Karyolysis, 266–267
Karyorrhexis, 266–267
Keratan sulfate, 310
Ketamine, induction of anesthesia by, 274
Ketorolac tromethamine, 169
Kiel bone, 252
Kinematics, 355
Knee
  joint ligaments, rupture of, 461
  kinematics, 187
  models, 417
Kruskal-Wallis test, 316
Kyphosis, 549

L

Labelling studies, 118
Laboratory animal(s)
  major sources of, 587
  technicians, 10
Laser
  -Doppler flowmetry, 99
  micrometer system, 178
Laws of mechanics, 139
Legg-Calve-Perthes disease, 272–273
Ligament(s)
  behavior shown by, 180
  biology, normal, 463
  biomechanical testing of, 188
  blood supply, 473

division, 464
fatigue of, 184
fixation of, 219
mechanical data of, 495
prosthesis, 41
reconstruction, 467
repair of, 234
research, species commonly used in, 462
response of to levels of stress, 182
substance, mechanical properties of, 176, 177
synthetic, 467
Ligament fixation, animal models of tendon and, 491–501
  commonly used animal models of ligament or tendon fixation, 494–498
    bone or absorbable plug, 498
    interference screw, 498
    screw/washer, 496–497
    spiked bushing, washer, or plate, 497
    staple, 496
    suture and suture anchor, 494–496
    Young's ligament anchor, 498
  evaluation methods, 499
Ligament repair, animal models of, 461–476
  evaluation of healing responses, 467–472
    biochemistry, 468
    biomechanics, 469–471
    histology, 468
    vascular physiology, 471–472
  overview of normal ligament biology, 463–464
  species commonly used in ligament research, 462–463
    dogs, 462
    goats and sheep, 462
    rabbits, 462
    rats, 463
  specific models of ligament healing, 464–467
    ligament division, 464–467
    ligament reconstruction and replacment, 467
    other models, 467
Ligaments and tendons, mechanical testing of, 175–193
  biological factors, 180–182
    effects of anatomical locomotion and functional role, 180–181
    effects of immobilization and exercise, 181–182
    effects of maturation and age, 181
  biomechanical properties of ligaments and tendons, 176–177
  determining functional role of ligaments/tendons, 184–188
  environmental factors, 183–184
    effects of dehydration, 183
    effects of freezing, storing, and thawing, 183–184
    effects of temperature, 183
    178–180
  experimental factors, clamping of testing specimens, 179–189
    determination of strain, 178–179
    effects of specimen orientation, 180

strain rate, 180
stress measurements, 178
viscoelastic properties of ligaments and tendons, 184
Light microscopy, 264
Limb function, monitoring of, 414
Linear variable displacement transducer (LVDT), 149
Link protein fragments, 360
Load application techniques, 167
Load displacement, 169
Long bone defect models, 247
Lordosis, 549
Lumbar intervertebral fusion, 508
Lumbar spine, 49
LVDT, *see* Linear variable displacement transducer
Lymphocyte, 543

M

Machine compliance, 149
Macrophage(s), 403
   activation, 431
   colony-stimulating factor, 380
Magnetic resonance imaging (MRI), 97, 261, 528
Mammary tumors, estrogen dependent, 284
Mann-Whitney U test, 31
MAR, *see* Mineral apposition rate
Mast cells, formalin-resistant, 358
Maximum torque capacity, 155
MBT, *see* Molecular biological technology
MCL, *see* Medial collateral ligament
Meachim's test, 334
Mechanical testing, of materials, 149
Medial collateral ligament (MCL), 42, 178, 462
Medial knee arthrotomy, 432
MedLine search, 33, 35
Meeting abstracts, 32
Meniscal repair, animal models of, 327–347
   animal selection, 330–332
   applications of models and future directions of research, 336–344
     evaluation of artificial substitute of meniscus, 337–341
     evaluation of biological graft, 341–344
     evaluation of pathophysiology and natural healing process of meniscal injury, 336–337
   commonly used models, 332–334
     meniscal repair, 332–333
     meniscal replacement, 333–334
   evaluation methods, 334–336
     gross morphology, 334
     histology, 334–335
     mechanical testing, 335
     others, 335–336
Meniscectomy, 327
MEP, *see* Motor evoked potentials
Methacrylate embedment, 119, 121
Methotrexate, 373, 374
Methoxyflurane, 65

Methyl methacrylate
   embedding, 120, 135
   solutions, preparation of, 135
Microsurgery, orthopædic animal models and, 567–581
   basic requirements for experimental microsurgery, 569
   basic techniques, 570–575
     nerve repair, 573–575
     vessel anastomosis, 570–573
   microsurgery and orthopædic animal models, 575–578
     reimplantation and transplantation of severed limbs, 575–576
     revascularization of avascularized bone, 577
     vascularized bone graft or whole joint transplantation, 576
     vascularized muscle transfer, 577
     vascularized nerve graft, 578
     vascularized periosteum graft, 576–577
     vascularized tendon graft, 577
   training of microsurgical techniques, 568–569
Microsuturing, 570
Microtrauma by proxy, 466
Midazolam, 62
Mineral apposition rate (MAR), 382
Mineralized bone, 94
Mitochondrial enzymes, 335
Model(s)
   Achilles tendon, 477
   anterior fusion animal, 517
   arthritis, 17, 293
   author-preferred, 352, 446
   bicortical plug, 23
   bilateral, 22
   bone defect, 44
   cancellous bone, 448
   canine, 273, 433, 479
   carrageenan injection, 384
   cartilage defect, 312–313
   controlled motion, 411
   diaphyseal bone, 447
   diaphyseal, 225
   Dowd, 434
   enzymatic, 352
   extensor tendon, 477
   fracture, 203
   heterotopic, 312
   hip replacement, 411
   implant, 398, 445
   in vitro, 332
   intramuscular implant, 399
   joint replacement, 416, 444, 448
   knee replacement, 412
   live animal, 505
   long bone defect, 247
   maxillofacial bone fracture, 227
   mechanical/surgical, 353
   nonhuman primate, 488
   non-rodent, 50
   nontraumatic, 276

nonweight-bearing gap, 408
nude mice, 312
osteoarthritis, 359
partial-weight bearing, 314
percutaneous implant, 399
porcine, 187, 511
posterior fusion, 516
prosthetic infection, 452
rat calvarial defect, 247
rupture, 465
scoliosis, 560
spontaneously hypertensive rat, 262
stress fracture, 204
stroke-prone spontaneously hypertensive rat, 264
subcutaneous, 311, 446
swine, 275
tendon, 486
total hip arthroplasty, 433
training, 568
traumatic, 276
unicortical, 23
unilateral, 22
weight-bearing, 411
Z-plasty apposition, 465
Molecular biological technology (MBT), 99
Molecular cloning, 100
Monocloncal antibodies, 335
Motion protocol, postoperative, 478
Motor evoked potentials (MEP), 528
MRI, see Magnetic resonance imaging
Mucopolysaccharides, 123
Multifactorial designs, 21
Murine osteosarcoma, 533
Muscle
  imbalance, 558
  transfer, 577
Musculoskeletal diseases, 99
Mycoplasmas, 370
Myeloid precursors, 381
Myofibroblasts, 395

N

Nasal bones, 227
Necropsy, 75, 382, see also Euthanasia and necropsy
Nembutal, 482
Nerve
  graft, vascularized, 578
  regeneration, 234
  repair, 24, 44, 233
Neurectomy, 289
Neuronal loss, 534
NHP, see Nonhuman primates
Nitrous oxide, 65
Nonbarbiturate agent, injectable, 74
Non-contact technology, 178
Nonhuman primates (NHP), 286
Non-osteonal healing, 198
Nonsteroidal antiinflammatory drugs (NSAIDS), 65, 169, 211

Nonweight-bearing gap model, 408
NSAIDS, see Nonsteroidal antiinflammatory drugs

O

OC, see Osteocalcin
OLF, see Ossification of ligamentum flavum
OM, see Osteomyelitis
ON, see Osteonecrosis
ONF, see Osteonecrosis of femoral head
Operating rooms, traffic flow in, 60
Operator performance, 27
OPLL, see Ossification of posterior longitudinal ligament
Orthopædic histology laboratory, equipment for, 137
Orthopædic surgery, 59
Osetotomy, radial, 204
Osseous spinal fusion, 521
Ossification of ligamentum flavum (OLF), 532, 535
Ossification of posterior longitudinal ligament (OPLL), 527, 535
Osteoarthritis, animal models of, 349–367
  animal models of mechanically/surgically-induced disease, 352
  applications of models, 359–360
    bone ingrowth to implants under osteoarthritic conditions, 360
    evaluation of pathophysiology of osteoarthritis, 359
    identification of markers of osteoarthritis, 359–360
    intervention and osteoarthritis models, 359
  author-preferred model, 352–355
  chemically-induced models, 352
  evaluation methods, 355–358
    biomechanical properties, 358
    clinical evaluation, 355–357
    macroscopic analysis, 357
    microscopic analysis, 357–358
    molecular biology methodology, 358
  general precautions, 355
  spontaneous models, 350–351
Osteoblasts, localization of alkaline phosphatase in, 125
Osteocalcin (OC), 101
Osteoclasts, 284
Osteogenesis, heterotopic models of, 242
Osteomyelitis (OM), 98, 444
Osteonecrosis (ONF), animal models of, 261–278
  applications of animal models, 276
  nontraumatic models of osteonecrosis, 262–271
    rabbit with hypersensitivity reactions, 267–269, 270–271
    rabbits with endotoxic reactions, 266–267
    spontaneously hypertensive rat model, 262–264
    steroid-treated rabbit model, 265–266
    stroke-prone spontaneously hypertensive rat model, 264–265

traumatic models of osteonecrosis, 271–276
  canine model by deep freezing, 273–274
  canine model by dislocation and vessel ligation, 274–275
  goat model, 272–273
  swine model, 275–276
Osteonecrosis of femoral head (ONF), 271, 276
Osteopenia, *see* Osteoporosis, animal models of osteoporosis or
Osteopetrosis genes, 282
Osteophyte formation, 531
Osteoporosis, animal models of osteoporosis or, 279–305
  animal models for osteoporosis, 282–287
    avian, guinea pig, rabbit, ferret, and cat, 285
    dog, 285–286
    mouse, 282–283
    nonhuman primate, 286–287
    pigs and sheep, 286
    rat, 284–285
  diffuse osteopenia, 287–290
    animals, 288–289
    common procedures, 289–290
    general, 287–288
  general consideration, 280–281
    bone loss and turnover rate rise after estrogen depletion, 281
    criteria, 280
    osteoporotic fractures and steady state osteopenia, 281
    remodeling, 281
    timeframe compression, 281
  glucocorticoid osteopenia, 290–291
    large animals, 291
    small animals, 290–291
  osteopenia associated with inflammation, 291–293
    animal models, 291–292
    fracture risk, 292–293
    general, 291
Osteosarcoma, 100, 242
Osteotomy, animal models of bone fracture or, 197–217
  animal models of delayed union and nonunion, 205
  animal models of diaphyseal fractures, 198–204
    author preferred diaphyseal models, 202–204
    models of diaphyseal fractures, 198–201
  animal models of epiphysometaphyseal models, 205
  evaluation methods, 205–209
    histology and histomorphometry, 208
    mechanical testing, 208
    other evaluation methods, 209
    radiography, 205–208
  fracture healing process, 198
  methods for enhancing fracture healing, 209–211
    biomaterials, 210
    biophysical stimulation, 210–211
    bone grafting, 209–210
    local use of growth promoting substances, 210
    systemic use of hormones or drugs, 210
  other factors affecting fracture healing, 211
    effects of drugs or medication, 211
    effects of fixation devices, 211
Ovariectomy (OVX), 47, 280
Over-decalcification, 89
OVX, *see* Ovariectomy
Oxytetracycline, 118

P

Paired designs, 22–23
Paraffin embedding, 132
Paralytic drugs, use of without anesthesia, 8
Partial-weight bearing models, 314
Particle toxicity, 6
Patient registries, computerized, 5
PCL reconstruction, 497
PCR, *see* Polymerase chain reaction
PDGF, *see* Platelet-derived growth factor
PE, *see* Polyethylene
Peak bone mass, 280, 286
Peer review authority, 10
Penicillamine, 373
Penicillin, 483
Pentobarbital, 482, 483
Perio-operative care, *see* Surgical techniques, surgical facilities, peri-operative care, anesthesia, and
Personal error, 26
Personnel
  categories of, 10
  performing surgery, 59
  qualifications, 8
PGA, *see* Polyglycolic acid
PGs, *see* Proteoglycans
Phenothiazine derivatives, 62
Phenylbutazone, 66
Philospohers, modern animal movement and, 4
Phosphotungstic acid hematoxylin (PTAH), 268
PHS, *see* Public Health Service
Pilot study, 19
Pin-on-disc
  friction apparatus, 170–171
  system, 168
Pinealectomy, in chickens, 554
Piroxicam, 383
PLA, *see* Polylactic acid
Plane-ended indenter, 167
Plasma cells, 403
Platelet-derived growth factor (PDGF), 253
Platelet-rich plasma (PRP), 270
PLSF, *see* Posterior lateral intertransverse process function
PMMA, *see* Polymethylmethacrylate
Poisson's ratio, 166
Polyethylene (PE), 428
Polyglycolic acid (PGA), 40, 219
Polyhydroxyacids, 220
Polylactic acid (PLA), 40, 219
Polymer(s)
  degradable, 229
  implantation of porous, 329

Polymerase chain reaction (PCR), 101, 487
Polymerization, 133
Polymethylmethacrylate (PMMA), 428
Polymorphonuclear granulocytes, 402
Polymyosystis, 370
Polytetrafluoroethylene (PTFE), 318
Polyvinyl alcohol hydrogen, 546
Porcine model, 187
Posterior lateral intertransverse process function (PLSF), 518, 520
Precision, 26
Prednisolone, 383
Preoperative agents, 62
Processing error, 27
Pro-inflammatory agents, 468
Prosthetic infection, animal models of orthopædic, 443–457
  applications of prosthetic infection models, 452–455
    effect of biomaterials on prosthetic infection rate, 454–455
    effect of infection on biomaterial surfaces, 455
    effect of prophylactic and therapeutic antibiotics, 454
    in vivo behavior of biofilm, 453–454
    pathogenesis of bacteria, 452–453
  commonly used evaluation methods and criteria, 448–452
    bacteriological study, 452
    clinical features, 448–451
    histological study, 452
    laboratory tests, 451–452
    radiography, 451
  design of animal model of prosthetic infection, 444–448
    animal selection, 444–446
    author preferred models, 446–448
    bacterial inoculation, 446
    implant fabrication, 446
    surgical and necropsy technique, 446
Protein adsorption, 39
Proteoglycans (PGs), 310, 463
Protocol
  preparation, 59
  review, 11
  submission, 11
PRP, *see* Platelet-rich plasma
PTAH, *see* Phosphotungstic acid hematoxylin
PTFE, *see* Polytetrafluoroethylene
Public Health Service (PHS), 6, 71
Publication, *see* Research ethics, experimental design, evaluation methods, data analysis, publication, and
Pullout tests, 157
Pushout tests, 157, 158
Pyridinium crosslinks, 95

Q

QCT, *see* Quantitative CT
QRD, *see* Quantitative roentgenographic densitometry
Quality assurance, 12
Quantitative CT (QCT), 97
Quantitative roentgenographic densitometry (QRD), 209
Quantitative trait linkage studies, 280
Quantitative ultrasound (QUS), 98
QUS, *see* Quantitative ultrasound

R

RA, *see* Rheumatoid arthritis
Rabbit
  bones, mechanical symmetry of, 16
  calvarial defect, 243, 246
  medullary nailing model, 447
  ulnar defect, 248
Radial fracture, 47
Radiographic scoring
  semi-quantitative, 87
  system, 250
Radiography, 205, 355
Radioimmunoassay (RIA), 95
Radiology facilities, 60
Randomization, 20
Range-of-motion measures, 506
Regression analysis, 30
Regulatory documents, 8
Removal torque, 159
Repair, types of, 574
Research ethics, experimental design, evaluation methods, data analysis, publication, and, 15–37
  data analysis, 27–31
    analysis of variance, 29–30
    correlation and regression analysis, 30–31
    descriptive statistics, 28–29
    nonparametric data, 31
    one sample analysis, 29
    paired comparisons, 29
    unpaired comparisons, 29
  evaluation methods, 24–27
    common sources of error, 26–27
    selection of, 26
  experimental design, 19–24
    controls, 21–24
    number of animals required, 19–20
    randomization and sampling error, 20–21
  publication, 31–33
    abstract, 31–32
    discussion, 33
    introduction, 32
    materials and methods, 32
    references, 33
    results, 32
    title page, 31
  question, hypothesis, and experimental purpose, 16–18
  research ethics, 33–35
    authorship and acknowledgment, 34
    ethical use of laboratory animals, 33

integrity, 35
intellectual property, 34
never duplicate publication, 35
obligations of researchers, 33–34
writing research proposal, 18–19
Revascularization, 491
Rheumatoid arthritis (RA) animal models of, 369–390
    applications of animal models, 382–386
        bone structure and strength, 386
        evaluation of pathophysiology of rheumatoid arthritis, 382–383
        fixation of prosthetic implants, 383–385
        observations of bone morphology and metabolism, 383
        soft tissue joint constraints in rheumatoid arthritis, 385–386
        synovectomy, 385
    evaluation methods, 380–382
        biochemical assessment of inflammatory disease, 381
        biochemical assessment of cartilage and bone turnover, 381
        clinical severity, 381
        histomorphometry, 382
        necropsy, 382
        roentgenographic methods, 381–382
    mechanistic approaches to models of arthritis, 380
        cytokine-induced arthritis and inhibition with anti-cytokine antibodies, 380
        mutation-based approaches, 380
    selection of, 371–380
        adjuvant arthritis, 372–373
        antigen induced arthritis, 377
        carrageenan induced arthritis, 378–380
        collagen induced arthritis, 373–376
        proteoglycan induced arthritis, 377
        streptococcal cell wall arthritis, 377–378
Rhizotomy, 554
RIA, see Radioimmunoassay
Rib resection, 555, 5578
Rigid plate fixation, 211
RNA nucleic acid sequences, localization of, 127
Robot/UFS testing system, 186, 187

S

Salmonella, 78
Sample size, 19
Sampling
    error, 20
    objective of, 21
Sawing-grinding, 90
Scanning electron microscopy (SEM), 93
Scanning frequency, 169
SCBF, see Spinal cord blood flow
SCID, see Severe combined immunodeficiency
Scientific paper, functions of, 31
Scintigraphy, 96, 357

Scoliosis, animal models of, 549–564
    animal selections, 550
    dietary methods, 551
    immobilization, 551–552
    local methods, 554–559
    systemic methods, 552–554
    use of scoliosis models in spinal research, 559
        exploration for new treatment, 559
        pathogenesis of scoliosis, 559
Scoliosis Research Society, 559
Screw pullout test, 155, 156, 158
Segmental bone defect, 24
SEM, see Scanning electron microscopy
SEP, see Somatosensory evoke potentials
Septic arthritis, 44
Severe combined immunodeficiency (SCID), 431
Sevoflurane, 64
Shear modulus, 166
Shigella, 78
Single-photon absorptiometry (SPA), 209
Sjögren's syndrome, 370
Skeletal disuse, conditions of, 287
Skeletal dysplasia, 116
Skin grafting, 319
Smoking, 261
Sodium pentobartbital, 274
Soft swelling, 86
Soft tissue biocompatibility of biomaterials, animal models for studying, 393–405
    common evaluation methods, 400–403
        histological and histomorphometrical evaluation, 402
        light microscopical preparation, 401–402
        other evaluation methods, 402–403
    commonly used implant models, 398–400
        intramuscular implant models, 399
        percutaneous implant models, 399–400
        subcutaneous implant models, 398–399
    factors affecting wound healing, 395–398
        animal considerations, 396
        material properties, 398
        surgical considerations, 397
        wound and repair process, 395
    general *in vivo* tests, 394–395
        commonly used animals, 394–395
        soft tissue biocompatibility assays, 394
Somatosensory evoke potentials (SEP), 528
SOP, see Standard operating procedure
SPA, see Single-photon absorptiometry
Spaceflight, 290
Species choice, 414
Specimen(s)
    decalcified, 119
    fixation of, 121
    -fixture interface, 150
    handling, 41
    harvesting, 118
    multiple thawing and freezing of, 145
    stiffness, overestimation of, 151
Spinal columns, radiographic examination of, 534

Spinal cord
  blood flow (SCBF), 528
  tumor, 536
Spinal cord compression, animal models of, 527–537
  commonly used models, 528–535
    classical models, 528–529
    herniated intervertebral disc, 529
    screw, 530
    spondylosis, 530–531
    tumor, 529–530
    vertebral hyperostosis, 532–535
  selection of animal models, 528
Spinal defects, 539
Spinal deformity, 549
Spinal fusion
  animal models of, 511
  implants, 514–515
  inhibitors to, 520
  osseous, 521
Spinal instability, animal models of spinal fusion and, 505–526
  animal models of spinal fusion and biology, 511–521
    anterior fusion animal models, 517–518
    inhibitors to spinal fusion, 520
    methods of evaluation, 520
    posterior fusion animal models, 516–517
    quadrupedal animal models, 519
    surgical approaches and techniques, 518–519
    vertebral segments, 519
  models of instability and internal fixation, 506–511
    biomechanical instability, 506
    ex vivo models of biomechanical instability and fixation, 506–509
    in vivo models of spinal instability and fixation, 509
Spinal instrumentation, load-sharing capacity of, 509
Spinal reconstruction, 540, 540
Spinal research, use of scoliosis models in, 560
Spinal surgery, 539
Spondylosis, 530
Spontaneous diseases, 42
Spontaneously hypertensive rat model, 262
Stain marking, 179
Standard operating procedure (SOP), 12, 75
Statistician, 19
Sterile sepsis, 234
Storing, effects of, 183
Strain gauge, 156
Streptobacillus moniliformis, 77
Stress
  fracture model, 204
  relaxation, 165, 169
Stroke-prone spontaneously hypertensive rat model, 264
Structural integrity, 316
Study protocol, 12
Substance P, 468
Support anvil, 202
Surgery, survival, 8

Surgical approach, description of, 32
Surgical facilities, *see* Surgical techniques, surgical facilities, peri-operative care, anesthesia, and
Surgical injury, 225
Surgical instruments, 60
Surgical procedure, type of, 60
Surgical suite, 61
Surgical techniques, surgical facilities, peri-operative care, anesthesia, and, 59–69
  anesthesia and analgesia, 62–65
    analgesia, 65
    inhalation anesthesia, 64–65
    injectable anesthetic protocols, 63–64
    preoperative agents, 62–63
    principles of analgesia, 62
  post-operative care, 67–68
  preoperative care, 61–62
  surgical facilities and equipment, 60–61
  surgical technique, 65–67
Surgical trauma, 394
Survival surgery, 8
Suture
  anchors, 496
  material, 67, 574
Swine model, 275
Synovectomy, 385
Synovitis, experimental, 46
Systemic lupus erythematosus, 261
Systemic sclerosis, 370

T

Tartrate-resistant acid phosphatase (TRAP), 123
TCP, *see* Tricalcium phosphate
TEM, *see* Transmission electron microscopy
Temperature, effects of, 183
Tendon(s)
  behavior shown by, 180
  biomechanical properties of, 175
  -bone interface, 499
  gliding function, 487
  healing, 478
  research, monkeys used in, 485
  specimens, 494
  synthetic materials used to construct, 122
Tendon repair, animal models of, 477–490
  animal and animal models selections, 479–480
  basic science of tendon, 478–479
    tendon biomechanics, 478–478
    tendon healing, 478
    tendon nutrition, 478
  commonly used models, 480–486
    achilles tendon, 486
    extensor tendons, 486
    flexor tendons, 480–485
    general tendon models, 486
  evaluation methods, 486–487
    biomechanical evaluation, 487
    morphological evaluation, 486–487
Tenosynovitis, 374

Index

Tenotomy, 289
Tensile test, 154, 499
Test(ing)
　bending, 151, 251
　biocompatibility, 398
　biodegradation, 220
　chi-square, 31, 316
　distribution-free, 31
　equipment, commercial, 167
　facility, 12
　Fisher exact, 316
　friction, 170
　high-strength, 169
　in vivo indentation, 172
　indentation, 145, 155
　Kruskal-Wallis, 316
　machine, 153
　Mann-Whitney U, 31
　Meachim's, 334
　mechanical, 173
　pullout, 157, 412
　pushout, 157–158, 412
　ranges, 142
　robot/UFS, 186
　screw pullout, 155–156, 158
　specimens
　　clamping of, 179
　　structural integrity of, 170
　systems, systemsic errors of, 27
　tensile, 154
　tensile, 499
　torsional, 155, 208
　toxicological, 39
　Wilcoxon Signed Rank, 31
Tetracycline
　incorporation, 117
　label, 125
Texas Scottish Rite Hospital (TSRH), 508
TGF, *see* Transforming growth factor
TGF-β, *see* Transforming growth factor-β
THR, *see* Total hip replacement
Thrombospondin, 311, 330
Thymidine kinase (TK), 101
Tibial osteotomy, 225
Tibial plating, 444
Tissue reaction, 129
TJR, *see* Total joint replacement
TK, *see* Thymidine kinase
TKR, *see* Total knee replacement
TNFI, *see* Tumor necrosis factor inhibitor
Tobial fracture, 201
Toe-to-hand transplantation, experimental, 576
Toluidine blue, 122
Torsional stiffness, 155
Torsional tests, 155, 208
Total hip replacement (THR), 408, 410
Total joint replacement (TJR), 427
　infection of, 453
　sepsis following, 443
Total knee replacement (TKR), 410
Toxicologcal tests, 39

Trabecular bone spatial connectivity, 93
Trabecular loss, 274
Training model, 568
Transarticular osteotomies, 24
Transcortical replacement, 409
Transforming growth factor (TGF), 253
Transforming growth factor-β (TGF-β), 413
Transmission electron microscopy (TEM), 76, 94, 358
Transverse processes, removal of, 557
TRAP, *see* Tartrate-resistant acid phosphatase, 124
Treatment control, 22
Tricalcium phosphate (TCP), 252
Trichrome, 122
TSRH, *see* Texas Scottish Rite Hospital
Tumor
　growth, 529
　necrosis factor, 376, 429
　necrosis factor inhibitor (TNFI), 102
Tyrosine hydroxylase, 91

U

UFS, *see* Universal Force Moment Sensor
UHMWPE, *see* Ultra high molecular weight polyethylene
Ultra high molecular weight polyethylene (UHMWPE), 430
Ultrasound, 157, 355
Unicortical plug models, 23
Unilateral models, 22
United States Department of Agriculture (USDA), 5–7
Universal Force Moment Sensor (UFS), 185
USDA, *see* United States Department of Agriculture
UV polymerization, 120

V

Vascular remodeling drugs, 473
Vascular repair, 93
VATS, *see* Video-assisted endoscopic technique
VDA, *see* Video dimension analyzer
Vendor health monitoring, 61
Vertebrae, histological changes of, 558
Vertebral column, animal models for reconstruction of invertebral disc and, 539–547
　animal models and animal selections, 540
　dynamic spinal reconstruction, 541–546
　　allograft, 514–544
　　artificial intervertebral discs, 545–546
　　autograft, 545
　spinal column reconstruction without invertebral disc, 540
Vertebral deformity, 552
Vertebral segments, fusion of destabilized, 505
Vessel
　anastomosis, 570
　regeneration, 233

repair, factors affecting success of, 573
Veterinary care, 7
Vibrio, 78
Video-assisted endoscopic technique (VATS), 519
Video dimension analyzer (VDA), 178
Video image file, 179
Video systemn, 126
Viral infections, diagnosis of, 79

W

Walking analysis, 336
Wear debris particles, effects of orthopædic, 439
Wilcoxon Signed Rank test, 31
World Health Organization, 281
Wound
   appearance of, 451
   healing, 396

X

Xenograft, 252
X ray
   absorptiometry, 98
   machine, high resolution, 205
   microanalysis (XRMA), 403
XRMA, *see* X ray microanalysis
Xylazine, 63

Y

Young's ligament anchor, 498
Young's modulus, 166–167, 177

Z

Zoledronate, 293
Zoonotic disease, species specific, 62
Zygomatic bone, 230